USDA Forest Service Experimental Forests and Ranges

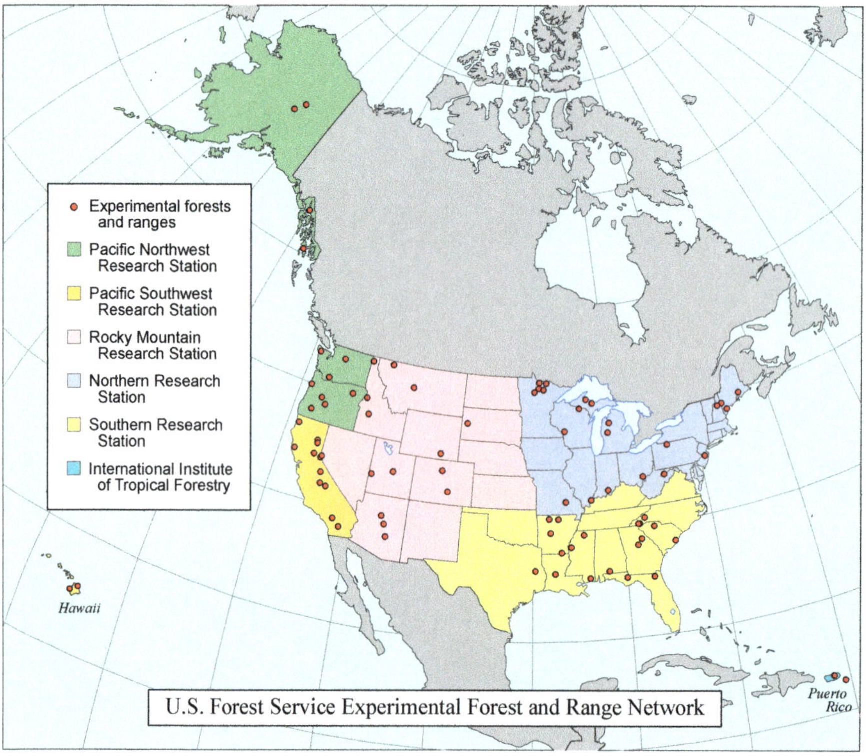

U.S. Forest Service Experimental Forest and Range Network

Deborah C. Hayes • Susan L. Stout
Ralph H. Crawford • Anne P. Hoover
Editors

USDA Forest Service Experimental Forests and Ranges

Research for the Long Term

Springer

Editors
Deborah C. Hayes
USDA Forest Service
Washington
District of Columbia
USA

Susan L. Stout
Northern Research Station
USDA Forest Service
Irvine
Pennsylvania
USA

Ralph H. Crawford
Northeastern Areas
USDA Forest Service
Newtown Square
Pennsylvania
USA

Anne P. Hoover
USDA Forest Service
Washington
District of Columbia
USA

ISBN 978-1-4614-1817-7 ISBN 978-1-4614-1818-4 (eBook)
DOI 10.1007/978-1-4614-1818-4
Springer New York Heidelberg Dordrecht London

Library of Congress Control Number: 2014940541

Printed on acid-free paper

Springer is part of Springer Science+Business Media (www.springer.com)

I am for this
with some changes.
GP

May 6, 1908.

<u>Memorandum for Mr. Pinchot</u>:

In accordance with your instructions of December 11, 1907, to work out a scheme for inaugurating forest experiments on the National Forests, I have the honor to submit the following plan for the creation of Forest Experiment Stations.

Respectfully,

Raphael Zon

Editors' Foreword

Research programs may move in new directions in response to changes in society's needs and values. Each chapter of this book reflects the relationship between the ecological results that emerge from a long-term research project and the social forces that influence questions asked and resources invested in ecological research. An example of this dynamic interplay is found in Chap. 21 (Graham), which describes the evolution of fire research in the Rocky Mountains. Another compelling example emerges from research on the Escambia Experimental Forest (EFR) (Chap. 4, Connor et al.), describing society's turn away from longleaf pine forests during a period when their regeneration processes were poorly understood. EFR research developed effective regeneration techniques and thus influenced the current move to restore this valuable natural system to the landscape.

Often, trends observed—or expected—in the early years of a research program are confounded or contradicted as the research record extends over decades. After a 30-year study of water chemistry at Coweeta Hydrological Laboratory (Chap. 17, Vose et al.) scientists realized that they had a more accurate picture of trends than suggested by earlier looks at 5-year increments. Several chapters provide examples of long-term research on EFRs that has provided credible data for questions not even imagined at the time the study was installed, such as the evidence for changes in structure and composition of forests with fire exclusion documented in Chap. 3 (Bragg). The development of long-term research on grey jays originated as a study of dwarf mistletoe (Chap. 11, Nicholls) and led to insights on the lifespan of Grey Jays and the movement of the West Nile Virus in the Rocky Mountains.

The natural resource problems now faced by the nation include questions of such regional and global scale that an unprecedented level of collaboration and large-scale focus is required of research programs. Six Forest Service EFRs were among the twenty-five sites first selected by the National Science Foundation as Long-Term Ecological Research sites. The research involves many disciplines of science, the application of the most up-to-date scientific methodology and data management, and long-term continuity of measurements on sites dedicated to experimental research. The long life of most forest communities creates an unusual need for long-term research; many studies will have to continue for additional centuries before they have followed even a single generation of trees. Similar efforts are needed on differing temporal scales for rangeland communities. In this book, scientists who work on EFRs step back and take the long view. Today, a network of experimental research sites is more relevant than ever.

Acknowledgements

We have many people to thank for contributions, ideas, and enthusiastic support that made this book a reality. We are grateful for the foresight of the USDA Forest Service in establishing and supporting a national network of Experimental Forests and Ranges. Foremost, we appreciate Raphael Zon, the first Deputy Chief of Forest Service Research. His vision and dedication in establishing the "Tree of Research" at Fort Valley Experimental Forest in 1908 has inspired generations of scientists. We acknowledge the authors for their contributions, dedication and patience throughout the writing process. We also appreciate the support of Forest Service Research and Development Deputy Chiefs Ann Bartuska and Jim Reaves for this project and for EFRs. We give special thanks to the peer reviewers of each chapter for their critical observations and comments. They have all improved this book well beyond what we originally envisioned.

The Editors

Foreword

On behalf of the USDA Forest Service Research & Development, I am proud and excited to share the science success stories made possible by the more than 100-year-old network of Experimental Forests and Ranges. These "EFRs" are living laboratories where Forest Service scientists and their colleagues make discoveries and demonstrate the importance and implications of research results. The entire network is recognized globally as valuable national asset for research, education, and conservation-related activities.

Since 1908, when the first experimental forest was established in Fort Valley, Arizona, the network has expanded to more than 80 research sites. The three most recent additions are the Sagehen Experimental Forest in northern California, the Hawaii Tropical Experimental Forest on the big island of Hawaii, and the Heen Letinee near Juneau, Alaska.

The network is the oldest and most extensive system of research sites in the USA dedicated to resolving the nation's natural resource problems through research and public education. They remain one of the few places where ecological research can be conducted over a long period of time and, if necessary, across a broad landscape. This feature of the network has resulted in some big datasets. In science, we know that if you've got good data, people will use it. In the past few decades, the Forest Service has moved a subset of its experimental forests and ranges into a system of ecological observatories by cooperating with other networks such as LTER and NASA and by collaborating with scientists from other agencies, many of whom are supported by the National Science Foundation.

I believe the philosophy that established the network of experimental forests and ranges is just as relevant now as it was at the beginning of the twentieth century, but tackling national and global issues at the proper scale will require the network to function as an integrated research platform. That's the main difference between the past and now. In the past we had a lot of great research projects happening on the ground, but they addressed specific problems on specific landscapes. The people who worked on one experimental forest didn't talk much to the scientists working on another experimental forest, especially if it was in another part of the country. Today, a single research project might involve every single experimental forest and range in the entire network.

That's because the network contains many gradients, like differences in climate, altitude, latitude, and biodiversity. It's through long-term comparative research across those gradients that scientists will unravel the consequences of climate change and study other global change. Developing an integrated program of long-term research that covers the entire continent, and Hawaii and Puerto Rico too, while continuing to do research that assists local communities is the next great challenge for Forest Service scientists entrusted with this network.

The complexity of the environmental challenges facing humanity in the new millennium requires a research focus that addresses environmental complexity of time and space on a transcontinental scale. Because of the foresight of Congress a century ago, scientists can build on past research results and the large datasets associated with EFRs to do their part in meeting those challenges.

Jim L. Reaves, Ph.D.
Deputy Chief, Research & Development
USDA Forest Service

Contents

List of Contributors

Mary Beth Adams Northern Research Station, USDA Forest Service, Parsons, USA

David H. Alban Northern Research Station, USDA Forest Service, Grand Rapids, MN, USA

Devendra Amatya Southern Research Station, USDA Forest Service, Cordesville, SC, USA

Peter E. Avers Avers Soil Surveys Inc., Snellville, GA, USA

Allan R. Bacon Nicholas School of the Environment, Duke University, Durham, NC, USA

Scott W. Bailey Northern Research Station, USDA Forest Service, North Woodstock, NH, USA

Sharon A. Billings University of Kansas, Lawrence, KS, USA

Dan Binkley Colorado State University, Fort Collins, CO, USA

William D. Boyer Southern Research Station, USDA Forest Service, Auburn, AL, USA

Don C. Bragg Southern Research Station, USDA Forest Service, Monticello, AR, USA

John C. Brissette Northern Research Station, USDA Forest Service, Durham, NH, USA

Dale G. Brockway Southern Research Station, USDA Forest Service, Auburn, AL, USA

Kenneth N. Brooks Department of Forest Resources, University of Minnesota, St. Paul, MN, USA

Marilyn Buford USDA Forest Service National Headquarters, Washington, DC, USA

Mac A. Callaham Southern Research Station, USDA Forest Service, Athens, GA, USA

John Campbell Northern Research Station, USDA Forest Service, Durham, NH, USA

Lee K. Cerveny Pacific Northwest Research Station, USDA Forest Service, Seattle, WA, USA

Susan Charnley Pacific Northwest Research Station, USDA Forest Service, Portland, OR, USA

Kenneth L. Clark Silas Little Experimental Forest, Northern Research Station, USDA Forest Service, New Lisbon, NJ, USA

Richard G. Cline Vienna, VA, USA

Kristina F. Connor Southern Research Station, USDA Forest Service, Auburn, AL, USA

Ralph H. Crawford Northeastern Area, USDA Forest Service, Newtown Square, Pennsylvania, USA,

Amy E. Curry Millersville University of Pennsylvania, Millersville, PA, USA

Robert Denner Rocky Mountain Research Station, USDA Forest Service, Moscow, ID, USA

Bov Eav Pacific Northwest Research Station (retired), USDA Forest Service, Portland, OR, USA

P. J. Edwards Northern Research Station, USDA Forest Service, Parsons, WV, USA

John D. Elioff Northern Research Station, USDA Forest Service, Grand Rapids, MN, USA

W. J. Elliot Rocky Mountain Research Station, USDA Forest Service, Moscow, ID, USA

Gus Erdmann Northern Research Station, USDA Forest Service, Grand Rapids, MN, USA

Peter F. Ffolliott School of Natural Resources and the Environment, University of Arizona, Tucson, AZ, USA

Gary O. Fiddler Pacific Southwest Research Station, USDA Forest Service, Redding, CA, USA

Ryan L. Fimmen Geosyntec Consultants, Columbus, OH, USA

R. B. Foltz Rocky Mountain Research Station, USDA Forest Service, Moscow, ID, USA

G. Sam Foster Rocky Mountain Research Station, USDA Forest Service, Fort Collins, CO, USA

Michael Gallagher Silas Little Experimental Forest, Northern Research Station, USDA Forest Service, New Lisbon, NJ, USA

Brian W. Geils Rocky Mountain Research Station, USDA Forest Service, Flagstaff, AZ, USA

Gerald J. Gottfried Rocky Mountain Research Station, USDA Forest Service, Phoenix, AZ, USA

Kathy L. Graham Rocky Mountain Research Station, USDA Forest Service, Moscow, ID, USA

Russell T. Graham Rocky Mountain Research Station, USDA Forest Service, Moscow, ID, USA

A. Stuart Grandy University of New Hampshire, Durham, NH, USA

J. Morgan Grove Northern Research Station, USDA Forest Service, Baltimore, MD, USA

Colin Hardy Rocky Mountain Research Station, USDA Forest Service, Missoula, MT, USA

Deborah C. Hayes USDA Forest Service, Washington, District of Columbia, USA

Anne P. Hoover USDA Forest Service, Washington, District of Columbia, USA

Tamara Heartsill Scalley International Institute of Tropical Forestry, USDA Forest Service, San Juan, PR, USA

Paul R. Heine Nicholas School of the Environment, Duke University, Durham, NC, USA

Michael Hofmockel Iowa State University, Ames, IO, USA

Nicholas School of the Environment, Duke University, Durham, NC, USA

Jason A. Jackson Southern Research Station, USDA Forest Service, Durham, NC, USA

Theresa B. Jain Rocky Mountain Research Station, USDA Forest Service, Moscow, ID, USA

Sherri Johnson Pacific Northwest Research Station, USDA Forest Service, Corvallis, OR, USA

Richard W. Judd Department of History, University of Maine, Orono, ME, USA

Laura Kenefic Northern Research Station, USDA Forest Service, Bradley, ME, USA

Laura S. Kenefic Northern Research Station, USDA Forest Service, Bradley, ME, USA

Christel Kern Northern Research Station, USDA Forest Service, Grand Rapids, MN, USA

Randall K. Kolka Northern Research Station, USDA Forest Service, Grand Rapids, MN, USA

Randy Kolka Northern Research Station, USDA Forest Service, Grand Rapids, MN, USA

Elizabeth LeMaster Francis Marion and Sumter National Forests, USDA Forest Service, Whitmire, SC, USA

Jianwei Li University of Kansas, Lawrence, KS, USA

Gene E. Likens Cary Institute of Ecosystem Studies, Millbrook, NY, USA

Department of Ecology and Environmental Biology, University of Connecticut, Storrs, CT, USA

Nelson S. Loftus Washington Office, USDA Forest Service, State College, PA, USA

Ariel E. Lugo International Institute of Tropical Forestry, USDA Forest Service, San Juan, PR, USA

Daniel A. Marion Southern Research Station, USDA Forest Service, Hot Springs, AR, USA

Daniel Markewitz University of Georgia, Athens, GA, USA

Gary W. Miller Northern Research Station, USDA Forest Service, Morgantown, WV, USA

Megan L. Mobley University of Wyoming, Laramie, WY, USA

Nicholas School of the Environment, Duke University, Durham, NC, USA

Mary W. Morrison Francis Marion and Sumter National Forests, USDA Forest Service, Columbia, SC, USA

J. E. Mulrooney Southern Research Station, USDA Forest Service, Starkville, MS, USA

Robert Musselman Rocky Mountain Research Station, USDA Forest Servcie, Fort Collins, USA

Daniel G. Neary Rocky Mountain Research Station, USDA Forest Service, Flagstaff, AZ, USA

Thomas H. Nicholls Northern Research Station, USDA Forest Service, St. Paul, MN, USA

Nature Education Center, Fifield, WI, USA

Susan D. Olberding Rocky Mountain Research Station, USDA Forest Service, Flagstaff, AZ, USA

Deborah Page-Dumroese Rocky Mountain Research Station, USDA Forest Service, Moscow, ID, USA

Brian Palik Northern Research Station, USDA Forest Service, Grand Rapids, MN, USA

C. J. Peterson Southern Research Station, USDA Forest Service, Starkville, MS, USA

Felix Ponder Northern Research Station, USDA Forest Service, Columbia, MO, USA

Robert F. Powers Pacific Southwest Research Station, USDA Forest Service, Redding, CA, USA

Michael Rains Northern Research Station, USDA Forest Service, Newton Square, PA, USA

Carol B. Raish Rocky Mountain Research Station, USDA Forest Service, Albuquerque, NM, USA

Jim Reaves Deputy Chief for Research, formerly Southern Research Station, USDA Forest Service, Washington, DC, USA

Charles Rhoades Rocky Mountain Research Station, USDA Forest Service, Fort Collins, CO, USA

Daniel deB. Richter Nicholas School of the Environment, Duke University, Durham, NC, USA

Martin W. Ritchie Pacific Southwest Research Station, USDA Forest Service, Redding, CA, USA

Douglas F. Ryan Pacific Northwest Research Station, USDA Forest Service, Olympia, WA, USA

Maria Schleidt Ouachita National Forest, USDA Forest Service, Hot Springs, AR, USA

Thomas M. Schuler Northern Research Station, USDA Forest Service, Parsons, WV, USA

Aaron Shapiro Department of History, University of North Carolina at Charlotte, Charlotte, NC, USA

Michael G. Shelton Southern Research Station, USDA Forest Service, Monticello, AR, USA

T. G. Shelton Southern Research Station, USDA Forest Service, Starkville, MS, USA

Carl N. Skinner Pacific Southwest Research Station, USDA Forest Service, Redding, CA, USA

Nicholas Skowronski Northern Research Station, USDA Forest Service, Morgantown, WV, USA

Peter Stine John Muir Institute of the Environment, University of California at Davis, Davis, CA, USA

Pacific Southwest Research Station, Research and Development, USDA Forest Service, Davis, CA, USA

Deanna J. Stouder Formerly Pacific Southwest Station, USDA Forest Service, Albany, CA, USA

Susan L. Stout Northern Research Station, USDA Forest Service, Irvine, Pennsylvania, USA

Michael S. Strickland Virginia Polytechnic and State University, Blacksburg, VA, USA

Terry Strong Northern Research Station, USDA Forest Service, Grand Rapids, MN, USA

Wayne T. Swank Southern Research Station (Emeritus), USDA Forest Service, Otto, USA

Frederick J. Swanson Pacific Northwest Research Station, USDA Forest Service, Blue River, OR, USA

Allan E. Tiarks Southern Research Station, USDA Forest Service, Council Bluffs, IA, USA

Donald Turton Natural Resource Ecology and Management, Oklahoma State University, Stillwater, OK, USA

Marty Vavra Pacific Northwest Research Station, USDA Forest Service, La Grande, OR, USA

James M. Vose Department of Forestry and Environmental Resources, College of Natural Resources, North Carolina State University, Raleigh, NC, USA

T. L. Wagner Southern Research Station, USDA Forest Service, Starkville, MS, USA

Thomas Waldrop Southern Research Station, USDA Forest Service Clemson University, Clemson, SC, USA

Carol G. Wells Southern Research Station, USDA Forest Service, NC, USA

Michael J. Wisdom Pacific Northwest Research Station, USDA Forest Service, La Grande, OR, USA

Part I
Introduction

Chapter 1
A Grand Experiment: USDA Forest Service Experimental Forests and Ranges

Aaron Shapiro

> *The greatest contribution of Forest Research is the spirit it has brought into the handling of national resources. Under the pressure of executive work, the technical ideas of the forester at times grow dim. It is Forest Research which has kept the sacred flame burning and has helped to raise Forestry to the level of the leading scientific professions.*
>
> Gifford Pinchot (1947)

Abstract The history of the US Department of Agriculture, Forest Service's network of experimental forests and ranges is traced starting in the late 1800s. This overview gives examples of forest research in the US predating establishment of the USDA Forest Service and features the leaders instrumental in subsequently planting the seeds for and nurturing experimental forests and ranges across the country in the twentieth century. Topics of research conducted at these locations reflect a variety of influences, such as controversy over livestock grazing in the West. This research has contributed extensively to greater understanding in areas from hydrology to forest products. It continues to inform and inspire the work of today's scientists and land managers, who increasingly must take into account the disparate needs and wants of multiple stakeholders.

Keywords Carlos Bates · Experimental forests and ranges · Forest experiment station · Fire research · Grazing · James Jardine · Gifford Pinchot · Arthur Sampson · USDA Forest Service history · Raphael Zon

1.1 Background

Reflecting on the importance of US Department of Agriculture (USDA) Forest Service research, Gifford Pinchot could undoubtedly recall the day in May 1908 when Raphael Zon provided him the initial plan for forest experiment stations. That summer, the first such station was established at Fort Valley near Flagstaff, Arizona, and it was officially organized in early 1909. According to Zon, the stations were to

A. Shapiro (✉)
Department of History, University of North Carolina at Charlotte, Charlotte, NC 28223, USA
e-mail: ashapi10@uncc.edu

D. C. Hayes et al. (eds.), *USDA Forest Service Experimental Forests and Ranges*,
DOI 10.1007/978-1-4614-1818-4_1,

conduct "experiments and studies leading to a full and exact knowledge of American silviculture, to the most economic utilization of the products of the forest, and to a fuller appreciation of the indirect benefits of the forest" (Zon 1908). Pinchot, the first chief of the USDA Forest Service, wrote of the plan, "I am for this with some changes" (Young 2010, p. 7). Building upon this foundation, Forest Service scientists have played a critical role in expanding human understanding of forests and improving the planet's welfare through sound science. Much of this work over the last century has taken place on the Forest Service's network of experimental forests and ranges, living laboratories where research engages with and responds to larger social, economic, and ecological questions.

1.2 Federal Forest Research Before the Forest Service

In many ways, Forest Service research predates the formation of a separate research branch in 1915, the first experiment station in 1908, and even the agency's establishment in 1905. In 1873, Franklin Hough, who would serve as the first chief of the Division of Forestry in the Department of Agriculture, presented *On the Duty of Governments in the Preservation of Forests* to the American Association for the Advancement of Science. After supervising the New York State census in 1855 and 1865, Hough compared the results and noticed a waning supply of timber. Fearing this decline could prove devastating to the nation over the long haul, he took action. His work ultimately helped establish the Division of Forestry. In 1876, the US government allocated US$ 2,000 for "some man of approved attainments," to be selected by the Commissioner of Agriculture, "to prosecute investigations and inquiries, with the view of ascertaining the annual amount of consumption, importation, and exportation of timber and other forest-products, the probable supply for future wants, the means best adapted to their preservation and renewal, the influence of forests upon climate, and the measures that have been successfully applied in foreign countries, or that may be deemed applicable in this country, for the preservation and restoration or planting of forests" (Fernow 1899; LaBau et al. 2007, p. 5). That person ended up being Hough, one of the few people qualified to undertake the work.

Hough embarked on the task with a sense of urgency and desire to gather information. He issued his multivolume *Report upon Forestry* (Hough 1884) over the ensuing years and in 1882 presented an idea for "Experimental Stations for Forest Culture." The 1883 annual report for the Division of Forestry mentioned the idea of forest experiment stations: "It would greatly aid in the dissemination of information in regard to trees and do much to encourage tree planting if there were established in different parts of the country forestry experiment stations, or test plantations" (USDA 1883, pp. 458–459). As a prodigious author and researcher and as head of the forestry division during its first 7 years, Hough provided the seedlings for what would later bloom into Forest Service research.

During these years, Hough was among the more preeminent forestry figures, but he surely was not alone in his interest in forest research. In 1884, Charles Sprague Sargent of Harvard University's Arnold Arboretum wrote *Report on the Forests of North America* for the US Department of Commerce, Bureau of the Census. This report included scientific information and warned of the need to reform destructive timber management policies. Sargent also began publishing *Garden and Forest* in 1887 and the publication stood at the vanguard of American forestry for the next decade. Historian Char Miller has labeled *Garden and Forest* the "most important late-nineteenth-century forum for discussing the role of science in human affairs" (Miller N.d.). During its decade-long run, more than 450 articles on forestry helped educate people about the growing field and broad significance of forest science, preparing the way for modern forest research in the twentieth century. Publications such as Gifford Pinchot and Henry Graves's *The White Pine* (Pinchot and Graves 1896) helped the cause as well.

At the same time, the nascent Division of Forestry under Bernhard Fernow began research on wood utilization and other investigations culminating in the 1897 Organic Act, which emphasized three major goals: (1) improve and protect forests within boundaries, (2) secure favorable water flow conditions, and (3) provide a continuous supply of timber for Americans. Fernow recognized the importance of research in achieving these tasks. Before stepping aside as chief of the division in 1898, Fernow (1899) wrote *Report upon the Forestry Investigations of the U.S. Department of Agriculture, 1877–1898*. Upon Pinchot's arrival as forester—Pinchot had requested the name of the position be changed from chief to forester—in 1898, he moved to establish a Special Investigations Section to undertake basic research that would aid forest management. By 1902, the section had become an agency division with 55 employees. Pinchot wrote of this unit in *Breaking New Ground,* reflecting on the years before management of the forest reserves was under the purview of the Forest Service, "Even before the transfer there was a little hall room in the Atlantic Building where technical forest facts and statistics were assembled. We called it compilation. Here was the first cradle and treasure house of forest research in America" (Pinchot 1947, p. 308). Out of that room at what was then Bureau of Forestry headquarters at 930 F Street NW in Washington, DC, would grow the Forest Service's research organization.

Things were also changing on the ground with new experiments emerging in the field. In 1903, the USDA's Bureau of Plant Industry established the Santa Rita Range Reserve in Arizona to protect native rangeland from grazing and to conduct research on livestock production. The Carnegie Institution established the Desert Laboratory in Tucson to conduct studies of native plants in arid regions. Santa Rita was expanded in 1907 and the Bureau of Plant Industry conducted additional studies on grazing capacity there before it was turned over to the Forest Service in 1915 and renamed the Santa Rita Experimental Range. Although the University of Arizona's College of Agriculture has administered the range in cooperation with the Forest Service and Arizona State Land Department since 1987, it remains the oldest experimental range in the nation and an important site for studying rangeland health and productivity (Gillespie 2003).

1.3 The Creation of Experimental Forests and the Forest Service Research Branch

In 1903, Pinchot established the Section of Silvics to coordinate data collection and tasked the Office of Forest Products with research in wood preservation, wood chemistry, wood utilization, forest measurement, and engineering. Raphael Zon, who had been with the agency since 1901, became head of the renamed Office of Silvics in the Branch of Silviculture in 1907. The following year, he traveled with Gus Pearson and Willard Drake to an area outside Flagstaff, where he proclaimed, "Here we shall plant the tree of research" (Pearson 1936). Fort Valley Experiment Station was established at the site with Pearson in charge of research focusing on ponderosa pine (*Pinus ponderosa* Dougl. ex Laws) ecology and management. Zon believed that restoration based on sound scientific research was the only way to heal the land after decades of overharvesting and overgrazing. Not only had Zon laid the foundation for experiment stations with the plan he had submitted to Pinchot but he had also chosen the first site.

While Zon was planting the "tree of research," President Theodore Roosevelt called the first Conference of Governors to address the nation's growing conservation problems. Since the beginning of his administration, the President had concerned himself with conservation policy and often discussed these issues with his good friend Pinchot. In 1903, Roosevelt established the Public Lands Commission to examine and report on "the condition, operation and effect of the present land laws, and on the use, condition, disposal and settlement of the public lands" (Roosevelt 1903). The Commission published its final report in 1905 with recommendations aimed at preventing abuse of existing land laws and assisting in the efficient development of natural resources. It also issued controversial recommendations regarding the classification and leasing of grazing lands. While opposition from western states contributed to a lack of implementation, grazing concerns called out for scientific research to better understand problems and offer solutions (US Public Lands Commission 1905).

The original proposal for a National Governors' Conference was made in May 1907 and was limited in its focus to the improvement of waterways and development of water resources. But W J McGee, whom historian Samuel Hays labeled the "chief theorist of the conservation movement" (Hays 1959, p. 102) and who served as one of conservation's greatest promoters, helped broaden the focus from solely water issues to all natural resource conservation concerns. Roosevelt and Pinchot supported this effort. Opening the conference in May 1908, Roosevelt informed those assembled that conservation was the "weightiest problem now before the Nation," and claimed that the nation's natural resources "are in danger of exhaustion if we permit the old wasteful methods of exploiting them longer to continue" (US Government Printing Office 1909, p. 3).

While the assembled governors delivered speeches, Pinchot and McGee wrote most of the conference presentations exploring water, forests, soil, minerals, wildlife, reclamation, grazing, land laws, navigation, and energy. The conference urged the adoption of practices to prevent the exhaustion of natural resources and provide for their more efficient use. Both Pinchot and McGee believed the conference, which was financed largely from Pinchot's personal income, should produce specific conservation recommendations that would garner public attention. Ultimately, the creation of a national conservation commission to inventory the nation's natural resources provided that opportunity and emerged, according to one historian, as "the singular and triumphant work of the Governors' Conference" (Hays 1959, pp. 139–140). The conference highlighted the Progressive Era's emphasis on expertise as essential for developing effective policy, which in the context of the period ranged from natural resources to urban reform. This need for expertise to develop scientific solutions to conservation problems helped the tree of research initially spread its roots throughout the national forests.

Within 5 years of the establishment of Fort Valley and the Governors' Conference, the Forest Service had field research stations or experimental forests in Idaho (Priest River), Washington (Wind River), Colorado (Fremont), Utah (Great Basin), and New Mexico (Jornada Experimental Range), among other locations, and the Forest Products Laboratory (FPL) in Wisconsin. At these sites, Forest Service researchers conducted experiments on fire, hydrology, and grazing on forests and rangelands. Their work provided the foundation for modern silviculture and improved understanding of forest ecosystems in the western USA (USDA Forest Service Undated d). In these early years, the stations were administered by districts (current regional offices) and were not separated from national forest administration. In a history of Forest Service Region 4, historian Thomas Alexander writes, "Forest Service experiment stations served as the backbone of the research effort. The goal of these stations was to establish a scientific basis for management policy" (Alexander 1987a, p. 90). To accomplish this goal, pioneering scientists undertook extensive research on these new experimental forests and ranges.

Lacking knowledge about the forest–water relationship and responding to growing concerns about the impact of erosion and unregulated logging on water flows, the Forest Service engaged in a new research effort. In cooperation with the US Weather Bureau at Wagon Wheel Gap on the Fremont Experimental Forest, Carlos Bates conducted the world's first scientific watershed experiments on the influence of deforestation on water yields. Bates's work proved influential in building support for the Weeks Act of 1911, which both established regulation of navigable streams as a way to manage stream flow and allowed for the purchase of watershed lands to preserve river navigability in the eastern USA, thus proving essential in expanding the national forest system. Bates' paired watershed studies laid a foundation for similar studies on other experimental forests, helping influence policy and transforming management of the nation's forests (Forest History Society N.d.; Hornbeck 2001; Ice and Stednick 2004).

While the water–forest relationship continued to be a key research area, scientists also responded to concerns raised by the initial Public Lands Commission report and began studying the relationship between grazing and forests. By the end of the nineteenth century, heavy grazing had substantially altered and deteriorated the western range. Drought hindered plant recovery and led to livestock loss. Native perennial grasses were destroyed. Competition for summer feed resulted in substantial forage depletion. A dire situation begged for answers. Pinchot viewed grazing as among the most significant issues facing the forest reserves. The new discipline of range science, which developed from research conducted on experimental forests and ranges, helped provide guidance and direction for managing range ecosystems.

Early range research began in the USDA Division of Botany under Frederick Coville. This botanist authored *Forest Growth and Sheep Grazing in the Cascade Mountains of Oregon* (1898), which determined that sheep would not harm the range if they were properly controlled. Coville elaborated what would become a basic principle of range science—that grazing livestock on rangelands can be destructive if not managed. He endorsed regulated grazing on the forest reserves, outlining a permit program and controlling the number of stock and season. His report received a warm reception from Fernow, Pinchot's predecessor as chief of the Bureau of Forestry: "I indorse [*sic*] fully Mr. Coville's conclusions that sheep grazing without proper restrictions and regulations, which have in view to prevent overstocking, is detrimental to the reproduction of forest growth and to soil conditions and water flow—in some localities more so than in others; hence, wherever forest growth is to be maintained and the washing of soils with consequent flood dangers is to be avoided, the greatest care and judgment must be exercised as to the manner in which sheep grazing may be carried out without detriment" (Coville 1898, p. 3).

In 1900, Albert Potter of the Arizona Woolgrowers Association invited Pinchot and Coville on a 3-week trip to investigate devastated rangelands (Miller 1999). Pinchot recalled in *Breaking New Ground* that the information from the trip regarding overgrazing proved instrumental in establishing Forest Service grazing policy as well as in the creation of national forests. In 1901, Pinchot, who expressed concerns with how range could best be protected and maintained, hired Potter to head the new Branch of Grazing in the Bureau of Forestry and establish a public lands grazing program (Pinchot 1947). Over the course of that first decade of work, in which the Bureau of Forestry became the USDA Forest Service, the necessity of scientific range studies became clear to Potter. To this end, he worked with Coville to establish a Forest Service Office of Grazing Studies. Like his western stockman colleague and future grazing chief Will Barnes, Potter wanted to ensure rangeland and ranching sustainability (Prevedel and Johnson 2005).

Range science developed in response to concerns about declining western range productivity and erosion from overgrazing. Even though the transfer of the forest reserves to the Forest Service in 1905 made research subordinate to management, the new agency needed to conduct research to address range concerns. In these early years, grazing rather than timber consumed forest managers. Range research on national forests began when Forest Service employees Arthur Sampson and James Jardine investigated degenerated mountain summer forage supply and grazing capacity in the Wallowa National Forest (Washington) in 1907. In 1910, Robert V R Reynolds began studying grazing in relation to erosion and floods in the Manti National Forest (Utah). Studies on other forests, including the Coconino (Arizona), Shasta (California), Malheur (Oregon), and Payette (Idaho), followed. That same year, the Forest Service established an Office of Grazing Studies headed by Jardine, who would later become chief of the Office of Experiment Stations for the Department of Agriculture, to study the effects of grazing on national forests. Thus began formal recognition of range research within the Forest Service, although the Office of Grazing Studies was not transferred to the Research Branch until 1926. Utah officials and residents urged scientific research into summertime floods originating on mountain watersheds that were damaging farms and rural communities. Overgrazing of sheep in the mountain ranges proved a leading cause. Flooding during the late nineteenth and early twentieth centuries led to the establishment of the Great Basin Experimental Range in the Manti National Forest in 1912 to conduct studies on range management, range erosion, and other grazing issues. Beginning with Reynolds' initial studies and continuing under the direction of Sampson and later Clarence Forsling, the Great Basin became the primary location for range management research in the nation (Mitchell et al. 2005; USDA Forest Service 1944).

At the Great Basin, Sampson and Jardine conducted reseeding studies and reported on growth habits and requirements of important range plants to help determine western range grazing capacity. Early research also examined the effect of grazing on plant species and streams, natural revegetation of overgrazed lands, soil acidity related to artificial range reseeding, and methods to restore rangeland vegetation and increase habitat. Sampson was the first range ecologist hired by the Forest Service, served as the first director of the Great Basin Experimental Range, and published *Range and Pasture Management,* the first comprehensive textbook in the field, in 1923. Known as the father of range science, Sampson promoted deferred and rotational grazing strategies and developed the indicator species concept to help managers evaluate range condition. Sampson also emphasized four stages of plant succession, using more simplistic classifications of "excellent, good, fair, and poor" to evaluate evolving range conditions. At the Great Basin, he established enclosures to study how having an area with no grazing impacted plants and soils. Sampson's research concerned the production of maximum forage through artificial and natural reseeding, forage utilization by livestock without undue damage to plant reproduction and watershed conditions, and securing the greatest grazing efficiency per unit (Klade 2006; Young 2000; Young and Clements 2001).

Other important individuals who contributed to early Forest Service range research included Jardine, who established and served as the first head of the Forest Service Office of Grazing Studies and later took charge of the USDA Office of Experiment Stations; Forsling, the first Forest Service employee in charge of range and cattle management research at Jornada Experimental Range; W R Chapline, who began his career as a grazing assistant under Sampson at the Great Basin and later served as director of range research for the Forest Service from 1920 to 1952; and Linc Ellison, a Forest Service ecologist who studied grazing effects on montane rangeland plant communities. Through their research, Jardine, Sampson, and Chapline sought to provide rangers with basic range management principles, including the proper number, kind, and distribution of livestock. In addition, the first generation of range scientists demonstrated that changes in species composition of plant assemblages provided the most biologically sensitive index of range condition. Ultimately, by studying the effects of grazing on various plant species, range scientists at the Great Basin determined appropriate methods for restoring rangeland vegetation, improving habitat, and rehabilitating watersheds. Professional range management emerged in the Forest Service, according to historian Thomas Alexander, largely because of the Intermountain Station's grazing research staff and especially those scientists who worked at the Great Basin Experimental Range (Alexander 1987a; Forest History Society 2009d; Keck 1972; Klade 2006).

The Forest Service began to produce scientific knowledge about range resources at a time of dramatic changes in plant science, such as Sampson and others' increased interest in plant succession. In 1910, the Forest Service focused on promoting range research and collecting and using technical range management information. In 1915, the Santa Rita and Jornada Experimental Ranges were transferred from the Bureau of Plant Industry to the Forest Service. At both locations, researchers sought to provide a scientific basis for managing rangelands in order to restore, improve, and maintain them at a sustained basis of productivity and to obtain the greatest returns on livestock. By 1915, the Forest Service Office of Grazing Studies had become the leading governmental organization in range research (USDA Forest Service 1944).

While scientists at experimental ranges conducted grazing research, experimental forests provided opportunities to study silviculture and fire. In 1910, 26-year-old Thornton Munger established growth plots on old burn sites in the Pacific Northwest, tagging and recording the size of every tree in these plots. Munger's plots showed how forests and individual trees within them grow over time, offering evidence that trees grow at a predictable rate (Herring and Greene 2007). Munger's work laid a foundation for future research, but it was another event in the Pacific Northwest in 1910 that captured the nation's attention and contributed to additional research programs on the experimental forests.

Particularly hot and dry conditions early in the 1910 fire season led the Forest Service to mobilize men for the fire season. On August 20, 1910, heavy winds across Idaho and Montana turned moderate fires into endless flames stretching for kilometers. Many firefighters were caught unprepared. Some took shelter in mine shafts or plunged into streams to soak themselves. Firefighter Edward Pulaski saved his crew of 45 men by forcing them into a shaft until the fire receded. Pulaski's heroic story was counterbalanced by others involving the loss of life as the fires blazed. More than 1.2 million ha of forest burned. The agency, turning to its scientists to help explain the conditions that had led to such devastation and tragedy, established Priest River Experimental Forest in northern Idaho in 1911. Scientists at Priest River studied all aspects of forestry in the Northern Rocky Mountains, including tree planting to help reforest burned areas and estimating fire danger and fuel flammability to help explain conditions that had contributed to the Big Blowup of 1910 (Graham 2004; Pyne 2001; USDA Forest Service Undated c). In the aftermath of the 1910 fires, the Forest Service conducted research on experimental forests to address questions about managing wildfire in the nation's forests. Fire protection and suppression efforts emerging from research at places like Priest River increasingly justified the agency's existence in the public's mind (Egan 2009; Graham 2004).

Back in the nation's capital, Chief Henry Graves established a Central Investigative Committee in 1912 to set priorities and supervise the agency's research program. The committee had three divisions—silviculture, grazing, and products—and its members included Zon, Jardine, and Carlisle P "Cap" Winslow, who would become director of the agency's FPL in 1917. Zon was appointed chief of forest investigations in 1914. The following year, Service Order 45 established a separate research branch in the Forest Service with Earle Clapp as chief of research and placed the branch on equal footing with the administrative side of the agency. The new branch initially assumed responsibility for studies of state forest conditions, the lumber industry, fire protection, silvicultural and statistical investigations, the Forest Service library, and FPL. Over the years, responsibilities expanded greatly (Steen 2004).

Seeds for FPL had been planted in 1907 when McGarvey Cline, head of the Forest Service's wood-use section, proposed that all wood products scientists be based in one location. Pinchot recognized the need for such a centralized entity and, after competition between the states of Wisconsin and Michigan, FPL was established at the University of Wisconsin-Madison in 1910. FPL scientists engaged in a wide variety of projects on wood and wood products, including testing wood fiber strength, examining the mechanical and chemical processes in pulping, and research on wood's chemical properties. FPL also served as the home of the Forest Service's first female scientist, Eloise Gerry, whose distinguished career at FPL ran from 1910 to 1954 (McBeath 1978). Unlike other areas of Forest Service research, wartime brought prosperity to FPL, increasing the workforce from less than 100 to nearly 450. During World War I, FPL research proved crucial in producing lightweight but strong airplanes. FPL scientists tested fuselage, wing, and propeller strength, and developed effective ways to use wood, cloth, and paint to strengthen airplane frames. Paper was in short supply during the war, so FPL scientists began research on tree species not commonly used for paper production. Scientific ingenuity and service to the nation coincided at the Forest Service's FPL (Forest History Society 2009a; Nelson 1971; Williams 2005).

In a 1917 *Journal of Forestry* article, Clapp claimed scientific research was "the foundation of permanent forest development in the United States." Surveying the scene, Clapp called for cooperation between universities, states, and the federal government in forest research. He saw a "need for a special force of well-trained men who shall be permitted to devote their entire time and efforts to the work" and wanted to ensure future Forest Service scientists received appropriate training (Clapp 1917). After World War I, these efforts increased. In 1919, the Forest Service issued "Range Management on the National Forests" and in the subsequent decade, Forest Service range science would move, as historian Alexander has elaborated, "from rule of thumb to scientific range management" (Alexander 1987b). The 1919 publication highlighted the need to regulate grazing in accordance with the needs of timber and water protection, wildlife, recreation, and the condition of the range itself. The agency began applying research findings from the Great Basin to managing the nation's rangelands (Jardine and Anderson 1919).

Forest research expanded during the Progressive Era as the application of scientific expertise proved crucial to effective management. In examining fire, hydrology, silviculture, and grazing on experimental forests and ranges, scientists pursued a basic understanding of underlying processes and forces in these areas. As custodian of the national forests, the Forest Service drew on research findings from experimental forests and ranges to help manage these public lands. By emphasizing professional management rooted in expertise and a new research program in which findings could be applied widely, the agency also encouraged applying this research to address concerns about practices in private forest and rangelands.

1.4 The Roaring 1920s

While flappers danced their way on to the national stage, Americans experimented with Prohibition and Hollywood studios moved from silent films to talkies, the responsibilities and facilities of the Forest Service Research Branch grew. By the end of the 1920s, the research program would receive formal Congressional recognition with the McSweeney–McNary Act. The 1920 "Capper Report," of which Research Chief Clapp was the primary author, reported on the extent of forest depletion. The following year, both the Appalachian and Southern Forest Experiment Stations were established and Clapp wrote *Forest Experiment Stations,* proposing ten additional regional experiment stations. Range research continued at the Great Basin while studies on the Tonto National Forest (Arizona) examined grazing in relation to watershed management. Clapp also claimed fire researchers would "become the leaders of the most important forest research activities in the country" (Hardy 1983, p. 2).

In the early 1920s, Harry Gisborne began his distinguished fire research career at Priest River Experimental Forest. Research on fire dated back to Pinchot's first year as chief when his staff examined 5,000 fires that had occurred since 1754. But Gisborne took fire research to a new level. He used a variety of tools, such as the Asman aspiration psychrometer, visibility meter, anemohygrograph, double tripod heliograph, and blinkometer, to gather information about forest fires. Much of Gisborne's groundbreaking research dealt with fire control and conditions causing severe wildland fires. Gisborne integrated weather, fuel type and conditions, ignition sources, and other variables to develop the first fire danger rating system. The agency used it to manage and assess conditions leading to wildland fires. Gisborne later became chief of the division of forest fire research for the Rocky Mountain Research Station in Missoula, Montana, and died doing what he loved while investigating the Mann Gulch fire in 1949 (Forest History Society 2009c; Graham 2004).

The expansion of forest research continued during the 1920s. Zon moved to St. Paul, Minnesota, to become the first director of the Lake States Forest Experiment Station in 1923. That year, he and colleague William Sparhawk published *Forest Resources of the World,* highlighting Forest Service scientists' research in the international realm. Thornton Munger assumed the directorship of the new Pacific

Northwest Forest Experiment Station in Portland, Oregon, in 1924. Munger had spent several years in the district office after his pioneering work at Wind River Experimental Forest, where he established a tree nursery and arboretum and conducted studies on tree heredity in the 1910s. Bent Creek Experimental Forest on North Carolina's Pisgah National Forest became the first experimental forest east of the Mississippi River in 1925. The decade also brought a new land use designation—Research Natural Area (RNA)—that was to be managed to maintain natural features and processes. The RNA system began with the Santa Catalina RNA on the Coronado National Forest in Arizona and Wind River RNA, the first of several hundred to be established in the ensuing years, including the Reynolds RNA on the Crossett Experimental Forest in Arkansas. In 1924, the Society of American Foresters established a special committee on forest research with Clapp as chairman and 2 years later, Clapp issued *A National Program of Forest Research,* which served as the basis of future government forest research. The report laid out the need for adequate scientific research on fire, timber growth, and wood utilization. Its release also coincided with the transfer of the Office of Grazing Studies to the Research Branch, allowing for improved coordination of range research with other ongoing agency research (Clapp and Society of American Foresters 1926).

In 1928, the McSweeney–McNary Act implemented Clapp's research program, which included a nationwide forest survey, legitimization of a formal network of experiment stations, and additional funds for research in forestry and range management. For two decades, the agency had carried out research without specific statutory authority, but the Act's passage authorized the research conducted at experiment stations and experimental forests. It contributed to improved coordination of the broader Forest Service research program of range, forest and watershed management, forest economics, and forest products. Each of these areas had a division within the Research Branch. Reflecting on the 10-year anniversary of McSweeney–McNary, Clapp called it the "second landmark in the recognition and development of forest research in the Forest Service" (Clapp 1938, p. 832). The first, as he saw it, was the initial creation of a separate Research Branch in 1915 (Clapp 1938).

1.5 Working Through the Depression

Despite an era of austerity amid devastating economic and social conditions during the 1930s, existing research facilities expanded and application of research findings at new experimental forests and ranges helped manage the nation's forests. McSweeney–McNary included provisions for continuing research at FPL and funds for a new building that was completed in 1931. The following year, FPL gained further public notoriety from its analysis of the wooden ladder used in the Lindbergh baby kidnapping. With the inauguration of President Franklin Roosevelt in 1933, New Deal work relief programs such as the Civilian Conservation Corps and Works Progress Administration provided labor and materials to construct research facilities. Cooperation with state agricultural experiment stations and other bureaus in the Department of Agriculture also increased. In 1933, Forest Service Research officials contributed to the "National Plan for American Forestry," also known as the Copeland Report, which advocated for federal–state control over forestry to protect the future national timber supply. Despite congressional rejection of the Copeland recommendations and political fallout over the control of private forestlands, the report did keep forest conservation in the public eye and highlighted the need for research (USDA Forest Service 1933).

By 1935, the Forest Service had 48 experimental forests and ranges, where scientists conducted research on forest, range, watershed, and fire. Among them were the Desert Experimental Range in Utah and the San Joaquin Experimental Range in California, as well as the Fernow Experimental Forest in West Virginia and the Cutfoot Sioux Experimental Forest in Minnesota, where Zon's ashes would be scattered upon his passing in 1956. The Desert Experimental Range was established in 1933 to demonstrate how "salt desert shrub zone could be managed to enhance sheep production and how different grazing strategies affected the vegetation" (Mitchell et al. 2005, p. 23). At Pringle Falls Experimental Forest in central Oregon, scientists began to examine the impact of grazing on timber growth. In the western slopes of the Sierra Nevada, the San Joaquin Experimental Range was established in 1934 "to ascertain the possibilities of sustainable livestock (cattle) husbandry in a transitional oak shrub community" (Mitchell et al. 2005, p. 22). Research at San Joaquin contributed to developing sustainable grazing systems in California (Mitchell et al. 2005). The establishment of these new experimental forests and ranges coincided with passage of the Taylor Grazing Act of 1934, which sought to address the problem of unrestricted grazing and reduced forage value on public lands. To that end, the act placed 32 million ha of public lands into grazing districts, ending previously free and unregulated grazing use of vast public areas and introducing broader federal protection and management. Two years later, "The Western Range" provided an assessment of the nation's rangelands based in part on Forest Service research. The study blamed the Department of the Interior for ineffective management and suggested the Taylor Grazing Act gave the livestock industry too much power. During an era of intense competition between the Interior and Agriculture departments, Interior Secretary Harold Ickes and the livestock industry challenged Forest Service research findings. Stockmen produced a study showing lack of rainfall, not overgrazing, caused range depletion (Steen 2004, p. 207; USDA Forest Service 1936).

Hydrologic and fire research continued at existing and new experimental forests. Paired watershed studies that built upon Carlos Bates' early work on the Fremont Experimental Forest in Colorado examined the effect of silvicultural treatments on water yield. In light of the differences in water characteristics by habitat, research at new sites could investigate specific questions and conditions. At Coweeta Hydrologic Laboratory, an experimental forest in western North Carolina, researchers began studying the relationship between forests and water. The Civilian Conservation Crew built roads and a network of rain gauges, weirs, and wells to support research activities. At Coweeta and other experimental forests, research addressed the protection of water quality, water quantity, and aquatic ecosystems as part of forest management (Forest History Society 2009b; Ice and Stednick 2004). While research at Priest River continued to contribute to knowledge about wildfire suppression, experimental forests in the South conducted studies beginning in the 1930s, including some that led to using prescribed burning as a forest management tool to maintain healthy southern forests. Started in 1937, studies on loblolly pine stands at coastal South Carolina's Santee Experimental Forest demonstrated fire was an important part of the ecosystem and helped maintain forest health. These studies, along with others at the Crossett Experimental Forest (Guldin 2009), also contributed to southern pine restoration (Adams et al. 2008). Furthermore, such studies provided benefits to the regional forest products economy. At Florida's Olustee Experimental Forest, for example, research helped transform inefficient gum naval stores extraction that destroyed trees into a more efficient and sustainable process using new methods that maintained trees (Forest History Society 2009e).

Wood products research continued at FPL, which produced the first laminated beam in 1935 and the first prefabricated house in 1937. Edward Munns, who previously served as chief of Forest Experiment Stations and chief of the Office (later Division) of Silvics and coauthored the watershed section of the Copeland Report, became chief of the Division of Forest Influences in 1937. Munns' office continued to direct research relating to watersheds, floods, and erosion. The McSweeney–McNary Act authorized a tropical forest experiment station and, in 1939, the International Institute of Tropical Forestry was established in Puerto Rico, expanding Forest Service research beyond the contiguous USA. The Luquillo Experimental Forest in Puerto Rico arose from these earlier efforts in the postwar years (Steen 1998).

Despite these gains and the work conducted over the previous decades, Earle Clapp sensed the tenuous nature of research within the agency. In a memo to Chief Silcox in 1936, he wrote: "The Forest Service in general has been indifferent to or has actively opposed practically every constructive move to develop research. For many years there have been periodic efforts to break it down… The recent decision to maintain the independent status of the forest experiment stations and the placing of Research in Washington on a par with national forest administration and State and private forestry are decidedly reassuring as far as they go" (Clapp 1936). Looking to insure a permanent place for research, he wrote Silcox to explain the branch's vital role within the agency. Research's first responsibility was to "supply the basic biological, social, economic, and other technical information which is necessary for the rapid and well-rounded-out progress of the whole forestry movement in the United States" (Clapp 1936). Second, sound research would make the Forest Service "a technical organization in spirit and in fact" And third, Clapp expressed the importance of research independence, writing of the need "to have at all times in the Forest Service a group not under administrative domination, idealistic from the very nature of its work, ready when occasion demands to supply the criticism which the Forest Service needs to keep it alive and forward-looking, and also to perform the same function for American forestry as a whole" (Clapp 1936). For Clapp, who guided the Research Branch for its first two decades and oversaw the initial growth and development of experimental forests and ranges, there was more work to be done to ensure the branch's permanent place within the agency.

1.6 From Wartime Research to Beyond Containment

While Americans fought in the European and Pacific theaters, women and men on the home front mobilized to support the war effort. Typical depictions highlight Rosie the Riveter working in an industrial plant, but Forest Service researchers also contributed to the cause. During World War II, research efforts included projects involving guayale (*Parthenium argentatum* Gray) and kok-saghyz (*Taraxacum kok-saghz*), commonly called the Russian dandelion, to address rubber shortages. Wartime also brought additional employees to FPL, where wood products research focused on wartime uses for airplanes, ships, containers, paper, and plywood. Forest Service scientists continued to conduct groundbreaking research on a variety of topics. Publications such as those by Gus Pearson on ponderosa pine at Fort Valley and Leo Isaac on Douglas fir (*Pseudotsuga menziesii*) at Wind River set standards for nomenclature, dendrology, and silvicultural methods. Range researchers contributed to the war effort by examining methods to obtain the greatest production of meat, hides, wool, and other products from livestock and game animals without damaging either future production or the range. After the war, research emphasized commodity production, studying the amount and quality of forage available for livestock. Forest Service range research provided sound science on how to best manage and educate people on proper management techniques and evolved into cooperative efforts with other public and private entities (Herring and Greene 2001; Klade 2006; Nelson 1971; Steen 1998).

As Winston Churchill warned of an iron curtain descending across Europe in 1946, diplomat George Kennan formulated the containment policy that would guide America during the Cold War. Historians have applied the containment idea to the domestic context in examining 1950s family and gender roles (May 2008), but postwar American consumption was anything but contained. Large numbers of returning GIs married and moved with their families to the suburbs. Responding to the growing demand for wood to support new housing construction in growing suburbs, researchers on experimental forests addressed timber production. Wood and wood products contributed to America's growing consumer economy, helping citizens live more comfortably with the products advertisers told them they needed and wanted (Cohen 2003). In 1959, Vice President Richard Nixon defended capitalism against communism by focusing on American appliances and ingenuity in the Kitchen Debate conversation with Soviet Premier Nikita Khrushchev in Moscow. Perhaps diplomatic in nature, it also highlighted an era of increased consumption and material wealth in which wood and other natural resources from forests played a key role (Cohen 2003).

The prosperous postwar economy captured in Nixon's remarks, along with funding increases in response to the Cold War, raised the profile of Forest Service research efforts. The baby boom, new housing and highways, and large-scale suburbanization contributed to more people moving into forested areas, which served as an impetus for additional wildfire research on experimental forests. As timber harvest from national forests increased in the 1950s to meet growing consumer demand, Priest River studies provided information on fire behavior, fuel flammability, and ecological effects. At the heart of postwar suburbanization in southern California, the San Dimas Experimental Forest was initially established at Los Angeles' request in 1933 to study hydrologic issues. Research in the postwar years also examined the ecological effects of wildfires as Americans faced greater danger from urban fires. This work contributed to the development of the Incident Command System used in firefighting and other emergency situations today (USDA Forest Service 2008; Wells 2009).

In 1954, most range research on the Great Plains was transferred from the Forest Service to the Agricultural Research Service although the Forest Service continued research on grazing management, range ecology, plant control, and range-related fire. During the late 1950s, range managers in Region 4 looked to reduce the numbers of sheep and cattle permitted on national forest grazing allotments because overuse was damaging vegetation and causing serious erosion, especially in streamside areas and canyon bottoms. While line officers were at the front of the "Range Wars," Forest Service range scientists played a supporting role. By the late 1950s, range research increasingly emphasized thinking about broader ecosystems and, in 1965, wildlife and range research were joined administratively (Klade 2006).

Before the 1960s, recreation specialists in regional offices conducted research as a side activity. Responding to growing concerns about decreasing outdoor recreation opportunities, Congress established the Outdoor Recreation Resources Review Commission in 1958 and its 1962 final report addressed the increased demand for recreation on America's public lands. In an expanding consumer society, people increasingly looked for places to play outdoors. Recreation research was needed to determine the current situation and future needs on the nation's forests. The 1962 McIntire–Stennis Act, which facilitated stronger research cooperation between land grant universities and the Forest Service, aided recreation research along with other areas (Williams 2005).

1.7 Experimental Forests and the Environmental Era

Rachel Carson's *Silent Spring* drew attention to the hazards of the pesticide DDT when it was published in 1962, contributing to growing public involvement in environmental issues. A year later, research on acid rain commenced at the Hubbard Brook Experimental Forest in New Hampshire. In 1970, the USA celebrated Earth Day. Environmental politics contributed to groundbreaking legislation of the 1960s and 1970s, including the Wilderness Act, Clean Air Act, National Environmental Policy Act, and Endangered Species Act, all of which required scientific examination of natural resource management issues. By the late 1970s, Sagebrush Rebels mobilized to demand the federal government turn over public lands to the states in the West. Nearly a decade later, the controversy over the spotted owl (*Strix occidentalis*) in the Pacific Northwest created a national political firestorm but also led scientists to new lines of research that focused on managing forests from an ecosystem perspective, accommodating wildlife and water quality on forest landscapes, and restoring degraded ecosystems.

Within this context, research on experimental forests and ranges increasingly examined questions of forest growth and sustainability, with particular attention to genetics, insects, and diseases. Priest River Experimental Forest and Deception Creek Experimental Forest in Idaho undertook experimental plantings of white pine blister rust-resistant stock and the Idaho Panhandle National Forest helped restore western white pine with knowledge gained from this research. Scientists also developed the nation's first insect risk-rating system based on research at Blacks Mountain Experimental Forest in California to determine which trees were most susceptible to insect attack. Studies also contributed to endangered species recovery. A chance encounter with running buffalo clover (*Trifolium stoloniferum* Muhl. ex. A. Eaton), which was believed to be extinct, on the Fernow Experimental Forest led to research that found periodic disturbance promotes long-term species survival (Madarish and Schuler 2002). This finding contributed to new management guidelines to increase the clover population. Studies on the Escambia Experimental Forest in Alabama have helped scientists understand the ecology and silviculture of longleaf pine (*Pinus palustris*) ecosystems, leading to policies that help manage for biodiversity. Efforts remain underway to restore the longleaf pine ecosystem, which provides habitat for the endangered red-cockaded woodpecker (*Picoides borealis*; Adams et al. 2008).

A growing sense of ecological awareness also meant that the concept of rangeland "improvement" shifted to "restoration." Research focused on providing information for sustainable use. In 1975, the third edition of *Range Management* defined the field as "the science and art of optimizing returns from rangelands in those combinations most desired by and suitable to society through the manipulation of range ecosystems" (Mitchell et al. 2005, p. 26). The following year, the Desert Experimental Range was designated a Biosphere Reserve by UNESCO, becoming one of few cold-desert biomes designated as such and the only one in the Western Hemisphere (USDA Forest Service Undated a). Research at the Great Basin Experimental Range helped establish basic principles of range management and provided a training ground for scientists to conduct ecologically based rangeland inventories. As a result, they have collectively impacted rangelands management (USDA Forest Service Undated b).

Finally, Smokey Bear debuted during the environmental era to deliver his message about fire safety to the American public. The National Fire Danger Rating System was developed from research conducted on experimental forests like Priest River. Fifty years after the McSweeney–McNary Act, the Forest and Rangeland Renewable Resources Research Act of 1978 laid out new possibilities for research on experimental forests and ranges that included fish and wildlife and their habitats, threatened and endangered species, and global natural resource issues (Forest and Rangeland Renewable Resources Research Act 1978).

1.8 From a Productive Past to a New Horizon of Scientific Possibilities

> In the past few decades, scientists have come to a better understanding of the ecosystem context of environmental problems. Accordingly, the research focus at experimental forests and ranges has shifted from local, narrow, applied topics to a wider range of topics with broader relevance—global climate change, watershed function, invasive plants, recovery after natural disturbances, and others. (USDA Forest Service 2008)

Across different eras, experimental forests and ranges have been living laboratories for studying issues involving complex interactions between forests and society. Today, these issues include the implications of climate change and energy development on forest and range ecosystems. Forests function in multiple ways to provide benefits to society. They can sequester significant amounts of carbon in a stable, long-term condition. Productive forests lead to steady streams of forest biomass, contributing to domestic supplies of energy. Demand for sustainably produced wood products continues to increase as America's population grows and the nation's reliance on wood for renewable energy and biofuels rises. At the same time, many Americans identify with and want to protect natural environments. Today's forest management practices aim to produce wood, increase plant and animal diversity, improve aesthetics, reduce the risk of destructive wildfires, protect water quality, and minimize and mitigate the effects of uncharacteristic outbreaks of destructive insects and diseases and anticipated changes in climate conditions. Experimental forests do not have all the answers for how to effectively balance these goals, but they offer a path rooted in a history of scientific discovery and knowledge that can help people manage forests to meet diverse societal needs in today's world.

Early research on experimental forests has proven useful in ways study originators could not have imagined. Daily weather data collected at Priest River Experimental Forest since 1911 have been used in climate forecasting models. Studies from the Kane Experimental Forest in Pennsylvania, Fraser Experimental Forest in Colorado, and Bartlett Experimental Forest in New Hampshire have helped evaluate carbon storage by managed forests. Historical silviculture research on timber management closely followed stand development. As such, these findings can be applied today to help meet forest management goals involving habitat management, ecosystem restoration, and climate change. Data compiled and lessons learned on the experimental forests generations ago continue to provide a foundation and ideas for today's scientists and land managers to address emerging issues (Adams et al. 2008).

Over the last century, experimental forests and ranges have developed as important sites to study fire, grazing, water, and climate change. Fire research initially focused on the science of fire behavior in forests and how to control wildfires. As a result of these early studies, scientists realized fire was an important part of the ecosystem and a tool for maintaining healthy forests. Much of the science of prescribed fire was developed on southern experimental forests. A lack of regular fire cycles can lead to devastating wildfires. Today, the risk of wildfire is exacerbated by suburban and exurban housing developments that were built during the postwar era and have increasingly extended into the wildland–urban interface. As the agency in its early years dealt with unregulated grazing, research established a scientific basis for managing range ecosystems. This research contributed to an understanding of how to manage rangeland and educate people on proper range management techniques and evolved into cooperative efforts to restore and sustain rangelands. In addition, early research on experimental forests helped scientists understand forest attributes that contribute to water quantity. From this fundamental information on the water process, forests could be managed within a watershed and ecosystem context that focused on water quality as well as water quantity. Water thus came to serve as an ecosystem health indicator (pers. commun. with the Experimental Forest and Range film working group, Mary Beth Adams 2008).

Forests are constantly changing. Some changes are gradual and others are abrupt. Forests change even when we do nothing. Experimental forests and ranges provide places to measure, monitor, study, and understand how forests, water, and wildlife change over time without disturbance, in response to management treatments and in response to unplanned disturbances such as wildfire (e-mail correspondence with Stephen R. Shifley, 9 January 2008; USDA Forest Service Undated c). Studies established decades ago on experimental forests and ranges continue to provide valuable information about the land and its resources. Scientists have built a body of research to support effective land management and further our understanding of natural processes. The historical record of accomplishment on experimental forests and ranges is vast, yet this national network of permanent outdoor laboratories is also uniquely positioned to respond to emerging issues. When the sites were established, officials could not predict the future, but studies were based on sound scientific principles. By drawing on historical data and integrating data with climate modeling, research can help address global climate change. As Douglas Ryan and Frederick Swanson highlight in this volume, today's experimental forests and ranges operate in what might be called a networked era, allowing scientists to more effectively share information within and outside the system to address global environmental issues. But for these forests and ranges to function as a fully integrated research network, they suggest existing data sharing must coalesce with the development of scientific social networks across the system. If individual experimental forests and ranges historically operated in relative isolation, the current networked era offers opportunities beyond the imagination of those early scientists.

In *Breaking New Ground* Pinchot wrote, "As research takes its rightful place in the vanguard of the forest movement, the early 'searchers for forest facts,' Raphael Zon perhaps foremost among them, will take their places alongside other pioneers who have helped to build the US Department of Agriculture Forest Service. Hand in hand with the development of scientific Forestry went the study of how to make the best use of the products of the forest" (Pinchot 1947, p. 310). Nearly 140 years after Franklin Hough delivered his report to the American Association for the Advancement of Science, these "searchers for forest facts" continue to contribute scientific knowledge to the nation and beyond through their research at experimental forests and ranges. Let us explore their journey across time and place in the following chapters.

References

Adams MB, Loughry L, Plaugher L (Comps.) (2008) Experimental forests and ranges of the USDA Forest Service. Revised edn. General Technical Report NE-321, USDA Forest Service, Northeastern Research Station, Newtown Square, PA (CD ROM: nrs.fs.fed.us/pubs/6741)

Alexander TG (1987a) The rise of multiple-use management in the Intermountain West: a history of Region 4 of the Forest Service (FS-399). USDA Forest Service, Washington, DC

Alexander TG (1987b) From rule of thumb to scientific range management: the case of the Intermountain Region of the Forest Service. West Hist Q 18(4 Oct 1987):409–428

Clapp EH (1917) The correlation of American forest research. J For 15(2):165–175

Clapp EH (1936) Letter to Ferdinand Silcox. 19 May. USDA Forest Service Headquarters History Collection, Forest History Society, Durham, NC

Clapp EH (1938) The decennial of the McSweeney-McNary Act. J For 36(9):832–836

Clapp E, Society of American Foresters (1926) A national program of forest research. American Tree Association, Washington, DC

Cohen L (2003) A consumers' republic: the politics of mass consumption in postwar America. Random House, New York

Coville FV (1898) Forest growth and sheep grazing in the Cascade Mountains of Oregon. Government Printing Office, Washington, DC

Egan T (2009) The Big Burn: Teddy Roosevelt and the fire that saved America. Houghton Mifflin Harcourt, Boston

Fernow BE (1899) Forestry investigations of the US Department of Agriculture, 1877–1898. Government Printing Office, Washington, DC

Forest and Rangeland Renewable Resources Research Act of 1978. P.L. 95–307

Forest History Society (2009a) Carlisle P. "Cap" Winslow. http:www.foresthistory.org/ASPNET/people/Scientists/Winslow.aspx. Accessed 4 Feb 2013

Forest History Society (2009b) Coweeta Hydrologic Laboratory. http://www.foresthistory.org/ASPNET/Places/ResearchSites/Coweeta.aspx. Accessed 4 Feb 2013

Forest History Society (2009c) Harry T. Gisborne. http://wwwforesthistory.org/ASPNET/people/Gisborne/Gisborne.aspx. Accessed 5 Feb 2013

Forest History Society (2009d) James T. Jardine. http://www.foresthistory.org/ASPNET/people/Scientists/Jardine.aspx. Accessed 4 Feb 2013

Forest History Society (2009e) Olustee Hydrologic Laboratory. http://www.foresthistory.org/ASPNET/Places/ResearchSites/Olustee.aspx. Accessed 4 Feb 2013

Forest History Society (N.d.) Carlos G. Bates. http://www.foresthistory.org/education/curriculum/Activity/activ11/BatesBio.htm. Accessed 4 Feb 2013

Gillespie WB (2003) Early years of the Santa Rita Experimental Range and Florida Station. http://www.gvrhc.org/Library/SantaRitaExperimental Range.html. Accessed 4 Feb 2013

Graham KL (2004) History of the Priest River Experiment Station. General Technical Report RMRS GTR–129, USDA Forest Service, Rocky Mountain Research Station, Fort Collins, CO

Guldin JM (2009) The Crossett Experimental Forest—72 years of science delivery in the silviculture of southern pines. In: Ashton SF, Hubbard WG, Rauscher HM (eds) A southern region conference on technology transfer and extension. General Technical Report SRS-116, USDA Forest Service, Southern Research Station, Asheville, NC, pp 203–209

Hardy CE (1983) The Gisborne era of forest fire research: legacy of a pioneer, FS-367. USDA Forest Service, Northern Forest Fire Laboratory, Missoula

Hays SP (1959) Conservation and the gospel of efficiency: the progressive conservation movement, 1890–1920. Harvard University Press, Cambridge

Herring MJ, Greene S (2001) Forest of time: research at the Wind River Experimental Forest 1908–1919. For Hist Today Spring/Fall:36–43

Herring MJ, Greene S (2007) Forest of time: a century of science at Wind River Experimental Forest. Oregon State University Press, Corvallis

Hornbeck J (2001) Events leading to establishment of the Hubbard Brook Experimental Forest. http://www.hubbardbrook.org/overview/HBEF_establishment.htm. Accessed 4 Feb 2013

Hough F (1884) Report on forestry. Government Printing Office, Washington, DC

Ice GG, Stednick JD (2004) Forest watershed research in the United States. For Hist Today Spring/Fall:16–26

Jardine JT, Anderson M (1919) Range management on the national forests. US Department of Agriculture, Washington, DC

Keck WM (1972) Great Basin Station—sixty years of progress in range and watershed research. USDA Forest Service, Intermountain Forest and Range Experiment Station, Ogden

Klade RJ (2006) Building a research legacy-the Intermountain Station, 1911–1997. RMRS-GTR-184, USDA Forest Service, Rocky Mountain Research Station, Fort Collins, CO

LaBau VJ, Bones JT, Kingsley NP, Lund HG, Smith WB (2007) A history of the forest survey in the United States: 1830–2004 (FS-S877). USDA Forest Service, Washington, DC

Madarish D, Schuler TM (2002) Effects of forest management practices on the federally endangered running buffalo clover (*Trifolium stoloniferum* Muhl. ex. A. Eaton). Nat Area J 22(2):120–128

May ET (2008) Homeward bound: American families in the Cold War era. 20th anniversary edn. Basic Books, New York

McBeath LW (1978) Eloise Gerry: a woman of forest science. J For Hist 22(July):128–135

Miller C (1999) Grazing Arizona: public land management in the Southwest. For Hist Today Fall:15–19

Miller C (N.d.) A high grade paper: "Garden & Forest" and nineteenth-century American forestry." http://www.loc.gov/preservation/about/prd/gardfor/essays/miller.html. Last modified April 2011. Accessed 4 Feb 2013

Mitchell JE, Ffolliott PF, Patton-Mallory M (2005) Back to the future: Forest Service rangeland research and management. Rangelands 27(June):19–28

Nelson CA (1971) History of the US Forest Products Laboratory, 1910–1963. USDA Forest Service, Forest Products Laboratory, Madison

Pearson GA (1936) The oldest forest experiment station. Research-Fort Valley. USDA Forest Service Headquarters History Collection, Forest History Society, Durham

Pinchot G (1947) Breaking new ground. Harcourt Brace, New York

Pinchot G, Graves HS (1896) The white pine. Century, New York

Prevedel DA, Johnson CM (2005) Beginnings of range management: Albert F. Potter, first chief of Grazing, U.S. Forest Service, and a photographic comparison of his 1902 Forest Reserve Survey in Utah with conditions 100 years later. R4-VM 2005–01, USDA Forest Service, Intermountain Region, Ogden, UT

Pyne S (2001) Year of the fires: the story of the great fires of 1910. Viking, New York

Roosevelt T (1903) Third annual message. 7 December. http://www.presidency.ucsb.edu/ws/index.php?pid=29544#axzz1P6OQUAIV. Accessed 4 Feb 2013

Steen HK (1998) Forest Service research: finding answers to conservation's questions. Forest History Society, Durham

Steen HK (2004). The U.S. Forest Service: a history. University of Washington Press, Seattle

US Department of Agriculture (1883) Report of the commissioner of agriculture for the year 1883. U.S. Government Printing Office, Washington, DC

US Department of Agriculture, Forest Service (1933) A national plan for American forestry. U.S. Government Printing Office, Washington, DC

US Department of Agriculture, Forest Service (1936) The western range. U.S. Government Printing Office, Washington, DC

US Department of Agriculture, Forest Service (1944) The history of western range research. Agric Hist 18(3):127–143

US Department of Agriculture, Forest Service (2008) Experimental forests and ranges. (Poster) USDA Forest Service, Pacific Northwest Research station, Portland, OR. http://www.fs.fed.us/pnw/pubs/experimental-forests-brochure.pdf. Accessed 4 Feb 2013

US Department of Agriculture, Forest Service (Undated a) Desert experimental range. http://www.fs.fed.us/rmrs/experimental-forests/desert-experimental-range/. Accessed 4 Feb 2013
US Department of Agriculture, Forest Service (Undated b) Experimental forest. http://www.srs.fs.usda.gov/organization/exforest/. Accessed 4 Feb 2013
US Department of Agriculture, Forest Service (Undated c) 1910 Fires: Edward Pulaski. www.fs.usda.gov/detail/r1/learning/history-culture/?cid+stelprdb5122876. Accessed 4 Feb 2013
US Department of Agriculture, Forest Service (Undated d) A retrospective look. http://www.fs.fed.us/rmrs/about/history. Accessed 4 Feb 2013
US Government Printing Office (1909) Proceedings of a conference of governors. Government Printing Office GPO, Washington, DC
US Public Lands Commission (1905) Report of the Public Lands Commission. U.S. Government Printing Office, Washington, DC
Wells G (2009) Experimental forests and ranges: 100 years of research success stories. USDA Forest Service, Forest Products Laboratory, FPL-GTR-R182, Madison
Williams GW (2005) The USDA Forest Service: the first century. USDA Forest Service, Washington, DC
Young JA (2000) Range research in the far western United States: the first generation. J Range Manage 53(January):2–11
Young JC (2010) Warrior of science: Raphael Zon and the origins of forest experiment stations. For Hist Today Spring/Fall:4–12
Young JA, Clements CD (2001) Range research: the second generation. J Range Manage 54(March):115–121
Zon R (1908) Forest Service-Research. Letter to G. Pinchot, 6 May. USDA Forest Service Headquarters History Collection, Forest History Society, Durham, NC

Chapter 2
Forest Service Experimental Forests and Long-term Data Sets: Stories of Their Meaning to Station Directors

Ariel E. Lugo, Bov Eav, G. Sam Foster, Michael Rains, Jim Reaves and Deanna J. Stouder

Abstract As Forest Service Research and Development worked to prepare this book reporting important results from long-term research conducted on U.S. Department of Agriculture Forest Service Experimental Forests and Ranges, the station directors added a chapter to highlight additional accounts of long-term research, its benefits to land managers and policy makers, and lessons learned from the first century of research on Experimental Forests and Ranges. The Northern Research Station described research on tree care and the opening it created to urban natural resource research. The Pacific Southwest Research Station described a series of studies on the relationships among logging, landslides, and water quality that began in 1963 and continues through the present. The International Institute of Tropical Forestry described pioneering work in measuring tree growth in tropical forests. The Pacific Northwest Research Station showed how conclusions vary with the length of an environmental record, and the ways in which their research has contributed to

A. E. Lugo (✉)
International Institute of Tropical Forestry, USDA Forest Service, 1201 Calle Ceiba,
Jardín Botánica Sur, San Juan, PR 00926-1119, USA
Phone: 787-766-5335
e-mail: alugo@fs.fed.us

B. Eav
Pacific Northwest Research Station (retired), USDA Forest Service, 1220 SW 3rd Ave.,
Portland, OR 97204, USA

G. S. Foster
Rocky Mountain Research Station, USDA Forest Service, 240 West Prospect Road,
Fort Collins, CO 80526, USA

M. Rains
Northern Research Station, USDA Forest Service, 11 Campus Boulevard, Suite 200,
Newton Square, PA 19073-3200, USA

J. Reaves
Deputy Chief for Research, formerly Southern Research Station, USDA Forest Service,
Washington, DC, USA

D. J. Stouder
Formerly Pacific Southwest Station, USDA Forest Service, 800 Buchanan Street,
West Annex Building, Albany, CA 94710-0011, USA

D. C. Hayes et al. (eds.), *USDA Forest Service Experimental Forests and Ranges*,
DOI 10.1007/978-1-4614-1818-4_2, © Springer New York 2014

understanding old-growth forests. The Southern Research Station highlighted the contributions of the Coweeta EFR to the science of forest ecosystem hydrology. The Rocky Mountain Research Station showed how data from even a single plot measured over many decades can enhance our understanding of forests and environmental change. Lessons learned include the importance of data quality, sampling intensity and consistency, scale, scientific creativity, and manipulative research. These stories also show us that sites on which long-term data have been collected can serve as settings for important conversations about important social and management questions.

Keywords Long-term research · Urban forestry · Landslides · Tropical forestry · Hydrological research · Old-growth forests · Forest change · Research policy

2.1 Introduction

The network of Experimental Forests and Ranges (EFRs) is a crown jewel of the Forest Service. There is no other country in the world with such treasures as the 80 individual EFRs. This is certainly evident by the number of scientists and students who visit from other countries and conduct research on a number of our EFRs. Since their establishment in 1908, the wealth and breadth of scientific knowledge gained from EFR research has provided both public and private land managers invaluable information on how to manage their forestland and has added to the very structure of natural resource science. In addition, seminal research on watershed issues has contributed to enhancing the quality and quantity of our nation's water resources. Research on EFRs ranges from development of new methods to study forests and ranges through studies whose longevity allows them to answer, in profound ways, questions not foreseen at the time the studies were initiated. Each chapter in this book tells a story that reflects these contributions and shows how EFR research enables Forest Service Research and Development to provide the science that managers and policy makers need to sustain forests, grasslands, and associated waters.

The research and development (R&D) branch of the U.S. Department of Agriculture (USDA) Forest Service is organized regionally. The seven research stations (Fig. 2.1) help the agency bridge the range of research needs—from questions specific to the ecological and social context of a local site through those appropriate to a single region to questions of national and international scale. In this chapter, the directors of each station highlight questions answered by research on the EFRs within their station. Each example illustrates the special contribution of EFR data sets to the mission of the Forest Service: "to sustain the health, diversity, and productivity of the Nation's forests and grasslands to meet the needs of present and future generations." Each research station is represented by at least one chapter elsewhere in this volume, but here, the directors place one or a few data sets or programs in the historical context of their respective parts of the country and highlight how the long-term focus has not only facilitated achieving the mission of the agency but also allowed Forest Service R&D programs to remain vibrant and relevant to the needs of society, even with the passage of time and changing research priorities.

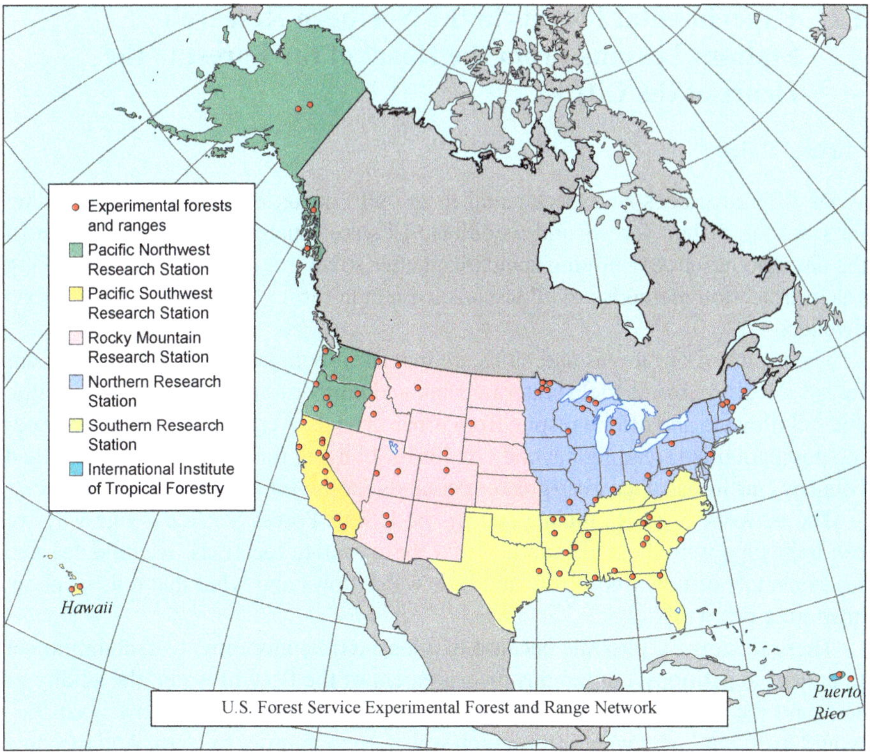

Fig. 2.1 The seven regional research stations of the USDA Forest Service, with locations of the 80 Experimental Forests and Ranges

The accounts span almost the full life of Forest Service R&D, from one of our most recent long-term ecological research sites in Baltimore to one of our very first experimental forests in northern Idaho.

The directors' questions range from how society values older forests through how the effects of forest management accumulate across space and time to how research in rural forests is shaping the environmental policies of cities today. The studies they report range from single plots, in Idaho and Puerto Rico, through watersheds in California and North Carolina to a city that is breaking the ground of urban EFRs. These examples demonstrate that the Forest Service's long-term approach to research gives the agency flexibility to address changing natural resource conservation questions and advance forest science methods while maintaining stable field research programs in the nation's premier experimental forest network, the EFRs of the USDA Forest Service (Lugo et al. 2006). In conclusion, the directors summarize the characteristics of this research approach that make it particularly powerful to address questions of sustainable management and global climate and environmental change.

2.2 Experimental Forests in the Northern Research Station: Lessons From the Heart of the Forest to the Heart of the City

Michael Rains

What difference does an experimental forest (EF) make, especially to those living far from the distant forests and woodlots? EFs are touching lives in the midst of the city. Modern tree trimming standards, better street tree placement, and drinking water protection methods are all lessons learned in rural settings that benefit urban dwellers.

The health of our towns and cities are inextricably linked with the health of our forests. Healthy forests yield water to drink and lumber to use; they shelter wildlife and wildflowers and offer a respite from urban crowds. Across the USA, an expanding population is spreading beyond city limits, blurring the lines between town and country. Our forests face ever-greater demands for the services they provide.

The network of EFRs maintained by the USDA Forest Service is one way we can help cope with those demands. From these remote locations, we have learned fundamentals of how trees grow and how water flows and what that tells us about how our cities work.

These lessons are possible because of a long-term commitment—a commitment to patient, repetitious, continuous measurement of the flow of water, the acidity of soil, and the growth of trees. In most cases, researchers did not know what they would learn but knew what information was needed in order to learn. And the slow and steady accumulation of data yields knowledge that could not be gained in any other way. Long-term data let us mark the changes occurring around us over the decades and help us anticipate what conditions we face in the future. The lessons we have learned so far surprise, excite, and encourage us.

As the baby boom was just underway in this country, a plant physiologist from Pennsylvania named Alex Shigo was wondering how trees lived and how they died. How does a tree make new wood? What keeps a branch growing? Why can an injury kill a tree?

He began exploring these questions through studies on EFRs in New England and around the Great Lakes. At first, his focus was economic, understanding the factors that degrade wood quality or kill the trees remaining after a harvest. Earlier researchers expected that wood products deteriorate in the same way that cut logs would. But Shigo's studies revolutionized how foresters and biologists thought about trees. They came to understand that, instead of being passive organisms acted upon by fungi, trees were living systems actively responding to injury, insect, drought, and changing soil conditions.

That knowledge changed the way we treat trees, especially those subject to the hazards of an urban environment, and launched modern arboriculture, the science and practice of tree care. Shigo's studies led the tree care industry to stop using tar to seal tree wounds, because trees naturally seal off an injury to resist the spread of infection. He shaped how utility crews and homeowners prune trees, based on a better understanding of how trees grow and how branches attach to trunks. Other research

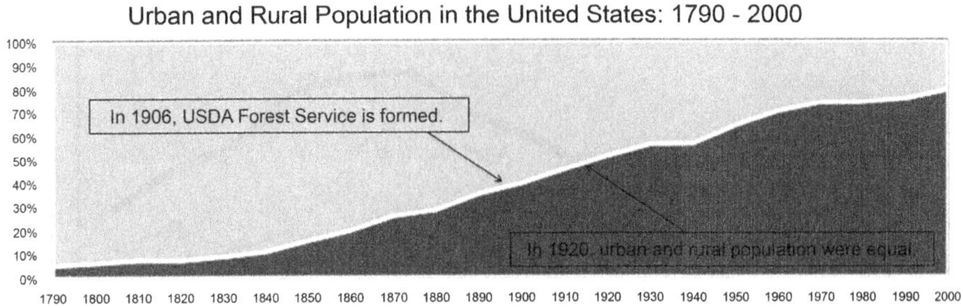

Fig. 2.2 Percentage of the US population that is urban (*dark area*) or rural (*clear area*) between 1790 and 2000. At 50%, the rural and urban population was equal

helped plant nurseries understand that soil organisms are vital to the health of trees, refocusing their industry from the biology of plants to the biology of the landscape.

Shigo succeeded in his work because he had access to much more data about trees and decay than he could have accumulated in his lifetime. By building on the long-term data from EFs, his base of knowledge could approach that of the lifespan of a tree rather than just the life of a single scientist.

EFs were also critical to his success because his research involved deliberately wounding trees and observing the results. Few forest managers willingly create damage to their resources. So a place for manipulative research was essential to fostering the health of trees elsewhere.

Shigo's work illustrates the importance of a place dedicated to understanding the linkage of forests and cities. As the significance of that linkage has grown, the Forest Service started a new set of EFs in urban settings: Urban Long Term Research Areas (ULTRAs). These fledging sites represent the next frontier of the EF legacy. Their importance is underscored by the predominance of urban population over rural population in the USA (Fig. 2.2) and the world. It is also a real tribute to the Shigo legacy that arboriculturists in Italy are creating an Alex Shigo Modern Arboriculture Park in Barasso, Italy.

The oldest of these urban sites, the Baltimore Ecosystem Study (BES), was launched just a decade ago on the shores of the Chesapeake Bay. Part of the National Science Foundation's Long-term Ecological Research Network, the BES explores a new field of knowledge, urban ecology, which looks at how organisms and environments in and around cities are affected by the buildings and paved surfaces, the things that people do (Fig. 2.3), and the new environments that cities create.

In many ways, the approach proven at older, rural EFs applies at the BES. Researchers here mirror the long-term watershed approach established at EFs such as Hubbard Brook, Coweeta, H.J. Andrews, and Luquillo. The demonstrated value of interdisciplinary research brings ecologists, soil scientists, hydrologists, and social scientists together to apply their traditional expertise to a nontraditional setting. Field trips and demonstrations are an important tool for sharing lessons learned.

But a unique focus of the BES is engagement of the community beyond the scientific circle. BES engages decision makers, from local communities to multistate policy makers, in the formation of questions, collection, and analysis of data,

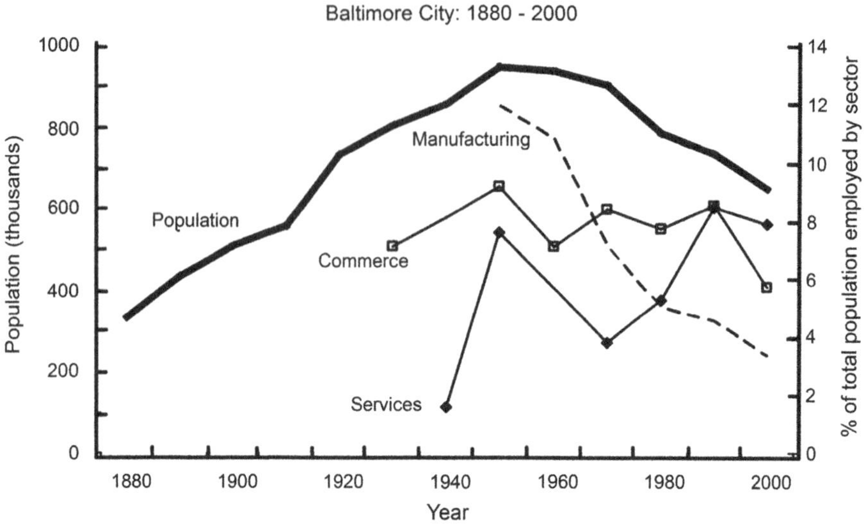

Fig. 2.3 The population of Baltimore between 1880 and 2000 and the proportion of the population involved in three economic activities

and the dissemination of findings. A core team of educators and researchers meets bimonthly to help others learn and teach about Baltimore's urban ecosystem. As a result of this engagement, the work here is already changing the face of Baltimore.

As a result of BES findings, regional planners at the Chesapeake Bay Program shifted their policy from planting trees in urban riparian buffers to increasing overall tree canopy in urban areas. In turn, the City of Baltimore established a goal of increasing its tree canopy from 20 to 40% by 2037. When achieved, that increase will mean a cooler city with less air and water pollution, increased property values, and a more pleasant and healthy setting for urban residents. And the circle of lessons widens. Following on Baltimore's success, New York City, Boston, and other metropolitan areas conducted similar analyses and have adopted urban tree canopy goals.

The legacy of EFs, new and old, is improvement of human communities from urban to rural settings. By maintaining this vital network for long-term research, we ensure that the learning continues. The more we learn about trees and forests, the more we understand how our health and that of our cities depends on them. From the heart of the forest to the heart of the city, EFs link us to those important lessons from the natural world.

2.3 EFs of the Pacific Southwest Research Station: Long-Term Research at the Caspar Creek Experimental Watersheds

Leslie M. Reid, Larry Rabin, and Deanna J. Stouder

Why study the same site for decades? Wouldn't more information be gained by studying many sites for short periods? The answers to these questions are at the

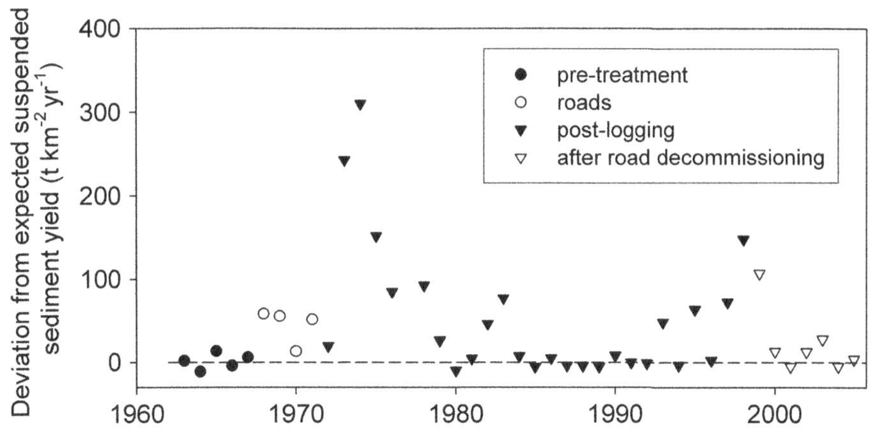

Fig. 2.4 Deviation of annual suspended sediment production in the South Fork Caspar Creek watershed from that expected for unlogged conditions. Roads were constructed in 1967, and the watershed was selectively logged between 1971 and 1973

same time simple and complex, like science itself. Consider the story of the Caspar Creek Experimental Watersheds, located in the forests of California's north coast. A century and a half ago, the North and South Fork watersheds supported an old-growth forest of towering coastal redwoods, some with diameters of 3 meters or more. But California's growing population required building materials, and by 1905 Caspar Creek's old growth had been replaced by young second-growth stands.

Over the next half century, an important fishing industry developed along the north coast, lowland communities grew, and logging continued at other sites. Conflicts between resource users raised new questions: Did logging reduce salmon populations by damaging spawning streams? Did it degrade the water supply by adding sediment? Did it aggravate the episodes of flooding and landsliding that the region had experienced in the 1950s? By the late 1950s, it was becoming clear that policy makers would need some solid information about the environmental impacts of logging if regulations were to be designed to reduce conflicts. In 1960, the USDA Forest Service and the California Department of Forestry and Fire Protection joined forces to address these issues by designating the North and South Forks of Caspar Creek as a site for studying the effects of logging on streamflow, sedimentation, and fish habitat in a rain-dominated forest.

The first study—the South Fork experiment—was developed to be a watershed-scale experiment quantifying the effects of tractor-yarded selective logging on streamflow and sediment yield. Construction of a small dam enabled monitoring at each watershed mouth, and measurements began in late 1962. After 5 years of pretreatment monitoring to define the baseline conditions at both forks, scientists left the 473-ha North Fork undisturbed as a control and roads were constructed in the 424-ha South Fork watershed. Logging then began on the South Fork in 1971, 4 years after road construction.

Data from the roading phase of the study showed that sediment inputs increased dramatically from road construction alone (Fig. 2.4). These preliminary results became evident just as a rising level of environmental concern was leading to passage of legislation at both federal and state levels. In 1973, the California legislature passed the Z'berg-Nejedly Forestry Practice Act, which authorized the California Board of Forestry to "...adopt rules for control of timber operations which will result or threaten to result in unreasonable effects on the beneficial uses of the waters of the state." The early Caspar Creek results underscored the need to develop methods for controlling road-related sediment, and later results quantified increases in runoff and sediment yield from tractor logging. Thus, Caspar Creek analyses and results contributed to the body of information that eventually resulted in restrictions on the use of tractor yarding on steep slopes in California.

By 1985, flow and sediment loads had returned to near pretreatment levels in the South Fork, and scientists initiated a new experiment in the North Fork watershed. By this time, controversy had begun to focus on the issue of the "cumulative impact," the "...impact on the environment which results from the incremental impact of the action when added to other past, present, and reasonably foreseeable future actions..." (40 CFR § 1508.7). Forest management plans governed by either Federal or California rules were required to assess cumulative impacts, but controversies—and litigation—persisted over what level of assessment was adequate and over what actually constituted the cumulative impact.

The North Fork experiment was designed to identify trends in the hydrologic and sediment response to clearcut logging along a downstream sequence of gauging stations. Comparison of proportional changes in flow and sediment as watershed size increased from 10 to 473 ha would reveal whether the cumulative downstream effects were additive, synergistic, or dampened with increasing scale. The design called for clearcutting five gauged tributary watersheds while leaving three unlogged watersheds as controls, allowing the direct effects of clearcutting to be evaluated. In addition, the overall changes at the downstream gauging weir could be compared with those observed during the previous long-term experiment to assess the relative effects of old and new logging practices. Information provided by the South Fork experiment allowed design of a North Fork sampling strategy that would provide the level of precision necessary to detect likely effects.

Monitoring for the experiment began with gauge installation in 1985, and 39% of the North Fork watershed was logged between 1989 and 1992. Results of monitoring through 1995 showed that runoff, low flows, peak flows, and sediment loads increased in most of the logged watersheds (Ziemer 1998). Hydrologic changes were additive over the range of watershed scales present, but sediment changes displayed no uniform downstream trend—other factors must be influencing the response. Additional information became necessary to explain the results.

That information became available because other kinds of research were also taking place during the North Fork experiment. Surveys of woody debris showed that riparian blowdown increased along the margins of clearcuts, and monitoring of channel cross sections revealed that this new wood trapped sediment. Further, data from a study of tributary channels showed that channels enlarged downstream of clearcut catchments, suggesting that the logging-related flow increases generated off-site erosion.

As results of the North Fork experiment were becoming available in the mid-1990s, a new issue emerged about 160 km north of Caspar Creek. Residents of several lowland communities in Humboldt County reported increased flood frequencies after intensive logging of upstream watersheds. Like studies elsewhere, the North Fork experiment had shown that logging greatly increased the small peak flows that occurred early in the wet season, when decreased transpiration after logging led to increased soil moisture and thus reduced the capacity for further water storage on hillslopes. But North Fork results also showed that the larger, midwinter peaks increased, and this result ran counter to expectations. Common wisdom had assumed that midwinter peaks are immune from increases because soils are near saturation in both logged and unlogged sites. Therefore, a rainstorm falling on saturated forest soils should affect stream flows in exactly the same way as it would falling on saturated logged soils.

But it did not. Clearly, changes in transpiration were not the only mechanism for hydrologic change after logging. Researchers conducted a brief study at Caspar Creek to determine whether the observed peak flow changes could be explained by differences in how much rain is trapped by foliage before and after logging. Comparison of rainfall under forest stands with that in adjacent clearcuts showed that the redwood forest canopy trapped and evaporated about 20% of even the largest rainstorms. When logging removed the canopy, more of a storm's rainfall hit the ground; so more runoff reached streams to increase peak flows. With the mechanism for change now explained, personnel from the California Department of Forestry used a flow model based on Caspar Creek results to calculate desired logging intensities for the Humboldt County watersheds.

Meanwhile, routine long-term monitoring continued at the South Fork weir. After 1985, flow and sediment loads remained at near pretreatment levels until the early 1990s, but then the sediment loads began to rise once again. Field surveys showed that culverts were rusting out along the main haul road, so the road was decommissioned in 1998 to deactivate potential future sediment sources. Even after decommissioning, sediment loads remained higher than prelogging levels, and the total excess sediment produced after 1993 rivals that contributed during the first 12 years after logging.

Research at the Caspar Creek experimental watersheds is nearing its 50th year. Work there has led to more than 100 publications describing research results, and hundreds of other publications have cited Caspar Creek results, demonstrating their relevance to research elsewhere. Caspar Creek studies have been cited in planning documents ranging in scale from environmental assessments for individual projects to the Northwest Forest Plan, which guides forest planning for multiple Federal agencies in the Pacific Northwest.

Today, new controversies are emerging in America's forestlands. Along the west coast, second- and third-growth forests are now being logged, this time using today's state-of-the-art practices. But concerns have been raised that the new practices are being superimposed on a landscape that still reflects effects from earlier logging, thus potentially generating a multi-cycle cumulative impact. We do not yet have the information needed to predict the outcome of multi-cycle logging, but that information awaits us in the South Fork watershed. South Fork now reflects the third-growth conditions present through much of the redwood region and is ready to be reentered. Here, the effects of earlier second-growth logging practices are known, and the nearby North Fork experiment quantified the effects of more modern practices. Because of the nearly 50 years of records that now exist, it will be possible to very quickly assess the short-term effects of multi-cycle logging. Baseline monitoring for the third watershed-scale experiment began in 2001.

So why study the same site for decades? First, long-term studies are essential for defining the full trajectories of impact and recovery. Had monitoring ended when the South Fork study formally ended in 1985, the portion of the management-related sediment that has been produced since 1990 would have gone unnoticed. In the North Fork, management treatment did not end with logging, and peak flow magnitudes and sediment yield continue to show the effects of pre-commercial thinning carried out in 2001. Only long-term monitoring will reveal when recovery has been achieved. Until that information is available, we will not know the extent to which impacts from sequential logging entries will be superimposed, thus contributing to cumulative impacts at the watershed scale.

Second, experimental responses are often subtle for watersheds, and variations in conditions through time may introduce "noise" that hinders detection of the response. A long period of pretreatment data helps researchers differentiate between signal and noise. At South Fork, 13 years had elapsed between initiation of monitoring and completion of the experimental treatment—only then did monitoring of experimental results begin. If multiple studies are conducted at a long-term research site, each new study can use some of the same baseline data, allowing new experiments to progress with shorter calibration periods.

Third, long-term study sites often allow quick response times when a new need for information becomes apparent. When logging-related flooding became controversial, preliminary results from Caspar Creek were already available to help address the issue. And when an additional study was needed to explain the observed changes, the necessary background work had already been completed at Caspar Creek, allowing the rainfall interception study to progress quickly.

Fourth, long-established research sites prove invaluable in attracting researchers across a range of disciplines, all of who can make use of the long-term data sets while themselves contributing to a more holistic understanding of process interactions and conditions at the site. Had the North Fork wood and channel morphology studies been carried out elsewhere, their relevance to the sediment load would have been less evident. Long-term research sites also provide the data needed to test hypotheses derived elsewhere, and can provide a long-term context for short-term measurements at other sites. Researchers from elsewhere often use Caspar Creek data to help determine how broadly their own results might be generalized.

And finally, long-term research is essential for understanding natural and managed systems that are continuously changing. For example, a 100-year-old redwood stand is changing in character as it continues to mature. Without data to define long-term trends, we would be unable to distinguish the effects of differences in experimental treatment from those of differences in initial condition. Recent human-caused climate change produces a similar challenge, increasing the need to identify causal factors that affect patterns observed in monitoring data. Long-term data sets allow us to describe a system's responses to a wide range of weather patterns, thus making it possible to distinguish the effects of current and future climate change from those of other influences.

A research watershed is an outdoor laboratory that provides the infrastructure to support many kinds of research, allows careful control of experimental conditions, and permits experimental treatments to be designed to most efficiently address particular problems. The product of the laboratory is knowledge. We rarely know beforehand how basic knowledge will be used, but each time a critical emerging problem demands an immediate response, the pool of existing research results provides the basis for the response. Knowledge gained from nearly 50 years of research at the Caspar Creek Experimental Watersheds is part of the edifice of understanding that guides science-based management of rain-dominated, temperate forests in the USA and elsewhere.

2.4 The EF in the International Institute for Tropical Forestry: Pioneering Tree Growth Measurements in the Tropics

Ariel E. Lugo

How to measure and document the effect of weather, management, and other disturbances on tropical forests where tree rings do not represent historical record in the same way that they do in temperate forests? When the USDA Forest Service began research in temperate and boreal forests, the record of growth found in tree rings gave scientists a leg up in understanding the forces that determine forest growth. In tropical forests, seasonality is less predictable and growth more constant year-round, challenging the ability of scientists to discern these patterns.

Before there was a USDA Forest Service, employees from the Division of Forestry of the USDA visited Puerto Rico and the Luquillo Mountains to assess the forestry situation and make land management recommendations (Hill 1899; Gifford 1905). The forest condition in Puerto Rico was dire. Only about 20% of the original forest cover remained and the situation was non-sustainable because "every year the people of Porto Rico consume over three times as much as the forests of the entire island produce" (Murphy 1916, p. 1).

Crown lands in the Luquillo Mountains were transferred to the USA after the Hispanic American War and the Luquillo National Forest (then named the Luquillo Forest Reserve) was established in 1903, 2 years before the USDA Forest Service was organized (the forest has undergone numerous name changes and is now the El Yunque National Forest). From the outset, forest managers faced many challenges, some unique to the tropical nature of the new National Forest. For example, the climate was different from familiar temperate and boreal climates. Rainfall and air temperatures remained high year-round and there was no frost. The vegetation was lush and diverse. The National Forest was later found to have 207 tree species in 133 genera and 55 plant families, all in the relatively small area of about 10,000 ha (Little 1970). Between 1913 and the 1940s, Puerto Rico and the National Forest were visited by prominent scientists under the leadership of Nathaniel Britton, who led a scientific expedition sponsored cooperatively by the New York Academy of Sciences, the University of Puerto Rico, and the Puerto Rico Legislature (Baatz 1996). Among the dozens of scientists who participated in the expedition was the prominent American ecologist H. A. Gleason, who with M. Cook described the vegetation of Puerto Rico (Gleason and Cook 1926) and had just published a paper that would revolutionize the field of ecological succession (Gleason 1926).

Efforts to manage the forest stands of the Luquillo National Forest became bogged down because familiar temperate forestry techniques led to failures, particularly the problem of land restoration and dealing with a high number of tree species for which silvicultural information was scant at best (Wadsworth 1995). By 1956, a tropical research station had been established in Puerto Rico (authorized by Congress in 1928 and operational since 1939) and the National Forest was designated the Luquillo Experimental Forest. Contrary to convention in the mainland, the entire National Forest was proclaimed an EF in recognition of the need for a close partnership between research and forest management. This close partnership was anticipatory of the model that would come to predominate in the Forest Service decades later.

Forest Service research solved the problems of reforestation in Puerto Rico through decades of research activity (see, Wadsworth 1995). But I will highlight the approach taken to address the challenges of assessing tree growth in tropical forests. I view this research as one of the most notable contributions by the Institute toward the understanding of the functioning of tropical forests.

In temperate and boreal forests, it is possible to assess the age and growth rate of trees by counting and measuring the width of the growth rings. Each tree carries its own history of growth in the width and number of radial rings of its woody parts. If a treatment is administered to a stand, the forester can assess its success by examining tree rings to determine the growth response of trees to the treatment compared to untreated trees. Tree ring analysis relies on a winter season (cold temperatures of near below or below freezing) with insignificant tree growth and a growing season (warm temperatures) where larger cells form. Combined, these define an annual cycle of growth. The width of the annual ring reflects the rate of growth during that year. In the tropics, tree ring analysis is not as simple.

Trees in the dry tropics experience a growing season that coincides with the rainy season, and they then grow more slowly during the dry season. However, the periodicity of the seasons and their length are not as neatly defined as they are in the temperate and boreal regions. The events that define tree growth in the tropics are not annual. Instead, the rainy season may or may not occur in a particular year or more than one rainy period may occur, which makes it very difficult to assign an age to the number of rings. In the moist and wet tropics, the situation is even more complicated because the growing season is year-round and trees may or may not produce rings, and for those that do, it is difficult to relate individual growth rings to particular time intervals.

Pioneer tropical foresters had no practical methods for determining age structure of stands, nor rates of tree growth. Forest Service research provided the solution and, in so doing, advanced the understanding of long-term processes in tropical forests.

The scientist involved was Frank H. Wadsworth, who in the early 1940s decided to assess tree growth in the Luquillo Mountains. His only alternative was to measure, mark, and remeasure trees over and over until rates of growth could be established by the differences in dimension between two known intervals of measurement. This approach is straightforward but it is time consuming, requires accurate record keeping, consistent measurements, long-term data management, and long-term institutional commitment. Dr. Wadsworth grouped his trees in plots, established plots in the major forest types of the Luquillo Mountains, and conducted experiments with basal area reductions to assess growth responses to change in spacing and tree sizes. The plots he established in the 1940s are the oldest known in the Neotropics. They continue to be studied today and were instrumental in the development of long-term ecological research in the Luquillo Mountains.

One of these plots is in the *Dacryodes excelsa* (tabonuco) forest, the tallest forest in the Luquillo Mountains. Known as El Verde 3, data for this plot extend from 1943 to 2005 and the 62-year record of the plot was just published by Drew et al. (2009). From that publication, I estimated rates of change to illustrate four short stories embodied in the long-term data set.

Before presenting the stories, it is important to keep in mind some of the events that characterize the 62-year record of tree measurements at El Verde 3. The first one happened in 1932, before the El Verde 3 plot was established. It was the passage of hurricane San Ciprián, one of the strongest hurricanes to pass over the Luquillo Mountains. Using data from El Verde 3, Crow (1980) attributed the behavior of the stand to that event, which thinned the forest and caused a growth pulse that was measurable up to 1951. This report by Crow, a USDA Forest Service scientist stationed at the Institute, was one of the first documentations of the long-term hurricane effects on Neotropical forests. After this event, the plot experienced a long period of uninterrupted development without any major interruptions by hurricanes. In 1989, Hurricane Hugo affected the Luquillo Mountains, including El Verde 3, and in 1998, Hurricane Georges passed south of the Luquillo Mountains and had some effects on the El Verde 3 plot. What do the El Verde 3 long-term data tell us about how these disturbances shaped the forest, and what do they tell us about today's management questions?

2.5 Hurricanes Cause Pulses of Tree Recruitment and Growth

After Hurricane Hugo in 1989, tree basal area growth (Fig. 2.5, closed circles) peaked sharply with the highest rate in the long-term record. At the same time, there was a peak in stem recruitment (Fig. 2.5, open circles). The highest rate of increase in stem density occurred at the beginning of the record, probably a residual effect from the 1932 hurricane. However, until the 1989 hurricane, stem density

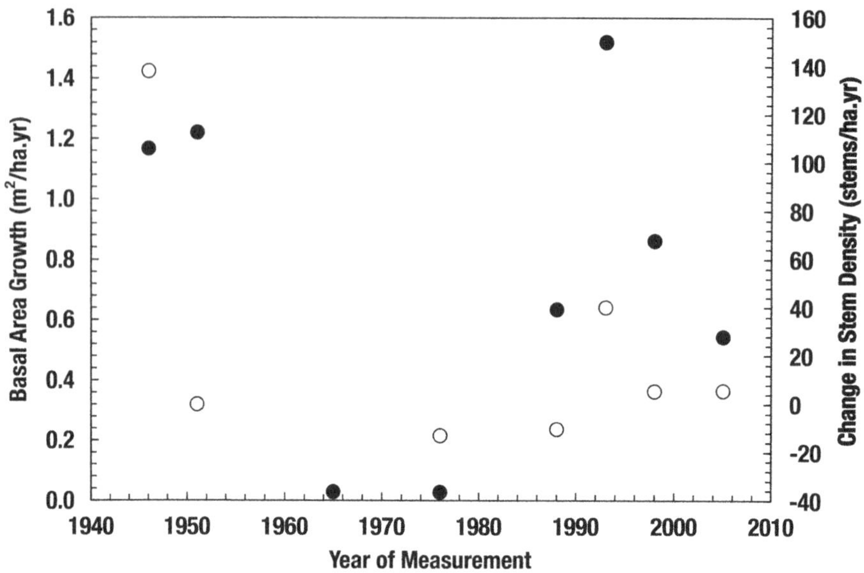

Fig. 2.5 Basal area growth rate (*solid circles*) and change in stem density (*open circles*) of a tabonuco forest stand (El Verde 3) in the Luquillo Experimental Forest. These rates were derived from data in Drew et al. (2009). The data points are plotted at the end of each interval of measurement, and the record extends from 1943 to 2005. (F. H. Wadsworth established the plot)

steadily declined, which provided more space for surviving trees. The basal area increment was low during the period of forest thinning and then increased significantly just before Hurricane Hugo passed over the forest. Again, the hurricane induced a peak in basal area growth and stem density was followed by reduced but positive rates.

2.5.1 The Bulletwood tree thrives after hurricanes

Manilkara bidentata, ausubo, or the bulletwood tree, is "one of the strongest and most attractive commercial woods in Puerto Rico" (Little and Wadsworth 1964, p. 444). It is also a primary forest species, with slow rates of growth and seedlings that remain on the forest floor for periods as long as 40 years before releasing in rapid bursts of growth toward canopy dominance (You and Petty 1991). Bulletwood growth benefits from the passage of hurricanes (Fig. 2.6). Bulletwood had peak recruitment and basal area increments after hurricane passages in 1932, 1989, and 1998 (Fig. 2.6). Bulletwood canopy trees benefited from the additional space allowed by stand thinning between the 1932 and 1989 hurricanes (1976 data point). However, the peaks in basal area accumulation after the hurricanes were higher than the 1976 basal area growth.

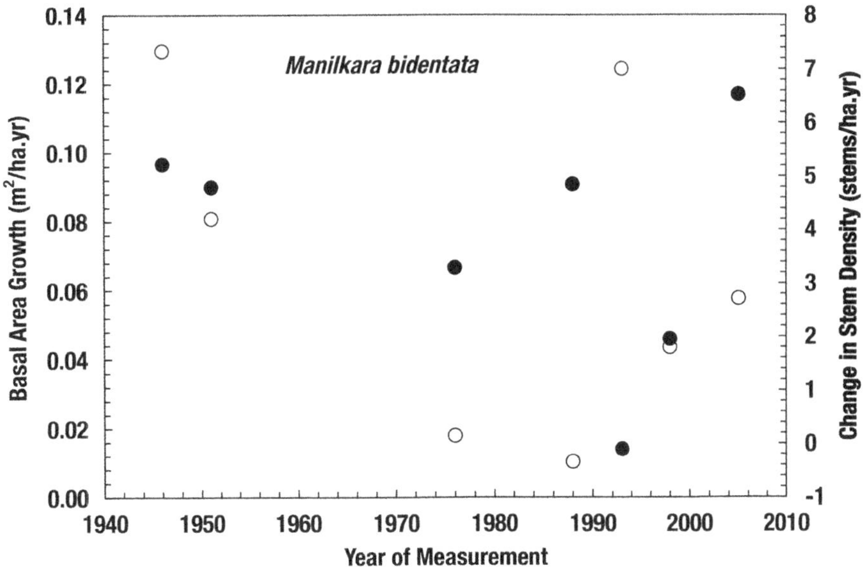

Fig. 2.6 Basal area growth rate (*solid circles*) and change in stem density (*open circles*) of *Manilkara bidentata* trees in a tabonuco forest stand (El Verde 3) in the Luquillo Experimental Forest. These rates were derived from data in Drew et al. (2009). The data points are plotted at the end of each interval of measurement, and the record extends from 1943 to 2005. (F.H. Wadsworth established the plot)

2.5.2 The Forest is a Carbon Sink

A steady accumulation of aboveground biomass occurred at El Verde 3 (Fig. 2.7). Accumulation rates were faster in the 1940s and 1950s than they were after 1976, but they were all positive with the exception of the period between 1951 and 1976. Others have interpreted short-term reductions in rates of biomass accumulation in tropical forests as a response to climate change (Phillips et al. 2005). However, our long-term record shows that the reduction is associated with forest maturation, i.e., as the forest approaches maximum biomass, its rate of biomass accumulation diminishes and this is compounded by increased tree mortality. Also of interest in this record is the continuing accumulation of biomass (a carbon sink) in spite of the passage of Hurricanes Hugo and Georges indicating rapid forest recovery following large-scale disturbance. The El Verde 3 plot reflects patterns documented elsewhere in the EF (Lugo 2008).

The 62-year record of tree growth at the El Verde 3 plot illustrates the complexity of tropical forest dynamics. Some species benefit from disturbance, while others fail to do so, and these responses seem to vary by site. Lacking tree ring data, the task of collecting, processing, and interpreting data is challenging and the results always seem far from conclusive. Each set of measurements initiates more questions given the variability of conditions and responses by so many species. Nevertheless,

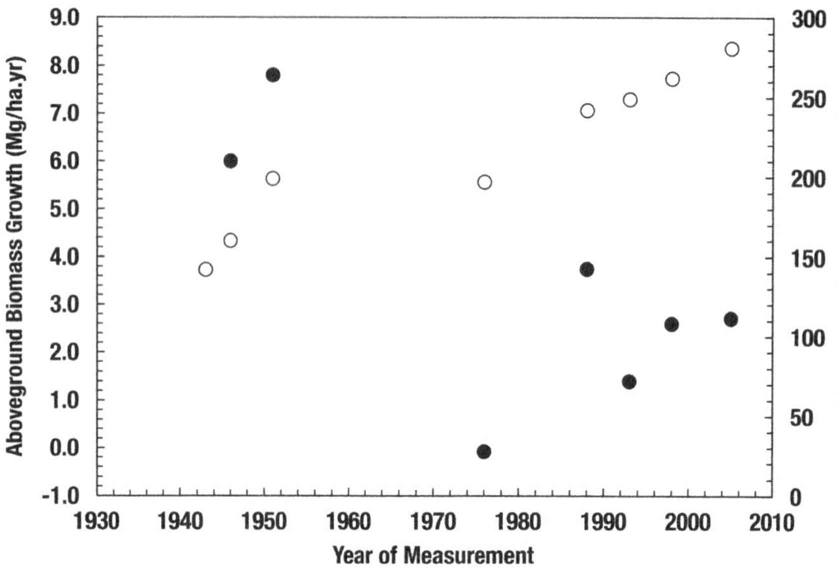

Fig. 2.7 Aboveground biomass growth rate (*solid circles*) and aboveground biomass (*open circles*) of a tabonuco forest stand (El Verde 3) in the Luquillo Experimental Forest. These rates were derived from data in Drew et al. (2009). The data points are plotted at the end of each interval of measurement, and the record extends from 1943 to 2005. (F. H. Wadsworth established the plot)

it behooves us to maintain and expand these plots as only through them will we be able to infer the long-term adjustments of tropical forests to climate and environmental change.

2.6 EFRs of the Pacific Northwest Research Station: From Water Yields to Old-Growth Forests

Bov Eav and Frederick J. Swanson

Are there patterns in the data from long-term records? Can we use them to inform our research program? Our experience with the EFRs of the Pacific Northwest Research Station provides examples of positive answers to both these questions. PNWS EFRs represent the programmatic and geographic diversity of Forest Service EFRs well. Sites range from several EFs on remote islands in southeast Alaska to the Wind River EF, a short drive from Portland, Oregon. On the Starkey Experimental Range, the presence of elk and cattle across the entire landscape integrates research that spans topics from grazing to forestry. Several studies use multiple EFs arrayed along environmental gradients or generally distributed across the region to assess broad-scale variation in peak streamflow in response to forest cutting and regrowth (Jones 2000), characteristics of forest communities (Acker et al. 1998), and ecological processes, such as wood decomposition (Chen et al. 2001). The historical

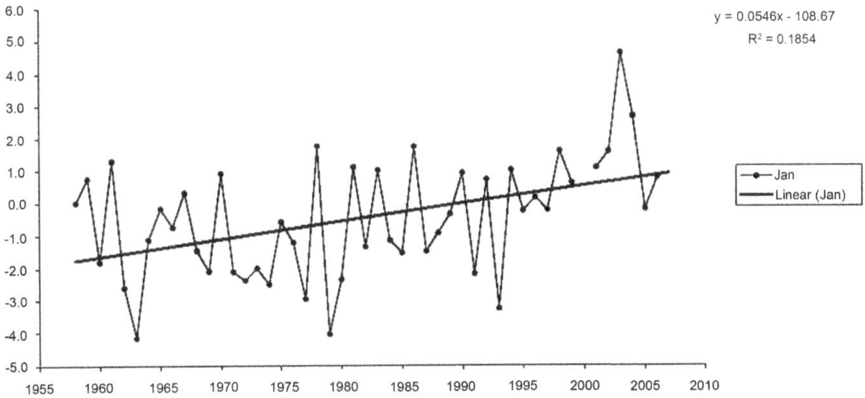

Fig. 2.8 Mean daily minimum air temperatures (°C) in January at the CS2 meteorological station in the H.J. Andrews Experimental Forest, 1958–2007. January is the month showing greatest warming at this meteorological station. January minimum air temperature has increased by 2.7 °C (36.86 °F) over the period 1958–2007. Monthly minimum air temperature is significantly correlated with the Pacific Decadal Oscillation (PDO), a measure of sea surface temperatures in the northern Pacific Ocean (r^2 = 0.19, 0.13, and 0.31 for January, March, April). When the effect of PDO is removed, January minimum air temperature has increased significantly by 2.0 °C (35.60 °F) from 1958 to 2007

scope of this work at several Pacific Northwest EFRs is captured in recent books (Luoma 2006, Geier 2007, Herring and Greene 2007, Joslin 2007).

2.6.1 Long-Term Records

Long-term records from EFRs have become exceptionally valuable scientific resources and common ground for intensive collaboration between the research and natural resource management communities. Some of the most valuable records have documented straightforward phenomena, such as air temperature and streamflow. Records initiated for one reason have gained unexpected value as science questions, research tools, and societal issues have moved in new directions over the years.

Records of air temperature and streamflow spanning about 55 years for the H.J. Andrew EF in the Cascade Range of Oregon offer excellent examples (Figs. 2.8 and 2.9), and the same could be said of many other EFRs. At the national level, EFRs have been the anchor points for these types of records. While the National Weather Service and U.S. Geological Survey sustain weather and gauging stations around the country, the EFRs are distinctive in combining these records with others, such as the chemistry of precipitation and streamflow and long-term vegetation change, and do so in mountain terrain and on small watersheds where other agencies seldom sample. This broad portfolio of ecosystem components sampled provides wonderful opportunities for interdisciplinary synergies and learning.

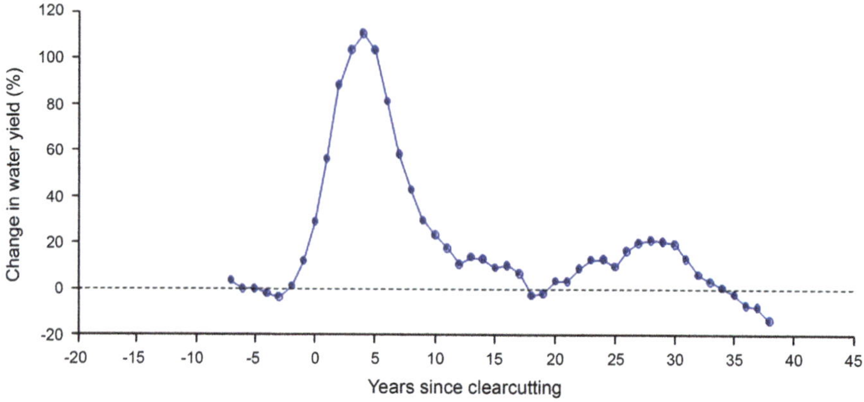

Fig. 2.9 Five-year running mean of percent changes in water yield relative to the pretreatment period in clearcut versus control watersheds at the H.J Andrews Experimental Forest for the water year (October–September). Maximum increases in annual water yield occurred in the first 5 years after clearcutting concluded in 1966, and were 110% at the Andrews (460-year-old evergreen conifer forest). Maximum decreases in annual water yield occurred in the most recent record, 30–40 years after clearcutting, when annual yields were 86% of pretreatment yields. (Adapted from Jones and Post 2004)

Uses of long-term environmental data have changed over time. Initially, data sets were used to characterize the environment of newly established EFRs. Record keeping commonly commenced with climate observations in support of studies in other fields, such as hydrology and forest growth. Initially, we thought that climate and streamflow were variable, but we did not expect them to exhibit long-term trends. However, as we now realize that climate change is underway, scientists return to the data sets with new questions; and interesting discoveries are emerging. For example, warming appears to be taking place in the H.J. Andrews EF, but the signal varies across the landscape and over the seasons of the year (most strongly in January—Fig. 2.8), which may have implications for biota and biological and geophysical processes. A part of the story now under investigation is the possibility that cold air drainage in this mountain landscape is ameliorating warming in valley floors during some seasons.

A focus of long-term streamflow studies is assessing annual water yield from clearcut watersheds blanketed with regrowing, young forest relative in comparison with runoff from unmanaged, control watersheds (Fig. 2.9). In the longest running experimental watershed study in H.J. Andrews EF, we observed greater runoff from the clearcut watershed for a period lasting more than 15 years. This is generally consistent with findings from similar studies in other sites.

Recently, we have made a surprising observation concerning streamflow from "control" watersheds (i.e., watersheds with no management actions). These have been exhibiting declining water yield for several decades. We do not yet have a clear explanation for this observation. Perhaps warmer air temperature is increasing evapotranspiration. Or perhaps vegetation succession over a period of several decades in the 150- and 500-year-old forests may be changing water use by vegetation. These observations are leading to new lines of research and fresh discussions of management implications.

Long-term environmental data have also become an important meeting ground for Forest Service researchers, land managers, and academic colleagues, thus fostering partnership efforts. Passions can run high on issues of public land management and policy, and long-term records ground discussions of these issues in objective measurements. Forest Service research has distinctive roles in the research–management partnerships involving these organizations collaborating at EFRs—maintaining long-term environmental records, guiding applied research projects, and helping bridge between research and land manager cultures. The partnerships extend well beyond the confines of the EFRs as scientists located elsewhere use long-term data for science and management purposes—in some cases never having to visit the field site.

An example of this role is found in the contributions that EFRs have made to our understanding and discussions of old-growth forests. Old-growth forests of the Pacific Northwest have meant many things to many people. In the 1930s, they were termed "large sawtimber"; in the 1960s, they were referred to as "decadent and overmature"; by the 1990s, those working for the preservation of these forests dubbed them "ancient" (Spies and Duncan 2009). Work at EFs brought important science into the picture. Highly interdisciplinary work in the International Biological Program at the H.J. Andrews EF during the 1970s revealed the great complexity of plant, animal, and fungal life of old forests and associated streams (Franklin et al. 1981). Further studies at Wind River, Cascade Head, and Pringle Falls EFs contributed to the picture of geographic variation in characteristics of old-growth forests. Findings from these studies have been used in conservation and restoration of old forests and also management of young plantations.

Over the 40 years history of public attention to old-growth forests, the EFs of the region have been a continuing source of new information from science, and also a meeting ground for public discussions of the future of the forest. This blending of the scientific work and public discourse is well represented in the recent book edited by Spies and Duncan (2009).

2.7 EFs of the Southern Research Station: Coweeta Hydrologic Laboratory: Long-Term Watershed Research in the Southern Appalachians

Jim Reaves, Katherine J. Elliott, and James M. Vose

How does our specific place within a larger landscape affect the way rain moves through forests and into soils and streams? How do changes in forest composition and structure affect these processes? How does our landscape affect the way we experience global weather patterns? The Coweeta Hydrologic Laboratory, located in the southern Appalachian mountains of western North Carolina, is one of 19

EFs across the southern USA within the Southern Research Station. Building on long-term climate, water quantity and quality, and vegetation data, the Coweeta Hydrologic Laboratory has advanced an interdisciplinary approach to understanding how watershed ecosystems respond to natural and human-caused disturbances. The basic philosophy is that if we understand how the watershed ecosystem works—the interconnections between climate, vegetation, soils, and water—then we can develop management practices to deal with the consequences of disturbances such as climate change, insects and disease, and extreme storm events. This approach requires integrating many scientific disciplines to understand the complex nature of both natural and managed forest ecosystems. Long-term data and experiments at Coweeta have been critical for separating treatment responses from natural variation; for testing hypotheses about vegetation, soil, and climatic controls on ecosystem processes; and for developing, testing, and validating predictive models.

2.7.1 Climate

Much of what is known today about mountain climatology and hydrology was determined from the extensive long-term climate network at Coweeta. For example, at the basin scale, precipitation consisted of frequent, small, low-intensity rains with occasional large storms at longer return intervals. In general, precipitation increases with elevation (about 5 % per 100 m) along the east–west axis of the Coweeta valley but changes little with elevation over north–south-facing side slopes (Swift et al. 1988). On an annual basis, precipitation exceeds evapotranspiration demand and streams flow perennially. Soil depth decreases and slope steepness increases with elevation. Both factors reduce the ability of the watershed to retain precipitation and thus increase the percentage that appears as streamflow. Upper elevation watersheds have lower precipitation minus runoff (P−RO) factors because they have less soil moisture storage capacity, a relatively higher percentage of precipitation as quickflow, and less evapotranspiration demand to create soil moisture storage opportunity before a rain (Swift et al. 1988). In addition to understanding rainfall–runoff relationships in managed and reference watersheds, accurate long-term climate measurements are critical for detecting variation in local climate. For example, climate trends at Coweeta reflect more extreme events in annual precipitation (Fig. 2.10) in the past two decades and an increasing temperature (Fig. 2.11) since the mid-1970s.

2.7.2 Water Quantity

Coweeta has been a leader in developing highly accurate measurement and data analysis procedures for gauged watersheds since 1934. Coweeta's 15 gauged watersheds provide some of the longest and best quality streamflow data in the world. The paired watershed approach—where one watershed is treated and one serves as a

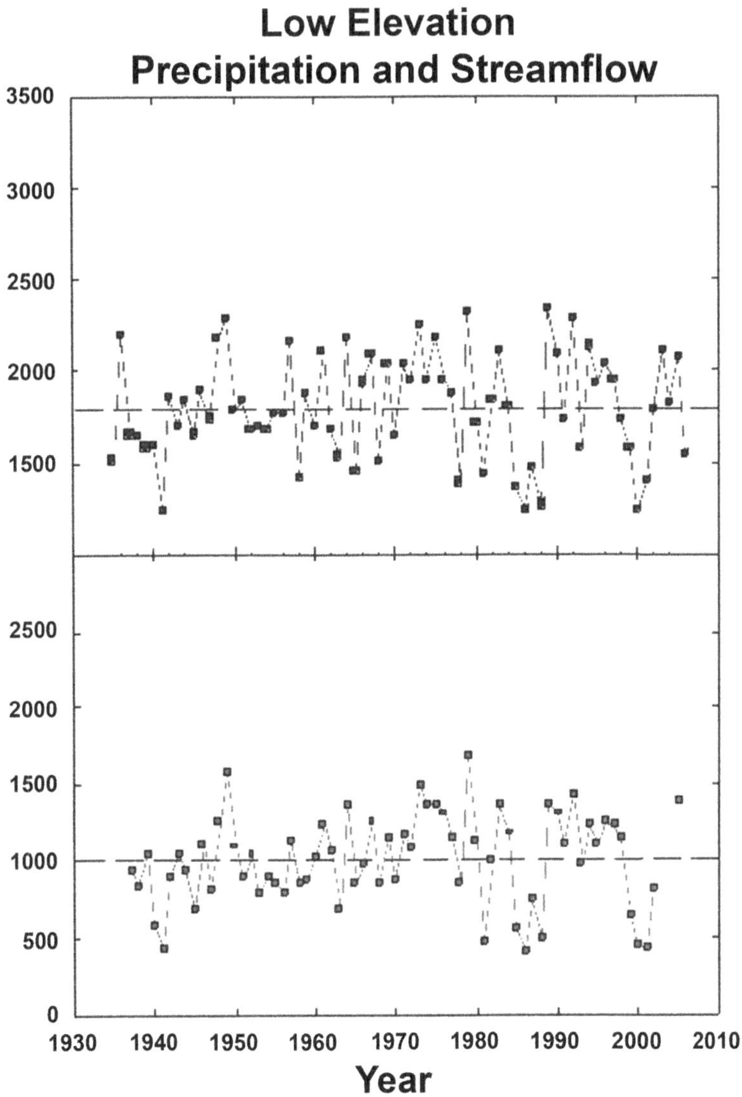

Fig. 2.10 Long-term annual precipitation for Coweeta Hydrologic Laboratory. Average annual rainfall is 180 cm (70.92 in.; *dashed line*) based on the 74-year record. Three significant droughts (2 or more consecutive years with ≥10 cm (3.94 in.) below average rainfall) have been recorded since the 1980s

reference—provides valuable information on the impacts of land management, unmanaged disturbances, and their interactions on the quantity and timing of streamflow from forested watersheds. The long-term data are also extremely valuable for developing and validating hydrologic models. Watershed-scale experiments include high-elevation (WS37) and low-elevation (WS7) clearcuts, pasture (WS6), white

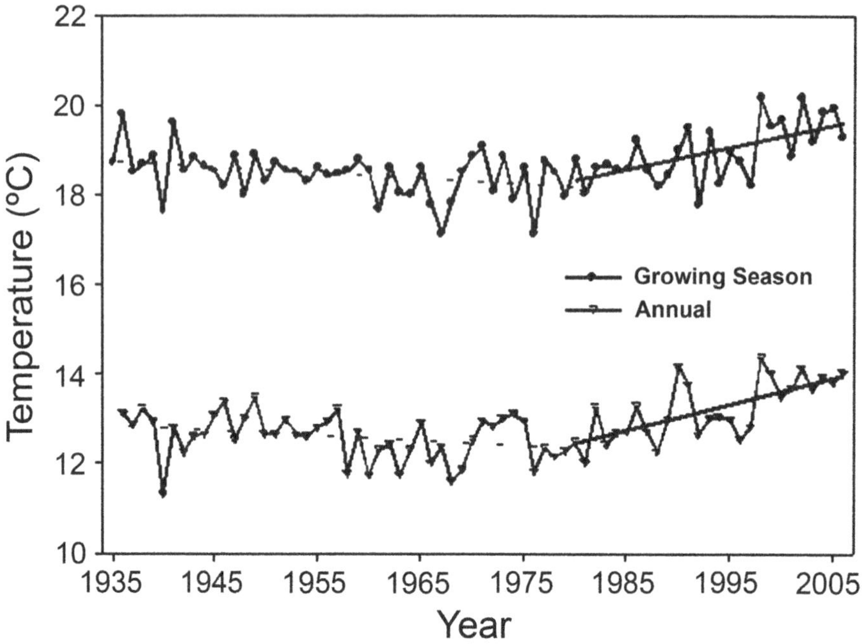

Fig. 2.11 Mean annual and growing season temperature for Coweeta Hydrologic Laboratory. Since 1976, a small but continuous increase in temperature has been recorded

pine (WS1 and WS17, Fig. 2.12), evergreen understory removal (WS19), 50% basal area removal (WS22), and multiple use (WS28). In addition to examining fundamental relationships among management, natural disturbance, and hydrologic processes, these long-term studies are being used to address important societal concerns both in the USA and internationally. For example, watershed experiments at Coweeta are being used to answer questions such as: (1) Does forest cutting increase flood risk? (2) Can forest cutting be used to augment streamflow to meet future water needs? (3) What are the hydrologic consequences of varying land cover types? (4) How does management interact with other stressors, such as extreme climatic events and native and nonnative invasive insects and diseases?

2.7.3 Water Quality

In the early days of Coweeta research, water quality research focused on sediment production as a result of land-use demonstration experiments (WS10 and WS3). In the 1970s, water quality research expanded to include water chemistry and nutrient transport, moving Coweeta into the area of ecosystem and nutrient cycling research (see Vose et al., Chap. 17). Coweeta now has a 36-year record of atmospheric deposition inputs and stream chemistry outputs. Coweeta scientists have used these

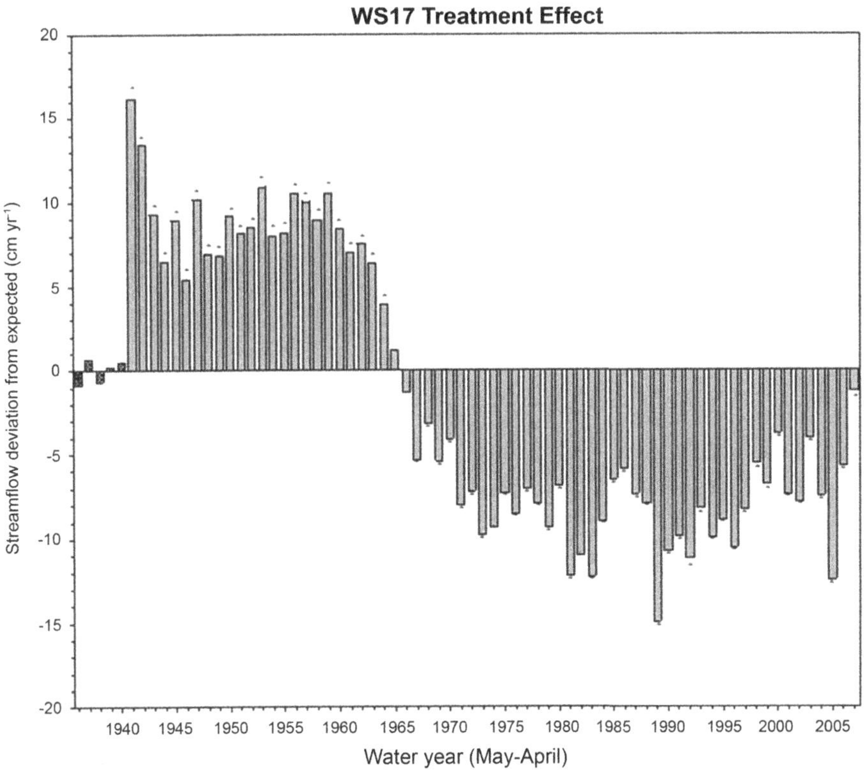

Fig. 2.12 Streamflow deviation from expected value for watershed 17, a deciduous forest that was converted to a white pine (*Pinusstrobus*) plantation in 1967. Reduced streamflow on the pine stand is attributable to greater interception and transpiration in the fall, winter, and spring

data to examine long-term changes in nutrient cycling patterns on reference watersheds (WS18 and WS27), effects of commercial clearcutting (Fig. 2.13, WS7), effects of prescribed burning, and effects of SO_4 deposition on class I wilderness areas. Knowledge gained from the long-term research program has had broad application. For example: Southern Appalachian forest watersheds retain nutrients—stream water has very low nutrient concentrations. Rapid vegetation recovery after disturbance retains nutrients on site. Long-term chronic acidic deposition may alter stream chemistry, particularly at high elevations. Losses of total site nitrogen after prescribed fire results from mass loss due to combustion not export via streams.

In addition to serving as signatures of ecosystem response to management and natural disturbances, stream chemistry studies provide guidance for forest management and streamside management and restoration. These long-term data are critical for understanding the relationships among disturbance, management, and water quality—and for developing the guidance to keep forests healthy and productive and water protected by the riparian zones that keep nutrients bound up in soils and vegetation. Further, these data have been important for validating computer-

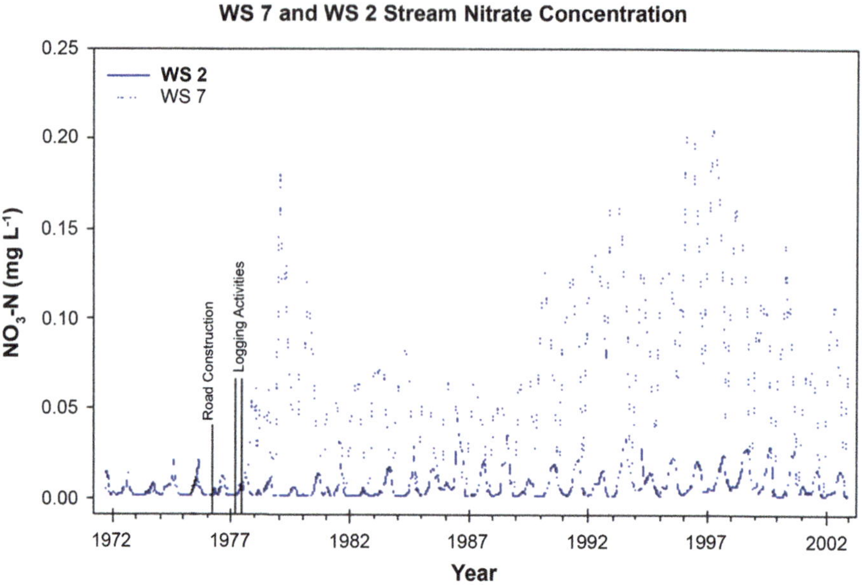

Fig. 2.13 Long-term changes (20 years) in water yield, the storm hydrograph, stream inorganic chemistry, and sediment yield were analyzed for a 59-ha (145.73 acres) mixed hardwood-covered catchment (Watershed 7; Swank et al. 2001). Stream chemistry has been measured on both WS7 (clearcut in 1977) and WS2 (reference) since late 1971. Nitrate (NO_3–N) concentrations began to increase on WS7 in early fall 1977, about 9 months after the initiation of logging and at the conclusion of site preparation cutting. Concentration increases remained low (0.02 mg L^{-1}) through the following summer and then peaked (0.18 mg L^{-1}) during the winter of 1978. A second peak also occurred the next summer. With forest regrowth, NO_3–N concentrations declined; peak summer values during the next 9 years were 0.07 to 0.12 mg L^{-1}. However, beginning in the summer of 1989, NO_3–N concentrations began to increase again, with peak values near 0.22 mg L^{-1}. In fact, in the summers of 1992 and 1995, stream water NO_3–N concentrations equaled or exceeded values observed in the first several years after clearcutting. Studies are currently being conducted to examine causal factors for the observed variation in stream NO_3 on WS7

based models that predict how Southern Appalachian watersheds might respond to changes in climate and air pollution.

2.7.4 *Vegetation*

A network of more than 900 permanent vegetation plots was first measured in 1934 and a subset has been remeasured in 1969–1972 and 1988–1993 (Elliot et al. 1999). Long-term vegetation plot surveys have allowed Coweeta scientists to evaluate the changes in forest structure, composition, and diversity from numerous disturbances, such as logging, drought, hurricanes, and invasive insects and pathogens. For example, American chestnut (*Castanea dentata*) was eliminated from the southern Appalachians and the eastern USA (Ellison et al. 2005) by the chestnut blight fungus

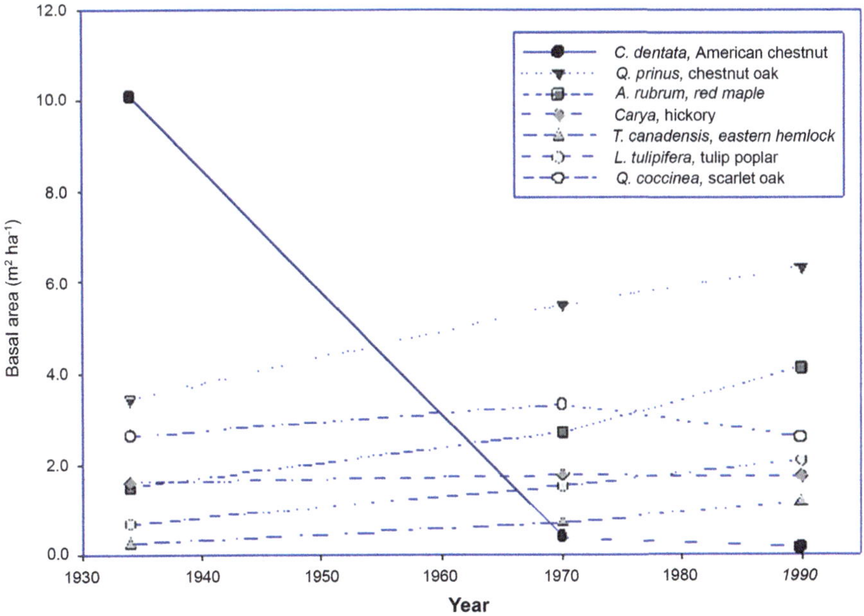

Fig. 2.14 Long-term changes in select tree species in the Coweeta Basin from inventories of permanent plots in 1934–1935, 1969–1972, and 1988–1993. American chestnut was the most abundant species in 1934, then it declined dramatically due to the chestnut blight fungus (*Endothiaparasitica*). Eastern hemlock increased in abundance and distribution, especially near streams across elevations. Tulip poplar replaced American chestnut in moist coves. Chestnut oak and red maple are ubiquitous, much like American chestnut before the chestnut blight, becoming dominant or codominant species across all environmental conditions. Red maple is now the second most abundant species in the Coweeta Basin. (Elliott and Swank 2008)

(*Endothiaparasitica*). In the 1934 survey, American chestnut was the most abundant species in the Coweeta Basin (Fig. 2.14). With the loss of American chestnut as the dominant species, other tree species replaced it in the forest canopy. Red maple (*Acer rubrum*) and chestnut oak (*Quercus prinus*) became the dominant species and tulip poplar (*Liriodendron tulipifera*) and eastern hemlock (*Tsuga canadensis*) were more abundant in coves and along riparian corridors (Elliott and Swank 2008). Hemlock, a species that increased following the loss of American chestnut, is threatened by another invasive species, hemlock woolly adelgid (HWA; *Adelgestsugae*; Ford et al. 2007; Nuckolls et al. 2009).

Long-term vegetation measurements at Coweeta have been used to understand the linkages between vegetation composition, structure, and watershed ecosystem processes. For example, ecosystem processes have changed with the demise of American chestnut and subsequent replacement of that species by others with different growth rates, litter qualities, and decomposition and nutrient cycling rates. Coweeta scientists are currently investigating the effects of the potential demise of eastern hemlock on ecosystem processes such as water (Ford and Vose 2007; Ford et al. 2007), carbon (Nuckolls et al. 2009), and nutrient cycling in riparian areas.

2.8 EFRs in the Rocky Mountain Research Station: Understanding Patterns of Forest Growth, Weather, and Disturbance

G. Sam Foster, Todd Mowrer, Russell Graham and Theresa B. Jain

How does forest growth integrate weather, insect and disease attach, management actions, and natural disturbance? Which of these has the most impact on forest growth, composition, structure, and change? These questions have animated the activities of scientists of the Rocky Mountain Research Station (RMRS) since its earliest days, and continue to animate our research today. RMRS is home to some of the first EFRs established in the Forest Service system: (1) Fort Valley Experiment Station was established in 1908 near Flagstaff, Arizona; (2) Fremont EF was established in 1909 on the Fremont and Pike National Forest near Wagon Wheel Gap Experimental Watershed (established 1911) and west of Colorado Springs, Colorado; (3) Priest River Experiment Station was established in 1911 at Priest River EF near Priest River, Idaho; and (4) Utah Experiment Station (now Great Basin Experimental Range) was established in 1912 near Ephraim, Utah. Perusal of the scientific and forest resource management literature, especially in the early to mid-twentieth century, reveals many examples of the research being conducted at least partially on EFRs (Daubenmire 1957; Davis 1942; Gisborne 1922). At one time, the current area of the RMRS contained at least 27 EFRs; the current number is 14.

The Priest River EF contains a long-term study that is representative of the ways we have answered the question about these interacting forces. The Priest River EF was established in the fall of 1911 and its northern Idaho location was selected because it contained the major forest types occurring in the northern Rocky Mountains and inland northwestern USA. The early researchers at Priest River EF recognized the value and importance of quantifying the response of forests to management actions (silviculture), disturbances, and weather. Within days of arriving at the forest, these scientists located a weather station in the compound at Priest River EF, at which temperature and precipitation have been continuously recorded since 1911 (see, Climate of the Priest River Experimental Forest 1983). After installing the weather station, they began establishing experiments on the forest to investigate how stands responded to silvicultural activities (cleanings, weedings, thinnings, and regeneration methods) and compared these results to how forests developed naturally. The resulting combination of vegetative growth data and weather data provides invaluable insights into patterns of forest change in managed and unmanaged stands, one example of which we provide here.

In 1914, a series of eight 0.2-ha plots were established adjacent to the weather station at Priest River EF to follow the development of both thinned and unthinned mixed conifer stands. These plots typified where western white pine (*Pinusmonticola* Douglas ex D.Don) forests grow as they receive at least 635 mm of precipitation a year and have a minimum of 25.4 cm of ash-capped soils underlying them (Haig et al. 1941; Jain et al. 2004). Western white pine and western larch (*Larixoccidentalis* Nutt.) were the dominant species with all other moist forest species present.

Initially, the plots were measured every 5 years; however, in 1954, the remeasurement cycle was extended to every 10 years. Each tree on the plots was tagged and heights and diameters were measured each time the plots were visited. As trees regenerated, they were added to the plot and the cause of death was noted for dead trees. Fire was excluded from the plots but weather (e.g., wind, snow ice), insects, and diseases impacted how the stands developed and the effects and evidence of these agents were recorded for each tree.

When established, the untreated plot contained a mix of naturally regenerated conifers about 30 years of age. The plot contained about 2,471 trees per hectare and 35.8 m^2/ha of basal area.

At Priest River EF, as in most of the western USA, the late 1920s and early 1930s was a period of low precipitation (Fig. 2.15c). Although this was the driest period in the history of Priest River EF, the reduced precipitation only minimally impacted tree growth as illustrated in Fig. 2.1 by the slight change in the slope of the basal area curve for the years 1933–1936. In addition, from 1921 through 1925 some low average minimum temperatures (Fig. 2.15b) occurred at Priest River EF. Growth of native forests on the Priest River EF was not impacted by this cold; however, the low temperatures killed a family of ponderosa pine trees of California origin growing at the forest. From 1940 through 1947, the average warmest high temperature ever recorded at the forest occurred and there was also a spike in the average minimum temperature observed (Fig. 2.15a and 2.15b). Again these climate fluctuations were not noticed in the growth of the native forests, as basal area exceeded 57.9 m^2/ha in 1948.

In 1985, the basal area occurring on the untreated plot peaked at 65 m^2/ha. A peak or culmination of volume and basal area around age 120 years is the norm as a western white pine forest ages or the stand becomes fully occupied (Haig 1932), not much different from the 110 years of age found on this plot. However, the decline exhibited in basal area in this undisturbed stand was the result of the mortality caused by disease in both western white pine and western larch. An introduced disease, white pine blister rust (*Cronartium ribicola*) killed the majority of the western white pines and red ring rot (Phellinuspini) significantly reduced the number of western larch. These tree species are being succeeded by grand fir (*Abiesgrandis* (Douglas ex D. Don) Lindl.), western red cedar (*Thujaplicata* Donn ex D. Don), and western hemlock (*Tsugaheterophylla* (Raf.) Sarg.); all shade-tolerant or late successional species, which in 2004 dominated the plot. This example showed that disease could be a major driver of forest development, while, at least in this case, the forests were very resilient to weather variation.

What we have illustrated here is only one of the hundreds of plots (replicated and non-replicated) for all types of vegetation, wildlife, insects, diseases and various other uses that have been established on the EFRs of the RMRS. This resource encompasses all of the vegetation types of the Rocky Mountains, desert and plains, and when combined with weather information the analytical and modeling possibilities are numerous. For example, data from these plots and many others located on RMRS EFRs are uniquely useful for validating models of forest growth, such as the Forest Vegetation Simulator (FVS; Dixon 2002). This forest growth model is used throughout North America and has utility at all levels of local to national forest management. Prediction of tree mortality, especially that induced by stresses and diseases, is greatly strengthened when data are available concerning trees and their growing conditions

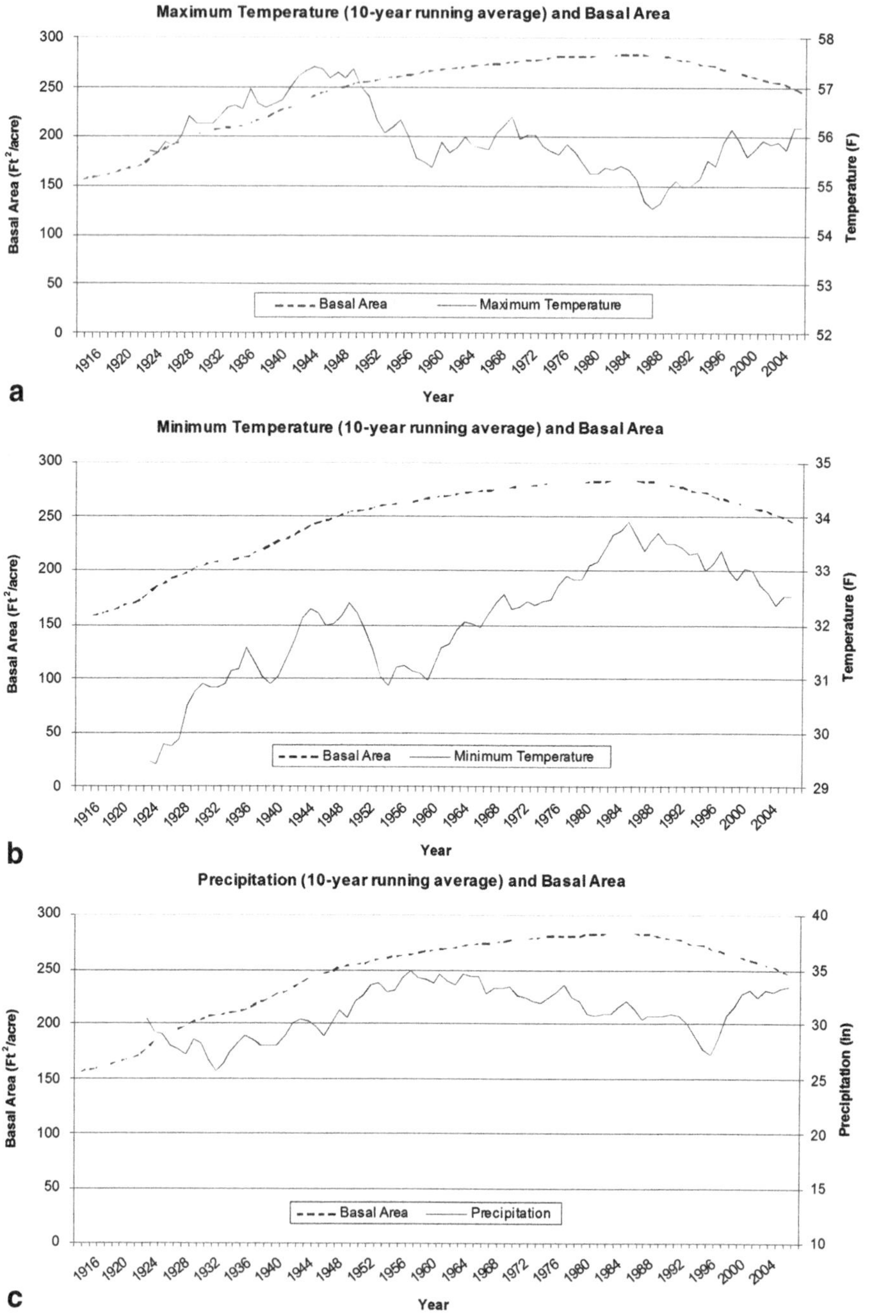

Fig. 2.15 The basal area growth of a natural mixed conifer stand located on the Priest River Experimental Forest in northern Idaho was minimally impacted by climate variations from 1914 through 2004. (**a**) Temperatures and precipitation (**b**) are displayed as 10-year running averages and basal area was estimated (**c**) from 5- and 10-year remeasurements of a fixed 0.2-ha (one-half acre) plot

over time. Permanent plots and associated long-term weather records such as those we have at Priest River EF are well suited to fulfilling this need, and will become more important as climate changes accelerate through the next century.

2.9 Conclusions

As directors of the seven regional research stations, each of us has found the long-term research conducted on EFRs a powerful tool for understanding patterns and processes of forests and forested watersheds at the deepest level. Scientists in Forest Service R&D have used—and continue to use—these data to answer critical societal questions, such as the cumulative effects of forest harvesting activities over time and space or how climate is changing. They have used EFRs to develop and test new methods of conducting natural resource science, from biogeochemistry to measurement of growth in tropical forests. What lessons emerge from our experience with these very valuable long-term data sets and studies on EFRs?

Data quality matters and reflects commitment throughout the R&D organization. The human dimension of long-term environmental research is as critical and challenging as the science itself. Long-term record keeping, even for seemingly simple measures of the environment, requires dedicated team effort. Field technicians and professionals maintain and calibrate instruments and collect the data over the years, punctuated by icy days with frozen fingers and sweaty days in the heat of late summer. Data managers carefully comb the data for glitches, store it, and make the hard-won data available to others for analysis. Scientists use the data in many ways—some planned, some serendipitous. A key to keeping the work moving forward is to balance persistence in record collecting with attention to the issue of the day to show the value of the records. Inattention either to the long-term persistence in record keeping or to addressing its relevance to current issues puts the enterprise at risk. All this takes a great deal of dedication of the entire workforce, which is motivated in part by respect for the forest and streams.

Data consistency matters. When measurements are skipped in one data series, our ability to capture the relationship among different elements of the system is compromised. Surprising patterns documented at the Caspar Creek EF, or in the long-term water yields from unmanaged watersheds on the H.J. Andrews EF, would be more difficult to detect if background stream measurements were suspended between formal studies. The comparison of insect and disease with weather on forest growth and development on the Priest River EF and the observations about species composition over time on the Luquillo EF are only possible because detailed measurements have been sustained through decades.

Scale and sampling intensity matter. Scientists at the Coweeta Hydrologic Laboratory are able to draw conclusions about the variability in hydrologic processes at different elevations and along different axes of the EF only because their sampling grid represents these differences adequately. Scientists at the Luquillo EF have gained better understanding of the site relationships of the hundreds of species in their forests because the network of plots across that forest encompasses a wide variety of sites.

Fig. 2.16 Scientists confer in an old-growth forest in the H.J. Andrews Experimental Forest. (Photographer: Michael Furniss, U.S. Forest Service)

Scientist creativity matters. Our public discourse often treats climate change in a fairly simplistic way, but scientists at the HJ Andrews have noticed seasonal differences in the pattern of warming, leading them to hypothesize and study patterns of the effects of cold air drainage on valley floor climate. Shigo, working with timber quality research data from northern EFRs, identified applications in urban and suburban tree care.

The ability to manipulate trees and forests matters. Shigo's research on tree quality and recovery from wounds depended upon the ability to damage trees, and Caspar Creek's ability to assess cumulative effects depended on the ability to superimpose disturbances on sites with long-term data records.

Sites with long-term data records are places in which important conversations can occur. Across the wide range of public opinion about sustaining forests and their values and benefits, passion for the forest itself is a common denominator. While EFRs are in no sense a panacea for the deep divisions in public opinion, their value for demonstrations and conversations cannot be overstated.

Data such as those acquired from permanent plots and long-term weather stations provide a more compelling and richer record than that acquired by sampling forest vegetation at a single point in time or from chronosequences of plots established in forests of different ages. Such sampling does not capture fine-scale disturbances, their interactions, and how they influence forest succession. In addition to the specific benefits illustrated by these examples from each station director, these data are valuable for addressing such issues as changes in wildlife habitat, vegetative responses to climate change, vegetative successional pathways, water use, sense of place, and timber production. Additionally, permanent plots provide great demonstrations of how forests develop and the treated plots readily show what different structures and compositions can be created and maintained through silvicultural treatments. Such demonstrations are invaluable to managers when making informed natural resource decisions and equally invaluable for communicating to policy makers, scientists, and the public.

The work at EFRs can lead to unexpected discoveries as well as findings based on careful hypothesis testing. Even rather simple observations can have great impact, if data are gathered for long periods of time, maintained well, and shared. The EFRs have a central role in sustaining observations at individual sites, and collectively the EFRs form a backbone of continental-scale sentinels for observing environmental change in the context of twenty-first century issues. But above all, the EFRs are meeting grounds (Fig. 2.16) for scientists, land managers, the public, and many others to discuss what we know about forests, rangelands, and watersheds, what else we need to know, and how to engage with these ecosystems in the future.

Much of what we currently know about forest and range ecosystems and their management has its roots in R&D conducted at least partially on EFRs.

References

Acker SA, McKee WA, Harmon ME et al (1998) Long-term research on forest dynamics in the Pacific Northwest: a network of permanent forest plots. In: Dallmeier F, Comiskey JA (eds) Forest biodiversity in North, Central, and South America and the Caribbean: research and monitoring. 1995 May 23–25. The Parthenon Publishing Group, Inc, Washington, DC. New York, NY, pp 93–106. (Jeffers JNR (ed) Man and the biosphere series, vol 21)

Baatz S (1996) Imperial science and metropolitan ambition: the scientific survey of Puerto Rico, 1913–1934. Ann N Y Acad Sci 776:1–16

Chen H, Harmon ME, Griffiths RP (2001) Decomposition and nitrogen release from decomposing woody roots in coniferous forests of the Pacific Northwest: a chronosequence approach. Can J For Res 31:246–260

Crow TR (1980) A rainforest chronicle: a 30-year record of change in structure and composition at El Verde, Puerto Rico. Biotropica 12:42–55

Daubenmire R (1957) Injury to plants from rapidly dropping temperature in Washington and northern Idaho. J For 55(8):581–585

Davis KP (1942) Economic management of western white pine forests. Technical Bulletin 830. U.S. Department of Agriculture, 77 p

Dixon GE (2002) Essential FVS: a user's guide to the Forest Vegetation Simulator. Internal report. U.S. Department of Agriculture, Forest Service, Forest Management Service Center, Fort Collins, CO, 189 p

Drew AP, Boley JD, Zhao Y (2009) Sixty-two years of change in subtropical wet forest structure and composition at El Verde, Puerto Rico. Interciencia 34(1):34–40

Elliott KJ, Swank WT (2008) Long-term changes in forest composition and diversity following early logging (1919–1923) and the decline of American chestnut (Castaneadentata (Marshall) Borkh.). Plant Ecol 197(2):155–172

Elliott KJ, Swank WT, Vose JM, Bolstad PV (1999) Long-term patterns in vegetation-site relationships in a southern Appalachian forest. J Torrey Bot Soc 126:320–334

Ellison AM, Bank MS, Clinton BD et al (2005) Loss of foundational species: consequences for the structure and dynamics of forested ecosystems. Front Ecol Environ 9:479–486

Ford CR, Vose JM (2007) *Tsuga canadensis* (L.) Carr. Mortality will affect hydrological processes in southern Appalachian forests. Ecol Appl 17(4):1156–1167

Ford CR, Hubbard RM, Kloeppel BD, Vose JM (2007) Acomparison of sap flux-based evapotranspiration estimates with catchment-scale water balance. Agric For Meteorol 145:176–185

Franklin AI (1983) Climate of the Priest River experimental forest, northern Idaho. Gen Tech Rep INT, 159

Franklin JF, Cromack K Jr, Denison W (1981) Ecological characteristics of old-growth Douglas-fir forests. General Technical Report PNW-118, U.S. Department of Agriculture, Forest Service, Pacific Northwest Forest and Range Experiment Station, Portland, OR, 48 p

Geier MG (2007) Necessary work: discovering old forests, new outlooks, and community on the H.J. Andrews Experimental Forest, 1948–2000. General Technical Report PNW-GTR-687, U.S. Department of Agriculture, Forest Service, Pacific Northwest Research Station, Portland, OR, 357 p

Gifford JC (1905) The Luquillo Forest Reserve, Porto Rico. Bulletin 54. U.S. Department of Agriculture, Division of Forestry, Washington, DC

Gisborne HT (1922) Weather records applied to the fire problem. USDA Forest Service, North. Rocky Mt. Forest and Range Experimental Station, Appl. For. Note 34, 4 p

Gleason HA (1926) The individualistic concept of the plant association. Bull Torrey Bot Club 53:7–26

Gleason HA, Cook MT (1926) Plant ecology of Porto Rico. In: Scientific Survey of Porto Rico and the Virgin Islands. New York Academy of Sciences, New York, pp 1–173

Haig IT (1932) Second-growth yield, stand, and volume tables for western white pine type. Technical Bulletin 323. U.S. Department of Agriculture, Washington, DC, 68 p

Haig IT, Davis KP, Weidman RH (1941) Natural regeneration in the western white pine type. Technical bulletin 767. U.S. Department of Agriculture, Washington, DC, 99 p

Herring M, Greene S (2007) Forest of time: a century of science at Wind River Experimental Forest. Oregon State University Press, Oregon, 188 p

Hill RT (1899) Notes on the forest conditions of Porto Rico. Bulletin 25. U.S. Department of Agriculture, Division of Forestry, Washington, DC

Jain TB, Graham RT, Morgan P (2004) Western white pine growth relative to forest openings. Can J For Res 34:2187–2197

Jones JA (2000) Hydrologic processes and peak discharge response to forest removal, regrowth, and roads in 10 small experimental basins, western Cascades, Oregon. Water Resour Res 36(9):2621–2642

Jones JA, Post DA (2004) Seasonal and successional streamflow response to forest cutting and regrowth in the northwest and eastern United States. Water Resour Res 40:W05203. doi:10.1029/2003WR002952

Joslin L (2007) Ponderosa promise: a history of U.S. Forest Service research in central Oregon. General Technical Report, PNW-GTR-711, U.S. Department of Agriculture, Forest Service, Pacific Northwest Research Station, Portland, OR, 121 p

Little EL (1970) Relationship of trees of the Luquillo Experimental Forest. In: Odum HT, Pigeon RF (eds) A tropical rain forest. National Technical Information Service, Springfield, pp B47–B58

Little EL, Wadsworth FH (1964) Common trees of Puerto Rico and the Virgin Islands. Agriculture Handbook 249. USDA Forest Service, Washington, DC

Lugo AE (2008) Visible and invisible effects of hurricanes on forest ecosystems: an international review. Austral Ecol 33:368–398

Lugo AE, Swanson FJ, Ramos González O (2006) Long-term research at USDA's Forest Service's Experimental Forests and Ranges. BioScience 56:39–48

Luoma JR (2006) The hidden forest: the biography of an ecosystem. Oregon State University Press, Corvallis, 228 p (Republished with new foreword by Jerry Franklin)

Murphy LS (1916) Forests of Puerto Rico: past, present, and future and their physical and economic environment. Bulletin 354. U.S. Department of Agriculture, Washington, DC

Nuckolls AE, Wurzburger N, Ford CR et al (2009) Hemlock declines rapidly with hemlock woolly adelgid infestation: impacts on the carbon cycle of southern Appalachian forests. Ecosystem 12:179–190

Phillips OL, Baker TR, Arroyo L (2005) Late twentieth-century patterns and trends in Amazon tree turnover. In: Mahli Y, Phillips OL (eds) Tropical forests & global atmospheric change. Oxford University Press, Oxford, pp 107–127

Spies TA, Duncan SL (2009) Old growth in a new world: a Pacific Northwest icon reexamined. Island Press, Washington, DC, 344 p

Swank WT, Vose JM, Elliott KJ (2001) Long-term hydrologic and water quality responses following commercial clearcutting of mixed hardwoods on a southern Appalachian catchment. For Ecol Manage 133:1–16

Swift LW Jr, Cunningham GB, Douglass JE (1988) Climatology and hydrology. In: Swank WT, Crossley JDA (eds) Forest hydrology and ecology at Coweeta: ecological studies, vol 66. Springer-Verlag, New York, pp 35–55

Wadsworth FH (1995) A forest research institution in the West Indies: the first 50 years. In: Lugo AE, Lowe C (eds) Tropical forests: management and ecology. Springer, New York, pp 33–56

You C, Petty WH (1991) Effects of Hurricane Hugo on Manilkarabidentata, a primary tree species in the Luquillo Experimental Forest of Puerto Rico. Biotropica 23:400–406

Ziemer RR (technical coordinator) (1998) Proceedings of the conference on coastal watersheds: the Caspar Creek Story. General Technical Report, PSW-GTR-168, U.S. Department of Agriculture, Forest Service, Pacific Southwest Research Station, Albany, CA, 149 p

Part II
Research Trajectories in Forest Ecology, Management, and Silviculture for Conifers

Chapter 3
The Value of Old Forests: Lessons from the Reynolds Research Natural Area

Don C. Bragg and Michael G. Shelton

Abstract In 1934, the Crossett Experimental Forest (CEF) opened to develop good forestry practices for the poorly stocked pine-hardwood stands that arose following the high-grading of the virgin forest. One CEF demonstration area has had no active silviculture other than fire protection since 1937; this 32.4-ha stand is now the Russell R. Reynolds Research Natural Area (Reynolds RNA). Periodic inventories of this tract provide a unique account of long-term stand development under minimal anthropogenic disturbance. For instance, successional change has been characterized by the slow conversion from pines to hardwoods. Gradually, as the dominant pines die, they are replaced by increasingly shade-tolerant hardwoods, resulting in a dense understory and midstory. Without concurrent fire to help prepare the seedbed, even a relatively severe bark beetle infestation in 1993–1994 failed to sufficiently disturb the site and permit the establishment of a new pine cohort. In addition to lessons learned on succession in this cover type, research associated with the Reynolds RNA has also helped develop old-growth restoration strategies, the ecological role of large dead wood in southern pine forests, the deleterious effect of dense midstory hardwoods on red-cockaded woodpecker habitat, the value of old forests in modeling tree allometry and carbon sequestration, and the unexpected benefits of preserving unique landscape features for future study. Clearly, the Reynolds RNA has demonstrated that there are opportunities to learn from passive stand management.

Keywords Crossett Experimental Forest · Disturbance · Red-cockaded woodpeckers · Southern pines · Succession

3.1 Background

Toward the end of the nineteenth century, timber supplies were nearly exhausted in the northern USA, and many companies began moving their logging operations to the south—a land of seemingly endless pine forests (Baker and Bishop 1986).

D. C. Bragg (✉) · M. G. Shelton
Southern Research Station, USDA Forest Service, P.O. Box 3516 UAM, Monticello, AR 71656, USA
Phone: (870) 367-3465
e-mail: dbragg@fs.fed.us

D. C. Hayes et al. (eds.), *USDA Forest Service Experimental Forests and Ranges*, DOI 10.1007/978-1-4614-1818-4_3, © Springer New York 2014

The early harvesting strategy was almost universally one of high-grading the virgin timber, in which only the best trees were taken (Della-Bianca 1983) and only the choicest parts of these trees were utilized, leaving huge quantities of usable timber in the woods to either burn or rot (Chapman 1913). The abundant supply of cheap, high-quality timber provided virtually no incentives to curb this reckless behavior. Large-scale removal of old-growth pine began in the 1890s in southern Arkansas, and was almost complete by 1930 (Reynolds 1980; Smith 1986). The Crossett Lumber Company, as well as numerous other operations, thoroughly cut the piney woods of Arkansas, Louisiana, and Texas, producing billions of cubic meters of valuable lumber and leaving behind logging slash, stumps, and scattered unmerchantable trees. Most of these companies then followed a "cut-out and get-out" strategy—they would either move their operations to the next area of virgin timber or close their business. By the late 1920s, though, the end was in sight for this free-for-all, and a few lumber companies such as Crossett resolved themselves to making sustainable, science-based forestry work on their cutover lands—they just needed help learning how to do it (Reynolds 1980; Darling and Bragg 2008).

At this time, ideas on the nature and value of the virgin forest had changed considerably, but were still far from the conservation of today. The science of forestry had yet to catch up with its practice, and the efficacy of these new techniques needed to be documented and demonstrated if the profession was to succeed. Indeed, the whole forest products industry required a revamp, and soon—the fast-growing second-growth timber that appeared after the "big cut" was thought to be substandard for lumber (Reynolds 1980) and the industry and local economy faced collapse. Experimental forests were vital to this learning process, and a number of them were established to provide proof of concept. In late 1933, Russell R. Reynolds of the USDA Forest Service scoured the lands of the Crossett Lumber Company for a suitable location for such an experimental forest and research center (Reynolds 1980).

And he found it! Opened in 1934, the Crossett Experimental Forest (CEF) is located on the Upper West Gulf Coastal Plain in Ashley County, Arkansas, 11 km south of the city of Crossett (Fig. 3.1). Over the decades, considerable research has been conducted on this experimental forest. Though the 680-ha CEF is renowned for its role in the development of uneven-aged silviculture, a small tract of unmanaged old timber, the Russell R. Reynolds Research Natural Area (Reynolds RNA), has also proven highly illustrative regarding ecosystem patterns and processes. The evolving role of the Reynolds RNA, first established as a control of active silviculture in pine-hardwood stands, now allows for the consideration of many contemporary resource issues and continues to provide new lessons for the future.

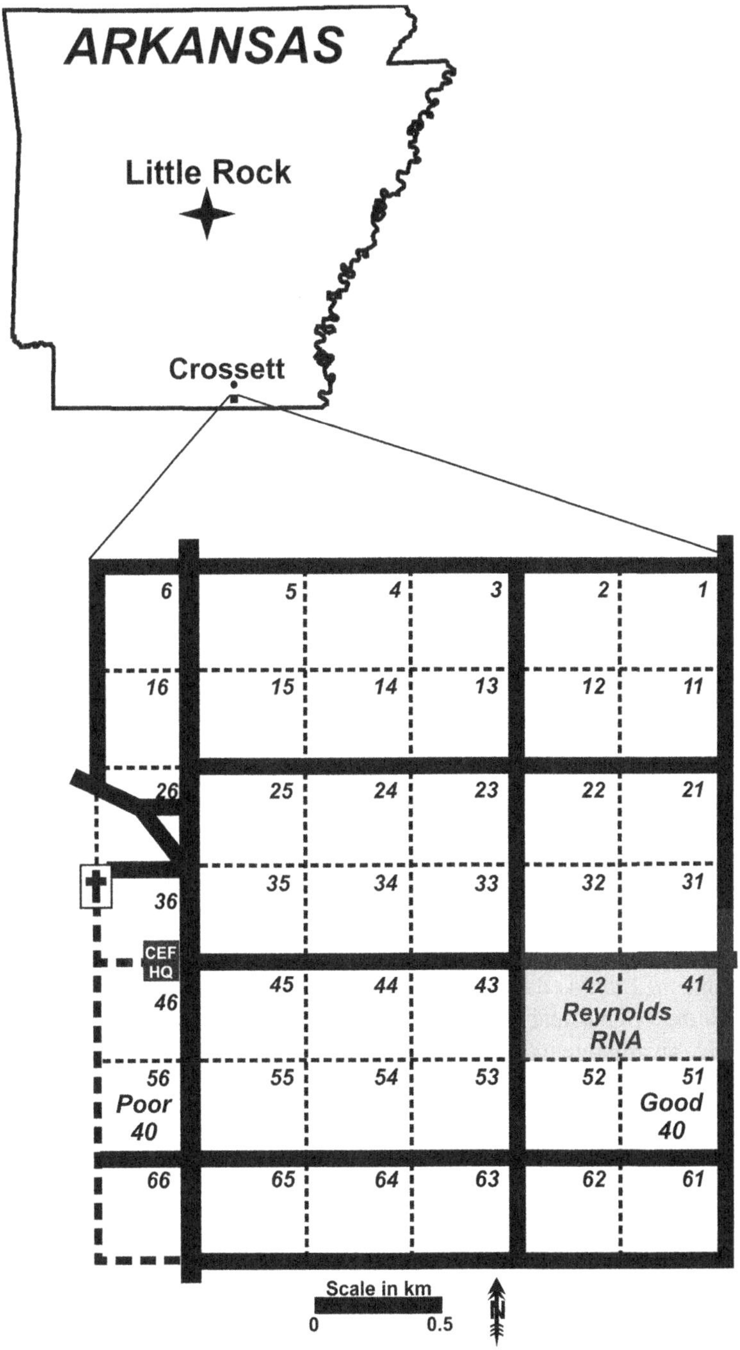

Fig. 3.1 Map of the location and layout of the CEF in Arkansas, including the Reynolds RNA

3.2 Reynolds RNA Description

3.2.1 Establishment History

Since hardwoods had little value during the original logging period, most were left among the surviving submerchantable pines. These legacy trees often grew at impressive rates once released and also provided a seed source for future stands. Periodic fires were common in the virgin forest; some fires were very intense and widespread, particularly in the heavy slash left after cutting (Reynolds 1980). Although these fires destroyed considerable quantities of trees, some fast growing individuals were able to reach fire-tolerant sizes between burns, and certain species—such as shortleaf pine (*Pinus echinata* Mill.) and most of the hardwoods—could resprout after being top-killed (Cain and Shelton 2000). These historic fires also prevented the buildup of litter on the forest floor and killed herbaceous vegetation, creating a receptive substrate for the small, wind-disseminated pine seeds. However, many areas burned too frequently and intensely to adequately reforest, resulting in poorly stocked, second- growth stands with a wide range of tree size classes—conditions viewed as too challenging for the practice of forestry (Anonymous 1981). Reynolds even credited local "woods burners" for helping to locate of the CEF—the Crossett Lumber Company was eager to have the Forest Service share responsibility in this arson-plagued portion of Ashley County (Reynolds 1980).

Most of the first year's activity on the CEF focused on building a modest headquarters, constructing 22 km of roads, establishing a system of 16.2-ha compartments, and inventorying the existing stands. By the summer of 1935, it was time to formulate a research strategy for the CEF, as Reynolds (1980, p. 12) recalled:

> ...[we] had a final meeting to agree on the assignment of research study areas on the Experimental Forest. The result was that 80 acres [32.4 ha] was to be left untouched as a 'Natural Area'...[another] 80 acres [32.4 ha] was for farm forestry studies.

This planning session established the three hallmark demonstration areas that have been maintained to this day—the compartments that would become the Reynolds RNA and the "Good" and "Poor" Farm Forestry Forties. The remaining area was allocated to an arboretum (planted in 1935), small plot research, large compartment studies, and administrative purposes.

The role of the natural area changed dramatically over the years. Initially, it served only as an unmanaged control to highlight the enhanced productivity of the Good and Poor Forties. However, this eventually changed. Following a rapid expansion and product diversification during the 1930s and 1940s, forest industry sought to intensify their timber management to ensure a large quantity of inexpensive raw materials (Heyward 1958). The rapid growth of naturally regenerated stands was not able to meet this increased demand. Furthermore, the implementation of effective fire control and the large-scale abandonment of marginal croplands spurred work on pine plantation management. Silvicultural research in southern pines had developed in tandem with the forest products industry, and had proven the efficacy of plantations of genetically superior seedlings, competition control, density control, and thinning to increase fiber production. During this period, the expansion of the southern pulp and paper industry also increased demand for fast-growing young pines (lower pitch content), which could be cut in mid-rotation thinnings of natural or planted stands (Heyward 1958).

By the 1960s, the industry was shifting from naturally regenerated southern pine stands to genetically improved pine plantations. To some, the selective timber management research at the CEF seemed dated and inadequate for a future silvicultural universe of artificially regenerated commercial forests (e.g., Wakeley 1964). Even after his retirement, Russ Reynolds maintained a spirited defense of this system (e.g., Reynolds 1974), but the growing interest in plantation silviculture was clearly one of several factors that led to the closing of the CEFin 1974. The CEF was reopened in 1979 after the Forest Service recognized that many landowners remained interested in silvicultural options other than intensive plantation culture. Hence, the focus of the CEF shifted toward low-cost management alternatives designed to appeal to small private landowners, as well as public agencies and large private owners interested in other uses of their timberlands. During this revisioning, the low productivity of the unmanaged natural area was still used to contrast the enhanced production of adjacent managed stands, but the non-timber attributes of what had become a mature, mixed pine-hardwood stand were also touted, especially aesthetic properties, wildlife benefits, and recreational potential (Baker and Bishop 1986). Also in the 1980s (about 70 years since the virgin forest was cut), the Reynolds RNA achieved a status of its own—it had become old and thus unique.

In 2005, this unmanaged parcel was officially designated as the Russell R. Reynolds RNA (USDA Forest Service 2005). The Forest Service developed their RNA program to preserve, as reference areas, examples of natural features and processes in ecosystems that can be contrasted with more human-influenced environments (Northern Research Station 2010). RNAs are considered to be natural laboratories and outdoor classrooms of historical and biological significance (Fountain and Sweeney 1987). Though some were established to protect small areas of old growth on national forest lands (Devall and Ramp 1992), many (such as the Reynolds RNA) were located in previously cutover stands and thus offered the opportunity to understand long-term vegetation dynamics and forest succession (Hemond et al. 1983).

3.2.2 Woody Vegetation, Past and Present

Prior to the early 1900s, records of the composition of virgin pine forests in the South are spotty at best (Eldredge 1952). Most accounts refer to pine and a handful of other commercial species, and are largely silent on the minor taxa that were present. However, the virgin forests are now known to be considerably more dynamic, complex, and robust than the ecological deserts they were once considered. A recent literature review of the region prior to lumbering found a wide range of pine dominance, with shortleaf being the most common pine on many upland sites, probably due to fire (Bragg 2008a). In southern Arkansas, the upland virgin forest was dominated by loblolly (*Pinus taeda* L.) and shortleaf pine in a roughly equal mixture, often with a significant hardwood component (Reynolds 1980; White 1984; Bragg 2004a; 2008a). The second-growth forests that appeared after lumbering differed in key ways. Where Yale Professor Herman Haupt Chapman had inventoried open, multi-cohort stands of old-growth pine-dominated forests in the uplands of the Crossett area (Chapman 1912, 1913), just two decades later Reynolds inherited a mix of variably stocked young pines and hardwoods that emerged following logging, overtopped by legacy trees.

Originally, the unmanaged CEF natural area was tallied with a 100% inventory by 2.5-cm-diameter at breast height (DBH) classes with a minimum measurement threshold of 9-cm DBH; eight inventories were conducted in this manner from 1937 to 1993. Only broad species groups, such as pine, oaks, and other trees, were recorded in these inventories. However, it became apparent that these coarse inventories provided an incomplete picture of the net changes that were occurring, and were unsuitable for determining stand dynamics, including survivor growth, ingrowth, and mortality. Thus, in 1989, 12 permanent 0.1-ha plots were established in the natural area where individual trees ≥9.0-cm DBH were numbered, identified by species, measured, and their location mapped. Two years later, eight additional plots were established for a total of 20 plots representing 6% of the area (Shelton and Cain 1999), and all plots are now measured about once every 5–10 years. Also, subplots within the permanent plots were established to collect information on seedlings and saplings of woody species.

Today, the overstory of the Reynolds RNA is still dominated by loblolly pine, with noticeably lower amounts of shortleaf pine. Of the canopy hardwoods present, white oak (*Quercus alba* L.), southern red oak (*Quercus falcata* Michx.), post oak (*Quercus stellata* Wang.), water oak (*Quercus nigra* L.), cherrybark oak (*Quercus pagoda* Raf.), sweetgum (*Liquidambar styraciflua* L.), and black gum (*Nyssa sylvatica* Marsh.) prevail (Cain et al. n.d.; Shelton and Cain 1999). The midstory is dominated by increasingly shade-tolerant hardwood species, including eastern hop hornbeam (*Ostrya virginiana* (Mill.) Koch.), elms (*Ulmus* L.), American holly (*Ilex opaca* Ait.), red maple (*Acer rubrum* L.), and flowering dogwood (*Cornus florida* L.). The understory is composed of numerous tree seedlings, although pines larger than recent germinants are conspicuously absent in this layer. A variety of shrubs and woody vines, including American beautyberry (*Callicarpa americana* L.), deciduous holly (*Ilex decidua* Walt.), hawthorns (*Crataegus* L.), huckleberries (*Vaccinium* L.), poison ivy (*Toxicodendron radicans* (L.) Kuntze), greenbrier (*Smilax* L.), grape (*Vitis* L.), Chinese privet (*Ligustrum sinense* Lour.), and sweetleaf (*Symplocos tinctoria* (L.) L'Hér) are abundant (Cain and Shelton 1995).

3.3 Current Research and Synthesis on the Reynolds RNA

Throughout much of human history, old forests were thought to be unproductive, stagnant, and even sterile environments. Rather than considering the value of old growth for the protection of its dependent species, scientists often dismissed the virgin forests as decadent timber stands with poor wildlife habitat scarcely fit for birds or rodents (e.g., Munger 1930). Over the past few decades, the role of old forests has been completely reevaluated. The following discussion focuses on two recent research outcomes from the Reynolds RNA that consider the value of old forests in contemporary landscapes, rather than its role as an unmanaged control for more conventional silvicultural research.

3.3.1 Structural and Functional Lessons

One of the first ecological lessons from the Reynolds RNA arose from the long-term dynamics of an unmanaged, relatively undisturbed second-growth pine-dominated stand. Stand developmental trajectories became of particular interest in the 1980s as ecological theories on forest succession matured, and the Reynolds RNA became the subject of publications describing the vegetative dynamics of its understory (Cain 1987) and overstory components (Guldin and Baker 1985). Their work was made possible by the long sheltered history of this stand, and has since been complemented by a number of follow-up studies that have further documented change to the structure and function of the Reynolds RNA (e.g., Cain and Shelton 1995, 1996; Shelton and Cain 1999; Bragg and Shelton 2011).

After the original logging of the area, most of the residual timber in the Reynolds RNA was loblolly and shortleaf pine, which destined the early stages of this development to be heavily pine-dominated for decades (Fig. 3.2). After lumbering, these legacy pines grew considerably larger than either the new crop of recently germinated pines or slower growing residual hardwoods, producing an irregular size distribution and multistoried pine canopy comparable to managed uneven-aged stands (Fig. 3.2, 1937 and 1942). This structure, however, was fleeting as the vigorous pine regeneration soon ascended into the overstory. Eventually, as the canopy closed, pine regeneration began to fail and a variety of hardwoods soon dominated the understory and midstory. The pine size distribution became broadly unimodal, with virtually all pines except the most suppressed individuals reaching the overstory (Fig. 3.2, 1952–1983). Over the past few decades, many of these pines died, interrupting what had largely been a continuous pine canopy across the Reynolds RNA (Fig. 3.2, 1993–2007). Mortality has claimed all of the overstory pines in some parts of this stand, leaving a pure hardwood overstory. In other areas, only a handful of super-canopy pines rise above a closed hardwood canopy. Pine density (stems >8.9-cm DBH) declined from 320 stems/ha in 1937 to 66 stems/ha in 2007, while hardwood density increased from 100 stems/ha to 383 stems/ha over the same period. Individual pine growth has ameliorated the impact of the loss of stems on relative stand density—pines accounted for an average of 66 % of the total basal area in 1937 and 54 % in 2007.

Coarse woody debris (CWD) has also accumulated differently in the Reynolds RNA than in managed stands in the same region. Although the ecological role of dead wood has been recognized for years (e.g., Lemon 1945; McMinn and Crossley 1996; Braccia and Batzer 2001), very little work has been done in assessing CWD

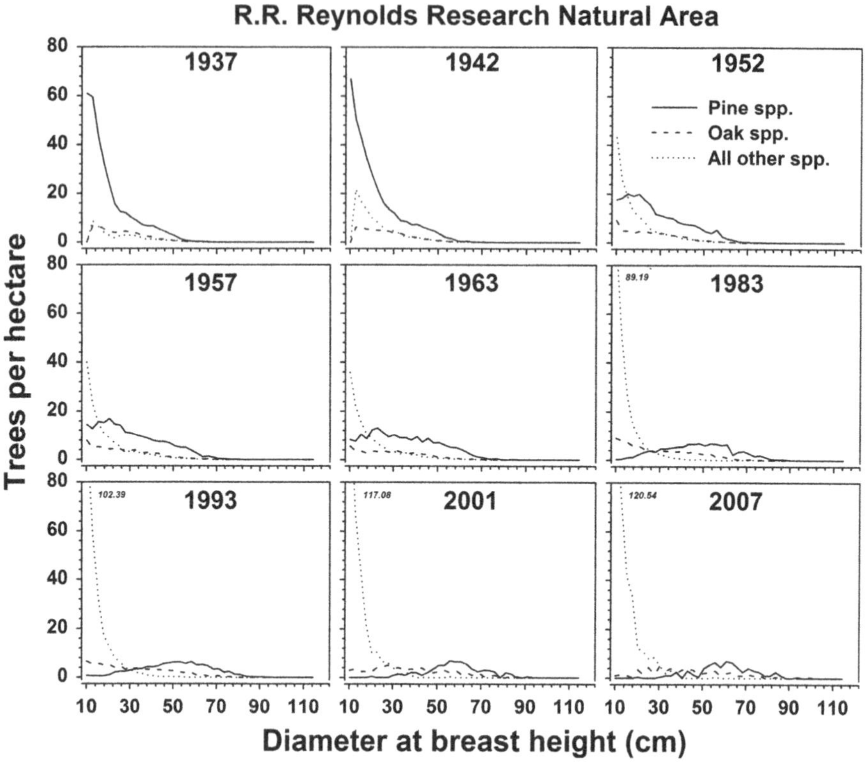

Fig. 3.2 Patterns in stem density (trees >8.9-cm DBH) over the 70-year period of observation on the Reynolds RNA. Small italicized numbers in the upper left corners of the graphs from 1983 to 2007 are the number stems in the smallest size class truncated by the density scale

volumes in southern forests, especially for mature stands of natural origin. From a silvicultural standpoint, dead wood was traditionally considered an undesirable attribute of any forest type, and was typically ignored or treated as a forest health problem. However, in recent years CWD has recognized as an important ecosystem metric associated with habitat quality (e.g., Braccia and Batzer 2001; Fan et al. 2005), carbon storage (Radtke et al. 2009), and other ecological functions.

Research in the Reynolds RNA has shown that this stand has accumulated considerably more dead wood than nearby examples of managed timber (Zhang 2000; Bragg 2004a). Table 3.1 provides estimates of CWD volume from a number of mature, pine-dominated forests from the region. The 90 to >300 m^3/ha totals in the Reynolds RNA are several times greater than mature managed second-growth stands, and comparable to that found in a nearby old-growth remnant that experiences periodic salvage (Bragg and Heitzman 2009). These accentuated CWD levels can be largely attributed to the senescence of many large pines due to windthrow, lightning strikes, and southern pine beetle (SPB) infestation (Cain and Shelton 1996; Zhang 2000). Intensively managed pine stands, such as the Good Forty

Table 3.1 Coarse woody debris volume in some mature pine-hardwood stands of southern Arkansas

Stand	Silvicultural regime	Volume (m^3/ha)	Source
Good Forty[a]	Managed second-growth	35.5	Zhang (2000)
Reynolds RNA[a]	Unmanaged second-growth (with some old-growth remnants)	93.7–309.7	Zhang (2000)
POWCNA[b]	Unmanaged second-growth	28.9	Bragg and Heitzman (2009)
Levi Wilcoxon DF[b]	Old-growth (some salvage of dead pine)	191.0	Bragg (2004a)
Hyatt's Woods[b]	Remnant old-growth (some salvage of dead trees)	19.8	D.C. Bragg, unpublished data

[a] Located on the Crossett Experimental Forest (AshleyCounty, AR) of the USDA Forest Service
[b] The Prisoner of War Camp Natural Area (POWCNA) is owned by the University of Arkansas at Monticello (Drew County, AR), the Levi Wilcoxon Demonstration Forest (DF) is owned by Plum Creek Lumber Company (Ashley County, AR), and Hyatt's Woods is a privately owned parcel in the southern portion of Drew County, AR

Demonstration Area on the CEF, typically have much less dead wood because of salvage, lower large tree density, and the higher vigor of the managed stands, which make them less vulnerable to certain mortality events such as SPB (Thatcher et al. 1980). Two other nearby unmanaged old stands (the POWCNA and Hyatt's Woods) have much less CWD than the Reynolds RNA because they have yet to reach the stage of extensive pine overstory mortality (Bragg and Heitzman 2009).

3.3.2 Forest Succession and Natural Disturbance

The development of a substantial hardwood component during the undisturbed natural succession of pine-dominated southern forests is well documented (e.g., Wahlenberg 1960; Quarterman and Keever 1962; Switzer et al. 1979; Glitzenstein et al. 1986; Huston and Smith 1987). This progression reflects differences in the autecology of the associated species in the southern forest community, which affect their establishment, development, and survival over long periods of time. Loblolly and shortleaf pines are opportunistic, shade-intolerant species which can rapidly establish and capture the resources of an unoccupied site (Shelton and Cain 2000). Although much harder to generalize, hardwoods typically tend to be less opportunistic, more shade-tolerant, and slower growing—at least on the upland sites they share with the pines. Under these circumstances, additional external factors are needed to ensure the long-term perpetuation of unmanaged pine-dominated ecosystems.

Conventional wisdom now holds natural disturbances capable of this, although this perspective was not always the case. For example, during the transition from unsustainable lumbering to scientific forestry, fire was almost always seen as a destroyer of timber and a limitation to forest productivity. Virtually all early published reports emphasized the losses that arose from uncontrolled fire (e.g., Barrett 1928; Forbes 1923; Bruner 1930; Garren 1941). At this time, most researchers were also confident that dominant pines would remain self-replacing without direct human intervention (e.g., Hall 1945; Eldredge 1952). Complete fire exclusion was thought to be the only reliable means to ensure adequate pine regeneration, especially in uneven-aged stands. Much of the early work at CEF consisted of fighting wildfires and trying to convince people not to burn their lands, even though this was a cherished local tradition (Bruner 1930; Shea 1940; Eldredge 1952). Of course, not everybody considered fire to be a problem. Reynolds and the Forest Service began a series of studies during the 1940s (Reynolds 1980, p. 38) in response to Herman Haupt Chapman's advocacy of the utility of certain types of controlled burning (e.g., Chapman 1916, 1932, 1942, 1952), work that demonstrated fire had both positive and negative silvicultural values.

Forest succession in the Reynolds RNA after the implementation of fire suppression was characterized by a 70+ year period relatively free of catastrophic disturbance. Rather, frequent, small-scale disturbances predominated, such as an individual tree or small groups of trees being killed by a lightning strike, severe winds, ice storms, insect infestations, or disease. Most of these minor disturbances went undetected by the early stand-level monitoring of the Reynolds RNA. Other than the broad umbrella of fire control, major anthropogenic disturbances have also been excluded from the Reynolds RNA area since its establishment. However, limited salvage was conducted to suppress a southern pine beetle (SPB) (*Dendroctonus frontalis* Zimm.) epidemic in southern Arkansas during the early 1970s (Ku et al. 1981). At that time, a 0.4-ha SPB "spot" was salvaged along the perimeter of the Reynolds RNA, and a cut-and-leave treatment was imposed on infested but isolated pines, affecting about 0.5 trees/ha.

Another SPB outbreak occurred in the 1990s when the Reynolds RNA was more closely monitored. Examination of this SPB infestation demonstrates how succession and disturbance interact to determine the composition of the forest community, and how multiple, small-intensity disturbances can have synergistic effects. The SPB infestations in 1993 followed an early spring windstorm that uprooted or damaged pines in a compartment immediately south of the Reynolds RNA. Although the pines were salvaged outside the perimeter, no suppression activity was conducted within the Reynolds RNA and the SPB remained active within the stand throughout the 1993 growing season. In February of 1994, the area was hit by an ice storm of historic proportions (Halverson and Guldin 1995). During 48 h, the accumulated ice broke tree limbs and even toppled a few old pines and hardwoods. Storm injuries further stressed the pines and exacerbated the SPB infestation, which intensified during the 1994 growing season. In 1995, however, the SPB activity stopped as abruptly as it started. Losses within the Reynolds RNA were not uniformly distributed. For example, one quarter of the permanent monitoring plots had pine mortality losses of about 50%, while losses were negligible on the other plots (Table 3.2). Four of the plots with SPB high activity were located in an infestation of about 4-ha in size located in the eastern part of the Reynolds RNA, while the remaining plot was located in a relatively isolated infestation of about 1 ha located

Table 3.2 Mean basal area in the Reynolds Research Natural Area of living pines and hardwoods over a 10-year period on 20 permanent, 0.1-ha plots—5 plots with high levels of southern pine beetle activity in 1993 and 1994 and 15 plots with low activity. (Adapted from, Shelton 2007)

Year[a]	Pines		Hardwoods	
	High activity	Low activity	High activity	Low activity
	m^2/ha		m^2/ha	
1990	24.5	22.1	12.2	15.0
1993	18.7	22.7	12.8	15.5
1994[b]	18.5	22.3	12.8	15.5
1994	9.0	22.0	10.6	14.6
1995	9.0	21.8	–[c]	–
1996	9.1	21.9	–	–
2000	9.2	22.1	13.0	16.0

[a] Inventoried (1990, 1993, and 2000) or visually inspected (1993–1996) for mortality after the growing season of the specified year. Survivor growth between inventories was interpolated

[b] Measurement after the February 1994 ice storm, but before the 1994 end of growing season inventory

[c] Not evaluated

in the western part of the Reynolds RNA. Although mortality losses were severe, even the locations with high activity still had live pine basal area averaging 9 m^2/ha after the infestation (Table 3.2).

A number of factors contributed to the patterns observed in this outbreak. On coastal plain sites in the South, overstocked stands of loblolly pine on good sites with reduced radial growth are most often attacked by SPB (Hicks 1980). In an assessment of SPB infestations in southern Arkansas, Ku et al. (1981) reported that high levels of pine basal area (>22 m^2/ha) increased the susceptibility of loblolly and shortleaf pines to attack. In addition, older pines are particularly susceptible to SPB infestations (USDA Forest Service 1993). Old, low-vigor pines in the Reynolds RNA provided the focal points for initial SPB attack, while the additional stress associated with the ice storm contributed to its expansion.

The primary effect of this SPB activity was to accelerate the successional transition of this stand to hardwoods (Fig. 3.3). This is especially apparent in the areas hardest hit by SPB, which amounted to about one sixth of Reynolds RNA's area. In these areas, pines currently make up less than 50% of the basal area (Table 3.2). As the dominant pines continue to die due to natural events, they are not being replaced by the next generation of pines because none currently exist (Fig. 3.2). The fairly intensive disturbance from the 1993–1994 SPB infestation was insufficient to permit a new cohort of pines to establish and recruit to the overstory. Furthermore, large dead pines tend to remain erect and gradually deteriorate as snags over many years rather than collapse and create gaps by knocking down nearby living trees (Jones et al. 1981). Such was the case following this SPB infestation—the canopy below dead and dying old-growth pines remained closed during the 1994 growing season because of the combined effects from understory, midstory, and overstory hardwoods. This was confirmed by measuring photosynthetically active radiation (PAR) at a height of 1.37 m. PAR averaged 7.9, 5.4, and 7.2% of full sunlight for

Fig. 3.3 View of the Reynolds RNA in 1959, showing how its original second-growth stands contained a prominent hardwood component, with pine regeneration occurring almost exclusively along the margins of the stand. (Photo from the USDA FS archives at the CEF)

areas where pine mortality was complete, partial, and none, respectively, and PAR was not significantly related ($P = 0.34$) to the intensity of the SPB activity (Cain and Shelton 1995). Thus, little direct sunlight penetrated to the forest floor under these closed canopy conditions.

Further evidence of the ineffectiveness of scattered SPB mortality on sustaining pine regeneration can be seen in the paucity of pine reproduction eight growing seasons after the onset of the SPB activity compared to the abundance of hardwood reproduction and shrubs (Table 3.3). Pines killed by SPB rarely disturb the ground surface enough to provide an adequate substrate for pine seedlings to germinate on, and existing hardwood and shrub competition is almost never reduced enough to provide pine seedling release opportunities. The handful of pine seedlings that were found were in the shortest (< 15-cm tall) height class recognized in the inventory procedure. According to Shelton and Cain (2000), an adequate stocking of pine regeneration to sustain a strongly pine-dominated overstory should exceed 500 stems/ha that are free to grow above competing non-pine vegetation and growing at least 15 cm in height per year.

Thus, the existing stand structure and long-term trends suggest that a pine-dominated overstory will not be sustained in the Reynolds RNA in the absence of some unforeseen large-scale disturbance or silvicultural intervention. In the past, fire, coupled with the high fuel loads associated with SPB activity, may have

Table 3.3 Mean density of seedlings (<1.3-cm DBH) and saplings (1.3–8.9-cm DBH) in the Reynolds RNA during the fall of 2000 (eight growing seasons after the onset of a southern pine beetle infestation) on 5 plots with high activity and 15 plots with low activity (there were five 8-m^2 subplots per plot)

Species group	High activity	Low activity
	Seedlings—stems/ha	
Pines	543	181
Oaks	5,286	4,907
Other canopy species	1,186	659
Midcanopy species	4,150	3,376
Shrubs	7,656	4,594
Total	18,822	13,717
	Saplings—stems/ha	
Pines	0	0
Oaks	99	16
Other canopy species	445	148
Midcanopy species	1,334	823
Shrubs	99	82
Total	1,977[a]	1,069[a]

[a] The difference between the high and low activity levels was significant at $P \leq 0.05$

created enough of a favorable environment for pine establishment and development (Waldron et al. 2007). Although it is tempting to apply silvicultural manipulations within the Reynolds RNA, the long history of "passive" management in this stand will be continued so that rates and direction of successional change can be determined. If nothing else, this approach will provide a certain environmental condition with specific ecological values. While this stay-the-course strategy leaves unanswered questions, two supplemental research studies (Bragg 2004b; Guldin 2005) have been implemented on other compartments in the CEF to evaluate the effectiveness of more intensive silvicultural treatments to produce old-growth-like characteristics in managed pine stands. These studies involve the use of fire either alone or in combination with herbicides, mowing, and selective harvesting to create an environment favorable to the development of new pine germinants into dominant overstory trees.

3.4 Future Research Opportunities

There are numerous research prospects in old forests, primarily because there are very few old, relatively undisturbed upland forests remaining in the southeastern USA, and the continuing intensification of silviculture across the region has further diminished their abundance. The uniqueness of these old stands gives them a particular value in the development and evaluation of certain concepts, especially those focusing on ecosystem goods and services as well as mensurational or modeling efforts. The following provide examples of some of these research opportunities.

Fig. 3.4 Unlike most managed forests, the Reynolds RNA still contains a number of examples of very large trees that can be very useful in the extension of tree allometric models. This specimen, a now-deceased loblolly pine, was 118 cm in diameter and 39.6 m tall when this photograph was taken in 1968. (Photo by James Burton, from the USDA FS archives at the CEF)

3.4.1 Carbon Sequestration in Old Forests

Carbon (C) sequestration has become an increasingly relevant aspect of forest management nationwide, and is of particular interest in southern forests, as this region has many timberland owners willing to consider alternative management opportunities. Currently, C credits are concentrated in the afforestation of non-timbered lands rather than the continued accumulation of C in existing stands (e.g., extended rotation silviculture). However, there is a good chance that other sequestration opportunities may arise if research can demonstrate that certain management practices can sustain C storage above and beyond that possible in business-as-usual silvicultural treatments. The well-documented stand history of the Reynolds RNA, coupled with other long-term research and demonstration projects on the CEF, lends itself to the description of C accumulation in mature pine-dominated forests, including what to expect following the transition from a pine- to hardwood-dominated overstory.

An under appreciated aspect of C sequestration linked to the study of old forests relates to the modeling of trees at the upper end of their physical dimensions (Fig. 3.4). Currently, most allometric relationships are developed using the typically small trees of managed landscapes—few specimens of considerable size are incorporated in these regressions. This in turn can have significant implications on the predictive models developed, which can be highly sensitive to the data used to derive them. For example, loblolly pines from the Reynolds RNA and a nearby privately owned old-growth remnant were used to predict height as a function of diameter (Bragg 2008b). If this model was fitted to the same data but with the upper diameters truncated, a different set of equations arose (Fig. 3.5). These new models had reasonably large sample sizes and were well fit (pseudo-$R^2 > 95\,\%$), with virtually no noticeable difference between them up to at least 40-cm DBH. However, removal of all pines > 70-cm DBH produced a height model that underestimated large tree heights, while removing pines > 40-cm DBH yielded a model that drastically overestimated height (Fig. 3.5b–d). It is clear that the addition of the big pines considerably improved large tree predictions while having virtually no impact on those for smaller diameter stems. Individually, the difference of a few percent of total tree height on a 40-m tall loblolly pine may not seem much, but the cumulative volume extrapolated over landscapes or regions is considerable, and could prove especially problematic if the model leads to inappropriate estimates of C storage.

3.4.2 Managing For Old-Growth-Like Attributes

Although left untreated for decades, compartments 41 and 42 (the lands eventually designated the Reynolds RNA) were not originally intended to protect old forests. Given their history, these compartments are not old growth, even though there were some trees in the stand that escaped the original lumbering period (Shelton and Cain 1999). The north–south fire break between compartments 41 and 42 intersects an east-west rail tram line built 90+ years ago to haul logs from this tract, which has long since been returned to forest (today, a number of large loblolly pines grow on the remains of tram line). Yet our modern-day sensibilities tell us that the towering trees and accumulation of dead wood in the Reynolds RNA differ from the most pine-dominated forests in the region. This suggests that we can learn from the Reynolds RNA to help frame management options in old, naturally regenerated, pine-dominated forests of the region. Even with our best science, we cannot recreate the virgin forest—the environment that now encompasses the region has changed too much for this to be practicable. However, it should be possible to encourage certain conditions in contemporary pine-dominated forests to satisfy the habitat requirements of at least some of the most threatened elements of these landscapes.

Managing for old-growth-like attributes in southern pine forests means different things to different people. Under some circumstances, simply retaining a number of larger-than-typical pines sufficiently improves desired habitat qualities. This could entail, for example, the permanent retention of seed trees to provide biological and structural legacies. Though not a seed tree system, the Reynolds RNA had a considerable number of pines left after the original high-grading. Most of these have since died, but during their lifespan they provided the mature tree structure that would have otherwise been absent in the developing stand. In death, these large pines have

$$HT = 1.37 + \frac{b_1}{1 + (1/b_2)DBH^{b_3}}$$

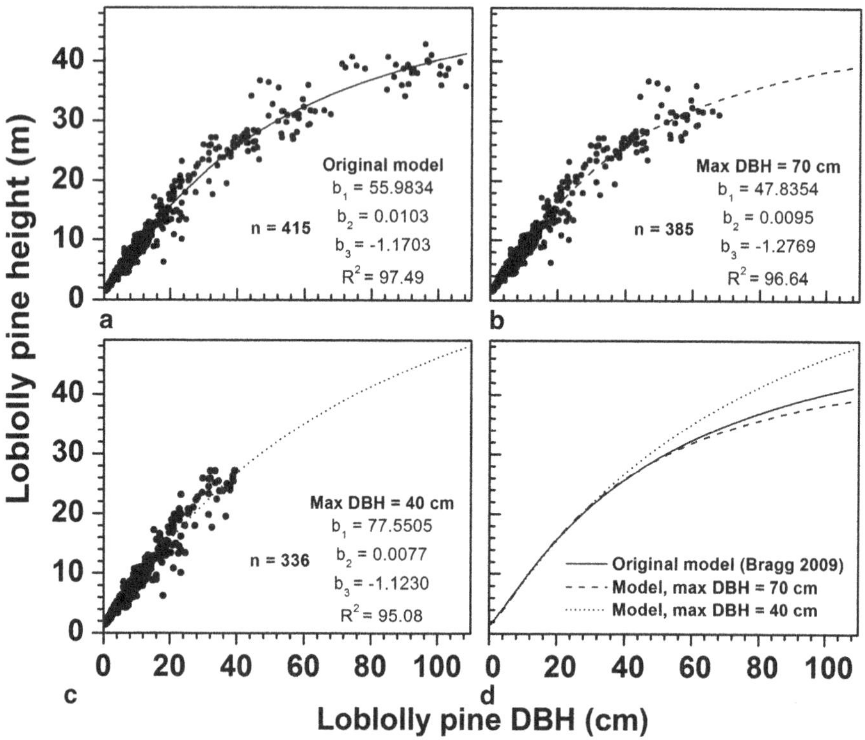

Fig. 3.5 Differences in total tree height models using truncated versions of the same data set. The original (**a**) modified logistic model is from Bragg (2008b), and included very large loblolly pine from the Reynolds RNA and a nearby privately owned old-growth stand. The next two graphs, (**b**) and (**c**), show the fit when this original data set was reduced to pines less than 70-cm DBH and 40-cm DBH, respectively. When compared, these models fit large pines poorly (**d**), suggesting that regression models based on limited data may substantially influence aggregated measures like total C sequestration in mature to old stands

contributed a considerable quantity of dead wood (see Table 3.1) that can act as stratum for a number of species.

Another objective may be to increase the proportion of certain taxa in an otherwise conventionally managed pine sawtimber stand—treatments can be developed to favor shortleaf pine over loblolly pine or hardwoods (Bragg 2004b; Bragg et al. 2008). Others may want to more closely emulate the open, grassy, large pine-dominated stand condition in which frequent fire and episodic pine recruitment drive the dynamics. The Reynolds RNA and numerous other protected areas have shown that without active intervention, conditions can quickly turn unfavorable for the red-cockaded woodpecker (RCW, *Picoides borealis* Vieillot; Saenz et al. 2001, Bragg et al. 2008). If the specific objective of this strategy is to improve upon RCW habitat, then removing the midstory and retaining live pines with heart rot (as opposed to culling them for more vigorous individuals) or preparing clusters for nest box inserts should prove more useful than hoping these conditions arise by chance (Saenz et al. 2001). Not surprisingly, the need to more intensively manage lands for non-timber attributes to ensure the persistence of certain elements (e.g., RCWs) has implications for public policy.

3.4.3 Ecological Implications of Public Land Management

Land management efforts have significant implications on the ecological response of the forests being treated. Over the years, the focus of management of public forestlands has shifted from timber production to a broader range of environmental services. This necessitates that a different perspective be taken on what constitutes appropriate and acceptable outcomes. For example, the success of a silvicultural approach that incorporates old-growth-like conditions depends on both intensive and extensive treatments to ensure that certain natural elements are retained. If burning is to be used to maintain stand conditions for fire-dependent species, is this treatment possible in a highly fragmented modern landscape, with its complicated mixture of public and private lands? What productivity, risk, and liability issues constrain this option?

For all of those interested in the active management of stands, there are others interested in the opposite—functionally, from a land management standpoint these people desire a version of the "set-aside" or "passive" management approach. From this perspective, no human intervention is acceptable, regardless of the intent, with the possible exception of fire protection to preserve a remnant condition. However, if an increasing fraction of the public ownership is removed from active management, a cascading series of responses will arise from both the biotic and socioeconomic communities associated with those lands. Although it is not of the appropriate scale to consider many impacts, the Reynolds RNA offers an excellent opportunity to evaluate the long-term outcome of such a passive strategy on a fine scale. Few examples highlight this issue more tellingly than rare woodpeckers in the South.

The most prominent woodpecker species to vanish from the forests of eastern USA was the ivory-billed woodpecker (*Campephilus principalis* L.), which relied upon large old-growth bottomland hardwood tracts. The ivory-bill was thought to have gone extinct when one of the last large remnants of virgin bottomland hardwoods in northeastern Louisiana was logged during World War II (Tanner 1942; Fitzpatrick et al. 2005). The apparent loss of this hallmark species helped galvanize portions of the environmental community to action in order to avoid such catastrophes in the future. During the next few decades, organizations dedicated to the acquisition and preservation of suitable habitats arose and federal, state, and local agencies began adapting to further protect species and habitats, culminating in the 1973 passage of the Endangered Species Act. Even most lumber companies and many private citizens changed at least some of their land management practices. In many places, preserves were established to protect the special conditions (e.g., old trees) needed for habitat for threatened species.

This transformation almost came too late for a different woodpecker. By the 1960s, bird-watchers and scientists across the southeastern USA began noticing a precipitous decline in the abundance of the RCW, a small and unassuming bird that once frequented the piney woodlands that had dominated this area. RCW's peculiar nesting requirements—a mature, live pine with extensive red heart (*Phellinus pini* (Thore) Fr.) disease growing in an open stand—coupled with changing forest structure and demographics resulted in the rapid collapse of RCW populations across the species range (Conner and O'Halloran 1987; Saenz et al. 2001). Forest management practices can be blamed for much of the decline suffered by the RCW. Even though the loss of the virgin pine forest, with its abundance of large, red heart-infected pines, was a devastating blow to the RCW, large areas of the region remained in mature, open second-growth stands—acceptable if not ideal habitat. During much of the mid-twentieth century, conventional silvicultural practices in naturally regenerated loblolly and shortleaf pine also helped to nurture RCW habitat, as both uneven-aged management and seed tree/shelter wood techniques of the time retained enough large, old, live individual pines to support the species.

However, pines with obvious signs of red heart were considered cull trees, treated as a loss, and often removed from stands to permit healthier trees more growing space. In addition, short-rotation (<35 years) intensively managed plantations proved more economical than naturally regenerated stands, and thus a large-scale conversion of southern upland forests to planted loblolly pine occurred (Schultz 1999; Conner and Hartsell 2002; Fox et al. 2004; Allen et al. 2005; Rousseau et al. 2005). While it is possible to grow pines of adequate diameter during this rotation length, RCW cavity trees are usually significantly older and slower growing, reflecting the gradual development of extensive red heart disease in the bole (Conner and O'Halloran 1987; Conner et al. 2004a, b).

The Reynolds RNA has a large number of old loblolly and shortleaf pine full of red heart disease, and would represent a good block of suitable RCW habitat if large, decaying live pines were all that mattered. However, there is not a single RCW nest cavity to be found on this tract, nor is there any evidence of a colony abandoned in the recent past. Pine age and overstory structure are only one part of the recent decline in RCW—the rest has to do with overall habitat quality in areas reserved for the perpetuation of this woodpecker. In Louisiana, Oklahoma, and Texas, Saenz et al. (2001) noted that unmanaged pine stands experienced severe RCW population declines relative to those in managed stands and attributed much of this drop to hardwood encroachment. The absence of periodic fire in the Reynolds RNA has permitted this stand to grow too dense, with too many hardwoods to provide suitable RCW nesting habitat. This was an unintended consequence of the passive management strategy employed on the RNA, and indicative of what has happened across much of the remaining RCW habitat in the South.

Another policy-related challenge lies in the determination of what to restore where, as this will help dictate priorities. For example, habitat conditions favorable for a particular rare species (e.g., RCW) may not be suitable for other taxa facing similar pressures. Aquilani (2006) noted that certain forest-obligate bird species (e.g., worm-eating warblers; *Helmitheros vermivorus* Gmelin) required interior areas with high amounts of shrubby understory coverage, a condition compatible with the modern-day structure of the Reynolds RNA but unsuitable if the stand is managed for more open conditions. This suggests that areas of older forest managed for ecosystem services other than fiber production be kept in a range of stand conditions—tree size or age are but some of the many components that affect habitat suitability.

3.4.4 Evidence of Climate Change

The relatively undisturbed soils of the Reynolds RNA also provide a good opportunity to study past climates. Recent research has suggested that the "pimple" or "prairie" mounds that dot the landscapes of the Midsouth may actually be "nebkha" or "coppice" dunes from much drier periods in the late Holocene (Seifert et al. 2009). These natural mounds are rapidly being destroyed by land leveling or the ripping and bedding practices that commonly precede pine plantation establishment. Hence, the mounds found in the Reynolds RNA and a few other protected old pine stands may be increasingly important records of prehistoric megadroughts.

Today, increasing atmospheric CO_2 threatens southern forests with global climate change, which is a possible catalyst for a number of other environmental concerns, including species migration, exotic species invasion, and the alteration of natural disturbance regimes. Researchers have begun to project species migration under a number of different climate scenarios using inferences from forest growth and yield plots located across the eastern USA (e.g., Iverson et al. 2004; Woodall et al. 2009). Long-term observations on a fixed location such as the Reynolds RNA can be used to directly observe the appearance or disappearance of trees as a function of climate change, the spread of exotic and endemic pests and pathogens, and natural successional tendencies. None of these purposes would have been anticipated in the 1930s when the Reynolds RNA was initially established, yet they are examples of important benefits of undisturbed natural areas in experimental forests.

3.5 Conclusions

Old forests offer opportunities to better understand the impacts of our management on the environment across a range of scales. Unfortunately, we have so few remaining examples of these stand conditions that it is increasingly difficult to study an ecosystem in the detail necessary to be able to predict outcomes under a variety of different scenarios. Natural areas, such as those found on experimental forests, offer a unique opportunity to observe the development of a particular stand over a long time period without worrying about how shifts in ownership or management strategy may affect the results. In the case of the Reynolds RNA, we have 70+ years of documentation on stand development under a fixed management regime—a record hardly equaled elsewhere in southern forests.

As can be seen in the climate change study opportunities, the lessons we can learn from long-term studies are not necessarily limited to those in place when the natural area was originally designated, nor do they need to be. Adaptive research policies that allow for refocusing of the analysis (if not the treatment) can supplement or extend contemporary investigations into problems not previously anticipated. For example, the study of C sequestration in southern forests would not be complete without the knowledge of how old forests such as the Reynolds RNA are organized and how they may respond to changes in atmospheric chemistry, precipitation patterns, or temperature regimes. Too many differences exist between the composition, structure, function, and genetics of 25-year-old pine plantations and old, natural origin pine-hardwood stands to extrapolate between these conditions. Furthermore, the ecological studies of the Reynolds RNA can document the impacts of disturbance exclusion over many years, thereby helping policymakers understand the consequences of certain decisions.

Long-term studies on experimental forests and ranges provide federal, academic, and even industrial research programs the flexibility and leverage they need to address future environmental issues in an efficient and predictable manner. It would be irresponsible to exclude unmanaged, protected old forests such as the Reynolds RNA from our toolbox, as we can sometimes learn as much from the unanticipated consequences of passive stand management as we do from direct treatments.

Acknowledgments We would like to thank the following people for their contributions to this chapter over the years—Michael Cain, Mike Chain, Jim Guldin, Kirby Sneed, and Bruce Walsh (all of the USDA Forest Service). In particular, we would like to give recognition to Russ Reynolds and other past station staff that helped to establish the long-term studies on the CEF. Without their vision, this work would not be possible.

References

Allen HL, Fox TR, Campbell RG (2005) What is ahead for intensive pine plantation silviculture in the South? South J Appl For 29(2):62–69

Anonymous (1981) "Dear Professor Chapman": letters from Yale forestry graduates, 1910–1912. J For Hist 25(4):197–208

Aquilani SM (2006) Bird communities in silvicultural fragments of Holly Springs National Forest, Mississippi. Southeast Nat 5(1):135–148

Baker JB, Bishop LM (1986) Crossett demonstration forest guide. General Report, R8-GR 6, U.S. Department of Agriculture, Forest Service, Southern Region, Atlanta, 55 p

Barrett LI (1928) Fire damage done to a mature loblolly pine stand. Savannah Wkly Nav Store Rev J Trade 38(18):11

Braccia A, Batzer DP (2001) Invertebrates associated with woody debris in a southeastern U.S. forest floodplain wetland. Wetlands 21(1):18–31

Bragg DC (2004a) Composition, structure, and dynamics of a pine-hardwood old-growth remnant in southern Arkansas. J Torrey Bot Soc 131(4):320–336
Bragg DC (2004b) A prescription for old-growth-like characteristics in southern pines. In: Shepperd WD, Eskew LD (comps.) Silviculture in special places: proceedings of the 2003 National Silviculture Workshop. RMRS-P-34, U.S. Department of Agriculture, Forest Service, Rocky Mountain Research Station, Ft. Collins, pp 80–92
Bragg DC (2008a) The prominence of pine in the Upper West Gulf Coastal Plain during historical times. In: Hardy LM (ed) Freeman and Custis Red River Expedition of 1806: two hundred years later. Bulletin of the Museum of Life Sciences Number 14, Louisiana State University-Shreveport, Shreveport, pp 29–54
Bragg DC (2008b) A comparison of pine height models for the Crossett Experimental Forest. J Ark Acad Sci 62:24–31
Bragg DC, Heitzman E (2009) Composition, structure, and dynamics of a mature, unmanaged, pine-dominated old field stand in southeastern Arkansas. Southeast Nat 8(3):445–470
Bragg DC, Shelton MG (2011) Lessons from 72 years of monitoring a once-cut pine-hardwood stand on the Crossett Experimental Forest, Arkansas, U.S.A. For Ecol Manage 261:911–922
Bragg DC, Shelton MG, Guldin JM (2008) Restoring old-growth southern pine ecosystems: strategic lessons from long-term silvicultural research. In: Deal RL (ed) Integrated restoration of forest ecosystems to achieve multiresource benefits: proceedings of the 2007 National Silviculture Workshop. General Technical Report, PNW-733, U.S. Department of Agriculture, Forest Service, Pacific Northwest Forest Experiment Station, Corvallis, pp 211–224
Bruner EM (1930) Forestry and forest fires in Arkansas. University of Arkansas Agricultural Extension Service Extension Circular 281, 30 p
Cain MD, Hestir JL, Shelton MG (n.d.) Establishment record for the R.R. Reynolds Research Natural Area within the Ouachita National Forest, Ashley County, Arkansas. Unpublished document on file at Monticello, AR: U.S. Department of Agriculture, Forest Service, Southern Research Station, 13 p
Cain MD, Shelton MG (1995) Thirty-eight years of autogenic, woody understory dynamics in a mature, temperate pine-oak forest. Can J For Res 25:1997–2009
Cain MD, Shelton MG (1996) The R.R. Reynolds Research Natural Area in southeastern Arkansas: a 56-year case study in pine-hardwood overstory sustainability. J Sustain For 3(4):59–73
Cain MD, Shelton MG (2000) Survival and growth of *Pinus echinata* and *Quercus* seedlings in response to simulated summer and winter prescribed burns. Can J For Res 30(11):1830–1836
Cain MD (1987) Survival patterns of understory woody species in a pine-hardwood forest during 28 years without timber management. In: Hay RL, Woods FW, DeSelm H (eds) Proceedings of the sixth central hardwood forest conference. The University of Tennessee, Knoxville, pp 141–147
Chapman HH (1912) A method of investigating yields per acre in many-aged stands. For Q 10:458–469
Chapman HH (1913) Prolonging the cut of southern pine. I. Possibilities of a second cut. Yale University Forestry School Bulletin 2:1–22
Chapman HH (1916) Organization of state forest protective systems. In: Proceedings of the Southern Forestry Congress. Seeman Printery, Durham, pp 81–97
Chapman HH (1932) Is the longleaf type a climax? Ecology 13(4):328–334
Chapman HH (1942) Management of loblolly pine in the pine-hardwood region in Arkansas and in Louisiana west of the Mississippi River. Forestry School Bulletin 49. Yale University, New Haven, 150 p
Chapman HH (1952) The place of fire in the ecology of pines. Bartonia 26:39–44
Conner RC, Hartsell AJ (2002) Forest area and conditions. In: Wear DN, Greis JG (eds) Southern forest resource assessment. General Technical Report SRS-53, U.S. Department of Agriculture, Forest Service, Southern Research Station, Asheville, pp 357–402
Conner RN, O'Halloran KA (1987) Cavity-tree selection by red-cockaded woodpeckers as related to growth dynamics of southern pines. Wilson Bull 99(3):398–412

Conner RN, Rudolph DC, Saenz D, Johnson RH (2004a) The red-cockaded woodpecker cavity tree: a very special pine. In: Costa R, Daniels SJ (eds) Red-cockaded woodpeckers: road to recovery. Hancock House Publishers, Blaine, pp 407–411

Conner RN, Saenz D, Rudolph DC, Schaefer RR (2004b) Extent of *Phellinus pini* decay in loblolly pines and red-cockaded woodpecker cavity trees in eastern Texas. In: Cripps CL (ed) Fungi in forest ecosystems: systematics, diversity, and ecology. New York Botanical Garden, Bronx, pp 315–321

Darling OH, Bragg DC (2008) The early mills, railroads, and logging camps of the Crossett Lumber Company. Ark Hist Q 67(2):107–140

Della-Bianca L (1983) Sixty years of stand development in a southern Appalachian cove-hardwood stand. For Ecol Manage 5:229–241

Devall MS, Ramp PF (1992) U.S. Forest Service Research Natural Areas and protection of old growth in the South. Nat Areas J 12:75–85

Eldredge IF (1952) Southern forests, then and now. J For 50(5):182–185

Fan Z, Shifley SR, Spetich MA, Thompson FR, Larsen DR (2005) Abundance and size distribution of cavity trees in second-growth and old-growth Central Hardwood forests. North J Appl For 22(3):162–169

Fitzpatrick JW, Lammertink M, Luneau MD, Gallagher TW, Harrison BR, Sparling GM, Rosenberg KV, Rohrbaugh RW, Swarthout ECH, Wrege PH, Swarthout SB, Dantzker MS, Charif RA, Barksdale TR, Remsen JV, Simon SD, Zollner D (2005) Ivory-billed Woodpecker (*Campephilus principalis*) persists in continental North America. Science 308:1460–1462

Forbes RD (1923) Research in forest fires. In: Proceedings of the fifth Southern Forestry Congress. A.W. Hyatt, Ltd, New Orleans, pp 41–46

Fountain MS, Sweeney JM (1987) Ecological assessment of the Lake Winona Research Natural Area. Research Paper SO-235, U.S. Department of Agriculture, Forest Service, Southern Forest Experiment Station, New Orleans, 14 p

Fox TR, Jokela EJ, Allen HL (2004) The evolution of pine plantation silviculture in the southern United States. In: Rauscher HM, Johnson K (eds) Southern forest science: past, present, and future. General Technical Report SRS-75, U.S. Department of Agriculture, Forest Service, Southern Research Station, Asheville, pp 63–82

Garren KH (1941) Fire wounds on loblolly pine and their relation to decay and cull. J For 39(1):16–22

Glitzenstein JS, Harcombe PA, Streng DR (1986) Disturbance, succession, and maintenance of species diversity in an east Texas forest. Ecol Monogr 56(3):243–258

Guldin JM (2005) Use of prescribed fire in restoration of old-growth loblolly-shortleaf pine stands on the Upper West Gulf Coastal Plain. Unpublished study plan for FS-SRS-4106-101 on file at Monticello, AR: U.S. Department of Agriculture, Forest Service, Southern Research Station, 15 p

Guldin JM, Baker JB (1985) Dynamics and development of a once-cutover, unmanaged loblolly pine stand in southeastern Arkansas. In: Shoulders E (ed) Proceedings of the third biennial southern silvicultural research conference. General Technical Report SO-54, U.S. Department of Agriculture, Forest Service, Southern Forest Experiment Station, New Orleans, pp 198–202

Hall WL (1945) Is pine coming or going in south Arkansas? J For 43:634–637

Halverson HG, Guldin JM (1995) Effects of a severe ice storm on mature loblolly pine stands in north Mississippi. In: Edwards MB (comp) Proceedings of the eighth biennial southern silvicultural research conference. General Technical Report SRS-1, U.S. Department of Agriculture, Forest Service, Southern Research Station, Asheville, pp 147–153

Hemond HF, Niering WA, Goodwin RH (1983) Two decades of vegetation change in the Connecticut Arboretum Natural Area. Bull Torrey Bot Club 110:184–194

Heyward F (1958) History of industrial forestry in the South. Col. William B. Greeley lectures in industrial forestry number 2. University of Washington College of Forestry, Seattle, 50 p

Hicks RR Jr (1980) Climate, site, and stand factors. In: Thatcher RC, Searcy JC, Coster JE, Hertel GO (eds) The southern pine beetle. Technical Bulletin, 1631. U.S. Department of Agriculture, Forest Service, Washington, DC, pp 55–68

Huston M, Smith T (1987) Plant succession: life history and competition. Am Nat 130:168–198
Iverson LR, Schwartz MK, Prasad AM (2004) How fast and far might tree species migrate in the eastern United States due to climate change? Glob Ecol Biogeogr 13:209–219
Jones SM, Van Lear DH, Cox SK (1981) Composition and density-diameter pattern of an old-growth forest stand of the Boiling Springs Natural Area, South Carolina. Bull Torr Bot Club 108:347–353
Ku TT, Sweeney JM, Shelburne VB (1981) Southern Arkansas. In: Coster JE, Searcy JL (eds) Site, stand, and host characteristics of southern pine beetle infestations. Technical Bulletin, No. 1612, U.S. Department of Agriculture, Washington, DC, pp 16–22
Lemon PC (1945) Wood as a substratum for perennial plants in the Southeast. Am Midl Nat 34(3):744–749
McMinn JW, Crossley DA Jr (eds) (1996) Proceedings of the workshop on coarse woody debris in southern forests: effects on biodiversity. General Technical Report SE-94, U.S. Department of Agriculture, Forest Service, Southern Research Station, Asheville, 146 p
Munger TT (1930) Ecological aspects of the transition from old forests to new. Science 72(1866):327–332
Northern Research Station (2010) About research natural areas. USDA Forest Service online statement. http://nrs.fs.fed.us/rna/about/. Accessed 30 Nov 2010
Quarterman E, Keever C (1962) Southern mixed hardwood forest: climax in the southeastern coastal plain, USA. Ecol Monogr 32:167–185
Radtke PJ, Amateis RL, Prisley SP, Copenheaver CA, Chojnacky DC, Pittman JR, Burkhart HE (2009) Modeling production and decay of coarse woody debris in loblolly pine plantations. For Ecol Manage 257:790–799
Reynolds RR (1974) Selective timber management—its role. For Farmer 33(5):6–7, 12
Reynolds RR (1980) The Crossett Story: the beginnings of forestry in southern Arkansas and northern Louisiana. General Technical Report SO-32, U.S. Department of Agriculture, Forest Service, Southern Forest Experiment Station, New Orleans, 40 p
Rousseau R, Kaczmarek D, Martin J (2005) A review of the biological, social, and regulatory constraints to intensive plantation culture. South J Appl For 29(2):105–109
Saenz D, Conner RN, Rudolph DC, Engstrom RT (2001) Is a "hands-off" approach appropriate for red-cockaded woodpecker conservation in twenty-first-century landscapes? Wildl Soc Bull 29(3):956–966
Schultz RP (1999) Loblolly---the pine for the twenty-first century. New For 17:71–88
Seifert CS, Cox RT, Foti TL, Forman SL, Wasklewicz TA, McColgan AT (2009) Relict nebkhas (pimple mounds) record late Holocene megadroughts in the forested region of South-Central United States. Quat Res 71(3):329–339
Shea JP (1940) Our pappies burned the woods. Am For 46(4):159–162, 174
Shelton MG (2007) Status of the litterfall, forest floor, and soils of the Good Farm Forestry Forty Demonstration Area and the R.R. Reynolds Research Natural Area on the Crossett Experimental Forest. Unpublished progress report for study FS-SRS-4106-66 on file at Monticello, AR, U.S. Department of Agriculture, Forest Service, Southern Research Station, 47 p
Shelton MG, Cain MD (1999) Structure and short-term dynamics of the tree component of a mature pine-hardwood forest in southeastern Arkansas. J Torr Bot Soc 126(1):32–48
Shelton MG, Cain MD (2000) Regenerating uneven-aged stands of loblolly and shortleaf pines: the current state of knowledge. For Ecol Manage 129(1–3):177–193
Smith KL (1986) Sawmill: the story of cutting the last great virgin forest east of the Rockies. University of Arkansas Press, Fayetteville, 246 p
Switzer GL, Shelton MG, Nelson LE (1979) Successional development of the forest floor and soil surface on upland sites of the East Gulf Coastal Plain. Ecology 60:1162–1171
Tanner JT (1942) The ivory-billed woodpecker. Research report 1, National Audubon Society, New York, NY, 111 p
Thatcher RC, Searcy JL, Coster JE, Hertel GD (eds) (1980) The southern pine beetle. Science and Education Administration, Technical Bulletin 1631, U.S. Department of Agriculture, Forest Service, Washington, DC, 266 p

USDA Forest Service (1993) Healthy forests for America's future. Miscellaneous publication 1513. U.S. Department of Agriculture, Forest Service, Washington, DC, 58 p

USDA Forest Service (2005) Record of decision: final environmental impact statement; revised land resource management plan. Ouachita National Forest, Arkansas and Oklahoma. Management Bulletin R8-MB 124D, U.S. Department of Agriculture, Forest Service, Southern Region, Atlanta, 19 p

Wahlenberg WG (1960) Loblolly pine. Duke University School of Forestry, Durham, 603 p

Waldron JD, Lafon CW, Carlson RN, Cairns DM, Tchakerian MD, Bist A, Klepzig KD (2007) Simulating the impacts of southern pine beetles and fire on the dynamics of xerophytic pine landscapes in the southern Appalachians. Appl Veg Sci 10:53–64

Wakeley PC (1964) A biassed history of the Southern Forest Experiment Station through fiscal year 1933. Unpublished document. http://www.treesearch.fs.fed.us/pubs/3261. Accessed 30 April 2014

White ZW (1984) Loblolly pine, with emphasis on its history. In: Karr BL, Baker JB, Monaghan T (eds) Proceedings of the Symposium on the Loblolly Pine Ecosystem (West Region). Mississippi Cooperative Extension Service, Mississippi State, pp 3–16

Woodall CW, Oswalt CM, Westfall JA, Perry CH, Nelson MD, Finley AO (2009) An indicator of tree migration in forests of the eastern United States. For Ecol Manage 257(5):1434–1444

Zhang M (2000) Quantification of snags and downed wood in the R.R. Reynolds Research Natural Area and Good Forty Demonstration Area in southeastern Arkansas. M.S. Thesis, University of Arkansas, Monticello, AR, 47 p

Chapter 4
Restoring a Legacy: Longleaf Pine Research at the Forest Service Escambia Experimental Forest

Kristina F. Connor, Dale G. Brockway and William D. Boyer

Abstract Longleaf pine ecosystems are a distinct part of the forest landscape in the southeastern USA. These biologically diverse ecosystems, the native habitat of numerous federally listed species, once dominated more than 36.4 million ha but now occupy only 1.4 million ha of forested land in the region. The Escambia Experimental Forest was established in 1947 through a 99-year lease with the T.R. Miller Mill Company of Brewton, AL, to explore all aspects of longleaf pine management. The 1,214-ha tract in southwest Alabama constitutes a unique example of longleaf pine ecosystems in all stages of development. Long-term studies and demonstrations include stand management alternatives, growth and yield of even-aged natural stands, cone production, and fire ecology.

Keywords Cone production · Forest management · Longleaf pine · Natural regeneration · Prescribed fire

4.1 Background

The Escambia Experimental Forest (Escambia), located in southern Alabama, was created to spur the revival and improved management of forests dominated by longleaf pine (*Pinus palustris* Mill.) Longleaf pine is a high-quality timber species that provided logs, poles, piling, posts, peelers, pulpwood, and naval stores for the building and transportation needs of early European settlers in the southeastern USA (Frost 2006). It was once the backbone of the early southern forest industry and a source of livelihood for southern communities. Referred to as "the most extensive forest ecosystem in North America dominated by a single tree" (Boyer 1997), longleaf pine encompassed 37.6 million ha from southeastern Virginia to eastern Texas. Longleaf pine trees grew in a range of habitats, from the dry sandy hills to the wet

Dr. Boyer is deceased.

D. G. Brockway (✉) · K. F. Connor · W. D. Boyer
Southern Research Station, USDA Forest Service, Auburn, AL 36849, USA
Phone: (334) 826-8700
e-mail: dbrockway@fs.fed.us

D. C. Hayes et al. (eds.), *USDA Forest Service Experimental Forests and Ranges*,
DOI 10.1007/978-1-4614-1818-4_4, © Springer New York 2014

Fig. 4.1 Young longleaf pine tree after a prescribed fire

flatwoods in the Carolinas, Georgia, Florida, Alabama, Mississippi, and Louisiana. From sea level along the Gulf Coast and the ridges and flats of central Florida, longleaf pine ecosystems reached up to an elevation of 607 m on the rocky hills of northern Alabama and northwestern Georgia (Burns and Honkala 1990).

The diversity and complexity of the understory in longleaf pine forests are indicative of the many habitats they occupy. Wiregrass (*Aristida* spp.) predominates in the sandy soils of the Atlantic and eastern Gulf coastal plains, while warm season grasses such as little bluestem (*Schizachyrium scoparium* (Michx.) Nash) and yellow indiangrass (*Sorghastrum nutans* (L.) Nash) are found on heavier soils in the flatwoods and western Gulf coastal plains (Jose et al. 2006). Fire was the common thread that united these habitats, creating the high understory diversity and making longleaf pine supreme. Frequent burning of these ecosystems ensured their survival. Longleaf pine trees, whether in the seedling (grass stage), sapling, or mature form, are resistant to flames as long as buds are intact and fire intensity is low (Fig. 4.1). Periodic surface fires, whether caused by lightning strikes or intentionally set by Native Americans or settlers, eliminated less fire-tolerant tree and shrub competitors and exposed the bare mineral soil for longleaf pine seed germination and seedling establishment.

Longleaf pine trees have many unique qualities. In addition to their fire tolerance, they are resistant to some of the insects and diseases that beset their southern pine competitors, including rusts (*Cronartium*), fungi (*Heterobasidion*), root rot (*Phytophthora*), pitch canker (*Fusarium subglutinans f. sp. pini* Wollenw. & Reink.), southern pine beetle (*Dendroctonus frontalis Zimmermann*), and tip moth (*Rhyacionia frustrana* (Comstock); see (1) http://edis.ifas.ufl.edu/pdffiles/FR/FR06400.PDF; (2) http://www.gfc.state.ga.us/forest-management/forest-health/pine-bark-beetles/index.cfm; (3) http://www.bugwood.org (all accessed 29 April 2014). They survive on poor-quality, sandy, droughty soils and are adapted to the hurricane zone, where they often rapidly recover from these catastrophic events (Hoyle 2009; Hughes 2006). However, unlike their competitor pines, longleaf pine seedlings resemble a clump of grass more than a tree and can remain in this stage for 1–10 years or more. Upon release, the seedlings grow rapidly, elevating their terminal bud beyond the reach of the next fire's flames and eventually up into the canopy of the forest if growing space is available. But sporadic seed production and this unique grass stage in the regeneration cycle often resulted in establishment failure after logging. That, along with fire suppression, feral hog predation on seeds and seedlings, and interest in short-term rotation forestry resulted in the fragmented, threatened longleaf pine forests we know today.

4.2 Decline of Longleaf Pine Ecosystems

The expansion of railroads into the southern forests in the late nineteenth and early twentieth centuries, along with the development of steam-engine logging, accelerated the harvest of longleaf pine trees (Jose et al. 2006). What were once isolated tracts in the vast longleaf pine interior forests were now within easy reach of railway spurs. Steam-powered skidders and other harvesting equipment represented the beginning of mechanized logging, and newly designed band saws increased mill efficiency. These industrial advances accelerated harvesting and reduced the possibility of successful longleaf pine regeneration.

Longleaf pine seed production was sporadic, with an average of 5–7 years between good seed crops (Boyer 1993; Boyer and White 1990). Removal of overstory trees, without regeneration already in place, resulted in decimation of these forests. Thus, at the peak of harvesting operations in the southern forests, there occurred, almost across the whole South, a failure of the longleaf pine to regenerate after logging.

Interruption of the fire cycle was another of the primary causes for longleaf pine regeneration failure. Burning southern forests was a tradition prior to the 1900s, and longleaf pine, a fire-dependent species, thrived. Native Americans and European settlers inhabiting the South set fires in the forests to promote habitat for particular game species, improve forage quality for domesticated animals, ease access by removing dense understory vegetation, fertilize the land (low intensity fires), facilitate travel through forested areas, and promote the growth of berry crops (Croker 1987;

Whitney 1996). However, in the 1930s, a battle raged between those advocating the benefits of controlled fires (i.e., prescribed burning) in longleaf pine forests and those completely opposed to all burning (Croker 1987). Many professional foresters vigorously fought against any use of fire, and a group begun in 1927, the Dixie Crusaders, toured the South spreading a fire prevention message. One newspaper editor, according to Croker (1987) "…condemned as unpatriotic those who would prescribe fire in the forest." In the 1940s, advertising campaigns featuring Smokey Bear, a black bear cub rescued from a wildfire, supported land management agency policies that viewed fire as the enemy, a destructive agent that must be suppressed at all costs. Thus, longleaf pine forests became overgrown with species less tolerant to fire, and longleaf pine either died out completely or became a minor component of the many ecosystems it once dominated.

Another contributing factor to regeneration failure was free-ranging packs of voracious feral hogs, so prevalent throughout the South that their populations at one time supported a meatpacking industry (Frost 2006). In addition to consuming mast crops of acorns and pine seeds, the hogs also showed a marked preference for the starchy roots of longleaf pine seedlings (Lipscomb 1989; Walkinshaw and Otrosina 2002). The few small longleaf pine seedlings that evaded consumption were usually overtopped by hardwoods and competing southern pines.

Other factors contributed to this regeneration failure but the end result was that many landowners abandoned longleaf pine trees as a cash crop. The species was falsely perceived as difficult to regenerate with highly variable establishment survival, slow early growth, and low initial productivity when compared to other southern pines. Millions of acres of forest land once dominated by longleaf pine were planted with loblolly pine (*P. taeda* L.) and slash pine (*P. elliottii* Engelm.), breaking the long southern cultural connection with longleaf pine. While the "piney woods" remained, the predominant tree species had changed. The forests that evoked literary poetry from its inhabitants (e.g. Ray 1999) were disappearing. The longleaf pine forests that now remain are among the most threatened ecosystems in the USA (Noss et al. 1995).

4.3 Establishing the Escambia Experimental Forest

The history of the area that now comprises the Escambia parallels that of the longleaf pine forests of the South. Extension of the railroads into southern forests accelerated harvesting, and, from 1900 to 1919, all merchantable longleaf pine trees on the land now occupied by the Escambia were cut (Croker 1987). Throughout the South, harvesting and loss of acreage to agriculture and development reduced longleaf pine forests from 38 million ha to the 1.2 million fragmented ha they occupy today.

The US Department of Agriculture (USDA) Forest Service began establishing research centers throughout the country at about the same time the Escambia forests were being cut. Seven centers built between 1934 and 1947 were located within the native range of longleaf pine, including one at Brewton, AL. On April 1, 1947, the

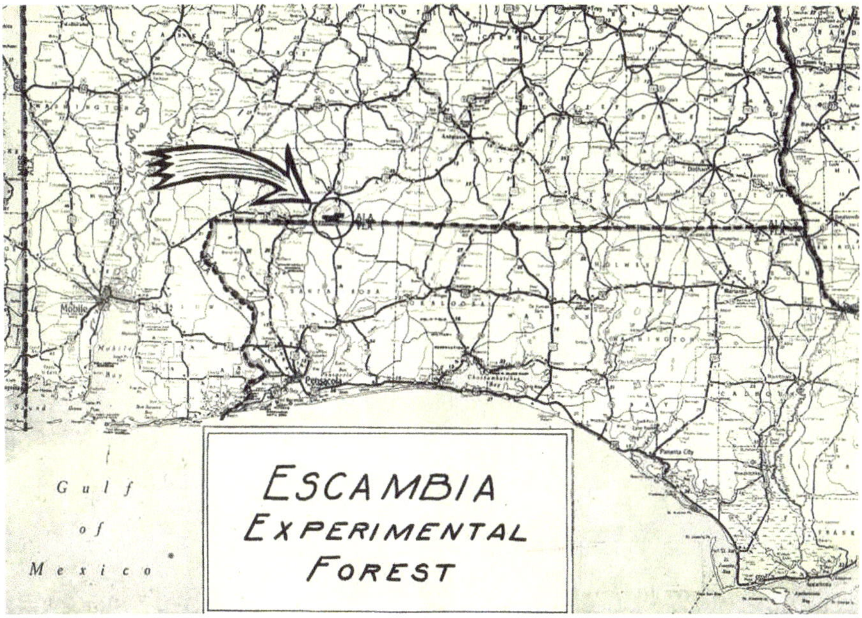

Fig. 4.2 Reconnaissance survey for a new experimental forest

Escambia was established 7 miles south of the center in Brewton, AL, and was later part of the East Gulf Coast Research Center, Southern Forest Experiment Station (Fig. 4.2). The T.R. Miller Mill Company of Brewton, AL, provided land for the Escambia at no cost to the Forest Service, through a 99-year lease. This 1,214-ha southwest Alabama tract, with trees then averaging 35–45 years of age, was selected because it typified the low-density, second-growth longleaf pine forests that then covered about 25.1 million ha in southern Alabama and northwestern Florida (Croker 1987), and it was centrally located within the species' natural range.

4.4 The First Half-Century of Progress

Under the leadership of professional forester Gifford Pinchot, the Bureau of Forestry evolved into the USDA Forest Service in 1905 with increased duties and responsibilities (Pinchot 1947). Unfortunately, while the Forest Service was still in its infancy, the longleaf pine forests were already falling under the axe. The southern timber industry and the communities dependent upon logging were booming. By the time professional foresters began gathering data on growth and yield, regeneration methods, and management techniques for American tree species, the vast longleaf pine forests were well on their way to becoming a historical footnote. Therefore, research at the Escambia was initially focused on solving basic management problems that plagued longleaf pine, such as natural regeneration, thinning regimes,

Fig. 4.3 Farm Forty Field Days: preparing the harvest for display (**a**); informing and advising the public (**b**, **c**)

and product rotation lengths (Croker 1952, 1958). The intent was to focus efforts on developing practical management alternatives in order to encourage retention of longleaf pine by the owners of the small farm forests that dominated the South. To that end, the Escambia was surveyed and divided into 16.2-ha compartments, and three studies were immediately installed:

a. *The Management Systems Study* The Management Systems Study was established in 24 of Escambia's 16.2-ha compartments, with 12 compartments randomly assigned to even-aged and 12 to uneven-aged management systems. The goal was to examine forest management and economic aspects of three rotation lengths for longleaf pine forests: short (40 years), medium (60 years), and long (80 years;) (Croker 1973). The study measured growth and yield, and management costs, and required a labor-intensive 100% inventory each year. Every tree harvested, whether log, pole, or pulpwood, was scaled in the woods and then again at the T.R. Miller company scale. Volume estimates made in the field were compared with company tickets. Croker (1987) reported truck miles, mule upkeep, labor rates, and equipment costs among the information recorded annually.
b. *The Farm Forty Study* The Farm Forty Study demonstrated management of a typical 16.2-ha longleaf pine farm woodlot. It was managed for logs and poles on an 80-year rotation. An annual field day was held to showcase the products harvested from the woodlot in an average year (Fig. 4.3a–c) (for a 30-year summary,

Fig. 4.4 An aerial view of an early seed-tree harvest on the Escambia Experimental Forest

see Boyer and Farrar 1981). Harvest was limited to two thirds of the computed annual growth on the woodlot (Croker 1987) and used the group shelterwood stand reproduction method, in which 5-acre groups of trees were periodically removed to encourage natural longleaf pine regeneration (Baker 1934).

c. *The Investment Forest Study* The Investment Forest Study was set up to simulate forest management of a typical investor. Records were kept of all activities on the 259-ha tract, such as timber marking, maintenance of roads and boundary lines, and prescribed burning (Croker 1973). It was managed on a 60-year rotation.

Two other significant events occurred in 1947, the consequences of which still resonate today. One was a bumper seed crop of longleaf pine, and the second was a decision to intentionally burn 10,522 ha of land on both the Escambia and the Conecuh National Forest (Croker 1987). H.O. Mills, a district ranger on the nearby Conecuh, noted a good crop of cones and with the support of Don Morris, an assistant forest supervisor, discussed performing a seedbed burn. They sought the help of Escambia employees Dave Bruce and Thomas Croker, who supported the decision to burn (Croker 1987). The resulting successful establishment of longleaf pine seedlings on burned areas was evidence of what could be achieved by noting patterns in nature and applying sound research to silvicultural problems. Fire, long considered a forest enemy, was now viewed as necessary for successful longleaf pine stand establishment. Croker (1987) wrote that almost all of the prolific advanced regeneration in his seed-tree study (Fig. 4.4) occurred before the study began and prior to the cutback to seed-tree densities. It was the 1947 seed crop that regenerated the site and not the harvest using the seed-tree method. Many of the third-growth forests at the Escambia were established from this one seed crop, and the seed-tree stand reproduction method for longleaf pine was abandoned at the Escambia. Not only did the seed-tree method result in poor regeneration, but it also provided insufficient quantities of pine needles and other fine fuels to carry prescribed fire across the study sites (Drs. Dale Brockway and William Boyer, U.S. Forest Service, pers. comm.).

In 1951, organizational changes in the research stations resulted in formation of the East Gulf Coast Branch, with Brewton as a subunit. Personnel changes and reduced research funds (US $15,000 for salaries and operating expenses; Croker 1987) resulted in simplification of the laborious Management Systems Study at the forest (Croker 1953). Other studies were put on a maintenance basis to conserve funds, and efforts were concentrated on the Farm Forty Study and Investment Forest Study. However, in 1955, with strong local support from the community and from a forest industry desirous of supporting science-based forestry, young foresters were hired to assist with research studies and management data. One of these, William D. (Bill) Boyer, later became project leader of the research unit responsible for managing the Escambia.

In 1956, Croker suggested that the shelterwood stand reproduction method should be used to regenerate longleaf pine forests (Croker 1956; Boyer 1963; Croker and Boyer 1975). His writing showcased this regeneration method as addressing longleaf pine's sporadic seed production and the need for advanced regeneration before release cuts. While the seed-tree method (with residual basal areas between 2 and 3.5 m^2/ha) resulted in understocked stands with low volume accretion and hardwood encroachment or severe competition from native grasses, the shelterwood method (with residual basal areas from 5.7 to 7 m^2/ha) retained considerable growing stock and waited until adequate reproduction was established before overstory removal. Croker's suggested use of this even-aged technique has benefited many longleaf pine growers throughout the South. Following the successful Farm Forty Study and Tom Croker's publication in the *Journal of Forestry* emphasizing the benefits of the shelterwood reproduction method, land managers began rethinking their approach to longleaf pine regeneration and stand management methods.

Recognizing that the clearcutting and seed-tree methods resulted in insufficient longleaf pine regeneration, Forest Service scientist Dr. Robert M. (Bob) Farrar initiated a study at the Escambia to examine the unexplored effects of uneven-aged management on longleaf pine stands (Farrar 1996). The objective of the study was to demonstrate and compare three uneven-aged management techniques with fixed basal area per acre (target BA = 12.7 m^2/ha). Plot sizes range from 12.1 to 16.2 ha. Fire is applied every 3 years, and the diameters of all trees on the study sites are measured every 5 years. All are techniques to maximize timber volume growth and may not always be well suited for achieving ecosystem management goals requiring variable stand structures. Phase 1 was established in 1977 using the volume-guiding diameter limit (V-GDL) technique for volume regulation. Phase 2 of the study was installed in 1981 employing the basal area-maximum diameter-diminution quotient (BDq) technique for structural regulation. Phase 3 was added in 1991, testing the diameter limit cutting (DLC) technique, in which a prescribed basal area is maintained by removing all trees over a specific diameter. Although easy to apply, it is not necessarily the best approach to use for improving a stand, and misapplication can easily degenerate into "high grading" that will seriously undermine the genetic diversity of a forest by removing only the best-quality trees. The diameter class

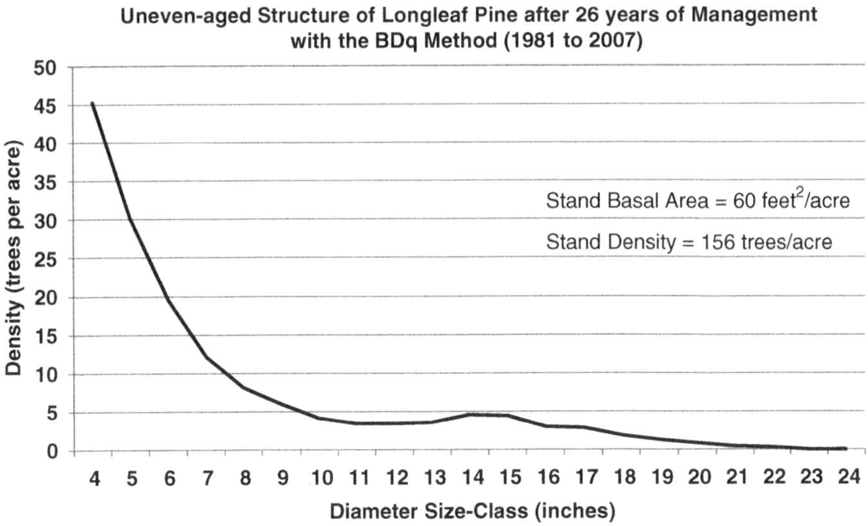

Fig. 4.5 Diameter distribution in an uneven-aged longleaf pine management demonstration, after 26 years of treatment with the basal area-maximum diameter-diminution quotient (BDq) method at the Escambia Experimental Forest. All data were collected in customary English units

distribution of pines from the BDq study, 26 years after installation, can be seen in Fig. 4.5. From such information, researchers discerned that this technique resulted in adequate regeneration to reproduce the stand and that growth could be sustained through periodic thinning on a 10-year cutting cycle.

In 1964, Farrar, working with partners in the Forest Service Regional Office in Atlanta and collaborators at Auburn University, established a much needed growth and yield study for naturally regenerated longleaf pine that was later expanded region-wide to several locations in Alabama, Mississippi, Florida, Georgia, and North Carolina (Kush and Tomczak 2007). Nearly half of the 305 plots in this study are located at the Escambia. While some plots on other study sites have been lost to development or otherwise destroyed, those at the Escambia remain protected and intact. The objective of the study, which continues today, is to quantify the growth and yield of natural, thinned longleaf pine forests spanning a range of ages, site types, and residual stand densities across the Southern Region. When the study began, site quality was measured by site index at base age 50 to be 15.2–27.4 m, and stand ages were determined to be 20–80 years. Study sites now are thinned to maintain the target basal area for each stand of 2.8–13.9 m^2/ha, and new stands in the 15-year age class are added every 10 years for temporal replication. All plots are remeasured every 5 years, with the 50-year remeasurement scheduled to take place in 2014.

Cone production studies at the Escambia date back to 1958 and, since 1966, these have been expanded to observe longleaf pine cone production throughout the region, as part of a shelterwood study entitled "Longleaf Regeneration Trials." Scientists still annually monitor mature longleaf pine trees from Louisiana to North

Table 4.1 Average total basal area (feet²/acre) and volume (feet³/acre) for different season and frequency of prescribed fire treatments on the Escambia Experimental Forest. Numbers in italic are significantly different than the other numbers in the column at a probability ≤0.05. All data were collected in customary English units (From Kush 2007)

	Basal Area						Volume					
	1984	1987	1990	1994	1999	2004	1984	1987	1990	1994	1999	2004
Treatment[a]												
Winter-2	11	25	39	62	86	111	74	256	546	1,155	1,869	2,840
Spring-2	11	23	38	61	80	104	76	236	524	1,125	1,750	2,639
Winter-3	11	26	41	63	87	112	78	275	595	1,208	1,969	2,966
Spring-3	12	27	44	68	*94*	114	80	268	612	1,300	2,096	2,987
Winter-5	21	25	40	65	89	116	86	262	573	1,230	1,969	3,022
Spring-5	13	26	42	*71*	*97*	*127*	94	284	608	1,366	*2,199*	*3,340*
No burn	11	27	45	*74*	*102*	*127*	72	274	638	1,468	*2,390*	*3,460*

[a] Treatments are prescribed burning during the winter or spring season at a 2-, 3-, or 5-year interval

Carolina, tallying the number of longleaf pine flowers, conelets, and cones to assess and predict longleaf pine seed production for the current and following years (Boyer 1974, 1987, 1998; Croker 1973). After many years of observing cone production in stands across a range of densities, optimum cone production appears to occur in shelterwood stands of 6.2 m²/ha. Scientists monitor 10–15 seed-bearing longleaf pine trees per study site and annually conduct pollen counts at the Escambia. Continuing Bill Boyer's earlier work, Dr. Dale Brockway annually prepares a report containing estimates for the regional cone crop. It is issued in June and is in high demand by southern forest managers.

As evidence supporting the role of fire in longleaf pine forests grew and appreciation for the importance of fire in successful seedling establishment increased, scientists formally began research on fire ecology in 1973 by establishing two continuing studies at the Escambia (Kush et al. 1999). To investigate the long-term effects of season of burn, plots are either burned once every 2 years in spring, summer, or winter, or left unburned as a control. In conjunction with the season of burn, some plots received an initial herbicide treatment while on others vegetation was periodically cleared away by hand. All pine height and diameters were measured, fire behavior documented every 3 years, and crown scorch recorded shortly after each burn. Understory species were also identified and measured. Still in progress, the study measurements are now repeated every 5 years.

A second fire study was established in 1985 to examine both fire season and the length of time between burns (i.e., one fire every 2, 3, or 5 years; Boyer 1990, 1994). Prescribed burning during the spring, rather than in summer or winter, enhanced longleaf pine seedling development and was also very effective in controlling hardwoods (Boyer 1990). Burning at 2-year intervals resulted in lower overstory basal area and volume growth than burning at either 3-year or 5-year intervals (Kush 2007; Whitaker et al. 2007; Table 4.1). The increased interval between burns places less stress on the trees and therefore increased growth.

4.5 Current and Future Directions

Through the decades of decline, longleaf pine forests retained supporters and admirers. In addition to those with memories of tapping pine trees for turpentine and collecting longleaf pine needles to weave baskets, remembering marvels of pitcher plant bogs and encounters with gopher tortoises, there were always those who recognized the timber production potential of this species and worked to promote and sustain research and practical management activities.

In 1974–1975, the headquarters for the research work unit managing the Escambia was relocated from Brewton, AL, to Auburn, AL. The future direction of forestry appeared to be moving away from the long rotations that produced high-value longleaf pine poles and sawlogs toward short rotations featuring loblolly pine and slash pine. Among the new objectives for the research unit was increased emphasis on herbicide control of undesirable weeds and hardwood vegetation (Dr. Bill Boyer, U.S. Forest Service, pers. comm.). The Escambia was scheduled for termination and only the combined efforts of Bill Boyer and Robert Farrar kept the land under Forest Service management. They argued that fire was an alternative means for natural hardwood control and that the fire studies on the Escambia would provide much needed information. Many of our existing databases and long-term records were maintained "unofficially" by Boyer who continued collecting data despite elimination of funds for longleaf pine research and the shift in research emphasis. Pleas for maintaining a research presence at the Escambia were still being made as late as 1987, stressing that the forest would be "…impossible to replace, quantitatively and qualitatively" (Boyer and Farrar 1987).

The persistent belief of Boyer and Farrar in the value of longleaf pine research to southern communities and to the southern timber industry was strengthened in 1995 with formation of the Longleaf Alliance (http://www.longleafalliance.org). This group was established to coordinate partnerships among the various parties—private landowners, educational institutions, state and federal governments, nongovernmental organizations, and industry groups—interested in longleaf pine. While interests in longleaf pine range from industrial forest management to complete ecosystem restoration, almost all of the 800 members of the Longleaf Alliance want to not only manage and restore longleaf pine forests but also promote increased interest in this species. Because of these individual and group efforts, research at the Escambia Experimental Forest has been continuous since its establishment in 1947. The long-term data sets collected by scientists from this forest are considered both invaluable and unique.

A little more than 80 % of the Escambia is now occupied by longleaf pine stands, with the remainder in slash pine and hardwood bottomlands. Tree ages range from young seedlings to 160-year-old trees with the second-growth timber approximately 95 years old. More than 485.6 ha of the forest have been naturally regenerated, and more than half of this is in stands ranging from 35 to 50 years in age. Stand densities vary widely, with some variation artificially created for the growth and yield studies started in 1964. Site index as a measure of site quality averages 21.3–22.9 m at age 50 years, with a range from 19.8 to 25.3 m. Very few locations in the South

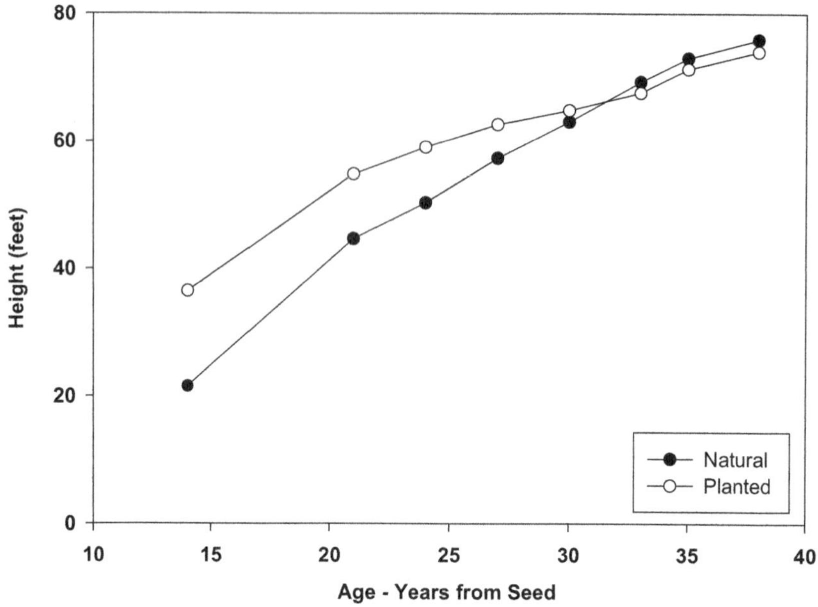

Fig. 4.6 Changes in longleaf pine total height: natural stands versus plantations on prepared sites. All data were collected in customary English units

can boast the combinations of stand ages, forest structures, and site conditions that are found at the Escambia.

The advantages of the long-term work in progress at the Escambia are exemplified by discoveries that could not have been made through short-term experiments. For instance, Boyer (1997) observed that natural longleaf pine regeneration catches and surpasses planted longleaf pine in height, even with understory control on the planted sites (Fig. 4.6). While planted longleaf pine trees on intensively prepared sites had a 4-m-height advantage at age 13 over natural regeneration on unburned sites, the difference had closed to 1.5 m by age 26. The height of naturally regenerated trees caught up to and surpassed planted trees by age 33. For a landowner or forest manager with longleaf pine already in place, natural regeneration methods are both effective and economical, eliminating the large sums needed for tree planting and intensive site preparation costs that, carried over years, diminish the economic benefit to the forest owner.

A study at the Harrison Experimental Forest near Gulfport, MS, has implications for the Escambia and for landowners throughout the South. In addition to exhibiting greater resistance to the many insect and disease pests that infect other southern pines, longleaf pine outgrows loblolly pine and is comparable in growth to slash pine by age 39 years (Harris et al. 2001; Table 4.2). Additionally, nearly 71 % of the longleaf pine trees in the study produced poles, while only 12 % of slash pine and 8 % of loblolly pine trees fell into that classification. Poles are consistently valued at about 150 % of sawtimber stumpage prices, making longleaf pine a far superior

Table 4.2 A comparison of basal area, diameter at breast height, total height, and volume of longleaf pine, slash pine, and loblolly pine at ages 25 and 39 years. Means followed by different letters indicate a significant difference between species within the same column. All data were collected in customary English units (Reproduced from Harris et al. 2001)

	Density (trees/acre)		Basal area ($feet^2$/acre)		DBH (inches)		Height (feet)		Volume ($feet^3$/acre)	
	1985	1999	1985	1999	1985	1999	1985	1999	1985	1999
Longleaf	154a	140a	49.51a	68.51a	7.44b	9.42ab	55.1a	68.7b	2,802ab	4,918a
Slash	151a	120b	43.06b	59.90a	7.83a	9.66a	56.7a	79.1a	2,908a	4,475a
Loblolly	132b	99c	47.95ab	50.02b	7.47b	9.17b	63.4b	63.4c	2,467b	3,580b

Table 4.3 Comparison of site index at base age 50 years (SI) for second-growth and third-growth longleaf pine forests at the Escambia Experimental Forest. All data were collected in customary English units (From Boyer 2001)

Compartment	2nd Growth				3rd Growth			
	No. of plots	Age	Height	SI	No. of plots	Age	Height	SI
74	8	60.8	72.5	67.9	7	41.5	72.7	81.2
75	8	63.4	72.4	65.8	7	38.8	75.0	84.8
81	5	59.4	74.1	68.7	10	39.6	74.3	84.2
83	9	55.3	69.9	66.9	6	39.2	76.8	87.2
102	7	50.6	65.5	65.9	14	40.3	71.1	80.6
103	8	55.8	64.4	61.6	14	40.8	68.9	77.1
107	9	46.0	67.3	70.1	6	40.0	74.7	84.6
115	11	55.0	64.2	61.8	7	40.7	73.1	82.6
125	9	51.1	58.8	60.6	7	39.7	75.2	85.8
Mean		*55.3*	*67.7*	*65.5*		*40.1*	*73.5*	*83.1*

investment for the forest landowner (for examples from two southern states, see Alabama Cooperative Extension System Reports and Mississippi State University Extension Notes at http://www.aces.edu/pubs/docs/A/ANR-0602/ and http://msucares.com/forestry/economics/reports/2009_harvest_report.pdf, respectively: accessed 8 April 2010). Longleaf pine is therefore a better economic choice than either loblolly or slash pine. Furthermore, a post-Hurricane Katrina survey and analysis showed that longleaf pine suffered substantially less damage and mortality than either slash pine or loblolly pine (Johnsen et al. 2009).

Boyer (2001) noted significant growth differences between second-growth and third-growth stands at the Escambia (Table 4.3). Examining data collected from two early studies, he noticed that estimates of site index for third-growth stands exceeded that obtained when second-growth stands were first inventoried. In 16 compartments, second-growth stands averaged 20.3 m in one study and 20.2 m in a second, while estimates of height growth in third-growth stands from studies in 17 compartments averaged 24.8 m. All of these stands are intermixed and cover a similar range of soil-site conditions. Additionally, less than 5% of second-growth trees in the study showed signs of early suppression followed by later release. In fact, based on early radial growth measurements of the first 25 rings, second-growth

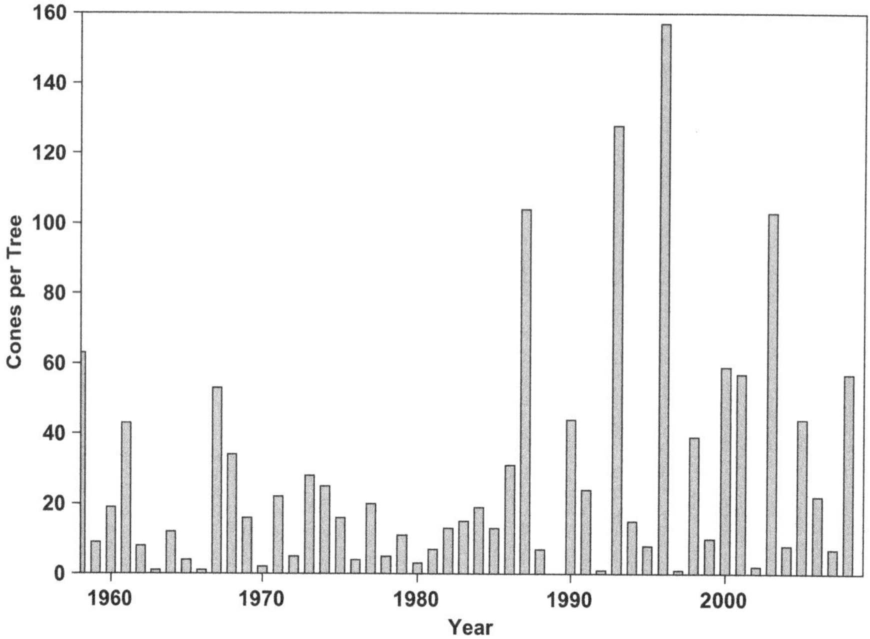

Fig. 4.7 Longleaf pine cone production on the Escambia Experimental Forest from 1958 to 2008

trees outgrew third-growth trees, suggesting that changes in growth are not due to differences in site, stand density, or early tree growth. Using 9 of the 16 compartments that included third-growth stands approximately 40 years of age, he made direct comparisons of the site index values and confirmed this significant site index shift. While further measurements are necessary, this study may have significant implications for climate change researchers.

Scientists are now examining data from the long-term growth and yield studies to determine whether the carbon storage capacity of longleaf pine could provide potential mitigation for climate change. Kush et al. (2004) noted that changes in the wood/fiber markets in the South resulted in a trend toward longer tree rotations. Since longleaf pine is a long-lived species with a low mortality rate and produces highly valued sawlogs and poles, it is an ideal candidate for long-term terrestrial carbon storage. Even after harvest, the wood products will continue to store carbon in the form of poles, building timbers, and veneers. There would also be both social and ecological benefits from increased restoration of longleaf pine acreage. Kush et al. (2004) also noted that longleaf pine outperforms other southern pines as the rotation lengthens and is tolerant of fire and many insects and diseases that decimate loblolly pine and slash pine.

Lastly, cone crop information that has been collected for 50 years on the Escambia and at many sites across the region in the "Longleaf Regeneration Trials" has produced interesting results. Because of this extensive long-term database, scientists have noted that cone production by longleaf pine trees on the Escambia has more than doubled during the period from 1986 to 2008 compared to the preceding 20-year average (Fig. 4.7). At this time, researchers are uncertain about the cause for this increasing frequency of good cone crops. While it may be a result of the same factors affecting site index of the third-growth stands, observers speculate that the increase may be related to advancing tree age, geographic origin of the species, or climate change. More research is needed to study these phenomena.

These important advances were possible only because of the long-term databases now available from experimental forests such as the Escambia, where studies have been actively maintained and protected for decades of information gathering. Because of research at the Escambia, we now know that the shelterwood reproduction method is a successful and cost-efficient means of regenerating longleaf pine forests, that fire is essential for longleaf pine regeneration and ecosystem maintenance, that height growth of naturally regenerated longleaf pine catches up to and surpasses planted seedlings after 33 years, and above all, that longleaf pine ecosystems are an integral and vital part of the southern culture.

The Escambia Experimental Forest is managed by the USDA Forest Service, Southern Research Station, Unit SRS-4158, headquartered in Auburn, AL, with scientists also stationed at Clemson, SC, and Pineville, LA.

References

Baker FS (1934) Theory and practice of silviculture. McGraw-Hill Book Company, Inc., New York, 502 p

Boyer WD (1963) Development of longleaf pine seedlings under parent trees. Research Paper SO-4, U.S. Department of Agriculture, Forest Service, Southern Forest Experiment Station, New Orleans, LA, 5 p

Boyer WD (1974) Longleaf pine cone production related to pollen density. In: Kraus J (ed) Proceedings of a colloquium: seed yield from southern pine seed orchards, Georgia Forest Research Council, pp 8–14

Boyer WD (1987) Annual and geographic variation in cone production by longleaf pine. In: Phillips DR (comp.) Proceedings of the fourth biennial southern silvicultural research conference, General Technical Report SE-424, U.S. Department of Agriculture, Forest Service, Southeastern Forest Experiment Station, Asheville, NC, pp 73–76

Boyer WD (1990) Growing-season burns for control of hardwoods in longleaf pine stands. Research Paper SO-256, U.S. Department of Agriculture, Forest Service, Southern Forest Experiment Station, New Orleans, LA, 7 p

Boyer WD (1993) Regenerating longleaf pine with natural seeding. Proc Tall Timbers Fire Ecol Conf 18: 299–309

Boyer WD (1994) Eighteen years of seasonal burning in longleaf pine: effects on overstory growth. In: Proceedings of the 12th international conference on fire and forest meteorology, Society of American Foresters, Bethesda, MD, pp 602–610

Boyer WD (1997) Longleaf pine can catch up. In: Kush JS (comp.) Proceedings of the first Longleaf Alliance conference. Longleaf pine: a regional perspective of challenges and opportunities, Longleaf Alliance Report No. 1, Auburn University, AL, pp 28–29

Boyer WD (1998) Long-term changes in flowering and cone production by longleaf pine. In: Waldrop TA (ed) Proceedings of the ninth biennial southern silvicultural research conference,

General Technical Report SRS-20, U.S. Department of Agriculture, Forest Service, Southern Research Station, Asheville, NC, pp 92–98

Boyer WD (2001) A generational change in site index for naturally established longleaf pine on a south Alabama coastal plain site. South J Appl For 25(2):88–92

Boyer WD, Farrar RM Jr (1981) Thirty years of management on a small longleaf pine forest. South J Appl For 5:73–77

Boyer WD, Farrar RM (1987) Justification for maintaining the Escambia Experimental Forest. In: Escambia Experimental Forest, 1987 Review, USDA Forest Service, Southern Forest Experiment Station, Internal document

Boyer WD, White JB (1990) Natural regeneration of longleaf pine. In: Farrar RE (ed) Proceedings of the symposium on the management of longleaf pine, General Technical Report SO-75, U.S. Department of Agriculture, Forest Service, Southern Forest Experiment Station, New Orleans, LA, pp 94–113

Burns RM, Honkala BH (tech. cords.) (1990) Silvics of North America, vol 1, Conifers, Agriculture Handbook 654, U.S. Department of Agriculture, Forest Service, Washington, DC, 675 p

Croker TC Jr (1952) Management plan, Escambia Experimental Forest, Internal document

Croker TC Jr (1953) Establishment and progress report, management systems, FS-SO-1103-1.1, Escambia Experimental Forest, East Gulf Coast Branch, Southern Forest Experiment Station, Internal document

Croker TC Jr (1956) Can the shelterwood method successfully regenerate longleaf pine? J For 54:258–260

Croker TC Jr (1958) Management Plan Revision, Escambia Experimental Forest Management, Internal document

Croker TC Jr (1973) Longleaf pine cone production in relation to site index, stand age, and stand density, Research Note SO-156, U.S. Department of Agriculture, Forest Service, Southern Forest Experiment Station, New Orleans, LA, 3 p

Croker TC Jr (1987) Longleaf pine, a history of man and a forest, Forestry Report R8-FR7, U.S. Department of Agriculture, Forest Service, Southern Region, Atlanta, GA, 37 p

Croker TC Jr, Boyer WD (1975) Regenerating longleaf pine naturally, Research Paper SO-105, U.S. Department of Agriculture, Forest Service, Southern Forest Experiment Station, New Orleans, LA, 21 p

Farrar RM Jr (1996) Fundamentals of uneven-aged management in southern pine, Tall Timbers Research Station Miscellaneous Publication No. 9, Tallahassee, FL, 63 p

Frost C (2006) History and future of the longleaf pine ecosystem, Chapter 2. In: Jose S, Jokela EJ, Miller DL (eds) The longleaf pine ecosystem. Ecology, silviculture, and restoration. Springer Science + Business Media LLC, New York, pp 9–48

Harris AH, Kush JS, Meldahl RS, Varner JM (2001) And the winner is…longleaf pine: response of planted pines to various cultural treatments 39 years later. In: Kush JS (comp.) Proceedings of the third Longleaf Alliance regional conference, Longleaf Alliance Report No. 5, pp 129–131

Hoyle, Z (2009) The Harrison: longleaf genetics come of age. Compass 2009(13):34–36

Hughes G (2006) Hurricane Katrina: impacts on different pine species and implications for landowners. Mississippi State University Extension Service. http://www.balmlandclearing.com/id67.html. Accessed 26 May 2009

Johnsen KH, Butnor JR, Kush JS, Schmidtling RC, Nelson CD (2009) Hurricane Katrina winds damaged longleaf pine less than loblolly pine. South J Appl For 33:178–181

Jose S, Jokela EK, Miller DL (2006) The longleaf pine ecosystem. An overview. In: Jose S, Jokela EJ, Miller DL (eds) The longleaf pine ecosystem. Ecology, silviculture, and restoration. Springer Science + Business Media LLC, New York, pp 3–8

Kush JS (2007) Fire effects on longleaf pine growth. In: Estes BL, Kush JS (comps.) Longleaf pine: seeing the forest through the trees, Proceedings of the sixth Longleaf Alliance regional conference, Longleaf Alliance Report No. 10, pp 33–36

Kush JS, Meldahl RS, Boyer WD (1999) Understory plant community response after 23 years of hardwood control treatments in natural longleaf pine (*Pinus palustris*) forests. Can J For Res 29:1047–1054

Kush JS, Meldahl RS, McMahon CK, Boyer WD (2004) Longleaf pine: a sustainable approach for increasing terrestrial carbon in the southern United States. Environ Manage 33(Suppl 1):S139–S147

Kush JS, Tomczak D (2007) The regional longleaf pine growth study—40 years old. In: Estes BL, Kush JS (comps.) Longleaf pine: seeing the forest through the trees, Proceedings of the sixth Longleaf Alliance regional conference, Longleaf Alliance Report No. 10, pp 102–103

Lipscomb DJ (1989) Impact of feral hogs on longleaf pine regeneration. South J Appl For 13(4):177–181

Noss R, LaRoe ET, Scott JM (1995) Endangered ecosystems of the United States: a preliminary assessment of loss and degradation, Biological Report 28, U.S. Department of Interior, National Biological Service, Washington, DC, 59 p

Pinchot G (1947) Breaking new ground. Harcourt, Brace, and Co., New York; Commemorative edition. 1998. Island Press, Washington, DC, 522 p

Ray J (1999) Ecology of a cracker childhood. Milkweed Press, Minneapolis, 200 p

Walkinshaw CH, Otrosina WJ (2002) Food reserves in mountain longleaf pine roots during shoot elongation. General Technical Report SRS-48, U.S. Department of Agriculture, Forest Service, Southern Research Station, Asheville, NC, pp 97–99

Whitaker B, Boyer WD, Kush JS (2007) A continued *Pinus palustris* burn study comparing frequency and season of fire to basal area growth loss. In: Estes BL, Kush JS (comps.) Longleaf pine: seeing the forest through the trees. Proceedings of the sixth Longleaf Alliance regional conference, Longleaf Alliance Report No. 10, pp 142–144

Whitney GG (1996) From coastal wilderness to fruited plain: a history of environmental change in temperate North America, 1500 to the present. Cambridge University Press, Cambridge, 488 p

Chapter 5
Northeastern Conifer Research: Multiple Species and Multiple Values

Laura S. Kenefic, John C. Brissette and Richard W. Judd

Abstract The northern conifer, or spruce-fir, forest of the northeastern USA and adjacent Canada has had a defining influence on the economy and culture of the region. The same can be said of the USDA Forest Service's research in this forest, which began more than 100 years ago. Forest Service research has evolved since that time in response to changes in the needs and prominence of the forest industry, and in public attitudes and concerns. Early studies of forest protection and rehabilitation first gave way to mid-century research on production forestry, then to twenty-first-century research on forest ecology. Though various lines of research have come and gone, long-term studies on the region's experimental forests continue to provide a unique perspective on the structure and dynamics of the forest, and the outcomes of silvicultural alternatives.

Keywords Acadian Forest · Penobscot Experimental Forest · Silviculture · Red spruce · Balsam fir

5.1 Background

A magnificent forest coupled with a complex of favorable environmental conditions made the Northeast a leader in lumber and pulpwood production in the 1800s and early 1900s (e.g., see Wilson 2005). Though forest industry provided an economic boon to the region, widespread cutting of progressively smaller trees resulted in forest degradation and raised concerns about resource sustainability (Judd 1997). At the time of his appointment as Chief of the US Division of Forestry (now the USDA Forest Service) in 1898, Gifford Pinchot observed that the heavy utilization of northeastern forests had "affect[ed] the forest to its injury" (Pinchot 1898, p. 31).

L. S. Kenefic (✉)
Northern Research Station, USDA Forest Service, Bradley, ME 04411, USA
Phone: (207) 581-2794
e-mail: lkenefic@fs.fed.us

J. C. Brissette
Northern Research Station, USDA Forest Service, Durham, NH, USA

R. W. Judd
Department of History, University of Maine, Orono, ME, USA

D. C. Hayes et al. (eds.), *USDA Forest Service Experimental Forests and Ranges*,
DOI 10.1007/978-1-4614-1818-4_5, © Springer New York 2014

In fact, by the end of the nineteenth century, the forests of the Northeast were a patchwork of brush, sprout-wood, relic old-growth, and second-growth timber.

Heavy cutting continued throughout the region in the early twentieth century, and opposition to exploitative practices intensified (Judd 1997). In response to industry pressure for continued production and public concerns about forest degradation, the Forest Service initiated research in the Northeast on forest rehabilitation, management, and protection. This research program was formalized by the creation of the Northeastern Forest Experiment Station (now the Northern Research Station) in 1923. Over the years, research by the Forest Service has evolved in response to changes in resource demands and the prominence of forest industry, and to changes in society's values and expectations. The focus of this chapter is the Forest Service's trajectory in northern conifer research over the following periods:

Resource assessment (1890s–1920s)
Forest improvement (1920s–1940s)
Changing directions (1940s–1950s)
Postwar prosperity (1950s–1990s)
Diversification (1990s–present)

5.2 Research Program

5.2.1 The Forest

Though northern hardwoods predominate in the Northeast, the northern conifer forest is locally abundant and was the primary source of wood for early industrial endeavors. This forest extends from the Adirondacks in New York through northern New England into maritime Canada (Fig. 5.1). Much of this forest is in the transitional zone between the eastern broadleaf and boreal forests and contains a diverse mixture of softwoods and hardwoods. Red spruce (*Picea rubens*) is largely confined to this region and is its signature species. Tolerant of shade, slow growing, and having excellent wood properties, this important commercial species is commonly known for its stiff and sharply pointed needles and woodsmen's confectionary "spruce gum." White (*P. glauca*) and black (*P. mariana*) spruces are also common, but the most prolific softwood in the region is balsam fir (*Abies balsamea*; Seymour 1995). Balsam fir is fast growing and competitive, but because it is more prone to insects, diseases, and decay, it is short lived relative to other shade-tolerant conifers. Consequently, it is less desirable than the spruces for most commercial products other than pulpwood.

Eastern hemlock (*Tsuga canadensis*) and eastern white pine (*Pinus strobus*) are abundant in the southern part of the forest; northern white-cedar (*Thuja occidentalis*) is more common in the north. Hardwoods include red maple (*Acer rubrum,* which leaf peepers love for its vibrant fall foliage and foresters grudgingly admire for its ubiquity and prolific sprouting), aspen (both *Populus grandidentata* and *P. tremu-*

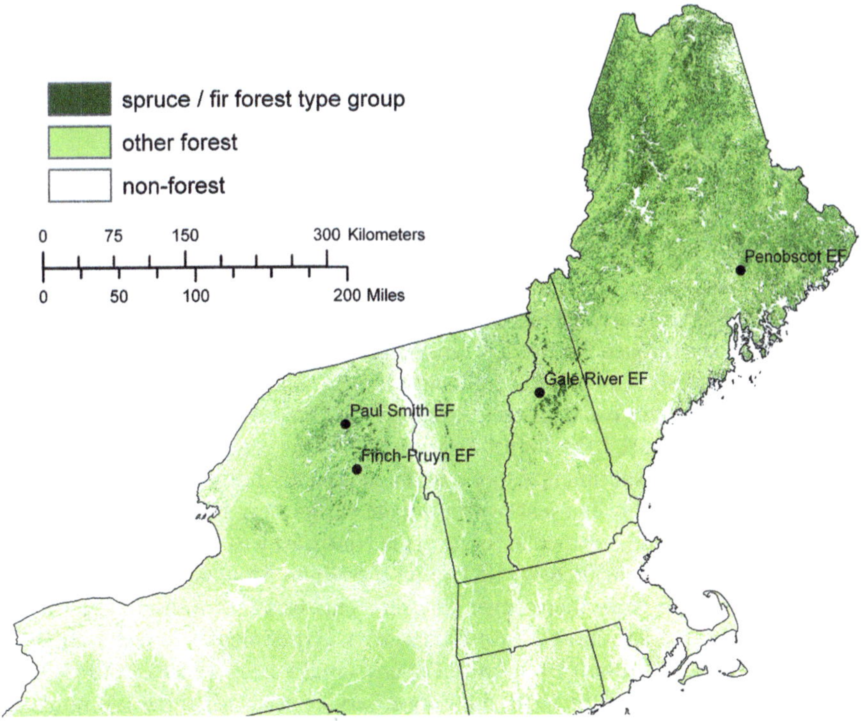

Fig. 5.1 Map of the northern conifer (previously call spruce-fir) forest in the northeastern USA. (Courtesy of B. Tyler Wilson, USDA Forest Service, Forest Inventory and Analysis)

loides), American beech (*Fagus grandifolia*), paper birch (*Betula papyrifera*) and gray birch (*B. populifolia*). Red oak (*Quercus rubra*), sugar maple (*A. saccharum*), and yellow birch (*B. alleghaniensis*) are found in better sites. Natural disturbances are primarily small scale, resulting from single-tree mortality and windthrow. Periodic larger-scale disturbances arise from insect infestation, such as the spruce budworm (*Choristoneura fumiferana*), and hurricanes (Seymour et al. 2002).The aspens, birches, oak, and pine are primarily associated with disturbances that open the forest canopy, while the other species commonly occur in closed forests. Though historically called spruce-fir, much of the forest is mixedwood; researchers in the region today call this the northern conifer forest. An important component of the northern conifer forest is the Acadian Forest (Rowe 1972), which includes most of the Canadian provinces of New Brunswick, Nova Scotia, and Prince Edward Island, as well as parts of eastern and northern Maine (Braun 1950).

One characteristic that sets the northern conifer forest apart from most other managed forests in North America is the tolerance to shade of many of the dominant species, and most of the important commercial tree species. In most regions, forest management focuses on shade-intolerant and mid-tolerant species, but northern conifer forest management emphasizes regenerating and favoring shade-tolerant trees.

The silvics of the species and predominantly small-scale natural disturbance regime result in a forest that is structurally and compositionally complex. These characteristics have important ramifications for stand dynamics and for the research that informs management of those stands.

5.2.2 The Experimental Forests

Though Forest Service researchers conducted a number of studies in the Northeast in the late 1800s and early 1900s, station leadership identified experimental forests as the preferred approach for forestry research beginning in the 1920s. Research on experimental forests was more secure and efficient than research on cooperators' lands, permitted intensive data collection, and facilitated correlation of different lines of investigation (USDA Forest Service 1935).This was a rather new idea; the first Forest Service experimental forest in the nation, the Fort Valley Experimental Forest, had been established just over a decade earlier near Flagstaff, AZ (Adams et al. 2008). The first experimental forest in the Northeast was the 754-ha (1,863-ac) Gale River Experimental Forest (GREF) on the White Mountain National Forest in New Hampshire; work there began in 1926. This was followed in the northern conifer type by the 252-ha (622-ac) Finch-Pruyn Experimental Forest (FPEF) in 1934 and 890-ha (2,200-ac) Paul Smith Experimental Forest (PSEF) in 1945; both were in the Adirondacks. In 1950, the Forest Service established the last northern conifer experimental forest, the 1,538-ha (3,800-ac) Penobscot Experimental Forest (PEF) in Maine.

5.2.3 Research Trajectory

5.2.3.1 Resource Assessment (1890s–1920s)

The earliest Forest Service research in northern conifers began in the late 1800s, predating the experimental forests. This work consisted of empirical studies aimed at characterizing the forest type and its component species, with an emphasis on red spruce. Very little was known about the condition and growth potential of the northern conifer forest, or about the impacts of logging. Early publications by Gifford Pinchot (1898), Henry Graves (1899), Ralph Hosmer (1902), Raphael Zon (1914), and Louis Murphy (1917) are characteristic of this period, which continued in the Northeast for about three decades. The objectives of these studies were to determine the condition of the forest (e.g., size- and age-class distribution, quality, vigor, and species composition), the silvical properties and growth potential of individual species and trees, and the influence of site on these characteristics. Most importantly, this work highlighted the consequences of uncontrolled logging and the growth potential of small, suppressed softwoods if protected from damage during harvesting

Fig. 5.2 Marinus Westveld (*left*), the Father of Spruce-Fir Silviculture

operations and free of hardwood competition. These observations led to recognition of the need for forest management and forest management research, and prompted the Forest Service to establish experimental forests and initiate manipulative studies in the Northeast.

5.2.3.2 Forest Improvement (1920s–1940s)

The first manipulative studies in the northern conifer type were motivated both by concerns about forest sustainability and by the desire to support economically important forest industries. In the 1920s, the northeastern pulp and paper industry manufactured more than half the nation's wood pulp and contributed substantially to the region's social and economic welfare (Meyer 1929; Westveld 1938). A rise in pulpwood prices earlier in the twentieth century had resulted in heavy cutting of spruce and fir, leaving understocked stands of poor form and defective softwoods and associated hardwoods (USDA Forest Service 1937, 1938). The forests and forest industries of the Northeast were under stress. Researchers, practitioners, industrial owners, and the public wanted to know how continuing demands for wood products could be met from the degraded forest, and how forest values could be protected. To that end, early studies on experimental forests focused almost exclusively on stand improvement and rehabilitation.

The first studies on the GREF were directed by silviculturist Marinus Westveld, now known as the Father of Spruce-Fir Silviculture (Berven et al. 2013; Fig. 5.2). Westveld's work investigated the use of partial cutting (then called "selective" cutting) to establish shade-tolerant softwood regeneration and accelerate the growth of crop trees. He also studied planting, weeding, release, and swamp drainage to improve production of spruce and fir. He put considerable effort into understanding how to secure shade-tolerant softwood regeneration in mixedwood stands with hardwood competition, and how to increase the proportion of sawtimber in managed stands. With Westveld's cooperation, similar research was initiated at the FPEF (Recknagel et al. 1933) and at the PSEF, where forester John Curry was investigating techniques for establishing regeneration and improving stand conditions.[1]

Forest Service researchers on the northern conifer experimental forests advocated the use of silviculture—the science of controlling forest establishment, growth, composition, and health—and generally focused their studies on improving forest composition and restoring production potential. Experiments that were started during this time were unreplicated and limited in scope, with a tendency to emphasize partial cutting. In this regard, research on the experimental forests established the potential of uneven-aged silviculture in the northern conifer type (e.g., Belotelkin et al. 1942; Westveld 1938), but did little to advance other types of silviculture such as even-aged systems. This emphasis was consistent with national trends in research; the period between 1925 and 1960 has been called the "selective cutting era" to reflect the near single-minded national focus on the selection system and other forms of partial cutting (Seymour et al. 2006; Smith 1972). Nevertheless, work on the experimental forests during this time demonstrated the feasibility of establishing spruce and fir seedlings prior to removing the overstory, using mechanical or chemical treatments to release overtopped softwoods (Westveld 1930, 1931) and retaining sawtimber in managed stands (Recknagel et al. 1933). This attention to advance regeneration, control of hardwood competition, and growing stock retention remain hallmarks of northern conifer silviculture today.

5.2.3.3 Changing Directions (1940s–1950s)

Despite a strong early start by station scientists working on northern conifer forest management research, change was in the wind. On September 21, 1938, the New England Hurricane made a landfall on Long Island, NY, USA with one of the highest forward speeds ever documented for a tropical cyclone (Grossi 2008). This fast-moving storm brought powerful rain and winds deep into New England, causing widespread flooding, wind damage, destruction of property and loss of forest growing stock. This hurricane was one of most destructive storms in the history of the USA, leveling nearly 7 million m^3 (250 million ft^3) of timber on more than 200,000 ha (almost 500,000 acres) of forest land. The winds were most destructive in central New England, where nearly one half of the total volume of softwood timber was destroyed, leaving behind "tangled heaps of splintered trunks and limbs piled like giant match sticks and waiting for sparks to turn a literal inferno loose" (Anonymous 1938, p. 252). Fifty percent of the station's northern conifer research plots in the hurricane zone were damaged (USDA Forest Service 1939). The partial cutting experiment and controls at the GREF were completely destroyed (Fig. 5.3). By the end of 1942, about 200,000 m^3 (7 million ft^3) of timber had been salvaged from the White Mountain National Forest alone (Spurr 1956).

The destruction of the GREF experiments was undoubtedly a great loss for Westveld, who had installed a 40-ha (100-acre) partial cutting experiment only 1 year

[1] Memorandum on file at the USDA Forest Service, Northern Research Station, Bradley, ME, USA.

Fig. 5.3 Damage on the Gale River Experimental Forest from the New England Hurricane in 1938. (USDA Forest Service photo courtesy of the Forest History Society)

before (Belotelkin et al. 1942). The 1938 hurricane was the beginning of the end for the GREF, which continued with limited work in the 1940s. In fact, most experimental forestry work was suspended during the 1940s as World War II shifted the national focus to meeting wartime demands for manpower and other resources. Staffing changes also occurred. Francis Rushmore took over the FPEF and PSEF, which were consolidated into the Adirondack Research Center (ARC). Though discussions to establish a new experimental forest in New England had been initiated after the hurricane, those efforts were put on hold until the disruption and turmoil of the 1940s gave way to new forest management and research paradigms.

One notable line of research by station scientists during this era was spruce budworm control through forest management. A spruce budworm outbreak in Canada had reached epidemic proportions by the late 1930s and was spreading toward the northeastern USA.[2] Westveld suggested a two-phase approach to silvicultural control: (1) pre-salvage cutting mature balsam fir to reduce imminent mortality and (2) converting mixed spruce-fir stands to predominantly spruce through selection cutting with removal of merchantable fir and poor vigor spruce (Westveld 1946). To determine the cost and feasibility of these practices, the station established large experimental cuttings on cooperators' lands throughout the northern conifer forest. In practice, treatments were a form of diameter-limit cutting with removal of all merchantable fir and sawtimber-sized spruce (McLintock and Westveld 1946). Marking was conducted during World War II by German prisoners of war. Though the study areas were not, in fact, infested by the budworm, they served as the basis for an evaluation of wind damage (McLintock 1954) and composition and growth in selectively cut stands (Hart 1956).

[2] Problem analysis on file at the USDA Forest Service, Northern Research Station, Bradley, ME, USA.

Fig. 5.4 Penobscot Experimental Forest owners, Forest Service staff, and cooperators at the forest's dedication in 1952

5.2.3.4 Postwar Prosperity (1950s–1990s)

The USA's entry into World War II challenged New England forest industries. Forest products were vital to the war effort, but the industry faced severe labor shortages, as rural workers left to join the military or seek defense jobs. The war emergency hastened mechanization of woods operations, while rationing resulted in pent-up demand for wood and paper products. After the war, forest industries in Maine reopened discussions with the station about a new northern conifer experimental forest. Nine pulp and paper and land-holding companies purchased the land in Maine that was leased to the Forest Service for the PEF (Fig. 5.4). A 1951 press release described this transaction as "the first instance in the annals of American forestry that a group of wood-using industries have united to purchase a timberland tract for lease to the government to do such work."

In 1950, the forest product industry was the largest industry in Maine in terms of product value, number of employees, and payroll. In a report of forest management research opportunities in the northern conifer type, Thomas McLintock, station silviculturist in Maine, noted that the foremost needs were to assess the financial feasibility and production potential of a range of treatments.[3] To that end, a large-scale compartment- (or stand-) level study was initiated on the PEF. Unlike earlier manipulative studies in northern conifers, this experiment was fully replicated, included an array of silvicultural systems, and has continued without interruption to the present day (Fig. 5.5). Similar work was also initiated at the PSEF in the

[3] Problem analysis on file at the USDA Forest Service, Northern Research Station, Bradley, ME, USA.

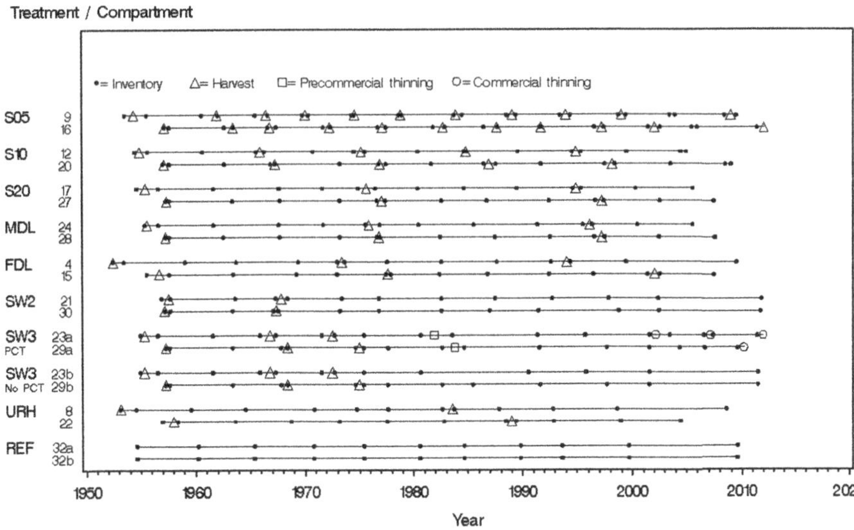

Fig. 5.5 Dates of treatment applications and inventories in the compartment study on the Penobscot Experimental Forest. Treatments are selection cutting on 5-, 10- and 20-year cycles (S05, S10, and S20), modified diameter-limit cutting (*MDL*), fixed diameter-limit cutting (*FDL*), shelterwood with two-stage overstory removal (*SW2*), shelterwood with three-stage overstory removal with and without pre-commercial thinning (*SW3 PCT* and *SW3 No PCT*), commercial clearcutting (*CC*) and no harvest (*NAT*). Numbers identify replicate compartments

Adirondacks, but was not completed. The GREF and ARC (including the FPEF and PSEF) were all closed by the early 1960s, resulting in consolidation of the station's northern conifer research program in Maine. Though the reasons for this decision were not clearly articulated, the declining relevance of earlier forest improvement research (which had been continued at the ARC; see Curry and Rushmore 1955; Rushmore 1956a, 1956b), hurricane damage at the GREF, reduced staffing, and emerging production-related research priorities were clearly contributing factors (Berven et al. 2013).

Arthur Hart, who had previously worked for Westveld on the GREF, took over the PEF research in the 1960s, and silviculturist Robert Frank Jr. joined the staff. Though research on the PEF had previously been restricted to sapling size and larger trees, Hart and Frank initiated studies of regeneration. Researchers quantified the competitive advantage of balsam fir over red spruce resulting from fir's larger and less palatable seed (Abbott and Hart 1960), more frequent seeding, deeper rooting, and faster growth (Hart 1963). It became clear that natural regeneration of northern conifers was prolific (Smith 1991), but questions remained about how to achieve desired species mixtures. Spruces were found to be less abundant than fir and hemlock under a range of selection and other partial cutting intensities, and hardwood-to-softwood ratios were higher in treatments with comparatively heavier removals (Brissette 1996).

Research on the PEF served as the basis for northern conifer silvicultural guidelines published during this time. These publications were typical of the Forest Service managers' handbooks and silvicultural guides that were common between the 1960s and 1980s, and which continue to serve as the basis for silvicultural prescription writing on managed lands, including national forests. Foresters throughout the northeastern USA used the *Silvicultural Guide for Spruce-Fir in the Northeast* (Frank and Bjorkbom 1973) and companion audio-slide program. This guide, which was based on PEF research results, is still being referenced today. In addition, management recommendations specific to uneven-aged silviculture were developed from the PEF selection treatments (Frank and Blum 1978). Findings after 20 years of treatment showed decreases in the amount of unmerchantable volume, increases in seedling density and proportions of spruce growing stock, and improved diameter distributions.

Though the uneven-aged (selection) system was emphasized on the PEF due to the most important commercial species' shade tolerance and Westveld's partial cutting research during the forest improvement era (e.g., Belotelkin et al. 1942; Recknagel et al. 1933; Westveld 1938, 1953), variants of even-aged systems were added to the experiment at the urging of cooperator David M. Smith from Yale University. Smith told McLintock that "...management and harvesting of spruce-fir types in this country would become pretty badly hog-tied in detailed refinements if an honest effort were made to superimpose the true selection principle...."[4] The experiments in even-aged silviculture became increasingly relevant to forest industries. Most industrial landowners regarded uneven-aged silviculture as unnecessarily complex, inefficient, prone to high-grading, and ill suited for maximizing pulpwood production. The national forestry paradigm shifted to even-aged silviculture, focusing on high-yield, low-cost wood production, around 1960 (Seymour et al. 2006). Smith's suggestion to include even-aged treatments in the PEF study proved to be an inspiration. Studies of planting, fertilization, thinning, strip clearcutting, and whole-tree harvesting were initiated on the PEF between the 1960s and 1980s in direct response to industrial needs. These studies were typical of Forest Service research nationwide during what is now known as the "production forestry era" (Seymour et al. 2006). The long-term silvicultural experiment on the PEF thus demonstrated that northern conifer stands can be managed effectively with both even- and uneven-aged silvicultural systems, giving managers a broad range of options.

The mid-twentieth century was a time of rapid change in harvesting technology. Retired PEF silviculturist Frank recalls that harvesting operations on the PEF transitioned from horses to modified farm tractors and bulldozers, skidders, and cut-to-length processors and forwarders during his tenure. While each of these technologies can be used for both even- and uneven-aged treatments, the physical size of early mechanical tree harvesters in Maine necessitated some form of clearcutting. Representatives of forest industries, many of whom were acting for the PEF landowners, requested that research be initiated to determine the effects

[4] Letter on file at the USDA Forest Service, Northern Research Station, Bradley, ME, USA.

of these machines on harvested sites. Because the preferred treatment at the time was strip clearcutting, Forest Service scientists established a large-scale study of this method on the PEF in the 1960s. The effects of slash disposal method, strip width, and whole-tree skidding were determined (Bjorkbom and Frank 1968), as was the nutrient status of soils in the residual stands (Czapowskyj et al. 1977). A companion study—also requested by forest industries—examined the effects of early mechanical tree harvesting on the survival of advance regeneration (Frank and Putnam 1972).

In the 1970s and 1980s, a severe spruce budworm infestation spread through southeastern Canada and northern New England. The name "spruce" budworm is misleading, because balsam fir is more susceptible than any of the spruce species. Periodic outbreaks of the budworm cause widespread defoliation, growth suppression, and mortality of fir, spruces, and even hemlock and pine at the peak of an epidemic (Blum and MacLean 1985). To salvage infested trees, reduce the amount of fir, and start a new cycle of regeneration, companies began clearcutting on a grand scale (Judd 1989). This activity created vast acreages of young, even-aged northern conifer stands and led to questions about precommercial thinning (PCT). Consequently, two PCT experiments were initiated on the PEF and would ultimately provide important information about the effectiveness of various operational methodologies and spacings. Analysis two decades after treatment revealed that crop trees in thinned stands had larger diameters, longer crowns, and higher volumes, and that thinned stands had more spruce and less fir (Brissette et al. 1999; Weiskittel et al. 2009).

Ironically, the most serious threat to long-term research on the PEF came just after this time of high research productivity. During the budworm years, Forest Service staffing had increased in Maine to include silviculturists, entomologists, soil scientists, and wildlife biologists. Despite continuing pressure from industry for forest management research, station leadership considered closing the research work unit in the late 1980s. The plan, which was intended to reduce costs, would have moved the scientific staff to New Hampshire. While the PEF research would have continued, it would have suffered without local management. After the announcement that the station might close the unit, an outpouring of support from the local forestry community convinced management to keep it open. While the overall number of staff in Maine was reduced, those assigned to the PEF were not affected. The most important outcome of the potential closing came several years later—in 1994—when the industrial landowners donated the PEF to the University of Maine Foundation.

5.2.3.5 Diversification (1990s–present)

In the 1980s and 1990s, forest industries in northern New England underwent profound changes involving consolidation, downsizing, and turnover in mill and forest ownerships. Early mills acquired large amounts of land and held it against wood shortages, but returns on this investment were low. As long as timberland was cheap, the mills retained their forest property, but with land values rising rapidly and demand for pulp declining in the 1980s and 1990s, many liquidated (Acheson 2000; Hagan et al. 2005). In some cases, the purchasers were organizations focused on preservation, conservation, or ecological forestry, such as the Nature Conservancy, which purchased 75,000 ha (185,000 acres) of International Paper Company's land along the St. John River in Maine (Acheson 2000; Irland 1999).

Increased emphasis on forest ecology had become common in forest management and research after the emergence of New Forestry, "a kinder and gentler forestry that better accommodates ecological values" (Franklin 1989, p. 38). In Maine, a number of unsuccessful citizens' referenda aimed at constraining forest management activities; the advent of forest certification and changing forest age-class structure motivated changes in the type and extent of silvicultural manipulations. On the PEF, the forest-type descriptor "spruce-fir" gave way to the more inclusive (and more accurate) "northern conifers." Frank began a new irregular shelterwood treatment with the objectives of creating multi-cohort stands and retaining large trees for cavities, snags, and downed woody material. The new project leader John Brissette added new measurement variables to the compartment study, including standing dead and downed trees, structural characteristics, such as tree location, height, crown radii, and crown length, and ground cover.[5] Laura Kenefic joined the project as a silviculturist and revised the stand prescriptions to exclude wildlife trees and some canopy emergents from cutting in the silvicultural treatments, and incorporated tree age into analysis of the selection treatments.

In 1994, the industrial owners of the PEF donated the property to the University of Maine Foundation, with the hope that faculty and graduate students would initiate new research. The station's earlier plan to close the research work unit played a part in the owners' decision to divest the property. In the donation document, they stated their expectation that the PEF would "afford a setting for long-term research conducted cooperatively among U.S. Forest Service scientists, University researchers and professional forest managers in Maine; to enhance forestry education of students and the public; and to demonstrate how the timber needs of society are met from a working forest."The Forest Service continued its research under a formal agreement, and encouraged collaboration between station and university scientists. In 1994, university scientists initiated the Acadian Forest Ecosystem Research Program, a long-term, multidisciplinary study on the effects of expanding-gap silviculture based on small-scale natural disturbance patterns (Saunders and Wagner 2005). This study was designed to complement the existing Forest Service experiment.

After the PEF became affiliated with the university, the number of short-term studies overlain on the Forest Service's long-term experiment increased. The earliest of these focused on ecophysiology, including photosynthesis, leaf morphology, leaf area, and growth efficiency (Day 2000; Kenefic and Seymour 1999; Maguire et al. 1998; Gilmore and Seymour 1996). The unique age structure of the PEF

[5] Study plan on file at the USDA Forest Service, Northern Research Station, Bradley, ME, USA.

selection stands, in which trees of the same species and age are in both suppressed understory and dominant overstory positions, allowed investigation of the effects of age on various growth parameters. Research by Forest Service staff and university collaborators revealed the negative effects of tree age—independent of tree size—on photosynthetic rate (Day et al. 2001) and the efficiency of stemwood volume growth (Seymour and Kenefic 2002). The influences of silvicultural treatments on tree-level growth dynamics were also investigated. The effect of PCT on root and height growth, for example, was found to be greater for balsam fir than for red spruce (Tian and Ostrofsky 2007; Weiskittel et al. 2009), suggesting that increases in the relative proportion of spruce in thinned stands are due more to preferential release of that species than to an advantage conferred by release.

Dead trees (and even living trees leaning more than 45°) were excluded from earlier inventories on the PEF, but snags and downed woody material became a focus of study during the 1990s. New research in the long-term compartment study revealed treatment differences in the amount, size, and decay class of standing dead and downed woody material (Garber et al. 2005; Weaver 2007). Decayed downed wood was found to be an important regeneration substrate for spruce and hemlock, but not for balsam fir and red maple, suggesting that management of dead wood can influence regeneration composition (Weaver et al. 2009). In addition, the tree numbers and mortality codes used on the permanent sample plots in the compartment study since the 1970s proved useful for relocating dead trees and identifying their dates of death. This allowed researchers to determine snag longevity and relate it to tree species and size, silvicultural treatment, and cause of death (Garber et al. 2005).

Though woody vegetation on the PEF had been inventoried in the 1960s (Safford et al. 1969), non-tree vegetation received little attention until the 1990s, when a study of understory plant associations with red spruce regeneration was conducted (Dibble et al. 1999). More recently, a complete inventory of understory plants on the sample plots in the Forest Service compartments was completed (Bryce et al. 2008). Understory species' richness and diversity generally declined with decreasing silvicultural intensity; differences in diversity and composition of understory plants were related to canopy composition and forest floor disturbance (Bryce 2009). Non-native invasive plants were uncommon in the experimental stands, but abundant in adjacent old-field stands, suggesting that mitigation is warranted to prevent future encroachment.

Other recent cooperative studies have covered a range of topics important to sustainable management of the region's forests, including wood decay and nutrient cycling (Smith et al. 2007), soil carbon (Hoover 2005), herbivory (Larouche et al. 2010), wildlife diversity and habitat suitability (Su and Woods 2001), and genetic implications of silvicultural treatments (Hawley et al. 2005). Such studies exemplify the potential of long-term silvicultural research to evolve and address emerging topics.

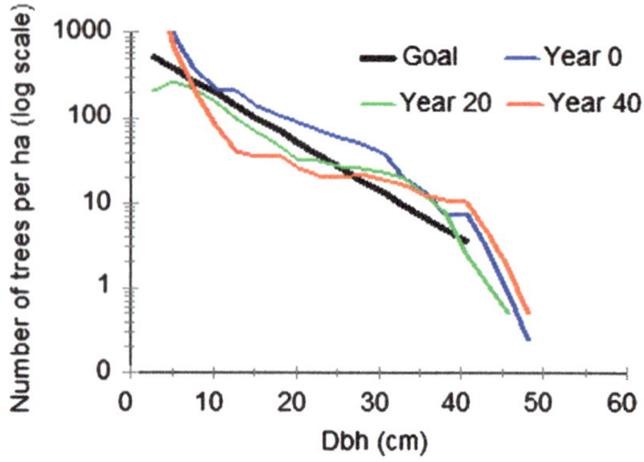

Fig. 5.6 Diameter distributions in a selection treatment on the Penobscot Experimental Forest reveal that the stands were closer to the target in year 20 than in year 40 of the experiment

5.3 Synthesis of the Research Trajectory

Forest Service research in northern conifers has evolved over the last century. Relevant research from earlier eras has continued to the present day (e.g., the PEF compartment study and PCT experiments), while other lines of research have been closed as industrial needs and societal expectations changed (e.g., the GREF swamp drainage and PSEF studies of chemical debarking). Though many of the individual trees studied by our predecessors are still living, management objectives and constraints differ from those when the first northeastern experimental forests were established in the 1920s. Early efforts related to forest protection and rehabilitation were supplanted by research on productivity and production, and later by studies of forest ecology and the impacts of manipulations on ecological processes.

5.3.1 New Perspectives

Despite these changes, long-term silvicultural studies like the one on the PEF continue to provide a unique perspective on forest dynamics. One of the advantages of building upon earlier work is that scientists can document treatment responses that vary over time. For example, the diameter distributions of the PEF selection treatments were close to their goals in the 1970s and researchers predicted that the stands would remain "essentially balanced" (Frank and Blum 1978, p. 11). However, later remeasurements revealed structural and compositional imbalances that were not apparent in earlier assessments (Kenefic and Brissette 2001; Seymour and Kenefic 1998; Fig. 5.6). In addition, though increases in the proportion of spruce growing stock led Frank and Blum (1978) to conclude that efforts to favor those species were successful, we now know that this result was a function of accretion rather than recruitment (Kenefic et al. 2007). Spruce trees in the selection treatments are almost all more than a century old (Seymour and Kenefic 1998). It can take more than 30 years for spruce seedlings in these stands to reach just 0.5 m (1.6 ft) in

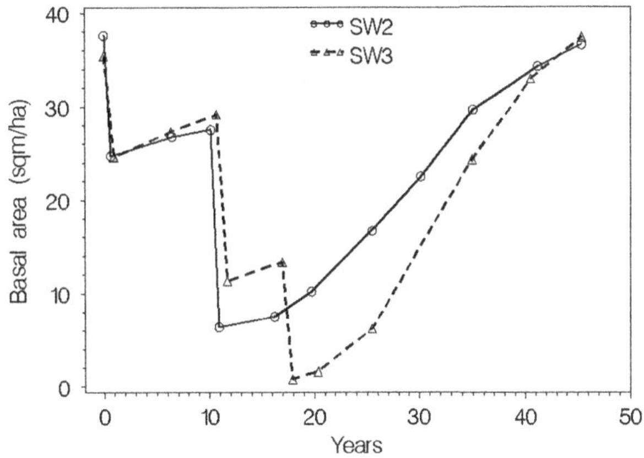

Fig. 5.7 Basal areas (trees ≥1.3 cm dbh) of the Penobscot Experimental Forest shelterwood treatments in which small and unmerchantable trees were retained (*SW2*) and removed (*SW3*). *SW3* had less growing stocking until the onset of residual mortality in *SW2*

height (Weaver 2007), and recruited saplings grow at a rate of about 2 cm (less than 1 in.) in diameter per decade.[6] Analysis of relationships between overstory stocking and growth of understory trees in the selection treatments revealed that there was no level of canopy closure favoring spruce over its competitors (Moores et al. 2007). These findings tell a much different story from those of the 1970s and raise concerns about long-term sustainability of structure and composition of the treatments.

The shelterwood treatments on the PEF have also revealed new findings over time. As an even-aged method, shelterwood has been more effective than clearcutting for regenerating shade-tolerant conifer-dominated stands. When the overstory is manipulated to provide partial shade, shade-tolerant and mid-tolerant species regenerate prolifically, while establishment of shade-intolerant species is limited. Unlike the selection treatments where sapling growth has been slow, development of shelterwood-regenerated stands on the PEF has been relatively rapid; stands grew to pretreatment levels of basal area (BA) in about 45 years (Fig. 5.7). The key difference between the selection and shelterwood treatments is that in the latter, the overstory is removed after a new cohort is established, providing full or nearly full sunlight.

The two variants of shelterwood in the long-term study on the PEF are two-stage (SW2) and three-stage (SW3) treatments; the stages are the number of harvests used to remove the overstory and release the new cohort. On the PEF, however, a partial overstory purposefully left in SW2 affected stand development more than the number of overstory removal cuts did. In SW2, more than 6 m^2/ha (approximately 28 ft^2/ac) of pole-timber and unmerchantable sawtimber were left, while in SW3, all trees ≥5 cm (2 in.) in diameter at breast height were felled (Fig. 5.7). Though SW2 had more residual BA for many years after treatment, the sparse overstory slowed development of the new cohort. By 45 years after the first (establishment) cut in both treatments, SW3 had as much BA as SW2, even though it was released later and started with almost no BA. The relatively slower growth observed in the

[6] Data on file at the USDA Forest Service, Northern Research Station, Bradley, ME, USA.

treatment with residual trees is consistent with the slow sapling growth in the selection treatments, and underscores the impact of overstory shade on the growth of young spruce and fir trees. Even though regeneration of tolerant conifers benefits from shady understory conditions for germination, seedling establishment, and early growth, full light is needed for rapid stand development.

5.4 Societal Impact

The trajectory of Forest Service research in northern conifers reveals a long history of research response to the needs of forest industry and society as a whole. Early manipulative studies were motivated not only by the desire to support the economically important forest products industries but also by public concern about degradation of the region's forests. Later investigations of production forestry were in response to the thriving economy of the postwar years and associated public demand for forest products. More recent studies of ecology and ecological forestry coincided with the regional transition of forest ownership from industry to other classes of landowners, and with society's demands for "a kinder and gentler forestry" (Franklin 1989). How, in turn, has Forest Service research affected the practice of forestry and the people of the northern conifer forest region?

Prior to the McIntire–Stennis Act of 1962, the capacity of university faculty to conduct forestry research was limited (Thompson 2004). Faculty members were concerned primarily with teaching, and had neither the funds nor the time to maintain substantial research programs. As a consequence, Forest Service studies were the sole source of information about many forest management topics. This was especially true in the northern conifer forest. Without the work of Westveld, Curry, Rushmore, McLintock, and Hart, northern conifer forest industries would have had little scientific basis for their management. Forest Service scientists developed all of the early spruce and fir silvicultural guides, yield tables, silvics manuals, and volume and site index equations. As such, the Forest Service can be credited for almost all of the early advances in northern conifer forest management, including the early twentieth-century transition from clearcutting to partial harvesting, the widespread practice of hardwood control in mixedwood stands via mechanical and chemical treatments, and the practice of establishing and protecting advance regeneration in stands managed for shade-tolerant conifers. Since that time, the Forest Service's collaboration with universities throughout the region has resulted in the frequent use of experimental forests, including the PEF, for laboratory exercises, student employment, and graduate research. In this way, Forest Service researchers on the experimental forests have contributed to the education and training of generations of natural resource professionals.

The impact of Forest Service research on northern conifer forest policy and forestry practice was particularly apparent during the years of spruce budworm outbreaks. Forest Service staffing in Maine was at an all-time high between the late 1970s and mid-1980s. Experts such as entomologist Gordon Mott were transferred

Fig. 5.8 Results from research on selection cutting (*left*) and diameter-limit cutting (*right*) on the Penobscot Experimental Forest led to regional changes in forest policy and practice

to Maine to tackle the issue of budworm control. Widespread insecticide spraying had triggered intense controversy over the potential hazards posed to the forest environment and human health (Irland 1980). Though spray programs had covered as many as 8 million ha (20 million acres) per year and slowed tree mortality, the need for repeated treatment led many to believe that they were ultimately futile. In response to these concerns, Forest Service researchers and cooperators developed an approach in which management objectives and stand conditions were used as guides for harvesting and targeted spraying (Dimond et al. 1984). Adoption of this practice by forest industries had a profound impact on northern conifer management at the time, and helped to reduce the exposure of northern forests and communities to potentially harmful chemicals.

More recently, assessments of long-term data from the PEF diameter-limit cutting and selection treatments resulted in changes in forest practices and policy throughout the Northeast and adjacent Canada (Fig. 5.8). Diameter-limit cutting as applied on the PEF (called fixed diameter-limit cutting, FDL) is the selective removal of large trees, without tending or deliberately establishing regeneration. Because FDL is easy to apply and results in high-value removals in the short term, it is widely practiced throughout North America. Though anecdotal evidence suggested that FDL was not sustainable, there were no data in the literature to support this claim. Analysis of the FDL and selection treatments on the PEF after 50 years of study revealed that repeated applications of diameter-limit cutting had resulted in lower stand volume and value, more unmerchantable timber, less medium-large sawtimber growth, and less desirable regeneration (Kenefic et al. 2005). This work clearly established the long-term degrading effects of FDL and led to a widely used primer for landowners and policy makers (Kenefic and Nyland 2005), a regional conference for practitioners (Kenefic and Nyland 2006), and numerous policy initiatives. Results have been used by the Bureau of Indian Affairs, the Northeastern Area State and Private Forestry Forest Stewardship Program, Maine Forest Service Landowner Assistance Program, Provincial Ministry of Natural Resources in Ontario, and numerous extension offices to promote sustainable forestry throughout the region.

5.5 Future Directions

5.5.1 Penobscot Experimental Forest

In recent years, data from the Forest Service's long-term study on the PEF have increasingly been used to answer large-scale questions about natural resource management. Carbon sequestration for climate change mitigation, for example, is one of the foremost challenges facing resource managers and policy makers today, but little is known about the long-term impacts of management alternatives in many forest types. With more than 60 years of data and a full array of silvicultural systems, the PEF is ideally suited to address the carbon sequestration implications of forest management in northern conifers. Ongoing research evaluating carbon sequestration across the treatments in the compartment study (e.g., Seymour et al. 2009) will provide information critical to forest management for climate change mitigation and carbon credit accounting.

Long-term tree growth records from Forest Service research on the PEF are also being used to develop a new individual-tree, distance-independent growth and yield model for the region (Weiskittel et al. 2010). Though practitioners and forest managers commonly use growth and yield models to predict future outcomes, existing models perform poorly in complex stands (Saunders et al. 2007). Data from the PEF are being used to parameterize equations to predict tree growth (Russell and Weiskittel 2010); these equations are key drivers for ongoing efforts to estimate stand productivity in mixed-species, even-, and uneven-aged stands. In addition, measurements from the PEF have recently been used to develop crown width equations for more than a dozen species (Russell and Weiskittel 2011); accurate estimates of crown width are increasingly important due to their use in fields employing light detection and ranging (LiDAR) and other novel remote sensing technologies. Modeling efforts such as these are expected to become more prevalent in the future.

Remote sensing is now commonly used in lieu of traditional, labor-intensive, on-the-ground inventory, and serves as the basis for harvest planning, habitat evaluation, forest health assessment, and estimation of biomass and carbon at large spatial scales. It is critical that the algorithms used in these analyses, and the resulting data, are strongly correlated with actual forest condition. The Forest Service's studies on the PEF are characterized by numerous stands with a range of structures, compositions, and treatment histories, all within a relatively small area and supported by intensive data collection. As such, the PEF has proven useful for ground truthing new technologies and new applications of existing technologies, such as the use of LiDAR and L-band radar to estimate standing biomass and carbon stocks (Cook et al. 2010).

As the research work unit responsible for the PEF has grown to include 14 experimental forests from Maine to Minnesota, the number and breadth of Forest Service staff at the PEF have diminished. As a consequence, research by cooperators makes up an ever-increasing proportion of the Forest Service's northern conifer research portfolio. One of the key requirements for the success of this collaborative model is readily available long-term data. To that end, all of the data from the PEF compartment study and a number of related studies were recently brought into a relational database (Russell et al. 2014). Because of those efforts, the PEF was identified as a high-priority location for national efforts to make Forest Service research data available online. Data, metadata, and supplemental files from the PEF are publicly available on the Internet (e.g., Brissette et al. 2012), resulting in increased awareness and use of the research.

5.5.2 Cross-Site Research

Today, there are more than 80 experimental forests and ranges and associated Forest Service research sites across the USA, representing important geographic, ecologic, atmospheric, and climatic systems. The continental distribution and longevity of these sites and their records are unmatched. In the past, research programs on experimental forests such as the PEF focused on concerns of local or regional importance. Though that function remains important, these sites are being increasingly valued as platforms for studies that cross landscape and continental gradients to address regional, national, and global problems. To accomplish this vision of an integrated network, it is critical that experimental forests and ranges are linked as collaborating research sites.

A number of Forest Service synthesis projects have been initiated to address large-scale questions; the PEF is involved in three such efforts. The first is a regional synthesis of long-term results from cutting practice level (CPL) studies.[7] Starting in the 1940s, researchers at experimental forests in the Northeast installed CPL plots to demonstrate a range of silvicultural practices rated "poor" to "high order." Though treatments were generally not replicated within individual experimental forests, they were similar across the station. Data from these plots were recently compiled and will be used to assess similarities and differences in treatment outcomes across forest types.

[7] Study plan on file at the USDA Forest Service, Northern Research Station, Bradley, ME, USA.

The second synthesis project to which the PEF is contributing is a national study of pollinating insects. This project, in collaboration with the US Geological Survey, originated from a multiagency federal task force formed in response to the international perception that populations of pollinating insects are in decline. Little information about the status or trends of native bees in the USA was available despite their key role in terrestrial environments. As a consequence, a federal pollinator network, including the PEF, was formed in 2010. During the first year of study, more than 3,500 bees were collected from these sites; future work will track changes in populations and relate them to local climate data.[8]

The PEF is also part of a regional study of long-term forest growth and yield across climatic and compositional gradients (Johnson 2012). This project involves a number of experimental forests and associated research sites in the Northeast and Lakes States. Response variables include stand-level growth and yield, i.e., stemwood volume and aboveground biomass accumulation and carbon storage. Researchers used a meta-analysis approach to compare variability within and between study sites in order to determine whether differences exist across the region. Climatic variables, such as minimum and maximum temperatures, growing degree days, and amount of precipitation, were assessed for their correlation to observed differences. Studies such as this one increase the relevance of experimental forest data by identifying and exploring the mechanisms for regional trends.

5.6 Conclusion

As we look to the future, it is essential that the Forest Service research on the PEF be continued. In the long-term silvicultural study, a comparison of productivity in even-and uneven-aged systems will not be possible until the even-aged stands reach rotation length, which will require another 30–40 years. The confidence associated with recommendations based on this research increases with each additional measurement cycle. The PEF research not only adds value to earlier studies but also serves as the basis for new studies on a wide range of emerging topics. It is our hope and expectation that Forest Service research in northern conifers will continue to evolve, and that Forest Service scientists and cooperators will continue to use the experimental forest to address critical science questions.

Acknowledgments We thank the many people who assisted us with our search for the historical information, images, and personal accounts presented in this chapter, particularly Robert Frank, Jr., Lawrence Safford, Frank Hagan, and Roger Collins (USDA Forest Service, retired); Michael Maguire, Peter McBride, Terry Miller, Mark Twery, Bill Leak, William McWilliams, B. Tyler Wilson, and Ralph Crawford (USDA Forest Service); Michael Rechlin (Principia College); Gould Hoyt (Paul Smith's College, deceased); Robert Seymour (University of Maine); and Eben Lehman (Forest History Society). Robert Frank, Jeremy Wilson (University of Maine), Susan

[8] Progress report on file at the U.S. Forest Service, Northern Research Station, Bradley, ME, USA.

Stout (USDA Forest Service), and an anonymous reviewer provided helpful reviews of an earlier version of this manuscript; this version was reviewed by Jeremy Wilson and Robert Seymour (University of Maine) and Thomas Schmidt (USDA Forest Service).

References

Abbott HG, Hart AC (1960) Mice and voles prefer spruce seed. Station Paper NE-153. U.S. Department of Agriculture, Forest Service, Northeastern Forest Experiment Station, Upper Darby, PA, 12 p

Acheson J (2000) Clearcutting Maine: implications for the theory of common property resources. Hum Ecol 28:145–169

Adams MB, Loughry L, Plaugher L (2008) Experimental forests and ranges of the USDA Forest Service. General Technical Report NE-321 Revised, U.S. Department of Agriculture, Forest Service, Northeastern Research Station, Newtown Square, PA, 178 p [CD ROM]

Anonymous (1938) Rebuilding of ruined forests confronts New England. Sci Newsl:252–253

Belotelkin KT, Reineke RH, Westveld M (1942) Spruce-fire selective logging costs. J For 40:326–336

Berven K, Kenefic L, Weiskittel A, Twery M, Wilson J (2013) The lost research of early northeastern spruce-fir experimental forests: a tale of lost opportunities. In: Camp AE, Irland LC, Carroll CJW (eds) Long-term silvicultural and ecological studies. Results for science and management, vol 2. Research Paper 013. Yale University, Global Institute of Sustainable Forestry, New Haven, CT: 103–114.

Bjorkbom JC, Frank RM (1968) Slash burning and whole-tree skidding fail to provide mineral-soil seedbeds for spruce-fir. North Logger Timber Process 16(7):20, 45

Blum BM, MacLean DA (1985) Potential silvicultural, harvesting and salvage practice in eastern North America. In: Sanders CJ, Stark RW, Mullins EJ, Murphy J (eds) Recent advances in spruce budworm research. Proceedings of the CANUSA spruce budworms research symposium, Canadian Forestry Service, Bangor, ME, 16–20 Sept 1984, pp 264–280

Braun EL (1950) Deciduous forests of eastern North America. Blakiston, Philadelphia, 596 p

Brissette JC (1996) Effects of intensity and frequency of harvesting on abundance, stocking and composition of natural regeneration in the Acadian Forest of eastern North America. Silva Fenn 30(2–3):301–314

Brissette JC, Frank RM, Stone TL, Skratt TA (1999) Precommercial thinning in a northern conifer stand: 18-year results. For Chron 75(6):967–972

Brissette JC, Kenefic LS, Russell MB, Puhlick JJ (2012) Overstory tree and regeneration data from the "Silvicultural Effects on Composition, Structure, and Growth" study at Penobscot Experimental Forest. Newtown Square, PA: USDA Forest Service, Northern Research Station. http://dx.doi.org/10.2737/RDS-2012-0008

Bryce E (2009) Relationships between nonnative invasive plant distribution, silvicultural treatment, and soil drainage in the northern conifer forest. M.S. Thesis, University of Maine, School of Forest Resources, Orono, ME, 123 p

Bryce E, Kenefic LS, Dibble AC, Brissette JC, Livingston WL (2008) Relationships between nonnative plant distribution, soil drainage and silviculture in the northern conifer forest. Eastern Canada—USA Forest Science Conference, University of Maine, Orono, ME, 17–18 Oct 2008, p 23

Cook BD, Sun G, Ranson KJ, Montesano PM, Luthcke SB, Blair JB (2010) Accuracy of DESDynI biomass estimates using LiDAR and data fusion methods. Eos Trans AGU 90(52), Fall Meeting Suppl., Abstract B31A-0318.

Curry JR, Rushmore JM (1955) Experiments in killing northern hardwoods with sodium arsenite and ammonium sulfamate. J For 53:575–580

Czapowskyj MM, Rourke RV, Frank RM (1977) Strip clearcutting did not degrade the site in a spruce-fir forest in central Maine. Research Paper NE-367. U.S. Department of Agriculture, Forest Service, Northeastern Forest Experiment Station, Upper Darby, PA, 8 p.

Day ME (2000) Influence of temperature and leaf-to-air vapor pressure deficit on net photosynthesis and stomatal conductance in red spruce (*Picea rubens*). Tree Physiol 20:57–63

Day ME, Greenwood MS, White AS (2001) Age-related changes in foliar morphology and physiology in red spruce and their influence on declining photosynthetic rates and productivity with tree age. Tree Physiol 21:1195–1204

Dibble AC, Brissette JC, Hunter ML Jr (1999) Putting community data to work: some understory plants indicate red spruce regeneration habitat. For Ecol Manag 114(2–3):275–291

Dimond JB, Seymour RS, Mott DG (1984) Planning insecticide application and timber harvesting in a spruce budworm epidemic. Agriculture handbook 618. U.S. Department of Agriculture, Forest Service, Cooperative State Research Service, 29 p

Frank RM, Bjorkbom JC (1973) A silvicultural guide for spruce-fir in the Northeast. General Technical Report NE-6. U.S. Department of Agriculture, Forest Service, Northeastern Forest Experiment Station, Upper Darby, PA, 29 p

Frank RM, Blum BM (1978) The selection system of silviculture in spruce-fir stands—procedures, early results, and comparisons with unmanaged stands. NE-425, U.S. Department of Agriculture, Forest Service, Northeastern Forest Experiment Station, Broomall, PA, 15 p

Frank RM, Putnam EL (1972) Seedling survival in spruce-fir after mechanical tree harvesting in strips. Research Paper NE-224, U.S. Department of Agriculture, Forest Service, Northeastern Forest Experiment Station, Upper Darby, PA, 16 p

Franklin J (1989) Toward a new forestry. Am For 95:37–44

Garber SM, Brown JP, Wilson DS, Maguire DA, Heath LS (2005) Snag longevity under alternative silvicultural regimes in mixed-species forests of central Maine. Can J For Res 35:787–796

Gilmore DW, Seymour RS (1996) Alternative measures of stem growth efficiency applied to Abies balsamea from four canopy positions in central Maine, USA. For Ecol Manag 84(1–3):209–218

Graves HS (1899) Practical forestry in the Adirondacks. Bulletin 26, U.S. Department of Agriculture, Division of Forestry, Washington, DC, 85 p

Grossi P (2008) The 1938 Great New England Hurricane: looking to the past to understand today's risk. http://www.rms.com/Reports/1938_Great_New_England_Hurricane.pdf. Accessed 26 Jan 2011

Hagan JM, Irland LC, Whitman AA (2005) Changing timberland ownership in the northern forest and implications for biodiversity. Report MCCS-FCP-2005-1, Manomet Center for Conservation Sciences, Brunswick, Maine, 25 p

Hart AC (1956) Changes after partial cutting of a spruce-fir stand in Maine Station Paper 86. U.S. Department of Agriculture, Forest Service, Northeastern Forest Experiment Station, Upper Darby, PA, 8 p

Hart AC (1963) Spruce-fir silviculture in northern New England. In: Proceedings of the Society of American Foresters National Convention, Boston, MA, Society of American Foresters, Bethesda, MD, pp 107–110

Hawley GJ, Schaberg PG, DeHayes DH, Brissette JC (2005) Silviculture alters the genetic structure of an eastern hemlock forest in Maine, USA. Can J For Res 35:143–150

Hoover C (2005) Carbon storage in northeastern U.S. forest soils. In: Kenefic LS, Twery MJ (eds) Changing forests—challenging times. Proceedings of the New England Society of American Foresters 85th Winter Meeting. 16–18 March 2005. General Technical Report NE-325, U.S. Department of Agriculture, Forest Service, Northeastern Research Station, Newtown Square, PA, p 17

Hosmer RS (1902) A study of Maine spruce. In: Fourth report of the Forest Commissioner of the State of Maine. Augusta, ME, pp 65–108

Irland LC (1980) Pulpwood, pesticides and people. Controlling spruce budworm in northeastern North America. Environ Manag 4(5):381–389

Irland LC (1999) The Northeast's changing forest. Harvard Forest and Harvard University Press, Petersham, 401 p

Johnson, SE (2012) Evaluating factors influencing forest growth across climatic and silvicultural gradients in northern forests of the United States. M.S. Thesis, University of Maine, School of Forest Resources, Orono, ME, 274 p

Judd RW (1989) Aroostook: a century of logging in northern Maine. University of Maine Press, Orono, 351 p

Judd RW (1997) Common lands, common people: the origins of conservation in northern New England. Harvard University, Cambridge, 335 p

Kenefic LS, Brissette JC (2001) Long-term dynamics of structure and composition in uneven-aged northern conifer stands. In: Forestry at the Great Divide—Proceedings of the 2001 Society of American Foresters National Convention. 13–17 Sept 2001, Denver, CO. Society of American Foresters, Bethesda, MD, pp 405–407

Kenefic LS, Nyland RD (2005) Diameter-limit cutting and silviculture in northeastern forests: a primer for landowners, practitioners, and policymakers. NA-TP-02-05, U.S. Department of Agriculture, Forest Service, Northeastern Area State and Private Forestry, Newtown Square, PA, 18 p

Kenefic LS, Nyland RD (2006) Facilitating a dialogue about diameter-limit cutting. In: Proceedings of the conference on diameter-limit cutting in northeastern forests. General Technical Report NE-342, U.S. Department of Agriculture, Forest Service, Northeastern Research Station, Newtown Square, PA, pp 1–2

Kenefic LS, Seymour RS (1999) Leaf area prediction models for Tsuga canadensis in Maine. Can J For Res 29:1574–1582

Kenefic LS, Sendak PE, Brissette JC (2005) Fixed diameter-limit and selection cutting in northern conifers. North J Appl For 22(2):77–84

Kenefic LS, Weaver J, Ray D, Moores A, Seymour RS, Brissette JC, (2007) Sustaining red spruce in uneven-aged Acadian Forest stands: results from long-term research. Proceedings of the Acadian Forest Science Conference. 10–13 Oct 2007. Atlantic Forestry Center, Canadian Forest Service, Fredericton, N.B., p 26

Larouche C, Kenefic LS, Ruel J-C (2010) Northern white-cedar regeneration dynamics on the Penobscot Experimental Forest in Maine: 40-year results. North J Appl For 27(1):5–12

Maguire DA, Brissette JC, Gu L (1998) Canopy structure and growth efficiency of red spruce in uneven-aged, mixed-species stands in Maine. Can J For Res 28:1233–1240

McLintock TF (1954) Factors affecting wind damage in selectively cut stands of spruce and fir in Maine and northern New Hampshire. Station Paper 70, U.S. Department of Agriculture, Forest Service, Northeastern Forest Experiment Station, Upper Darby, PA, 17 p

McLintock TF, Westveld M (1946) Some financial aspects of removal of overstory balsam fir as a budworm control measure. Forest Management Paper 1. U.S. Department of Agriculture, Forest Service, Northeastern Forest Experiment Station, Philadelphia, PA, 8 p

Meyer WH (1929) Yields of second-growth spruce and fir in the Northeast. Technical Bulletin No. 142, U.S. Department of Agriculture, Forest Service, Washington, D.C., 52 p

Moores AR, Seymour RS, Kenefic LS (2007) Height development of shade-tolerant conifer saplings in multiaged Acadian Forest stands. Can J For Res 37(12):2715–2723

Murphy LS (1917) The red spruce: its growth and management. Bulletin No. 544, U.S. Department of Agriculture, Forest Service, Washington, D.C., 100 p

Pinchot G (1898) The Adirondack spruce. Critic, New York, 157 p

Recknagel AB, Churchill HL, Heimberger C, Westveld M (1933) Experimental cutting of spruce and fir in the Adirondacks. J For 31(6):680–688

Rowe JS (1972) Forest regions of Canada. Publication 1300, Canadian Forestry Service, Department of the Environment, Ottawa, 172 p

Rushmore FM (1956a) The frill: a new technique in chemical debarking. J For 54(5):329–331

Rushmore FM (1956b) A small quantity of sodium arsenite will kill large cull hardwoods. Station Paper 83, U.S. Department of Agriculture, Forest Service, Northeastern Forest Experiment Station, Upper Darby, PA, 6 p

Russell MB, Weiskittel AR (2010) Assessing model prediction uncertainty in forecasting long-term tree diameter vs. basal area increment for the primary Acadian tree species. In: Proceedings of the ECANUSA Forest Science Conference, 12–14 October 2010. Universite de Moncton, Edmundston, NB, p 57

Russell MB, Weiskittel AR (2011) Maximum and largest crown width equations for fifteen tree species in Maine. North J Appl For 28(2):84–91

Russell MB, Meyer SR, Brissette JC, Kenefic LS (2014) Sixty years of research, 60 years of data: long-term USDA Forest Service data management on the Penobscot Experimental Forest. In: Kenefic LS, Brissette JC (comps). Penobscot Experimental Forest: 60 years of research and demonstration in Maine, 1950–2010. General Technical Report NRS-P-123. Newtown Square, PA: U.S. Department of Agriculture, Forest Service, Northern Research Station: 62–70.

Safford LO, Frank RM, Little EL Jr (1969) Trees and shrubs of the Penobscot Experimental Forest, Penobscot County, Maine. NE-128, U.S. Department of Agriculture, Forest Service, Northeastern Forest Experiment Station. 27 p

Saunders MR, Wagner RG (2005) Ten-year results of the Forest Ecosystem Research Program (FERP)—successes and challenges. In: Peterson CE, Maguire DA (eds) Balancing ecosystem values: innovative experiments for sustainable forestry. General Technical Report PNW-635, U.S. Department of Agriculture, Forest Service, Pacific Northwest Research Station, Portland, OR, pp 147–152

Saunders MR, Wagner RG, Seymour RS (2007) Thinning regimes for spruce-fir stands in the Northeastern United States and Eastern Canada. In: Cooperative Forestry Research Unit Annual Report. University of Maine, School of Forest Resources, Orono, ME, pp 27–32

Seymour RS (1995) The northeastern region. In: Barrett JW (ed) Regional silviculture of the United States. Wiley, New York, pp 31–79

Seymour RS, Kenefic LS (1998) Balance and sustainability in multi-aged stands: a northern conifer case study. J For 96:12–16

Seymour RS, Kenefic LS (2002) Influence of age on growth efficiency of Tsuga canadensis and Picea rubens trees in mixed-species, multi-aged northern conifer stands. Can J For Res 32:2032–2042

Seymour RS, White AS, de Maynadier PG (2002) Natural disturbance regimes in northeastern North America—evaluating silvicultural systems using natural scales and frequencies. For Ecol Manag 155:357–367

Seymour RS, Guldin J, Marshall D, Palik B (2006) Large-scale, long-term silvicultural experiments in the United States: historical overview and contemporary examples. Allg Forst-u J-Ztg 177(6/7):104–112

Seymour RS, Ray D, Scott N (2009) Linking forest structure to viable development pathways of diversely managed northern conifers: a pilot study exploring the potential of FIA data for modeling regional carbon dynamics. Northeastern States Research Cooperative, Final Report. http://nsrcforest.org/sites/default/files/uploads/seymour05full.pdf. Accessed 01 May 2014

Smith DM (1972) The continuing evolution of silvicultural practice. J For 70(2):89–92

Smith DM (1991) Natural regeneration from sprouts and advanced growth. In: Simpson CM (ed) Proceedings of the conference on natural regeneration management. Forestry Canada, Maritimes Region, Fredericton, N.B.:63–66

Smith KT, Shortle WC, Jellison J, Connolly J, Schilling J (2007) Concentrations of Ca and Mg in early stages of sapwood decay in red spruce, eastern hemlock, red maple, and paper birch. Can J For Res 37:957–965

Spurr SH (1956) Natural restocking of forests following the 1938 hurricane in central New England. Ecology 37:443–451

Su JC, Woods SA (2001) Importance of sampling along a vertical gradient to compare insect fauna in managed forests. Environ Entomol 30(2):400–408

Thompson DH (2004) Senator John C. Stennis: champion of forestry. For Hist Today (Spring/Fall):27–34.

Tian S, Ostrofsky WD (2007) Butt and root decay in precommercially thinned spruce-fir stands. North J Appl For 24(2):129–133

USDA Forest Service (1935) Annual report for fiscal year 1935. U.S. Department of Agriculture, Forest Service, Northeastern Forest Experiment Station, New Haven, CT, 34 p

USDA Forest Service (1937) Annual report for 1936. U.S. Department of Agriculture, Forest Service, Northeastern Forest Experiment Station, New Haven, CT, 34 p

USDA Forest Service (1938) Annual report for 1937. U.S. Department of Agriculture, Forest Service, Northeastern Forest Experiment Station, New Haven, CT, 61 p

USDA Forest Service (1939) Annual report for 1938. U.S. Department of Agriculture, Forest Service, Northeastern Forest Experiment Station, New Haven, CT, 34 p

Weaver JK (2007) Substrate availability and regeneration microsites of tolerant conifers in mixed-species stands in Maine. M.S. Thesis, University of Maine, School of Forest Resources, Orono, ME, 80 p

Weaver JK, Kenefic LS, Seymour RS, Brissette JC (2009) Decaying wood and tree regeneration in the Acadian Forest of Maine, USA. For Ecol Manag 257(7):1623–1628

Weiskittel AR, Wagner RG, Seymour RS (2010) Refinement of the forest vegetation simulator, northeastern variant growth and yield model: phase 1. In: Meyer SR (ed) 2009 Annual Report, University of Maine, Cooperative Forestry Research Unit, Orono, ME: 44–48

Weiskittel A, Kenefic LS, Seymour RS, Phillips LM (2009) Long-term effects of precommercial thinning on stem form, volume, and branch characteristics of red spruce and balsam fir crop trees. Silva Fenn 43(3):397–409

Westveld M (1930) Suggestions of management of spruce stands in the Northeast. Circular 134, U.S. Department of Agriculture, Washington, D.C., 24 p

Westveld M (1931) Reproduction on the pulpwood lands in the Northeast. Technical Bulletin 223, U.S. Department of Agriculture, Forest Service, Washington Office, Washington, D.C., 52 p

Westveld M (1938) Silvicultural treatment of spruce stands in northeastern United States. J For 36(10):944–950

Westveld M (1946) Forest management as a means of controlling spruce budworm. J For 44:949–953

Westveld M (1953) Ecology and silviculture of the spruce-fir forests of eastern North America. J For 51(6):422–430

Wilson J (2005) Nineteenth century lumber surveys for Bangor, Maine: implications for pre-European settlement forest characteristics in northern and eastern Maine. J For 103(5):218–223

Zon R (1914) Balsam fir. Bulletin 55, U.S. Department of Agriculture, Forest Service, Washington, D.C., 68 p

Chapter 6
Interdisciplinary Research on the Blacks Mountain Experimental Forest

Martin W. Ritchie and Carl N. Skinner

Abstract The US Department of Agriculture (USDA) Forest Service Blacks Mountain Experimental Forest is a 4,000 ha, pine-dominated tract in northeastern California, in the Lassen National Forest. Between 1934 and 1960, research at Blacks Mountain focused on various methods of harvest and bark beetle activity. These studies demonstrated the effectiveness of harvesting on increasing tree growth and reducing widespread mortality from bark beetles.

After a period of inactivity, research efforts accelerated in the early 1990s when a large-scale interdisciplinary research program was initiated, building on the foundation laid with the early harvest method studies. Although early in the life of the study, researchers have found that prescribed fire reduced surface fuel levels in stands with a pre-burn harvest. However, burning untreated areas of the forest resulted in high levels of mortality and elevated risk of high-severity fire in the short term. Thinning treatments in stands with a cohort of large-diameter trees increased survival and number of low-risk trees. In the short term, there were no direct effects observed on small mammals from either treatment. Changes in stand structural diversity, while having no apparent effect on species richness of birds, did appear to cause shifts among a few individual species.

In the aftermath of a recent fire, additional research is being conducted on the effects of fire salvage and artificial regeneration.

Keywords Fire · Thinning · Silviculture · Harvest methods · Sustainability · Bark beetle

6.1 Introduction

Blacks Mountain Experimental Forest is a pine-dominated forest in northeastern California (40° 40′ N latitude, 121° 10′ W longitude) about halfway between Susanville and Old Station. Although easily accessed by well-maintained gravel roads,

M. W. Ritchie (✉) · C. N. Skinner
Pacific Southwest Research Station, USDA Forest Service, 3644 Avtech Parkway, Redding, CA 96002, USA
Phone: 530-226-2551
e-mail: mritchie@fs.fed.us

D. C. Hayes et al. (eds.), *USDA Forest Service Experimental Forests and Ranges*, DOI 10.1007/978-1-4614-1818-4_6, it is rarely visited by passersby because of its remote location and status as a wildlife refuge (Fig. 6.1). The forest roads to Blacks Mountain are usually closed on account of snow from mid-November to late April.

The elevation at Blacks Mountain ranges from 1,700 to 2,100 m above sea level, and the light precipitation (approximately 460 mm annually) falls primarily as snow during the winter months. The warm, dry summers are interrupted by infrequent thunderstorms, but precipitation during these months is very light. Mean daily temperatures range from −9 in winter to 29°C in summer.

The primary tree species are ponderosa pine (*Pinus ponderosa* Laws)[1] and Jeffrey pine (*P. jeffreyi* Grev. & Balf.), with incense cedar (*Calocedrus decurrens,* Torrey; Florin) and white fir (*Abies concolor,* Gordon & Glend.; Lindley) increasing with elevational gain within the forest. The only other conifer present, sugar pine (*P. lambertiana* Douglas), is found rarely; only two specimens have been located. In addition to these coniferous species, mountain mahogany (*Cercocarpus ledifolius* Nutt.) may be found in rocky outcrops, and several small cluster of aspen (*Populus tremuloides* was recently located near the center of the forest Michaux).

6.2 Beginnings

Blacks Mountain Experimental Forest was formally established by F.A. Silcox, the Chief Forester of the USDA Forest Service, in March 1934. The roughly 4,000 ha tract of land was carved out of the Lassen National Forest on the Modoc Plateau in an area northeast of Lassen Volcanic National Park (Fig. 6.1). The selection followed 2 years of scoping work by A.E. Wieslander[2] of the California Forest and Range Experiment Station (the early name given to the current Pacific Southwest Research Station, PSW). Wieslander settled on an area just southeast of a small butte called Blacks Mountain, in an area of gentle topography dominated by ponderosa pine and Jeffrey pine fairly typical of the yellow pine forests of the Modoc Plateau (Fig. 6.2).

The Experimental Forest was intended to be the center for forest management research in lower-site-quality ponderosa pine, commonly found in northeastern California. The initial research plan[3] laid out the purpose for the Experimental Forest: "It is hoped there may be concentrated here a series of coordinated studies of management, economics, utilization and silviculture with a single objective: determination of the measures necessary to make sustained yield forestry possible on lands of this character."

[1] Nomenclature follows Hickman (1993).

[2] Wieslander was very familiar with the forests of northeastern California; his first job after graduating from the University of California in 1915 was on the Lassen National Forest. He went on to a long and productive career with the California Experiment Station and is best known for spearheading an effort in the 1920s and 1930s to develop a Vegetation Type Map for the state of California.

[3] Duncan Dunning, "Outline Plan of Development and Research, Blacks Mountain Experimental Forest." February 18, 1937. Memorandum on file, PSW Research Station, Redding, CA.

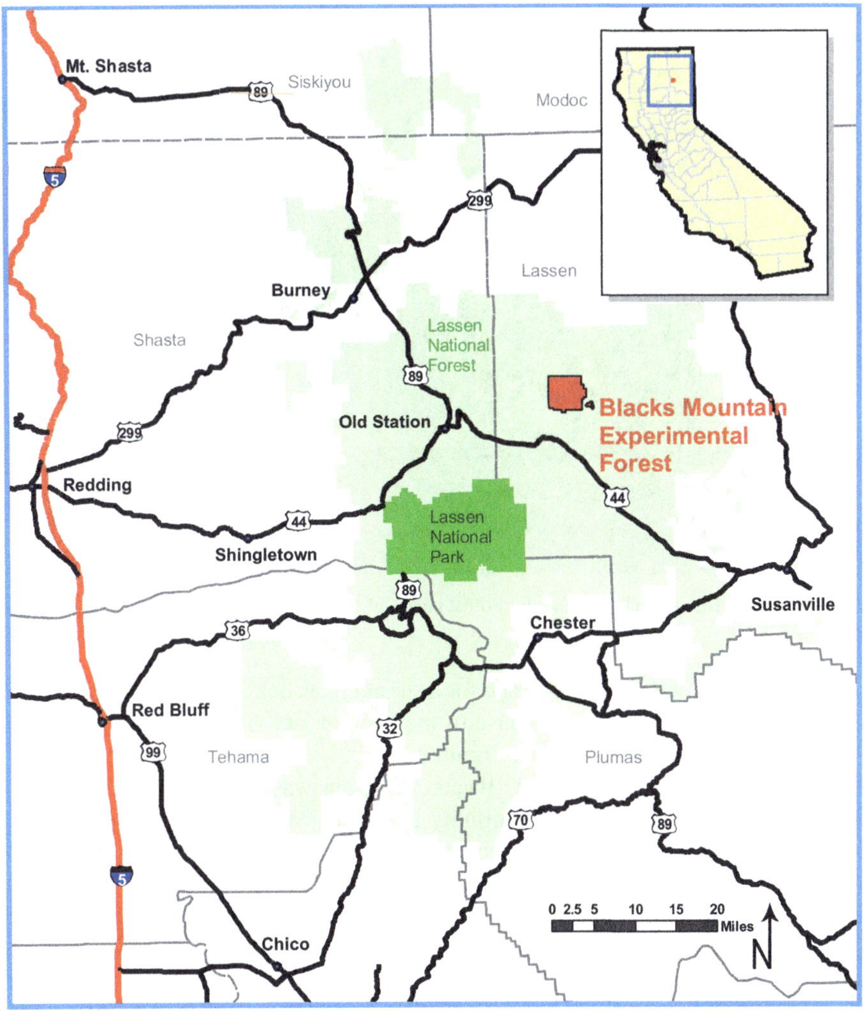

Fig. 6.1 Location of Blacks Mountain Experimental Forest in northeast California

Among the benefits of the location of Blacks Mountain Experimental Forest was its proximity to the Civilian Conservation Corps camp at nearby Halls Flat. The convenient labor supply allowed for the rapid development of an infrastructure needed to establish a research program in the forest. Researchers, led by Duncan Dunning, went to work quickly. In 1933 and 1934, they conducted a 100% inventory of all trees >9 cm in diameter, and shortly thereafter established a network of management units or compartments, each approximately 40 ha, and built 66 km of road designed to provide access to each compartment.

Dunning's philosophy was that research should support sound forestry practices rather than extractive "lumbering." He hoped to establish a sustained level of

Fig. 6.2 Blacks Mountain Experimental Forest near Halls Flat. (Photo by A.E. Wieslander 1934)

productivity of sawtimber and forage for deer and livestock in a regulated condition through a 140-year rotation. Because of the general lack of surface water flow in this area, watershed values are very limited.

Much of the early research at Blacks Mountain was influenced by the widespread bark beetle-induced tree mortality then being observed throughout the region (Fig. 6.3). At the time, the accepted method for dealing with this problem was to cut down beetle-killed trees and then strip and burn the bark, procedures that proved costly and produced only short-term benefits (Craighead et al. 1931). Furthermore, they treated symptoms rather than underlying causes of bark beetle outbreaks (Keen and Salman 1942). What managers needed was a way to anticipate the mortality so they could remove at-risk trees before they were killed. This would generate a financial return and reduce populations of bark beetles by denying them the stressed and susceptible trees needed to maintain populations. F.P. Keen (1936, 1943) developed and then refined an insect-risk rating system for interior ponderosa pine and then demonstrated the means by which the system could be applied to thin stands so as to reduce tree mortality and recover volume. Entomologists Kenneth Salman and J.W. Bongberg then set out to test a version of this risk-rating system at Blacks Mountain Experimental Forest (Salman and Bongberg 1942). After monitoring stands for 22 years, long-term results were published by Wickman and Eaton (1962). They found a substantial and sustained reduction in mortality that averaged approximately 80% for a 22-year period after the initiation of the sanitation/salvage program (Fig. 6.4). The application of the risk-rating system to selective logging was widely applied in California pine forests in the 1940s and 1950s and is regarded as the first risk-rating system of its kind to be developed and applied.

Fig. 6.3 Accelerated mortality at Blacks Mountain. (Photo by A.A. Hasel 1939)

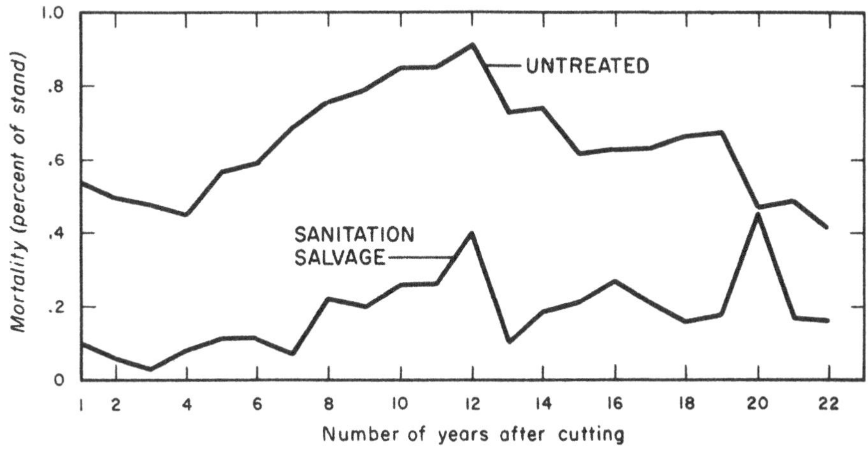

Fig. 6.4 Effect of sanitation harvesting at Blacks Mountain Experimental Forest between 1937 and 1959. (Source: Wickman and Eaton 1962)

Austin Hasel, drawing from an extensive time–motion study of the various harvest methods, concluded that light sanitation-salvage cutting described by Salman and Bongberg was economically feasible (Hasel 1946). The most important finding was that having trees of sufficient size for merchantability was the key factor, rather than intensity of cutting. His findings led to more widespread application of sanitation-salvage methods to reduce the threat posed by bark beetles on both Forest

Service and privately held timberland in California, and is still practiced today on some private commercial forestlands.

A key question for researchers at the time was: Could health be maintained, growth promoted, and forests regulated in an economically viable manner with intermediate harvests? Economic viability was an important element because if forest managers were to effectively move away from lumbering (the removal of sound merchantable material with little regard for future productivity) toward sustainable forestry, then it would be important to show that it was indeed economically feasible. A large-scale "methods of cutting" (MOC) study was established. In the MOC project, three different levels, or methods, of cutting would be compared with each other and with an uncut control.[4] The tree-cutting methods plus an untreated control were replicated ten times. The three harvest treatments included a heavy Forest Service harvest (removal of 80% of the volume in a single cut), a modified Forest Service harvest (removing 40% of the volume), and a silvicultural selection harvest, wherein a sanitation-salvage cut was applied and then followed some years later with an application of Dunning's unit area control. It was intended that over time, the silvicultural selection harvest would approach the retention level of the modified Forest Service harvest, but the initial harvest removed approximately 15% of the volume (Dolph et al. 1995). This management approach was similar to group selection with focus on the natural mosaic pattern in the forest in determining units for management, as opposed to individual tree selection. These research plots were phased in annually with a complete set of treatments installed in blocks annually beginning in 1938.

Effects of cutting on growth shifted over time (Hallin 1959). During the first 5 years after establishment, growth was greatest in the light sanitation-salvage units, but during the next 5 years, observed growth was greatest among the heavy Forest Service harvests. Annual net growth in the uncut control was slightly negative (−0.04% of standing volume[5]), while the heavy Forest Service harvest yielded the best growth rate (+1.5% of standing volume). Furthermore, Eaton (1959) showed that while insect-caused 10-year mortality was highest in the untreated controls (0.44%), there was little difference among the three primary treatments, where 10-year mortality ranged from 0.04 to 0.11%.

Some additional research on artificial regeneration, pruning, and precommercial thinning was conducted after establishment of the MOC study (Hallin 1959), but little was published. Between 1960 and 1990, the experimental forest was used very little for research. In the early 1990s, Forest Service research priorities shifted, with greater emphasis on restoration and enhancing forest ecosystem function (Brooks and Grant 1992; Kessler et al. 1992). Blacks Mountain provided an ideal location to conduct interdisciplinary ecosystem research and there was a rejuvenation of research efforts on the experimental forest.

[4] Austin Hasel, "Plan for Methods-of-Cutting Study, Blacks Mountain Sustained Yield Project." May 6, 1938. On File, PSW Research Station, Redding, CA.

[5] Volumes reported by Hallin (1959) were board-foot measure.

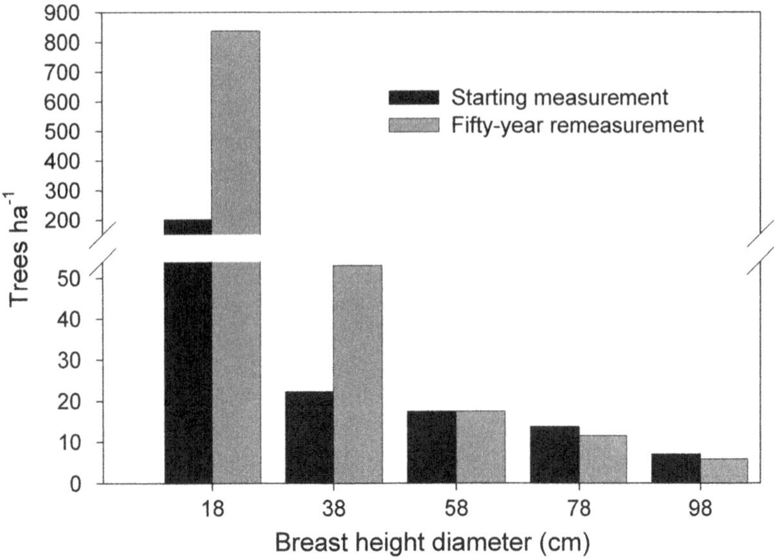

Fig. 6.5 Change in diameter distribution over 50 years at Blacks Mountain. (Source: Dolph et al. 1995)

6.3 Ecological Research at Blacks Mountain

In the late 1980s, Leroy Dolph and others at the Redding PSW Laboratory recognized that the MOC experiment provided an opportunity to quantify long-term changes in stand structure. After an extensive effort at re-monumenting and remeasuring the MOC study in 1990 and 1991, Dolph et al. (1995) reported substantial changes in the distribution of trees some 50 years after the initial establishment of the experiment (Fig. 6.5). Although there was some reduction in the number of large trees resulting from continued bark beetle mortality, the most striking change was in the influx of smaller-diameter trees in both harvested and unharvested stands. This is related to the effects of heavy grazing in the late nineteenth century followed by over 100 years of fire exclusion from these forests (Skinner and Taylor 2006). Fire exclusion as a general policy was instituted in the early twentieth century, resulting in removal of the periodic low-intensity fires that were common in interior pine stands before European settlement of the region and would have historically thinned young conifer regeneration (Norman and Taylor 2003). That the absence of frequent fires is likely the primary reason for the increased density of understory trees is evidence of similar stand structural changes from nearby Lassen Volcanic National Park in landscapes that have not experienced lumbering or other forms of timber harvesting (Taylor 2000). The increasing abundance of understory trees has contributed to an increase in the risk of high-intensity wildfire and an increase in competition for limited water and nutrients in these systems. Dolph et al. did not report a consistent increase in white fir, as expected, but did note that, in general, incense cedar increased in proportional representation, while pines filled in much of the remaining available space.

The early researchers did not understand the importance of fire as an ecological factor in the development of the forests at Blacks Mountain. The influence of fire in these forests was first altered in the late 1800s when widespread grazing was introduced to the area. Grazing was not managed until after the establishment of the Lassen National Forest in the early 1900s. Early grazing activity, through trampling and heavy consumption, removed much of the fine fuels (grasses, herbs, and pine needles) that carried frequent, low-intensity surface fires. This allowed for a flush of pine, incense cedar, and white fir regeneration that was not repeatedly thinned by fire (Norman and Taylor 2003, 2005). This grazing regime was followed by the introduction of systematic fire suppression early in the twentieth century, which effectively removed fire as an ecological process from the Blacks Mountain Experimental Forest area. The exclusion of fire thus promoted continued ingrowth and survival of young trees and created dense stands more susceptible to high-intensity fires than was typical of the original forests (Skinner and Taylor 2006).

These concerns, along with the growing questions about biological diversity, provided the emphasis for developing a long-term research project on the experimental forest. William Oliver, a research silviculturist with the research station, spearheaded the effort to bring together scientists specializing in fire, wildlife ecology, entomology, genetics, and forest mensuration to develop a single large-scale experiment that would allow an evaluation of many different ecological responses over time (Oliver 2000). Infrastructure such as the systematically located, spatially referenced grid points for recording baseline conditions and monitoring will facilitate future research efforts.

This experiment called for the creation of all combinations of two contrasting stand structures and two contrasting grazing treatments at a large scale (about 100 ha), replicated three times. Each of these 12 treatment units was then split so half could receive a prescribed fire treatment intended to represent the natural historical pattern and process of fire at Blacks Mountain. The strongly contrasting structures and the grazing treatments plus application of prescribed fire provided the strong response surfaces to test the hypothesis centered on biological diversity.

The experimental design, described in detail by Oliver (2000), is a randomized block experiment with split plots. Three blocks, related to slope position, have four large treatment areas each (about 100 ha each). The four treatments are a 2×2 factorial on grazing and structural diversity. The grazing treatments include grazing excluded (fenced) and grazing allowed (unfenced). The structural diversity treatments are high structural diversity (HD) and low structural diversity (LD). In the HD treatment, all canopy layers were maintained, and no trees in the dominant overstory were removed (almost all trees above 50 cm diameter at breast height, dbh, were retained). In the LD treatment, all of the large dominant trees over 50 cm were removed along with most of the smaller trees in the understory leaving mid-canopy trees with a quadratic mean diameter of about 25 cm in a fairly uniform density of approximately 280 trees ha^{-1} (Fig. 6.6). The large main plots were split in half, and prescribed fire was applied, post thinning, to one of the splits in each main plot.

Fig. 6.6 Treatments in the Blacks Mountain Ecological Research Project: **a** high diversity, burned; **b** high diversity, unburned; **c** low diversity, burned; **d** low diversity, unburned; and **e** untreated research natural area

Owing to the large scale of the treatments, there was insufficient experimental area for an additional untreated control. However, an established research natural area (RNA) was used as a qualitative control. The Blacks Mountain RNA (Cheng 2004) was established in the 1950s as a site representative of mature interior ponderosa pine. There are five separate compartments designated for the RNA. Four of these RNA compartments were used as a contrast to the treatments within the experiment, but because they were preexisting, the treatments were not randomly applied, and they are not a part of the formal experimental design.

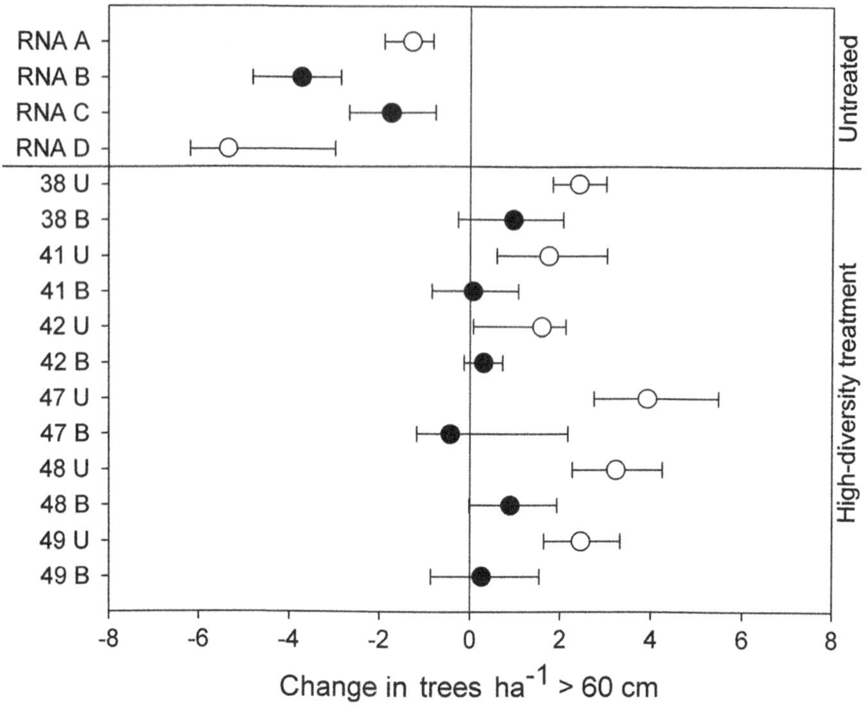

Fig. 6.7 Five-year median change (*first and third quartiles*) in large-tree (>60-cm dbh) density in high-diversity treatments and unthinned research natural areas at Blacks Mountain Experimental Forest (*closed circles* = burned, *open* = unburned)

Because of the scale of the project (about 1,300 ha treated and 160 ha in qualitative controls), planning and implementation of this project took a decade to complete. Treatments have now been in place for over 10 years, and the results are helping to shed more light on the earlier findings of the effects of thinning on tree health. The study was designed with a 50-year planning horizon, and we expect that the study will be fruitful for many years to come.

6.4 Results

A comparison of the untreated RNAs at Blacks Mountain with the HD showed that by thinning from below, ingrowth into the largest-diameter classes is increasing in the HD treatment areas (Fig. 6.7), while untreated "natural" stands are still losing the large-tree component at a rate not compensated by growth. Furthermore, treated stands actually show a reduction in the proportion of high-risk trees (Fig. 6.8), while the continued mortality in untreated "natural" stands indicates that there is little prospect for increasing the number of large-diameter trees without treatment (Ritchie et al. 2008).

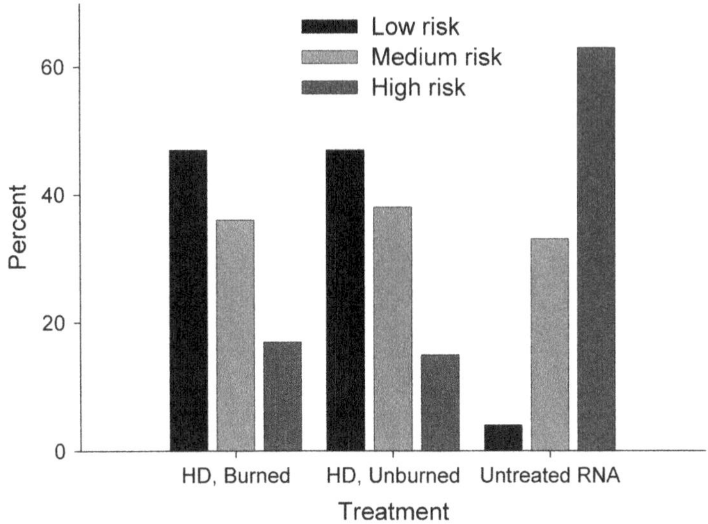

Fig. 6.8 Distribution of large trees (>60-cm dbh) by risk rating in high diversity and untreated units at Blacks Mountain Experimental Forest (observations made 5 years post treatment)

Some results from comparisons between stands with HD and LD are:

- Growth of both individual trees and stands (Fig. 6.9) is substantially higher in LD units (Zhang et al. 2008).
- The bird response to structural diversity has been subtle at Blacks Mountain. Species richness (the number of species present) showed no response to stand structure. However, four individual species showed differences in occupancy (preference). Occupancy was higher for American robins (*Turdus migratorius*) and chipping sparrows (*Spizella passerina*) in LD units, while occupancy was higher for the white-breasted nuthatch (*Sitta carolinensis*) and western tanager (*Piranga ludoviciana*) in HD units (George and Zack 2008).
- Small mammals showed minimal short-term response to stand structure and appeared to be more influenced by overstory density and, for some species, the presence of shrubs and coarse woody debris (CWD; Maguire et al. 2008).
- Surface CWD larger than 7.6 cm in diameter did not differ across treatment units regardless of the stand structure. There was less total CWD following treatments even in unburned plots. This reduction in CWD likely resulted from the harvest machinery breaking up CWD as the machinery moved about the units (Uzoh and Skinner 2009).
- Understory composition of shrubs is higher in terms of both cover and number of species in LD units (Zhang et al. 2008).

The prescribed fire treatments have been applied once, with re-burn planned as surface fuel levels rebuild over time. A number of researchers have published results of the effects of prescribed fire at Blacks Mountain.

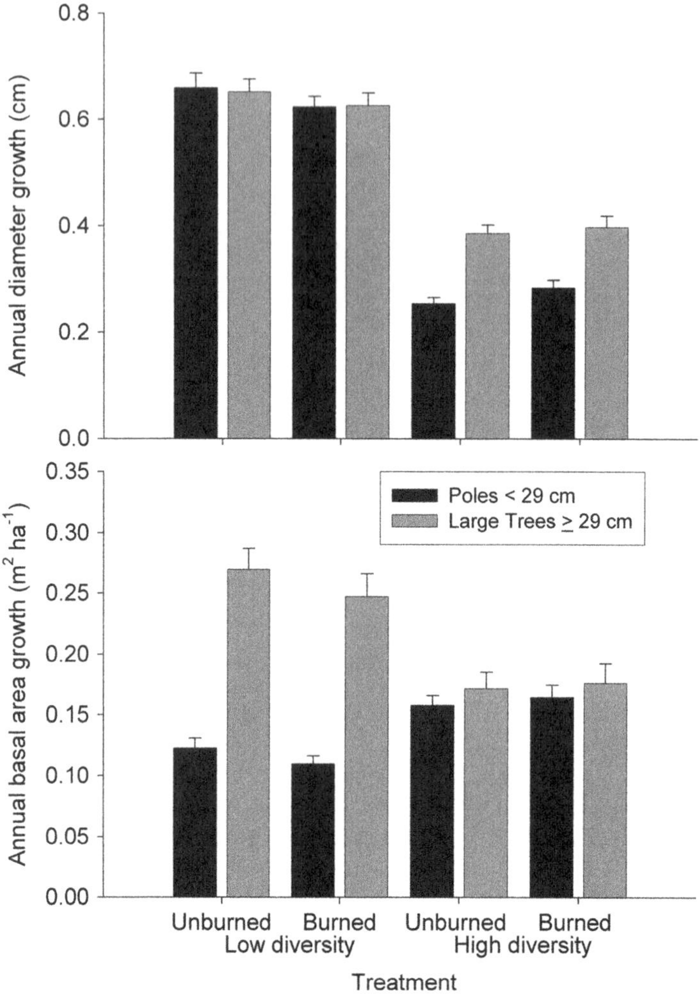

Fig. 6.9 Tree breast height diameter growth rates and stand basal area growth rates for treatments in the Ecological Research Project at Blacks Mountain. (Source: Zhang et al. 2008)

- The impact of fire on small mammals was difficult to determine with any precision. The effects of fire on small mammals appeared to be indirect in that some species are influenced by shrub cover and woody debris, both of which are influenced by prescribed fire (Maguire et al. 2008).
- Surface CWD decreased following the application of prescribed fire, with the most notable change in the CWD in an advanced state of decay. While CWD was significantly reduced in the LD following prescribed fire, the high variability in coverage of the burn treatment in HD led to an absolute but nonsignificant reduction in CWD in the latter areas (Uzoh and Skinner 2009).

Fig. 6.10 Aftermath of the Cone Fire, September 2002, at Blacks Mountain Experimental Forest. (Photo by M. Ritchie 2002)

- Prescribed fire had little influence on species richness of shrubs; however, the amount of shrub cover appears to be slightly higher in the more open LD burned units (Zhang et al. 2008).
- In a study focused on the RNAs, which have had no thinning, prescribed fire contributes to higher rates of mortality, fuel accumulation, and high-intensity fire risk in the short term, and it may take three applications of fire to achieve the reduction in high-intensity fire risk that can be obtained in one entry with a thinning treatment before fire (Skinner 2005).

6.5 The Cone Fire

More recently, a wildfire burned through part of the experimental forest in September 2002. The Cone Fire burned about 600 ha of the experimental forest including parts of three of the Ecological Research Project treatment units about 5 years after treatment implementation. At the time of ignition, 100-h fuel moisture and relative humidity were low (2 and 6%, respectively). Owing to these extreme fire weather and fuel conditions, much of the area burned with high severity, killing most or all of the trees (Fig. 6.10). While the fire did some damage to research plots, it was apparent the treatments at Blacks Mountain had considerably moderated fire intensity and the associated severity within treated areas (Fig. 6.11). This offered an opportunity to record the effects of various management treatments on subsequent wildfire behavior.

With funding from the Joint Fire Science Program beginning in 2003, researchers initiated a study of the effects of thinning and prescribed fire on wildfire behavior (Ritchie et al. 2007, Symons et al. 2008). Fire effects were generally much less severe in all treatment units. The combination of thinning and prescribed fire

Fig. 6.11 The Cone Fire underburned the thinned-only split of low diversity unit 46 (*left-center*), but failed to burn through the half with prescribed fire (*right-center*) in the Ecological Research Project at Blacks Mountain Experimental Forest, while the untreated area to the north burned with high severity. (Forest Service File Photo 2002)

essentially limited mortality to those trees along the edge of the stand (Fig. 6.12). The crown scorch and bole scorch was also highest outside the treated units and along their edges (Fig. 6.13).

Researchers have also initiated an experiment designed to study the effects of postfire (salvage) logging on fuel buildup, regeneration success, and understory species composition (Fig. 6.14). These plots are being monitored biannually to evaluate the changes over time. This experiment will provide a longer-term look at how salvage logging affects fuel loading, stand development, biodiversity, and wildlife habitat.

6.6 Conclusion

Blacks Mountain Experimental Forest has been a productive site for ponderosa pine research for 75 years. Each successive generation of research has drawn upon the previous work for inspiration, experimental material, and established research records. One lesson from these efforts is that long-term research can often provide unanticipated benefits. The experimental design with strongly contrasting treatments provides the opportunity for subsequent research questions, new management ideas, and new investigations. Such is the legacy of the modern Blacks Mountain Experimental Forest.

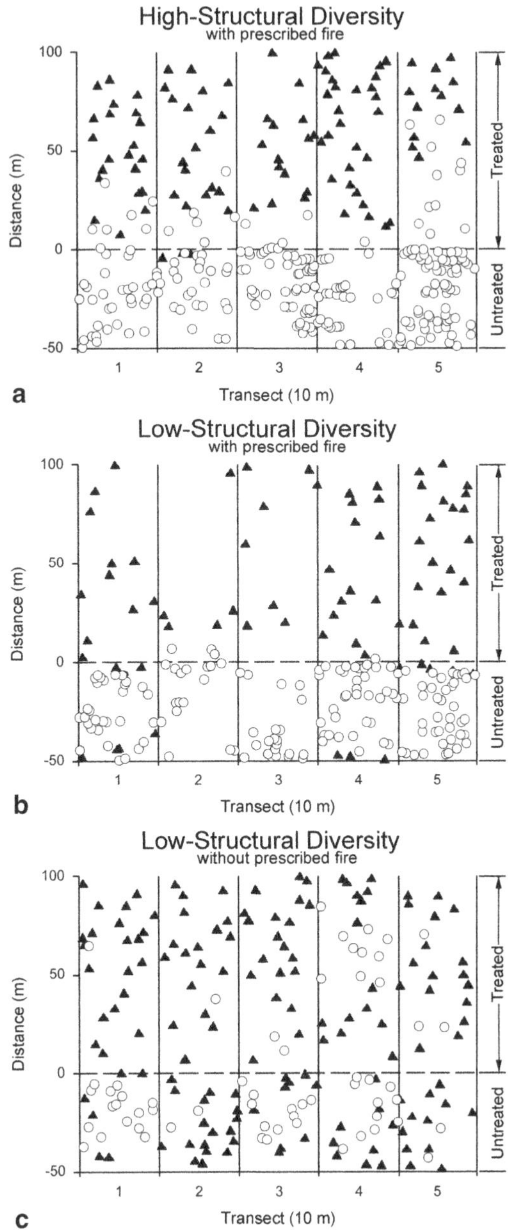

Fig. 6.12 Distribution of living (*solid triangles*) and dead (*open circles*) trees in transects on the border of treatment units within the Cone Fire at Blacks Mountain Experimental Forest shows the reduction in mortality in relation to distance from treatment boundary. (Source: Ritchie et al. 2007)

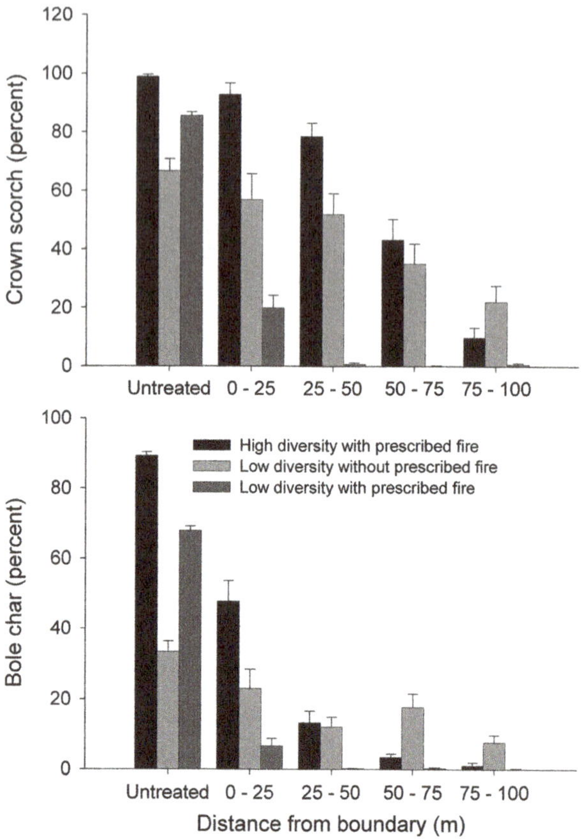

Fig. 6.13 Rates of crown scorch and bole char (with 1 standard error bar) decreased on the interior of treatment units at Blacks Mountain Experimental Forest; high levels of char were limited to the boundary. (Source: Ritchie et al. 2007)

Fig. 6.14 Variable-retention salvage study plots in the Cone Fire at Blacks Mountain Experimental Forest after the Cone Fire. (Photo by M. Ritchie 2005)

References

Brooks DJ, Grant GE (1992) New perspectives in forest management: background, science issues, and research agenda. Research Paper PNW-RP-456, Pacific Northwest Research Station, USDA Forest Service, 17 p

Cheng S (2004) Forest service research natural areas in California. General Technical Report PSW-GTR-188. Pacific Southwest Research Station, USDA Forest Service, 338 p

Craighead FC, Miller JM, Evenden JC, Keen FP (1931) Control work against bark beetles in western forests and appraisal of its results. J For 29:1001–1018

Dolph KL, Mori SR, Oliver WW (1995) Long term response of old-growth stands to varying levels of partial cutting in the eastside pine type. West J App For 10:101–108

Eaton CB (1959) Insect-caused mortality in relation to methods of cutting in ponderosa pine on the Blacks Mountain Experimental Forest. Technical Paper no 43, Pacific Southwest Forest and Range Experiment Station, USDA Forest Service, 33 p

George TL, Zack S (2008) Bird occupancy and richness in ponderosa pine forests with contrasting forest structure and fire history. Can J For Res 38:936–942

Hallin WE (1959) The application of unit area control in the management of ponderosa-Jeffrey pine at Blacks Mountain Experimental Forest. Technical Bulletin 1191, USDA Forest Service, 96 p

Hasel AA (1946) Logging cost as related to tree size and intensity of cutting in ponderosa pine. J For 44:552–560

Hickman JC (1993) The Jepson manual: higher plants of California. University of California Press, Berkeley, 1400 p

Keen FP (1936) Relative susceptibility of ponderosa pines to bark-beetle attack. J For 34:919–927

Keen FP (1943) Ponderosa pine tree classes redefined. J For 41:249–253

Keen FP, Salman KA (1942) Progress in pine beetle control through tree selection. J For 40:854–858

Kessler WB, Salwasser H, Cartwright CW Jr, Caplan JA (1992) New perspectives for sustainable natural resource management. Ecol Appl 2:221–225

Maguire CC, Maguire DA, Manning TE, Garber SM, Ritchie, MW (2008) Response of small mammals to alternative stand structures in the mixed-conifer forest of northeastern California. Can J For Res 38:943–955

Norman SP, Taylor AH (2003) Tropical and north Pacific teleconnections influence fire regimes in pine-dominated forests of north-eastern California, USA. J Biogeogr 30:1081–1092

Norman SP, Taylor AH (2005) Pine forest expansion along a forest-meadow ecotone in northeastern California, USA. For Ecol Manage 215:51–68

Oliver WW (2000) Ecological research at the Blacks Mountain Experimental Forest in northeastern California. General Technical Report PSW-GTR-179, Pacific Southwest Research Station, Forest Service, USDA Forest Service, 66 p

Ritchie MW, Skinner CN, Hamilton TA (2007). Probability of tree survival after wildfire in an interior pine forest of northern California: effect of thinning and prescribed fire. For Ecol Manage 247:200–208

Ritchie MW, Wing BM, Hamilton TA (2008) Stability of the large tree component in treated and untreated late-seral interior ponderosa pine stands. Can J For Res 38:919–923

Salman KA, Bongberg JW (1942) Logging high-risk trees to control insects in the pine stands of northeastern California. J For 40:533–539

Skinner CN (2005) Reintroducing fire into the Blacks Mountain Research Natural Area: effects on fire hazard. In: Proceedings of the symposium on ponderosa pine: issues, trends, and management. 2004, October 18–21; Klamath Falls Oregon. General Technical Report PSW-GTR-198, Pacific Southwest Research Station, USDA Forest Service, pp 245–258

Skinner CN, Taylor AH (2006) Southern Cascades bioregion. In: Sugihara NG, van Wagtendonk JW, Fites-Kaufmann J, Shaffer KE, Thode AE (eds) Fire in California's ecosystems. University of California Press, Berkeley, pp 195–224

Symons JN, Fairbanks DHK, Skinner CN (2008) Influences of stand structure and fuel treatments on wildfire severity at Blacks Mountain Experimental Forest, northeastern California. Calif Geogr 48:1–23

Taylor AH (2000) Fire regimes and forest changes in mid and upper montane forests of the southern Cascades, Lassen Volcanic National Park, California, USA. J Biogeogr 27:87–104

Uzoh FCC, Skinner CN (2009) Effects of creating two forest structures and using prescribed fire on coarse woody debris in northeastern California, USA. Fire Ecol 5(2):1–13

Wickman BE, Eaton CB (1962) The effects of sanitation/salvage cutting on insect-caused mortality at Blacks Mountain Experimental Forest. Technical Paper 66. Pacific Southwest Forest and Range Experiment Station, USDA Forest Service, 39 p

Zhang J, Ritchie MW, Oliver WW (2008) Vegetation responses to stand structure and prescribed fire in an interior ponderosa pine ecosystem. Can J For Res 38:909–918

Part III
Research Trajectories in Forest Ecology, Management, and Silviculture for Hardwoods

Chapter 7
Uneven-Aged Management After a Half-Century of Research on the Forest Service Fernow Experimental Forest in West Virginia

Thomas M. Schuler

Abstract Uneven-age management in forestry refers to a system of management that periodically selects individual trees or small groups of trees for harvest. In general, the concept of uneven-age management entails the sustained yield of forest products while maintaining continuous forest cover. In North America, interest in uneven-age management grew in the second half of the twentieth century after most of the old-growth forests had been harvested. In West Virginia, uneven-age management and its surrogates have been studied intensively since 1948 at the Forest Service Fernow Experimental Forest. The Fernow Experimental Forest is located in the Allegheny Mountains of the Central Appalachian Broadleaf Forest and has characteristics of both mixed-oak and northern hardwood forests, depending on site characteristics. In stands managed with the type of uneven-age management known as "single-tree selection," only those species capable of developing in reduced sunlight conditions have thrived. Concerns that the increased dominance of these so-called shade-tolerant species would lead to lower productivity have not been validated. It appears that managing the stocking level, or the size and number of trees after harvesting, has helped to sustain higher than expected levels of wood production and carbon storage. Economically, uneven-age-managed stands have produced periodic revenue that resulted in higher net present values than either clearcut or unharvested stands, and, unlike more exploitive procedures, did not deplete the residual stand value. Unexpected insights into forest ecology and management also were realized by carefully studying the response to treatments for more than a half century and the benefits of long-term forestry research are discussed.

Keywords Single · Tree selection · Diameter · Limit selection · Partial harvesting · Species diversity · Productivity · Mixed · Mesophytic · Long · Term forest research

T. M. Schuler (✉)
Northern Research Station, USDA Forest Service, Timber & Watershed Laboratory, Route 219 North – Nursery Bottom, P.O. Box 404, Parsons, WV 26287, USA
e-mail: tschuler@fs.fed.us

D. C. Hayes et al. (eds.), *USDA Forest Service Experimental Forests and Ranges*, DOI 10.1007/978-1-4614-1818-4_7, © Springer New York 2014

7.1 Introduction

7.1.1 Boom-and-Bust Era

Uneven-age forest management research on the Fernow Experimental Forest in West Virginia began after World War II. Initial interest in this form of management was in part a response to an earlier period of forest exploitation from about 1880 to 1920 that removed virtually all of the old-growth forests in the central Appalachian Mountains (Clarkson 1964). During this era, industrialists purchased vast tracts of land, or the timber rights to the land, and built mills that employed the newest technology capable of processing the timber from hectares of forestland each day. Temporary rail lines were constructed to reach the forests; and towns emerged or grew to provide the labor and services needed, leading to a generation-long economic boom in the region (Whitney 1994). For example, in Davis, WV, various mills processed roughly 2.4 million m^3 of timber during this period until the supply ran out (Clarkson 1964). Between the 1880s and 1920, harvesting proceeded across large tracts of land with little to no regard for any other resources. To make matters worse, wildfires of historic proportions often followed in cutover areas when sparks from steam-driven locomotives ignited the drying slash left behind after harvesting (Brose et al. 2001). Such fires burned with high intensity and degraded soils and water quality, destroyed valuable regeneration and wildlife habitat, and spread to adjacent uncut forests. As the timber supply ran out, rail lines were removed, mills were closed, unemployment jumped, and local populations declined. In some cases, small towns were depopulated entirely. In Tucker County, WV, where the Fernow Experimental Forest is now located, ten towns that existed during the timber boom era are gone today (Fansler 1962).

7.1.2 Early Twentieth-Century Forest Conservation Efforts

At the start of the twentieth century, the pattern of boom-and-bust forest exploitation in several regions and fears of looming timber shortages in the USA led to interest in forest conservation and the development of forestry as a science. The American Forestry Association and the National Academy of Science, supported by business interests, advocated for the federal purchase of forestlands in the eastern USA to serve as timber reserves (Robbins 1982). Eventually, the Weeks Act of 1911 authorized the federal purchase of private forestland in the eastern USA, which resulted in the establishment of the first eastern National Forests. Although the Weeks Act was controversial at the time, these new federal lands ultimately were created to protect the headwaters of navigable waterways and serve as examples of science-based forest management. Concurrently, the first forestry schools granting 4-year degrees were established in the USA to train the professionals needed to employ science-based forest management techniques. Concepts in these early forestry schools were taught by foresters trained in Europe such as Bernhard Fernow, a pioneer of professional forestry in North America and the founder of the first forestry schools to grant 4-year degrees in the USA and Canada (Rodgers 1991).

In Europe, forest management was already well established and professional foresters often managed estates for long-term objectives. Europeans had depended on their forests for centuries to supply many needs, including building materials, wood for fuel, and forage for livestock. Foresters in Europe often relied on partial harvesting and advanced the concept of uneven-age management, which involved removal of trees from all size classes to mimic the patterns of older, unmanaged forests (de Liocourt 1898). In theory, these stands would be kept at stocking levels (determined by the number and size of trees) low enough to allow for acceptable growth rates of existing trees and permit periodic harvests that could be sustained indefinitely. This theory was discussed in the USA early in the twentieth century (Graves 1910), and more fully articulated later in the century (Meyer 1943, 1952), when the social, political, and technological circumstances were more favorable.

7.1.3 The Fernow Experimental Forest and Late Twentieth-Century Forest Management

To establish the new Monongahela National Forest, as authorized by the Weeks Act, the first USDA Forest Service purchase of private land in West Virginia, referred to as the Arnold tract, occurred in 1915. In 1934, most of the Arnold tract became the 1,475-ha Fernow Experimental Forest (expanded to 1,900 ha in 1974) and was dedicated to forest research and forest management demonstration under the leadership of the Appalachian Forest Experiment Station. Roads, firebreaks, and a water reservoir were among the first projects on the Fernow and were built by the Civilian Conservation Corps in the 1930s. Research was just getting started in earnest when the national war effort forced the closing of operations in 1941. But in 1948, the facility was reopened under the administration of the Northeastern Forest Experiment Station (currently the Northern Research Station). The research mission at the time was focused on timber and watershed management (Trimble 1977). The Fernow was chosen as a research site because it is representative of the many ecological conditions found throughout the region, including those of the Allegheny Mountains of northern West Virginia, western Maryland, and central Pennsylvania, and to some extent those of the unglaciated Allegheny Plateau of western West Virginia, western Pennsylvania, and southwestern Ohio, and the Cumberland Mountains of eastern Kentucky.

At the beginning of the postwar era, Americans were still concerned about the exploitive harvesting practices seen earlier in the century. For example, when a Forest Service survey of forestlands across the USA was conducted in which harvest levels or "cutting practices" were evaluated as "excellent" to "poor" based on the amount of residual growing stock remaining after a harvest, "poor" was equated with the lowest levels (Harper and Rettie 1946). This terminology was further refined and implemented at the regional level and demonstrations of cutting practice levels (CPL) were established at several experimental forests throughout the region in the 1950s (Kenefic and Schuler 2008). In most cases, the "excellent" and "good" examples of forest management in these studies were implemented as uneven-age management using European-inspired forms of single-tree selection, which selects individual trees for harvesting and reserves others for continued growth. Poor management was most closely linked with heavier levels of removal, often implemented as a commercial clearcut without removal of trees with poor form or significant rot, usually referred to as cull trees (Weitzman 1949). Commercial clearcutting removes all trees of monetary value at one time and is similar to past exploitive practices, although on a smaller scale. In the 1950s, the lessons of the first half of the twentieth century were that exploitive harvesting, which looked similar to commercial clearcutting, compromised good forest stewardship and resulted in conditions that did not permit a sustainable flow of goods and services from the forest (Harper and Rettie 1946). By the postwar era, the capital-intensive logging that had occurred at the turn of the century in the Appalachians was no longer an option because virtually all of the forests had been exploitively logged just a few decades earlier. Many forest scientists and leaders in the field turned to uneven-age management as a possible alternative to foster better forest stewardship and enhance recovery of the region's cutover forests.

On the Fernow Experimental Forest, at least three early research initiatives addressed uneven-age management. One of the first studies was a variant of the regional CPL initiative that consisted of four treatments: two levels of uneven-age management implemented as single-tree selection, a diameter-limit cutting procedure, and a commercial clearcut harvest. In 1948, forest researchers designated these treatments as high-order, good, fair, and poor cutting practices, respectively. The commercial clearcut or poor cutting practice was considered the prevalent liquidation method of forest harvesting (Weitzman 1949). The CPL study was implemented as a small-scale case study and demonstration (10 ha, including the unmanaged reference area added in 1953).

A much larger experimental design involving three different site classes and three silvicultural practices, including uneven-age management, was started in 1950 on the Fernow as well. The larger study consisted of 21 physical research units, or compartments, on 280 ha. This study is referred to as the Large Area Comparison of Forest Management Practices, hereafter referred to as the LAMP study. The third prominent uneven-age type study was initiated in the 1970s and was designed to provide a silvicultural means for periodic partial harvests based on individual tree financial maturity and is referred to as financial maturity diameter-limit selection (FMDL; Trimble et al. 1974). The FMDL study was replicated on six research units or compartments on 97 ha. FMDL requires the removal of trees of poor vigor and quality as a first step and then provides guidelines for residual stocking based on financial maturity of individual trees and a desired financial rate of return.

Together, these three studies (CPL, LAMP, and FMDL) provide unparalleled insight into the ecological and economic qualities of uneven-age management in the central Appalachians after more than a half century of repeated harvests, forest responses, and continuous research and monitoring. These studies have been used by scientists to examine both ecological and management issues and have been the source of many scientific publications and management recommendations. The objective of this chapter is to provide an overview of these three studies and discuss some of the findings, management implications, and emerging issues related to using uneven-age forest management in the region today.

7.2 Fernow Experimental Forest

7.2.1 Characteristics and Layout

The Fernow Experimental Forest is located in the Allegheny Mountains of the Central Appalachian Broadleaf Forest, which forms the prominent uplift of the Appalachian Mountains in West Virginia. The Fernow is best characterized as a mixed-mesophytic forest type, which has characteristics of both mixed-oak and northern hardwood forests depending on aspect, elevation, and slope position (Braun 1950). The average growing season is 145 days (May–October) and the mean annual precipitation is about 142 cm, which is evenly distributed throughout the year (Pan et al. 1997). Growing-season temperatures are typically moderate and growing-season moisture deficits are uncommon (Leathers et al. 2000). The topography is mountainous and elevations range from 530 to 1,110 m above sea level. Common overstory species include northern red oak (*Quercus rubra*), sugar maple (*Acer saccharum*), and yellow-poplar (*Liriodendron tulipifera*) on the mesic sites, and chestnut oak (*Q. prinus*) and red maple (*A. rubrum*) emerging as dominants on the more xeric sites. Aspect and slope position are important determinants of site productivity and species composition. In all, more than 30 commercial species are found throughout the Experimental Forest (Madarish et al. 2002). Understory species composition is rich and often includes stinging nettle (*Laportea canadensis*), violets (*Viola* spp.), and several fern species; understory species vary with site and disturbance history (Gilliam et al. 1995).

The Elklick watershed (which later became the Fernow) was initially logged between 1903 and 1911 (Trimble 1977) during the railroad logging era (Fansler 1962). Horses and log slides were used to get harvested trees to the temporary rail lines. Trees close to the railroad were cut and used for a variety of purposes, including lower-value products such as mine timbers, but as distances and corresponding costs increased, the merchantability standards also increased. Some trees were left behind because their commercial values were less than the cost of removal. The landscape left behind had variable residual stocking and some older and larger trees with poor form or lower value.

At the beginning of the postwar era, the Fernow was compartmentalized into physical research units and small watersheds usually ranging in size from 10 to 30 ha. Sidney Weitzman was responsible for setting up the general research plan on the Experimental Forest, which emphasized silviculture research, large-scale demonstrations of various timber and watershed practices, and better understanding of

Fig. 7.1 Location of compartments on the Fernow Experimental Forest in West Virginia assigned to studies that incorporate uneven-age management concepts

the costs associated with a range of forest management options (Weitzman 1949). The research approach and some studies remain operative today and are a testament to the vision of Weitzman and his colleagues for setting up a plan that has persisted for more than 60 years. For example, the CPL, LAMP, and FMDL studies employ this approach and use 32 different compartments for various treatments, totaling nearly 400 ha (Fig. 7.1). By utilizing this approach, the research staff on the Fernow

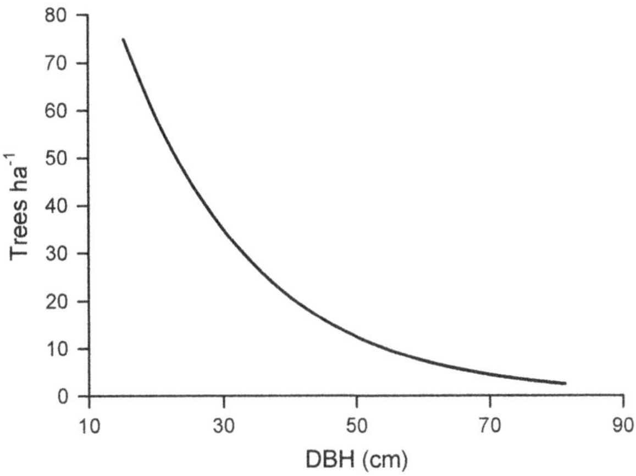

Fig. 7.2 Theoretical size-class distribution of an uneven-aged stand with a Q factor of 1.3

can examine the long-term consequences of managing forests in blocks of land that are large enough to be commercially operable, and typify the ownership size of private lands in the region. Moreover, long-term repetition of specific silvicultural treatments is critical when various forms of partial harvesting are being evaluated because understanding the full implications may require numerous harvests on 10- or 20-year cycles. For instance, the uneven-age management harvests in the CPL study were conducted for the seventh time in 2008, once in 1949 and once each decade thereafter, representing one of the longest-running examples of this type of uneven-age management in the eastern USA.

7.2.2 *Treatments and Experimental Design*

7.2.2.1 Demonstrations of CPLs

The CPL study was designed to demonstrate a range of residual stocking levels using different management intensities. Single-tree selection applying the balanced stand approach was used for uneven-age management. Residual stand goals were defined by three parameters: residual basal area (RBA), the largest tree to retain after each harvest in terms of diameter at breast height (dbh), and the ratio of trees in successively smaller size classes (Q), sometimes referred to as the BDq method of single-tree selection. The Q value results in a negative exponential size-class distribution (Fig. 7.2). Some forest scientists believed this type of stand structure to approximate the conditions that would develop naturally in older forests (Meyer 1952). Two levels of the BDq method were applied in the CPL study and both had a desired Q of 1.3 and a maximum dbh of 81 cm. The maximum dbh was deemed appropriate for the excellent growing conditions found at the site. In compartment 8D, the residual stand structure goals included both pole-size (minimum dbh of 18.0 cm) and sawlog-size (minimum dbh of 28 cm) trees and the targeted RBA was 20 m^2/ha. In subcompartment 8C, the residual stand structure goals included only sawlog-size trees, and the RBA for trees of that size and larger was 16 m^2/ha (Lamson and Smith 1991).

Both areas were first harvested after the 1948 growing season. Consistent with a 10-year cutting cycle, six additional harvests in each area have occurred, with the most recent in the fall of 2008. The diameter-limit harvest in the CPL study (8B) removed all trees larger than 39–cm dbh on a 20-year cutting cycle. After the first harvest in 1948, there have been three additional cutting cycles; the last harvest was in 2008. The commercial clearcut in the study (8A) removed all merchantable stems greater than 18.0 cm dbh in 1948 with no cull tree removal or additional silvicultural treatments to improve the next stand. This was not a true silvicultural clearcut but an exploitive type of harvest believed to be the "prevalent liquidation method of cutting" in 1948 (Weitzman 1949, p. 8). A second-rotation commercial clearcut is scheduled for 2018. A fifth area (8E) with similar site characteristics and disturbance history prior to the onset of experimental manipulations was also included as an unmanaged reference stand.

When the CPL study began, all areas were predominantly even-aged stands about 40 years in age with a few scattered older trees of poor form. All subcompartments were only 2 ha in size for demonstration purposes, and today the area remains easily accessible along a well-maintained and signed trail. Because of its easy accessibility, it is used for many guided tours to discuss the long-term implications of various types of forest management practices. Although demonstration was the primary purpose of the CPL study and therefore was not replicated for scientific testing, the treatments are similar to the LAMP study, which is replicated.

7.2.2.2 Studies in Large Area Comparison of Forest Management Practices

The LAMP study expanded the CPL concept to operationally sized units and featured three different site classes. The Fernow contains three broad site classes referred to as excellent, good, and fair and are estimated by northern red oak height growth. Excellent, good, and fair site classes relate to site index (SI) midpoints of 80, 70, and 60, respectively. SI has been used traditionally in terms of timber management to estimate potential wood volume growth and integrates several ecological factors such as soil type, aspect, landscape position, and precipitation, all of which are functionally related to the growth potential. SI also is a useful predictor of species associations and some ecological functions. Understanding how different sites shape the response to a silvicultural treatment such as single-tree selection greatly expands the geographic inferences that researchers can make from the results. An important benefit of the Fernow is that a range of sites exist and can be studied on a relatively small spatial scale (e.g., hectares $\times 10^3$); then research results can be extrapolated to a relatively large spatial scale (e.g., hectares $\times 10^7$).

Specific treatments in the LAMP study are uneven-age management, patch cutting in 0.16-ha openings, and a 43-cm diameter-limit cut. Unmanaged reference areas for each site class were also included and some treatments were modified for different site classes (Table 7.1). For example, larger trees were part of the residual stand

Table 7.1 Silvicultural treatments and compartments used for the Large Area Comparison of Forest Management Practices study initiated in 1950 on the Fernow Experimental Forest

Single-tree selection					
SI[a]	Cutting cycle (years)	RBA[b] ($m^2\ ha^{-1}$, $ft^2\ ac^{-1}$)	LDT[c] (cm, in.)	Q[d]	Compartment[e]
80	10	15 (65)	81 (32)	1.3	WS5A, 20A
70	10	12 (50)	66 (26)	1.3	7C, 16B
60	15	8 (35)	51 (20)	1.3	WS5B, 19B
Diameter-limit					
SI	Cutting cycle (years)	Harvest dbh[f] (cm, in.)			Compartment
80	15	43 (17)			WS2A, 9B
70	15	43 (17)			27A, 9A
60	20	43 (17)			WS2B, 20C
Patch cutting					
SI	Cutting cycle (years)	Rotation age			Compartment
80	10	65			18A, 17A
70	10	75			30, 18B
60	15	85			17C, 18C
Unmanaged					
SI					Compartment
80					WS4A
70					WS4B
60					WS4C

[a] Northern red oak site index, which is a measure of site quality (80 = excellent, 70 = good, and 60 = fair)
[b] Desired residual basal area of trees with dbh ≥ 28 cm (11 in.)
[c] Largest diameter-class tree to retain in the residual stand structure
[d] One measure of residual (postharvest) stand structure (see Fig. 7.2)
[e] See Fig. 7.1 for locations of compartments on the Fernow Experimental Forest
[f] Trees with dbh of 43 cm (17 in) or greater are harvested at each cutting cycle

goals on the more productive sites. The interested reader can find much more detailed information about this study in Schuler (2004). The patch cutting treatment in the LAMP study provides an opportunity to study stand dynamics in openings that are smaller than conventional clearcuts but large enough to establish groups of trees of the same age. In theory, patch cutting results in an uneven-aged stand structure made up of small even-aged patches. The LAMP diameter-limit harvest is not considered a silvicultural treatment, in part because there are no residual stand objectives. A simple diameter-limit harvest is considered an exploitive harvest, but it is one of the most common forms of timber harvesting in the central Appalachians (Fajvan et al. 1998) because it provides revenue with few internalized costs to the landowner.

7.2.2.3 FMDL Studies

The FMDL concept was conceived as a potential replacement for diameter-limit cutting in order to maintain quality and productivity in the residual stand, still provide an acceptable rate of return, and entail less complexity than other more conventional approaches to uneven-age management. George R. Trimble Jr., a Fernow scientist and project leader from 1950 to 1954 and 1957 to 1973, developed the procedure because he believed that good forest stewardship was achieved through sound ecological and economic principles, but too much complexity would deter acceptance and use. Trimble wrote that the predominantly second-growth forests created by the large-scale exploitive harvesting in the late 1800s and early 1900s created conditions where the traditional concepts of uneven-age management were difficult to apply (Trimble et al. 1974). The approach he developed estimates individual tree financial maturity and outlines when each tree should be harvested based on financial objectives and the growth potential of the site. Unlike diameter-limit cutting, FMDL directs the user to select residual stocking levels by simple guidelines referred to as rates of return. The Fernow FMDL study used three rate-of-return treatments (3, 4, and 6%) and each treatment was replicated in two compartments (Schuler and McGill 2007). Similar to more conventional approaches to uneven-age management, the first priority is removing all trees of poor quality and low vigor (Trimble et al. 1974). But unlike true single-tree selection, the marking procedures are much easier to use in the field once some guidelines have been selected. Balancing growth with removal is a fundamental principle of all uneven-age silviculture and an equally important component of the FMDL system.

7.3 Results and Discussion

7.3.1 Species Composition and Diversity

A common theme among all three studies has been the general decline of oaks and all other shade-intolerant species, including the commercially important black cherry (*Prunus serotina*), coupled with an increase in the abundance of shade-tolerant species, especially sugar and red maple. The proportion of sugar maple has risen from about 15% in 1950 to observed levels that exceed 50% recently in the overstory on many excellent growing sites (Schuler and Gillespie 2000). Current understory dominance of sugar maple is even greater, suggesting that the trend of increasing dominance by this shade-tolerant species will continue. Trimble (1965) predicted that cove hardwood stands in the central Appalachians that were subjected to some type of partial harvesting regime (e.g., single-tree selection or diameter-limit harvesting) would eventually be dominated by sugar maple.

The observed levels of sugar maple dominance today are nonetheless unprecedented. Early in the last century, Brooks (1911) estimated that sugar maple accounted for about 10% of the composition of the remaining old-growth hardwood forests in the vicinity of the Fernow. Looking back even further, researchers have studied pollen deposits that suggest sugar maple was not the most abundant species in the region at any time since the end of the Wisconsin glaciation about 17,000 years ago (Larabee 1986). Furthermore, no evidence suggests that current trends will abate or that species composition will shift back to the more diverse mixtures of the nineteenth and twentieth centuries (Schuler and Gillespie 2000). Loss of certain species means reduced food for wildlife and less resilience in the face of known and unknown perturbations yet to occur such as climate change, invasive plant species, and new insect pests. The reduction in diversity, clearly captured on the Fernow, is widespread throughout the central and eastern hardwood regions and has been referred to as the "mesophication" of the hardwood forest types in the eastern USA (Nowacki and Abrams 2008). Although patch cutting in openings about 0.16 ha in size has sustained some shade-intolerant species, none of the treatments has avoided the significant decline in oak abundance (Schuler 2004). Maintaining the compositional diversity of Appalachian forests and forest types is a high priority for the Monongahela National Forest (USDA Forest Service 2006) and other land management agencies in the eastern USA (Nowacki et al. 2009).

In the 1960s, as Trimble (1965) and researchers at the Vinton Furnace Experimental Forest in Ohio reported their findings regarding how only a few species were thriving following uneven-age management, USDA Forest Service scientists were advising National Forest managers to turn to even-age management, often in the form of clearcutting, to regenerate shade-intolerant species (Roach and Gingrich 1968). To achieve desired harvest levels and stay within budget for new road construction, clearcutting grew in size from tens to hundreds of hectares and sometimes were located in close proximity to each other over a period of years (Cravens 1975). Public opposition and controversy regarding how timber should be harvested from public land soon followed. When citizens objected to clearcutting, Forest Service managers cited research in West Virginia and Ohio (Trimble 1965; Roach and Gingrich 1968) to explain why they needed to move away from partial cutting or uneven-age management (H. Clay Smith, personal communication, 2009). But fear of a return to the vast exploitive harvesting that had ended just a half century earlier ultimately led to significant social conflict about whether clearcutting should be allowed. Finally, a legal decision about clearcutting (*West Virginia Division of the Izaak Walton League of America, Inc. v. Butz* 1973), sometimes referred to as the "Monongahela Decision," led to a temporary ban on all clearcutting in some eastern National Forests (Haines 1976) and ultimately to the passage of important federal legislation affecting the management of all National Forests (i.e., National Forest Management Act of 1976; Haines 1976).

As these historical events unfolded in the 1960s and 1970s, researchers continued to look for silvicultural options that would meet management objectives and be socially acceptable. The uneven-age management research on the Fernow continued to be evaluated as part of the research portfolio of alternative silvicultural options. Clay Smith, Fernow Research Forester and Project Leader from 1962 to 1967 and 1973 to 1994, stated that "the demonstration value of the uneven-age management studies during the clearcutting controversy was critical; everyone wanted to see examples of both even-age and uneven-age management first-hand and learn the pros and cons of each" (H. Clay Smith, personal communication, 2009). Legislators, members of the USDA Forest Service leadership team, and the general public took advantage of the Fernow's convenient location, approximately half a day's drive from Washington, DC, to see how these two contrasting practices looked on the ground (H. Clay Smith, personal communication, 2009). This period of controversy reinforced the need to have examples of a range of silvicultural practices that are easily accessible to the public, policy makers, and executive branch officials. Interest in seeing the treatments in person continues to this day and has become a major part of the technology transfer effort at the Fernow.

The interaction among scientists, managers, policy makers, and the public also produced a new silvicultural technique, sometimes referred to as "two-age" management, which addressed several of the issues. The first installations of two-age management were on the Fernow and other locations on the Monongahela National Forest. For a more detailed account of how this innovative technique emerged from the controversy surrounding the Monongahela Decision, see Miller (this volume). Knowledge and insight gained from this work are now based on more than 25 years of research and experience (Thomas-Van Gundy and Schuler 2008). Although no single system will ever satisfy every forest regeneration need, the collaboration among scientists and managers that developed two-age management, with input from concerned citizens, is a model of how future forest management issues can be addressed in a timely and effective manner.

As the long-term silvicultural experiments have continued on the Fernow and elsewhere, scientists have gained more insight into the effects of uneven-age management. In the Fernow studies, which included a range of site classes, sugar maple dominated the regeneration on the better sites and red maple dominated it on the poorer sites, especially in conjunction with any type of partial harvesting treatment. In general, of the 32 species identified in the overstory from 1951 to 2001, only American beech (*Fagus grandifolia*) and black birch (*Betula lenta*) increased in importance, apart from the maples already noted. All other species declined or remained minor components of the stand. Compartments managed with single-tree selection declined the most with respect to diversity and the decline was most pronounced on the better-quality sites (Schuler 2004). Fortunately for forest landowners interested in sustaining a periodic income, the commercial value of sugar maple is relatively high compared to that of other species. The commercial value of American beech, another shade-tolerant species that often increases following repeated partial harvests, is much less, however.

Responsible forest management embraces the concept of species diversity as a hedge against future changes in species' relative values and species- or genera-specific pathogens. For example, chestnut blight fungus, accidentally introduced into North America in the early twentieth century, effectively eliminated American chestnut (*Castanea dentata*) from eastern forests and urban settings in just a few decades (Anagnostakis 1987). Today in the eastern USA, up to ten nonnative insects or diseases threaten the forests of the eastern USA (Orwig 2002) and forest management strategies play an important role in minimizing their impacts (Waring and O'Hara 2005). Avoiding monocultures or forest stands that are dominated by one species reduces the risk of catastrophic mortality. Moreover, when management practices also reduce diversity, current forests are less reflective of historic (Schuler and Gillespie 2000) and prehistoric (Delcourt and Delcourt 1987) conditions and are less able to provide a full suite of forest products, wildlife needs, and ecological services.

The extent to which uneven-aged management (or no management) has contributed to the decline of many tree species suggests that the conditions created by single-tree canopy openings do not reflect conditions that permitted many species to persist in the past. Recent work suggests that oaks and other species had multiple canopy accession pathways before the twentieth century, including both large and small canopy openings (Rentch et al. 2003). It is therefore likely that the decline of oaks and sympatric species, which is associated with partial harvesting and uneven-age management, results from multiple factors. In the central Appalachians and elsewhere, increased deer abundance, loss of fire as a periodic understory disturbance, and changes in understory composition related to invasive species (both native and exotic) contribute to oak decline (Abrams 1992). In 1973, Fernow scientist George Trimble wrote that "deer browsing on the Fernow or in most of West Virginia is not a serious deterrent to obtaining satisfactory oak reproduction" and estimated the deer herd at six deer per square kilometer (Trimble 1973). But today, deer populations have increased dramatically and can be a limiting factor to obtaining adequate forest regeneration, regardless of the type of forest management used (Horsley et al. 2003). Foresters have used deer exclusion fencing, prescribed burning, and herbicides to overcome a variety of limitations to achieving adequate regeneration, but generally these techniques were associated with even-age management in the last century (Brose et al. 2008). The utility of these and other forest management practices in conjunction with uneven-age management has not yet been fully considered. New studies are underway at the Fernow to test how prescribed fire and deer influence regeneration following partial harvesting.

It should further be noted that uneven-age management is not always based on harvesting single trees and making small gaps in the forest overstory. Much larger gaps of one to several hectares could be included in one type of uneven-age management known as group selection. Group selection has been evaluated on the Fernow, and the LAMP study does include harvesting in 0.16-ha patches. In some cases, these patch openings have been satisfactory for regeneration of shade-intolerant species, but the oaks and hickories are not usually successful in such openings (Miller and Schuler 1995).

One of the unexpected benefits of uneven-age management and other long-term partial harvesting practices on the Fernow has been the apparent creation of a suitable habitat for the federally endangered running buffalo clover (*Trifolium stoloniferum*). Running buffalo clover prefers sunlight from small canopy gaps and canopy densities associated with various forms of partial harvesting (Madarish and Schuler 2002). This clover was once thought to be close to extinction and was listed as federally endangered in 1987. In 1993, it was discovered on the Fernow Experimental Forest, where it is now recognized as one of the largest occurrences of running buffalo clover known to exist. Nearly all of the running buffalo clover on the Fernow occurs in locations used for uneven-age management research and demonstration. Research continues to better understand why the conditions created by uneven-age management are so conducive to running buffalo clover persistence and how to fully recover this species (Burkhart 2010). It has also been shown that where running buffalo clover thrives, there is a much more diverse herbaceous community

(Burkhart 2010). Thus, while woody species diversity is declining in conjunction with uneven-age management, there appears to be a guild of herbaceous species that benefit from the repeated moderate level of disturbance associated with this type of harvesting and management.

7.3.2 Productivity

Surprisingly, unlike the loss of woody species diversity following decades of uneven-age management on the Fernow, productivity has met or exceeded levels expected for managed stands in all three studies and the importance of the effect of stocking on productivity has been demonstrated. After a half century of research in the LAMP study, growth in cubic wood volume as measured by mean periodic annual increment (PAI) ranged from 4.0 to 4.6 m^3/ha/year for all managed stands, but averaged just 2.5 m^3/ha/year for the unmanaged reference areas (Schuler et al. 2006). The documented growth rates illustrate that when managed properly, forest stands such as these are commercially important and are capable of growing high-quality wood products for a sustained period of time, as well as providing important ecological services.

Fernow scientist Trimble (1965) predicted declines in productivity as shade-tolerant species increased, but the effect of the documented changes in species composition appears to be less important than that of residual stocking. For example, despite increased maple dominance, productivity actually increased in some instances. Mean PAI for the diameter-limit-treated areas (SI = 70 and 80 combined) in the LAMP study increased from 4.3 to 4.8 m^3/ha/year from the first to last cutting cycles despite the loss of most shade-intolerant species. With respect to compartments managed with single-tree selection, declining trends in productivity are suggested graphically, but not statistically (Fig. 7.3; Schuler et al. 2006). Aside from environmental factors associated with site conditions, stocking is the most important factor associated with productivity. Uneven-age management conducted properly strives to maintain stocking at levels that will allow forest growth to remain vigorous for long periods of time.

The general relationship between growth and stocking is robust and an inherent part of many forest management guidelines for both commodity and non-commodity resources. In general, productivity is optimized at a stocking level that fully occupies the growth potential of a site and then declines as increased crowding reduces growth efficiency. This is the case as an even-aged stand develops through time and is clearly illustrated by the commercial clearcut and the unmanaged reference treatment areas in the CPL study (Fig. 7.4). In other words, growth continues, but the rate of growth declines as the stand gets more crowded. In contrast, with uneven-age management, residual stocking control is an inherent part of the process. By maintaining stocking in a more optimal range, productivity can theoretically be maintained at higher levels, such as the case with the single-tree selection and diameter-limit treatment areas in the CPL study (Fig. 7.4).

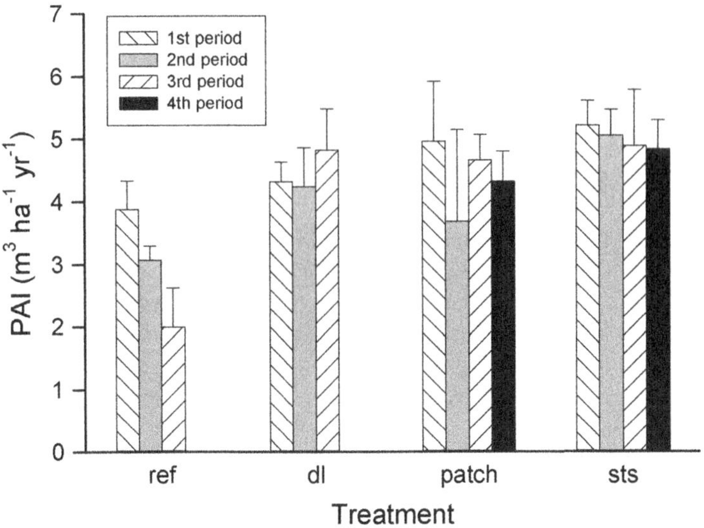

Fig. 7.3 Mean cubic volume net periodic annual increment (PAI) of merchantable trees (dbh ≥ 12.7 cm, 5.0 in.; vertical lines = 1 standard error) by measurement cycle and treatment category for good and excellent sites in the LAMP study (*ref* unmanaged reference compartments; *sts* single-tree selection; *dl* diameter-limit; and *patch* patch cutting)

Fig. 7.4 Cubic volume net periodic annual increment (PAI) of merchantable saw-log-size trees (dbh ≥ 27.9 cm, 11.0 in.) by measurement cycle and treatment in the CPL study (*ref* unmanaged reference compartment; *sts* single-tree selection (8C only); *cc* commercial clearcut; and *dl* diameter-limit. Note that cubic volumes here were converted directly from board feet, thus providing a conservative estimate of productivity

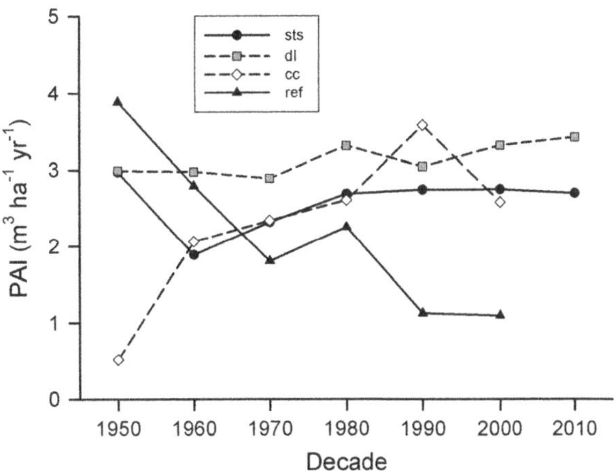

In the FMDL study, one of the opportunities was to evaluate the three rate-of-return levels to better understand their effects on the stocking and productivity relationship. After four periodic harvests, real distinctions in productivity have emerged, especially between the 3 and 6% rate-of-return treatments. The 6%

treatment yielded more than 29 m^3/ha/decade of cubic wood volume[1] in the second and third harvests, and more than 23 m^3/ha in the last decade (Schuler and McGill 2007). In contrast, the 3 % treatment is averaging a little more than 17 m^3/ha/decade and declined below that level in the last harvesting cycle. The 6 % treatment had the lowest RBA among treatment and time period interactions and was even below the recommended residual basal range of 13 m^2/ha (Trimble et al. 1974).

Some unexpectedly high growth rates associated with low RBAs were also documented in the CPL study. The diameter-limit treatment in this study resulted in a mean postharvest basal area after four cutting cycles of about 9 m^2/ha (dbh $\geq$ 12.7 cm). Yet the PAI resulted in the highest level of sustained productivity among all of the treatments (Fig. 7.4) and was well above the expected growth rate.

The unexpected level of productivity associated with low levels of residual stocking in all three long-term studies suggests our understanding of the stocking and productivity relationship is incomplete in the central Appalachians. Now that the carbon sequestered in forests is recognized as an important contributor to the global terrestrial carbon sink, it is more important to fully understand how growth rates are influenced by silvicultural prescriptions and management choices. Recent work on the Fernow using long-term measurements of water, carbon, and nitrogen in four different watersheds with different harvest histories has shown that forests managed with single-tree selection or diameter-limit harvesting stored about 37 % more carbon than did forests that were not harvested during the past half century (Davis et al. 2009). These findings are consistent with the higher growth rates documented in all three long-term studies on the Fernow where some form of partial cutting or uneven-age management has been employed. These findings have potentially important implications for managing greenhouse gases in the twenty-first century. The long-term measurements and forestry manipulations on the Fernow allow scientists to ask questions never anticipated 50 years ago, such as calibrating models of ecosystem productivity and understanding the state of carbon in forests managed using different scenarios. The partial cutting approaches on the Fernow may enhance the utility of forests as carbon sinks—at least for now. Understanding how forests and species respond to predicted changes in climate (Iverson and Prasad 1998) will be vitally important in the twenty-first century, and knowing how forests functioned prior to that period will be the benchmark for comparison.

7.3.3 Economic Considerations

Diameter-limit harvesting is by far the most common type of harvesting in eastern hardwoods (Fajvan et al. 1998). However, the repeated removal of commercial products with no investment in residual stand quality will lower residual stand values. Without regeneration of a commercial tree species, the residual for-

[1] Here, volume estimates were based on board feet which incorporate a loss in total volume due to sawing requirements. Cubic volume estimates are direct English to metric conversions (Miyata et al. 1981) without building the loss back into the estimate.

est stand may eventually have little to no commercial value. After six decades, the CPL study provides an excellent case study to examine this subject. Net present value (NPV) for each payment for each treatment was calculated using an internal rate of return (IRR) of 4% and a market rate of return (MRR) equal to the mean inflation rate for the period of interest plus the IRR. Residual stand values in 2008 also were calculated based on local 2008 timber sale bid rates for each species.

As expected, after six decades and four harvests, the diameter-limit residual stand value was the least among all of the treatments, but its NPV based on the MRR of all past timber-based revenues was the greatest (Table 7.2). The residual stand value of the unharvested stand (8E) was about US$ 19,760/ha and greater than any of the harvesting treatments, but its present value including the value of past payments was the lowest. All forms of management that included some form of harvesting resulted in a present value that was about three to five times greater than that of the unharvested stand. Accordingly, it is easy to understand why private forest landowners have been motivated to manage their forestland for periodic revenue. It appears that substantial revenue can be sustained about once per decade on managed land on excellent growing sites. Most recently, payments declined (Table 7.2), but this decline was due to a decrease in timber values caused by a historic downturn in the economy and not due to a decrease in wood quantity or quality.

It is also noteworthy that the residual stand values in 2008 for the single-tree selection compartments (8C and 8D) were about US$ 9,880/ha, or about half of the value of the unharvested compartment (8E), even though there had been at least seven previous harvests in these two compartments. Maximizing NPV is only one aspect to consider when managing forests and the value of past payments has little to no relevance to future revenue potential. Few forests are managed to optimize NPV, but our calculations and recorded payments illustrate the allure of partial cutting, especially diameter-limit cutting. Modifying partial harvests so that they can produce periodic income and maintain diversity and structural goals remains an important aspect of modern forest management research. Any management that favors the preponderance of one species over a diverse mixture of species provides less insurance against changes in consumer preferences, which could influence relative species values in the future. Moreover, some climate change scenarios predict sugar maple will become less suited to the region as temperatures warm (Iverson and Prasad 1998), so avoiding sugar maple dominance seems wise under those scenarios. However, when a desirable tolerant species can be regenerated at each periodic harvest, uneven-age management can be a realistic management alternative that can provide an acceptable rate of return (Miller 1993), while providing the benefits of continuous forest cover (Fig. 7.5).

7.4 Summary

Interest in uneven-age management on the Fernow Experimental Forest was in part a reaction to the forest exploitation era that occurred in the central Appalachian Mountains from about 1880 to 1920. After World War II, a more sustain-

Table 7.2 Fernow Experimental Forest Cutting Practice Level harvest payments and present values as of 2008 (USD ha^{-1}) using a risk-free rate of return (4%) and a market rate of return adjusted for inflation in the USA

ID	Year	Payment	Present value (4%)	Present value[a] (market rate)
8A	1949	694	7,022	57,077
	Residual timber in 2008		*14,600*	*14,600*
	Total		21,622	71,677
8B	1949	506	5,113	41,568
	1968	1,623	7,795	45,841
	1988	2,729	5,977	10,678
	2008	6,375	*6,373*	*6,375*
	Total payments		25,258	104,462
	Residual timber in 2008		*4,453*	*4,453*
	Total		29,711	108,915
8C	1949	321	3,243	26,370
	1958	128	914	6,375
	1968	277	1,334	7,842
	1978	889	2,885	9,537
	1988	946	2,075	3,705
	1998	4,881	7,225	9,248
	2008	2,505	*2,505*	*2,505*
	Total payments		20,181	65,582
	Residual timber in 2008		*9,166*	*9,166*
	Total		29,347	74,748
8D	1949	282	2,858	23,233
	1958	121	852	5,945
	1963	59	351	2,275
	1968	193	924	5,429
	1978	1,633	5,293	17,505
	1988	1,919	4,206	7,511
	1998	4,723	6,993	8,949
	2008	2,354	*2,354*	*2,354*
	Subtotal		23,831	73,201
	Residual timber in 2008		*9,934*	*9,934*
	Total		33,765	83,134
8E	2008		19,775	19,775

Compartment 8A was commercially clearcut in 1949; 8B was harvested using a diameter-limit (39.4 cm dbh) with a 20-year cutting cycle; 8C (includes trees 27.9 cm dbh and larger) and 8D (includes trees 12.7 cm dbh and larger) were harvested using single-tree selection once per decade; and 8E serves as an uncut reference stand of similar disturbance history prior to 1949

[a] Present values using market rates were computed using periodic rates of inflation. Annual inflation rates in the USA from 1949 through 2007 ranged from −0.95 to 13.58% and averaged 4.41%

able type of forest management was desired and interest in uneven-age management grew as a result. The Fernow became one of several Forest Service Experimental Forests in the eastern USA where this interest led to the establishment of long-term studies to better understand the consequences of a wide range of silvicultural options.

Fig. 7.5 A 60-year photo series from the CPL study (compartment 8C, camera point 4) managed with single-tree selection from 1948 to the present using 10-year cutting cycles. The sequence of images illustrates the decay of large coarse woody debris, the removal of trees through harvesting, and the growth of residual trees. Note the arrow pointing to the same tree in each image (a through d, 1948, 1958, 1979, 2008, respectively)

After more than a half century of uneven-age management research, much has been learned. The results illustrate that the mixed-mesophytic forest type, with its shade-tolerant commercial species, is capable of maintaining acceptable levels of productivity and stem quality, although concomitant reductions in woody species diversity are a byproduct and a serious management concern. Uneven-age management is shaping species composition in a way that seems to accelerate ongoing successional trends, and these trends may be unprecedented and counterproductive to some long-term management objectives. We have also demonstrated that simple diameter-limit harvesting may seem more profitable, at least for a few decades, but leaves the forest with less potential for future revenue. In so doing, however, we have also learned that lower levels of residual stocking, associated with more exploitive types of periodic partial harvesting, are not necessarily counterproductive to optimizing growth and carbon sequestration. Further research is needed to better understand the relationships among stocking, growth, and species recruitment, especially in the face of predicted climate change scenarios. Along with other Forest Service research units, we have learned that the original value-laden treatment names (e.g., excellent, good, fair, or poor cutting practice) associated with some early uneven-age management research were based on faulty assumptions, illustrating the need to test commonly held theories in forest stand dynamics. Both even-age and uneven-age management techniques can be used appropriately, depending on the circumstances, but the ecological amplitude of individual species must be recognized. For example, when individual tree canopy gaps constitute the largest planned opening in a forest, such as with single-tree selection, it is unreasonable to expect shade-intolerant species to be sustained.

Uneven-age management research is a long-term endeavor because the forest is changed gradually through time, at each harvest, and by unplanned perturbations. Determinations of sustainability are related to the simultaneous processes of growth, harvest and death, and regeneration, as well as social acceptance and economic viability. When the early researchers Sydney Weitzman and George Trimble Jr. planned the uneven-age management research on the Fernow, the objectives clearly involved the sustained yield of commercial forest products. They could not foresee the many changes ahead, such as the clearcutting controversy that came to a head in the 1970s; new federal laws pertaining to the protection of rare species, environmental assessments, and public involvement in the forest planning process; radically increased deer abundance affecting forest regeneration; exotic pests and invasive species displacing native ones; and the emerging issues of climate change and the related role of forests as carbon sinks or sources. Testing, developing, and refining uneven-age, even-age, two-age, and new forest management techniques remain a critical need and a vital part of the long-term mission of the Fernow Experimental Forest. Yet opportunities for exploring unanticipated issues will continue to be an important dimension of the research at the Fernow. Lindenmayer et al. (2010) refer to the discovery of unanticipated findings as "ecological surprises" and assert that such discoveries are more likely detected as part of long-term research because time is a major driver of change. In the absence of long-term forest research, it is difficult to identify changes in populations, ecosystems, or ecological processes because the reference conditions or processes are not well documented. We will continue to identify new forest research objectives as the world changes and new issues emerge. It is a virtual certainty, however, that healthy and productive forests will remain an essential societal objective and be part of a sustainable twenty-first century for everyone.

References

Abrams MD (1992) Fire and the development of oak forests. Bioscience 42(5):346–353

Anagnostakis SL (1987) Chestnut blight: the classical problem of an introduced pathogen. Mycologia 79(1):23–37

Braun EL (1950) Deciduous forests of eastern North America. Blakiston, Philadelphia

Brooks AB (1911) West Virginia geological survey, vol 5: Forestry and wood industries. Acme Publishing, Morgantown

Brose P, Schuler T, Van Lear D, Berst J (2001) Bringing fire back: the changing regimes of the Appalachian mixed oak forests. J For 99:30–35

Brose PH, Gottschalk KW, Horsley SB, Knopp PD, Kochenderfer JN, McGuinness BJ, Miller GW, Ristau TE, Stoleson SH, Stout SL (2008) Prescribing regeneration treatments for mixed-oak forests in the Mid-Atlantic region. General Technical Report NRS-33, USDA Forest Service, Northern Research Station, Newtown Square, PA

Burkhart JO (2010) Ecology of Trifolium stoloniferum (Muhl. Ex Eaton), a federally endangered vascular plant, at the Fernow Experimental Forest in West Virginia. M.S. thesis, West Virginia University

Clarkson RB (1964) Tumult on the mountains: lumbering in West Virginia 1770–1920. McClain, Parsons

Cravens JH (1975) The Monongahela decision. Unpublished document, on file at the Timber and Watershed Laboratory, Parsons

Davis SC, Hessl AE, Scott CJ, Adams MB, Thomas RB (2009) Forest carbon sequestration changes in response to timber harvests. For Ecol Manage 258:2101–2109

Delcourt PA, Delcourt HR (1987) Long-term forest dynamics of the temperate zone. Springer-Verlag, New York

de Liocourt F (1898) De l'amenagement des sapinieres (on the amelioration of fir forests) Société forestière di Franche-Comté et Belfort. Bulletin 6:396–409

Fajvan MA, Grushecky ST, Hassler CC (1998) The effects of harvesting practices on West Virginia's wood supply. J For 96:33–39

Fansler HF (1962) History of Tucker County West Virginia. McClain, Parsons

Gilliam FS, Turrill NL, Adams MB (1995) Herbaceous-layer and overstory species in clearcut and mature central Appalachian hardwood forests. Ecol Appl 5(4):947–955

Graves HS (1910) The selection system. Soc Am For Proc 5:1–17

Haines RW (1976) Monongahela and the National Forest Management Act of 1976. Environ Law 7:345

Harper VL, Rettie JC (1946) The management status of forest lands in the United States. USDA Forest Service, Washington, DC (Report 3)

Horsley SB, Stout SL, DeCalesta DS (2003) White-tailed deer impact on the vegetation dynamics of a northern hardwood forest. Ecol Appl 13(1):98–118

Iverson LR, Prasad AM (1998) Predicting abundance of 80 tree species following climate change in the eastern United States. Ecol Monogr 68:465–485

Kenefic LS, Schuler TM (2008) Regional synthesis of long-term findings from U.S. Forest Service management intensity demonstrations. In: Poster abstracts of the central hardwood forest conference; 2008 April 8–9; Lafayette, IN. Purdue University, West Lafayette, IN. FNR-402-W. Purdue Extension Publication: 9 (abstract)

Lamson NI, Smith HC (1991) Stand development and yields of Appalachian hardwood stands managed with single-tree selection for at least 30 years. Research Paper NE-655, USDA Forest Service, Northeastern Forest Experiment Station, Radnor, PA

Larabee PA (1986) Late-quaternary vegetational and geomorphic history of the Allegheny Plateau at Big Run Bog, Tucker County, West Virginia. MS thesis, University of Tennessee, Knoxville

Leathers DJ, Grundstein AJ, Andrew WE (2000) Growing season moisture deficits across the northeastern United States. Clim Res 14:43–55

Lindenmayer DB, Likens GE, Krebs CJ, Hobbs RJ (2010) Improved probability of detection of ecological "surprises". Proc Natl Acad Sci U S A 107:21957–21962

Madarish D, Schuler TM (2002) Effects of forest management practices on the federally endangered running buffalo clover (*Trifolium stoloniferum* Muhl. ex. A. Eaton). Nat Area J 22(2):120–128

Madarish DM, Rodrigue JL, Adams MB (2002) Vascular flora and macroscopic fauna on the Fernow Experimental Forest. General Technical Report NE-291, USDA Forest Service, Northeastern Research Station, Newtown Square, PA

Meyer HA (1943) Management without rotation. J For 41:126–132

Meyer HA (1952) Structure, growth, and drain in balanced uneven-aged forests. J For 50:85–92

Miyata ES, Steinhilb HM, Coyer LA (1981) Metric conversions for foresters. U.S. Department of Agriculture, Forest Service. North Central Forest Experiment Station, St. Paul, 7 p

Miller GW (1993) Financial aspects of partial cutting practices in central Appalachian hardwoods. Research Paper NE-673, USDA Forest Service, Northeastern Forest Experiment Station, Radnor, PA

Miller GW, Schuler TM (1995) Development and quality of reproduction in two-age central Appalachian hardwoods. In: Gottschalk KW, Fosbroke SLC (eds) Proceedings-10th central hardwood conference; 1995 March 5–8; Morgantown, WV. General Technical Report NE-197, USDA Forest Service, Northeastern Forest Experiment Station, Radnor, PA, pp 364–375

Nowacki GJ, Abrams MD (2008) The demise of fire and "mesophication" of forests in the Eastern United States. Bioscience 58:123–138

Nowacki G, Ablutz M, Yaussy D, Schuler T, Dey D (2009) Restoring oak ecosystems on National Forest system lands: an adaptive management approach. In: Hutchinson T (ed) Proceedings of the 3rd fire in eastern oak forests conference. General Technical Report NRS-P-46, USDA Forest Service, Northern Research Station, Newtown Square, PA, pp 133–139

Orwig DA (2002) Ecosystem to regional impacts of introduced pests and pathogens: historical context, questions and issues. J Biogeogr 29:1471–1474

Pan C, Tajchman SJ, Kochenderfer JN (1997) Dendroclimatological analysis of major forest species of the central Appalachians. For Ecol Manage 98:77–87

Rentch JS, Fajvan MA, Hicks RR (2003) Oak establishment and canopy accession strategies in five old-growth stands in the central hardwood forest region. For Ecol Manage 184:285–297

Roach BA, Gingrich SF (1968) Even-aged silviculture for upland central hardwoods. Agriculture Handbook 355. USDA Forest Service, Northeastern Forest Experiment Station, Washington, DC

Robbins WG (1982) Lumberjacks and legislators: political economy of the US lumber industry, 1890–1941. Texas A & M University Press, College Station

Rodgers AD (1991) Bernhard Eduard Fernow: a story of North American forestry. Forest History Society, Durham

Schuler TM (2004) Fifty years of partial harvesting in a mixed mesophytic forest: composition and productivity. Can J For Res 34:985–997

Schuler TM, Gillespie AR (2000) Temporal patterns of woody species diversity in a central Appalachian forest from 1856 to 1997. J Torrey Bot Soc 127:149–161

Schuler TM, McGill DW (2007) Long-term assessment of financial maturity, diameter-limit selection in the central Appalachians. Research Paper NRS-2, USDA Forest Service, Northern Research Station, Newtown Square, PA

Schuler TM, Ford WM, Adams MB, Kochenderfer JN, Edwards PJ (2006) Large area comparisons of forest management practices in West Virginia (1951-present). In: Irland LC, Camp AE, Brissette JC, Donohew ZR (eds) Long-term silvicultural and ecological studies: results for management and science. GISF Research Paper 005, Yale University, Global Institute of Sustainable Forestry, New Haven, CT, pp 94–103

Thomas-Van Gundy M, Schuler TM (2008) Deferred rotation harvests in central Appalachia: 20- and 25-year results. In: Jacobs DF, Michler CH (eds) Proceedings, 16th central hardwood forest conference; 2008 April 8–9; Lafayette, IN. General Technical Report NRS-P-24, USDA Forest Service, Northern Research Station, Newtown Square, PA, pp 423–435

Trimble GR Jr (1965) Species composition changes under individual tree selection cutting in cove hardwoods. Research Note NE-30, USDA Forest Service, Northeastern Forest Experiment Station, Upper Darby, PA

Trimble GR Jr (1973) The regeneration of central Appalachian hardwoods with emphasis on the effects of site quality and harvesting practice. Research Paper NE-282, USDA Forest Service, Northeastern Forest Experiment Station, Upper Darby, PA

Trimble GR Jr (1977) A history of the Fernow Experimental Forest and the Parsons Timber and Watershed Laboratory. General Technical Report NE-28, USDA Forest Service, Northeastern Forest Experiment Station, Upper Darby, PA

Trimble GR Jr, Mendel JJ, Kennell RA (1974) A procedure for selection marking hardwoods combining silvicultural considerations with economic guidelines. Research Paper NE-292, USDA Forest Service, Northeastern Forest Experiment Station, Upper Darby, PA

USDA Forest Service (2006) Monongahela National Forest land and resource management plan. USDA Forest Service, Eastern Region, Milwaukee

Waring KM, O'Hara KL (2005) Silvicultural strategies in forest ecosystems affected by introduced pests. For Ecol Manage 209:27–41
Weitzman S (1949) The Fernow Experimental Forest. Station Paper NE-61, USDA Forest Service, Northeastern Forest Experiment Station, Upper Darby, PA
West Virginia Division of the Izaak Walton League of America, Inc. v. Butz (1973) 367 F. Supp. 422- Dist. Court, ND West Virginia
Whitney GG (1994) From coastal wilderness to fruited plain: a history of environmental change in temperate North America, 1500 to the present. Cambridge University Press, Cambridge

Chapter 8
The Monongahela Clearcutting Controversy: Scientists and Land Managers Develop an Alternative Practice on the Fernow Experimental Forest

Gary W. Miller

Abstract The clearcutting controversy that erupted on the Monongahela National Forest (MNF) in West Virginia in the mid-1970s is best known for its public outcry and resulting litigation. However, there is more to this story than vocal opponents and lawsuits. This chapter reveals firsthand accounts of the partnerships between land managers on the MNF and USDA Forest Service scientists working on the Fernow Experimental Forest. It was this cooperation, formed over decades, which provided a scientific basis for responding to the controversy and led to a viable alternative to clearcutting. This chapter briefly details the history of the MNF and describes the reasoning behind using even-aged forest management in the 1960s. A review of the controversy illustrates how the concerns of vocal opponents and the ensuing lawsuits led to a more integrated approach to land management on the MNF as well as all national forests in the 1970s. Cooperative research between MNF land managers and Fernow scientists resulted in the expanded use of two-aged forest management in the 1990s. Finally, this chapter includes recommendations for building and maintaining similar partnerships.

Keywords Clearcutting controversy · Collaboration · Two-age management · Deferment cutting · Monongahela · Hardwood silviculture · West Virginia

8.1 Introduction

The clearcutting controversy on the Monongahela National Forest (MNF) in the mid-1970s illustrated how one relatively small, productive experimental forest can stand ready to offer solutions to an unforeseen and potentially explosive forest management challenge. Initially, hunters opposed the impact of clearcutting on small

---

G. W. Miller (✉)
Northern Research Station, USDA Forest Service,
180 Canfield Street, Morgantown, WV 26505, USA
Phone: 304-285-1521
e-mail: gwmiller@fs.fed.us

D. C. Hayes et al. (eds.), *USDA Forest Service Experimental Forests and Ranges*,
DOI 10.1007/978-1-4614-1818-4_8, © Springer New York 2014

game and turkey habitat, particularly when the harvests occurred on a favorite local hunting site. As the number and distribution of clearcut harvests increased across the MNF landscape in the late 1960s, the volume and number of opponents' voices also increased. In the years before and since the litigation proceedings that occurred from 1973 to 1975, critics also questioned the effect of clearcutting on soils, stream quality, and the appearance of forested landscapes.

Vegetation management research conducted on the Fernow Experimental Forest near Parsons, WV, provided a scientific basis for alternative timber harvesting methods on the MNF after the clearcutting controversy erupted. In particular, on-the-ground studies of a technique called deferment cutting by Fernow scientists provided a way for land managers to retain selected overstory trees, obtain desired natural tree regeneration after a harvest operation, and mitigate many of the negative impacts of clearcut harvests (Smith et al. 1989). Although many of the concerns about clearcutting have turned out to be unfounded, scientists continue to assess the long-term effects of both clearcutting and deferment cutting on songbird habitat, invertebrates, and mammals as the natural regeneration on harvested sites develops into a mature forest again.

This chapter describes how three factors—the close proximity of a laboratory and a national forest, well-established relationships between scientists and managers, and their mutual desire for science-based stewardship of the land—facilitated the development of a workable response to an urgent problem. Following a review of the underlying research questions brought to light by the clearcutting controversy, this chapter summarizes research results from previously published reports on deferment cutting and the resulting two-aged stand structures. Finally, this chapter describes current practices on the MNF to highlight the ongoing partnership between scientists and land managers, their innovative application of science-based tools, and their efforts to understand and sustain the Appalachian forest.

A brief history of the MNF and a review of the clearcutting controversy help explain how scientists on the Fernow and land managers on the MNF came together to find a workable solution to the problem. Historical information about timber management on the MNF and the resulting clearcutting controversy is available from numerous sources. Forest Service Chief John McGuire authorized an in-service analysis of events surrounding the Monongahela controversy to serve as a learning tool to avoid similar problems in the future (Weitzman 1977). Highlights were also reported for many years after the controversy in commentaries and scientific papers in peer-reviewed journals (Popovich 1976; Fairfax and Achterman 1977). These sources presented the viewpoints of both the plaintiffs and the defendants involved in the underlying lawsuit, as well as the resulting legislation that sought to resolve the major issues in the case (LeMaster and Popovich 1976).

Another compelling part of this story was gathered from retired Forest Service employees who worked on the MNF and the Fernow during those somewhat tense days of the clearcutting controversy. In February 2009, Clay Smith (now deceased) and James Kochenderfer (both retired scientists from the Fernow) met for a reunion of sorts with Harry Mahoney and Joe Tekel (retired district rangers in the MNF) to reflect on their experiences during the 1960s and 1970s when the call for alternative harvesting practices arose. Glen Juergens (now retired) and George Hudak, long-time employees on the MNF, also attended that meeting and recalled the resulting changes in management philosophy on the MNF from the early 1980s to the present.

I had the honor of presiding over their meeting to gather information for this chapter. The retirees were very energetic as they helped each other remember the days before, during, and after the clearcutting controversy. They explained the early management philosophy on the MNF and the chronology of events that led to the application of clearcutting in the first place. They also provided insight into the mood of the local public and the response of employees to the lawsuit and court decisions. Most importantly, they recalled with fondness their years of partnership in testing new harvest practices and developing two-aged forest management as a possible alternative to clearcutting. The 5-h meeting passed quickly, and it concluded with a discussion of how the clearcutting controversy had influenced plans for managing vegetation on the MNF in the twenty-first century.

8.2 Origins of the MNF

The MNF was established in 1920, but some of the land within its current boundaries was purchased by the US government much earlier. The first acquisition was the Arnold tract, comprising 2,888 ha formerly covered with mixed hardwoods and spruce stands growing on a variety of elevations, aspects, and soil conditions in Tucker County, WV (Weitzman 1949; Trimble 1977). The forest vegetation was typical of the mixed mesophytic hardwood region described by Lucy Braun (1950). The entire parcel had been logged between 1903 and 1911, and the cutover land was purchased for US$ 13.60 per ha by the Forest Service in 1915. In 1934, the Elklick Run watershed, a north-facing drainage covering 1,473 ha of the original parcel, was designated as an outdoor laboratory now known as the Fernow Experimental Forest.

The first scientific paper published from the Fernow resulted from data collected by young members of the Civilian Conservation Corps (CCC) stationed in nearby Parsons, WV. The report described fungal infections on yellow-poplar (*Liriodendron tulipifera* L.) and American basswood (*Tilia Americana* L.) whose branches had been damaged by snow in 1935 (Roth 1941). Early research on the Fernow focused on water resources and the impact of forest management on the growth and regeneration of trees and other native vegetation (Weitzman 1949).

Throughout eastern West Virginia in the 1920s and 1930s, timber companies and individuals readily sold their cutover tracts to the Forest Service due, in part, to depressed economic conditions and the absence of merchantable timber that promised any timber income in the near future. These cutover tracts were occupied by small residual trees, new natural regeneration stimulated by the logging disturbance, and scattered large, unmerchantable trees left by the early loggers. By 1940, the MNF comprised almost 324,000 ha.

Most of the disturbances that shaped the present species composition and structure of the MNF occurred when the land was still in private ownership, decades before the Forest Service began active management in the 1940s and 1950s. The earliest disturbances took the form of agriculture and fires applied by Native Americans (Brose et al. 2001), and later farming by European settlers (Carvell 1986). Tree harvesting by settlers in the mid-1800s focused on removing certain tree species for well-defined uses—white oak (*Quercus alba* L.) for cooperage, white pine (*Pinus strobus* L.) and red spruce (*Picea rubens* L.) for construction, and black walnut (*Juglans nigra* L.) and black cherry (*Prunus serotina* Ehrh.) for fine furniture.

In the late 1800s, railroads provided easier access to remote parts of the forest, new sawing technology increased productivity, and the demand for most hardwood species increased (Clarkson 1964). The advent of these factors led to broad-scale harvesting throughout the region in the early 1900s, often followed by accidental or incendiary forest fires. This disturbance pattern promoted a new forest dominated by shade-intolerant species that arose from stored seeds and sprouts from the previous forest.

Although most of the tree cover on the MNF dated back to the early 1900s, both older and younger age classes were represented. Some older stands of trees dated back to the 1860s when farms were abandoned after the Civil War. Unused fields and pastures reverted to tree cover as they were seeded by wind, birds, and rodents from surrounding forests (Carvell 1986). In much of Appalachia, the American chestnut (*Castanea dentata* (Marsh.) Borkh.) occupied 25% of the forest when it was ravaged by chestnut blight in the 1920s and 1930s (Brooks 1910). The loss of the American chestnut also stimulated natural regeneration, thus creating another age class of trees nearly two decades after the broad-scale logging.

By the time active forest management began on the MNF, the forest contained a wide range of tree species and several age classes. The most recognizable age classes included old residual trees left by the early loggers, mostly shade-intolerant species that regenerated after the broad-scale logging, and slightly younger patches of trees that filled voids formerly occupied by American chestnuts.

Early management activities on the MNF focused on fire protection, road construction related to fire control, and timber salvage operations to remove old residuals from the broad-scale railroad logging era and following the death of the American chestnut (Weitzman 1977). Both CCC and Works Progress Administration employees contributed to the road building effort on the MNF in the 1930s. The Forest Service employed local fire wardens and crews to suppress forest fires and patrol local forests, thus establishing personal relationships with the public living within the forest's boundaries. The Forest Service also hired local citizens to help with timber stand improvement projects, deaden old residual cull trees, and plant trees in reforestation projects.

During this early management period, the timber stands on the MNF were relatively young; most of the trees were less than 40 years old. Few stands had merchantable volumes that could support a timber sale because the preferred oaks (*Quercus* spp.), black cherry, yellow-poplar, and other commercial species that regenerated naturally after the early harvests had not yet accumulated sufficient size or grade to attract timber buyers.

During the 1940s, particularly after World War II, the demand for timber products increased nationwide, and timber harvesting increased accordingly. From 1940 to 1960, annual timber sales on the MNF increased from 31,320 to 76,560 m^3. Nearly all harvesting during that period took the form of a selection harvest or a partial overstory removal (Marquis 1994). Following the lead of their European counterparts, foresters in America applied periodic partial harvests with the expectation of a reliable, sustained yield of wood products. These harvests removed only the scattered old trees left after the broad-scale logging era along with some of the larger trees from the new, second-growth forest. The younger, smaller trees had no stumpage value and they were retained to grow for future partial harvests.

Forest managers had hoped to sustain the rich species diversity characteristic of the second-growth forest, but the residual overstory left in place after selection harvests usually prevented successful regeneration of the desirable shade-intolerant species like black cherry, yellow-poplar, and northern red oak (*Quercus rubra* L.; Trimble 1973; Schuler 2004). Instead, the partial harvests promoted the regeneration of shade-tolerant species like red maple (*Acer rubrum* L.), sugar maple (*Acer saccharum* Marsh.), black gum (*Nyssa sylvatica* Marsh.), and American beech (*Fagus grandifolia* Ehrh.) as well as a reduction in species diversity.

By the early 1960s, it became clear that widespread application of selection harvests or other partial harvest practices would not maintain the desired species diversity on the MNF. Decades of experiments conducted by the Forest Service and universities in the eastern hardwood region had shown that partial overstory harvests limited the regeneration of desirable shade-intolerant and mid-tolerant species (Korstian 1927; Averell 1929; Little 1938; Kuenzel and McGuire 1942; Liming and Johnston 1944; Sander and Williamson 1957; Weitzman and Trimble 1957; Bey 1964; Roach and Gingrich 1968). Shade from the residual overstory acted as an environmental filter (Oliver and Larson 1996), thus favoring the survival and dominance of mostly shade-tolerant species (Trimble 1973).

8.3 The Clearcutting Controversy

In 1962, Forest Service scientists at the Vinton Furnace Experimental Forest in Ohio prepared the *Timber Management Guide for Upland Central Hardwoods*. It recommended using even-aged management to regenerate and sustain a mix of species in eastern hardwoods (Roach and Gingrich 1962, 1968). As a result, the Forest Service adopted a new policy for managing eastern hardwoods, and even-aged management became the primary system for managing commercial forestland. In fact, foresters from national forests in the eastern region were required to visit Vinton Furnace research sites and learn about even-aged management prescriptions for managing hardwoods (Weitzman 1977).

These sites demonstrated that complete overstory removal (clearcutting) provided the sunlight and seedbed conditions that allowed all tree species to regenerate successfully. The Forest Service manual still allowed for partial-harvest management systems in locations where recreation or other uses required a continuous

Fig. 8.1 Two clearcuts in 60-year-old central Appalachian hardwoods on the Monongahela National Forest. The photo was taken from a helicopter in the late 1960s, approximately 3 years after logging. Note the road network along the contours for removing logs and the abundant natural reproduction covering the ground between the roads. (Photo by James N. Kochenderfer, USDA Forest Service)

forest cover, but most of the MNF at that time was managed as commercial timberland. The public's desire for less emphasis on timber production had not yet influenced planning and land management on its national forests.

As clearcutting on the MNF increased, there was growing local opposition from some turkey and squirrel hunters whose favorite hunting grounds were dramatically altered by logging activities (Fig. 8.1). Some clearcuts were quite large, covering hundreds of acres. Why were the early clearcut harvests in the 1960s so large? There are several factors. The MNF management policy in the 1950s involved partial harvests in relatively young forests that left some merchantable trees standing. Most of the MNF had been heavily logged at the turn of the century, so most of the new forests that occupied the MNF in the 1950s were only 40–50 years old. Because the merchantable trees were relatively small, the resulting sale volumes per acre were also small. As a result, harvests were bundled to offer enough marketable wood volume for local wood buyers.

Later, when it was determined that the tree regeneration in these large-sale units did not have a satisfactory mix of species, the management policy shifted to clearcut harvesting in the 1960s to initiate new, more diverse, tree regeneration. Managers were still required to meet expanding harvest volume targets, and budgets for new road construction were a constraint. Roads built for the large selection harvests in the 1950s were already in place, thus providing low-cost access for follow-up clearcuts in the 1960s. As a result, many of the large clearcut harvests in the 1960s were an attempt to rectify tree regeneration problems that followed large partial harvests in the 1950s. Good intentions did not always garner public support. Large clearcuts were also created in the 1960s because the underlying public opposition to clearcut harvesting was not fully appreciated by managers at the time.

Opposition to clearcutting eventually gained the attention of the West Virginia Legislature. The House of Delegates passed three resolutions to study clearcutting. The first report found no technical basis for halting small clearcuts, and the second report essentially asked that the Forest Service return to the use of selection cutting. A third report in 1970 made 15 recommendations to the Forest Service regarding management practices on the MNF. All but one recommendation, which requested that "uneven-aged be the dominant system of management," were eventually adopted in concept by the Forest Service.

The Forest Service also issued an internal report *Even-aged Management on the Monongahela National Forest* that was prepared by agency specialists of various disciplines. Weitzman (1977) summarized the report, which called for less emphasis on timber management and a more balanced approach to multiple use concerns. Even-aged management continued as the "primary" system of management until 1971, when the Chief testified before the Senate Committee on Public Lands that "a new plan of action for the Monongahela has been put into effect."

Clearcutting on the MNF continued into the mid-1970s. Although the Chief announced a change in timber management policy, clearcutting was still considered to be a viable silvicultural alternative, along with various other practices, needed to meet multiple use objectives. At the end of 1970, proposed timber sales were halted so they could be reconsidered. However, previous sale contracts involving clearcut harvests were honored by the Forest Service. In some cases, there was a 3- to 7-year lag between the contractual agreement with the timber buyer and the resulting logging operation on the ground. To the public, and especially the vocal opponents of clearcutting, the continuation of such harvests made it appear that the Forest Service had not changed its "primary" focus on the even-aged management system.

As clearcutting expanded from a local issue for hunters in West Virginia, several national conservation and environmental interest groups helped seek a resolution in the courts. On May 14, 1973, a lawsuit was filed in Federal District Court in Elkins, WV, by the Natural Resources Defense Council, Sierra Club, Izaak Walton League, West Virginia Highlands Conservancy, and one individual against the Forest Service. The lawsuit charged that three proposed timber sales on the MNF violated provisions of the Organic Act of 1897, which authorized the agency to sell timber within national forests. Judge Robert E. Maxwell ruled in favor of the plaintiffs on November 6, 1973, based on the meaning intended by Congress of the phrases "mature," "large growth," "marked and designated," and "cut and removed" specified in the law under review (Fairfax and Achterman 1977). The court later granted a summary judgment on December 21 that prohibited the Forest Service from selling trees that were not dead, mature, or of large growth. The judgment also required that trees included in timber sales be individually marked and then removed after felling (Schirck 1976).

The Forest Service appealed the decision, and on August 21, 1975, the Fourth US Circuit Court of Appeals in Richmond, VA, upheld the lower court ruling (*Izaak Walton League vs. Butz*). The appeals court decision also suggested that the three-judge panel recognized that "their reading of the Organic Act would have serious and far-reaching consequences… that the law may be an anachronism that no longer serves the public interest… and that the appropriate forum to resolve the issue was not in the courts but the Congress" (LeMaster and Popovich 1976).

After Judge Maxwell's decision was upheld, the Forest Service called a temporary halt to new timber sales in the jurisdiction of the Fourth Circuit Court of Appeals, which included Virginia, West Virginia, North Carolina, and South Carolina. Timber sales on the MNF reached their lowest ebb during that period (Fig. 8.2), and

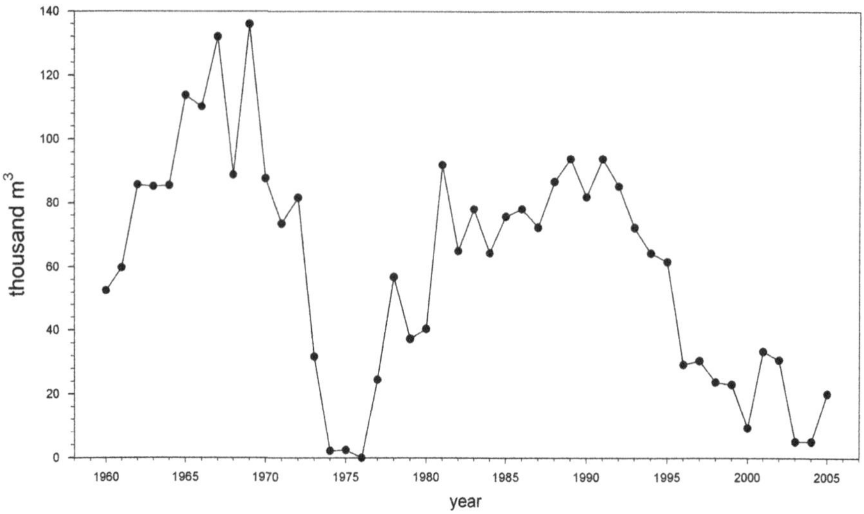

Fig. 8.2 Volume of commercial timber sold on the Monongahela National Forest from 1960 through 2005

the "Monongahela Decision" jeopardized almost 75% of the ensuing 1976 timber sales program, some 41,760,000 m^3 of timber, on all national forests (Popovich 1975). By extension, the court decision jeopardized timber sales and local economies as far away as the western Douglas fir region in the Pacific Northwest.

As a result of the clearcutting controversy on both the Bitterroot and MNFs, President Ford signed the National Forest Management Act (NFMA) into law on October 22, 1976. The new law directed the Secretary of Agriculture to establish guidelines for the use of even-aged management, thus permitting continued applications of clearcutting and relaxing the strict constraints of the Organic Act of 1897 (Williams 2000). It also emphasized the planning process in achieving multiple-use objectives and increased public involvement as an amendment to the Renewable Resources Planning Act of 1974.

The NFMA contained a provision for a committee of scientists to provide advice to the Secretary on how to implement the intent of the law, a process that took 3 years to complete. As the Forest Service moved forward in the late 1970s, NFMA provided enough flexibility so that land managers could continue to apply professional judgment and new science as it became available. The focus of planning activities progressed beyond timber management to a much broader concern for multiple uses on the MNF and other national forests.

Many changes in timber harvest practices that occurred on the MNF during and after the clearcutting controversy were the result of an ongoing partnership between land managers on the MNF and scientists working at the Fernow Experimental Forest. Their partnership was based on both the necessity to apply new scientific information and cordial relationships formed over several decades. This partnership is best described here in five parts:

1. Background information on two-aged management practices to clarify terminology
2. Review of the history of two-aged management on the MNF
3. Summary of the personal relationships between land managers and scientists as a possible model for others to use in developing innovative management practices
4. Description of the communication tools employed by land managers on the MNF and Fernow scientists
5. Assessment of two-aged management practices over a 20-year period to illustrate the current status of this practice and how it continues to be applied as an alternative to clearcutting

8.3.1 Background on Two-Aged Management

Public opposition to clearcutting led land managers on the MNF to seek alternative harvest practices that would provide adequate sunlight and seedbed conditions to regenerate a diverse community of tree species, enhance vertical structure for wildlife habitat, and provide a more appealing visual impact after harvest. Early efforts to develop suitable alternatives to clearcutting included consultations with scientists at the Vinton Furnace, Bent Creek, and Fernow Experimental Forests to evaluate the progress of experimental applications of group selection, patch cutting, and strip clearcutting. All of these practices resulted in a postharvest appearance that simply looked like smaller clearcuts.

A new approach was needed that resulted in a different postharvest visual impact when compared to clearcutting. Although Smith (1962) described a similar practice much earlier, MNF managers seriously considered two-aged management for the first time because it had the potential to mitigate several aspects of clearcutting, most notably its negative visual impact. While the problem began with complaints about hunting sites and wildlife habitat, a major part of the solution called for retention of overstory trees so that harvested areas covered less acreage and did not "look like clearcuts."

Two-aged stands resemble those harvested using what foresters call a seed-tree practice (Fig. 8.3). In the seed-tree practice, all but a few mature overstory trees are harvested to make sure there is an adequate seed source left behind to regenerate a new forest. Once seeds fall and new seedlings become established, the seed trees are removed, usually within 10 years after the initial harvest. Unlike seed-tree practices, however, the reserve trees in two-aged stands are not harvested once the reproduction becomes established. Instead, the harvest of the reserve trees is deferred until the new reproduction is between 40 and 80 years old, generally one-half to a full sawtimber rotation. The resulting stand takes on a two-aged structure, with the scattered reserve trees being taller and much older than the new reproduction that develops beneath it. The result is a two-aged stand structure for many years.

Once the two-aged stand structure is established, similar practices can be repeated at certain intervals to maintain this condition indefinitely. A small, 2-ha trial

Fig. 8.3 A comparison of deferment cutting and clearcutting in 80-year-old central Appalachian hardwoods near Elkins, WV. The photo on the *left* was taken in 2004, one year after deferment cutting that left 37 overstory reserve trees per hectare and an average residual basal area of 4.1 m^2/ha. The edited photo on the *right* simulates the appearance of clearcutting in the same stand. The post-logging appearance is temporary. In both cases, rapid development of natural tree reproduction will cover the area and reach a height of 7.6–9.1 m within 10 years. (Photo by Gary W. Miller, USDA Forest Service)

of this approach was made on the Bent Creek Experimental Forest in 1965, where the residual stand contained 30 mature northern red oaks per hectare and all other trees were removed or deadened with herbicides (Beck 1986).

Several terms are used to describe regeneration harvest methods that effectively create a two-aged stand structure. These terms include deferment cutting, clearcut with reserves, shelterwood with reserves, and irregular shelterwood. For example, Smith (1962) suggested an irregular shelterwood or reserve seed-tree practice where a few overstory trees are retained after harvest to provide seed production, improve esthetics, protect soil, or meet a variety of objectives. More recently, the retention system (Mitchell and Beese 2002) and continuous cover forestry (Pommerening and Murphy 2004) have emerged as similar approaches to regenerating and sustaining desired species after logging without ever removing the entire overstory. Deferment cutting and clearcut with reserves refer to practices in which reserve trees are retained to achieve goals other than reproduction for an entire sawtimber rotation (Smith et al. 1989). The other terms refer to practices in which reserve trees are retained for several years, but not necessarily for an entire rotation. In practice, two-aged regeneration methods generally describe a host of harvest practices designed to create and maintain a stand composed of two distinct age classes.

Harvest practices that promote a two-aged stand structure have been initiated on national forests, state forests, industrial forests, and to a lesser degree on non-industrial private forests in many eastern states. A similar practice called insurance silviculture was applied in Pennsylvania to retain a seed source in the event of a regeneration failure in Allegheny hardwoods composed of black cherry and maples (Bennett and Armstrong 1981). In most cases, two-aged regeneration methods were employed on the MNF to regenerate a variety of species, particularly those that are shade intolerant, and to mitigate the negative impacts of clearcutting on esthetics and habitat. Similar to clearcut stands, natural reproduction that follows two-aged regeneration methods includes a variety of both shade-tolerant and shade-intolerant commercial hardwoods (Trimble 1973; Miller and Schuler 1995; Miller and Kochenderfer 1998).

Unlike clearcutting, however, the presence of reserve trees improves esthetics (Pings and Hollenhorst 1993) and maintains a more diverse vertical stand structure that may benefit certain wildlife species (McDermott and Wood 2009). As the new cohort develops beneath the reserve trees, the stand has two distinct height strata. These strata provide a diverse habitat for songbirds that forage in high canopy trees (Duguay et al. 2001), as well as those that require a brushy cover characteristic of a young even-aged stand (DeGraaf et al. 1991). Two-aged regeneration methods also provide an opportunity to retain species that produce hard mast for wildlife food.

8.3.2 History of Two-Aged Management on the Monongahela

On a visit to Germany in the midst of the clearcutting controversy, Robert Schirck, timber staff officer on the MNF, was shown a harvest practice known as deferment cutting. Troup (1928) and Kostler (1956) had described a "two-storied high forest" practice used to manage pine, larch, and oak-beech stands in Europe. The harvest removed most of the existing mature stand, while retaining a low-density residual overstory to improve the post-logging appearance and perhaps meet a range of other objectives. The removal of the residual overstory trees was deferred until the new regeneration was well developed many years later, hence the name "deferment cutting."

When Robert Schirck returned from Germany, the concept of deferment cutting made a rather circuitous journey before making its way into a test application on the ground. According to some accounts, the idea went up the chain of command from the MNF Supervisor's Office in Elkins, WV, to the Forest Service Region 9 Headquarters in Milwaukee, WI. From there, it went to Forest Service Headquarters in Washington, DC, where it moved from the National Forest System Deputy Chief's office to the Research and Development Deputy Chief's office. The idea then moved down the chain of command to the Northeastern Forest Experiment Station Headquarters in Philadelphia, PA, and continued on down to the Timber and Watershed Laboratory, office of the Fernow Experimental Forest in Parsons, WV. Elkins is only 35 km from Parsons, so the idea could have traveled more efficiently via a local telephone call or perhaps a half-hour drive by car.

According to other sources, Bob Schirck contacted Clay Smith, project leader on the Fernow, soon after his return from Germany. Although the general concept of deferment cutting traveled through proper administrative channels over a period of weeks and months, the longstanding personal relationships between employees on the MNF and the Fernow allowed a local exchange of ideas to begin almost immediately.

Land managers and scientists needed a little time after the clearcutting controversy erupted to devise a workable alternative, even though silviculturists in Europe had described deferment cutting, a possible solution, decades earlier. Research was needed to determine if forests in the Eastern United States would regenerate and respond as they had in Europe. Such research would require reviewing the published literature, developing a study plan and a defensible study design, selecting research sites, installing monitoring devices, preparing the harvest treatment, and finally conducting an experimental logging operation to test deferment cutting as a possible alternative to clearcutting.

The first experimental application of deferment cutting on the MNF occurred 6 years after Judge Maxwell's "Monongahela Decision" in 1973 and 3 years after the NFMA in 1976 defined clearcutting as an acceptable forest management tool. In 1979, an experimental deferment cut was applied on the Greenbrier Ranger District of the MNF in cooperation with scientists at the Fernow Experimental Forest. Additional experimental deferment cuts were applied on the Fernow and four other sites on the Cheat, Greenbrier, and Marlinton Ranger Districts by 1983. These study sites, which ranged from 4 to 6 ha, provided valuable preliminary information on the implications of two-aged regeneration methods for forest managers (Smith 1988; Smith et al. 1989). In addition, forest managers on the MNF visited these stands over the course of several growing seasons to monitor the progress of the new reproduction and to evaluate the potential of this technique as an alternative to clearcutting.

The first operational deferment cuts were applied on the Potomac and Greenbrier Ranger Districts of the MNF in 1991, about 10 years after the experimental sites were installed. A brief review of basic forest ecology explains why managers waited so long to implement operational deferment cuts. Hardwood forests on the MNF require about 10 years after a harvest disturbance for tree seedlings and sprouts to form a new forest cover where the species composition is set for the future. The new trees are about 7.6–9.1-m tall in this age class, and they form a closed canopy of tree crowns that will persist for decades.

After 10 years, the experimental deferment cuts had begun to answer many of the key research questions initially posed by managers and scientists. Would deferment cutting promote the same desired diversity of tree regeneration as that provided by clearcutting? After 10 years, the answer was "yes." Would the residual overstory trees remain standing for long periods, resistant to wind and the risks of sudden exposure? After 10 years, the answer was "yes." In short, preliminary research results indicated that managers could begin to prescribe deferment cutting as a workable alternative to clearcutting.

After operational deferment cutting began, the evaluation of two-aged forest structures on the MNF and Fernow continued. The early research questions about deferment cutting focused on the fate of the residual overstory trees and the development of the new age class of trees. Would logging operations damage the residual trees? Would the residual trees grow faster? Would the market value of the residual trees decline due to branches and knots that usually form when the main stem is suddenly exposed? Would the new reproduction develop slower under the influence of the residual trees? Would the new reproduction include shade-intolerant species like yellow-poplar and black cherry, and mid-tolerant species like northern red oak, chestnut oak (*Q. montana* Wild.), and white oak?

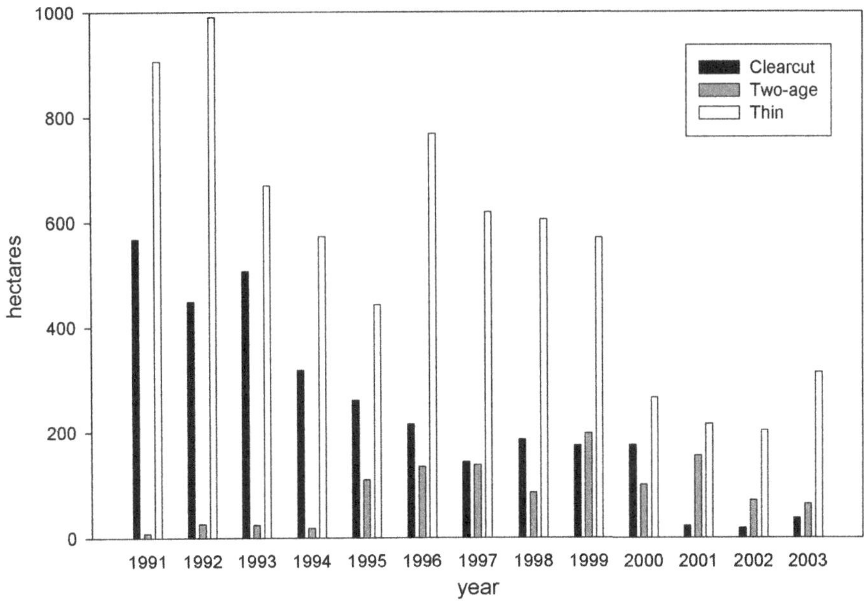

Fig. 8.4 Distribution of hectares harvested on the Monongahela National Forest from 1991 through 2003. Major harvest prescriptions included clearcutting, deferment cutting (two-aged), or commercial thinning

After preliminary results were known and study sites with two-aged structures became available, more pointed research questions about wildlife habitat and esthetics could be studied. How would songbird habitat and nesting behavior in two-aged forests differ from young, even-aged forests resulting from clearcuts? The esthetics of two-aged forests was compared to that resulting from other silvicultural methods (Pings and Hollenhorst 1993). A study of songbird density and nesting success in two-aged stands was conducted from 1992 to 1996 (Miller et al. 1995; Duguay et al. 2001). Logging economics and product market options were also investigated (Miller and Baumgras 1994; Baumgras et al. 1995).

Clearcutting has since been prescribed on a decreasing number of acres each year, while the application of two-aged harvests increased from 1991 through 1997 (Fig. 8.4). The number of acres harvested by any harvesting practice on the MNF has generally declined for two decades, but two-aged harvests are now viewed as a viable alternative to clearcutting in specific cases.

8.3.3 Relationships Between Land Managers and Scientists

When the clearcutting controversy emerged, personnel on the MNF and Fernow had been partners for decades in building roads, clearing firebreaks, marking property boundaries, and erecting buildings that facilitated the operation of a working forest. They had also been partners in applying the latest scientific information in management plans for the MNF as well as research plans for the Fernow. As early as 1934, a water-supply reservoir for the town of Parsons was built on the Fernow on a special-use permit from the MNF. Both groups of employees worked together to meet the needs of the public while preserving the integrity of a valuable outdoor laboratory.

When research funding increased after WW II, the Forest Service established a branch research unit called the Mountain State Research Center in 1948. The main office for the research unit and scientists working on the Fernow was located within the Supervisor's Office of the MNF in Elkins, 35 km from the Fernow. For the next 16 years, scientists and land managers shared office space and had daily contact that encouraged a free exchange of ideas about issues affecting forest management.

In 1964, the Forest Service built the Timber and Watershed Laboratory in Parsons, which became the new headquarters for personnel working on the Fernow. Still, strong working relationships between the two branches of the agency continued, mainly because the culture of cooperation had been handed down from one generation of employees to the next. It was the close ties between scientists and land managers that allowed them to quickly search for science-based answers to the clearcutting issue.

An important first step in dealing with the clearcutting controversy on the MNF was to narrowly define the problem and clarify the land management objectives so that specialists from a variety of disciplines could work toward a solution. Initially, the problem was broadly defined as the need for practical alternatives to clearcutting. More specific objectives were then established: To identify harvest practices that provided adequate tree regeneration, improved both esthetics and habitat compared to clearcutting, and yielded wood volume that allowed feasible timber sales.

A range of regeneration methods were evaluated based on how effectively they provided the light and seedbed conditions necessary to regenerate desirable shade-intolerant species. Closer examination of potential alternatives to clearcutting resulted in additional management objectives. For example, group selection harvests provided desirable tree regeneration and improved esthetics, but this alternative would require a dramatic increase in the timber sale acreage to achieve timber volume targets. The concentration of deer damage on regeneration in harvested gaps was also recognized as a shortcoming of group selection. The search for alternatives to clearcutting therefore disclosed two new objectives: a harvest practice that would keep sale acres within practical limits and mitigate deer damage to new reproduction.

The implications for wildlife habitat were also considered in evaluating potential alternatives to clearcutting. Several options such as seed tree, shelterwood, and two-aged practices, which entail the retention of residual trees, were recognized as improvements over clearcutting because residual trees provide mast, dens, perches, and habitat for the foraging and nesting needs of many wildlife species. Two-aged practices were considered to be superior in the sense that residual trees are retained for a relatively long time compared to seed tree and shelterwood practices. As a result, improved wildlife habitat also emerged as one of the management objectives used to define suitable alternative harvest methods.

As land managers and scientists worked toward defining a suitable alternative to clearcutting, a variety of silvicultural options were also evaluated in terms of their potential effect on Forest Plan goals and management prescriptions. For example, each alternative was evaluated in light of its effect on recreation, wildlife, esthetics, harvest volumes and acreages, and economics. The interaction among resource specialists, scientists, and administrators that occurred during this process resulted in a valuable exchange of ideas and a more integrated approach to land management that continues today. Through this process, land managers at the district level became better prepared to prescribe stand treatments that satisfy many management objectives involving multiple forest resources.

8.3.4 Communication Tools for Land Managers and Scientists

Land managers and scientists utilized several communication tools over the years to develop and improve two-aged harvest prescriptions (Miller et al. 1997). The following suggestions are presented as four possible strategies that might be helpful to other groups who are considering the use of unconventional silvicultural methods: using demonstration areas, holding meetings in the field, conducting forest-wide staff training sessions, and sharing preliminary information.

The stands in which experimental two-aged harvests were applied served as valuable demonstration areas for sharing information. Initially, forest managers had the opportunity to evaluate the general appearance of a two-aged stand and ask important questions about the long-term implications of such treatments. The initial visits prompted numerous questions about the fate of residual trees and the new reproduction. As a result, the demonstration areas challenged land managers and scientists to better understand the long-term implications of two-aged stands. These areas also provided a medium for presenting and interpreting preliminary results.

The demonstration areas were located throughout the MNF on three ranger districts, so forest management personnel had a sense of ownership and were able to visit these areas and monitor their development with minimal investment of time and expense. Later, these areas served as a place to hold meetings, train personnel, and educate the public about the objectives of two-aged regeneration methods. It is important that demonstration areas are accessible and inexpensive to visit to maximize the effectiveness of communications.

Another successful communication strategy entailed holding meetings among resource specialists and the forest leadership team in the field, against the backdrop of one or more of the demonstration areas. These sessions were two-way exchanges of questions and answers for personnel at different levels of management. The outdoor forum promoted an open dialog of ideas and reduced confusion about the appearance, objectives, and dynamic nature of two-aged stand structures. In particular, such meetings were an efficient use of time because participants could address tangible issues such as residual tree quality, road conditions, or stand dynamics that are visible on the site. Scientists often attended such meetings to provide updated research results and to stay abreast of new research needs of the managers.

Forest-wide staff training helped ensure that two-aged regeneration methods were more consistently applied across the forest. Marking crews and silviculturists from all districts had the opportunity to interact with each other through hands-on marking exercises in the field. Participants practiced selecting residual trees, asked questions of each other, and resolved confusion about management objectives associated with two-aged management. These activities clarified the criteria for selecting residual trees for a range of objectives. Moreover, the training sessions prepared each marking crew to better implement a stand prescription as intended by the silviculturist.

Scientists also shared preliminary research results with land managers as they became available using preliminary written reports (prior to the formal publication of results), telephone conversations, and field visits to the demonstration areas. Obtaining preliminary research results at frequent intervals helped land managers develop appropriate stand prescriptions more quickly. Similarly, presenting such information and obtaining feedback helped scientists focus on the high-priority research needs of the land manager.

There are also opportunities to communicate the implications of innovative silvicultural practices through existing channels within the Forest Service. For example, two-aged regeneration methods are now part of the National Advanced Silviculture Program, formerly known as the Program of Advanced Studies in Silviculture throughout Regions 8 and 9. Another formal communication tool that grew out of the NFMA was the planning process at the local forest level. Planning staffs benefit from the involvement of outside groups as they fine-tune the desired results and consider the likely impacts of proposed actions, particularly timber harvest projects.

8.3.5 Assessment of Two-Aged Management Practices after 20 Years

In general, reserve trees after two-aged harvests exhibited faster diameter growth compared to similar trees in uncut stands. The reserve trees were free to grow with an average crown growing space of 6.1 m to adjacent reserve tree crowns after treatment (Miller and Schuler 1995). For black cherry reserve trees, the average growth at diameter breast height (d.b.h.) of released trees did not increase compared to unreleased controls, probably due to the sudden shock of exposure. For all other species tested, released trees had greater average d.b.h. growth compared to control trees. This result supported the notion that the reserve trees were biologically young, capable of responding to additional growing space, and likely to survive a very long time.

A total of 667 reserve trees were monitored for survival and quality following two-aged harvests (Miller 1996). After 10 years, 89 % of reserve trees had survived. Only six trees (1 %) were destroyed or removed due to inadvertent damage during logging. After logging, 22 trees (3 %) died within 2 years, and an additional 38 trees (6 %) died between the 2nd and 5th year. Mortality after the 5th year was greatly reduced; only an additional seven trees (1 %) died by the end of the 10th year. Mortality was greatest for black cherry (more than 20 %) and least for yellow-poplar (less than 5 %).

The risk of epicormic branches developing from dormant buds on the boles of reserve trees increases as stand density is reduced by harvesting (Trimble and Seegrist 1970). Such branches diminish the market value of hardwood species because they result in knots that reduce the amount of clear boards or veneer within the tree. Epicormic branching on the boles of reserve trees increased for all species within 2 years after two-aged harvests compared to pretreatment levels. Between the 2nd and 10th years, there was no significant increase in the number of epicormic branches on the bottom 5-m log section. The net effect on quality was that 12 % of residual trees exhibited a reduction in log quality due to new epicormic branches. Of the few grade reductions observed, white oak, northern red oak, and black cherry were most susceptible, while less than 1 % of yellow-poplar trees lost quality due to epicormic branching.

After 10 years, the canopy of the tree reproduction developing beneath the overstory reserve trees was nearly closed and codominant trees on an average were 10.7 m tall. The new reproduction included nearly 2,500 trees/ha ≥2.5-cm d.b.h. with 1,000 trees/ha exhibiting the potential to become crop trees in the future. On excellent growing sites, northern red oak reproduction was relatively sparse, with only 25 potential crop trees per hectare. Codominant reproduction of other species included a variety of both shade-tolerant and shade-intolerant species distributed over 74 % of the stand area. The diversity of commercial species in a codominant position after 10 years was similar to that observed 12 years after clearcutting on similar growing sites.

Songbird density and nesting success in even-aged and two-aged stands were compared 10 years after clearcutting and deferment cutting, respectively (Miller et al. 1995; Duguay et al. 2001). Songbird density was greater in the two-aged stands compared to the even-aged stands. Interior-edge and edge songbird species were more abundant in two-aged stands while the abundance of interior species was about the same for both stand structures. Nest survival (151 nests) was not significantly different between the two-aged and even-aged stands. Predation by mammals was the most common cause of nest failure.

During the second decade of study, managers and scientists began to notice a divergence between the development of reproduction in two-aged stands and that observed after clearcutting. They suspected that the reserve trees were causing a somewhat different pattern of development due to the expansion of their large crowns and an obvious increase in overhead shading on the new age class of trees. Long-term results from Fernow research and examples of both practices on the MNF allowed scientists to test that hypothesis and help managers incorporate new information into their management practice. Ongoing research examined the impact of the 100-year-old reserve trees on the growth, composition, and distribution of the reproduction 20 years after the experimental two-aged harvests.

Measurements were collected on the 20-year-old reproduction adjacent to randomly selected northern red oak and yellow-poplar overstory reserve trees in four

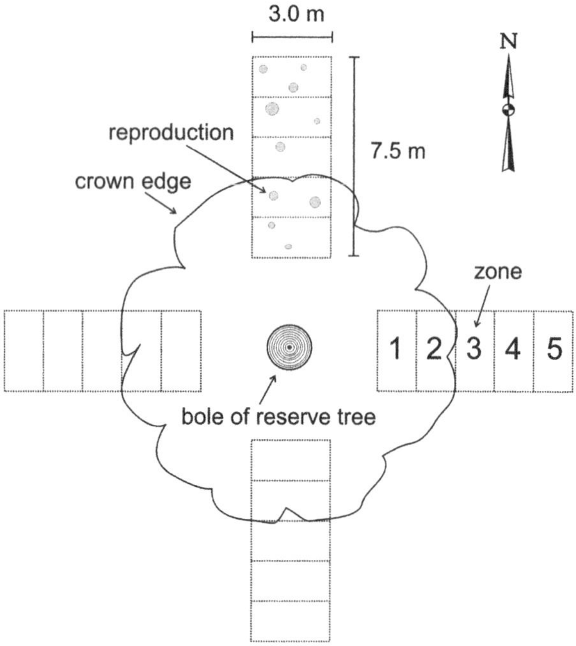

Fig. 8.5 Schematic of sampling transects used to collect data on natural tree reproduction under and adjacent to reserve trees 20 years after deferment cutting

experimental stands (Miller et al. 2006). Species, d.b.h., crown class, height, and distance to the reserve trees were recorded for all reproduction ≥2.5-cm d.b.h. within transects aligned along the four cardinal directions from each reserve tree (Fig. 8.5). D.b.h., height, and crown radius along each transect were also recorded for each of the reserve trees. Crown class of the reproduction was based on the status of each tree relative to neighboring trees of the same age, without regard for the reserve trees. In addition, each transect was divided into five equal zones, thus providing a basis for assessing the density and species distribution of reproduction in relation to the distance from the reserve trees. Zones 1 and 2 were located under the reserve tree crown, and zones 3, 4, and 5 were located farther away from the reserve tree. For reproduction within each zone, tree density was computed for each of three species groups according to shade tolerance defined by Trimble (1975).

For the reserve trees, the d.b.h. growth of northern red oaks exceeded that of yellow-poplars by an average of 27%, even though yellow-poplar was slightly larger when harvests occurred 20 years earlier. Yellow-poplar reserve trees were slightly taller than northern red oak in each stand when the study began, but overall the two species exhibited similar annual height growth. Northern red oak reserve trees also had faster crown radial growth than yellow-poplars in each stand. The crown radius of northern red oak and yellow-poplar reserve trees increased by an average of 1.7 and 0.7 m, respectively. As a result, the area covered by the reserve tree crowns increased by approximately 88% for northern red oak and 44% for yellow-poplar in the 20 years since the harvests. There was an average of 32 reserve trees per hectare after 20 years, down from 37 reserve trees per hectare immediately after harvest.

For the 20-year-old reproduction, the average d.b.h. and height of dominant and codominant stems increased from zone 1 to zone 5 (Fig. 8.5). These findings support the general conclusion that the growth of reproduction was strongly related to distance from the reserve trees. Because the density of reproduction, particularly intolerant species, also increased from zone 1 to zone 5, the results indicated that the reserve trees also influenced the species composition and distribution of the reproduction. In addition, the development of reproduction in the two-aged stands appeared to lag behind that observed in even-aged stands on similar growing sites. Applying the appropriate proportion of stand area and tree density observed in each zone, stand-wide basal area of reproduction averaged 14.9 m^2/ha. In contrast, basal area in 11 nearby even-aged stands averaged 22.0 m^2/ha when 20 years old (Miller et al. 2001). While basal area growth of reproduction was less than that observed after clearcut harvests, the difference was essentially offset by 13.1 m^2/ha of basal area growth on the surviving reserve trees since the harvest.

Songbird populations and nesting success were observed in two-aged stands resulting from deferment cutting and even-aged stands resulting from clearcutting when tree reproduction in both practices was about 15 years old. Results were compared to similar observations in 80-year-old undisturbed second-growth stands on the MNF. Abundance and nesting success did not differ among treatments for the wood thrush, rose-breasted grosbeak, Acadian flycatcher, and red-eyed vireo. Veerys were more abundant in two-aged and unharvested stands compared to even-aged stands (Duguay et al. 2001). A follow-up study further indicated that two-aged stand structures are an advantageous choice over either even-aged or mature forest structures because they can provide habitat for both mature-forest and early successional bird species over an extended period of time on the same site (McDermott and Wood 2009). Two-aged structures provide for greater bird species diversity as they mature, so they also offer a sustainable alternative to clearcutting in locations where brown-headed cowbird parasitism is uncommon.

8.4 Discussion

Although the lawsuit against the Forest Service was based on legal issues associated with clearcutting, the size of harvested areas and the complete removal of overstory trees drove the litigation. The plaintiffs strongly objected to the dramatic change in habitat characteristics and appearance of the land after clearcutting. The land managers responded by considering alternative practices that retained a low-density overstory cover for many years, even as long as an 80-year rotation (Smith et al. 1989). Deferment cutting and other two-aged harvest practices provided immediate relief from the problem of postharvest esthetics. Deferment cuts simply look better than clearcuts because the land has a continuous cover of big trees (Fig. 8.6). In addition, early evaluations of two-aged stand structures indicated that the species diversity of reproduction was similar to that observed after clearcutting, and two-aged stands actually provided better songbird habitat than young even-aged stands.

Fig. 8.6 Crowns of overstory reserve trees immediately after deferment cutting in 80-year-old central Appalachian hardwoods. The space between the reserve trees diminished as their crowns expanded, thus affecting the development of reproduction beneath. (Photo by H. Clay Smith, USDA Forest Service)

Further study of two-aged practices revealed some trade-offs compared to even-aged practices. The 20-year research results from the experimental two-aged harvests on the MNF and Fernow suggested that retaining reserve trees for more than 20 years had a measurable effect on reproduction (Miller et al. 2006). Site resources captured by the overstory reserve trees led to slower growth of reproduction compared to even-aged stands. Expansion of the reserve tree crowns also influenced the species composition and distribution of trees in the new age class. In light of these findings, land managers have begun to weigh competing management objectives over the long term when deciding how many trees to retain and how long to retain them.

Zinke (1962) suggested that the composition and distribution of vegetation within a forest reflect the aggregated effect of many single-tree "influence circles." The crown and roots of a given tree intercept sunlight and draw moisture and nutrients from the soil such that the vigor of other nearby plants is affected (Fig. 8.7). This phenomenon was evident in the experimental two-aged stands studied on the MNF and Fernow (Miller et al. 2006). Results from other experiments with canopy retention systems also indicated that individual reserve trees influence the development of new reproduction (Logan 1965, 1973; Boettcher and Kalisz 1990; Packer and Clay 2003; Joys et al. 2004). The observed influence of reserve trees on reproduction is similar to that observed in several studies of gap dynamics in eastern hard-

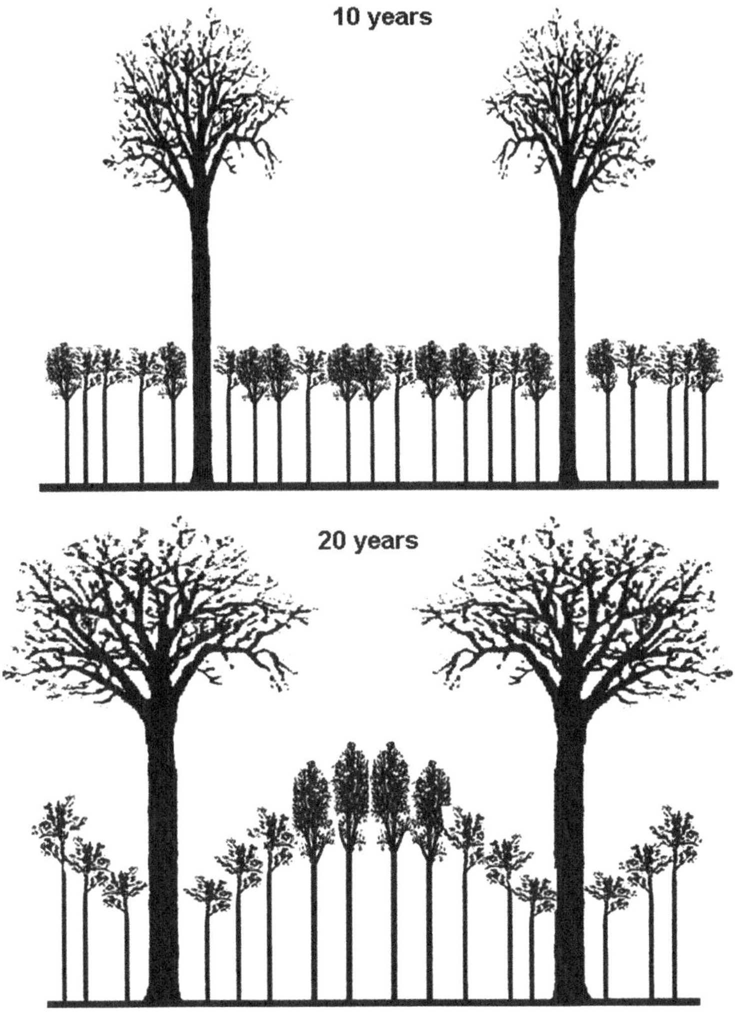

Fig. 8.7 Influence of reserve trees on tree reproduction in two-aged central Appalachian hardwood stands. After 10 years, new reproduction from a variety of tree species formed a closed canopy in a distinct new age class beneath the reserve trees. After 20 years, the reserve trees had expanded their crowns and exerted a noticeable effect on the growth and species composition of the new age class (Smith 1962). Shade-intolerant species remained codominant in the spaces between the reserve trees, while only shade-tolerant species survived beneath the reserve trees

woods, whereby distance from an individual reserve tree is analogous to distance from a gap edge (Dale et al. 1995; Walters and Nyland 1989; Runkle 1990; McClure and Lee 1993; Fownes and Harrington 2004).

In the initial applications of two-aged harvests on the MNF, the long-term impact of reserve trees on reproduction was not a serious concern, because preliminary results on regeneration were favorable. To be clear, however, land managers and scientists had concerns about reproduction issues as early as 1979 when the first experimental two-aged harvest was applied on the Greenbrier Ranger District. As new information became available, forest managers on the MNF adapted prescriptions for two-aged harvests to better define the combination of species and spacing of reserve trees that favors desired reproduction.

Scientists and land managers on the MNF have identified possible alternative strategies for managing two-aged stands if the influence of reserve trees on reproduction is a concern:

1. Retain fewer reserve trees per ha in the initial harvest operation
2. Leave species with relatively smaller crowns and less aggressive crown expansion
3. Leave clumps of residual trees to increase open space and reduce the effect of reserve tree crown expansion
4. Start with a given number of reserve trees per ha and later reduce the residual stand density by chemical or mechanical means
5. Plan to harvest some of the reserve trees once the new age class is established

Forest managers also weigh the risks and benefits of harvesting some of the reserve trees in two-aged stands once the reproduction is well established. Falling and skidding the relatively large reserve trees can cause significant damage to young trees in the new age class. Combining a thinning treatment in the new age class with the harvest of reserve trees in the older age class would provide corridors in which to fall the reserve trees, but additional care is needed to minimize damage from skidding logs and the movement of logging equipment.

Reproduction in operational two-aged stands on the MNF is approaching 20 years old, and there are a couple of lessons that can enhance the success of future applications of this practice. First, as the area under the crowns of the reserve trees expands, the remaining area where shade-intolerant and mid-tolerant species can reach maturity in the upper canopy steadily declines. Although the space under the reserve tree is not really needed for new regeneration (because there is already a mature tree occupying that space), care must be taken to select reserve trees so that the desired long-term species composition is achieved. For example, if the reserve trees are predominantly shade-tolerant American beech, and the space between their crowns is later dominated by shade-tolerant red maple and American beech, then the eventual species composition may lack the desired diversity. However, if the reserve trees include a mix of desired species, then the composition of reproduction between the reserve trees is less critical and can be managed to promote the reserve trees of the future.

Second, opportunities to successfully release intolerant and mid-tolerant species developing in the new reproduction may be limited to areas located between the reserve trees. Crown release treatments free selected trees from nearby competition, thus accelerating their growth and improving their long-term survival (Miller et al. 2007). There is an active program to release young crop trees within two-aged stands to enhance the development of desired species. It is noteworthy that the MNF and Fernow staffs have partnered to study crop tree management since the 1960s, well before the clearcutting controversy, and they continue to do so.

8.5 Recent Trends and Future Directions in Vegetation Management on the MNF

Scientists at the Fernow, other experimental forests, and within the general science community continue to produce scientific discoveries about the development of Appalachian forest plant communities and offer published guidelines for improving silvicultural prescriptions. Likewise, the professional staff on the MNF continues to incorporate new science products into their management plans and stand-level prescriptions. Shelterwood cutting, clearcutting, deferment cutting, and selection cutting are all viable options available to land managers depending on management objectives and conditions on the ground.

MNF forest managers are aware of several new factors that must be considered in developing long-range vegetation management plans. First, the second-growth forests of the MNF are nearly 100 years old. Large commercial volumes are available and management units can be smaller than in the past. Second, fire has been absent from the forest for many decades, and there has been an accumulation of understory trees and shrubs that interferes with the development of advance reproduction vital to sustaining the desired tree species composition. Third, sprout-origin reproduction is not as reliable after harvests today as it was in a younger forest. Fourth, emerging risk factors such as deer browsing and invasive plants add a level of complexity to management that requires a more deliberate sequence of treatments to achieve the desired outcome.

The current focus in vegetation management on the MNF is to apply a sequence of preparatory treatments, such as controlling interfering plants coupled with shelterwood cutting, to increase the probability of successful regeneration well before the final removal of the overstory. Their responsive approach stems from the ongoing partnership with scientists at the Fernow, Kane, Vinton-Furnace, and other experimental forests in the eastern hardwood region. Most recently, field workshops located on the Fernow and the MNF as well as published guidelines for prescribing regeneration treatments in the Mid-Atlantic region (Brose et al. 2008) have provided up-to-date summaries of scientific findings for sustaining Appalachian forests. Land managers on the MNF participated in both the underlying scientific studies and the technology transfer efforts to fill gaps in the literature and make sure that new findings are put into practice on the ground.

The partnership between the MNF and the Fernow will most certainly continue and grow in the future. In fact, new research sites needed to study emerging problems will most likely be located on the Fernow and on suitable research sites on the MNF. Scientists will benefit from monitoring operational activities on the MNF to shed light on research problems on a broader spatial and temporal scale than that found on the Fernow. Indeed, networking research data from multiple experimental forests and ranges, including the Fernow, will certainly enhance the scientific basis for managing the MNF and other national forests. Still, the accumulation of long-term data on both the Fernow and MNF sites is analogous to fine wine; its value will steadily increase with age until it becomes a unique treasure for both scientists and land managers.

Managers on the MNF recently identified four priorities for new science and technology transfer efforts in vegetation management: (1) more information about the impact of management activities on herbaceous species in the plant community, (2) guidelines for mitigating the impact of high deer populations on both herbaceous and woody plants, (3) guidelines for preventing or mitigating the impact of nonnative invasive plants, and (4) assistance in developing demonstration sites and other communication tools for increasing public awareness of their management efforts. Indeed, the partnership between Forest Service scientists and land managers will continue because there seems to be a clear purpose and desire to journey forward together.

References

Averell JL (1929) Factors affecting the reproduction of hardwood forests in southern Connecticut. J For 27:55–61

Baumgras JE, Miller GW, LeDoux CB (1995) Economics of hardwood silviculture using skyline and conventional logging. In: Lowery G, Meyer D (eds) Advances in hardwood utilization: following profitability from the woods through rough dimension. Proceedings of the 23rd annual hardwood symposium. 1995 May 17–20; Cashiers, NC. National Hardwood Lumber Association, Memphis, TN, pp 5–17

Beck DE (1986) Management options for southern Appalachian hardwoods: the two-aged stand. In: Phillips DR, comp. Proceedings of the fourth biennial southern silvicultural research conference. 1986 November 4–6; Atlanta, GA. General Technical Report SE-42, U.S. Department of Agriculture, Forest Service, Southeastern Forest Experiment Station, Asheville, NC, pp 451–454

Bennett AL, Armstrong MM (1981) Insurance silviculture in the black cherry-maple type. J For 79(146–149):154

Bey CF (1964) Advance oak reproduction grows fast after clearcutting. J For 62:339–340

Boettcher SE, Kalisz PJ (1990) Single-tree influence on soil properties in the mountains of eastern Kentucky. Ecology 71:1365–1372

Braun EL (1950) Deciduous forests of eastern North America. Hafner Publishing Company, New York, 596 p

Brooks AB (1910) West Virginia geological survey, vol 5. Forestry and wood industries. Acme Publishing Company, Morgantown, 481 p

Brose PH, Schuler TM, Van Lear DH, Berst J (2001) Bringing fire back: the changing regimes of the Appalachian mixed-oak forests. J For 99(11):30–35

Brose PH, Gottschalk KW, Horsley SB, Knopp PD, Kochenderfer JN, McGuinness BJ, Miller GW, Ristau TE, Stoleson SH, Stout SL (2008) Prescribing regeneration treatments for mixed-oak forests in the Mid-Atlantic region. General Technical Report NRS-33, U.S. Department of Agriculture, Forest Service, Northern Research Station, Newtown Square, PA, 90 p

Carvell KL (1986) Effect of past history on present stand composition and condition. In: Smith HC, Eye MC (eds) Proceedings, guidelines for managing immature Appalachian hardwood stands. 1986 May 28–30; Morgantown, WV. SAF Publ. 86-02. West Virginia University Books, Morgantown, WV, pp 1–7

Clarkson RB (1964) Tumult on the mountains—lumbering in West Virginia 1770–1920. McClain, Parsons, p 210

Dale ME, Smith HC, Pearcy JN (1995) Size of clearcut opening affects species composition, growth rate, and stand characteristics. Research Paper NE-698, U.S. Department of Agriculture, Forest Service, Northeastern Forest Experiment Station, 21 p

DeGraaf RM, Scott VE, Hamre RH, Ernst L, Anderson SH (1991) Forest and rangeland birds of the United States: agricultural handbook 688. U.S. Department of Agriculture, Washington, DC, 625 p

Duguay JP, Wood PB, Nichols JV (2001) Songbird abundance and avian nest survival rates in forests fragmented by different silvicultural treatments. Conserv Biol 15(5):1405–1415

Fairfax SK, Achterman GL (1977) The Monongahela controversy and the political process. J For 75:485–487

Fownes JH, Harrington RA (2004) Seedling response to gaps: separating the effects of light and nitrogen. For Ecol Manage 203:297–310

Joys AC, Fuller RJ, Dolman PM (2004) Influences of deer browsing, coppice history, and standard trees on the growth and development of vegetation structure in coppiced woods in lowland England. For Ecol Manage 202:23–37

Korstian CF (1927) Factors affecting germination and early survival in oaks. Yale University School of Forestry Bulletin 19. Yale University, New Haven

Kostler JN (1956) Silviculture. Oliver and Boyd Ltd., Aylesbury, 416 p (M.L. Anderson, trans)

Kuenzel JF, McGuire JR (1942) Response of chestnut oak reproduction to clear and partial cutting of overstory. J For 40:238–243

LeMaster DC, Popovich L (1976) Development of the National Forest Management Act. J For 74:806–808

Liming FG, Johnston JP (1944) Reproduction in oak-hickory forest stands of the Missouri Ozarks. J For 42:175–180

Little S (1938) Relationships between vigor of resprouting and intensity of cutting in coppice stands. J For 36:1216–1223

Logan KT (1965) Growth of tree seedlings as affected by light intensity. I. White birch, yellow birch, sugar maple and silver maple. Department of Forestry Publication No. 1121, Ottawa, Canada

Logan KT (1973) Growth of tree seedlings as affected by light intensity. V. White ash, eastern hemlock, and general conclusions. Canadian Forestry Service Publication No. 1323. Department of the Environment, Ottawa, Canada

Marquis DA (1994) Thinning principles and practices: control of stand density, structure, and species composition during thinning in hardwood stands in the Alleghenies. In: Marquis DA (ed) Quantitative silviculture of hardwood forests in the Alleghenies. General Technical Report NE-183, U.S. Department of Agriculture, Forest Service, Northeastern Forest Experiment Station, Radnor, PA, pp 247–287

McClure JW, Lee TD (1993) Small-scale disturbance in a northern hardwoods forest: effects on trees species abundance and distribution. Can J For Res 23:1347–1360

McDermott ME, Wood PB (2009) Short- and long-term implications of clearcut and two-age silviculture for conservation of breeding forest birds in the central Appalachians, USA. Biol Conserv 142:212–220

Miller GW (1996) Epicormic branching on central Appalachian hardwoods 10 years after deferment cutting. Research Paper NE-702, U.S. Department of Agriculture, Forest Service, Northeastern Forest Experiment Station, Radnor, PA, 9 p

Miller GW, Baumgras JE (1994) Efficient silvicultural practices for eastern hardwood management. In: Opportunities for the hardwood industry to address environmental challenges: Proceedings of the 22nd annual hardwood symposium of the hardwood research council; 1994 May 12–15; Cashiers, NC. Hardwood Research Council, Memphis, TN, pp 23–35

Miller GW, Kochenderfer JN (1998) Maintaining species diversity in the central Appalachians. J For 96(7):28–33

Miller GW, Schuler TM (1995) Development and quality of reproduction in two-age central Appalachian hardwoods—10-year results. In: Gottschalk KW, Fosbroke SLC (eds) Proceedings, 10th central hardwood forest conference; 1995 March 5–8; Morgantown, WV. General Technical Report NE-197, U.S. Department of Agriculture, Forest Service, Northeastern Forest Experiment Station, Radnor, PA, pp 364–374

Miller GW, Wood PB, Nichols JV (1995) Two-age silviculture: an innovative tool for enhancing species diversity and vertical structure in Appalachian hardwoods. In: Forest health through silviculture: Proceedings of the 1995 national silviculture workshop. Eskew LG, comp. General Technical Report RM-GTR-67, U.S. Department of Agriculture, Forest Service, Rocky Mountain Range and Forest Experiment Station, Fort Collins, CO, pp 175–182

Miller GW, Johnson JE, Baumgras JE, Bustamente RG (1997) Two-age silviculture on the Monongahela National Forest—managers and scientists assess 17 years of communication. In: Proceedings of the national silviculture workshop; 1997 May 19–22; Warren PA. General Technical Report NE-238, U.S. Department of Agriculture, Forest Service, Northeastern Forest Experiment Station, Radnor, PA, pp 123–133

Miller GW, Kochenderfer JN, Knibbs J, Baumgras JE (2001) Vegetative conditions and management options in even-age stands on the Monongahela National Forest. In: Barras SJ (ed) Proceedings, national silviculture workshop. 1999 October 5–7; Kalispell, MT. Proceedings RMRS-P-19. U.S. Department of Agriculture, Forest Service, Rocky Mountain Research Station, pp 40–47

Miller GW, Kochenderfer JN, Fekedulegn DB (2006) Influence of individual reserve trees on nearby reproduction in two-aged Appalachian hardwood stands. For Ecol Manage 224:241–251

Miller GW, Stringer JW, Mercker DC (2007) Technical guide to crop tree release in hardwood forests. Southern Regional Extension Forestry Publication SREF-FM-011. http://www.sref.info/publications/online_pubs. 24 p. Accessed 25 June 2013

Mitchell SJ, Beese WJ (2002) The retention system: reconciling variable retention with the principles of silvicultural systems. For Chron 79(3):397–403

Oliver CD, Larson BC (1996) Forest stand dynamics. Update edition. Wiley, New York, 520 p

Packer A, Clay K (2003) Soil pathogens and *Prunus serotina* seedling and sapling growth near conspecific trees. Ecology 84:108–119

Pings P, Hollenhorst S (1993) Managing eastern hardwoods for visual quality. In: Vander Stoep GA (ed) Proceedings of the 1993 northeastern forest recreation symposium. 1993 April 18–20; Sarasota Springs, N.Y. General Technical Report NE-185, U.S. Department of Agriculture, Forest Service, Northeastern Forest Experiment Station, Radnor, PA, pp 89–93

Pommerening A, Murphy ST (2004) A review of the history, definitions and methods of continuous cover forestry with special attention to afforestation and restocking. Forestry 77(1):27–44

Popovich L (1975) Monongahela and the Forest Service—decisions, decisions. J For 73:668–669

Popovich L (1976) A monster on the Monongahela. J For 74:169–170

Roach BA, Gingrich SF (1962) Timber management guide for upland central hardwoods. U.S. Department of Agriculture, Forest Service, Central States Forest Experiment Station and North Central Region, Columbus, 33 p

Roach BA, Gingrich SF (1968) Even-aged silviculture for upland central hardwoods. Agricultural Handbook 355. U.S. Department of Agriculture, Forest Service, Washington, DC, 39 p

Roth ER (1941) Top rot in snow-damaged yellow-poplar and basswood. J For 39:60–62

Runkle JR (1990) Gap dynamics in an Ohio *Acer-Fagus* forest and speculations on the geography of disturbance. Can J For Res 20:632–641

Sander IL, Williamson MJ (1957) Response of a mixed hardwood stand in eastern Kentucky to a harvest cutting. J For 55:291–293

Schirck RA (1976) Timber management on the Monongahela. J For 74:346–352

Schuler TM (2004) Fifty years of partial harvesting in a mixed mesophytic forest: composition and productivity. Can J For Res 34:985–997

Smith DM (1962) The practice of silviculture. Wiley, New York, 578 p

Smith HC (1988) Possible alternatives to clearcutting and selection harvesting practices. In: Smith HC, Perkey AW, Kidd WE (eds) Proceedings: guidelines for regenerating Appalachian hardwood stands; 1988 May 24–26; Morgantown, WV. SAF Publication 88-03. West Virginia University Books and U.S. Department of Agriculture, Forest Service, Morgantown, WV, pp 276–289

Smith HC, Lanson NI, Miller GW (1989) An esthetic alternative to clearcutting? J For 87(3):14–18

Trimble GR Jr (1973) The regeneration of central Appalachian hardwoods with emphasis on the effects of site quality and harvesting practice. Research Paper NE-282, U.S. Department of Agriculture, Forest Service, Northeastern Forest Experiment Station, Upper Darby, PA, 14 p

Trimble GR Jr (1975) Summaries of some silvical characteristics of several Appalachian hardwood species. General Technical Report NE-16, U.S. Department of Agriculture, Forest Service, Northeastern Forest Experiment Station, Upper Darby, PA, 5 p

Trimble GR Jr (1977) A history of the Fernow Experimental Forest and the Parsons Timber and Watershed Laboratory. General Technical Report NE-28, U.S. Department of Agriculture, Forest Service, Northeastern Forest Experiment Station, Upper Darby, PA, 46 p

Trimble GR Jr, Seegrist DW (1970) Distribution of diameter growth rates and clear stem lengths as a basis for selecting superior phenotypes. Research Paper NE-160, U.S. Department of Agriculture, Forest Service, Northeastern Forest Experiment Station, Upper Darby, PA, 11 p

Troup RS (1928) Silvicultural systems. Oxford University Press, Oxford, 199 p

Walters RS, Nyland RD (1989) Clearcutting central New York northern hardwood stands. North J Appl For 6:75–78

Weitzman S (1949) The Fernow Experimental Forest. Misc. Publication. U.S. Department of Agriculture, Forest Service, Northeastern Forest Experiment Station, Upper Darby, PA, 16 p

Weitzman S (1977) Lessons from the Monongahela experience. In-service briefing paper authorized by Chief John R. McGuire, U.S. Department of Agriculture, Forest Service, Washington, DC, 62 p

Weitzman S, Trimble GR, Jr (1957) Some natural factors that govern the management of oak. Station Paper 88, U.S. Department of Agriculture, Forest Service, Northeastern Forest Experiment Station, Upper Darby, PA, 40 p

Williams GW (2000) The USDA Forest Service—the first century. FS-650, U.S. Department of Agriculture, Forest Servicem, Washington, DC, 154 p

Zinke PJ (1962) The pattern of influence of individual forest trees on soil properties. Ecology 43:120–133

Chapter 9
Development of the Selection System in Northern Hardwood Forests of the Lake States: An 80-Year Silviculture Research Legacy

Christel Kern, Gus Erdmann, Laura Kenefic, Brian Palik and Terry Strong

Abstract The northern hardwood research program at the Dukes Experimental Forest in Michigan and Argonne Experimental Forest in Wisconsin has been adapting to changing management and social objectives for more than 80 years. In 1926, the first northern hardwood silviculture study was established in old-growth stands at the Dukes Experimental Forest. In response to social demands for more "natural" forestry, the study included then-contemporary practices (e.g., liquidation of old-growth forest) and new approaches (e.g., partial cuttings). By 1953, the partial cutting treatments were deemed most sustainable (Eyre and Zillgitt, Partial cuttings in northern hardwoods of the Lake States: twenty-year experimental results. Technical Bulletin LS-1076, 1953), and led to the creation of an uneven-aged stand structural guide that is still widely used today: the famed "Arbogast Guide" (Marking guides for northern hardwoods under the selection system. Station Paper 56, 1957). Charismatic figures such as Raphael Zon, Windy Eyre, William Zillgitt, and Carl Arbogast Jr. were important to establishing this research and its early application in the Lake States region. Since then, research at the Dukes and Argonne Experimental Forests has expanded to evaluate a range of management alternatives for northern hardwood forests, including approaches designed to sustain biodiversity, habitat, and timber production. In addition, the long-term studies provide new opportunities for larger-scale applications and research unforeseen at the studies' establishment. The lessons learned from the 80 years of research on northern hardwood ecosystems at the Dukes and Argonne Experimental Forests have led to numerous publications and management guides and have impacted thousands of forestry professionals and millions of hectares of land.

C. Kern (✉) · G. Erdmann (Retired) · B. Palik · T. Strong (Retired)
Northern Research Station, USDA Forest Service, 1831 Hwy 169 E, Grand Rapids, MN 55744, USA
Phone: 218-326-1734
e-mail: cckern@fs.fed.us

L. Kenefic
Northern Research Station, USDA Forest Service, 686 Government Rd., Bradley, ME 04411, USA

D. C. Hayes et al. (eds.), *USDA Forest Service Experimental Forests and Ranges*,
DOI 10.1007/978-1-4614-1818-4_9,

Keywords Uneven-age silviculture · Long-term research · Permanent plots · Technology transfer · Applied science · Upper Peninsula (Dukes) Experimental Forest · Argonne Experimental Forest · Arbogast structure

9.1 Introduction

Forest managers face increasingly diverse and exciting land management challenges, including the call for socially acceptable and ecologically sustainable practices. Over the past few decades, the values that society has associated with forests have changed fundamentally. The previously dominant paradigm of economic growth, control of nature, and the view that natural resources were abundantly available has been replaced by one that focuses on sustainable development, harmony with nature, belief that the supply of natural resources is finite, and a strong emphasis on public involvement in land management decision making (Bengston 1994). Moreover, management of forestland used to be the domain of government agencies, forest product companies, and private landowners; now conservation organizations such as The Nature Conservancy are taking an active role in managing forestland. In addition, users of forestland are diverse in their characteristics and expectations and demand a broad range of high-quality recreational experiences. At the same time, managers are facing an expanding range of threats (e.g., climate change, nonnative insects and diseases, and landscape fragmentation) that endanger the ecological integrity of forests. This uncertainty has heightened the need for creative, innovative silvicultural prescriptions to increase forest heterogeneity and complexity (Coates and Burton 1997; Lindenmayer et al. 2006; Puettmann et al. 2009), and to help insure the long-term sustainability of forested ecosystems and the goods and services they provide. Addressing today's challenges in forest management requires both basic and applied research, short- and long-term perspectives, and insight into ways to apply findings at both small and large spatial scales. The amount of knowledge needed to address current challenges appears daunting, but long-term research on USDA Forest Service experimental forests (EFs) already provides some of that information.

The EFs were established to study problems at scales ranging from individual trees to forest stands, forests, and watersheds. These forests are outdoor laboratories where scientists observe natural phenomena and test hypotheses about ecosystem functioning. They enable scientists to test stand-level silvicultural practices and examine tree-level processes. Today, the EFs are incorporated into national networks to address basic and applied watershed- and landscape-scale questions.

In 1926, the then cutting-edge research on uneven-age silviculture, an alternative to high-grading and exploitive cutting in the region, was established by the Lake States Forest Experiment Station (LSFES, now part of the USDA Forest Service, Northern Research Station) on the Dukes (also known as Upper Peninsula) EF. This experiment, designed to sustain timber yield by increasing diversity in tree age and size, triggered eight decades of research that expanded over time to include basic and applied science questions. Research results have been applied across large regions and diverse forest types.

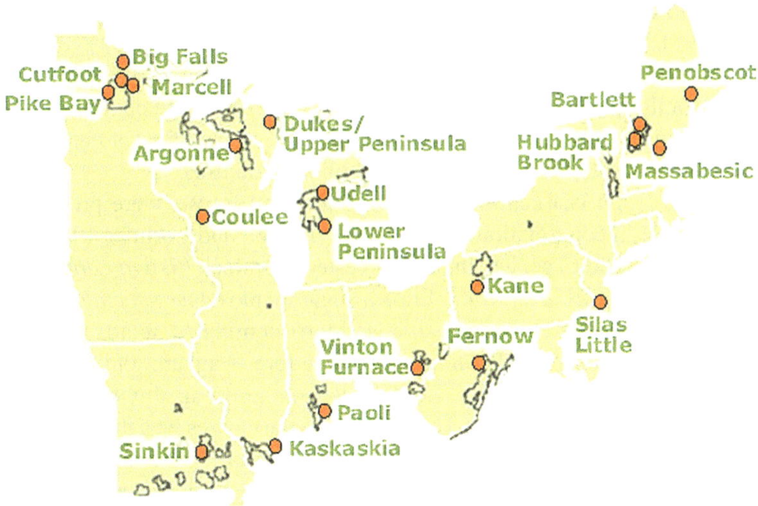

Fig. 9.1 Experimental Forests of the USDA Forest Service, Northern Research Station

9.1.1 *The Dukes and Argonne EFs*

This chapter focuses on the Dukes and Argonne EFs and the long-term uneven-age silviculture studies conducted on them. Research began on the Dukes EF in 1923 on a half section (142 ha) of land donated by Cleveland Cliffs, Inc., about 30 km southeast of Marquette, Michigan (Fig. 9.1). In 1938, the Dukes EF was officially established and incorporated into the Hiawatha National Forest; it has since expanded to 2,225 ha. A small amount of eastern white pine (*Pinus strobus* L.) and American elm (*Ulmus americana* L.) was logged in the early 1900s but the EF areas discussed in this chapter were otherwise considered old-growth northern hardwoods before research began (Adams et al. 2008). The 2,656-ha Argonne EF was established in 1947 and is located 200 km southwest of Dukes within the Chequamegon-Nicolet National Forest in northeast Wisconsin (Fig. 9.1). When silvicultural research at the Argonne EF began around 1950, the areas discussed in this chapter were second-growth northern hardwoods (Adams et al. 2008). The study stands had been commercial, clearcut, and were dominated by trees that established about half a century earlier, but also contained a component of older residuals (Erdmann and Oberg 1973). The site and stand differences between the Dukes and Argonne EFs provided opportunities to include more of the conditions encountered by managers in the region, and to expand and replicate research on northern hardwood silviculture.

9.1.2 *The Lake States Northern Hardwood Forest*

By the late 1890s and early 1900s, the white pine forests of northern Michigan and Wisconsin had been cut over, and many timber operators sold their holdings and moved west to California, Oregon, and Washington or south to Texas and Louisiana (Stearns 1997). Some of the larger operators in the Lake States, however, were familiar with the region's hardwood resources. They bought more land and converted old mills or built new ones to take advantage of this resource. In contrast to the pine forests of the region, the hardwood forests contained many different hardwood species that grew together, each with its own attributes and uses.

At the turn of the last century, northern hardwood forests were predominantly late successional in composition and structure (Frelich and Lorimer 1991). Shade-tolerant to mid-tolerant species, including eastern hemlock (*Tsuga canadensis* L.), American beech (*Fagus grandifolia* Ehrh.), sugar maple (*Acer saccharum* Marsh.), and yellow birch (*Betula alleghaniensis* Britton) dominated stands. Shade-intolerant species such as aspen (*Populus* spp.) were rare (Canham and Loucks 1984; Frothingham 1915; Stearns 1949). The forests were multi-aged with both very old and young trees growing together. This variety of age classes was the result of varying intensities of wind, the primary disturbance agent for this forest type. Frequent, low-intensity wind events result in small canopy gaps that support small saplings of shade-tolerant species (e.g., sugar maple, the dominant species in this forest type) within a stand of large trees. Rare, large-scale wind events maintain complexity at the forest scale and provide opportunities for less shade-tolerant species to establish (Frelich 2002; Hanson and Lorimer 2007).

9.1.3 A Need for Silvicultural Research

In the early twentieth century, some companies in the Lake States were experimenting with partial cutting, and, to facilitate regeneration, started tree nurseries. Many questions arose regarding the management and use of hardwood forests, and several companies hired professional foresters to manage their lands. In addition, the number of tax-delinquent land holdings grew in the 1920s, resulting in the establishment of county forests, state reserves, and national forests throughout the region. Much of this land had been cut over and presented significant management challenges, raising the need for more and better information to manage these public lands.

With this social context, the LSFES was established in 1923 to address problems of reforestation, forest fires, and local economy. By 1926, the Station had initiated partial-cutting studies on the Dukes EF, beginning an 80-year research legacy in uneven-age silviculture that expanded to the Argonne EF and across the northern hardwood range. In particular, the single-tree selection method has been studied extensively at the Dukes and Argonne EFs (Erdmann 1986; Godman and Books 1971; Strong et al. 1995). Though even-age silvicultural systems have also been studied at both EFs with successful results (Godman and Tubbs 1973; Tubbs 1977) and a research natural area at Dukes EF adjacent to the silvicultural research has provided stand dynamics information (Woods 2004, 2000), this chapter focuses on the uneven-aged silviculture research. The studies and staff at the Dukes and Argonne EFs were key in developing these systems, which were unknown and untested in the USA in the early twentieth century. Research results and applications from the

Dukes and Argonne EFs have crossed state and national borders, generated new areas of research, and influenced forest management throughout the region. The long-term silviculture studies also provide a platform for addressing current challenges that were not anticipated at the initiation of the early research program.

9.2 Beginnings of the Selection Paradigm at the Dukes EF (1920s–1950s)

The "Arbogast Guide," a structural guide for an uneven-age silvicultural system called single-tree selection, is one of the most influential products of research at the Dukes EF. The Arbogast Guide was developed during the Selective Cutting Era (1925–1960), an early stage of development in the profession of American forestry (Seymour 2004). The forestry profession was focusing on sustained yield, a concept developed in Europe (the origin of most US silviculture). The sustained yield concept suggested that a negative exponential diameter distribution would result in sustained timber yield at the forest level (de Liocourt 1989). Meyer (1952, 1943) recommended negative exponential diameter distributions (q-structures, mathematically derived structures in which the ratios between numbers of trees in successive diameter classes are constant) for single-tree selection in the USA, thus applying the sustained yield concept at the stand level. At the time, the discipline of ecology was in its infancy and Clements' (1916) theories on stable-state climax forests, where disturbances were viewed as unnatural, were shaping public opinions on forest management. Most people strongly opposed clearcutting and pressed for "near natural" forestry practices. In response to this cultural paradigm, newly formed USDA Forest Service Experiment Stations began research on uneven-aged silvicultural systems, viewed as more "natural" than even-age cutting practices. "Partial cutting" experiments were installed nationwide (Seymour 2004).

A Partial Cutting Study was installed at the Dukes EF in 1926. It consisted of eight experimental cutting treatments of varying intensities: reserve (uncut), diameter-limit cutting, light and heavy improvement cutting, overmature and defective (OMD) cutting, and clearcutting. The objective of the study was to devise a method of partial cutting that would prolong utilization of old-growth northern hardwood resources while gradually converting them into managed forests. This approach would sustain existing mature forests and support local economies until second-growth forests reached merchantable size (Eyre and Zillgitt 1950).

The first LSFES Station Director (1923–1944), Raphael Zon, designed the Partial Cutting Study. Zon was a charismatic figure in American forestry. He was a Russian immigrant who studied under Bernhard Fernow at Cornell University. His leadership resulted in the establishment of the USDA Forest Service research branch and the Journal of Forestry, the first professional periodical in American forestry (Ross 2008). Zon had keen interest in the selection system as an alternative to exploitive logging, an interest that was evident in the design of the Partial Cutting Study. As a strong proponent of the selection system, his aggressive promotion of

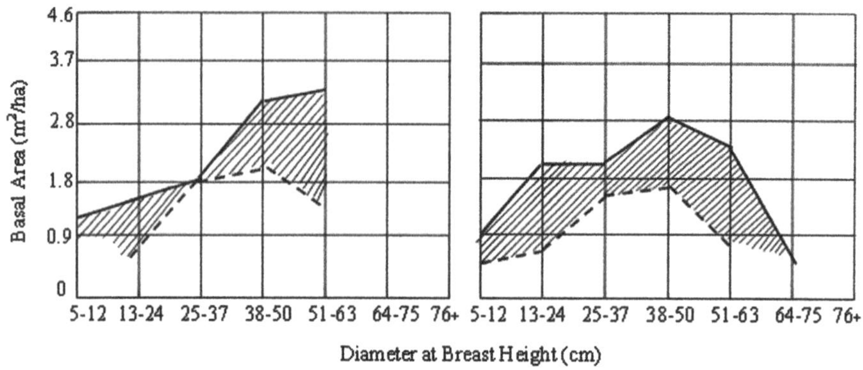

Fig. 9.2 Basal area after harvest, basal area 20 years later, and resulting cutting cycle growth in the overmature and defective No. 2 (*right*) and No. 1 (*left*) treatments. (Eyre and Zillgitt 1953)

this concept made him an important figure in the uneven-age silviculture legacy at the Dukes and Argonne EFs.

After the first 20 years of observation in the Partial Cutting Study, silviculturists F.H. "Windy" Eyre and William M. Zillgitt developed a monograph that explored the ecology, management, and special concerns of northern hardwood forests; it included the first structural guide for northern hardwood silviculture in the USA (Eyre and Zillgitt 1953). Carl Arbogast Jr., silviculturist, later presented this guide as a field forester's technical publication. Consequently, the structure that Eyre and Zillgitt (1953) originally proposed is better known today as the Arbogast Guide (1957).

9.2.1 Early Results, an Empirically Derived Structure

Arbogast (1957) recommended creating and maintaining uneven-aged stands with the empirically derived, reverse-J diameter distribution and residual stocking suggested by Eyre and Zillgitt (1953). Silviculturists observed this distribution, which was based on the OMD treatments of the Partial Cutting Study at the Dukes EF, to provide good growth (Fig. 9.2), continuous ingrowth, and adequate seedling reproduction. The OMD treatments removed all OMD trees regardless of the diameter (Eyre and Zillgitt 1953). Because residual stand growth could be harvested periodically without depleting the base structure, stands with the Arbogast structure (1957) were believed to provide stand-level sustained yield. Eyre and Zillgitt (1950) judged that the distribution of diameter classes "improved upon nature" by bringing in more size classes and more closely emulating a reverse-J distribution than the uncut, reserve treatment. In practice, structural changes are monitored and diameter classes with a surplus of trees are reduced to the recommended level of stocking at regular cutting cycles. In addition, mature trees are removed to regenerate new trees in their place. In balanced stands, the goal is to harvest only the growth gained in a cutting cycle without compromising the base structure of the stand (Arbogast 1957; Eyre and Zillgitt 1953, 1950; Meyer 1952; O'Hara 2002).

9.2.2 *Acceptance and Application of Single-Tree Selection*

An early emphasis on outreach and training by LSFES silviculturists working on the Dukes EF facilitated acceptance and application of the selection system as developed by Eyre and Zillgitt (1953). Raphael Zon promoted publication of early findings in the form of one-page "notes" as well as journal articles explaining detailed results, to get research information into the hands of practitioners and other user groups. His assistant director, Windy Eyre, encouraged outreach and training to share results with forest administrators even before data were published (Rudolf 1985).

This outreach and publication of early results from the Partial Cutting Study at the Dukes EF led to widespread adoption of single-tree selection in the Lake States. Memos in Station archives indicate that major landowners were already starting to apply the selection system instead of clearcutting by the mid-1930s. The regional forester required national forests in the Lake States to use single-tree selection in northern hardwood forests, and the Menominee Indian Reservation adopted selection-marking guides based on results from the Partial Cutting Study. Moreover, many large, private landowners, such as Goodman Lumber in northern Wisconsin and American Can Co. and US Steel in the Upper Peninsula of Michigan, began to use single-tree selection in northern hardwoods instead of clearcutting practices. This widespread application of research results is an excellent example of the successful outreach and training that affected agency and company policies and management of the region's forests.

9.3 Refinement of Single-Tree Selection (1950s–1990s)

During the 1950s, forestry began moving into the Production Forestry Era (1960–1990); research and management emphasized high forest productivity to meet post-World War II economic demands and forecasts of timber scarcity. Even-age silviculture became popular again for many forest types, and practices started to follow agricultural paradigms, including intensive site preparation, vegetation control, and thinning (Seymour 2004). In the northern hardwoods of the Lake States, USDA Forest Service research had always conducted both even- and uneven-aged silvicultural research. Even-age methods were shown to be feasible and productive for the northern hardwood forest (e.g., Godman and Tubbs 1973), unlike the turn-of-the-century exploitive cuttings. However, the success of the selection system and strong support for it by Station leadership and Station silviculturist Carl Arbogast Jr. resulted in its continued importance in the Lake States during the Production Forestry Era.

9.3.1 *Results at Dukes*

The unreplicated Partial Cutting Study motivated scientists to pursue replicated research at both the Dukes and Argonne EFs, especially with Joseph Stoeckler, Ph.D., trained in experimental design and statistics, on board. At the Dukes EF, a replicated Stocking and Cutting Cycle Study was installed in 1951 to test Eyre and Zillgitt's (1953) findings. The design was created by Stoeckeler, a strong proponent of replication, and Arbogast, a strong advocate for single-tree selection. The blocked design included different levels of stocking and different cutting cycles. After 20 years, the results supported Eyre and Zillgitt's (1953) recommendations of 16-m^2/ha residual sawlog stocking (3.5 m^2/ha in poles) and a cutting cycle of about 10 years, with growth rates around 0.5 m^2/ha after single-tree selection cutting. The recommended stocking was a result of high residual tree growth and low mortality. Because differences in net growth were negligible between cutting cycles of different lengths, practical considerations were recommended to guide management decisions about cutting cycle lengths (Crow et al. 1981). This replicated study provided robust, empirical evidence for using the selection system in northern hardwoods.

Single-tree selection increased the dominance of shade-tolerant sugar maple in long-term studies at the Dukes and Argonne EFs (Eyre and Zillgitt 1953; Metzger and Tubbs 1971); this led to the development of special cutting practices for mid-tolerant species. In even-aged stands, shelterwood cutting with release and thinning were effective for regenerating less shade-tolerant species and increasing tree species richness (Erdmann 1986; Erdmann and Peterson 1971; Godman and Tubbs 1973; Metzger and Tubbs 1971). In uneven-aged stands, experiments on the Dukes EF demonstrated that group openings (0.04 ha) near seed trees, with scarification, successfully regenerated yellow birch. Consequently, when appropriate, adding some group-selection openings to single-tree selection stands was recommended, to accelerate recruitment and increase tree species richness (Eyre and Zillgitt 1953; Arbogast 1957; Erdmann 1986).

Before the end of the Production Era, research activity at the Dukes EF, including the Partial Cutting Study and the Stocking and Cutting Cycle Study, ceased (temporarily) in 1981 when the field office in Marquette, MI, was closed and site administration was transferred to the Station's laboratory in Grand Rapids, MN (Adams et al. 2008; unpublished memoranda on file with the USDA Forest Service, Grand Rapids, MN). A letter from the deputy station director to the chief of the Forest Service in 1985 stated that some study areas on the Dukes EF were converted to demonstration areas for professional training and public education purposes.

9.3.2 Results at Argonne

When research started at the Dukes EF in 1926, one third of the forests on the Upper Peninsula of Michigan were still old growth; research focused on sustaining that resource base. By the 1950s, however, the Lake States landscape had changed and most forests were cutover. Yet, silvicultural systems for the second-growth northern hardwoods of the Lake States were still largely untested in 1950. As a consequence, the Station started to focus northern hardwood silvicultural research on second-growth forests at the Argonne EF.

A "Cutting Methods" study designed by Stoeckeler and Arbogast was installed on the Argonne EF in 1951, and included replicated even- and uneven-age treatments. The even-age results highlighted the effectiveness of the shelterwood method for northern hardwood regeneration (Godman and Tubbs 1973) and the value of crop-tree release to stimulate growth and quality development (Erdmann 1986; Erdmann and Oberg 1973). Research showed that tree quality could be increased by controlling tree density to reduce main stem forking and epicormic branching, which tend to degrade sawlog quality (Godman 1992). The work on uneven-age silviculture evaluated applicability of the diameter distribution devised by Eyre and Zillgitt (1953) to second-growth forests. Results from the selection cuts showed that higher residual basal areas (20.7 m^2/ha) increased stand quality (Godman and Books 1971), while lower residual basal areas increased growth (13.8 m^2/ha). The intermediate residual basal area, 17.3 m^2/ha, provided the best balance between stimulating growth and developing quality, which are both important considerations in managing northern hardwoods (Erdmann and Oberg 1973).

With most of the landscape in second-growth forests, a second important outcome from the Argonne EF research was the development of guidelines for conversion of even- to uneven-aged structures. Findings suggested that the first selection cuts in cutover stands could be heavier than previously thought; early annual growth of dominant and codominant trees did not differ between residual basal areas in the Cutting Methods Study. The heavier cuts increased early returns and extended the time until the next entry to 20 years. In addition, Arbogast (1957) had hypothesized that during conversion, pole-sized stands could be cut back to 20 m^2/ha, but experiments at the Argonne EF showed that this level of reduction was not economical. Finally, the second-growth stands had irregular diameter distributions, with few trees in the large diameter classes. Crop-tree release was found to accelerate growth into larger size classes in early harvests, and helped to hasten the development of reverse-J diameter distributions during conversion to an uneven-aged structure (Erdmann and Oberg 1973). The studies at the Argonne EF were maintained through the Production Era.

9.3.3 Outreach Traditions Continued

During the Production Forestry Era, the early emphasis on outreach and training at the Dukes EF was extended to the Argonne EF, where research focused on regional issues of managing second-growth northern hardwoods. Moreover, industry was interested in better utilization of second-growth forests and the work at Argonne addressed this need. At this time, the North Central Forest Experiment Station (formerly, the LSFES) began developing user-friendly management guides for major forest types. The first of these guides since the Arbogast Guide (1957) was for northern hardwoods (Tubbs 1977), and dealt with both even- and uneven-age management options and special issues associated with second-growth forests.

In addition to a new management guide, Richard (Dick) Godman developed a user-friendly series called the "Northern Hardwood Notes" (Godman 1992). The series was presented as a Station-published three-ring binder with one-page summaries of silvicultural systems for northern hardwoods, stand dynamics, regeneration, economics, and wildlife. The Notes could be augmented with additional pages as new information became available. Godman was an influential mentor to other Station scientists and a strong advocate for outreach. Though not trained in graduate school, Godman was knowledgeable about forest stand dynamics and the silviculture of northern hardwood forests. He was gifted in communicating technical information in terms that were meaningful to managers, foresters, and the public. He devoted his work time to outreach in order to get information into practitioners' hands rather than publishing journal articles for the scientific community. The regional forestry network respected him highly. The scientists working at the Argonne EF also disseminated information via field tours. Carl Arbogast Jr. continued outreach and tours about the single-tree selection system at the Dukes and Argonne EFs. In his opinion, single-tree selection was the "only way" to manage northern hardwoods and other options, such as even-aged systems, were not productive. He was known for expressing his opinions and highlighting his ideas in his tours. Silviculturists Dick Godman and Gayne (Gus) Erdmann designed a marking course to be incorporated into Argonne EF tours. The course trained field foresters in single-tree selection as well as even-age marking practices such as shelterwood cuttings, thinnings, and crop-tree release in second-growth northern hardwoods. Training sessions began in the 1970s and continue today. Virtually all forest management agencies in the Lake States have personnel trained at the Argonne and Dukes EFs, and thousands of foresters, forestry students, loggers, and small woodland owners have been trained at or have toured the Dukes and Argonne EFs.

9.3.4 Widespread Application to the Land

The emphasis on outreach and training at the Dukes and Argonne EFs greatly increased the application of single-tree selection in the Lake States and elsewhere, even though the Production Forestry Era emphasized even-age practices nationally. Major landowners within the Lake States based their silvicultural handbooks on the Dukes and Argonne EFs' study results. These landowners included state departments of natural resources (Wisconsin and Michigan), Wisconsin counties, Native American natural resource agencies (Menominee Tribal Enterprises, Lac Du Flambeau, Keweenaw Bay, and Bureau of Indian Affairs), large industrial landowners (American Can Co., Owens-Illinois, Keweenaw Land Co., Mead Paper Co., Champion International, Consolidated Papers, Inc.), national forests (Chequamegon-Nicolet, Ottawa, and Hiawatha), and even provincial lands (Quebec and Ontario). Estimates by Jacob (1987) suggest single-tree selection is the most widely used silvicultural system in the northern hardwood forests of the Lake States.

9.4 Tests of the Empirical Structure Outside the Lake States

Tests of the research results from the Dukes and Argonne EF experiments at other locations were important because northern hardwoods are found from the Lake States to the northeastern USA and eastern Canada. In Michigan and Wisconsin alone, more than 4 million ha are in northern hardwood forest (Jacobs 1987). Silvicultural research elsewhere in the forest type highlights the range of forest outcomes possible.

9.4.1 Lake States

The Ford Center, a property of Michigan Technological University, installed studies in its lower-quality sites and high-graded, second-growth northern hardwood stands to test the structure developed by Eyre and Zillgitt (1953) and recommended in the Arbogast Guide (1957). Eric A. Buordo designed the study in consultation with Carl Arbogast and used selection treatments similar to those at the Dukes and Argonne EFs. Although the study treatments are not replicated, the study has been maintained for more than 50 years and is unique because the cuttings were initiated when the stands were at low stocking levels relative to recommendations by Arbogast (1957). One study at the Ford Center showed that after 40 years of single-tree selection cutting, stand structure varied and only one of five stands had a reverse-J diameter distribution (Neuendorff et al. 2007). Another study at the Ford Center and in the surrounding region demonstrated that commonly practiced uneven-age methods, which do not adhere to strict stocking guides, resulted in compositional and structural variability similar to that observed in selection stands (Schwartz et al. 2005). The recent research at the Ford Center illustrates the variability of individual stands, and the possibility that applying the same treatment to other locations might not result in the anticipated outcomes.

Economic modeling was used to compare alternative diameter distributions to the Arbogast empirical structure. Analyzing data from northern hardwood forests of central Wisconsin, Adams and Ek (1974) showed that more volume in smaller size classes resulted in higher economic gain. This study was one of the first of many to use modeling to test the empirical structure derived at the Dukes EF.

9.4.2 Northeastern States

Long-term studies of selection cutting in northern hardwood forests in the Northeast included similar experiments. Unlike the work at the Ford Center, they did not involve collaboration with silviculturists working on the Dukes or Argonne EFs. In general, studies of the selection system in the northeastern USA have shorter temporal records and little or no replication.

Table 9.1 Basal area distribution of two independent, empirical stand structures for northern hardwood forests by diameter at breast height (dbh) class

	Bartlett EF[a]	Dukes EF[b]
Dbh (cm)	Basal area (m^2/ha)	Basal area (m^2/ha)
15–30	4.6	4.8
30–45	6.4	6.7
45+	7.3	7.3

[a] Gilbert and Jensen (1958), Leak et al. (1969, 1987)
[b] Eyre and Zillgitt (1953), Arbogast (1957)

Gilbert and Jensen (1958) reported 25-year results from a trial of single-tree selection cutting in northern hardwoods on the Bartlett EF in New Hampshire. Findings were later updated by Leak et al. (1987, 1969). Though developed independently of research at the Dukes EF, this empirically derived structure has a similar size class distribution to that of its counterpart in Michigan (Nyland 2002; Table 9.1).

Additional trials of selection cutting explicitly using the structure and guidelines suggested by Eyre and Zillgitt (1953) and Arbogast (1957) were initiated in the 1970s and 1980s in New York (Mader and Nyland 1984; Nyland 1987). Stand structures have remained stable and growth rates (Nyland 2002) have been similar to those observed on the Dukes (Eyre and Zillgitt 1953) and Bartlett EFs (Leak et al. 1969). Hansen and Nyland (1987) also used computer simulation to test Eyre and Zillgitt's (1953) suggestion that a lower residual basal area and longer (15- to 20-year) cutting cycle could be maintained over a 30-year period (Nyland 2002). Growth rates were comparable if managers matched the residual density with an appropriate cutting cycle (Hansen and Nyland 1987). Nyland (2002, 1998, 1987) has recommended the Arbogast (1957) guidelines for selection system silviculture of northern hardwoods in the Northeast.

Studies of the selection system in the Northeast highlighted the dominance of shade-tolerant species in the regeneration (Seymour 1995). American beech, an economically low-value species because of the widespread effects of the beech bark disease (a complex between the scale insect *Cryptococcus fagisuga* and fungi of the genus *Nectria*), was favored by single-tree selection cutting (Leak and Wilson 1958). Group (or patch) selection was shown to be more effective for recruitment of mid-tolerant species such as yellow birch (Leak 2005; Leak and Wilson 1958; Nyland 1998).

9.4.3 Eastern Canada

Long-term studies of the selection system started in Quebec in the late 1980s. Diameter-limit cutting had commonly been used to remove high-value species. The selection system was regarded as the best silvicultural option to restore stand structure and sustain yield of high-quality products (Bedard and Huot 2006). In 1987, the first long-term study of the selection system was installed to test a theoretical

q-structure and compare findings with empirical structures of Eyre and Zillgitt (1953), Crow et al. (1981), and other researchers. Ten years after cutting, net annual growth rates were similar to those reported by Crow et al. (1981) and Eyre and Zillgitt (1953). In addition, residual densities of 16.8–21.2 m^2/ha (similar to Eyre and Zillgitt 1953) were effective at increasing numbers of sugar maple and yellow birch saplings (Bedard and Majcen 2001).

9.5 Criticisms and Alternatives to the Empirical Structure

Even though single-tree selection using the target diameter distribution suggested by Arbogast (1957) has been widely applied, the system has had its critics. Criticisms of the Arbogast Guide and other theoretically balanced stand structures come from the profession of forestry and discipline of ecology. The Arbogast Guide assumes that stand-level sustained yield is both necessary and attainable and its complexity has created some "obsession" to impose the structure without considering the ecological applicability of doing so (Guldin 1991; O'Hara 2002). This type of application appears to be in contrast to a tenet of silviculture, which is expressed in one textbook as follows: "The practice of silviculture does not consist of rigid adherence to any set of simple or detailed rules of procedure" (Smith et al. 1997, p. 17). The idea of balanced uneven-aged distributions was based on observations at the forest level in Europe, not at the stand level (de Liocourt 1898), which contributes to the problem of attaining stand-level sustained yield. Even today, in European forests, where silviculture has been practiced for centuries, selection systems are still evolving to address sustainability (O'Hara et al. 2007) and sustained yield at the stand level remains an idyllic concept (Smith et al. 1997).

Selection cuttings manipulate diameter distributions under the assumption that size is a surrogate for age; in other words, balanced size structures are balanced age structures. Yet studies of tree age–size relationships have revealed that diameter is not equivalent to age, particularly for shade-tolerant species (e.g.,Seymour and Kenefic 1998). In fact, mixed-species and even-aged stands may develop a diameter distribution similar to the reverse J, due to crown class differentiation within a single age class, or stratification among species (Oliver and Larson 1996; Smith et al. 1997). This pattern of development will influence the outcomes of silvicultural treatments because trees in lower crown classes or of slower-growing species may not respond to release as expected.

The single-tree selection system has no specific provision for regeneration aside from residual stand basal area. Seedlings of desirable species are assumed to establish and recruit in a timely manner if stand-level stocking is controlled. In practice, insufficient regeneration or regeneration of undesirable species (e.g., beech) may occur (e.g., Donoso and Nyland 2006), or seedlings and saplings may experience slow growth and competition-induced mortality.

Studies of old-growth forests show that diameter distributions other than reverse-J shapes are common and could be used as guides for management (Frelich and Lorimer 1991; Janowiak et al. 2010; Leak 1964; Tyrrell and Crow 1994). For instance, the rotated sigmoid distribution has been observed in old-growth northern hardwoods, and may be a more realistic model for managed stands. A rotated sigmoid distribution reflects high mortality of small-diameter trees from self-thinning, vigorous growth, and low mortality in the middle size classes, and increased mortality due to senescence in the largest size classes (Goff and West 1975). Past work on the Argonne and Dukes EFs has focused on reverse-J diameter distributions.

Silvicultural systems may not duplicate the effects of natural disturbance. The single-tree selection system, originally perceived as more "natural" (O'Hara 2002), does not in reality emulate the complete natural disturbance regimes of northern hardwood forests. Frequent light harvests perpetuate sugar maple, reduce tree species richness (Neuendorff et al. 2007; Strong et al. 1995), and do not emulate intermediate-intensity wind events. Intermediate disturbances, such as thunderstorm downbursts, remove 10–50% of the overstory trees in localized stands and tend to occur at least once over the life span of a cohort of trees (Frelich and Lorimer 1991; Hanson and Lorimer 2007; Seymour et al. 2002). Patchiness in regeneration perpetuates mid-tolerant species and creates irregularly uneven-aged rather than "balanced" stands (Frelich and Lorimer 1991; Woods 2004).

Because the balanced reverse-J structures have received criticism from both inside and outside of the profession of forestry, alternative methods for uneven-aged silviculture have been proposed. Beginning in the 1970s, alternative structures were suggested for different management objectives (e.g., Leak and Gottsacker 1985), economic factors (e.g., Buongiorno and Michie 1980), and optimization of production (e.g., Adams and Ek 1974). More recently, research has focused on group selection as an option to maintain mid-tolerant species in uneven-aged stands (Leak 1999; Shields et al. 2007; Webster and Lorimer 2005, 2002). In addition, multicohort approaches, which use area control to create spatially distinct cohorts, have been recommended (Lorimer and Frelich 1994; O'Hara 1998; Seymour and Hunter 1999). These approaches have been suggested for maintaining landscape patches of mid-tolerant seed sources, creating heterogeneity in the landscape for habitat, and more closely mimicking intermediate disturbance events (Hanson and Lorimer 2007). Furthermore, newer concepts highlight techniques to support ecological function in managed stands by increasing large woody debris, maintaining decadent trees, creating microtography, designating permanent no-cut areas, and extending cutting cycles (Franklin et al. 2007).

9.6 Maintenance of Legacy Studies and New Directions (1990s–present)

As societal values related to environmental management began to shift in the later part of the twentieth century, forestry began experiencing a corresponding shift to a new resource management paradigm (Bengston 1994). The old "multiple-use sustained-yield" forest management model had guided management for decades; the sustained-yield philosophy can be traced back to eighteenth- and nineteenth-century central European traditions of forest management. By the late 1990s, forestry began to embrace the concept of ecological forestry (Seymour and Hunter 1999), a concept that also influenced work on the Dukes and Argonne EFs. The Ecological Forestry Era (1990–present) entails a landscape triad approach, where production forests and ecological reserves are embedded in a forest matrix managed for a broad range of ecosystem goods and services. Ecological forestry uses natural variability in structure and composition as a way to manage for most species in a forest, a coarse filter approach in biodiversity conservation (Seymour et al. 2007). With the profession focusing on more than just production of wood in the 1990s, the silvicultural research vision at the Dukes and Argonne EFs expanded to include ecological forestry, while still maintaining much of the legacy work of previous eras.

During the Ecological Forestry Era, Terry Strong, silviculturist, maintained the Cutting Methods Study, along with several other studies at the Argonne EF. His primary emphasis was on outreach with more than 1,800 people trained from 20 different organizations between 1993 and 2003. He also helped to integrate Station silviculture research results into new editions of silvicultural handbooks for state, federal, tribal, and industrial landowners, many of whom were already applying selection practices. In addition, some new research was added on the effects of management on a wider array of ecosystem goods and services. For example, the Cutting Methods Study at the Argonne EF was used to compare economic and tree diversity responses among the study's treatments. Long-term data (40 years) showed that heavy to medium single-tree selection was more economical and provided the benefit of higher regeneration diversity than diameter-limit cutting, which initially appeared more economical (Niese and Strong 1992).

9.6.1 Visions for the Future at the Dukes and Argonne EFs

With the full support of Station leadership, research activity at the Dukes EF was revitalized in 2002 when new silviculture project leader Brian Palik reopened the Stocking and Cutting Cycle Study. Palik saw value in extending the existing long-term databases and recognized that analysis of data from historical experiments could offer insights into forest sustainability. Reinvigorating research at Dukes EF also offered opportunities to look at old work, at Dukes and Argonne EFs, in new ways. Under Palik's direction, permanent plots were remeasured and harvests were scheduled to reapply the original treatments at the Dukes and Argonne EFs. In addition, several new projects were started at both EFs, reflecting three new visions: (1) maintenance of long-term studies for their original purposes, (2) new research and unforeseen opportunities, and (3) increasing the scale of inference for various studies.

9.6.2 *Vision 1: Maintenance of Long-Term Studies for Their Original Purposes*

9.6.2.1 More than 80 Years of Balanced Stand Structure Research

With renewed interest in the historical experiments at the Dukes EF, Palik scheduled the original Partial Cutting Study established in 1926 to be re-treated following the original study plan. The study had been periodically measured until 1966 and had been treated by the Hiawatha National Forest for demonstration purposes in 1985. No analysis of long-term findings had been published since Eyre and Zillgitt's (1953) monograph, even though the Arbogast Guide (1957) remains the most widely used structural goal in uneven-age northern hardwood silviculture. In 2007, a remeasurement of the study provided an 80-year record which scientists can use to support or challenge the predictions by Eyre in Zillgitt (1953) and evaluate the sustainability of their empirical structure.

9.6.2.2 Stand Structure in Managed Old Forests

While the Stocking Levels and Cutting Cycles Study was conceived with other objectives in mind (Crow et al. 1981), this study presented an opportunity to examine tree quality and structural characteristics after more than 50 years of selection system silviculture. Extending rotation ages in even-aged stands or growing and maintaining older cohorts of trees in uneven-aged stands is of increasing interest in many different forest types (Curtis 1997; North and Keeton 2008). One rationale for retaining older trees is that such stands will contain more of the structural characteristics of old-growth stands such as large-diameter trees with deeply fissured bark and large crowns, snags, and large deadwood on the ground that are important to some plant and animal species. Empirical data on tree growth and stand structure from long-term experiments are needed to inform managers about what to expect when managing for older tree ages. Structural data indicate that snag density is not greatly affected by cutting cycle (5, 10, or 15 years), but does vary widely by stocking level (Gronewold et al. 2010). Density of large snags (>30-cm dbh) is consistently higher at high stocking levels (20.7 m^2/ha), regardless of cutting cycle, and these densities are similar to unmanaged old growth. Additionally, density of small snags (<30-cm dbh) is higher at lower stocking levels (Gronewold et al. 2010). The work emphasizes the flexibility that managers have in using different selection system options to address the management concerns of the Ecological Forestry Era.

9.6.2.3 Informing Simulation Models Using Data from Long-Term Studies

Empirical data from long-term studies are invaluable for answering questions about management effects on forests. Recent advances in computer modeling provide an additional avenue for addressing these questions. Yet these models are only as good as the data used to calibrate and verify them, and depend on long-term studies on EFs as a valuable source of that information. One such effort is the CANOPY model developed by Craig Lorimer and students at the University of Wisconsin (Choi et al. 2001). CANOPY can simulate a wide variety of treatments in northern hardwood stands over long time periods. It generates an array of response variables related to forest composition, diameter distributions, volume growth, and large deadwood abundance. CANOPY was calibrated and verified using data sets from regional northern hardwood silvicultural studies, including long-term studies on the Dukes and Argonne EFs. It simply would not have been possible to bring the CANOPY model to its current level of accuracy and precision without use of the long-term data sets from the EF studies.

9.6.3 Vision 2: New Research and Unforeseen Opportunities

9.6.3.1 Incorporating Ecological Forestry

Managing forests to restore or sustain structural and compositional characteristics of old unmanaged stands has emerged as an issue of concern for northern hardwood ecosystems in the Ecological Forestry Era (Crow et al. 2002; Keeton 2006). The goal generally is not to create old-growth stands per se, but to create greater structural complexity and maintain older cohorts of trees within stands managed for wood products, perhaps with silvicultural approaches that better emulate natural canopy disturbances. Recent research in the Great Lakes region has found that mesoscale canopy disturbances that remove 30–60 % of the overstory basal area in a patchy distribution are fairly common and important for establishment of mid-tolerant species, and for creating multi-cohort age structure (Hanson and Lorimer 2007). A silvicultural analog for this disturbance regime is being tested operationally on the Argonne EF in collaboration with the Wisconsin Department of Natural Resources and the University of Wisconsin. The treatment consists of single-tree selection cutting with interspersed 0.8-ha shelterwood cuts and deliberately creating large deadwood (both snags and logs) to enhance wildlife habitat. The concept is rooted in basic philosophies of ecological forestry (Franklin et al. 2007) and supplements structural attributes missing in traditional single-tree selection system cutting. This work may suggest new empirical structures and alternate approaches for management to address needs of Ecological Forestry Era.

9.6.3.2 Simulation Modeling of Ecological Objectives in Northern Hardwood Management

Currently, the CANOPY model is used to ask questions about long-term effects of traditional management, as well as some newer approaches, on stand structure and composition. For example, simulations examine how long-term application of single-tree selection based on the Arbogast Guide influences the composition of mid-tolerant tree species and compares the predictions to outcomes from Menominee Tribal Forestry group selection cutting tailored to maintain these species. Another question is how permanent retention (for the life of the tree) of selected canopy trees influences wood volume production in stands managed with the selection system. This guideline is now being implemented on the Chequamegon-Nicolet National Forest (USDA Forest Service 2004) to enhance wildlife habitat and maintain sources of large deadwood.

9.6.3.3 Sustaining Productivity and Carbon Storage

The long-term data from the Dukes and Argonne EFs provide opportunities for addressing emerging science questions that could have not been foreseen when the studies were established. Responses to increased threats to forests and the impending changes, whether from climate, introduced pathogens and pests, or market demands, can be informed by long-term data on stand dynamics and response to different disturbances. The long-term research in the EFs provides this type of information, including detailed data on level of harvest or disturbance intensity. For example, recent work is examining how soil physical attributes affect ecosystem productivity varies across a gradient of cutting intensities over long-time periods at the Argonne EF and other EFs (Tarpey et al. 2008).

9.6.4 Vision 3: Increasing Scale of Inference for Various Studies

A third vision for the Dukes and Argonne EFs is to expand their scale of inference from stands to landscapes. Silvicultural and ecological research has generally moved to studies larger in scale than the small, stand-scale studies established on the Dukes and Argonne EFs. Understanding research results at larger scales provides context and applicability of the findings. Increasing scale entails both installation of operational-scale studies and creatively using already existing long-term study data at larger scales.

Operational-scale research, with treatment units similar in size to managed stands, provides realistic examples of innovative silvicultural systems and a training ground for managers. Large-scale treatments were initially used in silvicultural research in the 1920s. However, later study installations were small in scale because maintaining large-scale studies was expensive and replication was difficult within limited land bases for research use (Seymour et al. 2007). Today, new silvicultural studies are operational in scale, like the new ecological forestry study at the Argonne EF.

Recently, Forest Service research and development has encouraged and financially supported the development of an Experimental Forest and Range Network encompassing more than 80 EFs and Ranges across the USA. Other chapters in this book demonstrate that the research histories of many of these EFs have been long and similar to those of the Dukes and Argonne EFs. This support from the Washington office has led to unprecedented opportunities for EF managers to exchange ideas and findings with each other at national EF and Range workshops (e.g., Adams et al. 2010). One goal of this group is to link the long-term databases across EFs (Vavra and Mitchell 2010). The Dukes and Argonne EFs will contribute to this network, opening opportunities to address questions that could not be answered with individual EF databases. Research topics could range from basic silviculture to climate change.

9.7 Summary

The Dukes and Argonne EFs have rich research histories that have inspired unique research trajectories and widespread application of uneven-age management to the northern hardwood resource of the Lake States, northeastern USA, and eastern Canada. In response to social and professional needs in silviculture, innovative research on partial cuttings for sustaining old-growth forest resources was established early in the twentieth century and resulted in the empirically derived Arbogast Guide. The research, key staff, and extensive outreach influenced policy and management across the range of northern hardwood forest; they also helped to educate managers, students, and private landowners about appropriate methods for hardwood management. The Dukes and Argonne EFs research has transformed the northern hardwood landscape by informing managers and changing regional practices from exploitive cutting to more sustainable and socially acceptable practices. The Argonne and Dukes EFs also have served as sites for some of the longest replicated silvicultural research in the northern hardwood forest type. Today, these studies and databases offer opportunities to test long-term responses of forests to management and to address new questions unanticipated at the time of study establishment. The future of the Dukes and Argonne EFs includes maintenance of those long-term studies, addressing needs of the current Ecological Forestry Era, capitalizing on the long-term databases for emerging science needs, and expanding the scale of inference.

Acknowledgments The authors want to recognize Station scientists and administration (former and current) for decades of hard work and leadership. Support staff has always been critical to maintaining this unique legacy over the years. In particular, we want to acknowledge Wilho "Bill" Salimen on the Dukes EF and Gib Mattson on the Argonne EF. We thank Tom Crow for nominating this chapter for the book, Pam Jakes for assisting us with social context, and Ralph Nyland, Cynthia Moser, Tom Schmidt, and two anonymous reviewers for improving the chapter.

References

Adams DM, Ek AR (1974) Optimizing the management of uneven-aged forest stands. Can J For Res 4:274–287

Adams MB, Loughry L, Plaugher L (2008) Experimental forests and ranges of the USDA Forest Service. General Technical Report NE-321, USDA Forest Service, Northeastern Research Station, Newtown Square, PA

Adams MB, McNeel J, Rodriguez-Franco C (2010) Meeting current and future conservation challenges through the synthesis of long-term silviculture and rande management research. General Technical Report WO-84, USDA Forest Service, Washington Office, Washington, DC
Arbogast CJ (1957) Marking guides for northern hardwoods under the selection system. Station Paper 56, USDA Forest Service, North Central Forest Experiment Station, St. Paul, MN
Bedard S, Huot M (2006) Long-term silviculture research on northern hardwoods in Quebec. In: Long-term silvicultural & ecological studies: results for science and management. Global Institute Sustainable Forestry Research Paper 005, Yale University, School of Forestry and Environmental Studies, New Haven, CT
Bedard S, Majcen Z (2001) Ten-year response of sugar maple-yellow birch-beech stands to selection cuttings in Quebec. North J Appl For 18:119–126
Bengston DN (1994) Changing forest values and ecosystem management. Soc Nat Resour 7:515–533
Buongiorno J, Michie BR (1980) A matrix model of uneven-aged forest management. For Sci 26:609–625
Canham CD, Loucks OL (1984) Catastrophic windthrow in the presettlement forests of Wisconsin. Ecology 65:803–809
Choi J, Lorimer CG, Vanderwerker J, Cole WG, Martin GL (2001) A crown model for simulating long-term stand and gap dynamics in northern hardwood forests. For Ecol Manage 152:235–258
Clements FE (1916) Plant succession: an analysis of the development of vegetation. Carnegie Institution of Washington, Washington, DC
Coates KD, Burton PJ (1997) A gap-based approach for development of silvicultural systems to address ecosystem management objectives. For Ecol Manage 99:337–354
Crow TR, Buckley DS, Nauertz EA, Zasada JC (2002) Effects of management on the composition and structure of northern hardwood forests in Upper Michigan. For Sci 48:129–145
Crow TR, Tubbs CH, Jacobs RD, Oberg RR (1981) Stocking and structure for maximum growth in sugar maple selection stands. Research Paper NC-199, USDA Forest Service, North Central Forest Experiment Station, St. Paul, MN
Curtis RO (1997) The role of extended rotations creating a forestry for the 21st century: the science of ecosystem management. Island Press, Washington, DC, pp 165–170
Donoso PJ, Nyland RD (2006) Interference to hardwood regeneration in northeastern North America: the effects of raspberries (*Rubus* sp.) following clearcutting and shelterwood methods. North J Appl For 23:288–296
Erdmann GG (1986) Developing quality in second-growth stands. In: The northern hardwood resource: management and potential, Houghton, MI, August 18–20, 1986. Michigan Technological University, Houghton, MI, pp 206–222
Erdmann GG, Oberg RR (1973) Fifteen-year results from six cutting methods in second-growth northern hardwoods. Research Paper NC-100, USDA Forest Service, North Central Forest Experiment Station, St. Paul, MN
Erdmann GG, Peterson RMJ (1971) Crown release increases diameter growth and bole sprouting of pole-size yellow birch. Research Note NC-130, USDA Forest Service, North Central Forest Experiment Station, St. Paul, MN
Eyre FH, Zillgitt WM (1950) Size-class distribution in old-growth northern hardwoods twenty years after cutting. Station Paper No. 21, USDA Forest Service, Lake States Forest Experiment Station, St. Paul, MN
Eyre FH, Zillgitt WM (1953) Partial cuttings in northern hardwoods of the Lake States: twenty-year experimental results. Technical Bulletin LS-1076, USDA Forest Service, Lake States Forest Experiment Station, Bromall, PA
Franklin JF, Mitchell RJ, Palik BJ (2007) Natural disturbance and stand development principles for ecological forestry. General Technical Report USDA FS NRS-19, USDA Forest Service, Northern Research Station, Newtown Square, PA
Frelich LE (2002) Forest dynamics and disturbance regimes. Cambridge University, Cambridge
Frelich LE, Lorimer CG (1991) Natural disturbance regimes in hemlock-hardwood forests of the upper Great Lakes region. Ecol Monogr 61:145–164

Frothingham EH (1915) The northern hardwood forest: it's composition, growth, and management. Technical Bulletin No. 285, USDA, Washington, DC

Gilbert AM, Jensen VS (1958) A management guide for northern hardwoods in New England. Station Paper NE-112, USDA Forest Service, Northeastern Forest Experiment Station, Upper Darby, PA

Godman RM (1992) Northern hardwood notes. USDA Forest Service, North Central Research Station, St. Paul, MN

Godman RM, Books DJ (1971) Influence of stand density on stem quality in pole-size northern hardwoods. Research Paper NC-54, USDA Forest Service, North Central Forest Experiment Station, St. Paul, MN

Godman RM, Tubbs CH (1973) Establishing even-age northern hardwood regeneration by the shelterwood method—a preliminary guide. Research Paper NC-99, USDA Forest Service, North Central Forest Experiment Station, St. Paul, MN

Goff GF, West D (1975) Canopy-understory interaction effects on forest population structure. For Sci 21:98–108

Gronewold CA, D'Amato AW, Palik BJ (2010) The influence of cutting cycle and stocking level on the structure and composition of managed old-growth northern hardwoods. For Ecol Manage 259:1151

Guldin JM (1991) Uneven-aged BDq regulation of Sierra Nevada mixed conifers. J For 89:89–36

Hanson J, Lorimer C (2007) Forest structure and light regimes following moderate wind storms: implications for multi-cohort management. Ecol Appl 17:1325–1340

Hansen GD, Nyland RD (1987) Effects of diameter distribution on the growth of simulated uneven-aged sugar maple stands. Can J For Res 17:1–18

Jacobs RD (1987) Silvicultural systems for northern hardwoods in the mid-west and Lake States. In: Proceedings of the symposium on managing northern hardwoods, Syracuse, NY, SUNY College of Environmental Resources and Forestry, Syracuse, NY, pp 383–388

Janowiak M, Nagel LM, Webster C (2010) Minimum tree size and interpretation of stand structure in uneven-aged northern hardwoods. North J Appl For 27:34–37

Keeton WS (2006) Managing for late-successional/old-growth characteristics in northern hardwood-conifer forests. For Ecol Manage 235:129–142

Leak WB (1964) An expression of diameter distributions for unbalanced, uneven-aged stands and forests. For Sci 10:39–50

Leak WB (1999) Species composition and structure of a northern hardwood stand after 61 years of group/patch selection. Northern. J For 16:151–153

Leak WB (2005) Effects of small patch cutting on sugar maple regeneration in New Hampshire northern hardwoods. North J Appl For 22:68–70

Leak WB, Gottsacker JH (1985) New approaches to uneven-age management in New England. North J Appl For 2:28–31

Leak WB, Wilson R Jr (1958) Regeneration after cutting of old-growth northern hardwoods in New Hampshire. Research Paper NE-310, USDA Forest Service, Northeastern Forest Experiment Station, Upper Darby, PA

Leak WB, Solomon DS, Filip SM (1969) A silvicultural guide for northern hardwoods in the Northeast. Research Paper NE-143, USDA Forest Service, Northeastern Forest Experiment Station, Upper Darby, PA

Leak WB, Solomon DS, DeBald PS (1987) Silvicultural guide for northern hardwood types in the Northeast (revised). Research Paper NE-603, USDA Forest Service, Northeastern Forest Experiment Station, Broomall, PA

Lindenmayer DB, Franklin JF, Fischer J (2006) General management principles and a checklist of strategies to guide forest biodiversity conservation. Biol Conserv 131:433–445

de Liocourt F (1989) De l'amenagement des sapinieres, English translation: (2001) On the amelioration of fir forests, translated by Maria Nygren. Bulletin trimestriel, Societe forestiere de Franche-Comte et Belfort, juillet, pp 396–409

Lorimer CG, Frelich LE (1994) Natural disturbance regimes in old-growth northern hardwoods. J For 92:33–38

Mader SF, Nyland RD (1984) Six-year responses of northern hardwoods to selection system. North J Appl For 1:87–91
Metzger FT, Tubbs CH (1971) Influence of cutting methods on regeneration of second-growth northern hardwoods. J For 69:559–564
Meyer HA (1943) Management without rotation. J For 41:126–132
Meyer HA (1952) Structure, growth, and drain in balanced uneven-aged forests. J For 50:85–92
Neuendorff JK, Nagel LM, Webster C R, Janowiak MK (2007) Stand structure and composition in a northern hardwood forest after 40 years of single-tree selection. North J Appl For 24:197–202
Niese JN, Strong TF (1992) Economic and tree diversity trade-offs in managed northern hardwoods. Can J For Res 22:1807–1813
North MP, Keeton WS (2008) Emulating natural disturbance regimes: an emerging approach for sustainable forest management. In: Lafortezza R, Sanesi G, Chen J, Crow TR (eds) Patterns and processes in forest landscapes: multiple use and sustainable management. Springer, Netherlands, pp 341–372
Nyland RD (1987) Selection system and its application to uneven-aged northern hardwoods. In: Nyland RD (ed) Managing northern hardwoods. SUNY College of Environmental Science and Forestry, Syracuse. (Misc. Publ. 13 (ESF 87-002))
Nyland RD (1998) Selection system in northern hardwoods. J For 96:18–21
Nyland RD (2002) Silviculture: concepts and applications, 2nd edn. McGraw-Hill, Boston
O'Hara KL (1998) Silviculture for structural diversity: a new look at multi-aged stands. J For 96:4–10
O'Hara KL (2002) The historical development of uneven-aged silviculture in North America. Forestry 75:339–346
O'Hara KL, Hasenauer H, Kindermann G (2007) Sustainability in multi-aged stands: an analysis of long-term plenter systems. Forestry 80:163–181
Oliver CD, Larson BC (1996) Forest stand dynamics, update edition. Wiley, New York
Puettmann KJ, Coates KD, Messier C (2009) A critique of silviculture: managing for complexity. Island Press, Washington, DC
Ross AT (2008) Exploring the collections, acquisitions, and treasures of the Forest History Society. Forest History Society, Durhan
Rudolf PO (1985) History of the lake states forest experiment station. USDA Forest Service, North Central Forest Experiment Station, St. Paul
Schwartz JW, Nagel LM, Webster CR (2005) Effects of uneven-aged management on diameter distribution and species composition of northern hardwoods in Upper Michigan. For Ecol Manage 211:356–370
Seymour RS (1995) The Northeast region. In: Barrett JW (ed) Regional silviculture in the United States, 3rd edn. Wiley, New York, pp 31–79
Seymour RS (2004) Silviculture: lessons from our past, thoughts about our future. In: Palik B, Levy L (eds) Proceedings of the Great Lake silviculture summit. General Technical Report NC-254, USDA Forest Service, North Central Research Station, St. Paul, MN
Seymour RS, Hunter M (1999) Principles of ecological forestry Maintaining biodiversity in forest ecosystems. Cambridge University Press, Cambridge, pp 22–61
Seymour RS, Kenefic LS (1998) Balance and sustainability in multiaged stands: a northern conifer case study. J For 96:12–17
Seymour RS, White AS, deMaynadier PG (2002) Natural disturbance regimes in northeastern North America—evaluating silvicultural systems using natural scales and frequencies. For Ecol Manage 155:357–367
Seymour RS, Guldin J, Marshall D, Palik B (2007) Large-scale, long-term silvicultural experiments in the United States: historical overview and contemporary examples. Allg Forst Jagdztg 177:104–112
Shields JM, Webster CR, Nagel LM (2007) Factors influencing tree species diversity and *Betula alleghaniensis* establishment in silvicultural openings. Forestry 80:293–307
Smith DM, Larson BC, Kelty MJ, Ashton PMS (1997) The practice of silviculture: applied forest ecology, 9th edn. Wiley, New York

Stearns FW (1949) Ninety years of change in a northern hardwood forest in Wisconsin. Ecology 30:350–358

Stearns FW (1997) History of the Lake States forests: natural and human impacts. In: Vasievich JM, Webster HH (eds) Lake States regional forest resources assessment. General Technical Report NC-189, USDA Forest Service, North Central Forest Experiment Station, St. Paul, MN, pp 8–29

Strong TF, Erdmann GG, Niese JN (1995) Forty years of alternative management practices in second-growth, pole-size northern hardwoods. I. Tree quality development. Can J For Res 25:1173–1179

Tarpey RA, Jurgensen MF, Palik BJ, Kolka RK (2008) The long-term effects of silvicultural thinning and partial cutting on soil compaction in red pine (*Pinus resinosa* Ait.) and northern hardwood stands in the northern Great Lakes Region of the United States. Can J Soil Sci 88:849–857

Tubbs CH (1977) Manager's handbook for northern hardwoods in the North Central States. General Technical Report NC-39, USDA Forest Service, North Central Forest Experiment Station, St. Paul, MN

Tyrrell LE, Crow TR (1994) Structural characteristics of old-growth hemlock-hardwood forests in relation to age. Ecology 75:370–386

USDA Forest Service (2004) Land and resource management plan. Chequamegon-Nicolet National Forest, Park Falls and Rhinelander, WI

Vavra M, Mitchell J (2010) Experimental forests and ranges as a network for long-term data. In: Adams MB, McNeel J, Rodriguez-Franco C (eds) Meeting current and future conservation challenges through the synthesis of long-term silviculture and range management research. General Technical Report WO-84, USDA Forest Service, Washington Office, Washington, DC, pp 15–23

Webster CR, Lorimer CG (2002) Single-tree versus group selection in hemlock-hardwood forests: are smaller openings less productive? Can J For Res 32:591

Webster CR, Lorimer CG (2005) Minimum opening sizes for canopy recruitment of midtolerant tree species: a retrospective approach. Ecol Appl 15:1245–1262

Woods KD (2000) Long-term change and spatial pattern in a late-successional hemlock-northern hardwood forest. J Ecol 88:267–282

Woods KD (2004) Intermediate disturbance in a late-successional hemlock-northern hardwood forest. J Ecol 92:464–476

Part IV
Research Trajectories in Wildlife, Fauna, Insects

Chapter 10
The Starkey Experimental Forest and Range: Long-Term Research to Meet the Needs of Managers

Marty Vavra and Michael J. Wisdom

Abstract As the current concern over climate change illustrates, there is a need, sometimes unforeseen, for the development and maintenance of long-term data sets. Coupling long-term vegetation data sets with long-term wildlife demographics allows scientists to evaluate potential cause–effect relations between changing landscapes and wildlife populations. Vegetation data sets contain important variables often included in habitat effectiveness models that predict the value of a landscape for a particular wildlife species. In this chapter, we focus on Starkey's research history and long-term research activities on wildlife and rangeland resources as case examples of applied research that has addressed societal needs for ecosystem services. In 1939, Pickford wrote a letter that initiated discussions on the Starkey Cattle and Horse Allotment in the Whitman National Forest as suitable for a research station to develop information on the management of cattle summer range within the ponderosa pine (*Pinus ponderosa* Dougl. ex. Loud) type. In 1940, the Starkey Experimental Forest became a reality. The major research problem to be addressed was the overgrazing of mountain ranges. The current research trajectory was initiated in 1989 with the building of a game-proof fence and development of a telemetry system to monitor ungulate movements. The initial studies focused on perceived conflicts between deer and elk, and timber harvest and livestock grazing. Current research focuses on the role of ungulates as chronic ecosystem disturbance agents and their interaction with episodic disturbance to alter successional trajectories. Also included is the confounding role of human disturbance in modifying ungulate distribution across landscapes.

Keywords Livestock grazing · Deer · Elk · Wildlife habitat · Forage quality · Herbivory

M. Vavra (✉) · M. J. Wisdom
Pacific Northwest Research Station, USDA Forest Service, 1401 Gekeler Lane, La Grande, OR 97850, USA
Phone: 541-420-2120
e-mail: martinvavra@fs.fed.us

D. C. Hayes et al. (eds.), *USDA Forest Service Experimental Forests and Ranges*, DOI 10.1007/978-1-4614-1818-4_10, © Springer New York 2014

10.1 Introduction

National leaders and policy makers are facing profound issues regarding people and the environment. Among these are impacts of climate change, biodiversity, and the value of ecosystem services. Managers are considering these developing issues while addressing immediate problems such as increasing wildfires from fuels accumulation and exotic weed invasions. Fire suppression, wildfires, livestock grazing, logging, road construction, recreation, and many other human disturbances have singly and interactively caused extreme alterations in plant community composition of forests, in turn, causing a cascading effect on wildlife. As management scenarios are developed to deal with these present and emerging problems, managers need better knowledge about the impact of those practices on nontarget species and their habitats. As the current concern over climate change illustrates, there is a need, sometimes unforeseen, for the development and maintenance of long-term data sets. Cody and Smallwood (1996) cite two reasons for long-term data collection: (1) The opportunity occurs to observe important but rare events, e.g., episodic disturbance (fire, insect outbreaks), before and after and (2) detection of processes that evolve/unfold slowly, e.g., climate change.

Vegetation structure and composition are the basic ingredients of wildlife habitat. All wildlife use vegetation for shelter, either for the security aspects (hiding) or for thermal regulation. Herbivores are, of course, dependent on vegetation for food and the carnivores are dependent on the vegetation to produce their prey species. Coupling long-term vegetation data sets with long-term wildlife demographics allows scientists to evaluate potential cause–effect relations between changing landscapes and wildlife populations. Vegetation data sets contain important variables used in habitat effectiveness (HE) models to predict the use or other values of a landscape for a particular wildlife species. These models are effective planning tools to identify and evaluate landscape options for managing plant communities, associated habitats, and other important ecosystem properties. Swetnam et al. (1999) have shown benefits from the analysis of long-term vegetation data sets. As an example, they cited fire scar chronologies being widely used to justify and guide fuel reduction and natural fire reintroduction in forests. Climate change and the need to account for its potential effects in land-use planning have particularly brought sharp focus to the lack of available data (Powledge 2008).

Experimental forests and rangelands, managed by the USDA Forest Service, form a network of locations amenable to long-term data collection across many of the major ecosystems of the continental USA, Alaska, and Hawaii (Lugo et al. 2006). Historically, long-term understory/rangeland data collection on these sites encompassed a 10–15-year period and typically focused on different intensities of livestock grazing or timber management. Common variables included those for weather, structure, and composition of merchantable timber and associated silvicultural regimes; production and composition of livestock forage and changes related to grazing intensity (including exclusion of grazing); and livestock gains (Klipple and Costello 1960; Martin and Cable 1974; Skovlin et al. 1976). In some cases, the effects of livestock management on wildlife were reported (Skovlin et al. 1976).

Truly long-term data sets (20+ years) for wildlife and understory and range vegetation are rare but do exist. Linkhart and Reynolds (2004, 2007) studied return rate, fidelity, dispersal, and longevity of flammulated owls on the Manitou Experimental Forest in Colorado. The H.J. Andrews Experimental Forest served as one of the important study areas for long-term research on the biology and ecology of the spotted owl (Anthony et al. 2006). Studies examining understory/rangeland vegetation successional trends have been carried out on a few experimental ranges (Buffington and Herbel 1965; Angel and McClaran 2001). Unfortunately, the status of many older studies or supporting data collections, in terms of longevity and variables monitored, are incompletely documented. These records often are boxed and archived in warehouses and other locations generally inaccessible to scientists (personal communication, Richard Oakes, Site Manager, Manitou Experimental Forest, Woodland Park, CO).

Pearson et al. (2008) reported on grazing and excluded plots at the Fort Valley Experimental Forest in Arizona that were sampled from 1912 to 1941 and rediscovered and resampled from 1996 to 2007. Johnson (2003) illustrates the value of long-term data collection on vegetation change in his description of 90 years of plant succession on green fescue (*Festuca viridula* Vasey) grasslands in the subalpine ecological zone of the Wallowa Mountains in Oregon, where changes in sheep grazing over time resulted in dramatic changes in plant community composition. Gibbons and Beck (1988) reported on long-term plots that were established between 1915 and 1932, continuously read until 1947, and only a portion read annually until 1979 when the collection was discontinued.

A number of reasons have likely contributed to the scarcity of long-term data sets. Personnel turnovers and a lack of ownership of long-term studies that may only be sampled every 5 or 10 years, coupled with limited funds available to hire personnel for data collection, are likely causes. Universities, which undertake much of the natural resource-related research, tend to receive support for short-term studies because the university reward system promotes research leading to frequent publications (Fribourg 2005; Vavra 2005). Federal research agencies, alternatively, commonly focus their work on longer-term projects that are tied to experimental areas (Klade 2006). Another shortcoming of some long-term data sets with infrequent monitoring is the lack of concomitant weather and soils statistics (Mueggler 1992). Sharp et al. (1990) pictorially demonstrated the link that growing season and annual precipitation have on species composition and productivity on a research site in southern Idaho.

The objective of our study is to review research on the Starkey Experimental Forest and Range that has been driven by the needs of land and wildlife managers and to discuss some of the management implications. We focus on Starkey's research history and long-term research activities on wildlife and rangeland resources as case examples of applied research that has addressed societal needs for ecosystem services.

10.2 Research Program

10.2.1 The Starkey Experimental Forest and Range: The Early Years

The Pacific Northwest Research Station (PNWRS) was established in Portland, OR in 1924 (Skovlin 1991). This coincided with a national movement to have a Forest Service research organization present in each major forested region of the country. Even previous to this time, consideration had been given to establishing a western range experiment station in the Blue Mountains. In 1936, the US Senate Document 199 focused national attention on the deteriorating state of the western range. The problem of overgrazing mountain ranges was added to the list of concerns that the PNWRS was to address. Gerald Pickford came to the PNWRS in 1936 and was joined in 1937 by Elbert "Bert" Reid. Much of their summer field seasons were spent east of the Cascades visiting problem sites and assessing research needs. These men were among the first to identify excessive elk use of mountain meadows in the Elkhorn Mountains (Pickford and Reid 1942). In 1939, Pickford wrote a letter that initiated discussions on the Starkey Cattle and Horse Allotment in the Whitman National Forest as suitable for a research station to develop information on the management of cattle summer range within the ponderosa pine (*Pinus ponderosa* Dougl. ex. Loud) type. In 1940, the Starkey Experimental Forest became a reality, but World War II interrupted research at Starkey. After the war, research focused on improved forages for livestock production. Joseph Pehanec became the chief of range research for the PNWRS in 1945. His efforts led to the addition of "Range" to the Starkey Experimental Forest title. Today Starkey remains the only experimental forest and range.

Pehanec later served as the first president of the Society for Range Management. Robert Harris joined the range research staff in 1946. He would go on to be the deputy chief of research for the Forest Service.

In the early 1950s, Jon Skovlin, rangeland scientist, and Gerald Strickler, ecologist (Fig. 10.1), came to Starkey and conducted important research there for the next three decades. They would build on the earlier research on overgrazing. In 1953, the first long-term study on cattle grazing methods and stocking rates was initiated (Skovlin et al. 1976). Cattle were grazed under deferred rotation and season-long grazing systems and at three different stocking rates for each method for 11 years. The study also included a no-grazing control. Prior to the study, most cattle grazing was season-long because there was no scientific support for a rotational system. Deferred rotation at moderate stocking rate provided better protection of the forage base and soil resources. This was the first study to show the impacts of cattle grazing on the forage resource in forests and how that in turn affected deer and elk distribution and food habits. Cattle stocking rates and the associated deferred rotation system described in this study are still in use by National Forest range managers.

Fig. 10.1 Range Ecologist Gerald Strickler clipping standing crop during an early grazing study on the Starkey Experimental Forest and Range

10.2.2 The Starkey Experimental Forest and Range: Recent

Long-term wildlife research was initiated by Evelyn Bull in 1973 with her study of pileated woodpeckers. This species was selected as a management indicator species for older forests by some regions in the USDA Forest Service because of its use of large-diameter hollow trees and snags for nesting and roosting. Large, hollow trees and snags often are felled for safety reasons during forest management activities or are felled for firewood use. Consequently, detailed research regarding the species' ecology and habitat requirements was needed. The initial study lasted 10 years (Bull 1987), with resampling done in 1989–1990 (Bull and Holthausen 1993) and 2003–2005 (Bull et al. 2007). The resampling efforts allowed the evaluation of long-term forest management practices on the woodpecker and its habitat. These studies revealed the importance of snags and down logs to the ecology of pileated woodpeckers. Forest Service woodcutting permit and logging regulations were modified to maintain large snags for nesting habitat. Recommendations for forest stand composition for optimum pileated woodpecker habitat were made. Landscapes with extensive tree mortality from insect outbreaks continued to be acceptable pileated woodpecker habitat if they were not extensively harvested. Timber harvesting that provides retention of coarse woody debris (snags and down wood) will maintain acceptable pileated woodpecker habitat.

Beginning in 1976, intensive studies of livestock performance and seasonal forage nutritional quality were initiated in cooperation with Eastern Oregon Agricultural Research Center, Oregon State University, to evaluate existing grazing systems and provide information for the development of new systems for riparian zone improvement (Walburger 2002). This research was driven by the emergence of concern for the condition of streams and associated riparian corridors across the western USA. In northeastern Oregon, many streams are spawning and/or rearing habitat for endangered salmon and steelhead stocks, which made the research more critical.

As part of this research, Holechek et al. (1982a, b) discovered a nutritional disparity between the forage on north-facing versus south-facing grassland slopes as the grazing season progressed. Forage quality deteriorated more rapidly on grasslands than in forests, but the grasslands responded with nutritionally better regrowth to late summer and fall precipitation than did the forests. Utilizing each slope at its nutritional optimum improved livestock weight gain over cattle allowed free choice of both slopes. This increased production provided some offset to the increased cost associated with fencing if streams required total protection. If livestock permittees were required to initiate rest rotation or deferred rotation systems over season-long grazing to improve riparian conditions, the question arose on whether weight gains would be similar between systems or if the forced movements required of a rotation system would reduce overall summer weight gain. Research at Starkey indicated no detrimental effects on animal production with the use of rest rotation or deferred rotation grazing systems (Holechek et al. 1987).

The 25 years of study (Walburger 2002) revealed to managers and livestock producers that the use of nutritional calendars in the design of grazing systems could increase beef weight gains. At the same time, prudent use of nutritional calendars reduces cattle use of riparian zones while maintaining the desired stocking rate and grazing area.

10.2.3 The Starkey Project

In 1973, Jack Ward Thomas arrived in La Grande, OR, as a research wildlife biologist and project leader at the Pacific Northwest Research Station's Range and Wildlife Habitat Laboratory. Thomas's duties included oversight and coordination of research at Starkey.

Earlier that year, eastern Oregon's dry forests experienced an extensive outbreak of the tussock moth. Inch-long tussock moth caterpillars had voraciously defoliated thousands of acres of Douglas fir (*Pseudotsuga menziesii,* Mirbel, Franco) and grand fir (*Abies grandis,* Dougl., Lindl) trees, leaving forests brown, ailing, and, at worst, bare in their wake. Intending to clear the landscape of dead and dying trees, National Forest managers found their efforts challenged by the recently enacted National Environmental Policy Act, which required that environmental effects of major actions on public lands be examined and disclosed before actions commence.

At the same time, eastern Oregon's elk population was increasing, raising fears among livestock producers that their cattle would face competition for forage. Wildlife managers were equally concerned about the availability of forage, but their fear was that livestock posed a greater threat to the browse that elk and deer needed to survive. Moreover, increases in timber harvest and intensive building of roads to facilitate these harvests worried managers that elk and deer were losing thermal and hiding cover. These roads had the secondary effect of opening up tracts of deer and elk habitat to hunters, offering unprecedented access. By the end of eastern Oregon's 2-week-long elk hunting season, managers were startled to find as few as 1 or 2 bulls remaining for every 100 cows. To managers, declining herd productivity, not forage, emerged as the greatest threat to the region's elk population.

Donavin Leckenby, a research project leader with the Oregon Department of Fish and Wildlife, worked in the same building as Thomas and was studying elk and their habitat in northeastern Oregon. Leckenby was a seasoned biologist who had previously completed an extensive study of mule deer behavior and habitat use in central Oregon. Leckenby and Thomas—two scientists representing two different and diverse agencies and research styles—found common ground for research on the emerging issues of elk habitat, timber harvest, and potential competition among elk, mule deer, and cattle. Leckenby brought a naturalist's perspective, which he obtained from his direct observation studies of elk. These studies also generated a sense of frustration in him—he would watch elk but did not have the experimental controls necessary to determine if their behavior was based on the preference for a habitat or if that habitat was a requirement for the well-being of the elk.

In the 1980s, a dramatic intersection of management and policy, wildlife and livestock interactions, and multiple resource use occurred as part of major changes in National Forest management. State and Federal agencies, livestock growers, and sportsmen were at odds over various issues pertaining to the management of National Forest lands. Thomas and Leckenby took stakeholder and agency input from a series of regional meetings as well as from informal correspondence. Martin Vavra, then with the Eastern Oregon Agricultural Research Center, Oregon State University, was contracted to develop the problem analysis "Forage allocation for big game and livestock in northeastern Oregon." A significant portion of the problem analysis was a compilation of elk, mule deer, and cattle issues identified by ranchers, county extension agents, and personnel from Oregon Department of Fish and Wildlife, Forest Service, Bureau of Land Management, and Isaac Walton League.

In response to the cumulative input, Thomas and Leckenby formulated major research questions to address both the sustainability of elk and deer herds and the perceived conflicts with timber harvest and livestock grazing. The research questions centered on the effects of timber harvest and roads on elk, deer, and cattle productivity and distribution; on how these large animals use landscapes and food sources; and on the effect of breeding bull elk on herd productivity (Rowland et al. 1997). Controlled experiments were needed to separate animals' preferences from requirements. Controlled experiments could only be conducted in a controlled environment; and the idea of a game-proof fence around Starkey was initiated. In July 1986, then Forest Service Chief Max Peterson approved the project. With that, the Starkey Project was born.

With the support of his colleagues, Thomas turned his attention from being a scientist to serving as the Starkey Project's promoter. First, he needed to persuade his supervisors and the National Forest System (NFS) that developing a 40-square-mile fenced research laboratory together with an automated telemetry system to monitor animal movement—as he proposed to do with the Starkey Project—was a worthwhile investment. As Thomas saw it, this field laboratory would yield data that could help managers make the inherently difficult decisions that encompass management of landscapes, elk, deer, and cattle. Ultimately, the NFS agreed and

Fig. 10.2 Elk fitted with a global positioning system (GPS)-based transmitting collar. GPS has replaced the original telemetry system

provided funding from regional and national levels. The land that Thomas proposed to fence off and change human activities as part of the project was a long-time hunting and camping area. Convincing local residents of the importance of this project was, clearly, a tough sell.

As funds were being allocated to the Starkey Project and its studies, Larry Bryant, a wildlife biologist at the La Grande Laboratory, worked with the La Grande District Forest Service to develop a timber sale around the site's perimeter. It was the longest clearcut in Forest Service history, 26 miles in length. By removing the trees, Bryant was able to install a game-proof perimeter fence down the middle of the cleared corridor, which helped to guard against wind-thrown trees damaging the fencing structure. Bryant then secured a track machine that was able to dig hundreds of post holes and navigate the rough landscape at the Starkey Project site.

The Eastern Oregon Agricultural Research Center, branch station of Oregon State University, already a long-term cooperator, provided 100 cows to the allotment on Starkey with a portion of those fitted with radio telemetry collars. Center staff provided both scientific and technical support to the project with leadership in cattle-related studies.

Over the course of the first 20 years of the project, 40 cooperators participated in the research. Included were universities, state wildlife and parks agencies, NFS, private industry, Native American tribes, federal agencies, and the Oregon Cattlemen's Association.

With the fence complete, the next challenge was to invent an automated telemetry system that could monitor animal movement frequently and accurately. The Starkey Project's success hinged on the ability to remotely and efficiently monitor exactly how elk, deer, and cattle moved and interacted across the vast project landscape in response to manipulative experiments. Bryant worked with a Texas company to develop a telemetry system that was based on a government-operated navigational aide to aircraft and ships at sea (Rowland et al. 1997). The system permitted the Starkey Project researchers to gather almost continuous movement data of large animals that had been fitted with transmitting collars (Fig. 10.2). With the fence and a reliable and advanced system for capturing data in place, the Starkey Project was operational.

The original four research problems, the influence of roads and traffic on deer and elk; the influence of intensive timber management on deer, elk, and cattle; exploring forage allocation among cattle, deer, and elk; and the influence of bull age on breeding success and calf crop of elk, were addressed in the work conducted during the 1990s (Rowland et al. 1997). The enclosure fence and the automated telemetry system were critical to the success of the research (Thomas and Wisdom 2005).

10.2.3.1 The Effect of Roads and Traffic

Fitting female elk with radio collars and recording their location several times each day allowed scientists to record more than 100,000 elk locations in relation to distance from roads. Data gathered were used (1) to test the hypothesis that elk use increased as distance from a road increased; (2) to evaluate the standard elk HE model that used a road density variable as one of the predictors of elk distribution; and (3) to examine potentially confounding effects of different spatial patterns of roads on HE model performance (Rowland et al. 2000).

The road density component and other variables in the HE model had undergone little validation even though the model had been used extensively by NFS planning and management. In the HE model, there was no component for addressing the variability in the spatial pattern of roads across a landscape. Female elk consistently selected areas away from open roads in both spring and summer. In spite of this relationship, little or no significant relations existed between the number of elk locations and HE scores based on road density. From this study, the formulation of distance bands was developed, offering managers a more spatially appropriate scale for predicting road effects on elk distribution. This new approach gave managers a better estimation of the true landscape scale habitat available to elk.

Roads per se are not really the issue in moving elk. In fact, it is the traffic on those roads that initiates movement (Wisdom et al. 2005b). Within Starkey, traffic counters were placed in selected locations on roads open to traffic (Rowland et al. 1997, 1998). Counts of traffic were used to characterize the rate of traffic on each road segment. Animal movements were again recorded with the automated telemetry system. Traffic rates as low as one vehicle per 12 h caused elk to move farther from the roads, and elk avoidance of roads increased with increasing rates of traffic (Wisdom et al. 2005b). Deer responded in the opposite manner, with increased selection toward roads of higher traffic rates, which was shown to be related strongly to elk avoidance rather than a reaction to roads or traffic (Johnson et al. 2000; Wisdom et al. 2005a).

This study reinforced the findings of Rowland et al. (2000, 2005) and also illustrated that mule deer habitat selection was strongly influenced by elk distribution. The findings of these two studies would strongly influence management decisions on road management in Forest Plans across the western USA and lead to new research dealing with the influence of human disturbance on distribution and behavior of elk and mule deer.

10.2.3.2 Intensive Timber Harvest

At the inception of the Starkey Project, intensive timber harvest was practiced on much of the Forest Service lands in the western USA. Wildlife managers raised concerns over the loss of thermal and hiding cover and the potential negative impacts on elk productivity. The intensive timber harvest study was initiated on 1,416 ha acres enclosed by a game-proof fence so that elk and cattle utilizing the study area could not leave and were forced to reside within the area being logged. This allowed scientists to directly measure elk and cattle response to the timber harvest treatment (Wisdom et al. 2005c). The following summary of methods and results provided in this study are from Wisdom et al. (2005c). Radio-collared animals utilized the study pasture before harvest (1989), during timber sale planning and layout (1990–1991), during harvest (1992), and during postharvest management (1993–1996). Timber harvest occurred on 50% of the forest lands in the study area and was primarily shelterwood and seed tree regeneration cuts. Cattle and elk distribution before, during, and after timber harvest was monitored and data were utilized to map spatial distributions. Both elk and cattle in the study area were weighed annually and those weights were compared to those of elk and cattle grazing the main study area.

Elk distribution changed substantially during timber harvest with the use concentrated on the study area's outer boundaries. Elk distribution also became more diffuse during harvest; nearly twice as much of the study area was within the 50% use volume compared to the period before harvest. Elk decreased their use near roads during timber harvest when traffic rates increased substantially. They also utilized steeper slopes during harvest indicating a selection for greater security. After harvest, the elk not only returned to the western half of the study area but also increased their use in the interior portion, the area which was largely unused during harvest. Elk distribution was more diffuse after harvest than before but less diffuse than that during harvest.

In contrast to elk, cattle showed little change in distribution during all periods of study. The areas of highest concentration remained consistent before, during, and after timber harvest. Unlike elk, cattle showed no evidence of selection of areas with characteristics of greater security from humans.

If timber harvest has a negative impact on elk or cattle then one metric to use as an indicator would be decreased weight gains. Cattle, (cows and calves) were weighed on and off Starkey in both the main study area and the intensive timber study area each year. Likewise, elk (cows in spring, cows and calves in fall) were weighed before turn out onto the study areas and again in the late fall when they entered the feedground. The main study area served as a control to the intensive timber harvest treatment. If timber harvest caused a treatment effect in weight gains, then within a given year weight gains in the intensive timber harvest treatment would be expected to decrease or increase compared to those in the main study area. The direction and degree of variability in annual weight gains was generally consistent between the two study areas for both species. Weight gains then were largely affected by weather patterns that affect annual changes in forage biomass and nutritional quality.

An important aspect of intensive timber harvest is the development of a more open landscape and the opportunity for hunters to harvest more elk with less effort. Elk vulnerability to hunter harvest increased with timber harvest. For the years before timber harvest, hunter success averaged 22% and required an average of 19 days to achieve that level of success. During timber harvest, success increased to 35% with only 9 days required to achieve that success. After timber harvest, hunter success was 32% and was achieved in 14 days. The data just reported were for hunter entry to the study area on foot only. When vehicle use was allowed, the success increased to 54% with only 14 days required to reach that. For landscapes with extensive timber harvest or fuels treatments that greatly reduce hiding cover, a combination of road closures and limited hunter entry is likely to be effective in preventing overharvest.

This research, in conjunction with other studies that evaluated the physiological response of elk to varying levels of thermal cover (closed forest canopy; Cook et al. 1998), revealed that elk productivity will not be impaired by intensive timber harvest. This research illustrated the importance of the Starkey Project because enclosure fencing provided the opportunity to differentiate between preference and requirement in regard to canopy cover and elk. At times, elk prefer habitats that provide security cover in the form of a dense multistoried canopy but do not require this stand structure for homeostasis. The study shifted potential management emphasis from thermal cover retention to management that reduced elk vulnerability to hunter harvest. Reduced vulnerability is achieved through the retention of security areas and restrictions on motorized access. These goals could be reached by planning timber harvest activities in time and space such that a mosaic of seral stages is maintained to provide a variety of foraging conditions and security cover areas (Wisdom et al. 2005c).

10.2.3.3 Forage Allocation

Range managers have long attempted to develop proper animal stocking of western rangelands. Where multiple ungulates graze common landscapes, the questions of forage allocation and competition for limited forage among species have loomed large. Managers have used animal unit equivalencies based on body weight ratios to allocate forage and develop stocking rates; a cow is five times larger than a deer, so five deer equal one cow. The goal of this phase of the Starkey Project was to develop a forage allocation model to evaluate different grazing management strategies on interior forest summer ranges and test various hypotheses about the effects of alternative stocking rates for ungulates. Trying to model forage removal and animal performance for multiple species of ungulates across large heterogeneous landscapes is difficult due to the temporal and spatial variability in animal distributions, forage production, and forage nutritional value.

At Starkey, animal distribution was estimated with the use of resource selection functions (RSF). An RSF represents the probability that an animal will select or avoid resources over space and time in relation to available resources. RSFs can then be integrated within a larger forage allocation model. Coe et al. (2005) developed RSFs for elk, mule deer, and cattle on the summer range at Starkey. During early and midsummer, the presence of cattle affected elk distributions, as shown clearly by the RSFs. Specifically, elk either moved to areas of the pasture where cattle did not go or elk left the pasture entirely. Likewise, elk affected where mule deer were located in the study pasture. Mule deer moved away from elk. These interactions indicate cascading effects of the larger herbivore displacing the smaller one. Management decisions that change cattle distribution or season of use change elk distribution which in turn changes mule deer distribution.

Late summer and fall are typically warm and dry in the interior Pacific Northwest. Forage for ungulates has matured and dried at this time causing a decline in their nutritive quality. Earlier in the summer when forage is both abundant and nutritious, animals separate themselves across the landscape. However, once nutrients become limiting, overlap in animal distribution increases. RSFs become more similar in relation to the scarcity of nutritional resources, forcing coexistence among ungulate species and increasing the potential for competition.

The traditional animal unit equivalency calculation based on animal weights requires that animals overlap 100 % in their distribution and diets. RSFs developed for elk, mule deer, and cattle indicate that animals vary seasonally in their distributional overlap and that cascading effects occur whereby larger animals displace smaller species.

Dietary overlap is the other major issue to be addressed in developing a forage allocation model. Findholt et al. (2005) designed a manipulative study on Starkey that addressed that issue by utilizing tame elk, mule deer, and cattle in enclosure studies with the specific intent of estimating diets and overlap among the species.

In ungrazed pastures, dietary overlap between cattle and mule deer was lowest but increased in response to previous cattle grazing. Nutritional analysis of the diets indicated no change in nutrient densities suggesting that competition was not occurring. With dietary overlap as an indicator, the greatest potential for competition was between mule deer and elk. However, if competition does occur, it may be interference competition as mule deer move away from elk (Coe et al. 2001). Since mule deer and elk tend to have similar RSFs in late summer when resources are scarce, it would appear that this early summer movement is preference related. It would seem that if in moving, mule deer encountered scarce nutritional resources, they would in fact move back near elk as they do in late summer when nutritional resources are scarce. Cattle and elk have similar diets particularly on pastures previously grazed by cattle. However, when cattle and elk grazed pastures previously grazed by cattle there is no compromise in either species' diet nutritional quality. Over time cattle nutrition decreases on pastures previously grazed by cattle, indicating intraspecific competition. The lack of 100 % dietary overlap among the ungulate species again negates the use of animal unit equivalencies based on animal weight.

From the data accumulated in the Starkey Project, development of a foraging model was initiated (Ager et al. 2005). The model predicts animal weight dynamics and animal distributions in response to stocking rates, grazing systems, and the influences of other human activities such as road use. The major challenge to refine the current model is to determine what mechanisms in the foraging process are the most important determinants of landscape scale foraging behavior and animal performance. An objective is to identify the existence of key stocking thresholds that correspond to changes in animal performance at the species level. Such a tool is currently not available (Ager et al. 2005).

10.2.3.4 Age of Breeding Bulls

Elk herds in northeastern Oregon were declining when the Starkey Project began. Potential causes were identified as nutritional condition of females on conception dates, pregnancy rates, and age of male sires. In many elk management units in northeastern Oregon, only one or two bulls remained at the end of hunting season. That meant that most breeding in the following fall rut was by yearling bulls. Wildlife managers and sportsmen were concerned over the future of elk herds given the potential for declining productivity. Therefore, a study was developed to assess the effects of male age and female nutritional condition on conception dates and pregnancy rates of female elk (Noyes et al. 1996, 2002, 2005). Again, an enclosed study area was required to conduct the research so that manipulation of the male population could be accomplished. The elk herd during the study was managed so that a single cohort of males functioned as principal herd sires as they matured from 1 to 5 years of age.

Breeding by mature males achieved early and synchronous conception of females. Conception dates for females were strongly influenced by the age of sires; conception dates became progressively earlier as sires matured from yearlings to 5-year olds. Earlier birth dates are linked to increased neonate survival. In fall, calves born earlier are larger than calves born later. Calf body size is an important factor in surviving the first winter. Synchronous births also influence calf survival. A critical period for calf predation is the first 2 weeks after birth when the calf has limited mobility. Typically, calves remain hidden and motionless only getting up when the dam arrives for nursing. During this time, calves are particularly susceptible to predation. Synchronous breeding means synchronous calving over a short time frame. Calves born during a short time span flood the system during that 2-week susceptibility period and predation is reduced in contrast to a calving season that continues over an extended period, the latter of which allows for increased opportunities for predation.

Elk hunting seasons in Oregon and other western states were drastically modified based on the results from the breeding bull study at Starkey. Restrictions were placed on mature bull harvest and limited entry to specific game management units was implemented. Now, more mature bulls remain in the herds after the hunting season. Some of these bulls are available to hunters the next year. Use of these results in management enhanced both herd productivity and hunter opportunity to harvest a mature bull.

10.3 Synthesis of Research Trajectory

Over the past 22 years, results from the Starkey Project have been presented to managers in a variety of venues. These results have, in turn, been used by state and federal managers to formulate hunting seasons designed for a more sustainable harvest and in the forest planning process. Three workshops have been conducted to target different audiences. The workshop "Elk, mule deer, and cattle in forests: A workshop for managers" was held in La Grande, Oregon on September 5 and 6, 2001. Registration was 160 people from 6 states representing 15 National Forests, 3 Native American tribes, 2 universities, Department of Army, Bureau of land Management, and 5 state fish and game agencies. On February 26, 2003, a wildlife professionals' workshop "The Starkey Deer and Elk Project: Results of long-term research for management" was sponsored by and conducted at the Rocky Mountain Elk Foundation Annual Meeting. Another workshop for wildlife professionals was held on March 22, 2004, at the North American Wildlife and Natural Resources Conference. Papers from this last workshop were published in the Conference Proceedings and as a book titled *The Starkey Project: A Synthesis of Long-term Studies of Elk and Mule Deer* (Wisdom 2005). The book is in its third printing.

Roads and traffic management studies produced findings about mule deer and elk response to open roads and traffic rates that are now being used in the current round of forest planning. Elk selected habitats away from roads (Rowland et al. 2000; Rowland et al. 2005; Wisdom et al. 2005b) while mule deer selected habitats away from elk and closer to roads (Johnson et al. 2000). From this work, research was developed to evaluate and compare the effects of off-road vehicle use (ORV), horse riding, mountain biking, and hiking on elk and mule deer. All four forms of recreation disturbed elk and elicited a flight response. ORV and mountain biking impacts were greater than the other two forms of recreation (Wisdom et al. 2005a; Naylor et al. 2009). Results of the roads and traffic study and the recreation study are being used across the western USA by National Forests in forest and travel management planning.

Timber harvest has little negative effect on mule deer, elk, and cattle, based on Starkey studies, as long as the increased vulnerability of elk and mule deer is recognized and road access is managed (Wisdom et al. 2005c). These results are easily extrapolated to include mechanical fuels reduction. Additionally, the perceived physiological requirement of elk for thermal cover was not supported by research (Cook et al. 1998, 2005).

New tools for innovative and equitable allocation of stocking rates among elk, deer, and cattle are being refined from Starkey Project research (Ager et al. 2005; Coe et al. 2001, 2005; Findholt et al. 2005). This phase of the original studies is still ongoing. The goal is to develop a HE model for cattle that is useful in predicting cattle distribution across seasons and is user friendly for managers.

The results of the breeding age of bulls study have been utilized by state fish and wildlife agencies to establish new elk hunting regulations that insures the carryover of mature bulls (Noyes et al. 1996, 2005). Prior to the study, bull elk hunting seasons in Oregon allowed unlimited entry into all management units for the entire season and did not have antler (age) restrictions on bull harvest. Study results were used to modify hunting seasons so that entry to specific game management units was restricted and the number of mature bull tags was restricted. Washington and Idaho also modified seasons in a similar fashion for some of their game management units. Results have been very positive in that now there are more mature bulls available for breeding and harvest which has led to increased hunter satisfaction. Elk hunting tag application is, however, more complicated and hunting opportunities for mature bulls do not occur for an individual every year.

Starkey has yielded rich data that addressed the fundamental research questions on which it was founded (Thomas and Wisdom 2005). But, after more than 20 years of active research, the project was in a state of uneasy transition. The original research problems Thomas and his colleagues developed earlier had been addressed, countless publications outlining their findings had been prepared, and a great deal of findings had been disseminated. The Starkey Project had lived up to its expectations, so its researchers, frankly, were unsure of how they should proceed with potential future research.

An external review team was commissioned and met in June 1998 to review accomplishments and recommend research direction for the future. The review team was composed of an individual each from Wildlife Management Institute, Colorado Division of Wildlife (representing state wildlife agencies), US Geological Survey, Forest and Rangeland Ecosystem Service Center, and NFS. The team praised the project for developing a "world-class research facility," fostering interagency cooperation in research, the high level of professionalism and productivity, and the depth of the data collected. They went on to report that the facility and the tractable elk herd made the Starkey Project uniquely suited to become a world-leading center for long-term, large-scale research on ecosystem processes relating the dynamics of herbivores and plants. Suggestions for future research included:

1. Large-scale influences of ungulate densities on behavior of individuals, reproduction dynamics, and spatial relations among species of herbivores
2. The effects of domestic and native herbivores on the composition and structure of vegetation, ecosystem processes, and long-term productivity
3. More thorough analyses of existing data on animal distribution and forest landscape characteristics
4. Utilization of the tractable elk to study the grazing preferences and behavior as related to silvicultural prescriptions, cover characteristics, and experimental grazing regimes
5. Research related to human disturbance and animal distribution

During this time, Tom Quigley, first as the program manager of the Managing Disturbance Regimes research program and then as PNW station director, provided key leadership and funding support during critical periods. As the program manager, Quigley established positions for two new scientists at Starkey and secured funding for the initiation of new research directions. Then, as the station director, he provided funds for the development and construction of a replacement telemetry system based on global positioning system technology. Thanks to Quigley's support, those funds became a permanent part of the Starkey Project's research budget. With this renewed support, a new era of research began at Starkey and continues today—one that addresses contemporary issues, including fuels reduction impacts on wildlife; ungulate impacts on ecosystems; human disturbance impacts on elk and deer; and the ecology of invasive plants.

10.4 Societal Impacts

Results of Starkey research have had direct benefits to both state and federal land and wildlife management agencies for land use and population management. Environmental, wildlife, hunting, and nongovernment organizations have benefited from the many applications of results in land and wildlife management. Results have also been applicable to the operations of private landowners and managers, forest product industries, and Native American tribes.

Research findings on the effects of roads traffic and off-road recreation on mule deer and elk now constitute part of the foundation for the national roads policy established by the US Department of Agriculture, Forest Service, thus affecting road and recreation management on all National Forests. Public use of National Forests will be affected by these new travel policies.

Data on elk and mule deer distribution as affected by physical features of the landscape, management, and seasonal nutritional value collected during the Starkey Project are now being used to develop new HE models for eastern Oregon forests. The models will have general application for most forests in the interior West. At minimum, model design can serve as a template for developing more local models. Models will provide information to managers developing forest plans.

Opportunities for wildlife viewing and hunting have benefited from Starkey research on the age of breeding bulls. More mature bulls are now present in herds in the Northwest, providing enhanced recreational opportunities for viewing and hunting.

Improved cattle management in forests in eastern Oregon should occur as further refinements are developed for cattle distribution models. Better cattle management should improve forage availability for mule deer and elk, improve vegetation diversity, and improve riparian vegetation conditions.

Native American tribes are increasingly active in the management of elk and deer on tribal lands, most also graze cattle, and they regulate tribal harvest on public lands where they have hunting rights.

Results from the Starkey Project have enhanced ecosystem services in the form of ecological functions that sustain and improve human life in association with National Forests and their management. Ecosystem services provided by the Starkey Project have future implications for mediating climate change and sustaining associated forest resources. For example, better management of ungulates could enhance the regulating ecosystem services of carbon sequestration and water management through more sustainable use of plant communities. Supporting ecosystem services, primary production and nutrient cycling, likewise, will benefit from better ungulate management. Cultural services like recreation will also be improved.

10.5 Future Direction

Upon completion of the original four research thrusts, Starkey scientists utilized recommendations from the 1998 review, input from public and industrial forest biologists, and a review of literature to develop new directions for the project. A problem analysis "The role of ungulate herbivory and management on ecosystem patterns and processes" was completed and signed off by the PNWRS in 2003. That document now provides the general research direction for the Starkey Project.

10.5.1 Herbivory

At the close of the first four studies, new directions for research at Starkey were needed. Years of forest management in regard to timber harvest methods, livestock grazing, and fire suppression, while considered to be good practices at the time, have left National Forests out of alignment with what is considered a more ecologically based range of variability (Hann et al. 1997). Changes in the composition and density of forest canopies in the interior West had resulted in forests that were more prone to catastrophic wildfire. Recent management directions utilizing both mechanical and prescribed fire methods are being instituted to reduce fuels present on forestlands with the intention of reducing fire risk. Little was known on how these efforts might impact the behavior and distribution of mule deer, elk, and cattle. Research on Starkey involving dietary overlap of elk, mule deer, and cattle (Skovlin et al. 1976; Findholt et al. 2005) revealed that these animals would consume a wide range of forage species and could potentially alter the composition of plant communities through the herbivory process. Riggs et al. (2000) reported on long-term data taken from grazing exclosures in northeastern Oregon confirming that ungulate herbivory should be considered a chronic disturbance to plant communities and that plant succession following episodic disturbance, e.g., logging, fire, could be altered by ungulate herbivory.

A new direction for Starkey research was initiated based on the new emphasis of fuels reduction on forests and the potential interaction with ungulate herbivory and the possibility for alternative plant successional trajectories (Fig. 10.3). During 2001, 2002, and 2003, a total of 2,000 acres at various locations on Starkey was treated mechanically for fuels reduction and then followed with prescribed fire. Long-term research is focused on the impacts of ungulate herbivory on plant succession following the episodic disturbance of fuels reduction (Vavra et al. 2005). As Riggs et al. (2000) pointed out, the knowledge generated to date is rudimentary. Studies must be designed to integrate herbivory into disturbance

Fig. 10.3 An illustration of different trajectories of secondary succession following wildfire as influenced by herbivory or a lack thereof (inside fence)

research at scales meaningful to management. The Starkey study is designed to follow successional trajectories on plant communities subjected to mechanical fuels reduction treatments followed by prescribed fire and on similar untreated control plant communities. Replicated study sites each containing three levels of elk or cattle herbivory, low, medium, and high, have been established and a sampling protocol developed to follow successional trajectories (Vavra et al. 2005).

10.5.2 Fuels Reduction and Animal Behavior

As part of the fuels reduction research over a 2-year period, the influence of fuels reduction on mule deer and elk distribution was evaluated (Long et al. 2008a, b). Elk did not alter their behavior within a home range but rather altered their home ranges as a result of fuels reduction. Plant communities that had fuels reduction were used by elk in spring and early summer but avoided in late summer because of a decline in forage quality. Elk used non-treated controls during that time. Other influences on elk distribution noted in earlier studies also pertained here. Elk use of fuels-treated communities was affected by the presence of cattle and distance to roads. Mule deer use of fuels-treated communities was dependent on elk distribution as had been noted in previous studies. Fuels reduction opened up forest canopies and allowed more sunlight to reach understory vegetation, causing forage to mature earlier in the summer and lose its nutritive quality faster than in untreated controls. Elk responded to this array of forage by using treated stands early in the year and controls later. If elk are a consideration on forest landscapes being treated for excessive fuels, a mosaic of treated and untreated units would provide elk with an array of foraging choices across spring, summer, and fall that provide high-quality forage early in the growing season (fuels-treated units) as well as late in the season (untreated units).

Fig. 10.4 National Forest recreationists often encounter wildlife. Such encounters may lead to altered distribution of the impacted species

10.5.3 Human disturbance

Human access to National Forest lands has been facilitated by years of timber harvest and road construction. Human disturbance via road traffic is a potential modifier of elk distribution as noted by previous Starkey research. Recreation use has changed drastically on National Forests. The leisurely pursuits of a camping trip have been replaced by recreationists that use more of the landscape. Use of motorized off-road vehicles, motorcycles and all terrain vehicles (ATVs), mountain bikes, horses, and even hiking have put more people into the habitats occupied by mule deer and elk (Fig. 10.4). In the past, many National Forests have had few restrictions on road and off-road travel. The current forest planning cycle, however, has seen an increased awareness by managers that restrictions on off-road travel and road closures are needed to maintain ecological integrity of landscapes. Motorized users of public lands have been very vocal in their opposition to regulation and restriction of access.

Initial research at Starkey found that mountain biking, ATV use, hiking, and horse riding all disrupted elk behavior (Wisdom et al. 2005a; Naylor et al. 2009). Mountain biking and ATV use were the most disruptive. Elk spent less time foraging, more time in security cover, moved farther from use areas, and expended more energy when exposed to human disturbance than when none was present.

Results from this study are being used by forest planners to develop off-road recreation management plans that provide suitable refugia for wildlife. Results from this study were also used to develop an ongoing study on the use of ATVs during hunting season and that influence on the behavior of mule deer and elk.

10.6 Conclusions

Early research at Starkey focused on the problems of the day. Improper cattle grazing and the decline in the integrity of the forage base were addressed. This research was supported by Forest Service managers and livestock owners alike.

Many of the management recommendations that came from that research are still in use today.

Later, research dealing with wildlife and forest management interactions provided managers with information critical to retention of species of interest. Concerns over the impacts of grazing on stream and riparian zone integrity lead researchers to initiate studies on novel grazing systems to alleviate the season-long use of riparian corridors. These studies were the first long-term data sets evaluating cattle performance seasonally on eastern Oregon forest ranges. The research provides information to managers attempting to blend environmentally sustainable grazing systems with sustainable beef production for the permittees.

The Starkey Project—now in its 26th year—owes a great deal to the team of visionary researchers who established it and supported its research over the years. The implications of Starkey's work are far-reaching, with forest managers in the Interior Northwest and beyond routinely using findings yielded from Starkey's research to manage natural resources. The foundation of the Starkey Project is solid, significant, and resilient, meaning the site will continue to provide important contributions to natural resource management in the Interior Northwest for many years to come.

Acknowledgments The authors would like to thank Yasmeen Sands, writer and technology transfer specialist, Pacific Northwest Research Station, for assistance in the development of portions of the narrative. The authors would also like to thank anonymous reviewers for their constructive input.

References

Ager AA, Johnson BK, Coe PK, Wisdom MJ (2005) Landscape simulation of foraging by elk, mule deer, and cattle on summer range. Pages 170–184 in Wisdom MJ, technical editor, The Starkey Project: a synthesis of long-term studies of elk and mule deer. Reprinted from the 2004 Transactions of the North American Wildlife and natural Resources Conference, Alliance Communications Group, Lawrence, Kansas, USA

Angel DL, McClaran MP (2001) Long-term influences of livestock management and a non-native grass on grass dynamics on the desert grassland. J Arid Environ 49:507–520

Anthony RG, Forsman ED, Franklin AB et al (2006) Status and trends in the demography of northern spotted owls, 1982–2003. Wildl Mongr 163:48 (The Wildlife Society)

Bull EL (1987) Ecology of the pileated woodpecker in Northeastern Oregon. J Wildl Manage 51:472–481

Bull EL, Holthausen RS (1993) Habitat use and management of pileated woodpeckers in Northeastern Oregon. J Wildl Manage 57:335–345

Bull EL, Nielsen-Pincus N, Wales BC, Hayes JL (2007) The influence of disturbance events on pileated woodpeckers in Northeastern Oregon. For Ecol Manage 243:320–329

Buffington LC, Herbel CH (1965) Vegetational changes on a semidesert grassland range. Ecol Monogr 35:139–164

Cody ML, Smallwood JA (eds) (1996) Long-term studies of vertebrate communities. Academic Press, San Diego, 597 p

Coe PK, Johnson BK, Kern JW, Findholt SL, Kie JG, and MJW (2001) Responses of elk and mule deer to cattle in summer. J Range Manage 54:A51–A76

Coe PK, Johnson BK, Stewart KM, Kie JG (2005) Spatial and temporal interactions of elk, mule deer, and cattle. Pages 150–158 in Wisdom MJ, technical editor, The Starkey Project: a synthesis of long-term studies of elk and mule deer. Reprinted from the 2004 Transactions of the North American Wildlife and Natural Resources Conference, Alliance Communications Group, Lawrence, Kansas, USA

Cook JG, Irwin LL, Bryant LD, Riggs RA, Thomas JW (1998) Relations of forest cover and condition of elk: a test of the thermal cover hypothesis in summer and winter. Wildl Monogr 141:61

Cook JG, Irwin LL, Bryant LD, Riggs RA, Thomas JW (2005) Thermal cover needs of large ungulates: a review of hypothesis tests. Pages 185–196 in Wisdom MJ, technical editor, The Starkey Project: a synthesis of long-term studies of elk and mule deer. Reprinted from the 2004 Transactions of the North American Wildlife and Natural Resources Conference, Alliance Communications Group, Lawrence, Kansas, USA

Findholt SL, Johnson BK, Damiran D, DelCurto T, Kie JG (2005) Diet composition, dry matter intake, and diet overlap of mule deer, elk, and cattle. Pages 159–169 in Wisdom MJ, technical editor, The Starkey Project: a synthesis of long-term studies of elk and mule deer. Reprinted from the 2004 Transactions of the North American Wildlife and Natural Resources Conference, Alliance Communications Group, Lawrence, Kansas, USA

Fribourg H (2005) Where are the land-grant colleges headed. J Nat Resour Life Sci Educ 34:40–43

Gibbons RP, Beck RF (1988) Changes in grass basal area and forb densities over a 64-year period on grassland types of the Jornada Experimental Range. J Range Manag 41:186–192

Hann WJ, Jones JL, Karl MG et al (1997) Landscape dynamics of the basin. In: Quigley TM, Arbelbide SJ (eds) An assessment of ecosystem components in the interior Columbia Basin and portions of the Klamath and Great Basins. General Technical Report PNW-GTR-405. USDA Forest Service Pacific Northwest Research Sta. Portland, Oregon

Holechek JL, Vavra M, Skovlin J, Krueger WC (1982a) Cattle diets in the Blue Mountains of Oregon I: grasslands. J Range Manage 35(1):109–112

Holechek JL, Vavra M, Skovlin J, Krueger WC (1982b) Cattle diets in the Blue Mountains of Oregon II: forests. J Range Manage 35(2):239–242

Holechek JL, Berry TJ, Vavra M (1987) Grazing system influences on cattle performance on mountain range. J Range Manage 40(1):55

Johnson CG Jr (2003) Green fescue rangelands: changes over time in the Wallowa Mountains. General Technical Report PNW-GTR-569. USDA Forest Service, Pacific Northwest Research Station, 41 p

Johnson BK, Kern JW, Wisdom MJ, Findholt SL, Kie JG (2000) Resource selection and spatial separation of mule deer and elk during spring. J Wildl Manage 64:685–697

Klade RJ (2006) Building a research legacy: the intermountain station, 1911–1997. General Technical Report RMRS-GTR-184 USDA Forest Service, 259 p

Klipple GE, Costello DF (1960) Vegetation and cattle responses to different intensities of grazing on short-grass ranges on the central Great Plains. USDA Technical Bulletin 1216. USDA Washington, DC, 82 p

Linkhart BD, Reynolds RT (2004) Longevity of flammulated owls: additional records and comparisons to other North American strigiforms. J Field Ornithol 75(192):195

Linkhart BD, Reynolds RT (2007) Return rate, fidelity, and dispersal in a breeding population of flammulated owls (*Otus flammeolus*). Auk 124:264–275

Long RA, Rachlow JL, Kie JG (2008a) Effects of season and scale on response of elk and mule deer to habitat manipulation. J Wildl Manage 72:1133–1142

Long RA, Rachlow JL, Kie JG, Vavra M (2008b) Fuels reduction in a western coniferous forest: effects on quantity and quality of forage for elk. Rangel Ecol Manage 61:302–313

Lugo AE, Swanson FJ, Gonzalez OM, Adams MB, and others (2006) Long-term research at the USDA Forest Service's experimental forest and ranges. Bioscience 56:39–48

Martin SC, Cable DR (1974) Managing semidesert grass-shrub ranges Vegetation responses to precipitation, grazing, soil texture, and mesquite control. USDA Technical Bulletin 1480. USDA Washington, DC, 45 p

Mueggler WF (1992) Cliff Lake bench research natural area: problems encountered in monitoring vegetation change on grasslands. Research Paper INT-454. Intermountain Research Station, Ogden, UT, 13 p

Naylor LM, Wisdom MJ, Anthony RG (2009) Behavioral responses of North American elk to recreational activity. J Wildl Manage 73:328–338

Noyes JH, Johnson BK, Bryant LD, Findholt LL, Thomas JW (1996) Effects of bull age on conception dates and pregnancy rates of cow elk. J Wildl Manage 60:508–517

Noyes JH, Johnson BK, Dick BL, Kie JG (2002) Effects of male age and female nutritional condition on elk reproduction. J Wildl Manage 66:1301–1307

Noyes JH, Johnson BK, Dick BL, Kie JG (2005) Influence of age of males and nutritional condition on short- and long-term reproductive success of elk. Pages 94–101 in Wisdom MJ, technical editor, The Starkey Project: a synthesis of long-term studies of elk and mule deer. Reprinted from the 2004 Transactions of the North American Wildlife and Natural Resources Conference, Alliance Communications Group, Lawrence, Kansas, USA

Pearson HA, Clary WP, Moore MM, Sieg CH (2008) Range management research, Fort Valley experimental forest. In: Olberding, SD, Moore MM (tech coords). Fort Valley experimental forest—a century of research 1908–2008. Proceedings RMRS-P-53CD. Fort Collins, CO: USDA, Forest Service, Rocky Mountain Research Station, 408 p

Pickford GD, Reid EH (1942) Competition of elk and domestic livestock for summer range forage. J Wildl Manage 7:328–332

Powledge F (2008) Climate change and public lands. Bioscience 58:912–918

Riggs RA, Tiedemann AR, Cook JG, Ballard TM et al (2000) Modification of mixed-conifer forests by ruminant herbivores in the Blue Mountains ecological province. Research Paper PNW-RP-527 USDA Forest Service Pacific Northwest Research Station, 77 p

Rowland MM, Bryant LD, Johnson BK et al (1997) The Starkey Project: history, facilities, and data collections methods for ungulate research. PNW-GTR-396 USDA Forest Service Pacific Northwest Research Station, 62 p

Rowland MM, Coe PC, Stussy RJ et al (1998) The Starkey habitat data base for ungulate research: construction, documentation, and use. PNW-GTR-430 USDA Forest Service Pacific Northwest Research Station, 48 p

Rowland MM, Wisdom MJ, Johnson BK, Kie JG (2000) Elk distribution and modeling in relation to roads. J Wildl Manage 64:672–684

Rowland MM, Wisdom MJ, Johnson BK, Penninger MA (2005) Effects of roads on elk: Implications for management in forested ecosystems. Pages 42–52 in Wisdom MJ, technical editor, The Starkey Project: a synthesis of long-term studies of elk and mule deer. Reprinted from the 2004 Transactions of the North American Wildlife and Natural Resources Conference, Alliance Communications Group, Lawrence, Kansas, USA

Sharp LA, Sanders K, Rimbey N (1990) Forty years of change in a shadscale stand in Idaho. Rangelands 12:313–328

Skovlin JM (1991) Fifty years of research progress: a historical document on the Starkey Experimental Forest and Range Gen Tech Rep PNW-GTR-266. USDA, Forest Service, Pacific Northwest Research Station, 58 p

Skovlin JM, Harris RW, Strickler GS, Garrison GA (1976) Effects of cattle grazing methods on ponderosa pine-bunchgrass range in the Pacific Northwest. USDA Forest Service Technical Bulletin 1531. USDA Washington, DC, 40 p

Swetnam TW, Allen CD, Betancourt J (1999) Applied historical ecology: using the past to manage the future. Ecol Appl 9:1189–1206

Thomas JW, Wisdom MJ (2005) Has the Starkey Project delivered on its commitments? Pages 240–248 in Wisdom MJ, technical editor, The Starkey Project: a synthesis of long-term studies of elk and mule deer. Reprinted from the 2004 Transactions of the North American Wildlife and Natural Resources Conference, Alliance Communications Group, Lawrence, Kansas, USA

Vavra M (2005) Random musings on land grant universities: their relevance and future. Proc West Sect Am Soc Anim Sci 56:3–6

Vavra M, Wisdom MJ, Kie JG, Cook JG, Riggs RA (2005) The role of ungulate herbivory and management on ecosystem patterns and processes: future direction of the Starkey Project. Pages 233–239 in Wisdom MJ, technical editor, The Starkey Project: a synthesis of long-term studies of elk and mule deer. Reprinted from the 2004 Transactions of the North American Wildlife and Natural Resources Conference, Alliance Communications Group, Lawrence, Kansas, USA

Walburger K (2002) Twenty-five years of grazing research at Meadow Creek in the Starkey Experimental Forest and Range. M.S. Thesis Oregon State University, Corvallis, OR, 130 p

Wisdom MJ, Ager AA, Preisler HK, Cimon NJ, Johnson BK (2005a) Effects of off-road recreation on mule deer and elk. Pages 67–80 in Wisdom MJ, technical editor, The Starkey Project: a synthesis of long-term studies of elk and mule deer. Reprinted from the 2004 Transactions of the North American Wildlife and Natural Resources Conference, Alliance Communications Group, Lawrence, Kansas, USA

Wisdom MJ, Cimon NJ, Johnson BK, Garton EO, Thomas JW (2005b) Spatial partitioning by mule deer and elk in relation to traffic. Pages 53–66 in Wisdom MJ, technical editor, The Starkey Project: a synthesis of long-term studies of elk and mule deer. Reprinted from the 2004 Transactions of the North American Wildlife and Natural Resources Conference, Alliance Communications Group, Lawrence, Kansas, USA

Wisdom MJ, Johnson BK, Vavra M et al (2005c) Cattle and elk response to intensive timber harvest. Pages 197–216 in Wisdom MJ, technical editor, The Starkey Project: a synthesis of long-term studies of elk and mule deer. Reprinted from the 2004 Transactions of the North American Wildlife and Natural Resources Conference, Alliance Communications Group, Lawrence, Kansas, USA

Chapter 11
Evolution of a Short-Term Study of Lodgepole Pine Dwarf Mistletoe Vectors that Turned Into a Long-Term Study of the Remarkable Gray Jay on the Fraser Experimental Forest, Colorado, 1982–2009

Thomas H. Nicholls

Abstract This is a summary of a 5-year short-term study that evolved into 28 years of long-term research on the US Department of Agriculture, Forest Service's Fraser Experimental Forest in Colorado. The study was begun in 1982 by Forest Service Research Scientists Thomas H. Nicholls and Frank G. Hawksworth to determine the importance of mammal and bird vectors in the long-distance dissemination of lodgepole pine (*Pinus contorta*) dwarf mistletoe (*Arceuthobium americanum*), an economically devastating forest disease of western forests. The original vector study evolved into a study of a plant growth regulator using ethephon for controlling small pockets of lodgepole pine dwarf mistletoe initiated by vector-disseminated seed. An in-depth study of the ecology of the most common vector of dwarf mistletoe, the gray jay (*Persisoreus canadensis*), followed. The gray jay study evolved into sub-studies on: radio-tracking gray jays between infected and healthy stands in 1983, the exotic West Nile virus in 2003, the development of a method to predict populations and habitat carrying capacities for gray jays by using modeling and geographic information systems techniques in 2005, the use of the polymerase chain reaction technique to sex gray jays for the first time in 2005, and determining the impact of the native mountain pine beetle (*Dendroctonus ponderosae*) outbreak on gray jay habitat and population in 2009.

Keywords Gray jay · Fraser experimental forest · Lodgepole pine · Dwarf mistletoe · Ethephon · Frank G. Hawksworth · Mountain pine beetle · Polymerase chain reaction · West Nile virus

T. H. Nicholls (✉)
Northern Research Station, USDA Forest Service, 1561 Lindig Ave., St. Paul, MN 55108, USA

Nature Education Center, W7283 Walnut St., P.O. Box 63, Fifield, WI 54524, USA
Phone: 715 762 3076
e-mail: nicho002@umn.edu

D. C. Hayes et al. (eds.), *USDA Forest Service Experimental Forests and Ranges*,
DOI 10.1007/978-1-4614-1818-4_11,

11.1 Introduction

11.1.1 The Fraser Experimental Forest Study Area

The 9,300-ha Fraser Experimental Forest (EF), in the Arapaho-Roosevelt National Forest, is located 9.6 km southwest of Fraser, in Grand County, CO. It is administered by the Rocky Mountain Research Station headquartered in Fort Collins, CO. The elevation of the Forest ranges from 2,682 to 3,903 m.

The USDA Forest Service dedicated the Fraser EF in 1937 as an outdoor laboratory to research subalpine forests representative of much of the central and southern Rocky Mountains. The Fraser EF is particularly noted for its long-term climate and hydrological research. In 1978, the United Nations designated the Fraser EF a World Biosphere Reserve, one of many worldwide dedicated to the study and conservation of the diversity and integrity of plant and animal communities within natural ecosystems. Forest Service EFs and Biosphere Reserves by their very nature are strategic places for carrying out long-term ecological and environmental studies with minimal human disturbance.

11.1.2 Research with Frank Hawksworth

An animal vector study on lodgepole pine (*Pinus contorta*) dwarf mistletoe (*Arceuthobium americanum*) was initiated in 1982 on the Fraser EF by this author (Fig. 11.1) and Frank Hawksworth (Fig. 11.2). At the time, I was a Forest Service Research plant pathologist and project leader of the Forest Disease Research Project at the US Department of Agriculture, Forest Service's North Central Forest Experiment Station (NCFES), now the Northern Research Station, St. Paul, MN. Hawksworth held similar positions at the Rocky Mountain Forest and Range Experiment Station, now the Rocky Mountain Research Station.

In the western USA, dwarf mistletoe species have a greater impact on forests than any other pathogen, decreasing growth rates, distorting tree form, reducing wood quality, and killing trees. It has been estimated that more than 11 million ha are infested in western forests and Alaska with 4.64 million m^3 of wood lost annually (Hawksworth and Shaw 1984). Lodgepole pine dwarf mistletoe affects about 40 % of the lodgepole pine type, causing major economic timber production losses through tree deformity and mortality (Drummond 1982). Assessments of the effects of lodgepole pine dwarf mistletoe in Montana, Colorado, and Wyoming indicated the annual loss exceeds 1.1 million m^3/year (Hawksworth and Dooling 1984).

The major economic impact of lodgepole pine dwarf mistletoe inspired scientists to learn more about its biology and management. Their studies are summarized in Hawksworth and Johnson (1989). Some of these studies were initiated on the Fraser EF as follows.

Fig. 11.1 Project Leader, Research Plant Pathologist, and Wildlife Biologist Thomas H. Nicholls from the North Central Forest Experiment Station, St. Paul, MN holding a banded 11-year-old gray jay in 1993 (**a**, *left*) and four gray jays in 2009 (**b**, *right*), Fraser Experimental Forest, CO

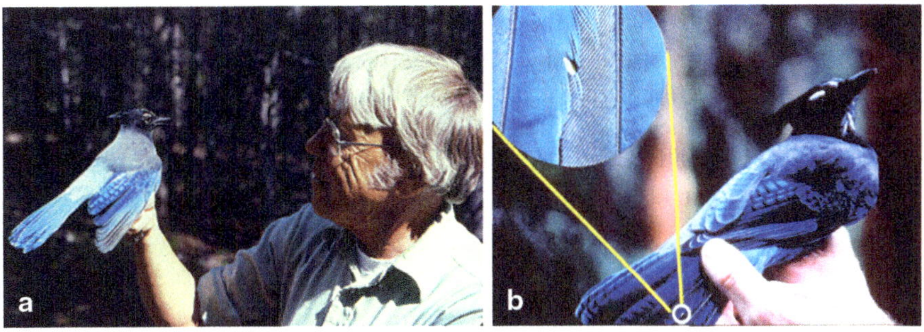

Fig. 11.2 Project Leader and Research Plant Pathologist Frank G. Hawksworth from the Rocky Mountain Forest and Range Experiment Station, Fort Collins, CO, holding a Steller's jay in 1982 (**a**, *left*) with a dwarf mistletoe seed on its tail feather (**b**, *right*), Fraser Experimental Forest, CO

In 1982–1983, Frank Hawksworth and I decided to work together because we were both interested in studying how dwarf mistletoes spread long distances and how isolated pockets of infection become established in otherwise healthy forest stands. My experience and academic training were in both plant pathology and wildlife ecology and I held a Federal Bird Banding Permit needed to capture, band, and release birds after examining them for various pathogens. I was interested in how plant and animal pathogens were picked up and disseminated by vectors and Frank was a world authority on dwarf mistletoes. Frank was also incredibly productive as a scientist—writing more than 275 papers—while having significant administrative duties as well. He had initiated several long-term dwarf mistletoe studies and had several of them ongoing in various western states when we worked together. Frank was also an avid birder, keeping life lists even of birds seen in his own backyard. His expertise and mine made a good match to conduct a study of vectors of lodgepole pine dwarf mistletoe on the Fraser EF, where long-term studies were encouraged.

Frank turned out to be an important mentor for me. I found him to be an amazing friend, teacher, scientist, forest pathologist, administrator, and always a pleasure to work with and be with. He had supreme attention to scientific detail, a necessary scientific skill I wanted to learn. He taught me that skill through his patient guidance and his subdued, good-natured humor, often displayed when least expected. He also taught me to be curious about almost everything biological and to see how everything is tied together in the web of life.

I also learned why Frank was such a productive scientist. In those days, Forest Service project leaders were supposed to spend about 30% of their time on administrative and leadership duties and 70% on scientific research. I was spending too much time on administrative duties and not enough time on scientific research and leadership where my real interests were. In Frank's office, I noticed a big pile of papers in his in-box. I asked him about it one time. He claimed that he let all administrative requests for what he thought was "nonessential" information generated by others to pile up in his in-box until someone asked for the information a second—or sometimes a third—time, after which he promptly responded. More often than not, however, no one ever asked him again for that information, giving him more time for his research. Another reason Frank was so productive was that he worked long hours, often on weekends. He was efficient and he hired good people who were willing to work as hard as he did.

11.2 Lodgepole Pine Dwarf Mistletoe Studies on the Fraser Experimental Forest

11.2.1 Background

Early studies revealed that dwarf mistletoe spreads short distances by sticky seed shot up to 15 m from ripening explosive fruits. The seeds stick to anything they hit, including birds and mammals.

It was not known how long-distance spread occurred in lodgepole pine until our vector study. However, bird and mammal vectors were suspected as a result of previous studies by former students and technicians who worked on black spruce (*Picea mariana*) eastern dwarf mistletoe (*Arceuthobium pusillum*) in Minnesota (Hudler et al. 1974, Ostry et al. 1983). Hudler et al. (1979) went on to discover birds also disseminated seeds of ponderosa pine (*Pinus ponderosa*) dwarf mistletoe (*Arceuthobium vaginatum*).

The objective of our original 1982 study was to identify animal vectors of dwarf mistletoe on lodgepole pine and to determine their importance in the establishment of new infection centers that could not be explained by normal short-distance spread of sticky mistletoe seeds. The term vector as used here is defined as any animal able to transmit a pathogen. Cell traps, mammal ear tags, mist nets, bird bands, and radio telemetry were used to trap mammals and birds and to mark them individually to document their movements and to learn how mistletoe seeds were carried beyond the normal seed dispersal range of infected trees.

11.2.2 Bird and Mammal Vectors

The study identified the following ten bird and four mammal vectors of lodgepole pine dwarf mistletoe seed: the gray jay (Fig. 11.1), Steller's jay (*Cyanocitta stelleri*; Fig. 11.2), mountain chickadee (*Parus gambeli*), dark-eyed junco (*Junco hyemalis*), hermit thrush (*Catharus guttatus*), American robin (*Turdus migratorius*), yellow-rumped warbler (*Dendroica coronata*), northern saw-whet owl (*Aegolius acadicus*), Townsend's solitaire (*Myadestes townsendi*), three-toed woodpecker (*Picoides tridactylus*), least chipmunk (*Eutamius minimus*), golden-mantled ground squirrel (*Spermophilus lateralis*), red squirrel (*Tamiasciurus hudsonicus*), and pine marten (*Martes americana*). The study found that birds and mammals foraging in infected lodgepole pine become inadvertent targets of explosive, sticky mistletoe seeds that stick to feathers or fur. Although such events are rare, a sufficient proportion of birds (27 %) carried dwarf mistletoe seeds to make some dispersal probable. As animals move about the forest, or clean their bodies, mistletoe seeds can be deposited on healthy lodgepole pine where they sometimes stick, germinate, and cause new infections.

Previous studies on the Fraser EF had found 59 species of birds using the Forest. During this study, we added 14 species to the Forest bird list that we either banded or observed. Detailed results are reported in Hawksworth and Weins (1996), Hawksworth et al. (1987), Nicholls and Hawksworth (1983), and Nicholls et al. (1984a, b, 1987c, 1989).

11.2.3 Control of Dwarf Mistletoes

11.2.3.1 Chemical Control

A chemical control method was tested on the Fraser EF in the 1980s in cooperation with Forest Pathologists David Johnson, USDA Forest Service, Forest Pest Management, Lakewood, CO, and Kathy Robbins of NCFES, both now retired. This work was inspired by William Livingston et al. (1985), who showed that ethylene-releasing agents could help prevent the spread of black spruce eastern dwarf mistletoe in Minnesota.

The objective of this work was to determine whether small pockets of infected trees originating from vector-disseminated seed could be saved and adjacent

Fig. 11.3 Lodgepole pine dwarf mistletoe before (**a**, *left*) and several days after (**b**, *right*) being sprayed with an ethephon growth regulator, Fraser Experimental Forest, CO

healthy trees protected by treating infected trees with ethephon, a growth regulator. It acts by releasing ethylene, a plant hormone, which when absorbed by the plant interferes with the growth process. Ethephon is found in nature and, among other things, causes tree leaves to abscise at the end of a growing season. We found that it did the same to mistletoe shoots (Nicholls 1989; Nicholls et al. 1987a, b).

Ethephon at 2,500 ppm in water with a spreader applied by ground sprayers was effective in causing mistletoe shoots, flowers, and fruits to drop off trees (Fig. 11.3), thereby significantly reducing seed dispersal for up to 4 years after treatment. However, it did not kill the parasite's endophytic system in the host tissue, so shoots often resprouted in 3–5 years. Many of the sprouts eventually produced seeds. As a result, frequent ethephon treatments would be required to effectively manage this disease, making it economically unfeasible for use under forest conditions.

Ethephon treatments using ground sprayers can be used to slow the development and spread of dwarf mistletoe in high-value trees located in campgrounds, small parks and golf courses, and around buildings. Control of dwarf mistletoe shoots presumably can also reduce the drain of nutrients from the host tree to the parasite. Because of these benefits, ethephon was registered for these uses with the US Environmental Protection Agency. Aerial applications of ethephon by helicopter, however, were not effective in controlling lodgepole pine dwarf mistletoe under forest conditions (Robbins et al. 1989). These research results spawned additional studies of ethephon control of other western dwarf mistletoe species, which achieved similar results.

11.2.3.2 Silvicultural Control

Vector study results explained how new satellite infection centers become established in healthy stands far removed from main infection centers. Dwarf mistletoe plants are dioecious, so a female and male plant would have to become established within pollination range to develop a satellite infection. Although satellite infection centers are relatively scarce, the explosive mechanism of seed dispersal utilized by dwarf mistletoes enables them to intensify and spread rapidly once a new infection center is established.

The most practical management plan recommended for controlling small infection centers in otherwise healthy stands is to find them through periodic, systematic land or aerial surveys and remove them. Removal can be done by cutting infected trees and a 40-m-wide buffer strip of adjacent trees that may be harboring latent infections yet not showing signs and symptoms. Dwarf mistletoe is an obligate parasite; once the host is killed, so is the parasite. Follow-up surveys 5–10 years after eradication efforts will determine whether all infected trees were removed, or whether more eradication of infected trees is needed. This management action is recommended only in stands being managed for forest products. When used with other dwarf mistletoe management strategies, this approach effectively prevents, or slows, the spread of dwarf mistletoe (Hawksworth and Johnson 1989).

11.2.3.3 Benefits of Dwarf Mistletoes to Wildlife

In the previous study, I had been looking through the eyes of a forest pathologist. Later, having switched disciplines by becoming a wildlife project leader and scientist in 1986, I found that control of dwarf mistletoe may not be the best management action to take in all situations, especially in terms of overall ecosystem health and diversity of wildlife.

Dwarf mistletoes are a major cause of tree deformity and mortality in affected coniferous stands throughout the Northern Hemisphere and lead to significant economic damage. However, the resulting dead and declining trees have a positive effect on many wildlife species that use infected trees for nesting, shelter, and food (Hawksworth and Weins 1996; Nicholls et al. 1984a; Ostry and Nicholls 1991). Mistletoes create canopy openings providing conditions suitable for plant species not ordinarily found in dense, healthy stands, which in turn attract a variety of animal species. Mistletoe shoots are eaten by some mammal and bird species. Insects are abundant in affected trees and thus attract a wide variety of insect-eating birds such as woodpeckers, nuthatches, and warblers, as observed in the Fraser EF. In addition, several species of raptors and songbirds nest and forage in the witches' brooms produced by various dwarf mistletoe species.

The Fraser EF study and related studies have shown that mistletoes play a positive role in creating more compositional (both plant and animal), structural, and functional diversity in forests. Whether this result is good or bad depends upon management objectives. For example, if timber management for tree health and forest products is the primary objective, dwarf mistletoe control is essential when economically feasible. If forest health, wildlife management, species diversity, and wildlife viewing are the primary stand objectives, mistletoe control is not essential and may even be detrimental to some wildlife species, especially in marginal and noncommercial stands.

Fig. 11.4 Forest Service Volunteer Mary Lou Nicholls preparing suet bait in 2007 for trapping gray jays in mist nets, Fraser Experimental Forest, CO (**a**), and gray jays responding to the bait (**b**)

11.3 Gray Jay Studies on the Fraser Experimental Forest

11.3.1 Background

I found the ecology of the gray jay so interesting in its own right that I continued research on this species after my retirement in 1994. My wife, Mary Lou, and I continued to work on the project as Forest Service volunteers (Fig. 11.4). Mary Lou's assistance was essential to the research effort as she took on duties such as making suet bait balls and preparing meals for all the researchers, working dawn to dusk along with the rest of us.

11.3.1.1 Natural History of the Gray Jay

Although the gray jay is a vector of lodgepole pine dwarf mistletoe, it became clear in our subsequent studies that the gray jay is an integral part of a healthy ecosystem and an important indicator of forest health where it lives. We found the gray jay to be a remarkable, highly intelligent, long-lived, territorial bird with an amazing memory and skills in survival, foraging, and communication, as detailed in the following summary of its natural history gathered from our research, general sources, and material in *The Birds of North America* (Strickland and Ouellet 1993).

The gray jay is in the order of perching birds (Passeriformes) and belongs to the family Corvidae, which includes crows, ravens, magpies, and jays, all highly intelligent birds. Some of its common names are camp robber, Canada jay, meatbird, and whiskey-jack. The species is found only in North American boreal and subalpine forests. Preferred habitats include black spruce, tamarack (*Larix laricina*), white spruce (*Picea glauca*), Engelmann spruce (*Picea engelmannii*), subalpine fir (*Abies lasiocarpa*), and lodgepole pine.

The gray jay is well known for its bold, almost tame, behavior around humans. A frequent visitor at backwoods cabins, campsites, logging camps, or isolated resorts, it will often raid human food caches, or even take food from human hands. Despite its association with humans, this jay does not live in towns or civilized developments. It is exclusively a bird of remote forests and thus a good indicator species of healthy, pristine forests. In fact, gray jays disappear as soon as an area becomes overly developed, or its habitat destroyed.

Male and female gray jays have similar plumage, making it difficult to sex in the field. Their average weight is 73 g. Adults generally pair for life, but will take a new mate upon the loss of one. They are nonmigratory, long-lived birds (up to 17 years) with an adult annual mortality rate of about 20 %, as compared to about 50 % mortality in migratory birds. Being nonmigratory, it avoids many migration perils, which helps to explain its higher survival rate. Adult gray jays maintain and defend a permanent year-round territory ranging from about 25 to 150 ha based on various studies.

Gray jays have an omnivorous diet that includes insects, berries, mushrooms, carrion, and nestling birds. During non-winter months, especially in late summer and fall, the jay shapes some of its food into oval pellets, or boli, with its tongue. It permeates the bolus with saliva produced by two large mandibular salivary glands located on each side of the head just behind the beak. The sticky saliva is used to glue boli to vegetation or within bark crevices, where they dry to form a hard protective covering around the food. With its remarkable memory, the gray jay finds most of its hidden food, allowing this species to survive often harsh, winter weather conditions. This food-stashing behavior eliminates the need to migrate south for winter to find food, as many bird species do. Stored food caches also allow gray jays to nest earlier than most birds as some of the food is used to feed nestlings before spring food supplies become plentiful.

Gray jays build a new nest every year in late February or early March on a branch or in an upright crotch of a conifer, commonly 2 m off the ground. It is a bulky, compact structure often containing strips of bark, twigs, grasses, mosses, spider webbing, cocoons, and catkins. It is heavily lined with fine grasses, feathers, fur, and hair, providing thick insulating walls that help to keep eggs warm during freezing temperatures. The male chooses the nest site, initiates the nest building, and is joined later by the female. The nest takes about 3 weeks to build. Only one brood is produced annually. The female lays two to five greenish, speckled eggs, which she incubates. The male will feed the female on the nest. Eggs hatch in 18–20 days and young fledge in 20–23 days, well before migrant birds return. Fledging birds are sooty gray all over and look completely different from their parents. They achieve their adult plumage by late fall after molting juvenile feathers.

Juvenile gray jays start a remarkable deadly struggle for dominance in June. The winning dominant juvenile bird expels its rival siblings from its parents' territory (Strickland 1991). About 80 % of expelled siblings die by fall. The other 20 % move into a failed nesting territory occupied by two adults. This behavior explains the trios of gray jays commonly seen in fall and winter. The winner has the advantage of learning the locations of its parents' food caches; using this food source allows it to overwinter successfully. This nonbreeding bird does not help parents with nest building or feeding of young in the nest. Nest activity, like nest predation, are thus reduced. However, it will help feed the young after they fledge. The dominant offspring will often stay with its parents for 2–3 years, or until a nearby territory becomes available, which it then occupies.

Gray jays have a complex communication system that uses many different vocalizations to keep in touch with each other or to defend their territories. They can also easily imitate other birds such as ravens and Steller's jays.

According to Breeding Bird Survey and Christmas Bird Count long-term data, the gray jay overall is doing well. However, there are concerns of possible declining local populations, especially in Alberta, Manitoba, southern Ontario, northeastern Minnesota, and northern Wisconsin. Although the International Union for the Conservation of Nature classifies the conservation status of the gray jay as a species of least concern, there are threats to the species on the horizon. These threats are loss of habitat due to extensive forest fires; insect outbreaks; logging of the boreal forest and oil, gas, and mining exploration and development; West Nile virus (WNV); and global climate change.

Waite and Strickland (2006) recently reported that climate change has contributed to a rapidly declining gray jay population, studied for about 50 years, at the southern edge of its range in Ontario. Recent warmer autumns apparently rot stored perishable food caches the bird depends upon to survive the winter. Reduced food supplies lead to delayed breeding, reduced reproductive success, and sometimes abandonment of territories. As global warming heats up their natural "refrigerator," allowing food caches to rot before they can be used, a range contraction to cooler climates is expected. This remarkable species requires more research into its natural history as it faces new environmental threats to its survival.

11.3.1.2 The Gray Jay as an Indicator Species

Our early study results found that not only is the gray jay one of the most common vectors of lodgepole pine dwarf mistletoe, but it is also an important indicator of forest health. As a result, our later research objectives began to focus on gray jay movements, longevity, health, site fidelity, and ecology so that we could better understand this species. The baseline information obtained proved valuable to our future work on the Fraser EF.

Leanne Egeland (Fig. 11.5), NCFES biological technician now with the USDA Forest Service, Forest Health Management Unit, Gunnison, CO, formerly retired, helped on all aspects of the dwarf mistletoe and gray jay work for most of the 28 years of research on the Fraser EF. We believe this to be the longest continuous study of gray jays in the USA. It was started at a time when the study of nongame species like the gray jay was moving more to the forefront of wildlife conservation research efforts.

Fig. 11.5 Biological Technician Leanne Egeland using an Osborne Fire Finder to radio-track a gray jay (**a**). The tracked gray jay showing a dwarf mistletoe seed on its wing feather in 1983 (**b**). The bird is wearing a radio transmitter used to track its movements between infected and healthy stands of lodgepole pine on the Fraser Experimental Forest, CO

11.3.2 Gray Jay-Dwarf Mistletoe Seed Study

The objective of this study was to find out if gray jays moved back and forth between dwarf mistletoe-infected and healthy stands of lodgepole pine while carrying dwarf mistletoe seeds. Five gray jays were fitted with 3.7-g transmitters and radio-tracked from three fixed radio-tracking receiving stations located 800–1,000 m apart in the Fraser EF (Fig. 11.5). By obtaining degree bearings and triangulating, bird locations could be determined similar to the way forest fires are found.

A 58.7-ha stand of 70-year-old lodgepole pines within the radio-tracking study area was examined for satellite infections centers (Hawksworth et al. 1987). Radio tracking showed that gray jays frequently flew between dwarf mistletoe-infected and healthy parts of the stand. Some birds were known to have mistletoe seed on their feathers when retrapped and released during that study (Fig. 11.5).

Locations from five radio-tracked birds over a period of 4–12 days showed a home range average of 85 ha (200 ac) based upon 50 data days. The home range of one pair of jays overlapped, but the other birds maintained distinct boundaries between each other. The pair of jays exhibited territorial behavior similar to what we found in banding studies.

11.3.3 Gray Jay Life Span and Habit Study

Since 1982, we banded and released 704 individual gray jays in the Fraser EF. To capture birds, we annually used up to 50 fixed mist-net trapping locations about 800 m apart by road centered in the Fraser EF headquarters. The average weight of these birds was 73 g. We handled 1,968 new and retrapped gray jays over the course of our study. Many were retrapped and released numerous times, often at or near the very location where they were originally banded—showing an extremely high site fidelity.

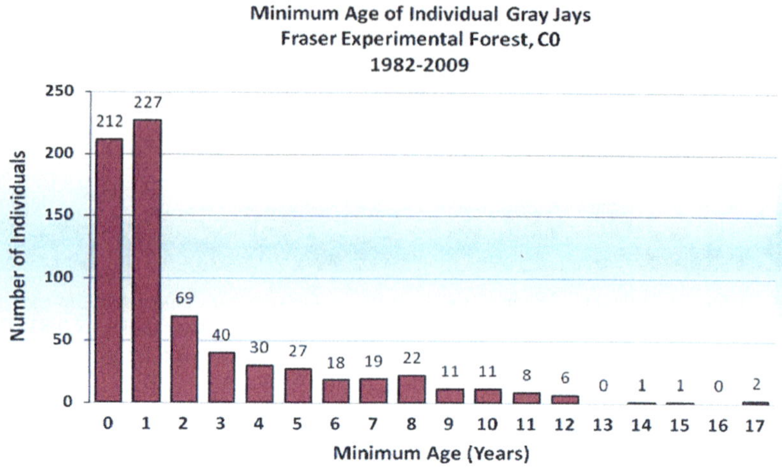

Fig. 11.6 Minimum age of 704 individual gray jays trapped, banded, and released from 1982 to 2009, Fraser Experimental Forest, CO. Zero birds are hatch year birds less than 1 year old

Study results revealed a relatively long life span for resident adult gray jays, which live in permanent, year-round, all-purpose territories in the subalpine forests in the Fraser EF. They often live under extreme winter weather conditions at high elevations.

Of the 704 gray jays we banded, 29 lived 10 or more years and 2 as long as 17 years (Fig. 11.6). Our first 17-year-old gray jay was first banded August 16, 1985, as a young hatch-year bird. It was last captured and released on Aug. 30, 2002. This bird holds the US record for gray jay longevity as confirmed by the US Geological Survey Bird Banding Laboratory (Fig. 11.7). This bird dispersed about 3.2 km away from its original banding location at Fraser EF headquarters. As an adult, it subsequently maintained a territory near the headquarters, where it was retrapped and released four times over the years (1998, 1999, 2000, and 2002). The bird's weight remained normal over the years, ranging between 69 and 73 g. Our second 17-year-old bird was last captured at the King/Elk Junction on September 6, 2009. It was first trapped and banded at that same location as an adult bird on August 25, 1993. It was caught and released nine times (1993, 1994, 1995, 1996, 1997, 1999, 2005, 2007, 2009) on its territory over the 17 years.

11.3.4 Study of the Threat of West Nile Virus to Gray Jays

WNV is an emerging, infectious, exotic disease that was first discovered in the African country of Uganda in 1937. In recent years, it has spread beyond its traditional boundaries, causing illness in birds, horses, and humans in Europe and now the USA. It was first discovered in the USA in 1999 in New York City. Since that time, WNV has been detected in humans, animals, and mosquitoes across the USA as it spread like a wave from east to west over a period of 5–6 years.

Fig. 11.7 Fraser Forest Technician Manuel Martinez holding a banded 17-year-old gray jay in 2002 at the Fraser Experimental Forest, CO. This bird was the oldest gray jay recorded in the wild in the USA according to the US Geological Survey Bird Banding Laboratory

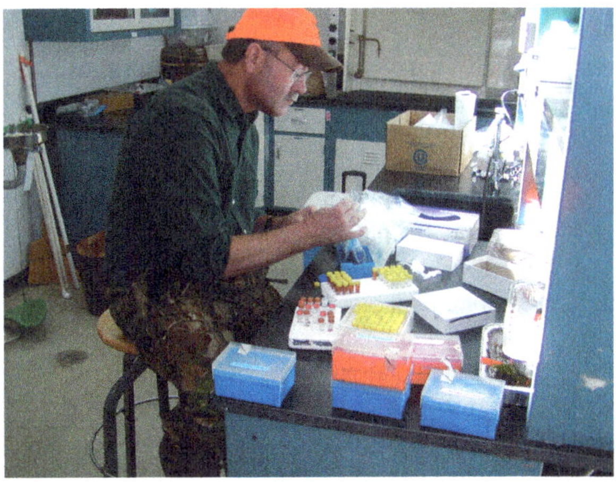

Fig. 11.8 Zoonotic Pathologist and Microbiologist Kurt Reed, M.D., preparing gray jay blood samples in 2003 for analysis of WNV and other blood parasites in the Fraser Experimental Forest field laboratory in Colorado

WNV is spread by the bite of an infected mosquito. Mosquitoes become infected when they feed on infected birds. Infected mosquitoes can then spread WNV to humans and other susceptible animals when they bite them. Animals and people die when they cannot produce enough antibodies to fight off the disease, especially if they have a compromised or underdeveloped immune system. A particularly serious outbreak of WNV occurred in Colorado in 2003. It posed a serious threat to the susceptible gray jay population in the Fraser EF as well as other susceptible bird populations and humans throughout Colorado. In 2003, more humans (2,947) were infected in Colorado than in any other state, and 63 people died.

Zoonotic Pathologist Kurt Reed, M.D. (Fig. 11.8), previously with the Marshfield Clinic Research Foundation and now Professor, Department of Pathology and

Laboratory Medicine, University of Wisconsin-Madison, joined our research team in 2003 to study the impact of WNV on gray jays. He took blood samples from 296 gray jays from 2003 to 2007. Of 236 blood samples tested to date, all were negative for WNV antibodies except for two gray jays (bird band nos. 9822-51941 and 9822-52036) in 2003 that tested positive, with high WNV antibody titers of $\cong$1:20 and $\cong$1:40, respectively.

Jays are highly susceptible to the virus and many individuals may have died before they could develop protective antibodies and before we could find them in the field, where they can quickly be scavenged. However, circumstantial evidence as follows showed a major decline in the Fraser EF jay population that we hypothesized was caused by WNV.

The population of gray jays in the Fraser EF had been healthy for at least 21 years, but something was different about the birds in 2003. They were more difficult to trap and far fewer birds were trapped than in the previous 4 years despite similar trapping efforts. Their social structure seemed to be affected as some family groups appeared to be disorganized. There was a 37% decrease in the number of birds netted in 2003 (N=74) compared to 2002 (N=117). The 2003 capture rate declined 33% compared to the average capture rate of 110 birds for the previous 4 years, 1999–2002. We are not sure whether WNV mortality was the entire cause of this decline, but our WNV blood test results and field observations strongly hinted that WNV might have been the cause.

WNV waned in Colorado after 2003. By 2009, there were only 102 human WNV cases and three deaths, and only three WNV-infected birds were reported in the state. The gray jay population in the Fraser EF started to rebound in 2005 and 2006, but declined again by 2009, when an insect outbreak and salvage logging began to destroy gray jay habitat.

WNV outbreaks, like other arboviruses, are cyclical in nature. We might be able to further strengthen our hypothesis that WNV was the cause of the population decline in 2003 by making more observations during the next WNV outbreak in Colorado. If and when that will happen is an unknown. That uncertainty is why places for long-term studies are valuable for ecological and disease research.

11.3.5 Habitat Use Study

In 2005, 39 of our gray jays in the Fraser EF were color-banded to determine habitat use. The birds were studied by Jennifer Berg (now Jennifer Lansing), a master's student at the University of Colorado-Denver under the direction of Prof. Diana Tomback. The objective of this study was to create a method to predict carrying capacities for birds by using geographic information systems techniques with our long-term data on gray jay distribution, habitat use, and territorial sizes (Berg 2006). Based upon this work, Berg and Tomback estimated the population of gray jays in the Fraser EF to be 682. Their models and field observations showed that the jays primarily used the more extensive lodgepole pine habitat although the Engelmann spruce/alpine fir habitat was used as well.

Fig. 11.9 Effects of the mountain pine beetle infestation of 2005 through 2009 on the Fraser Experimental Forest, CO. The beetle infested much of the older lodgepole pine, killing thousands of trees in prime gray jay habitat. Byers Peak, the tallest mountain (3,903 m or 12,805 ft) in the Forest, is seen in the background

11.3.6 Use of Polymerase Chain Reaction Technique to Sex Gray Jays

The plumage of female and male gray jays is identical, so we were unable to sex the birds in the field. The objective of this study was to use for the first time in 2005, DNA in blood samples to identify the sex of gray jays using polymerase chain reaction (PCR). Avian-sexing PCR uses the fact that male and female birds have different genes or chromosomes, much like mammals. We were able to determine the sex of 74 gray jays in 2005, 38 males and 36 females. As a control, the PCR method was used to correctly sex four red-shafted flickers and one hairy woodpecker that can be sexed by plumage in the field. As this database develops, it can serve as another spin-off study of genetic parentage, relationships, and dispersal of related gray jays in the Fraser EF.

11.3.7 Threat of Native Mountain Pine Beetle to Gray Jay Habitat

Lodgepole pine is the predominant cover type in the Fraser EF covering about half of the Forest. According to Forest Service Forest Health Management aerial surveys, a few 3-ha pockets of the native mountain pine beetle (MPB) were detected in the Fraser EF in 2002. By 2003, three areas of 80 ha or more were detected in our gray jay study area, and by 2009 much of the older lodgepole pine in the Fraser EF had been hit hard by the beetle (Fig. 11.9).

In 2009, the gray jay in the Fraser EF faced destruction of much of its prime lodgepole pine habitat because of the unprecedented MPB outbreak. Thousands of older trees in the Forest, including many infected with dwarf mistletoe, were killed. Salvage clearcutting of some stands starting in 2008 added to the loss of

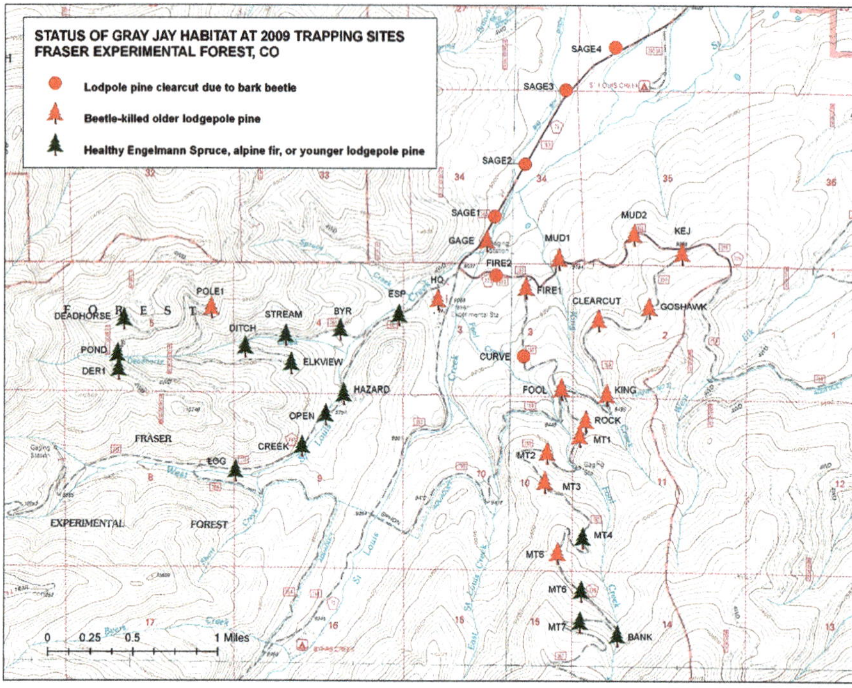

Fig. 11.10 The 22 (in *red*; out of 38) gray jay trapping sites and their associated lodgepole pine habitats that were seriously impacted by the mountain pine beetle in 2009, Fraser Experimental Forest, CO

the gray jay's habitat. Twenty two (58 %) of 38 gray jay trapping sites and adjacent gray jay habitat were seriously impacted by 2009 (Fig. 11.10). Only 67 gray jays were captured and released in 2009, a 39 % decrease in the average capture rate for 1999–2002 based upon a similar trapping effort. Providing nothing happens to the remaining Engelmann spruce/subalpine fir habitat in the meantime, it may be an important refuge for the gray jay as the lodgepole pine forest regenerates to once again provide suitable habitat.

The unprecedented MPB outbreak is thought to be caused by a combination of prolonged drought, warmer winters that allowed beetles to survive over winter, lack of timely forest thinning management in some areas, and fire control over past years that allowed development of large concentrations of fuel and overmature trees to develop. This combination of events apparently set the perfect stage for a massive MPB attack on stressed trees too weak to resist the beetle attack. The current situation is part of an unprecedented multispecies beetle outbreak and has caused significant and widespread tree mortality from Mexico to British Columbia (Rhoades et al. 2013). Millions of hectares of older lodgepole pine have been affected.

The gray jay is a forest bird that prefers to forage and nest in mature forests. It cannot successfully live outside a forested habitat that does not provide it with essential resources. These resources include food, water, shelter, and space required to successfully carry out its life cycle with a minimum of stress and loss of life. Losing millions of hectares of one of its preferred habitats to the bark beetle along with subsequent salvage logging could have a significant negative impact upon this species, according to our preliminary research results. Potential impacts of its habitat loss include lack of breeding, nesting failures, fewer fledglings, reduced life span, abandonment of territories, or even death.

11.4 Summary

11.4.1 A Lifetime of Collaboration with Frank Hawksworth

Decades ago, Frank Hawksworth started our collaboration on dwarf mistletoe research. I assisted Frank in resurveying some long-term dwarf mistletoe plots located along the rim of Grand Canyon National Park in Arizona and in the Mascelero Apache Indian Reservation in New Mexico. We were working 6 days a week to get the resurveying done. We also made trips to look for new species of dwarf mistletoe in such places as the San Juan Mountains in Colorado. On similar trips, Frank did discover new species of mistletoe and was honored for his efforts by having one named after him, *Arceuthobium hawksworthii.*

These field trips showed me that discovery was truly in Frank's nature. It was always an adventure traveling with him while discovering and learning new things about the biological world around us. Anything biological was "official business" and of interest to him. And he was always willing to share what he learned with others in a quiet, non-assuming way. He left a legacy of scientific accomplishment second to none, summarized in his most important publication coauthored with Delbert Weins entitled *Dwarf Mistletoes: Biology, Pathology and Systematics* (1996).

After Frank's retirement and up until a couple of years before his death in 1993, he remained as a Forest Service volunteer working on his beloved mistletoe research. As of 2009, the research continues partly in memory of Frank and partly because of new, important scientific research spin-offs resulting from the original lodgepole pine dwarf mistletoe vector study. Over time, these spin-offs evolved into a series of smaller studies as reported here as new knowledge was gained, new research questions emerged, and new technologies developed.

11.4.2 The Value of Long-term Research in Addressing New Questions

When we initiated our study in 1982, we were sure that dwarf mistletoe was the worst possible pest threat to lodgepole pine and that its major economic impact warranted significant research and management. Now, more than 28 years later, our research begs answers to many emerging questions not envisioned at the beginning of our study. For example, which pest, dwarf mistletoe, or MPB is more important in the long run to the health of the lodgepole pine ecosystem and those animals that depend upon it? Could timelier forest management have minimized economic and aesthetic losses caused by dwarf mistletoe and MPB? If our forests are not managed in a timely manner, will massive outbreaks of insects, diseases, and fires continue to inflict enormous economic losses, as seems to be happening in the western USA in recent years? How will habitat loss from such events affect gray jay populations? Are global climate changes predisposing forests to pest and fire outbreaks? Only long-term research can answer many of these and other emerging questions, questions that might never have been asked during a short-term study.

11.5 Conclusions

Forest Service EFs are essential for long-term ecological and pest research such as that reported here. Along with Natural Areas and National Parks, they are often the only places where such research can be done without undue artificial disturbance. This unique characteristic is a compelling reason for continued support and protection of our Forest Service EFs and other protected areas.

These areas will remain important as researchers continue to draw upon their long-term databases to answer emerging science questions. The following suggested future studies on the Fraser EF would build upon or complement the long-term research findings as reported in this paper:

- Determine survival, spread, distribution, and cumulative impact of lodgepole pine dwarf mistletoe as a new lodgepole pine forest regenerates after a MPB outbreak
- Document increased woodpecker and other cavity-nesting bird activity resulting from thousands of dead and dying trees killed by dwarf mistletoe and MPB
- Document what happens to other birds and animals dependent on lodgepole pine habitat during and after the MPB outbreak and salvage logging
- Determine impacts of MPB tree mortality and salvage logging on gray jay habitat and populations
- Determine whether the Engelmann spruce/alpine fir habitat will provide a refuge for gray jays and other forest birds, and whether it will provide a source of birds to repopulate a regenerating lodgepole pine forest
- Identify potential zoonotic (affecting both humans and animals) diseases and blood parasites affecting gray jays and other wildlife species stressed by the loss of their primary habitat
- Determine genetic parentage, gene dispersal, and movements of banded gray jays using DNA and PCR techniques
- Determine potential impact of global climate change on gray jay habitat, populations, reproduction, and perishable winter food caches

Acknowledgments Thanks to all of the following individuals who made this research possible: Frank Hawksworth, Leanne Egeland, Mary Lou Nicholls, Cory Nicholls, Kurt Reed, Manuel Martinez, Jennifer Meece, Mark Dixon, Kelly Elder, Jennifer Lansing, Mike Ostry, Patrick Stockwell, Laura Merrill, David Johnson, Charles Troendle, Doug McClain, Terry Shaw, Robert Alexander, William Barnard, Virgil Scott, Kathy Robbins, and many others who volunteered their time! We also thank the USDA Forest Service, Forest Health Management, Gunnison, CO, Service Center and the Forest Disease Research Project, Northern Research Station, St. Paul, MN, for their support.

Thanks are also given to reviewers of this paper who provided helpful comments for improvement, Drs. Douglas Boyce, Mike Ostry, and Kurt Reed.

References

Berg J (2006) Estimating population sizes of the gray jay (*Perisoreus canadensis*) using geographic information systems and habitat preferences. M.S. thesis, University of Colorado—Denver, 69 p

Drummond DB (1982) Timber loss estimates for the coniferous forest of the United States due to dwarf mistletoes. Report 83-2, US Department of Agriculture, Forest Service, Forest Pest Management, Methods Applications Group, Fort Collins, CO, 24 p

Hawksworth FG, Dooling OJ (1984) Lodgepole pine dwarf mistletoe. Forest Insect and Disease Leaflet 18. US Department of Agriculture, Forest Service, Washington, DC, 11 p

Hawksworth FG, Johnson DW (1989) Biology and management of dwarf mistletoe in lodgepole pine in the Rocky Mountains. General Technical Report RM-169, US Department of Agriculture, Forest Service, Rocky Mountain Forest and Range Experiment Station, Fort Collins, CO, 38 p

Hawksworth FG, Shaw CG III (1984) Damage and loss caused by dwarf mistletoes in coniferous forests of Western North America. In: Wood RKS, Jellis GJ (eds) Plant diseases: infection, damage and loss. Blackwell Scientific Publications, Oxford, 327 p

Hawksworth FG, Weins D (1996) Dwarf mistletoes: biology, pathology, and systematics. Agricultural Handbook 709, US Department of Agriculture, Forest Service, Washington, DC, 410 p

Hawksworth FG, Nicholls TH, Merrill LM (1987) Long-distance dispersal of lodgepole pine dwarf mistletoe. In: Troendle CA, Kaufmann MR, Hamre RH, Winokur RP (eds) Management of subalpine forests: building on 50 years of research: proceedings of a technical conference; 1987 July 6–9; Silver Creek, CO. General Technical Report RM-149, US Department of Agriculture Forest Service Rocky Mountain Forest and Range Experiment Station, Fort Collins, CO, pp 220–226

Hudler G, Nicholls T, French DW, Warner G (1974) Dissemination of seeds of the eastern dwarf mistletoe by birds. Can J For Res 4:409–412

Hudler G, Oshima N, Hawksworth FG (1979) Bird dissemination of dwarf mistletoe on ponderosa pine in Colorado. Am Midl Nat 102(2):273–280

Livingston WH, Blanchette RA, Brenner ML, Zuzek KJ (1985) Effective use of ethylene-releasing agents to prevent spread of eastern dwarf mistletoe on black spruce. Can J For Res 15:872–876

Nicholls TH (1989) Ethephon tests for control of lodgepole pine dwarf mistletoe in Colorado. In: van der Kamp B J (Comp) Proceedings of the 36th Western International Forest Disease Work Conference, 1988 September 19–23, Park City, UT. University of British Columbia, Department of Forest Science, Vancouver, BC, pp 34–37

Nicholls TH, Hawksworth FG (1983) Animal vectors of lodgepole pine dwarf mistletoe. Abstract ASSI. Phytopathology 73(5):836

Nicholls TH, Hawksworth FG, Merrill LM (1984a) Animal vectors of dwarf mistletoe, with special reference to *Aceuthobium americanum* on lodgepole pine. In: Hawksworth FG, Scharpf RF (eds) Biology of dwarf mistletoes: proceedings of the symposium, 1984 August 8, Fort

Collins, CO. General Technical Report RM-111. US Department of Agriculture, Forest Service, Rocky Mountain Forest and Range Experiment Station, Fort Collins, CO, pp 102–110

Nicholls TH, Hawksworth FG, Merrill LM (1984b) A preliminary look at animal vectors of lodgepole pine dwarf mistletoe. In: Dubreuil SH (Comp) Proceedings of the 31st Western International Forest Disease Work Conference, 1983 August 22–26, Coeur d'Alene, ID. US Department of Agriculture, Forest Service, Northern Region, Missoula, MT, pp 54–56

Nicholls TH, Egeland L, Hawksworth FG, Johnson DW (1987a) Control of dwarf mistletoe with ethephon. In: Cooley SJ (Comp) Proceedings, 34th Annual Western International Forest Disease Work Conference, 1986 September 8–12, Juneau, AK. US Department of Agriculture, Forest Service, Forest Pest Management, Portland, OR, pp 78–85

Nicholls TH, Egeland L, Hawksworth FG, Johnson DW, Robbins MK (1987b) Control of dwarf mistletoes with a plant growth regulator. In: Troendle CA, Kaufmann MR, Hamre RH, Winokur RP (eds) Management of subalpine forests: building on 50 years of research: proceedings of a technical conference, 1987 July 6–9, Silver Creek, CO. General Technical Report RM-149, US Department of Agriculture, Forest Service, Rocky Mountain Forest and Range Experiment Station, Fort Collins, CO, pp 154–156

Nicholls TH, Hawksworth FG, Egeland L (1987c) Animal vectors of *Arceuthobium americanum* on lodgepole pine. In: Cooley SJ (Comp) Proceedings of the 34th Annual Western International Forest Disease Work Conference, 1986 September 8–12, Juneau, AK. US Department of Agriculture, Forest Service, Forest Pest Management, Portland, OR, pp 48–50

Nicholls TH, Egeland L, Hawksworth FG (1989) Birds of the Fraser Experimental Forest, Colorado, and their role in dispersing lodgepole pine dwarf mistletoe. Colo Field Ornithol 23:3–12

Ostry ME, Nicholls TH (1991) The forest beyond the trees: beneficial roles of disease in forest health. In: Proceedings of the 1997 Society of American Foresters National Convention, Meeting in the Middle, 1997 October 4–8; Memphis, TN. Society of American Foresters, Bethesda, MD, pp 167–172

Ostry ME, Nicholls TH, French DW (1983) Animal vectors of eastern dwarf mistletoe on black spruce. Research Paper NC-232, US Department of Agriculture, Forest Service, North Central Forest Experiment Station, St. Paul, MN, 16 p

Rhoades CC, McCutchan JH Jr, Cooper LA, Clow D, Detmer TM, Briggs JS, Stednick JD, Veblen TT, Ertz RM, Likens GE, Lewis WM Jr (2013) Biogeochemistry of beetle-killed forests: explaining a weak nitrate response. PNAS 110(5):1756–1760

Robbins K, Johnson DW, Hawksworth FG, Nicholls TH (1989) Aerial application of ethephon is ineffective in controlling lodgepole pine dwarf mistletoe. West J Appl For 4(1):27–28

Strickland D (1991) Juvenile dispersal in Gray Jays: dominant brood member expels siblings from natal territory. Can J Zool 69:2935–2945

Strickland D, Ouellet H (1993) Gray Jay (*Perisoreus canadensis*). In: Poole A, Gill F (eds) The birds of North America, No. 40. The Birds of North America, Inc., Philadelphia

Waite TA, Strickland D (2006) Climate change and the demographic demise of a hoarding bird living on the edge. Proc R Soc Lond Ser B 273:2809–2813

Chapter 12
Subterranean Termite Control Examinations on Current and Former Experimental Forests and Ranges

T. G. Shelton, T. L. Wagner, C. J. Peterson and J. E. Mulrooney

Abstract For more than 70 years, the USDA Forest Service's Termite Team has engaged in research to extend the life of wood in service by studying chemical (and a few nonchemical) subterranean termite control products. These efficacy data are produced in distinct field trials on experimental forests across the USA, and are used by industry cooperators to register their products with the Environmental Protection Agency. Experimental forests and ranges allow long-term undisturbed efficacy examinations of termiticides in preventing subterranean termite attack on wood. This chapter provides historical information on the development of these efficacy studies over the years and the places where these data are collected.

Keywords Termite · Termiticide · Pesticide · Urban entomology · Subterranean pests

12.1 Introduction

Of all the types of pest management, few involve such an intimate relationship between affected persons and pests than urban pest management. Urban entomology involves pests affecting the three basic needs of people (health, food, and shelter) and includes medical, veterinary, structural, and stored products pests. As with many other pests, some urban pests may best be understood and controlled by examining their natural habitats as opposed to the urban environment. It surprises many people to find that the USDA Forest Service works in urban pest management by studying control technology for wood products insects.

The term "pest" has been used to describe a species that is found in an area where it is unwanted (Bennett et al. 1997). Termites fit this definition perfectly, as they are not considered pest species in forests. In fact, termites benefit forests by removing downed timber and recycling nutrients back into the soil. In the southern

T. G. Shelton (✉) · T. L. Wagner · C. J. Peterson · J. E. Mulrooney
Southern Research Station, USDA Forest Service, Wood Products Insect Research,
201 Lincoln Green, Starkville, MS 39759, USA
e-mail: tshelton@fs.fed.us

T. G. Shelton
Phone: (662) 338-3108

D. C. Hayes et al. (eds.), *USDA Forest Service Experimental Forests and Ranges*,
DOI 10.1007/978-1-4614-1818-4_12, © Springer New York 2014

Fig. 12.1 Subterranean termites and their damage to a pine bolt

USA, forests are being converted into urban areas quickly, due to the phenomenon of "urban sprawl" resulting in increased development in formerly forested areas (Cordell and Overdevest 2001). Subterranean termite colonies (Fig. 12.1) are fairly long lived (as long as 16 years for some species; Grace et al. 1995) and some may feed on downed timber 1 year and on a newly built home the following year. Proper preventive control measures can protect structures from this occurrence. One of the most common subterranean termite control methods over the past century has been the use of pesticides applied to soil to provide a barrier that prevents entry of termites into structural wood (Kofoid 1934; Mallis 1997).

Residences are one of the largest investments made by Americans, estimated at US$ 12.5 trillion nationwide for the year 2000 (Peterson et al. 2008). Obviously, protecting that investment is a key goal for homeowners, part of which is the prevention of termite damage (Fig. 12.2). The economic cost of termite control and repair of their damage is quite large, estimated at roughly US$ 5 billion annually (Su and Scheffrahn 2000; Peterson et al. 2008). From a resource preservation perspective, improving termite control is important. By protecting structures, wood in use lasts longer, decreasing the need for harvesting replacement wood.

The USA has not always regulated pesticides through efficacy and nontarget toxicity testing (Ware 1994; Peterson et al. 2008). Currently, all pesticides undergo some level of efficacy testing for registration (Pedigo 1989), but only two groups of pesticides require independent efficacy studies: termiticides, because of the investment loss resulting from heavy termite infestations, and public health pesticides, such as mosquito control products. For termiticides to be registered, the US Environmental Protection Agency (EPA) guidelines (Office of Prevention, Pesticides, and Toxic Substances, OPPTS 810.3600, and Pesticide Registration Notice 96–7) state that candidate product rates should achieve 100% control at four US locations (detailed later) for a period of 5 years. After federal registration, state regulatory officials also review the product before it can be sold or applied legally in that state. Just over half of the states (26 plus Puerto Rico) belong to an organization

Fig. 12.2 Subterranean termite damage to a 2.5 × 10 cm (1 × 4 in.) board

that makes these decisions as a group—the Association of Structural Pest Control Regulatory Officials (ASPCRO). It is important to consider that the EPA acceptance requirements are guidelines rather than hard-and-fast rules; thus, the EPA can register products that do not meet the efficacy standards perfectly if the product has other attractive properties, such as low nontarget (including mammalian) toxicity (Shelton and Wagner 2005). This chapter discusses the involvement of the USDA Forest Service in the protection of wood in use from subterranean termite damage.

12.2 The Termite Team, Past and Present

The termite team of the Insects, Diseases, and Invasive Plants Research Work Unit (SRS-4552) has been working in the Southeast for 75 years. The team was started by Thomas E. Snyder in 1934 in New Orleans to address problems with insect pests of forest products and forests. Snyder was one of the most respected termite authorities of his time.

In 1938, Snyder hired Harmon R. "Johnny" Johnston as a research entomologist (Mauldin 1989; Kard 2000). He was stationed at the Harrison Experimental Forest (HEF) in Saucier, MS. The HEF is part of the DeSoto National Forest in southern Mississippi. Johnston initiated work investigating means of protecting wooden structures and crates from termites. Over the years, Johnston was also recognized for his work on controlling wood-destroying beetles, such as ambrosia beetles (Kard 2000). In the 1940s, the US military funded research on soil-applied termiticides, allowing the work to expand into other areas (Kowal and St. George 1948; Mauldin 1989). Some areas of the world have more aggressive termite fauna than the USA, so any reliance on wooden containers or structures required protection from damage.

Johnston's work involved the development of an appropriate field test for examining large numbers of replicates of candidate termiticides in the field. Due to

Fig. 12.3 Ground board and modified ground board plots on the Santa Rita Experimental Range

Fig. 12.4 Termiticide application to a ground board plot

the numbers involved with simultaneous testing of various formulations and chemistries, a simple evaluation method was needed to speed the process. The termite team's scientists eventually decided on a pair of simple protocols for measuring termite damage. One of these, the ground board test, was designed to simulate the use of a chemical applied to the soil as a barrier protecting wood on top of the soil (Figs. 12.3 and 12.4). The other test, the stake test, was also designed to work with soil applications of chemicals but measured the protection of stakes driven into treated soil. Both tests used a simple damage rating scale allowing rapid evaluation of test wood.

As a side note, soil applications of termiticides have changed in importance over the past century. In the early 1930s, soil pesticides for termite control were considered experimental and generally not recommended due to questions about efficacy and persistence of these applications (Peterson et al. 2008). The recommendations at the time were focused on treated wood products (brush or dip treatments; Peterson et al. 2008).

In 1960, the termite team moved into a new laboratory in Gulfport, MS, where it remained until 1996 when the team was moved to Starkville, MS. Work within the team is currently divided into more basic research (behavior, forest ecology), toxicology and environmental fate, and the termiticide testing program. The termiticide testing program is responsible for evaluating the performance of subterranean termite control materials in the laboratory and field, producing data that are used in registration packages submitted to EPA. In a sense, the termiticide testing program is the part of the unit that is most familiar to outsiders. The efficacy data are eventually made public following the successful registration of a termiticide in an annual report published in a trade journal for the pest management industry (most recently, Wagner et al. 2011).

Through the annual report and various appearances at meetings, the efficacy data are widely distributed as are topics surrounding the regulation of such materials. Thus, the termiticide testing program data are available for both the industry and the homeowners to whom the services are sold. This allows the consumers to educate themselves on the performance of the products before agreeing to purchase a pest control application, a very costly endeavor depending on treatment type and foundation of the structure. While some consumers access the testing program data through the published annual reports, many come to the termite unit staff directly via e-mail, phone calls, or occasionally in person.

12.3 The Termiticide Testing Program Locations

Originally, the Forest Service conducted termiticide research on the HEF. The HEF is a prime habitat for native southeastern subterranean termite species belonging to the genus *Reticulitermes* Holmgren. Over the years, the termite team has added (and dropped) termiticide research sites from use. For example, by 1942, termiticides were being tested in the Panama Canal Zone and at Beltsville, MD (Kowal and St George 1948). Military installations in several locations have also been used as testing sites, including Fort Dix, NJ, Fort Sill, OK, Puerca Point in Puerto Rico, and the Panama Canal Zone. Midway Island, HI, was also used as a location for the termiticide testing program. A further expansion occurred in 1965 with termiticides installed in seven locations, including Arizona, Florida, Hawaii, Maryland, Missouri, Oregon, and South Carolina (Carter and Stringer 1970a; Carter et al. 1970). This study was also used for an examination of termiticide persistence under differing rainfall and soil conditions (Carter and Stringer 1970a; Carter et al. 1970). Bait studies (using mirex) were performed near the city of Lake Charles, LA, in 1968 against the Formosan subterranean termite, *Coptotermes formosanus* Shiraki.

Many of these locations were on experimental forests, as long-term access to those areas could be secured. Private lands often changed hands, or were harvested from time to time, making a stable, consistent outdoor laboratory nearly impossible. Experimental forests filled this need nicely, and by working with the managers who maintained those forests (or ranges), the termiticide testing program could work in

Table 12.1 Soil characteristics of the current termiticide testing program locations

Soil type	Harrison	Chipola	Calhoun	Santa Rita
	Rumford sandy loam	Lakeland sand	Cataula loamy sand	Continental gravelly loamy sand
pH	5.1	4.8	5.8	6.9
Clay	4.9%	2.8%	7.0%	7.5%
Silt	25.2%	2.7%	10.0%	15.1%
Sand	69.9%	94.5%	83.0%	77.4%
Mean rainfall	170 cm/yr	163 cm/yr	127 cm/yr	35.5 cm/yr

areas where there was little chance of man-made disturbance. However, that does not prevent nature from disturbing plots, as seen in 2005 on the HEF with Hurricane Katrina (Wagner et al. 2006, 2007).

While additional locations have been used from time to time, Beal (1986) states that from 1972 onward all candidate products considered for registration were studied on the HEF, Arizona, Florida, and South Carolina (detailed below). These decisions were based in part on variations observed in the initial penetration of organochlorine termiticides in plots of different soil types and moisture content (Table 12.1; Carter and Stringer 1970a, 1970b). Studies in Maryland were discontinued due to low termite attack on control plots (Kard et al. 1989), and the studies in Panama were terminated when control of the Canal Zone was returned to the government of Panama (Kard 2000).

Harrison Experimental Forest The value that this experimental forest has had on the multibillion dollar pest management industry is overwhelming. Virtually all soil-applied termiticides over the past 70 years have been tested there. This site tends to have the greatest termite activity (Mulrooney et al. 2007), and thus tends to be the most challenging site for candidate termiticides.

Chipola Experimental Forest The Chipola Experimental Forest (Calhoun County, FL) was established in 1952 on 1,116.93 hectares of land provided by corporate donors. This private property was leased to the Forest Service for research purposes, with the lease to be renegotiated after 50 years. The original research projects were aimed at establishing optimal procedures for growing pines in "sandhill" areas (Hopkins and Hebb 1954). Termiticide testing began there in 1965 (Carter et al. 1970). Over the years, the properties changed hands numerous times, moving from corporate to private (individual) holdings, and the Forest Service has periodically reduced its size. In 2004, the Southern Research Station purchased the property. The total area of the Chipola Experimental Forest is currently 380.4 hectares, and the scientists responsible for the testing program continue to annually install and evaluate termiticide products on this site.

Calhoun Experimental Forest The South Carolina site is located on the Calhoun Experimental Forest on the Sumter National Forest. The Calhoun occupies 4,451.54

hectares in the western end of South Carolina, near the town of Union. This area is known for soil erosion and water holding problems, and the Calhoun was originally set up in 1947 to research soil stabilization and forest productivity for similar Piedmont areas (Metz 1958). Although the testing program has been executed at two locations within the Calhoun, this forest has provided a stable location since 1965 (Carter et al. 1970). The Calhoun site has been one of the four major test locations used for termiticide efficacy testing (Beal 1986), with new termiticide products installed and evaluated annually.

Santa Rita Experimental Range This range is located south of Tucson, near the towns of Green Valley and Sahuarita in Pima County, AZ. The Santa Rita Experimental Range (SRER) was the first experimental range in the USA. Sayre's (2003) extensive history of research on the SRER provides much of the detail of the numerous property transfers affecting the SRER. A land transfer between the US Department of the Interior and the State of Arizona was finalized in 1991, with Arizona taking possession of the Santa Rita. It is currently managed by the University of Arizona (Sayre 2003).

Candidate termiticides have been installed and evaluated annually on the SRER since 1965 (Carter et al. 1970). It is an important part of the registration process as it provides access to two subterranean termite species not encountered on the southeastern sites, *Heterotermes aureus* (Snyder) and *Reticulitermes tibialis* (Banks; Table 12.2; Kofoid 1934; Mallis 1997). Efficacy data against *H. aureus* are a required part of the registration packages for candidate termiticides under current EPA guidelines (PR notice 96–7; OPPTS 810.3600). Thus, the efficacy data taken from this location provide information for western American homeowners and pest management companies on prevention and control of their local subterranean termites. It also provides information to chemical companies on the influence of high temperature and low moisture conditions on the persistence of their termiticides.

Dorman Lake An additional site has been established near Starkville, MS. It is used mainly for generic termiticides seeking Florida registration. This location is in a pine forest near Dorman Lake and is part of the John W. Starr Memorial Forest owned and managed by the Department of Forestry, Mississippi State University. This site is not used for new products; it is used primarily for additional data that are sometimes required by states, often after an active ingredient is no longer protected by patent.

12.4 Changes in Chemistry Lead to Changes in Testing

Advances in chemistry over the years have given rise to new products. As with previous termiticides, the new products have been tested by members of the termite team. Some products have required deviations from standard protocols. Additionally, new test methods were added to account for changes in American construction practices, most importantly the increased prevalence of concrete slab-on-grade con-

Table 12.2 Termite species present at the current termiticide testing program locations

Termites	Harrison	Chipola	Calhoun	Santa Rita
Reticulitermes flavipes (Kollar)	×	×	×	–
Reticulitermes virginicus (Banks)	×	×	×	–
Reticulitermes hageni (Banks)	×	×	–	–
Reticulitermes tibialis (Banks)	–	–	–	×
Heterotermes aureus (Snyder)	–	–	–	×
Gnathamitermes perplexus (Banks)[a]	–	–	–	×

[a]*G. perplexus* is present, but not a pest species

struction occurring during the 1950s and 1960s (Peterson et al. 2008).The previous ground board test, that had been used, was appropriate for testing open (uncovered) weathering conditions, and was a worst-case scenario in testing termiticides. The modified ground board test (concrete slab test; Fig. 12.3) was used beginning in 1967 for a variety of organophosphate and carbamate products (Beal and Smith 1972). This test provided data on protected termiticide treatments (i.e., termiticide persistence beneath a slab foundation), and, to date, new studies are still being installed using both methods.

The termiticide testing program has also responded to requests for other types of termite control products developed over the years. Physical barriers (such as stone particles, metal meshes, plastic barriers, etc.) are nonchemical methods used for keeping termites out of structures, and have various benefits and drawbacks compared with conventional termiticide applications (Su 2002). When these products are impregnated with insecticides, the manufacturers must submit efficacy data similar to those for soil-applied termiticides for registration. The termite team also tests these products providing the necessary data for registration, as well as examining some of the non-impregnated products from a scientific perspective.

In the early 1970s, studies were performed on the HEF to investigate the feasibility of termite baiting (Esenther and Beal 1974). The objective of these studies was to examine the possibility of area-wide management of large numbers of colonies of subterranean termites (Esenther and Beal 1974). In the past decade, the use of termiticidal baits has become fairly common in the control of subterranean termites, although on a smaller scale (few or single colonies; Grace and Su 2000).

New chemistries such as the delayed action, non-repellent termiticides (DANR) have required modification of experimental layouts. Prior to the use of DANR products, test layouts were simple, completely randomized block designs with plots spaced on five-foot centers. Many products could be tested together in a design like this because plots of repellent compounds should not influence one another. However, the new DANR termiticides required changes to the layout because now there was a possibility for some products/concentrations to influence other nearby products. Several layouts were tried as scientists with the termite team looked for a reasonable accommodation for these products, balancing available space against the need for product separation (Peterson et al. 2006). Currently, a variety of layouts is being used to accommodate the DANR products (Peterson et al. 2006), using distances supported by colony foraging territory size and distribution data in the literature (Howard et al. 1982; Vargo 2003).

During the past decade, a slight change in the evaluation of products has also been implemented. Previously, when a block from a treated plot was damaged for the first time, it was recorded and removed from further consideration. Now, when a plot block is damaged it is recorded and provided with a new test block and evaluated again on future visits. Thus, additional data are collected on the fate of previously damaged plot blocks (are they damaged again and how badly), information that is needed for registration in Florida.

Changes in product evaluation have also been required in the past 10 years to give cooperators and regulators additional information regarding the performance of candidate products. Florida has implemented its own regulations on the efficacy of products that must also be satisfied by registrants before a product is registered in the state (Wagner et al. 2004). Most importantly, the Florida rule does not use 100% control as a registration criterion (used by EPA). This leads to an increase in the number of years of efficacy data (compared to those used by EPA) for products when used as a standard (Wagner et al. 2004, 2006, 2009, 2011). EPA has initiated the process to change OPPTS 810.3600 (Shelton and Wagner 2005), and the methods of installation or evaluation of the testing program will be modified accordingly.

Clearly, the termiticide testing program benefits both the general public and those who determine public policy. The program's consistent studies on experimental forests produce unbiased efficacy testing of termite control products for the American consumer and also aids efforts by the EPA to register these products.

References

Beal RH (1986) Field testing of soil insecticides as termiticides. In: Proceedings, 17th annual meeting of the International Research Group on Wood Preservation—Working Group IB: biological problems. IRG Secretariat, Stockholm, Sweden, Paper 1294, pp 1–9

Beal RH, Smith VK (1972) Carbamate or phosphate insecticides for subterranean termite control? Pest Control 40:20, 22, 43

Bennett GW, Owens JM, Corrigan RM (1997) Truman's scientific guide to pest control operations, 5th edn. Advanstar Communications, Cleveland, 520 p

Carter FL, Stringer CA (1970a) Soil moisture and soil type influence initial penetration by organochlorine insecticides. Bull Environ Contam Toxicol 5:422–428

Carter FL, Stringer CA (1970b) Residues and degradation products of technical heptachlor in various soil types. J Econ Entomol 63:625–628

Carter FL, Stringer CA, Beal RH (1970) Penetration and persistence of soil insecticides used for termite control. Pest Control 38:18–21, 24, 62

Cordell HK, Overdevest C (2001) Footprints on the land: an assessment of demographic trends and the future of natural lands in the United States. Sagamore, Champaign, 314 p

Esenther GR, Beal RH (1974) Attractant-mirex bait suppresses activity of *Reticulitermes* spp. J Econ Entomol 67:85–88

Grace JK, Su N-Y (2000) Evidence supporting the use of termite baiting systems for long-term structural protection (Isoptera). Sociobiology 37:301–310

Grace JK, Yamamoto RT, Tamashiro M (1995). Relationship of individual worker mass and population decline in a Formosan subterranean termite colony (Isoptera: Rhinotermitidae). Environ Entomol 24(5):1258–1262

Hopkins W, Hebb EA (1954) Guide to Chipola experimental forest. Southern Forest Experiment Station, Asheville, 24 p

Howard RW, Jones SC, Mauldin JK, Beal RH (1982) Abundance, distribution and colony size estimates for *Reticulitermes* spp. (Isoptera: Rhinotermitidae) in southern Mississippi. Environ Entomol 11:1290–1293

Kard BM (2000) Professional biography of Harmon R "Johnny" Johnston, the first entomologist assigned to the USDA Forest Service, wood products insect research work unit, Gulfport, Mississippi (Project Leader and Research Entomologist; service: 1934–1971). USDA Forest Service Occasional Paper No. RWU-4502-1-2000

Kard BM, Mauldin JK, Jones SC (1989) Evaluation of soil termiticides for control of subterranean termites (Isoptera). Sociobiology 15:285–297

Kofoid CA (1934) Termites and termite control, 2nd edn. University of California Press, Berkeley, 795 p

Kowal RJ, St George RA (1948) Preliminary results of termite soil-poisoning tests. J Econ Entomol 41:112–113

Mauldin JK (1989) Southern Forest Experiment Station. In: Fulton H, Smith JW, Morgan ER, Young DF, Wilson CA, Parencia CR, Murphree LC, Bennett AG, Martin DF, Hepner LW, Lloyd EP (eds) A history of entomology in Mississippi. Mississippi Entomological Association, Mississippi, pp 47–48

Mallis A (1997) Handbook of pest control, 8th edn. Mallis Handbook and Technical Training Company. Cleveland, 1453 p

Metz LJ (1958) The Calhoun experimental forest. Southeastern Forest Experiment Station, Asheville, 25 p

Mulrooney JE, Wagner TL, Shelton TG, Peterson CJ, Gerard PD (2007) Historical review of termite activity at forest service test sites from 1971–2004. J Econ Entomol 100(2):488–494

Pedigo LP (1989) Entomology and pest management. Macmillan, New York 646 p

Peterson CJ, Wagner TL, Shelton TG, Mulrooney JE (2006) New termiticides necessitate changes in efficacy testing: a case study of fipronil. In: Lyga JW, Theodoridis G (eds) Synthesis and chemistry of agrochemicals VII, Chap 16. American Chemical Society Symposium Series 948, pp 179–193

Peterson CJ, Wagner TL, Shelton TG, Mulrooney JE (2008) A historic pest: 90 years of USDA publications reveal changes in termite management. Pest Manag Prof 76(2):25, 28–30, 32

Sayre NF (2003) Recognizing history in range ecology: 100 years of science and management on the Santa Rite Experimental Range. In: McClaran MP, Ffolliott PF, Edminster CB (eds) Santa Rita Experimental Range: 100 years (1903–2003) of accomplishments and contributions. Rocky Mountain Research Station, Ogden pp 1–15

Shelton TG, Wagner TL (2005) Brief overview of subterranean termite issues in the southern U.S. Proceedings of the American Wood-Preserver's Association 101:104–106.

Su N-Y (2002) Novel technologies for subterranean termite control. Sociobiology 40:95–101

Su N-Y, Scheffrahn RH (2000) Termites as pests of buildings. In: Abe T, Bignell DE, Higashi M (eds) Termites: evolution, sociality, symbioses, ecology. Kluwer, Boston, pp 437–453

Vargo EL (2003) Hierarchical analysis of colony and population genetic structure of the eastern subterranean termite, *Reticulitermes flavipes*, using two classes of molecular markers. Evolution 57:2805–2818

Wagner TL, Shelton TG, Peterson CJ, Mulrooney JE (2004) 2003 termiticide report: test results and efficacy standards. Pest Control 72(2):23–29

Wagner TL, Mulrooney JE, Shelton TG, Peterson CJ (2006) 2005 termiticide report: testing demand continues; USDA-FS's Mississippi site hit hard by Hurricane Katrina. Pest Control 74(2):55–56, 58, 60–65

Wagner TL, Shelton TG, Peterson CJ, Mulrooney JE (2007) 2006 termiticide report: reorganization of the southern research station and impact of Hurricane Katrina on Mississippi test plots. Pest Control 75(2):58–69

Wagner TL, Mulrooney JE, Shelton TG, Peterson CJ (2009) 2008 USDA-FS termiticide report. Pest Manag Prof 77(2):26–27, 30, 32, 34

Wagner TL, Peterson CJ, Shelton TG (2011) USDA-FS termiticide report, termiticide efficacy results for 2010. Pest Manag Prof 79(2):28–34

Ware GW (1994) The pesticide book, 4th edn. Thomson, Fresno, 384 p

Part V
Research Trajectories in Forest and Range Hydrology, Biogeochemistry, Soils and Ecosystem Science

Chapter 13
Cycles of Research at Fort Valley, Arizona, Our First Forest Experiment Station

Brian W. Geils and Susan D. Olberding

Abstract We present a history of the origin and development of long-term, place-based research at Fort Valley, Arizona, site of the first experiment station established by the USDA Forest Service. Initially named the Coconino Experiment Station, this field laboratory was founded to solve a crucial problem with ponderosa pine reforestation and to institute a federal research program in forest science. The laboratory was soon renamed the Fort Valley Experiment Station and later designated along with its forest tracts as the Fort Valley Experimental Forest (FVEF). The science program at Fort Valley has proceeded through three cycles of research, each consisting of problem recognition, investigation, resolution, and incorporation into subsequent work. Fundamental investigations in silvics began in the first research cycle and broadened in the second cycle to integrate practical knowledge in silviculture, ecology, and related fields for improving resource management. The principal research objective in the present third cycle is to provide the scientific basis of restoration and forest health projects for managing ecosystems at a landscape scale. A common theme across these cycles has been ponderosa pine and its ecological interactions with climate and disturbance. From cycle to cycle, however, the scope of that theme has expanded from regenerating trees, to interacting biotic communities, to functioning ecosystems. The next cycle of research at FVEF will need to develop adaptive and mitigation strategies for responding to new environmental and resource crises. This research will require better integration of ecological and evolutionary perspectives through such approaches as ecosystem genomics. FVEF could contribute to this research by conducting innovative, manipulative experiments, enhancing physical and management infrastructures, and compiling long-term records of climate, disturbance, and vegetation responses.

Keywords Fort Valley Experimental Forest · Ponderosa pine · Silviculture · Forest management

---

B. W. Geils (✉) · S. D. Olberding
Rocky Mountain Research Station, USDA Forest Service,
2500 South Pine Knoll Drive, Flagstaff, AZ, USA
Phone: 928-522-3642
e-mail: bgeils@npgcable.com

D. C. Hayes et al. (eds.), *USDA Forest Service Experimental Forests and Ranges*,
DOI 10.1007/978-1-4614-1818-4_13, © Springer New York 2014

13.1 Introduction

A popular conception of science is that it seeks objective truth of universal relevance. Livingstone (2003) observed, however, that the personal interests of scientists and their sponsors determine what research is conducted and how it is applied. The geography and history of a place determine whether a line of inquiry and its resulting technologies are relevant for solving the contemporary problems perceived by these scientists and sponsors.

The more than 100-year history of forest research at Fort Valley, AZ, is marked by establishment of the first Forest Service experiment station, which continues today as the Fort Valley Experimental Forest (FVEF). That history well illustrates Livingstone's thesis on the importance of place to science. The early scientists working in the Fort Valley area, their administrators, and stakeholders were motivated by practical concerns of the time and were strongly influenced by the prevailing natural and societal environment of America and the Southwest (Olberding 2002, 2000). Over many decades, researchers at the Fort Valley field station developed the scientific basis of management for improving forests and rangelands across Arizona, New Mexico, and other western states (Olberding and Moore 2008). Although issues and environments have changed, the present research program at FVEF still contributes to adaptive learning for sustaining diverse, healthy, and productive forests (Geils 2008). Strengths of FVEF are an opportunity for conducting innovative experiments in forest stands, a research infrastructure, and detailed records of past investigations. The challenge is to relate results of historical, site-specific studies to emerging landscape-scale ecosystem management problems.

In this chapter, we discuss the motivations, methods, and influences on forest research at Fort Valley by addressing several key questions. Why did the Forest Service locate its first forest experiment station in an unpopulated territory known more for its deserts than forests? What were the critical original and subsequent forestry issues and their resolution? Who organized and conducted this research? How did resolution of one problem lead to a new crisis that began another cycle of research? Can long-term, place-based research at FVEF be made relevant for addressing present and future concerns over southwestern forests in changing environments?

13.2 Beginnings: A Resource Problem and the Scientific Method

13.2.1 Natural Geography

Fort Valley is a natural park or forest opening amidst the southern Colorado Plateau which is a large pine-covered area set in a topographically complex, warm, and arid landscape (Brown and Lowe 1980). The volcanic San Francisco Peaks rise

Fig. 13.1 A graphical rendition of the San Francisco Peaks (circa 1890) illustrates that forest stands were composed of several size classes arranged into tree groups and openings. Because several non-forested alpine slopes in the composition closely resemble what is seen today, this artwork is considered a faithful representation. (Source: Merriam 1890)

from the northern side of Fort Valley to an alpine zone above tree line (Fig. 13.1). Rocky Mountain ponderosa pine (*Pinus ponderosa* var. *scopulorum*) dominates the montane forests below the subalpine forests of spruce, firs, and white pines (Merriam 1890; Springer et al. 2009). Ponderosa pine, known locally as southwestern ponderosa pine, is also called "yellow pine" from the appearance of its platy, old-growth bark or "blackjack pine" from the dark, furrowed bark of young trees. The sparse vegetation between ponderosa pine trees consists of bunch grasses, herbs, and scattered, low shrubs. Pinyon-juniper woodlands encircling the montane forest extend from north-central Arizona into western New Mexico. These characteristically southwestern conifer forests and woodlands are surrounded by chaparral, grassland, and desert. Vegetation varies at multiple scales because of differences in landform, biogeography, climate, and disturbance history.

Compared to other regions in the American West, the southwestern climate is more variable and more prone to severe, multiyear droughts (Pearson 1951; van Mantgem et al. 2009). A precipitation pattern of winter-storm snows and summer-monsoon rains is characteristic of southwestern highlands. Meager annual precipitation limits plant growth, but a favorable distribution allows forest trees to establish and grow at highland elevations (> 1,500 m). Climate controls fire regimes by influencing the abundance of fuels and timing of ignitions (Grissino-Mayer and Swetnam 2000; Pearson 1922b). Soils and natural disturbances (especially wind and mistletoe) influence tree species composition and age-size structure, but climate and fire are most widely responsible for the naturally open pine stands typical of Fort Valley and southwestern forests (Dahms and Geils 1997; Pearson 1950).

13.2.2 Establishment History

The territory of northern Arizona was sparsely populated in the late 1800s, but its rich physical and biotic diversity was well known in Washington, DC, to scientists of the US Department of Agriculture. They were keenly interested in the roles climate and elevation played in determining which natural and domestic crops could be cultivated in a region. C. Hart Merriam recognized that Arizona's San Francisco Peaks provided an ideal location for testing his hypotheses on plant and animal distributions (Maienschein 1994). Merriam (1890) identified the southwestern flora and fauna, which he observed on the Peaks and surrounding landscapes, as a distinctive mixture derived from the adjacent Pacific, Mexican, and Rocky Mountain bioregions. He demonstrated his life zone concept by relating forest types and their associated animals to landform (elevation and aspect) and climate (temperature and moisture). He later returned to northern Arizona with Bernhard Fernow, chief of the Division of Forestry. In addition to spectacular deserts, mesas, and canyons, Fernow found large open forests of sawtimber-sized pines (Fernow 1897; Rodgers 1951). Gifford Pinchot, first chief of the Forest Service, also knew these forests from his own travels across the region (Moir et al. 1997; Pinchot 1947).

Pinchot's chief of silvics, Raphael Zon (Young 2008), and Fernow were familiar with the state-level administration of forest management and research practiced in Germany. However, America's vast virgin forests, commercial interests in timber "mining" instead of tree farming, and few educational or research institutions dedicated to forestry warranted a different approach. The alternative Zon and Fernow envisioned was a well-administered, federal system of national forests which managed the lands and experiment stations which scientifically investigated forestry practices.

The seed, which led to locating the first experimental field station in northern Arizona, was a letter sent in 1903 by Flagstaff lumberman T. A. Riordan to forester Gifford Pinchot (Schmaltz 1980; on file, FVEF archives). Ponderosa pine had become commercially valuable when a railroad linked the abundant timber supply in northern Arizona to expanding markets across the country. Both lumbermen and foresters, however, were concerned the supply was rapidly declining and regeneration would not sustain commercial harvest (Woolsey 1911). Riordan asked Pinchot to investigate the reforestation issue and offered a pasture for conducting experiments. Pinchot sent Zon in 1908 to locate a suitable site for studying climate effects on ponderosa pine regeneration and on silvicultural methods for improving forest growth and yield. Because of its proximity to the Flagstaff rail station, access to Leroux Springs, and biodiversity, Zon determined to "plant the tree of research" on the Coconino National Forest near the small community of Fort Valley (Olberding and Moore 2008).

The immediate and specific purpose for an experiment station on the Coconino National Forest was to resolve regeneration and silviculture issues unique to ponderosa pine in the Southwest (Sánchez Meador and Olberding 2008). Nationally, the station's objectives were to develop the principles of scientific forestry and to establish a professional, federal research organization (Clapp 1917; Moore 1916; Pearson 1914). Soon after facilities were built and studies began, an education program was initiated. During 2-week summer sessions, newly recruited scientists and rangers received specialized forestry training in field methods and office procedures (Woolsey 1909). These sessions continued until World War II by which time state and private colleges had developed education programs in forestry.

13.2.3 G. A. Pearson and the Roots of Scientific Forestry in the Forest Service

Gustaf A. Pearson was the first scientist and director of the Coconino Experiment Station, which was renamed the Fort Valley Experiment Station in 1911 (Olberding 2008, 2000). Pearson was outstandingly qualified for the position because of his education, leadership, work ethic, and social conscience. His academic training in plant physiology and forestry experience facilitated the establishment of experiments on reproduction, forest types, and silviculture. His professionalism gained him support from supervisors, esteem from scientists, and respect from community leaders. Pearson was an early and important contributor to the new American scientific journals of *Ecology* (Pearson 1920a, b) and *Journal of Forestry* (Pearson 1918, 1923). Pearson used his position and publications to advocate that scientifically based ponderosa pine silviculture in the Southwest could make important economic and societal contributions to a region and nation needing wood products and jobs (Pearson 1940).

The condition of southwestern forests demonstrated Meinecke's (1917) assertion that the prevailing issue for American forestry was converting unmanaged, old-growth forests into productive, young stands regulated to sustain yield and minimize waste. Forest conservation of the time was understood as "wise use" through silviculture implemented on public lands by trusted professionals of the Forest Service. For Pearson and others to develop a scientific silviculture for the American situation, they first needed to understand the physical factors (climate, weather, soil, and fire) and biotic agents (herbivores and parasites) that reduced the regeneration and growth of crop trees (Pearson 1928, Sánchez Meador and Olberding 2008).

Pearson (1914) organized research at the Fort Valley station to facilitate application of the scientific method, which uses observation, hypothesis, experiment, analysis, and interpretation. Specific research objectives were to explain the basic natural processes of tree growth and stand development and to test alternative silvicultural systems of harvesting and regenerating ponderosa pine. Seedlings were grown in the relatively controlled environments of forest nurseries. Experiments with local and nonlocal plants were conducted in research studies that could be completed in a few years. Mature forest trees, however, are large and long-lived, and respond slowly to treatment. They are also affected by factors and agents that are difficult to control within experiments. Investigations on forest stands therefore required township-sized forest tracts and research studies extending over many decades. Since naturally growing trees of various ages and conditions can be found in stands with different histories and site attributes, comparative studies could serve as observational experiments. Planned experiments that implemented alternative treatments, such as modified silvicultural systems, needed large areas dedicated to research and long-term intensive monitoring—permanent sample plots. Studies of the physical and biotic influences on pine reproduction and growth were accompanied with the development of silvicultural concepts and mensuration techniques (see Gaines and Shaw 1958).

Zon, Pearson, and the early investigative committees (reports on file, FVEF archives) also established the administrative structure for managing research in the Forest Service. National policy, budget, and institutional decisions came from the Washington office. Management and coordination with the National Forests in Arizona and New Mexico were conducted by District 3 until the research functions were assigned to the Southwestern Forest and Range Experiment Station after 1927. Research sites included field laboratories and, later, experimental forests, which were under the jurisdiction of a station director and managed in cooperation with a Forest Supervisor. Forest Service researchers were organized into research projects chartered under 5-year plans and funded by annual appropriations. The documents which planned and tracked research consisted of a problem analysis and various study plans, internal progress and final reports, and journal articles. Results and recommendations were usually presented in bulletins, notes, and papers published by the US Department of Agriculture or Forest Service.

13.2.4 Early Studies in Forest Meteorology

Zon (1908) and others understood that climate and forest development were related and fundamental for improving silvicultural practices. Several early studies provided starting points for a century-long record of meteorological observation at the Fort Valley station (Huebner et al. 2008). Mattoon (1909) noted an inverse relationship between forest cover and date of snowmelt, a key event in a region where water was, and continues to be, the limiting factor. Pearson (1922b) described how the region's distinctive features of weather, wildfire, and forest growth were strongly controlled by monsoon rains (cf. Pearson 1930 on climate and soils). In a 1916 study inspired by Merriam (1890), five stations were established along an elevational transect on the San Francisco Peaks to represent each of the major forest types. Meteorological instruments were installed, and trees typically found at other elevations were transplanted to each station (see Crouse et al. 2008). This was one of the first uses of a reciprocal transplant (or "common garden") approach for studying gene-by-environment interactions. Haasis (1921, 1923) studied the effects of soil type and frost heaving on the survival of ponderosa pine seedlings (also see Heidmann 2008).

13.2.5 Resolution of the Reforestation Problem

It was not regeneration failure, but success, that ultimately proved ponderosa pine's reproductive capability and reconfigured the forest age structure in northern Arizona (Savage et al. 1996). In the early 1900s, Pearson and other forest scientists were studying local climates, weather patterns, and forest disturbances. They learned about the regeneration requirements of ponderosa pine and that many of these pines lived for more than 200 years. In 1919, the scientists observed a series of events, which had been uncommon in preceding decades but sufficiently frequent in the previous two centuries that ponderosa pine had become the dominant species in montane forests. These events followed the sequence: (1) disturbance-caused preparation of a seedbed, (2) bountiful cone crop, (3) sufficiently wet spring to sustain seedlings from snowmelt to monsoon, and (4) several years without weather or fire events lethal to young pines. Before the late 1800s and early 1900s, disturbances such as windthrow and fire had occurred frequently among groups of one to several trees and allowed multiple age cohorts at this tree-group scale to develop across the Southwest (White 1985; see also Swetnam and Brown 2011). However, early settlements, especially in proximity to a railroad, led to significant changes in disturbance regimes. So much of the forest had been heavily logged, burned, and grazed (see Leopold 1924) that the patchy mosaic of diverse ages and sizes was replaced with a uniform forest dominated by the 1919 cohort. Although establishment of these pines resolved the original reforestation problem, these trees and their management presented researchers with new questions. The principal subject of research shifted from old-growth, cutover stands to two-story stands with few large, old yellow pines and many small, young blackjack pines (Fig. 13.2).

13.2.6 Basic Tools for Observation

Several primary tools of forest and range science were developed by early researchers at the Fort Valley station. Recording changes in plant distribution at a meter scale was important in many range and ecology studies. Hill (1920) described a mechanical pantograph for mapping plants on small quadrats. The average diameter of a tree with a noncircular bole had been routinely estimated by multiple measurements using large (and cumbersome) calipers. Krauch (1924) demonstrated that a properly graduated tape wrapped around the bole provided a quick and accurate measure of tree diameter. Although it was well known that temperature affected plant growth, Pearson (1924b) was an early proponent of the "degree-days" concept that accumulated exposure to warmth or chill could be used to time phenological events. Pearson (1937) also identified the relevant and fundamental physical factors

Fig. 13.2 As ponderosa pine trees age beyond their first century, the general appearance of the bark changes from furrowed and black to having large, yellowish to orange plates. Younger trees, especially common in dense thickets, are called "blackjack pine" and older trees, as solitary or open groups, are called "yellow pine" (or "big pumpkins"). (Photo by B. Geils at Fort Valley Experimental Forest, AZ)

that control tree-ring formation. He encouraged Glock (1933) in development of the science of dendrochronology. These and other simple, practical technologies developed by Fort Valley researchers made forest science more precise and predictive.

13.2.7 Fort Valley, Proof of Concept for an American Forest Experiment Station

Pearson and other Fort Valley scientists contributed significantly to the development of three concepts that moved forest research from practical tinkering to scientific methodology:

- *Permanent sample plots.* For long-term tracking of survival and growth by large trees responding to various cutting methods, permanent sample plots were designed and data requirements were specified (Woolsey 1912). Because ponderosa pine stands, especially after cutting, were composed of scattered, irregular groups of trees, the size of these permanent sample plots ranged from tens to hundreds of acres. Every mature tree was tagged, and its size (bole and crown) and condition (live or dead) were monitored over many decades.

- *Experimental forests and research natural areas.* To ensure continuity in silvicultural studies, forest tracts were withdrawn for primarily scientific study and experimentation with alternative and innovative methods (designated experimental forests). Although most permanent sample plots were cut according to silvicultural prescriptions, Pearson also established sample plots in uncut areas as treatment controls and advocated for the establishment of research natural areas (Pearson 1922a). A portion of one permanent sample plot became the G. A. Pearson Natural Area (GAP-NA) in 1950 and continued to be remeasured every 5–10 years, so it now has a 90-year record of growth and survival for individually tagged trees.
- *Education and support.* Experimental forests provide instructive, comparative demonstrations of silvicultural practices. Pearson extended the education role of the Fort Valley station from those who made management, policy, and investment decisions to include those who just wanted to know what tree to plant in their yard. He understood that the scientific prestige of the Forest Service was essential for public support of the agency's program. Pearson (1914,1924a) described effective research as requiring well-educated and inquisitive investigators, modern facilities, adequate administrative and technical support staff, satisfying living accommodations, and professional and public recognition.

By the 1930s, the Coconino National Forest had adequate pine regeneration, and the Fort Valley scientists had hundreds of acres of research plots for studying growth and yield under various silvicultural and disturbance regimes. National Forest lands specifically withdrawn for research were designated as the FVEF in 1931 and the Long Valley Experimental Forest (LVEF) in 1936. To understand better how results from one site with its unique environment and climate could be generalized, sample plots and studies were established across New Mexico and Arizona. Several decades of observation yielded sufficient, reliable results to produce the initial long-term syntheses (e.g., Krauch 1930; Pearson 1930).

Silvics of ponderosa pine has always been a central focus of research at the Fort Valley station. Station scientists, however, also investigated range, wildlife, animal husbandry, soils, hydrology, entomology, plant pathology, fire ecology, genetics, botany, and other sciences (see subject reviews in Olberding and Moore 2008). An experiment station that began with a simple cabin at Fort Valley where Pearson spent the winter of 1908–1909 (Olberding 2008) evolved by 1935 into a multistate research organization administered as the Southwestern Forest and Range Experiment Station. This station organized research in Arizona and New Mexico and included the nation's first experimental range at Santa Rita, Arizona. Administration moved to Tucson, increased cooperation with state agricultural schools, and added a focus in watershed research (Gottfried et al., Chap. 14). Research at FVEF continued with expertise in silviculture and extended research in range science and other disciplines which supported national forest management.

From the beginning, a common theme of research at the Fort Valley station has been that *significant interactions of climate and ecological disturbances with the biota affect important ecosystem functions and values*. What began at Fort Valley as a local, silvicultural problem of regenerating ponderosa pine and sustaining timber harvests expanded into developing tools for managing young stands across the Southwest for multiple uses. In recent years, that research has matured into a pursuit for understanding how to restore and sustain healthy ecosystems (Geils 2008). To this ecological imperative, the future requirement is to add an evolutionary perspective, which can together maintain genetically fit and adaptable populations in a dynamic environment responding to climate change.

13.3 First Results: Useful Information and Tools

13.3.1 Long-Term and Place-Based Research

Pearson (1935) expressed a common sentiment of the time that livestock and timber were the primary societal values for which public resources in the Southwest should be managed. The question was which range and forest management practices best sustained financial return and employment. Because disturbances and demographic processes such as tree regeneration and death were episodic and varied with stand age and condition, addressing this question required long-term observation and corrections when initial hypotheses proved wrong.

For silviculture, the important questions were how to convert virgin forests into regulated stands, how to minimize animal damage, and how to improve timber stand quality. At first, changes in tree density were determined by repeated counts of live trees per unit area; but Krauch (1930) recommended that monitoring individual trees and their condition was more informative. Many old-growth yellow pines were dying in cutover stands. This mortality could be "captured" in a periodic harvest if the forester knew which trees were at risk from specific hazards. Pearson (1939) extended monitoring to a 30-year period on several cutover permanent sample plots at FVEF. He concluded that: (1) most of the stand volume losses were from mortality of trees more than 2 ft (60 cm) in diameter; (2) mortality varied greatly from year to year; and (3) the most common mortality agents were ranked in the order of wind, lightning, mistletoe, and bark beetles.

Early evidence pointed to serious losses of ponderosa pine saplings by domestic livestock grazing (Pearson 1931). Although national forests were established to protect timber and water as well as to provide livestock forage, the interests of foresters and ranchers often conflicted. Even Grazing Examiner Robert Hill (1917) acknowledged that sheep and cattle damaged young ponderosa pine. After 17 years of study, however, Pearson was so impressed by the recuperative capacity of ponderosa pine that he eventually moderated his strong opposition to grazing. Initially, rodents and other animals were also held responsible for seedling destruction. Results with differential exclosures suggested to Taylor and Gorsuch (1932) that with a good seed crop and favorable weather, ponderosa pine could successfully regenerate in spite of seed predation. Birds and rodents could even increase pine regeneration by distributing and planting seeds, but Taylor and Gorsuch (1932) cautioned that their observations were limited. Seedling success was also apparently affected by disposal of logging slash, and the "pulled tops" of harvested trees were thought to provide shelter from damage. But Pearson (1921) observed that seedlings were most abundant on bare spots where brush had been burned, were moderately abundant on grassy spots, and were least abundant under heavy brush.

One of the earliest studies on controlling dwarf mistletoe was established in a forest stand near the Fort Valley station (Korstian and Long 1922). Foresters initially hoped that removing only some infected trees would be sufficient to reduce mistletoe infestation. Heidmann (1983) concluded after 27 years of observation, however, that satisfactory silvicultural control would require either multiple entries or complete stand replacement. According to the first strategy, a stand was entered every 5–10 years to remove the most severely infected trees and to prune any remaining infected branches. With the second strategy, a stand was cut by seed-tree or shelterwood methods. After regeneration was established but before it was infested, the remaining overstory trees were removed, many of which would be infected. These and subsequent studies (e.g., Myers et al. 1972) revealed the initially slow but accelerating increase in mistletoe infestation over many decades. A few infections in a young stand could spread mistletoe to most trees within several decades, while causing little immediate damage. But, after several decades of intensification, severe disease effects on growth and survival would render the stand commercially worthless.

A long-term perspective in forestry is necessary because environments, technologies, economics, and politics change. Railroad logging had required harvesting enough timber to pay for the expense of laying track; but, as Pearson (1938) realized, harvest cuts with truck logging could be lighter and more frequent. Data from permanent sample plots were initially summarized for growth and yield under silvicultural systems of heavy and infrequent cutting. These data were reanalyzed with new statistical methods to test new hypotheses for different silvicultural regimes (Lexen 1939). Several detailed studies documented stand development for the permanent sample plots (e.g., Pearson 1933). Pearson and Wadsworth (1941) tracked the 30-year history of a twice-cut unit at Wing Mountain that was severely infested by mistletoe. They concluded that potential timber income could exceed production and protection costs. Sánchez Meador and Moore (2008) presented a 93-year case history of harvest, regeneration, and grazing at Coulter Ranch. The authors reported that harvesting reduced the number of trees under most treatments but density greatly rebounded for most treatments—up to nine times the preharvest level under seed-tree silviculture.

13.3.2 A Spatial Context for Research

The Fort Valley researchers recognized both that forests change over time and that the silvics of widely distributed species, such as ponderosa pine, varied by region and site. For example, Pearson (1943) identified the primary mortality agents of ponderosa pine as wind, lightning, and mistletoe in the Southwest but recognized that bark beetles were the most important in California and the Pacific Northwest (Oregon and Washington). Pearson (1951) reported that differences in the climates of Arizona, California, the Pacific Northwest, and Black Hills required modification of ponderosa pine silviculture for each region. Pearson (1943) had earlier asserted that "silviculture is local," but what was local? How far could results from a permanent sample plot be extended? Because Arizona and New Mexico have a similar monsoon climate (Krauch 1939), ponderosa pine in these states might show similar silvicultural responses. Pearson (1944) concluded from 35 years of observation that New Mexico sample plots agreed well enough with those from Arizona that common lessons could be applied to similar sites across the Southwest. Because all the permanent sample plots on the FVEF were located on basaltic soils, the station scientists also established studies on limestone soils of the LVEF (80 km south of Fort Valley). Like Pearson's regional analyses, comparisons of finer-scale effects (e.g., an edaphic factor) helped determine how similar a site must be to a permanent sample plot for research results to be applicable there.

As crowding or competition from trees of the 1919 cohort increased, research on density management became more important. The questions of interest were which trees should be removed to concentrate growth on high-quality fast-growing crop trees and when should removals occur. Pearson (1936) analyzed 10-year data from the Prescott National Forest for the first research thinning plots in the Southwest. He had initially focused on older, multi-aged, cutover stands; this work in second-growth stands helped develop and refine Pearson's concept of timber stand improvement. Thinning was integrated as part of a comprehensive silvicultural prescription that included crop tree section, spacing, tending, and regeneration (De Blois et al. 2008). Gaines and Kotok (1954) later reviewed stocking and scheduling studies and observations from the Prescott thinning plots.

13.3.3 Synthesis and Admonishment

Pearson successfully accomplished his two primary assignments as the first director of a Forest Service experiment station: (1) discover the constraints on silviculture of southwestern ponderosa pine and (2) demonstrate the utility of experimental and observational science to forest management. Pearson published frequently in several forestry outlets during and after his career. For example, his compilation of work on vegetation as a factor controlling regeneration of ponderosa pine was produced as an *Ecological Monograph* (Pearson 1942). His book on management of ponderosa pine was published posthumously by the US Department of Agriculture (Pearson 1950). This classic work was used in forestry classes for the next 50 years. In addition to promoting reliable, practical research, Pearson was also an outspoken advocate for ponderosa pine, silviculture, and the application of science and experimental forests to resource management (Pearson 1946). He advised scientists and managers to learn from past mistakes but held adamantly to several strong sentiments. These included his contentions that: "research [administration] has not exerted leadership...experimental management has lagged...the goal of a well-balanced, fully productive selection forest is still far from being realized...the West needs experimental forests but it should have more active ones...[not just] show places lacking suitable forest stands and an adequate research setting" (Pearson 1946).

With 50 years of investigations completed at the Fort Valley station, scientists were able to produce seminal papers on ponderosa pine, dwarf mistletoe, and other work conducted by the Southwestern Forest and Range Experiment Station (Gaines and Shaw 1958). Studies on ponderosa pine regeneration were especially important in the initial years at the Fort Valley station and continued for another 50 years on its permanent sample plots. Meagher (1950) provided a contemporary review of the silvics learned there, and Ffolliott (2008) presented a more comprehensive summary of that work. The Fort Valley Experiment Station was a realization of the vision by Zon, Pearson, and others that a Forest Service research organization could provide the scientific foundation for conservation of ponderosa pine in the Southwest. Although transformed from an experiment station to an experimental forest, Fort Valley research continued at both the headquarters campus (including labs and a greenhouse) and in the nearby forest stands.

13.4 Reorganization: New Perspectives

13.4.1 Rocky Mountain Forest and Range Experiment Station

In 1953, the Southwestern Forest and Range Experiment Station was merged into the Rocky Mountain Forest and Range Experiment Station (RMFRES) with headquarters in Fort Collins, CO (Price 1976). This reorganization gave FVEF new administrators, scientists, partners, and clients with broader geographic perspectives, different interests, and alternative approaches to science and public resource management. For example, Pearson (1940) considered "multiple use" as a fundamental tenant of Forest Service philosophy that applied to the forest as a whole rather than each individual acre of ground. By his interpretation, timber production should be optimized using intensive silviculture on those sites most productive for tree growth. Range, wildlife, and other resources would be given preference where they constitute a site's best use. After Pearson's departure and station reorganization, the research emphasis shifted to greater integration of multiple uses such as timber and grazing (Clary et al. 1975). More attention was given to wildlife biology, plant pathology, and fire ecology. More research was conducted through university researchers funded by Forest Service cooperative agreements. A half century of silviculture had converted uneven-aged, virgin ponderosa pine stands to even-aged, managed stands (Myers and Martin 1963). By time of the merger, the 1919 cohort was more than 30 years old. Contemporary stands appeared as thickets of full-crown blackjack pine filling between the dwindling groups of yellow pines.

13.4.2 Headquarters Renovation

The early residents of Fort Valley endured long, cold winters and short, cool summers. Since they usually did not stay long, when Zon and Pearson first visited the area in the summer of 1908, there were few neighbors. Flagstaff was about 2 hours away by mule-drawn wagon, and visits to town were infrequent. The remote setting was ideal to focus on research. The scientists and their families living at the station built a root cellar to keep perishables cold and relied on hunting to supplement their diets. After 19 years, the Fort Valley station headquarters contained only four buildings. With Civilian Conservation Corps construction, the headquarters eventually expanded to 12 buildings including offices, laboratories, shops, sheds, greenhouse, residences, and even a schoolhouse. For many decades, the headquarters campus provided the satisfying living accommodations and suitable research setting which Pearson (1946) stipulated as necessary for an effective field station.

After the 1953 reorganization, scientists were assigned to offices and laboratories on the state college campus in Flagstaff, and much of FVEF was abandoned. For several decades, some research in tree physiology, entomology, and geology continued at the Fort Valley headquarters, and a few employees resided there. After 1990, research in Arizona was consolidated to a new laboratory on the Northern Arizona University (NAU) campus. Then, Rocky Mountain and Intermountain Research Stations merged (RMRS) and set up new headquarters in Fort Collins, Colorado. With RMRS interest directed elsewhere, maintenance of the FVEF buildings was repeatedly deferred until it became critical. Serious attention to renovation began when the site was added to the National Register of Historic Places. Most of the original buildings remain, but only a few meet occupancy standards. Rehabilitated buildings now house the historical archives and provide office and meeting space for small groups. Potential uses of the remaining buildings include shops and laboratories, short-term residences, and an environmental education center.

With the diverse interests of larger stations (RMFRES and RMRS), Fort Valley researchers developed new concepts and methods for studying the biology, ecology, and management of southwestern ponderosa pine and associated species. Laboratories and equipment were more sophisticated. New inventions designed included an automated, greenhouse lighting system and specialty containers for nursery-grown seedlings (Landis et al. 1989). Descriptions of forest types were refined from Merriam's broad life zones to detailed classifications of forest vegetation (Layser and Schubert 1979). Transportation and communication improvements allowed researchers from distant locations to conduct studies at Fort Valley. For example, a forest pathologist stationed in Fort Collins might visit Fort Valley plots to study mistletoe biology over an entire season (Hawksworth 1961). Research of station scientists expanded to study the human dimensions or sociology of resource management and use (Raish et al. 1997). The factors of interest to researchers and managers still included natural effects of site, climate, weather, fire, disease, and insects, but anthropogenic effects such as urbanization and recreation were added.

Since 1953, long-term, cooperative research at FVEF has been especially prominent in the fields of genetics, silviculture, ecophysiology, and fire ecology. Varietal differences in ponderosa pine across its distribution were studied by reciprocal transplant experiments begun in 1910. The importance of long-term, place-based, cooperative research is well illustrated by this provenance trial. Some extra-regional varieties survived the Fort Valley climate for several decades; but after a few more decades, the southwestern varieties proved best adapted to the site (DeWald and Mahalovich 2008). The young, even-aged stand at Taylor Woods (on FVEF) was included in range-wide levels of growing stock study (Bailey 2008; Myers 1967). Data from these and other FVEF permanent sample plots significantly contributed to early development of a computer program for projecting stand growth and yield (Myers et al. 1972). Later versions of this program (Edminster et al. 1991) provided the core functions for simulation models currently used in preparing silvicultural prescriptions and forest plans. The Taylor Woods plots have also been monitored by Kolb and McDowell (2008) to study the ecophysiology of trees growing under different stand densities. A pioneering study of fire history by Dieterich (1980) was conducted at Chimney Spring (on FVEF) where one of the longest investigations on intervals of burning is still active (Sackett and Haase 2008). Although these and other studies were originally conceived or established by Fort Valley scientists, their continuation has increasingly relied on work by other Forest Service units, university professors, and university students. Local environmental organizations other than the Forest Service have adapted elements of the forest experiment station model. They conduct research and education at the Centennial Forest (NAU), the Arboretum at Flagstaff, and the Merriam-Powell Environmental Research Center (NAU).

The summary and application of research produced from the plots, laboratories, scientists, and cooperators associated with FVEF has been presented in numerous proceedings and technical reports. Schubert (1974) compiled a state-of-knowledge review for silviculture of southwestern ponderosa pine. Clary (1975) did likewise for range science in the Southwest. Bakker et al. (2008) and Laughlin and Moore (2008) reviewed range science research first established at the Hill plots and Wild Bill plots by Fort Valley scientists and continued by university cooperators. Research on the biology and ecology of the Abert squirrel as a wildlife species rather than a damaging pest was compiled by Patton (2008). Aspen is a less common but valuable forest species for special wood products, wildlife, biotic diversity, and recreation (fall color). Aspen research shifted over the decades from silviculture for producing excelsior (used in evaporative coolers to make life tolerable in Arizona desert cities) to protection from browsing damage (Shepperd and Fairweather

1994). Shepperd et al. (2006) suggested that sustaining aspen in northern Arizona involved not just restoration but also promoting resilient ecosystems. Sackett et al. (1996) reviewed the lessons from decades of prescribed burning on FVEF. Much of the information synthesized by Moir et al. (1997) on the ecology of southwestern ponderosa pine was derived from Fort Valley research. Changes in the objectives of forest management in the Southwest have required a reexamination of values, approaches, and participation of partners (see Hayes et al. 1991). The most comprehensive review of research, its application, and history was published in the *Fort Valley Experimental Forest—A Century of Research 1908–2008 Conference Proceedings* (Olberding and Moore 2008).

13.5 Renewal: Adapting for Change

After 100 years of research, FVEF is set for yet another renewal. Pearson and his Fort Valley scientists saw research in terms of resource management, forest stands, and recurring cycles of harvest and regeneration. Scientific leadership now requires an expansion of those terms to include ecosystem management, urban-wildland landscapes, and global changes in environments and societies. The present generation has inherited several legacies to steward for their scientific and cultural values. First, there is "the land," by which Aldo Leopold (1949) meant the air, water, soil, plants, and animals of a place. FVEF is especially valuable because the study plots are documented in the archives (with study plans, data, and publications), giving new research-specific context and continuity. The historic buildings and artifacts (including forestry and scientific tools) provide physical evidence of how and why research was conducted so differently from the present. Past research was the work of a few resident, federal scientists and assistants supported mostly by direct appropriations to the station for designated studies. Research in the present, highly connected world is broadly collaborative and funded from multiple public and private sources through competitive grants. Science and education are more integrated. Larger, more complex studies involve numerous researchers, professionals, technicians, students, and volunteers working on experimental forest sites or in distant laboratories, offices, and schools.

The first principle for guiding a new cycle of research at FVEF is that human activities (including resource extraction and pollution) have increased so much they profoundly affect the ecosystems upon which our civilization depends (Millennium Ecosystem Assessment 2003). A corollary to this principle is that the primary objective of forest management should be to sustain desirable, resilient ecosystems with their diverse biotic and abiotic components, processes, and patterns (Dahms and Geils 1997; Kaufmann et al. 1994). How this could be accomplished for the ponderosa pine ecosystem was explored in a Flagstaff conference on the three critical aspects of biology, sociology, and economics (Vance et al. 2000).

The concept of renewal applied to forest research and management is demonstrated at FVEF by two approaches. The first is *restoration ecology,* whereby cutting and burning treatments re-create the forest structure and pattern of an historical reference condition (Covington et al. 1997). Restoration of ecosystem functions consequently follows (cf. Wagner et al. 2000). The second approach is *adaptive management,* which complements restoration with the principle that management is an experiment or learning opportunity involving diverse stakeholders and dealing with uncertain changes in both natural and human systems (see Geils 2008).

13.5.1 Restoration

Reference conditions have been described for various forest areas, including many applicable to FVEF. White (1985) provided the first description of presettlement forest structure on the GAP-NA. Huffman et al. (2001) identified reference conditions for several FVEF units (Chimney Spring, Wing Mountain, and Coulter Ranch). Bell et al. (2008) reexamined the permanent sample plots established by Woolsey (1912). A typical reconstruction uses dendrochronology, but Kerns et al. (2008) analyzed phytoliths (hydrated silica formed in living plants) to infer past distributions of parks in the montane forest. Molecular genetic techniques enable scientists to investigate community genetics or ecosystem genomics and to discern how species assemblages have evolved and adapted.

Current ponderosa pine restoration treatments may or may not lead to return of presettlement stand structure, fire regime, and ecosystem health. As experimental forests, many areas on FVEF and LVEF were protected from harvest and other activities that would have removed old-growth yellow pine and increased blackjack pine. These experimental forests offer opportunities to determine the appearance of natural stand structures in terms of age cohorts and patterns of distribution. Current restoration treatments focus as much on reduction of severe fire risk through fuels reduction as on broader goals of ecological restoration. Experimental forests offer the opportunity to test assumptions that reestablishment of stand structure and fire regime leads to a restoration of ecosystem function.

Because treatment experiments are prohibited in research natural areas, a portion of the GAP-NA was decommissioned and rededicated to study restoration methods and effects (Moore et al. 2008). This small research plot is located next to the historic FVEF headquarters and is the subject of intensive study by researchers of the Ecological Restoration Institute (NAU). An added benefit of this plot is that it provides a firebreak between the densely stocked GAP-NA and the historic headquarters compound. The consequences of restoring the presettlement pattern and size distribution of yellow pine (mostly by removing trees of the 1919 cohort) have been well investigated in numerous studies. These include research on tree physiology, mortality and regeneration, soil attributes, potential fire behavior, and plant diversity (Bailey and Covington 2002; Feeney et al. 1998; Fulé et al. 2001; Kaye and Hart 1998; Kolb et al. 2001; Laughlin et al. 2008). Success of restoration as a scientific paradigm and its popular appeal as a means to forest health ensures this approach will remain an important component of future FVEF research.

13.5.2 Adaptive Management

Based on several regional case histories, Holling (1978) described the fundamental concepts of adaptive management as a practical response to environmental crises in complex natural and social systems. Although these concepts were not developed at FVEF, their utility is well demonstrated in the Fort Valley history. These concepts also offer an approach for design of the next cycle of FVEF research. A complex system such as forest management in the Southwest has both natural and societal aspects. Forest responses to environmental changes and management interventions are often accompanied with unanticipated surprises. Rather than managing ecosystems as engineering projects with little uncertainty, treatments and policies can be considered as an experiment or learning experience where the expected response is a hypothesis and system behavior is tracked by monitoring (Holling and Meffe 1996). Broad participation by diverse stakeholders increases the range of options considered and builds public acceptance.

From Merriam to Pearson to the present, research has repeatedly confirmed the important lesson that natural and social systems cycle though stages of establishment, expansion, collapse, and renewal (see Holling 2004). For example, consider the Quaternary biogeography of pines in the Southwest. Betancourt et al. (1990) reconstructed the region's vegetation during the late Wisconsin Glacial Episode (40,000 years BP) using data from packrat (*Neotoma*) middens. They concluded that the widespread and dominant pine was not ponderosa pine but limber pine (*Pinus flexilis*), which is now replaced by pinyons (*Cembroides*) in the woodlands and persists as a minor species in mixed-conifer forests of the montane and subalpine zones. In the relatively wetter past 10,000 years, Holocene populations of ponderosa pine have expanded in area and number. Although specific triggers cannot be predicted, a variable dry climate and other consequences of human activities could cause rapid and profound ecological and genetic changes in local communities and populations (Millar et al. 2007). The biological system that establishes and develops after a collapse depends on legacy retention (for example, ecological structures, genetic variability, and refugia), genetic evolution, and migration. Millar et al. (2007) described a mix of adaptive and mitigation strategies and priority-setting approaches for sustaining ecosystem values at risk of inevitable change.

Concern over the threat of extreme wildfire near Flagstaff led a number of partners to conduct landscape-scale fuel reduction projects in the Fort Valley area. Moseley and KenCarin (2001) described the conflicts and eventual collaborations that developed in such partnerships. The Fort Valley Forest Health Restoration Project examined the economics of wood product recovery (Larson and Mirth 2001; Lowell and Green 2001). DeWald and Springer (2001) described a project on the San Francisco Peaks to rehabilitate an ecologically sensitive and threatened willow community. Olberding et al. (2008) reviewed the history of cooperation between the FVEF and the Coconino National Forest.

Silvicultural research for enhanced water flow was conducted by Fort Valley researchers primarily at the nearby Beaver Creek watershed (Gottfried et al., Chap. 14). Flow from the Little Leroux Spring, one of the very few springs on the San Francisco Peaks, is a delayed response to precipitation (Martin 1969) and has served as an arboretum water source. A present concern over this water that might be addressed in an adaptive management framework is whether to restore overland flow and study the return of transient riparian vegetation or to use the water to maintain arboretum trees for genetics studies.

In relation to its population, the relatively small community of northern Arizona has a large number of scientific institutions and residents with advanced education. Along with the Forest Service (Coconino and Kaibab National Forests and RMRS), there are two national astronomical observatories, a US Geological Survey Laboratory, the Arboretum at Flagstaff, the internationally acclaimed Museum of Northern Arizona, several National Parks, NAU with its many research institutes, the Arizona Game and Fish Department Research Lab, and several citizen conservation organizations (e.g., The Natural Conservancy, Hart Prairie Preserve, and Willow Bend Environmental Education Center). Civic support and pride is manifested each autumn in a 2-week Festival of Science when scientific institutions are open to visitors with special programs, hikes, and events. FVEF hosts an Open House during this time, drawing visitors especially interested in environmental education and historic preservation. Throughout the year, these institutions work together to present programs that promote collaboration. The Forest Service cooperates with public–private partnerships to address forest health issues critical to the local community (see Moseley and KenCairn 2001). Partnerships and volunteers contribute to rehabilitation of FVEF facilities, an activity which also generates opportunities for conservation education and builds support for a Forest Service role in research.

13.5.3 Conclusion

Research at FVEF over the past century has completed several cycles of problem definition, investigation, synthesis, and application. For the first 50 years, research focused on pine harvesting and regeneration with the aim of conversion to regulated stands. For the next 25 years, the focus was on integrated resource management for multiple-use productivity. The current period of research at FVEF is focused on restoration of historical forest structure and fire regimes to sustain healthy forests, diverse communities, and threatened species. However, by Pearson's observation, ponderosa pine trees can live more than 200 years, so another century is required to study just one complete rotation. The most substantial challenges for ponderosa pine and FVEF in the next century are to accommodate and adapt to rapid and severe environmental and social changes. Visionary leadership, prudent stewardship, and collaborative partnership provide the keys to success.

References

Bailey JD (2008) Forty years later at Taylor Woods: merging the old and new. In: Olberding SD, Moore MM (tech coords) Fort Valley Experimental Forest—a century of research 1908–2008, 2008 August 7–9, Flagstaff, AZ. Proceedings RMRS-P-55, U.S. Department of Agriculture, Forest Service, Rocky Mountain Research Station, Fort Collins, CO, pp 100–105

Bailey JD, Covington WW (2002) Evaluating ponderosa pine regeneration rates following ecological restoration treatments in northern Arizona, USA. For Ecol Manage 155:271–278

Bakker JD, Moore MM, Laughlin DC (2008) The Hill plots: a rare long-term vegetation study. In: Olberding SD, Moore MM (tech coords) Fort Valley Experimental Forest—a century of research 1908–2008, 2008 August 7–9, Flagstaff, AZ. Proceedings RMRS-P-55, U.S. Department of Agriculture, Forest Service, Rocky Mountain Research Station, Fort Collins, CO, pp 116–123

Bell DM, Parysow PF, Moore MM (2008) Assessing the representativeness of the oldest permanent inventory plots in northern Arizona ponderosa pine forests. Restor Ecol 17:369–377

Betancourt JL, Van Devender TR, Martin PS (1990) Synthesis and prospectus. In: Betancourt JL, Van Devender TR, Martin PS (eds) Packrat Middens: the last 40,000 years of biotic change. University of Arizona Press, Tucson, pp 259–292

Brown DE, Lowe CH (1980) Biotic communities of the Southwest. General Technical Report RM-78, U.S. Department of Agriculture, Forest Service, Rocky Mountain Forest and Range Experiment Station, Fort Collins, CO

Clapp EH (1917) The correlation of American forest research. J For 15(2):165–175

Clary WP (1975) Range management and its ecological basis in the ponderosa pine type of Arizona: the status of our knowledge. Research Paper RM-158, U.S. Department of Agriculture, Forest Service, Rocky Mountain Forest and Range Experiment Station, Fort Collins, CO, 35 p

Clary WP, Kruse WH, Larson FR (1975) Cattle grazing and wood production with different basal areas of ponderosa pine. J Range Manage 28(6):434–437

Covington WW, Fulé PZ, Moore MM, Hart SC, Kolb TE, Mast JN, Sackett SS, Wagner MR (1997) Restoring ecosystem health in ponderosa pine forests of the Southwest. J For 95(4):23–29

Crouse JE, Moore MM, Fulé PZ (2008) Revisiting Pearson's climate and forest type studies on the Fort Valley Experimental Forest. In: Olberding SD, Moore MM (tech coords) Fort Valley Experimental Forest—a century of research 1908–2008, 2008 August 7–9, Flagstaff, AZ. Proceedings RMRS-P-55, U.S. Department of Agriculture, Forest Service, Rocky Mountain Research Station, Fort Collins, CO, pp 135–144

Dahms CW, Geils BW (tech eds) (1997) An assessment of forest ecosystem health in the Southwest. General Technical Report RM-295, U.S. Department of Agriculture, Forest Service, Rocky Mountain Forest and Range Experiment Station, Fort Collins, CO, 97 p

De Blois BP, Finkral AJ, Sánchez Meador AJ (2008) Early thinning experiments established by the Fort Valley Experimental Forest. In: Olberding SD, Moore MM (tech coords) Fort Valley Experimental Forest—a century of research 1908–2008, 2008 August 7–9, Flagstaff, AZ. Proceedings RMRS-P-55, U.S. Department of Agriculture, Forest Service, Rocky Mountain Research Station, Fort Collins, CO, pp 145–149

DeWald LE, Mahalovich MF (2008) Historical and contemporary lessons from ponderosa pine genetics studies at the Fort Valley Experimental Forest, Arizona. In: Olberding SD, Moore MM (tech coords) Fort Valley Experimental Forest—a century of research 1908–2008, 2008 August 7–9, Flagstaff, AZ. Proceedings RMRS-P-55, U.S. Department of Agriculture, Forest Service, Rocky Mountain Research Station, Fort Collins, CO, pp 150–155

DeWald LE, Springer AE (2001) Incorporating ecological and nonecological concerns in the restoration of a rare, high-elevation Bebb willow riparian community. In: Vance RK, Edminster CB, Covington WW, Blake JA (comps) Ponderosa pine ecosystems restoration and conservation: steps towards stewardship, 2000 April 25–27, Flagstaff, AZ. Proceedings RMRS-P-22, U.S. Department of Agriculture, Forest Service, Rocky Mountain Research Station, Ogden, UT, pp 134–140

Dieterich JH (1980) Chimney Spring forest fire history. Research Paper RM-220, U.S. Department of Agriculture, Forest Service, Rocky Mountain Forest and Range Experiment Station, Fort Collins, CO, 8 p

Edminster CB, Mower HT, Mathiasen RL, Schuler TM, Olsen WK, Hawksworth FG (1991) GENGYM: a variable density stand table projection system calibrated for mixed conifer and ponderosa pine stands in the Southwest. Research Paper RM-297, U.S. Department of Agriculture, Forest Service, Rocky Mountain Forest and Range Experiment Station, Fort Collins, CO, 32 p

Feeney SR, Kolb TE, Wagner MR, Covington WW (1998) Influence of thinning and burning restoration treatments on presettlement ponderosa pines at the Gus Pearson Natural Area. Can J For Res 28(9):1295–1306

Fernow BE (1897) The forests and deserts of Arizona. Nat Geogr 8:203–226

Ffolliott PF (2008) Historical review of Fort Valley studies on stand management. In: Olberding SD, Moore MM (tech coords) Fort Valley Experimental Forest—a century of research 1908–2008, 2008 August 7–9, Flagstaff, AZ. Proceedings RMRS-P-55, U.S. Department of Agriculture, Forest Service, Rocky Mountain Research Station, Fort Collins, CO, pp 18–24

Fulé PZ, McHugh C, Heinlein TA, Covington WW (2001) Potential fire behavior is reduced following forest restoration treatments. In: Vance RK, Edminster CB, Covington WW, Blake JA (comps) Ponderosa pine ecosystems restoration and conservation: steps towards stewardship, 2000 April 25–27, Flagstaff, AZ. Proceedings RMRS-P-22, U.S. Department of Agriculture, Forest Service, Rocky Mountain Research Station, Ogden, UT, pp 28–35

Gaines EM, Kotok ES (1954) Thinning ponderosa pine in the Southwest. Station Paper RM-17, U.S. Department of Agriculture, Forest Service, Rocky Mountain Forest and Range Experiment Station, Fort Collins, CO, 20 p

Gaines EM, Shaw EW (1958) Half a century of research, Fort Valley Experimental Forest 1908–1958. Station Paper RM-38, U.S. Department of Agriculture, Forest Service, Rocky Mountain Forest and Range Experiment Station, Fort Collins, CO, 17 p

Geils BW (2008) Fort Valley: a natural laboratory for research and education. In: Olberding SD, Moore MM (tech coords) Fort Valley Experimental Forest—a century of research 1908–2008, 2008 August 7–9, Flagstaff, AZ. Proceedings RMRS-P-55, U.S. Department of Agriculture, Forest Service, Rocky Mountain Research Station, Fort Collins, CO, pp 1–2

Glock WS (1933) Tree-ring analysis on Douglass System. Pan Am Geol 60:14

Grissino-Mayer HD, Swetnam TW (2000) Century-scale climate forcing of fire regimes in the American Southwest. Holocene 10(2):207–214

Haasis FW (1921) Relations between soil type and root form of western yellow pine seedlings. Ecology 2(4):292–303

Haasis FW (1923) Frost heaving of western yellow pine seedlings. Ecology 4(4):378–390

Hayes DC, Bumstead JS, Richards MT (eds) (1991) A Southwestern Mosaic: Proceedings of the Southwestern Region New Perspectives University Colloquium, 1991, October/November, New Mexico and Arizona. General Technical Report RM-216, U.S. Department of Agriculture, Forest Service, Rocky Mountain Forest and Range Experiment Station, Fort Collins, CO, 52 p

Hawksworth FG (1961) Dwarfmistletoe of ponderosa pine in the Southwest. Technical Bulletin, 1246, U.S. Department of Agriculture, Forest Service, Washington, DC, 112 p

Heidmann LJ (1983) Silvicultural control of dwarf mistletoe in Southwestern ponderosa pine. Research Note RM-433, U.S. Department of Agriculture, Forest Service, Rocky Mountain Forest and Range Experiment Station, Fort Collins, CO, 4 p

Heidmann LJ (2008) Forest regeneration research at Fort Valley. In: Olberding SD, Moore MM (tech coords) Fort Valley Experimental Forest—a century of research 1908–2008, 2008 August 7–9, Flagstaff, AZ. Proceedings RMRS-P-55, U.S. Department of Agriculture, Forest Service, Rocky Mountain Research Station, Fort Collins, CO, pp 25–37

Hill RR (1917) Effects of grazing upon western yellow-pine reproduction in the national forests of Arizona and New Mexico. Bulletin 580, U.S. Department of Agriculture, Forest Service, Washington, DC, 27 p

Hill RR (1920) Charting quadrats with a pantograph. Ecology 1(3):270–273

Holling CS (ed) (1978) Adaptive environmental assessment and management. Blackburn, Caldwell, 377 p

Holling CS (2004) From complex regions to complex worlds. Ecol Sci 9(1):10

Holling CS, Meffe GK (1996) Command and control and the pathology of natural resource management. Conserv Biol 10:328–337

Huebner DP, Olberding SD, Peterson B, DeSimone D (2008) A century of meteorological observations at Fort Valley Experimental Forest: a cooperative observer program success story. In: Olberding SD, Moore MM (tech coords) Fort Valley Experimental Forest—a century of research 1908–2008, 2008 August 7–9, Flagstaff, AZ. Proceedings RMRS-P-55, U.S. Department of Agriculture, Forest Service, Rocky Mountain Research Station, Fort Collins, CO, pp 183–185

Huffman DW, Moore MM, Covington WW, Crouse JE, Fulé PZ (2001) Ponderosa pine forest reconstruction: comparisons with historical data. In: Vance RK, Edminster CB, Covington WW, Blake JA (comps) Ponderosa pine ecosystems restoration and conservation: steps towards stewardship, 2000 April 25–27, Flagstaff, AZ. Proceedings RMRS-P-22, U.S. Department of Agriculture, Forest Service, Rocky Mountain Research Station, Ogden, UT, pp 3–8

Kaufmann MR, Graham RT, Boyce DA Jr, Moir WH, Perry L, Reynolds RT, Bassett RL, Mehlhop P, Edminster CB, Block WH, Corn PS (1994) An ecological basis for ecosystem management. General Technical Report RM-246, U.S. Department of Agriculture, Forest Service, Rocky Mountain Forest and Range Experiment Station, Fort Collins, CO, 22 p

Kaye JP, Hart SC (1998) Ecological restoration alters nitrogen transformations in a ponderosa pine-bunchgrass ecosystem. Ecol Appl 8(4):1052–1060

Kerns BK, Moore MM, Hart SC (2008) Understanding ponderosa pine forest-grassland vegetation dynamics at Fort Valley Experimental Forest using phytolith analysis. In: Olberding SD, Moore MM (tech coords) Fort Valley Experimental Forest—a century of research 1908–2008, 2008 August 7–9, Flagstaff, AZ. Proceedings RMRS-P-55, U.S. Department of Agriculture, Forest Service, Rocky Mountain Research Station, Fort Collins, CO, pp 191–195

Kolb TE, McDowell NG (2008) Tree ecophysiology research at Taylor Woods. In: Olberding SD, Moore MM (tech coords) Fort Valley Experimental Forest—a century of research 1908–2008, 2008 August 7–9, Flagstaff, AZ. Proceedings RMRS-P-55, U.S. Department of Agriculture, Forest Service, Rocky Mountain Research Station, Fort Collins, CO, pp 196–202

Kolb TE, Fulé PZ, Wagner MR, Covington WW (2001) Six-year changes in mortality and crown condition of old-growth ponderosa pines in different ecological restoration treatments at the G.A. Pearson Natural Area. In: Vance RK, Edminster CB, Covington WW, Blake JA (comps) Ponderosa pine ecosystems restoration and conservation: steps towards stewardship, 2000 April 25–27, Flagstaff, AZ. Proceedings RMRS-P-22, U.S. Department of Agriculture, Forest Service, Rocky Mountain Research Station, Ogden, UT, pp 61–66

Korstian CF, Long WH (1922) The western yellow pine mistletoe. Bulletin 1112, U.S. Department of Agriculture, Washington, DC, 36 p

Krauch H (1924) Comparison of tape and caliper measurements. J For 22(5):537–539

Krauch H (1930) Mortality in cut-over stands of western yellow pine in the Southwest. J For 28(12):1085–1097

Krauch H (1939) Analysis of a 36-year record of precipitation at Cloudcroft, NM. Research Note SW-60, U.S. Department of Agriculture, Forest Service, Southwestern Forest and Range Experiment, Tucson, AZ, 5 p

Landis TD, Tinus RW, McDonald SE, Barnett JP (1989) Container tree nursery manual. Agriculture Handbook 672, U.S. Department of Agriculture, Forest Service, Washington, DC

Larson D, Mirth R (2001) Projected economic impacts of a 16-inch tree cutting cap for ponderosa pine forests within the Greater Flagstaff urban-wildlands. In: Vance RK, Edminster CB, Covington WW, Blake JA (comps) Ponderosa pine ecosystems restoration and conservation: steps towards stewardship, 2000 April 25–27, Flagstaff, AZ. Proceedings RMRS-P-22, U.S. Department of Agriculture, Forest Service, Rocky Mountain Research Station, Ogden, UT, pp 154–160

Laughlin DC, Moore MM (2008) Forest and range research on the "Wild Bill Plots" (1927-2007). In: Olberding SD, Moore MM (tech coords) Fort Valley Experimental Forest—a century of research 1908–2008, 2008 August 7–9, Flagstaff, AZ. Proceedings RMRS-P-55, U.S. Department of Agriculture, Forest Service, Rocky Mountain Research Station, Fort Collins, CO, pp 203–208

Laughlin DC, Bakker JD, Daniels ML, Moore MM, Casey CA, Springer JD (2008) Restoring plant species diversity and community composition in a ponderosa pine-bunchgrass ecosystem. Plant Ecol 97:139–151

Layser EF, Schubert GH (1979) Preliminary classification for the coniferous forest and woodland series of Arizona and New Mexico. Research Paper RM-208, U.S. Department of Agriculture, Forest Service, Rocky Mountain Forest and Range Experiment Station, Fort Collins, CO, 27 p

Leopold A (1924) Grass, brush, timber and fire in southern Arizona. J For 22(6):1–10

Leopold A (1949) A Sand County almanac and sketches here and there. Oxford University Press, New York, 228 p

Lexen BR (1939) Growth following partial cutting in ponderosa pine. J For 37(12):943–946

Livingstone DN (2003) Putting science in its place: geographies of scientific knowledge. University of Chicago Press, Chicago, 244 p

Lowell EC, Green DW (2001) Lumber recovery from small-diameter ponderosa pine from Flagstaff, Arizona. In: Vance RK, Edminster CB, Covington WW, Blake JA (comps) Ponderosa pine ecosystems restoration and conservation: steps towards stewardship, 2000 April 25–27; Flagstaff, AZ. Proceedings RMRS-P-22, U.S. Department of Agriculture, Forest Service, Rocky Mountain Research Station, Ogden, UT, pp 161–165

Maienschein J (1994) Pattern and process in early studies of Arizona's San Francisco Peaks. Bioscience 44(7):479–485

Martin EC (1969) The relation of precipitation to flow from Little Leroux Spring on the San Francisco Peaks, Arizona. Research Note RM-129, U.S. Department of Agriculture, Forest Service, Rocky Mountain Forest and Range Experiment Station, Fort Collins, CO, 4 p

Mattoon WR (1909) Measurements of the effects of forest cover upon the conservation of snow waters. For Q 7(3):245–248

Meagher GS (1950) Reproduction of ponderosa pine. J For 48(3):189–191

Meinecke EP (1917) Basic problems in forest pathology. J For 15(2):215–224

Merriam CH (1890) Results of a biological survey of the San Francisco Mountain region and desert of the Little Colorado, Arizona. North American Fauna. 3. U.S. Department of Agriculture, Division of Ornithology and Mammalogy, Washington, DC, 128 p

Millar CI, Stephenson NL, Stephens SL (2007) Climate change and forest of the future: managing in the faces of uncertainty. Ecol Appl 17(8):2145–2151

Millennium Ecosystem Assessment (2003) Ecosystems and human well-being: a framework for assessment. Island Press, Washington, DC, 212 p. http://www.maweb.org/en/Framework.aspx

Moir WH, Geils BW, Benoit MA, Scurlock D (1997) Ecology of southwestern ponderosa pine forests. In: Block WM, Finch DM (eds) Songbird ecology in southwestern ponderosa pine forests: a literature review. General Technical Report RM-292, U.S. Department of Agriculture, Forest Service, Rocky Mountain Forest and Range Experiment Station, Fort Collins, CO, pp 3–27

Moore B (1916) The relationship of forestry to science. For Q 14(3):375–379

Moore MM, Covington WW, Fulé PZ, Hart SC, Kolb TE, Mast JN, Sackett SS, Wagner MR (2008) Ecological restoration experiments (1992–2007) at the G.A. Pearson Natural Area, Fort Valley Experimental Forest. In: Olberding SD, Moore MM (tech coords) Fort Valley Experimental Forest—a century of research 1908–2008, 2008 August 7–9, Flagstaff, AZ. Proceedings RMRS-P-55, U.S. Department of Agriculture, Forest Service, Rocky Mountain Station, Fort Collins, CO, pp 209–218

Moseley C, KenCairn B (2001) Problem solving or social change? The Applegate and Grand Canyon forest partnerships. In: Vance RK, Edminster CB, Covington WW, Blake JA (comps) Ponderosa pine ecosystems restoration and conservation: steps towards stewardship, 2000 April 25–27, Flagstaff, AZ. Proceedings RMRS-P-22, U.S. Department of Agriculture, Forest Service, Rocky Mountain Research Station, Ogden, UT, pp 121–129

Myers CA (1967) Growing stock levels in even-aged ponderosa pine. Research Papers RM-33, U.S. Department of Agriculture, Forest Service, Rocky Mountain Range and Forest Experiment Station, Fort Collins, CO, 8 p

Myers CA, Martin EC (1963) Fifty years progress in converting virgin southwestern ponderosa pine to managed stands. J For 61(8):583–586

Myers CA, Hawksworth FG, Lightle PC (1972) Simulating yields of southwestern ponderosa pine stands, including effects of dwarf mistletoe. Research Paper RM-87, U.S. Department of Agriculture, Forest Service, Rocky Mountain Range and Forest Experiment Station, Fort Collins, CO, 16 p

Olberding SD (2000) Fort Valley: the beginnings of forest research. For Hist Today (Spring):9–15

Olberding SD (2002) Fort Valley: then and now. Fort Valley Publishing, Flagstaff, 90 p

Olberding SD (2008) "It was a young man's life": G. A. Pearson. In: Olberding SD, Moore MM (tech coords) Fort Valley Experimental Forest—a century of research 1908–2008, 2008 August 7–9, Flagstaff, AZ. Proceedings RMRS-P-55, U.S. Department of Agriculture, Forest Service, Rocky Mountain Range and Forest Experiment Station, Fort Collins, CO, pp 3–17

Olberding SD, Moore MM (tech coords) (2008) Fort Valley Experimental Forest—a century of research 1908–2008, 2008 August 7–9, Flagstaff, AZ. Proceedings RMRS-P-55, U.S. Department of Agriculture, Forest Service, Rocky Mountain Research Station, Fort Collins, CO, 282 p

Olberding SD, Malis-Clark K, Pilles PJ Jr, Lund D (2008) A century of cooperation: the Fort Valley Experimental Forest and the Coconino national forest in Flagstaff. In: Olberding SD, Moore MM (tech coords) Fort Valley Experimental Forest—a century of research 1908–2008, 2008 August 7–9, Flagstaff, AZ. Proceedings RMRS-P-55, U.S. Department of Agriculture, Forest Service, Rocky Mountain Range and Forest Experiment Station, Fort Collins, CO, pp 224–229

Patton DR (2008) The Fort Valley Experimental Forest, ponderosa pine and wildlife habitat research. In: Olberding SD, Moore MM (tech coords) Fort Valley Experimental Forest—a century of research 1908–2008, 2008 August 7–9, Flagstaff, AZ. Proceedings RMRS-P-55, U.S. Department of Agriculture, Forest Service, Rocky Mountain Research Station, Fort Collins, CO, pp 81–88

Pearson GA (1914) The administration of a forest experiment station. For Q 12:211–222

Pearson GA (1918) Studies of yield and reproduction of western yellow pine in Arizona and New Mexico. J For 16(3):273–291

Pearson GA (1920a) Factors controlling the distribution of forest types, Part 1. Ecology 1(3):139–159

Pearson GA (1920b) Factors controlling the distribution of forest types, Part 2. Ecology 1(4):289–308

Pearson GA (1921) Brush disposal in western yellow pine. J For 19(1):36–38

Pearson GA (1922a) Preservation of natural areas in the National Forests. Ecology 3(4):284–287

Pearson GA (1922b) Seasonal distribution of rainfall as related to forest fires in Arizona and New Mexico. J For 20(7):776–779

Pearson GA (1923) Natural reproduction of western yellow pine in the Southwest. Bulletin 1105, U.S. Department of Agriculture, Washington, DC, 143 p

Pearson GA (1924a) Some conditions for effective research. Science 60(1543):71–73

Pearson GA (1924b) Temperature summations with reference to plant life. Mon Weather Rev 52(4):218–220

Pearson GA (1928) Measurement of physical factors in silviculture. Ecology 9(4):404–411

Pearson GA (1930) Studies of climate and soil in relation to forest management in the southwestern United States. J Ecol 18(1):139–144

Pearson GA (1931) Recovery of western yellow pine seedlings from injury by grazing animals. J For 29(6):876–894

Pearson GA (1933) A twenty year record of changes in an Arizona pine forest. Ecology 14(3):272–285

Pearson GA (1935) Public values in cutover timber lands. J For 33(6):590–594

Pearson GA (1936) A thinning experiment applied to timber stand improvement. J For 34(9):855–861

Pearson GA (1937) Factors influencing the growth of trees. In: Glock WS (ed) Principles and methods of tree-ring analysis. Carnegie Institution, Washington, DC, pp 65–68

Pearson GA (1938) Lighter cuts and larger yields in ponderosa pine. J For 36(8):779–789

Pearson GA (1939) Mortality in cutover stands of ponderosa pine. J For 37(5):383–387

Pearson GA (1940) Forest land use. J For 38(3):261–270

Pearson GA (1942) Herbaceous vegetation a factor in natural regeneration of ponderosa pine in the Southwest. Ecol Monogr 12(3):315–338

Pearson GA (1943) How local is silviculture? J For 41(12):917–919

Pearson GA (1944) Growth, mortality, and cutting cycles in New Mexico ponderosa pine. J For 43(1):901–905

Pearson GA (1946) A plea for applied silviculture in forest research. J For 44(11):958–961

Pearson GA (1950) Management of ponderosa pine in the Southwest. Agricultural Monograph 6, U.S. Department of Agriculture, Forest Service, Washington, DC, 218 p

Pearson GA (1951) A comparison of the climate in four ponderosa pine regions. J For 49(4):256–258

Pearson GA, Wadsworth FH (1941) An example of timber management in the Southwest. J For 39(5):434–452

Pinchot G (1947) Breaking new ground. Harcourt, New York

Price R (1976) History of Forest Service research in the central and southern Rocky Mountain regions 1908–1975. General Technical Report RM-27, U.S. Department of Agriculture, Forest Service, Rocky Mountain Forest and Range Experiment Station, Fort Collins, CO, 100 p

Raish C, Yong W, Marzluff J (1997) Contemporary human use of southwestern ponderosa pine forests. In: Block WM, Finch DM (eds) Songbird Ecology in southwestern ponderosa pine forests: a literature review. General Technical Report RM-292, U.S. Department of Agriculture, Forest Service, Rocky Mountain Forest and Range Experiment Station, Fort Collins, CO, pp 3–27

Rodgers AD III (1951) Bernhard Eduard Fernow: a story of North American forestry. Princeton University Press, Princeton, 623 p

Sackett SS, Haase SM (2008) Fire and fuels research at Fort Valley and Long Valley Experimental Forests. In: Olberding SD, Moore MM (tech coords) Fort Valley Experimental Forest—a century of research 1908–2008, 2008 August 7–9, Flagstaff, AZ. Proceeedings RMRS-P-55, U.S. Department of Agriculture, Forest Service, Rocky Mountain Research Station, Fort Collins, CO, pp 38–45

Sackett SS, Haase SM, Harrington MG (1996) Lessons learned from fire use for restoring southwestern ponderosa pine ecosystems. In: Covington WW, Wagner PK (tech. coords) Conference on adaptive ecosystem restoration and management restoration of Cordilleran conifer landscapes of North America, 1996 June 6–8, Flagstaff, AZ. General Technical Report RM-278, U.S. Department of Agriculture, Forest Service, Rocky Mountain Forest and Range Experiment Station, Fort Collins, CO, pp 54–61

Sánchez Meador AJ, Moore MM (2008) 93 years of stand density and land-use legacy research at the Coulter Ranch study site. In: Olberding SD, Moore MM (tech coords) Fort Valley Experimental Forest—a century of research 1908–2008, 2008 August 7–9, Flagstaff, AZ. Proceedings RMRS-P-55, U.S. Department of Agriculture, Forest Service, Rocky Mountain Research Station, Fort Collins, CO, pp 230–236

Sánchez Meador AJ, Olberding SD (2008) Fort Valley's early scientists: a legacy of distinction. In: Olberding SD, Moore MM (tech coords) Fort Valley Experimental Forest—a century of research 1908–2008, 2008 August 7–9, Flagstaff, AZ. Proceedings RMRS-P-55, U.S. Department of Agriculture, Forest Service, Rocky Mountain Research Station, Fort Collins, CO, pp 237–241

Savage M, Brown PM, Feddema J (1996) The role of climate in a pine forest regeneration pulse in the Southwestern United States. Ecoscience 3:310–318

Schmaltz NJ (1980) Raphael Zon: forest researcher. J For Hist 24(1, 2):24–39, 86–97

Schubert GH (1974) Silviculture of southwestern ponderosa pine: the status of our knowledge. Research Paper RM-123, U.S. Department of Agriculture, Forest Service, Rocky Mountain Forest and Range Experiment Station, Fort Collins, CO, 71 p

Shepperd WD, Fairweather ML (1994) Impact of large ungulates in restoration of aspen communities in a southwestern ponderosa pine ecosystem. In: Covington WW, DeBano LF (tech coords) Sustainable ecological systems: implementing an ecological approach to land management, 1993 July 12–15, Flagstaff, AZ. General Technical Report RM-247, U.S. Department of Agriculture, Forest Service, Rocky Mountain Forest and Range Experiment Station, Fort Collins, CO, pp 344–347

Shepperd WD, Rogers PC, Burton D, Bartos DL (2006) Ecology, biodiversity, management, and restoration of aspen in the Sierra Nevada. General Technical Report RMRS-178, U.S. Department of Agriculture, Forest Service, Rocky Mountain Research Station, Fort Collins, CO, 122 p

Springer JD, Daniels ML, Nazaire M (2009) Field guide to forest and mountain plants of Northern Arizona. Ecological Restoration Institute, Flagstaff, 649 p

Swetnam TW, Brown PM (2011) Climatic inferences from dendrochronological reconstructions. In: Hughes MK, Swetnam TW, Diaz HF (eds) Dendroclimatology: progress and prospects. Springer, New York, pp 263–296

Taylor WP, Gorsuch DM (1932) A test of some rodent and bird influences on western yellow pine reproduction at Fort Valley, Flagstaff, Arizona. J Mamm 13(3):218–223

van Mantgem PJ, Stephenson NL, Byrne JC, Daniels LD, Franklin JF, Fulé PZ, Harmon ME, Larson AJ, Smith JM, Taylor AH, Veblen TT (2009) Widespread increase of tree mortality rates in the western United States. Science 323(23):521–524

Vance RK, Edminster CB, Covington WW, Blake JA (comps) (2000) Ponderosa pine ecosystems restoration and conservation: steps toward stewardship, 2000 April 25–27, Flagstaff, AZ. Proceedings RMRS-P-22, U.S. Department of Agriculture, Forest Service, Rocky Mountain Research Station, Ogden, UT, 188 p

Wagner MR, Block WM, Geils BW, Wenger KF (2000) Restoration ecology: a new forest management approach or another merit badge for foresters. J For 98(10):22–27

White AS (1985) Presettlement regeneration patterns in a Southwestern ponderosa pine stand. Ecology 66(2):589–594

Woolsey TS Jr (1909) The Coconino Ranger School. For Q 7(3):243–244

Woolsey TS Jr (1911) Western yellow pine in Arizona and New Mexico. Bulletin 101, U.S. Department of Agriculture, Forest Service, Washington, DC, 64 p

Woolsey TS Jr (1912) Permanent sample plots. For Q 10:38–44

Young JC (2008) Roots of research: Raphael Zon and the origins of forest experiment stations. In: Olberding SD, Moore MM (tech coords) Fort Valley Experimental Forest—a century of research 1908–2008, 2008 August 7–9, Flagstaff, AZ. Proceedings RMRS-P-55, U.S. Department of Agriculture, Forest Service, Rocky Mountain Research Station, Fort Collins, CO, pp 257–261

Zon R (1908) Principles involved in determining forest types. For Q 6(3):263–271

Chapter 14
Contributions of Studies on Experimental Forests to Hydrology and Watershed Management

Gerald J. Gottfried, Peter F. Ffolliott, Kenneth N. Brooks, Randall K. Kolka, Carol B. Raish and Daniel G. Neary

Abstract The link between healthy forests and watersheds and healthy streamflow and quality water is universally recognized. The major rivers of the USA originate in the forested mountains of the western and eastern USA and the glaciated regions of the Lake States and Great Plains and produce almost two-thirds of the nation's clean water supply. Original logging and mismanagement of upstream forested watersheds often resulted in degradation of land and water resources and adversely impacted aquatic and human populations. During the 30-year period, between the 1930s and 1960s, experimental forests, ranges, and watersheds were established on national forests and adjacent lands to study the impacts of land conditions on water yield, stormflow, water quality, and nutrient cycling. While the impact of sustained timber production was an original research focus, current efforts include research on nutrient cycling, carbon sequestration, climate change, fire effects, and the impacts of insects and diseases. The experimental forest network of long-term meteorological and biological records is invaluable for evaluations of potential climate change and its consequences on the forests and water resources. This chapter reviews hydrology

G. J. Gottfried (✉)
Rocky Mountain Research Station, USDA Forest Service, 2324 East McDowell Rd., Phoenix, AZ 85006-2496, USA
Phone: 602-225-5357
e-mail: ggottfried@fs.fed.us

P. F. Ffolliott
School of Natural Resources and the Environment, University of Arizona, Tucson, AZ, USA

K. N. Brooks
Department of Forest Resources, University of Minnesota, St. Paul, MN, USA

R. K. Kolka
Northern Research Station, USDA Forest Service, 1831 Hwy 169 East, Grand Rapids, MN 55744, USA

C. B. Raish
Rocky Mountain Research Station, USDA Forest Service, 333 Broadway Boulevard Southeast, Albuquerque, NM 87102, USA

D. G. Neary
Rocky Mountain Research Station, USDA Forest Service, 2500 South Pine Knoll Drive, Flagstaff, AZ 86001, USA

D. C. Hayes et al. (eds.), *USDA Forest Service Experimental Forests and Ranges*,
DOI 10.1007/978-1-4614-1818-4_14, © Springer New York 2014

and watershed management research at a sample of experimental forests and areas from throughout the USA. The chapter includes studies conducted in the forests near the Fort Valley Experimental Forest and on the Sierra Ancha Experimental Forest in Arizona, on the Fraser Experimental Forest in Colorado, and the H.J. Andrews Experimental Forest in Oregon. In the East, research at the Coweeta Hydrologic Laboratory in North Carolina, the Hubbard Brook Experimental Forest in New Hampshire, and the Marcell Experimental Forest in Minnesota is reviewed.

Keywords Forest hydrology · Watershed management · Western and eastern experimental forests · Long-term studies and records · Climate change

14.1 Introduction

Forests and rangelands produce almost two-thirds of the nation's clean water supply. The Weeks Act of 1911 acknowledged the important link between forests and watersheds and the protection of navigable rivers. Efforts to ensure quality water and flows that sustain stream and floodplain habitats were a significant reason for the establishment of the National Forest System. A century of research and monitoring in experimental forests and rangelands, largely by the USDA Forest Service, provides a rich body of knowledge and long-term data to guide land management stewardship that preserves the integrity of our water resources (Ice and Stednick 2004).

This chapter reviews research on selected experimental forests in the western and eastern USA to highlight their contributions to forest hydrology and watershed management and the link between good silviculture and good watershed management. Many studies were initiated because of local concerns that forestland management could have an adverse effect on water resources or produce peak stormflows that would negatively impact downstream infrastructures. In the southwestern USA, there were concerns that increased tree densities were causing declines in streamflow volumes. In the Pacific Northwest, concerns centered on large-scale clearcutting in the forested headwaters and the impacts of higher stormflows on water quality, stream habitat, and on endangered or sensitive wildlife species. Impacts of changes in summer streamflows on salmon fisheries were also important concerns in the Pacific Northwest. While timber and water yield augmentation are no longer a paramount concern, the knowledge gained from the early research on experimental forests is vital to current planning and interpretation of multiple resource management. Experimental forests and ranges can play an important part in answering current questions concerning the role, management, and effects of fire. Hydrologic and meteorological records, which are fundamental to watershed experiments, are invaluable for setting baselines to evaluate current and future climate change and new forestland management options. Some records are more than 100 years old. The value of long-term experimental watershed studies and monitoring for addressing future forest water issues has gained considerable attention in the last decade (National Research Council (NRC) 2008; Buttle et al.

2005; Stednick et al. 2004). However, concerns associated with changing demographics and climate, and issues associated with expanding temporal and spatial scales, pose new questions that may not be adequately addressed from past research (NRC 2008). The uncertainty of land use and climate change suggests that a new and expanded forest hydrology research and monitoring program is necessary to support the development of predictive models that complement, but do not replace watershed studies (Stednick et al. 2004). This chapter discusses the importance of continuing monitoring programs in many existing experimental watersheds and an expanded vision of watershed research for the twenty-first century.

14.2 Experimental Watersheds of the West

Watershed studies from four western experimental forests and research watershed areas are considered representative of watershed research in the West. Fort Valley, the oldest experimental forest in the country (Olberding and Moore 2008), does not contain gauged watersheds, but the silvicultural research provides the basis for treatments on the adjacent Beaver Creek watersheds in central Arizona and higher-elevation forests in the Sierra Ancha Experimental Forest, north of Globe, Arizona, and its satellite, Castle Creek, in eastern Arizona. The Fraser Experimental Forest in the Front Range of Colorado emphasizes the link between forest conditions and water in high elevation, snow-dominated watersheds. Finally, the H.J. Andrews Experimental Forest in the Cascade Mountains of Oregon represents watershed research in rain-dominated forests of the Pacific Northwest.

14.2.1 Mixed Conifer and Ponderosa Pine Experimental Program in Arizona

Water has always been a critical issue in the arid Southwest with its periodic cycles of drought and abundant precipitation. Roosevelt Dam, on the Salt River in central Arizona, was completed in 1911 to ensure that the farms and developing cities in the Phoenix Area had adequate water supplies. The Tonto National Forest was established to protect and manage the lake's watershed. Watershed research has been an important part of activities to protect and, where possible, to augment streamflows. The Sierra Ancha Experimental Forest and the Beaver Creek watersheds are centers for watershed studies in mixed conifer and ponderosa pine forests. However, because of the large variety of ecosystem habitats in the Southwest and the need to show that results from the experimental forests were valid in other areas, additional experimental sites were instrumented and treated on adjacent national forestlands (Baker 1999).

14.2.2 Sierra Ancha Experimental Forest

Forest Service watershed research began near Roosevelt Lake in 1925 to ascertain sedimentation rates from the granite soils surrounding the lake. The Parker Creek Experimental Forest was dedicated in 1932 and enlarged in 1938 to form the Sierra Ancha Experimental Forest in the mountains northeast of Roosevelt Lake (Gottfried et al. 1999a). Some of the initial efforts were to gauge several of the creeks to determine the hydrology of streams originating in the higher elevation forests. The three forks of Workman Creek, at 2,075–2,355 m in elevation, that support mixed conifer forests of Rocky Mountain Douglas-fir (*Pseudotsuga menziesii* (Mirb.) Franco var. *glauca* (Beissn.) Franco), white fir (*Abies concolor* var. *concolor (Gord.* and *Glend.*)), and ponderosa pine (*Pinus ponderosa* var. *scopulorum* Engelm.) and forests of pure ponderosa pine were part of this effort. Annual precipitation is about 840 mm. One objective at Workman Creek was to study the effects of watershed vegetation on streamflow, erosion, floods, and sedimentation. Treatments were applied to North Fork and South Fork, while Middle Fork was designated as the hydrologic control. The experimental approach used in most watershed studies throughout the USA is based on statistical relationships between paired watersheds (Hewlett 1971). Streamflow or other characteristics from two, and occasionally more, watersheds are compared during a calibration period and then one watershed is treated, while one remains as a hydrological control to account for changing weather from year to year. The pre- and posttreatment relationships are compared to determine if significant differences have occurred.

The first treatments on North Fork were implemented in 1953, and designed to evaluate the impacts of removing vegetation in a series of steps to determine the effects on water yields (Fletcher and Rich 1955). The North Fork treatments were experimental and not designed as potential management options. The first step was to remove riparian trees, mainly Arizona alder (*Alnus oblongifolia* Torr.) and bigtooth maple (*Acer grandidentatum* Nutt.). This treatment, which only covered a small area, did not result in a significant increase in water yields. The second treatment cleared the mixed conifer forest from 32.4 ha and seeded grasses. The third treatment removed the adjacent ponderosa pine forest from 40.5 ha. These two combined treatments resulted in significant water yield increases of 84 % or 69 mm when compared to original conditions (Rich and Gottfried 1976). The increases were still being measured until the weirs were closed in 1983.

The South Fork was harvested according to a standard single-tree selection prescription starting in 1953 (Rich and Gottfried 1976; Gottfried et al. 1999a). A slight but statistically significant increase was detected. The second treatment in 1966 was designed to convert the mixed conifer forest into a ponderosa pine forest and to maintain the stand at 9.2 m^3/ha to optimize both water and tree production. The stocking levels were not achieved and ponderosa pine regeneration efforts were not successful. However, the treatment did result in significant increases in runoff of 111 % or 93 mm with little increase in sedimentation.

Fig. 14.1 Castle Creek East Fork was monitored with 120-degree v-notch weir. The objective on the two Castle Creek watersheds was to start management of the ponderosa pine forest to a system using a 120-year rotation with 20-year cutting cycles. East Fork was the original control watershed but a preharvest watershed prescribed fire treatment was evaluated after 1981. (Photo by G. Gottfried)

The Workman Creek studies indicated that removal of forest vegetation in significant areas and amounts could increase runoff (Rich and Gottfried 1976). They suggested that even-aged management could enhance both tree and water resources and that a system of patch clearcuts scheduled at 20-year intervals over a 120-year rotation could meet these criteria. Current silvicultural prescriptions favor uneven-aged management although small group selection is a component of many prescriptions.

The three watersheds burned in the Coon Creek Wildfire in April 2000. The Middle Fork, with its uncut, old-growth forest suffered high-severity fire damage, the South Fork suffered moderate-severity fire damage, and North Fork had low-severity damage. Postfire runoff and sedimentation are being monitored to determine the hydrologic effects related to different fire severities (Neary et al. 2006).

14.2.3 Castle Creek Timber Harvesting and Thinning Tree Overstories

The two Castle Creek watersheds were established in the White Mountains of eastern Arizona, to study the effects of commercial timber harvesting on streamflow regimes (Fig. 14.1). The watersheds were an extension of the research results from Workman Creek to evaluate the impacts of scattered small patch clearcut openings and harvesting of adjacent trees on water yields (Gottfried et al. 1999b). The Castle Creek watersheds are between 2,380 and 2,620 m in elevation. L. R. Rich, who was based at the Forest Hydrology Laboratory in Tempe, AZ, was instrumental in establishing the Castle Creek watershed experiment.

A silvicultural treatment was applied to the 364-ha West Fork watershed at Castle Creek in 1965–1967 to obtain timber products and place the remaining tree overstory into the "best growing condition possible" by starting the transition from an uneven-aged to an even-aged stand condition (Rich and Thompson 1974). Timber was cleared from one-sixth of the watershed in small openings fitted to the existing stands. The remaining five-sixths were thinned to a basal area of 13.7 m^2/ha to initiate a shelterwood system of "commercial timber management" over a 120-year rotation using a 20-year cutting cycle based on studies at Fort Valley (Schubert 1974). About 50% of the original basal area of 31 m^2/ha was removed by the harvesting and thinning treatment.

Increases in streamflow remained stable at 13 mm or about 30%, for more than 20 years after treatment, with these streamflow increases attributed to increased snowpack accumulations and reduced evapotranspiration (ET) rates (Gottfried et al. 1999b). More specifically, the increase in streamflow was presumed to be a result of increased snowpacks in openings because of aerodynamic conditions created by the surrounding mature trees and because of lower transpiration by the new trees in the openings.

The treatment achieved its purpose of initiating a schedule of timber harvesting and management of ponderosa pine in a shelterwood system (Gottfried et al. 1999b). Silvicultural studies at Fort Valley and elsewhere in the region suggested that an even-aged stand structure has a greater "timber-production potential" than an unbalanced stand (Schubert 1974). However, because timber production is no longer a management emphasis in the region, the management focus has shifted to obtaining ecosystem-based, multiple-use benefits. This shift reflects the increasing environmental concerns of the 1960s and 1970s, many of which focused on logging and USDA Forest Service timber management programs. In 1972, the agency produced an action plan, National Forests in a Quality Environment (USDA Forest Service 1972), in response to the National Environmental Policy Act. The plan discussed a variety of issues including problems with logging practices, such as clearcutting. Additional research coming out of this effort recommended increased efforts in multiple-resource management, understanding human interactions with the forest environment, and developing timber management programs that enhance or have minimal environmental impact and are compatible with other uses (Steen 1998). The research at Beaver Creek and Castle Creek anticipated the shift in emphasis as indicated by the bibliography prepared by Baker and Ffolliott (1998) listing research articles covering the entire range of forest resource values.

14.2.4 Prescribed Fire and Watershed Values

In the 1980s, land managers were becoming more concerned about the dangers of large, devastating wildfires. Prescribed fire treatments had not been attempted in forests at that time, although fire had been used in chaparral communities (DeBano et al. 1999, 1998). More recent efforts have evaluated prescribed burning with and without thinning in forest environments. Dieterich (1983) of the Rocky Mountain Research Station suggested that a prescribed burn on an area scheduled for harvesting would reduce the possibilities of a wildfire and increase streamflow because of the reduced forest floor depths and stand densities. Castle Creek was selected for the experiment. The results from the original treatments were stable, and it was decided to reverse the watersheds and treat East Fork and hold the previously treated West Fork as the control. The burn was ignited in November 1981, but only covered 43 % of East Fork, mostly adjacent to the main channel (Gottfried and DeBano 1988). Surface fuels were removed, middle forest floor layers were slightly charred, and tree reductions were minor. The burn resulted in a slight but statistically nonsignificant increase in runoff of 8 % or 8 mm. Some significant changes occurred in nutrient concentrations, but the changes were very small and of little consequence in terms of site productivity or water quality.

14.2.5 The Beaver Creek Watersheds

Much of the forest watershed research in the West was initiated in the 1950s because of concerns that increasing forest and woodland stand densities were resulting in declines in surface runoff for local and downstream users during a severe drought period. Results from the watershed experiments influenced forest management into the mid-1980s. The long history of silvicultural studies conducted at Fort Valley, located 17 km northwest of Flagstaff, AZ, provided a basis for testing clearcutting and thinning treatments in ponderosa pine forests on Beaver Creek (Gottfried et al. 2008). Management of ponderosa pine forests in the region has changed since the Beaver Creek and Sierra Ancha treatments were initiated from a timber production emphasis to a more holistic, ecosystem-based perspective. While the importance of land management for water yield augmentation also declined as management for other ecosystem services increased, it remains an important consideration. The interest in managing forests to improve water yields and to regulate snowmelt timing may again become important as the West's population continues to grow.

The 20 Beaver Creek watersheds were established 83 km south of Flagstaff to investigate the potential for increasing streamflow from ponderosa pine forests and pinyon-juniper woodlands in central Arizona (Brown et al. 1974). The watersheds are within the Coconino National Forest and located between 1,980 and 2,440 m in elevation. The Beaver Creek watersheds were established as a result of a meeting of ranchers, the Salt River Valley Water Users Association, which provides water to much of the Phoenix Metropolitan Area, and the Forest Service. Participants were concerned that the dense forests and woodlands were reducing streamflows during a period of extended drought. Watershed management research, which began in 1960, evolved into the evaluation of multiple resources. Impacts of silvicultural treatments on herbage production for livestock and wildlife habitats were important aspects in all watershed studies. Economic models were developed based on research findings. The Beaver Creek watersheds were not formally designated as an experimental forest; however, treatments were implemented cooperatively by personnel from the Coconino National Forest and scientists from Rocky Mountain Forest and Range Experiment Station. H. Brown and D. Worley were the first project leaders, and key scientists included M. Baker, Jr, W. Clary, P. Ffolliott, F. Larson,

Fig. 14.2 The clearcut watershed will not produce ponderosa pine crops for a long time, but treatment has resulted in increases in forage production for livestock; it is valuable as wildlife habitat, and the clearing created a more diverse landscape by breaking up the continuous forest cover. The watershed is also becoming a source for firewood for neighboring communities. (Photo by P. Ffolliott)

and J. Rodgers. The Beaver Creek watersheds are a biosphere reserve in UNESCO's Man and the Biosphere Program. They provide a valuable outdoor classroom and laboratory for university students and visiting scientists.

14.2.5.1 Complete Clearcutting of Tree Overstories

Silvicultural options to meet hydrology and watershed management goals include complete or partial clearcutting tree overstories, thinning tree stands, or combinations of these options. One of the first objectives was to determine the maximum potential of increasing runoff from ponderosa pine forests. The tree overstory of ponderosa pine and intermingling Gambel oak (*Quercus gambelii* Nutt.) and alligator juniper (*Juniperus deppeana* Steud.) on a 172-ha watershed was completely clearcut in 1966–1967 to evaluate the effects of this "most drastic" treatment on streamflow (Fig. 14.2). Merchantable trees were felled and removed from the watershed, and smaller trees and slash were piled in windrows aligned perpendicular to the stream to facilitate transport of overland flow into the channel. The clearcut harvest was not intended as a possible land management treatment.

Streamflow increases for 7 years following the treatment averaged 43 mm or about a 30% increase. This increase was caused by reductions in tree transpiration and more overland flow from melting snowpacks, the primary source of annual streamflow from the watershed (Baker and Ffolliott 1999). Windrowed slash trapped snow and delayed melting on the lee sides of the windrows until the ambient temperature rose, resulting in more of the overland flow reaching the stream channels. Vegetation and the transpiration rates recovered sufficiently within 7 years for the soil-water depletion to be the same as under the original forest.

The treatment removed the watershed from commercial timber production. Stocking of ponderosa pine reproduction declined from 65 to 15% after the clearing and remained constant for the following 23 years (Ffolliott and Gottfried 1991).

Fig. 14.3 Heavy thinning was applied to this watershed reducing the basal area to 5.7 m^2/ha. The treatment resulted in significant increases in water yields, but it is unlikely that timber production can be sustained. Forage production and wildlife habitat conditions have improved. (Photo by P. Ffolliott)

It is unlikely that the watershed can be managed for future timber production without site preparation and artificial regeneration. However, research at Fort Valley showed that artificial regeneration must be initiated soon after clearcutting to minimize problems with competing vegetation (Schubert 1974). Management goals other than "commercial timber production" included increases in forage production for livestock grazing and abundant posttreatment oak sprouts, which are beneficial to indigenous ungulates, and firewood for local consumption. The clearing created a more diverse landscape by breaking up the continuous ponderosa pine forest cover.

14.2.5.2 Thinning Tree Stands

A subsequent study at Beaver Creek evaluated the impacts of thinning trees on water production. The ponderosa pine overstory on a 121-ha watershed was commercially harvested by group selection and the residual trees uniformly thinned in 1969 (Fig. 14.3), leaving stands in even-aged groups at a basal area level of about 5.7 m^2/ha. The prescribed basal area level, while relatively low, was above the level where windthrow of ponderosa pine trees occurs. Seventy-five percent of the original basal area was removed. Slash was piled perpendicular to the stream channel (similar to the clearcut treatment).

Increased streamflow following the thinning treatment persisted for 10 years and averaged 41 mm annually—ranging from 10 to 30% above the predicted streamflow if the watershed had not been treated (Baker and Ffolliott 1999). The streamflow response was attributed to reduced transpiration losses and increased efficiency of winter overland flow to stream channels because of the influence of the windrowed slash.

Stocking of regeneration was reduced from over 50% before thinning to 2% after treatment. However, nearly 40% of the watershed became restocked with natural regeneration within 10 years of the treatment (Ffolliott et al. 2000). It is unlikely that timber production will be sustained, but managing for other resources is possible. Heavy thinning and slash treatments could influence fire behavior and fuel management in the wildland–urbaninterface.

14.2.5.3 Partial Clearcutting and Thinning Tree Overstories

A combination of thinning and clearing trees in strips was carried out on a 546-ha watershed on Beaver Creek in 1970–1971 (Ffolliott and Baker 2001). Trees were cleared in irregularly shaped strips 18 m wide with intervening leave strips 36 m wide. The harvested strips were oriented downslope for more efficient overland flow of water. Ponderosa pine in the leave strips were thinned to 18.4 m^2/ha. The treatment removed 40% of the basal area on the watershed. Slash was piled in the center of the stripcuts and burned.

The hypothesis was that streamflow increases would occur following treatment because of increased efficiency in transporting overland flow to the stream channel due to the uphill–downhill orientation of the stripcuts and decreased transpiration. Annual streamflow increased 25 mm, but the increase only lasted 4 years (Baker and Ffolliott 1999) because of the rapid recovery of vegetation in the cut strips.

The treatment removed most of the trees in the cut strips and left a mosaic of even-aged tree stands, mostly of trees 20–46 cm in diameter at breast height (d.b.h.) in the intervening leave strips. Stocking of trees in the leave strips increased in the 25-year posttreatment evaluation period (Ffolliott and Baker 2001). Stocking of ponderosa pine seedlings on the watershed declined after treatment and did not fully recover during the subsequent 25 years.

Research results based on existing data or computer simulation models continue to be published providing land and water managers with new knowledge and tools. More information about Beaver Creek and Castle Creek, including a searchable bibliography, can be found at a Rocky Mountain Research Station website: http://www.fs.fed.us/rm/boise/AWAE/labs/awae_flagstaff/watersheds/ and information about Sierra Ancha is presently available at http://www.fs.fed.us/rm/boise/AWAE/labs/awae_flagstaff/Hot_Topics/SierraAnchaExperimentalForest.html. Current research efforts at Sierra Ancha continue to monitor the impacts of the Coon Creek Fire. Although the Beaver Creek installations were closed in the mid-1980s, some of the watersheds have been re-instrumented to study a variety of forestry and ecological questions including the effects of forest restoration treatments, prescribed fire effects, evaluations of ponderosa pine growing stock levels, tree ecophysiology, and long-term climate change.

14.2.6 Fraser Experimental Forest

The Fraser Experimental Forest is situated in the Central Rocky Mountains about 105 km northwest of Denver. Its location is suited for investigations of water yields in high-elevation, snow-dominated forest areas (Alexander and Watkins 1977). Three-fourths of the subalpine forests of lodgepole pine (*Pinus contorta* Dougl. ex Loud.), Engelmann spruce (*Picea engelmannii* Parry), subalpine fir forest (*Abies lasiocarpa* (Hook.) Nutt.), Douglas-fir (*Pseudotsuga menziesii* var.*glauca*), and aspen (*Populus tremuloides* Michx.) are within the "snow zone," above 3,100 m. Snowpack dynamics, therefore, have been studied since the establishment of the experimental forest. Observations were made in natural stands initially to determine how forest types and canopy patterns affect snowpack accumulation. Plot studies of timber harvesting and thinning with measurements of snow accumulation, melt rates, and ET changes came next. The final step in this sequence of research applied timber harvesting and thinning combinations to whole watersheds and measured the effects on streamflow.

14.2.6.1 Plot Studies

Plot studies in mature lodgepole pine provided initial information on how management affected the water cycle, primarily the snow component. The first plots were harvested in 1940 after initial snowpack measurements, with the cuttings ranging from a complete (commercial) clearcut of all trees to partial cuttings retaining a specified reserve of timber volumes (Leaf 1975). An unharvested natural area was retained for comparison. Initial studies determined how different methods and intensities of cutting affected snowpack accumulation and tree reproduction, growth, and mortality. The greatest snowpack accumulation was observed in clearcut plots. Lesser snow accumulations occurred in plots with intermediate cuttings, and the least accumulations occurred in the unharvested natural stands. Differences in snowpack accumulations were caused by the elimination of sublimation from snow intercepted by tree crowns. Sublimation from tree crowns was greater than sublimation from snowpacks in the cut openings where shade and wind were less than on the tree crowns.

14.2.6.2 Watershed Studies

Partial Clearcutting

Results from the plot study were tested at the watershed level on the Fool Creek Watersheds at the Fraser Experimental Forest in stands of lodgepole pine, Engelmann spruce, and subalpine fir. Forty percent of the 289-ha watershed was partially cut in 1954–1956 leaving alternating cut and leave strips varying from 20 m to nearly 121 m wide and 151–182 m long. Cut strips were oriented perpendicular to the contour and changes in streamflow, other hydrologic variables and the residual tree overstory conditions were measured. No cutting was allowed within 27 m of the stream to minimize direct damage to the channel. Logging slash was lopped and scattered. Fifty percent of the merchantable timber was removed from the watershed area. Snow courses were established to study the effects of the treatment on snow accumulation patterns.

Average annual streamflow increased 81 mm or nearly 40 % in the first 28 years following treatment. This increase was attributed largely to an average annual increase in snowpack water equivalent of about 10 % because of reduced interception and sublimation of snow and increases in soil storage (Troendle and King 1985). However, the increase in streamflow has been declining along with decreasing snowpack water equivalents. The streamflow increase 50 years after the treatment averages 58 mm or about 30 % (Elder et al. (2006) in preparation). As vegetation continues to regrow, ET losses will increase, resulting in the diminishing increased streamflow. However, analysis of streamflow records since treatment suggests that treatment effects might last 70 years (Stednick and Troendle 2004).

Sufficient advanced regeneration (seedlings and saplings) survived the treatment to restock all clearcut strips, with greater stocking levels on the wider strips. The number of seeds dispersed from standing trees into the cleared strips in the initial 10 years following treatment was greater in years with the heaviest seed production (Alexander and Watkins 1977). Windfall of trees adjacent to the strips was the major cause of mortality in this period, with about two-thirds of the mortality occurring in the first 2 posttreatment years. According to Elder et al. (2006), there was "adequate regeneration" and "conservative growth" of lodgepole pine 50 years after treatment, while Engelmann spruce exhibited "poor" regeneration and growth. Current constraints on reducing the cover of subalpine forests to increase streamflow include environmental concerns about sustaining commercial timber production practices, providing wildlife habitats, and maintaining outdoor recreational opportunities.

Patch and Shelterwood Cuttings

Effects of silvicultural cuttings on streamflow volumes were subsequently tested on Deadhorse Creek, a 270-ha watershed within the Fraser Experimental Forest that supported spruce-fir along stream bottoms and northern and upper slopes, lodgepole pine on lower and middle-southern slopes, and alpine vegetation above timberline. Patch and shelterwood cutting prescriptions were initiated on Deadhorse Creek based on guidelines from earlier studies on Fraser and elsewhere in the region (Alexander 1969, 1971, 1973; Noble and Ronco 1978; and others). Twelve circular clearcuts of 5H (i.e., 5 tree heights) in diameter—each about 1.2 ha—were cut on the 40-ha North Fork of Deadhorse Creek in 1977–1978 to investigate harvest effects on snow accumulation and resultant streamflow. Trees were harvested on approximately one-third of the watershed in this treatment, which represented an initial effort at creating a stand that could sustain timber management. In a second treatment on Deadhorse Creek, 40 % of the original forest basal area on the 40-ha North Slope was uniformly removed in the first of a three-step shelterwood cutting in 1980–1981. A third treatment on Deadhorse Creek affected about 25 % of the

78-ha Upper Basin, which was harvested in irregularly shaped clearcut openings varying from 1 to 10 ha in size in 1983–1984.

Changes in streamflow regimes following the treatments on Deadhorse Creek have been summarized by Troendle and Kaufmann (1987), Troendle and King (1985), and others. An average annual increase of 61 mm (30 %) in streamflow was observed on North Fork in the first 7 posttreatment years. These yearly increases were correlated with snowpack water equivalents; wetter years had larger increases than drier years. Streamflow response to the treatment on the ungaged North Slope was estimated from the differences between the gaged streamflow on the main stem of Deadhorse Creek and that from the streamflow from the North Fork and Upper Basin. Although partitioning of streamflow in this way decreases the reliability of analyzing the effects of the treatment on this watershed, an analysis of covariance showed no significant increase in streamflow in the first 3 years after the North Slope treatment. Increases in streamflow following treatment on the Upper Basin have occurred only periodically. Clearcuts on this part of the watershed were apparently not as "effective" as those on the North Fork in increasing snowpack accumulation, and consequentially, streamflow. While streamflow significantly increased over the initial evaluation period, covariance analysis indicated that changes in streamflow were not detectable in drier years.

Silvicultural impacts of the timber harvesting and silvicultural practices implemented on Deadhorse Creek—and the Fraser Experimental Forest generally—have been mostly consistent with those expected in subalpine forests. Partial clearcutting in strips or patches (circular openings) and thinning to initiate a shelterwood system of management represent viable and appropriate silvicultural options. Studies at Fraser and in the region have provided guidelines to prescribe reproduction methods for lodgepole pine, Engelmann spruce, and subalpine fir relative to stand structure, windthrow risks, and management goals (Smith 1987). Furthermore, simulation techniques to estimate stand growth and volumes are also available (Alexander and Edminster 1980, 1981) to evaluate management alternatives in these forests. Current research at Fraser is concerned with linkages between forests, riparian areas, and streams to understand nutrient cycling, carbon storage, and snow hydrology. More than 30 studies currently are in progress in addition to long-term monitoring efforts. Such studies demonstrate a changing research focus favoring topics described by Steen (1998) as environmentally sensitive. As always, they are also responsive to current issues of concern. Ecologically based forest management was mandated in a range of legislation including the Clean Water Act of 1972, the Endangered Species Act of 1973, and the National Forest Management Act of 1976, with its mandate to sustain both biological diversity and soil productivity. As a result of these legislative mandates, the USDA Forest Service has shifted the way it manages the land, moving away from operating along functional lines toward an ecosystem management mode of operation (Steen 1998).

Additional information about the Fraser Experimental Forest, including a partial list of publications, can be found at a Rocky Mountain Research Station website: http://www.fs.usda.gov/efr/fraser/.

14.2.7 H.J. Andrews Experimental Forest

H.J. Andrews Experimental Forest is located on the western slope of the Cascade Mountains of Oregon about 80 km east of Eugene. It is administered jointly by the USDA Forest Service's Pacific Northwest Research Station, Oregon State University, and the Willamette National Forest. The experimental forest, which was originally designated as the Blue River Experimental Forest, was renamed in 1953 to honor H. J. Andrews, a past regional forester, who was instrumental in establishing the research area. The climate on the experimental forest is influenced by the Pacific Ocean and is characterized by wet winters and dry summers (Rothacher 1970). Average annual precipitation is about 2,286 mm mostly occurring as rain, except at the highest elevations. Approximately 56–63 % of the annual precipitation runs off as streamflow (Harr 1976). Before timber cutting began in 1950, about two-thirds of the Experimental Forest was cloaked with old-growth forests comprised mainly of 450-year-old Douglas-fir, with the remainder largely in stands that regenerated following wildfires in the mid-1800s. Upper elevations of the Experimental Forest support noble fir (*Abies procera* Rehd.), Pacific silver fir (*Abies amabilis* Dougl. ex Forbes), Douglas-fir, (*Pseudotsuga menziesii* (Mirb) Franco var. *menziesii*), and western hemlock (*Tsuga heterophylla* (Raf.) Sarg.), while lower elevations are dominated by Douglas-fir, western hemlock, and western redcedar (*Thuja plicata* Donn ex D.Don). Ongoing research projects include experimental watersheds to study the effects of alternative logging systems and road construction on streamflow regimes, sedimentation and nutrient cycling, and the development of posttreatment vegetation and biological diversity. Early concerns focused on how widespread clearcutting in headwater forested watersheds might impact flooding, downstream roads, and developments.

The H.J. Andrews Experimental Forest is a component of the National Science Foundation's Long-Term Ecological Research (LTER) Program. It is also a Biosphere Reserve in UNESCO's Man and the Biosphere Program.

14.2.7.1 Watershed Studies in Old-Growth Forests

H.J. Andrews contains three instrumented watersheds in dense stands of 450-year old-growth Douglas-fir forest (Rothacher 1970) and three in 130-year second-growth Douglas-fir stands (Harr et al. 1982). The watersheds occur on steep, northwest-facing topography. The first study in the old-growth stands evaluated complete clearcutting, a common practice in the forests of the Pacific Northwest, using a skyline system on a 96-ha watershed. Roads were not constructed in this watershed. Slash was broadcast burned. The treatment on the second watershed, approximately 101 ha in size, followed a patch-cutting prescription and included a full road system. A third watershed of about 61 ha was reserved as the hydrologic control.

The initial streamflow increases in the clearcut watershed were roughly proportionate to the area harvested (Rothacher 1970). Statistically significant increases did not occur until 40 % of the watershed was cleared, however. Water yields in 1966 and 1967 (the first 2 years after harvest), which were representative of average climatic conditions, were between 259 and 457 mm greater than the amount that would have been expected prior to treatment.

Approximately 8 % of the patch-cut watershed was cleared during road construction prior to the harvesting operation (Rothacher 1970). Logging began 34 months later, in 1962 and 1963, and cleared an additional 22 % of the watershed in three clearcut units. Slash was burned. Statistical analyses did not indicate a significant difference in streamflow related to road construction. Although greater yields of 170 mm were measured after 25–30 % of the watershed was cleared, the differences were not significant. Two major storms occurred after harvesting that resulted in major stream channel scouring.

The results from the two treatments indicate that streamflow responses are proportional to the reduction in forest cover. Increases are related to declines in ET of the new herbaceous and tree cover relative to the old-growth forest. Streamflow increases should decline in the future as the replacement forest stand develops and transpires more water. Eighty percent of the increases occurred during the wet October–March season. However, summer increases were proportionally greater and possibly more important to downstream users. Rothacher (1970) points out that large streamflow increases on headwater watersheds may not be detectable within the larger watersheds.

Low streamflows in August have implications for regional important salmonid fisheries because it affects the availability of rearing habitats and levels of competition with other species. There are also issues related to water temperatures and oxygen depletion (Hicks et al. 1991). These authors examined the long-term effects of logging the old-growth watersheds on low summer (August) streamflows. Annual streamflows increased on the clearcut watershed relative to the control watershed for the first 8 years following harvesting, but significant increases declined in the subsequent 10 years. Water yield increases for August were eliminated within 3 years of treatment because of vegetation recovery; however, annual water yields remained higher than predicted throughout the study period.

Summer streamflow increased from the patch-cut watershed and remained higher than expected for the 1963–1978 posttreatment period. However, summer and annual water yields from the 1979–1988 posttreatment period were not significantly different from the preharvesting period. The difference between the longevity of increased summer streamflows between the two treated areas was related to the nature of the riparian vegetation along the respective creeks. The riparian corridor in the clearcut area had a vigorous stand of mainly red alder (*Alnus rubra* Bong.). The riparian stand development was related to relatively flat geomorphology along the creek, while the patch-cut area, with a narrow steep valley, did not have a similar well-developed zone (Hicks et al. 1991). The authors assume that the influence of the riparian zone on the decline in summer runoff will be less as conifers begin to dominate the sites and the shade-intolerant broadleaved species become less vigorous.

14.2.7.2 Watershed Studies in Second-Growth Forests

In the second-growth stand, one watershed was clearcut, one harvested according to a shelterwood prescription, and the third was maintained as a control (Harr et al. 1982). The clearcut area contained 383 trees/ha and the shelterwood area contained 334 trees/ha prior to harvesting. High-lead cable logging, where logs are suspended above the ground during skidding operations, was used for 90 % of the clearcutting operation. The watershed was planted with Douglas-fir seedlings after slash was burned. The shelterwood method was used because of the difficulty of obtaining satisfactory tree regeneration on south-facing slopes. Approximately 44 m^2/ha or 60 % of the total basal area was removed from the shelterwood watershed using tractor and high-lead techniques. Harvesting significantly increased annual water yields for both treatments (Harr et al. 1982). Average increases were 380 mm (30 %) for the first 4 years after clearcutting and 200 mm (22 %) for the shelterwood. Water yields were declining by the 5th year.

The annual runoff results for clearcutting on this set of watersheds and those harvested earlier (Rothacher 1970) were similar. The results from the shelterwood and earlier patch cutting also were similar. Logging resulted in fewer low-flow days and greater summer volumes on the treated watersheds relative to the control. An analysis of peak flows and peak flow timing did not indicate changes related to logging. Streamflow increases from the second-growth watersheds also were attributed to reduced ET of the residual vegetation following harvesting (Harr et al. 1982).

Current work on H. J. Andrews has moved away from a focus on commercially logged watersheds toward an interdisciplinary program of ecosystem science (USDA Forest Service 2009), reflecting the general direction of forest research and management in recent years. Researchers on the forest have conducted research on dead wood in old-growth forest streams, probing the role of wood jams in providing calm pools where fish can rest, gravel bars for spawning, and cover from predators. Their work has led to changes in stream management practices on forestlands. The ecological value of dead wood on land has also been examined, yielding valuable information on the role of dead wood in wildlife habitat, carbon dynamics, and nutrient cycling (Geier 2007; Luoma 2006). Geier (2007) describes the forest as becoming a natural and human environment that links people, place, and community with an emerging vision of ecosystem management. More information about H.J. Andrews and a bibliography can be found at http://andrewsforest.oregonstate.edu/.

14.2.8 Watershed Research in the West: A Summary

Research studies conducted at the Sierra Ancha Experimental Forest, Beaver Creek Watersheds, the Fraser Experimental Forest, and the H.J. Andrews Experimental Forest were initially designed to evaluate treatments to achieve a compromise between water yield augmentation and sound silvicultural practices to sustain timber production. The first treatments determined the maximum amounts of water available after clearcutting. Subsequent treatments evaluated silvicultural options to enhance and sustain both water and timber resources. While neither timber nor water is the prime resource in many areas of the West today, this situation is changing. The demands for water will surely increase as the West's population increases and climate change-related droughts expand. Research at the experimental forests and at other locations will alert managers about the potentials and limitations of managing forestlands for water production. There is renewed interest in silvicultural practices as overly dense stands are viewed as potential sources for dangerous and damaging wildfires especially in the wildland–urbaninterface. The experimental forests provide important information for land management and its impact on wildlife and recreation. Determining and quantifying the role of forests as carbon sinks is an important new area of research on many experimental forests. These sites continue to provide sources of long-term forest monitoring that are of local, national, and international importance to scientists and land managers studying forest influences over a wide range of forest environments (i.e., arid, high elevation, high precipitation areas).

14.3 Experimental Forests of the East

The hydrologic role of forests in watershed management has received considerable research attention east of the Rocky Mountains in the last century. Experimental forests and watersheds across this vast region have focused attention on how forest management and land-use changes affect the water resource. Examples of experimental forests and watersheds in the Southeast, Northeast, and Upper Midwest are presented to illustrate the contributions of their research programs to each respective region. With the exception of the Wagon Wheel Gap study in Colorado (Bates and Henry 1928), much of the early research concerning forest influences on water flow began in the Southeast at the Coweeta Hydrologic Laboratory (Jackson et al. 2004).

14.3.1 Coweeta Hydrologic Laboratory

Questions about forest influences on streamflow and flooding in the southeast USA led to the establishment of the Coweeta Hydrologic Laboratory in 1934 (Hornbeck and Kochenderfer 2004; Jackson et al. 2004; Swank and Crossley 1988). The Coweeta Hydrologic Laboratory is located in western North Carolina in the Nantahala Mountain Range of the Blue Ridge Physiographic Province. Coweeta occupies 2,185 ha and is composed of two east-facing watersheds (Coweeta Creek and Dryman Fork) that range in elevation from 670 to 1,580 m. Multiple sub-catchments (Fig. 14.4) have been used for manipulative studies in the main treatment watershed (Coweeta Creek). Coweeta is one of the longest continuous watershed and

Fig. 14.4 Streamflow on Coweeta Watershed 17 is monitored by a 90-degree v-notch weir. The mixed mature hardwood forest on this 13.5 ha watershed was cut in 1942 and replaced by planted eastern white pine (*Pinus strobus* L.) to evaluate the impacts of type conversion on water yields (Swank and Miner 1968). Water yields decreased relative to the hardwood forest as the pine crowns closed and evapotranspiration and interception increased. (Photo courtesy of the Coweeta Hydrologic Laboratory.)

ecosystem studies in North America, guided by the philosophy that "…the quantity, timing, and quality of streamflow provide an integrated measure of the success or failure of land management practices" (Jackson et al. 2004).

14.3.1.1 Studies of Water Flows

In the early 1960s, research focused attention on developing a better understanding of how forests influence runoff, streamflow, and flooding in the Southeast. Hewlett (1961), Hewlett and Hibbert (1963), and Betson (1964) discovered that stormflow and floods in southeastern forested catchments are the product of variable source areas. Importantly, they determined that subsurface flow in forested catchments was the major pathway of stormflow to channels, not "Hortonian" overland flow that was widely thought to be the cause of flooding from all landscapes at that time. High infiltration capacities of the deep forest soils explained why subsurface flow was the dominant component of stormflow in stream channels. Hewlett and Hibbert (1967) expanded the Coweeta research, finding that the variable source area concept explained flows during rainstorms and for extended inter-storm periods in 24 southeastern and eastern streams. Hewlett and Troendle (1975) subsequently developed one of the early variable source area models that incorporated flow paths with topography and landscape position. This relationship provided a better understanding of forestland use and water quality responses in numerous long-term, paired watershed studies.

14.3.1.2 Watershed Studies

Jackson et al. (2004) summarized the key findings that reducing forest cover increases water yields and reforestation decreases water yields. The magnitudes of these changes were proportional to the changes in forest density and solar isolation values. The conversion of hardwoods to pine forests reduced water yields by 30–60 %. However, the conversion of forest cover to grass had little effect on water yields when production of grass was high, but as grass production declined, water yields increased by 15 % and baseflows also increased. Stormflow volumes and peak flow rates increased 11 and 7 %, respectively, following the clearcutting of hardwood forests.

The long-term research contributions of the Coweeta Hydrologic Laboratory enhanced the science of forest hydrology and provided information to better manage forested watersheds of the Southeast. Importantly, this research is as relevant today as in the early years when the focus was on timber harvesting and water yield. This solid foundation illuminates the hydrologic role of forests allowing us to discern the impact of forest fragmentation on impaired water quality, and ascertain how climate change affects both our forests and water. Additional information on the Coweeta Hydrologic Laboratory can be found at http://www.srs.fs.usda.gov/coweeta/.

14.3.2 Hubbard Brook Experimental Forest

The 3,138-ha Hubbard Brook Experimental Forest was established by the United States Department of Agriculture (USDA) Forest Service in the White Mountains of New Hampshire in 1955 as a center for forest hydrology research in the Northeast. Through a cooperative agreement, the Hubbard Brook Ecosystem Study was initiated in 1963 and in 1988 the Hubbard Brook Experimental Forest was designated a Long-Term Ecological Research (LTER) site by the National Science Foundation. Today, it represents one of the most comprehensive and longest continuous ecological studies in North America.

14.3.2.1 Watershed Studies

Paired watershed research began in the 1950s with the aim of better understanding the hydrologic cycle of mature forests in the Northeast (Hornbeck and Kochenderfer 2004). Forest effects on flooding were of particular interest initially, given the wet environment and uniform precipitation of the Northeast. Annual precipitation at Hubbard Brooks is 1,400 mm of which nearly one-third is snow (Federer et al. 1990).

In the Northeast, city-owned watersheds are a primary source for municipal water; the relationship of forest management to water quantity and quality has been a primary interest to municipal watershed managers since the mid-1970s (Hornbeck and Federer 1975). Studies of nutrient cycling and biogeochemical processes were emphasized in the 1960s through the 1990s with the development of cooperative projects and the LTER designation (Hornbeck and Kochenderfer 2004).

One of the most well-known paired watershed experiments at Hubbard Brook is a forest-clearing experiment to determine how complete forest removal influences ET and water yield (Hornbeck and Federer 1975; Hornbeck et al. 1970). In 1965, all of the woody vegetation on one watershed was cut and left where it fell. In the following three summers, herbicides were applied to the watershed to prevent regrowth of vegetation (Likens et al. 1970). For each of the first 3 years following the initial clearing, water yield increased by an average of 290 mm/year with the majority of the increased flow occurring in late summer–early fall. Given that the woody vegetation was not removed from the site, some level of interception by the stems and dead biomass occurred, but most of the increased flow was due to the elimination of transpiration.

A stripcutting experiment to promote regeneration of birch (*Betula* spp.), a valuable timber and aesthetic species, was conducted in 1970. In this experiment, one-third of a watershed was harvested in 24.4 m strips every other year until the entire watershed was harvested. Following the first year in which one third of the watershed was stripcut, water yield increased by 33 mm with the majority of the increase occurring in late summer. This increase in water yield following the strip cut was only about 10 % of the increase observed from the cleared watershed, although 33 % of the watershed was harvested. The difference was explained by both rapid regrowth of vegetation within the strips and the high interception and transpiration rates of trees along the border of uncut strips with roots extending into the moist soil in the harvested strips.

14.3.2.2 Relevance of Results

The relevance of the two watershed experiments became apparent to municipal watershed managers in the Northeast. The potential for manipulating forest cover to increase water yield was minimal and could not be sustained when using an accepted silvicultural practice. By clearcutting, annual water yield could be increased by 100–250 mm, but the increases diminish rapidly with forest regrowth (Hornbeck and Kochenderfer 2004). The prolonged water yield increases observed in other regions of the country, such as at Coweeta and Fraser Experimental Forests, could not be duplicated in the Northeast because of the combination of shallow soils and rapid regeneration. The roots of woody and herbaceous regeneration can fully occupy a shallow soil mantel and utilize all the available soil water including any surpluses created by removing the original cover. Analysis of postharvesting response further indicated that increases in water yield were lower during dry years than wet years. Although clearing and using herbicide to eliminate regrowth increased water yield by nearly 300 mm, the combination treatment had serious water quality and nutrient cycling impacts. Nitrate concentrations increased from 2 to 80 ppm following the clearing and herbicide treatment (Hornbeck and Federer 1975; Likens and Bormann 1995). This level of nitrate is far above drinking water standards of 45 ppm.

Increases in water yield following forest removal raised questions about possible increases in stormflow and flooding. However, 13 major rainstorms following forest clearing were found to increase stormflow by an average of 11 mm per storm, with a maximum increase of 30 mm (Hornbeck and Federer 1975). Since rarely would more than 25 % of any watershed be cleared at any point in time, the effects of accepted forest harvesting practices on stormflow and flooding would be even less than what was observed in the paired watershed experiment. Increases in flood-producing snowmelt following clearing was also considered negligible, given the limited areas undergoing clearing at any one time and the natural desynchronization of snowmelt in the mountainous areas of the Northeast (Federer et al. 1972). Again, a small soil moisture reservoir quickly depleted by transpiration from new vegetation negates the impact of forest harvesting on water yield and peak stormflows in comparison to results from other experimental forest sites with deeper soils, long vegetation recovery periods or both. A recent publication has summarized long-term trends developed from ecosystem research at Hubbard Brook (Campbell et al. 2007). Additional information on Hubbard Brook can be found at the following website: http://nrs.fs.fed.us/ef/locations/nh/hubbard-brook/.

14.3.3 Marcell Experimental Forest

The 1,141-ha Marcell Experimental Forest (MEF) was established in 1960 in the Upper Midwest to study the physical, silvicultural, and hydrological aspects of peatlands and how peatlands affect forest productivity, water availability, and water chemistry (Bay 1966, 1962; Boelter and Verry 1977). The MEF was established to fill a geographic and ecological void in the USDA Forest Service Experimental Forest and Range network of sites. Located approximately 42 km north of Grand Rapids, MN, USA, the area was considered to be representative of upland/peatland forests in the Upper Midwest. Bay (1966) emphasized the need for fundamental hydrologic studies, given the prominence of the 6.1 million ha of organic soils in headwater watersheds of the northern Lake States. Research at Marcell was directed by R. Bay and S. Verry, and is currently directed by R. Kolka and S. Sebestyen. The following summary represents 50 years of research conducted at Marcell that is described in a book edited by Kolka et al. (2011).

14.3.3.1 Process Studies

Fundamental studies were initiated in the 1960s to determine the hydrologic characteristics of organic soils and ET from peatlands. The hydrologic characteristics of organic soils in peatlands were studied by Boelter (1966, 1965, 1964) and Nicholes and Boelter (1984); they developed relationships among fiber content, bulk density, ash content, water storage capacity and water retention, hydraulic conductivity, drainable porosity (specific yield), and von Post degree of humification for organic soils at MEF. These studies provided the foundation for understanding and explaining peatland responses to rainfall and snowmelt in subsequent experimental watershed studies (Boelter and Verry 1977; Verry 1997, 1975).

Studies of ET using lysimeters, energy budgets, and water budgets demonstrated the dominance of evaporative processes in peatland systems. Nichols and Brown (1980) conducted growth chamber studies and found that evaporation rates from sphagnum moss ranged from 0.291 mm/h) at 8.9 °C to 0.558 mm/h at 25.3 °C and were from 1.3 to nearly 2 times the evaporation rates of free water surfaces. Subsequent field studies using eddy correlation methods reported ET rates in a sphagnum bog (May to early October) to average 3.6 mm/day, but ranged from 0.9 to 6.0 mm/day (Verma et al. 1993). Water budget studies of two watersheds with forested peatlands and upland aspen forests found that ET from May 1 to November 1 ranged from 87 to 121 % of Thornthwaite potential evapotranspiration (PET; Bay 1967a, b). Verry and Timmons (1982) reported that ET of a black spruce (*Picea mariana*) forested sphagnum bog, determined as the residual of the water balance, equaled the Thornthwaite PET. From these studies, we can conclude that ET in peatlands is at or near the PET rate as long as the water table is within 30 cm of the soil surface. As a result, ET from peatlands in northern Minnesota averages about 500 mm/year with annual average precipitation of 775 mm/year (Verry 1997).

14.3.3.2 Watershed Studies

Studies of silvicultural treatments on upland and peatland forests provided insight into the hydrologic characteristics and water budgets of these ecosystems (Fig. 14.5). Verry (1986) summarized the hydrologic response of forest harvesting in the Northern Lake States, based largely on the research at the MEFs. Among important results from silvicultural treatments was the finding that clearcutting strips of black spruce and complete clearcutting in a peatland did not change annual water yield (Brown 1972). However, harvesting caused water tables and streamflow to rise during the wet spring period, but it resulted in water tables and streamflow falling below predicted values during the summer dry period. The increase in flow during the spring was explained by the reduced interception losses due to removal of the black spruce. The lower flow rates in the summer following forest harvesting were due to the changes in the vegetative conditions at the forest floor. Following harvesting, the grass-sedge cover on the forest floor was exposed to greater solar energy and wind that promoted higher ET. In response to the changed microclimatic

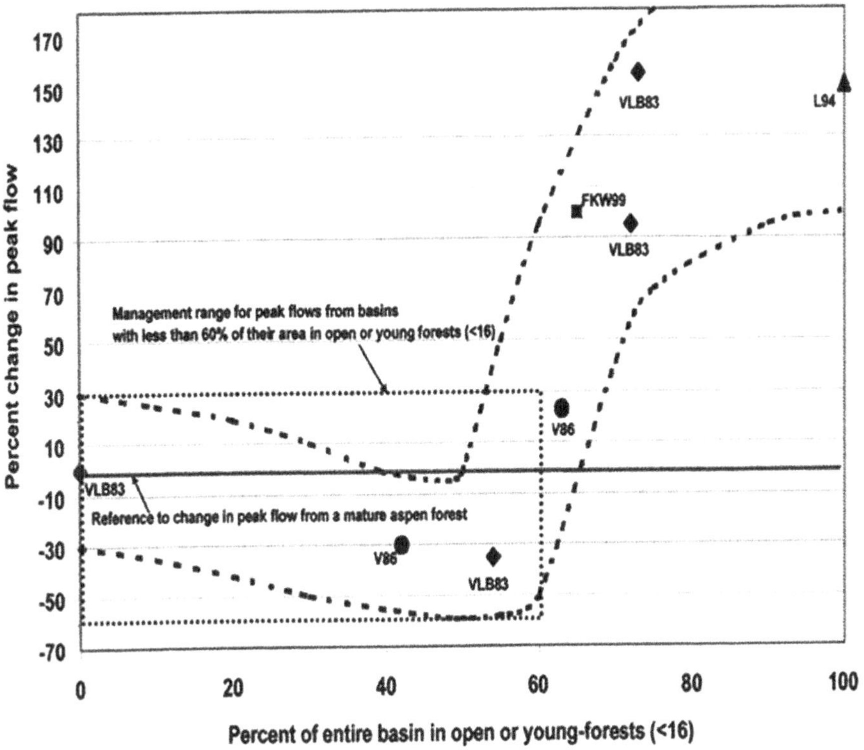

Fig. 14.5 Relationship between peak discharge response to percentage of basin in open land or young forests in the Midwest (Verry et al. 2004). Data points are from individual paired watershed experiments (Verry et al. 1983), double mass curve historical evaluations (Verry 1986), historical geomorphic and modeling analyses (Fitzpatrick et al. 1999), and modeling of upland/peatland watersheds (Lu 1994)

conditions, the grass-sedge biomass in the clearcut strips increased fivefold in the first growing season following tree removal. Furthermore, these species are apparently more physiologically active and have higher transpiration rates in the summer period than did the black spruce.

In contrast to the black spruce watersheds, harvesting of aspen (*Populus* spp.) in the upland portion of a peatland-upland watershed increased annual water yield initially by an average of 90 mm/year (Verry 1972). The increase in water yield was reduced to zero in 12–15 years as a result of aspen regeneration. Snowmelt produced stormflow peaks and stormflow volumes that doubled for 2 years following upland aspen clearcutting. For snow and rain periods, the average annual peak flow (recurrence interval of 2 years) increased 71 % after clearcutting (Verry et al. 1983).

Snowmelt peak response to forest harvesting varied according to the percentage of the watershed that was clearcut (Verry et al. 1983). Clearcutting an entire watershed will double snowmelt peaks, whereas clearcutting 30–50 % of the watershed will reduce snowmelt peaks by as much as 30 % compared to a watershed with a mature mixed aspen forest. Partial cutting desynchronizes snowmelt, and, thus, snowmelt runoff at the watershed outlet. This finding was particularly useful for land-use planners; by tracking the percentage of any watershed that is non-forested through development or timber harvesting, the goal of maintaining at least 50% forest cover will not increase snowmelt peaks. Computer modeling of streamflow response to timber harvesting, using the paired watershed data at Marcell and long-term precipitation records in the region, showed that clearcutting has a minimal effect on large floods with recurrence intervals in excess of 20–25 years (Lu 1994; Verry 2000). Additional information about the MEF can be found on the website: http://nrs.fs.fed.us/ef/marcell/.

14.3.4 Watershed Research in the East: A Summary

Watershed research in the eastern and midwestern USA also has focused on how forest management and land-use changes have affected water yields and water quality. These issues are important particularly because of the population growth in these regions. Many urban areas depend on their forested watersheds for high quality water that does not require intensive purification. The research in the East has been a combination of basic hydrologic process studies and traditional paired watershed experiments. Research results, such as the relationship between forest treatments and water yield impacts over time, support the general conclusions from research in the West. Pioneering work at Coweeta showed how common land-use practices affected water resources. Subsequent research explained how water moves through the watershed's soil and geological mantle and that all areas are not equally productive sources for water. The importance of transpiration by postharvest replacement vegetation, especially where soils are shallow, was demonstrated at Hubbard Brook. Research at Marcell developed a basic understanding of the hydrology of peatlands and the impacts of land management options. Information from the experiments is the basis for the development of computer models used to explain and predict the impacts of present or future management on water resources.

14.4 Conclusions

A controversy in the early twentieth century concerned the relationship between forests and streamflow. Many foresters maintained that forests reduce surface runoff and help to regulate streamflow. Others believed that only precipitation was important. The evidence in favor of the importance of forests resulted in the Weeks Act of 1911 that allowed for the purchase of forested lands to protect navigable streams and rivers (Dana 1956). Experimental forests were established to answer many questions about forestry and forest resource management. Common key questions concerned the natural role of forests and how various common silvicultural prescriptions and land-use options affect streamflow quantities, peak flows, and timing. In the West, a goal was to clarify the interconnection between forests and snowpacks and how management impacts these resources. Concerns about soil erosion and general site deterioration were also important in the West and Southeast.

Land managers and the public raised questions usually driven by local or regional interests. Maintenance and enhancement of adequate water supplies were, and still are, important concerns throughout the USA. One objective of research on experimental areas surrounding Fort Valley and at Hubbard Brook was to improve water yields and to clarify the forest role in providing water for downstream users. A common question is: "Will harvesting during a drought increase water yields?" Adequate water supplies are also an issue in the humid Northeast where periodic droughts have caused some water restrictions in downstream metropolitan areas such as New York City. Snowpacks are an important source of water throughout the Mountain West, and their management is vital to the City of Denver, Phoenix Metropolitan Area, Salt Lake City, Las Vegas, Los Angeles, Portland, Seattle, and to other western urban centers (Dissmeyer 2000). The impacts of common forestry practices on streamflow quantities and seasonal timing are important issues in the Pacific Northwest with its important fisheries resources, and in the peatlands of the Upper Midwest.

Researchers and staffs assigned to the experimental forests discussed above and to other research locations throughout the USA attempt to provide governmental and private land managers and the general public with information for planning forest treatments and interpreting the impacts on water. Climate goes through cycles and water is a critical resource when there is too much or too little. People look to the forests with hope when there is a drought and when there are floods. Scientists working in the field of watershed management attempt to provide answers for feasible and relevant options. While timber production and water yield augmentation are no longer the prime forest resources in many regions, knowledge gained from USDA Forest Service experimental forests has provided information for more holistic management that includes wildlife populations, recreational opportunities, and livestock management. Increasing demands for water in the future could once more increase interests in managing forestlands for water production.

Land and water managers are currently planning for the consequences of climate change on water supply and flood flows. One of the most important aspects of the Experimental Forest System is the availability of long-term hydrologic and meteorological records. It is particularly relevant now as society considers a future of climate change. These records can be used in simulation models to determine how the climate is changing and at what rate. Linking climate data with long-term vegetation trends provides insight to what our future landscape could look like. The 1950s drought in the Southwest caused large-scale changes in the region's forests and woodlands. Such droughts are a current concern in the region. Will these changes be duplicated by the current drought or a generally more arid environment? Changes in climate and vegetation will affect the regional hydrology. Watershed mangers can modify silvicultural treatments to adjust to these changes, depending on the rate of change. One option is to maintain more open stands of healthier trees that could survive without the excessive competition from their neighbors, but markets for intermediate harvests and mature tree harvests need to be redeveloped. Treatments to mitigate the potential for wildfires are being implemented and will become more common. How do they affect water resources from the forests? Experimental forests will become more important as outdoor laboratories providing scientific answers to questions about climate change treatment options, especially in view of wildfire concerns, landscape dynamics and management, carbon sequestration, and hydrology.

References

Alexander RR (1969) Seedfall and establishment of Engelmann spruce in clearcut openings: a case study. Research Paper RM-53, USDA Forest Service

Alexander RR (1971) Initial partial cutting of old-growth spruce-fire. Research Paper RM-76, USDA Forest Service

Alexander RR (1973) Partial cutting in old-growth spruce-fir. Research Paper RM-110, USDA Forest Service

Alexander RR, Edminster CB (1980) Management of spruce-fir in even-aged stands in the Central Rocky Mountains. Research Paper RM-217, USDA Forest Service

Alexander RR, Edminster CB (1981) Management of lodgepole pine in even-aged stands in the Central Rocky Mountains. Research Paper RM-229, USDA Forest Service

Alexander RR, Watkins RK (1977) The Fraser Experimental Forest, Colorado. General Technical Report RM-40, USDA Forest Service

Baker MB Jr (Comp) (1999) History of watershed research in the Central Arizona Highlands. General Technical Report RMRS-GTR-29, USDA Forest Service

Baker MB Jr, Ffolliott PF (1998) Multiple resource evaluations on the Beaver Creek watersheds: an annotated bibliography (1956–1996). General Technical Report RMRS-GTR-13, USDA Forest Service

Baker MB Jr, Ffolliott PF (1999) Interdisciplinary land use along the Mogollon Rim. In: Baker MB Jr (comp) History of watershed research in the Central Arizona Highlands. General Technical Report RMRS-GTR-29, USDA Forest Service, pp 27–34

Bates CG, Henry AJ (1928) Forest and streamflow experiment at Wagon Wheel Gap, Colorado. U.S. Weather Bureau monthly weather review, Suppl 30

Bay RR (1962) Establishment report, Marcell Experimental Forest. USDA Forest Service, (Chippewa National forest and adjacent private, Itasca County, and State of Minnesota lands), p 26

Bay RR (1966) Evaluation of an evapotranspirometer for peat bogs. Water Resour Res 2:437–441

Bay RR (1967a) Ground water and vegetation in two peat bogs in northern Minnesota. Ecology 48(2):308–310

Bay RR (1967b) Factors influencing soil-moisture relationships in undrained forested bogs. In: Sopper WE, Lull HW (eds) Forest hydrology. Pergamon Press, New York, pp 335–342

Betson RP (1964) What is watershed runoff? J Geophys Res 69(8):1541–1552

Boelter DH (1964) Water storage characteristics of several peats in situ. Soil Sci Soc Am Proc 33:433–435

Boelter DH (1965) Hydraulic conductivity of peats. Soil Sci 100:227–231

Boelter DH (1966) Hydrologic characteristics of organic soils in Lakes States watersheds. J Soil Water Conserv 21(2):50–53

Boelter DH, Verry ES (1977) Peatland and water in the northern Lake States. General Technical Report NC–31, USDA Forest Service

Brown JM (1972) Effect of clearcutting a black spruce bog on net radiation. For Sci 18:273–277
Brown HE, Baker MB Jr, Rogers JJ, Clary WP, Kovner JL, Larson FR, Avery CC, Campbell RE (1974) Opportunities for increasing water yields and other multiple use values on ponderosa pine forest lands. Research Paper RM–129, USDA Forest Service
Buttle JM, Creed IF, Moore RD (2005) Advances in Canadian forest hydrology, 1999–2003. Hydrol Process 19:169–200
Campbell JL, Driscoll CT, Eager C, Likens GE, Siccama TG, Johnson CE, Fahey TJ, Hamburg SP, Holmes RT, Bailey AS, Buso DC (2007) Long-term trends from ecosystem research at the Hubbard Brook Experimental Forest. General Technical Report NRS–17, U.S. Forest Service, Northern Research Station
Dana ST (1956) Forest and range policy its development in the United States. McGraw-Hill, New York
DeBano LF, Neary DG, Ffolliott PF (1998) Fire's effects on ecosystems. Wiley, New York, 333 p
DeBano LF, Baker MB Jr, Gottfried GJ (1999) Providing water and forage in the Salt-Verde River Basin. In: Baker MB Jr (comp) History of watershed research in the Central Arizona Highlands. U.S. Forest Service General Technical Report RMRS-GTR-29, pp 13–18
Dieterich JH (1983) Fire history of southwestern mixed conifer: a case history. For Ecol Manage 6:13–31
Dissmeyer GE (2000) Drinking water from forest and grasslands: a synthesis of the scientific literature. General Technical Report SRS-39, U.S. Department of Agriculture, Forest Service, Southern Research Station, Ashville, NC, 246 p
Elder K, Porth L, Troendle C (2006) The effect of timber harvest on the Fool Creek watershed after five decades. Abstract B21F-01, American Geophysical Union
Federer CA, Pierce RS, Hornbeck JW (1972) Snow management seems unlikely in the Northeast. In: American Water Resources Association national symposium on watersheds in transition, Urbana, Illinois, pp 212–219
Federer CA, Flynn LD, Martin CW, Hornbeck JW, Peirce RS (1990) Thirty years of hydrometeorologic data at the Hubbard Brook Experimental Forest, New Hampshire. General Technical Report NE-141, Northeastern Forest Experiment Station, USDA Forest Service, Radnor, PA
Ffolliott PF, Baker MB Jr (2001) Stripcut-thinning of ponderosa pine stands: an Arizona case study. Research Paper RMRS-RP-34, USDA Forest Service
Ffolliott PF, Gottfried GJ (1991) Natural regeneration after clearcutting in Arizona's ponderosa pine forests: two long-term case studies. Research Note RM-507, USDA Forest Service
Ffolliott PF, Baker MB Jr, Gottfried GJ (2000) Heavy thinning of ponderosa pine stands: an Arizona case study. Research Paper RMRS-RP-22, USDA Forest Service
Fitzpatrick FA, Knox JC, Williams HE (1999) Effects of historical land-cover changes on flooding and sedimentation, North Fish Creek, Wisconsin. USGS Water Resource Investigations Report 99-4083
Fletcher HC, Rich LR (1955) Classifying southwestern watersheds on the basis of water yields. J For 53:196–202
Geier M (2007) Necessary work: discovering old forests, new outlooks, and community on the H. J. Andrews Experimental Forest, 1948–2000. General Technical Report PNW-GTR-687, USDA Forest Service Pacific Northwest Research Station, Portland, OR
Gottfried GJ, DeBano LF (1988) Streamflow and water quality responses to preharvest prescribed burning in an undisturbed ponderosa pine watershed. In: Krammes JS (comp) Effects of fire management of southwestern natural resources. General Technical Report RMFRES-GTR-191, USDA Forest Service, pp 222–228
Gottfried GJ, DeBano LF, Baker MB Jr (1999a) Beginning of water studies in the Central Arizona Highlands. In: Baker MB Jr (comp) History of watershed research in the Central Arizona Highlands. General Technical Report RMRS-GTR-29, USDA Forest Service, pp 7–12
Gottfried GJ, DeBano LF, Ffolliott PF (1999b) Creating a basis for watershed management in high elevation forests. In: Baker MB Jr (comp) History of watershed research in the Central Arizona Highlands. General Technical Report RMRS-GTR-29, USDA Forest Service, pp 35–41

Gottfried GJ, Folliott PF, Neary DG (2008) Contributions of silvicultural studies at Fort Valley to watershed management of Arizona's ponderosa pine forests. In: Olberding SD, Moore MM (tech cords) Fort Valley-a century of research 1908–2008. Proceedings RMRS-P-53CD, U.S. Department of Agriculture, Forest Service, Rocky Mountain Research Station, Fort Collins, CO, pp 86–97

Harr RD (1976) Forest practices and stream flow in western Oregon. General Technical Report PNW-49, USDA Forest Service, 18 p

Harr RD, Levno A, Mersereau R (1982) Streamflow changes after logging 130-year-old Douglas fir in two small watersheds. Water Resour Res 18:637–644

Hewlett JD (1961) Soil moisture as a source of baseflow from steep mountain watersheds. Station Paper 132, Southeastern Forest Experiment Station, USDA Forest Service, Asheville, NC

Hewlett JD (1971) Comments on the catchment experiment to determine vegetal effects on water yield. Water Resour Res 7:376–381

Hewlett JD, Hibbert AR (1963) Moisture and energy conditions within a sloping soil mass during drainage. J Geophys Res 68(4):1081–1087

Hewlett JD, Hibbert AR (1967) Factors affecting the response of small watersheds to precipitation in humid areas. In: Sopper WE, Lull HW (eds) Forest hydrology. Pergamon Press, New York, pp 275–290

Hewlett JD, Troendle CA (1975) Non-point and diffused water sources: a variable source area problem. In: Watershed management, American Society of Civil Engineers, Logan, UT, pp 21–45

Hicks BJ, Beschta RL, Harr RD (1991) Long-term changes in streamflow following logging in western Oregon and associated fisheries implications. Water Resour Bull 27:217–226

Hornbeck JW, Federer CA (1975) Effects of management practices on water quality and quantity: Hubbard Brook Experimental Forest, New Hampshire. In: Municipal watershed symposium proceedings, General Technical Report NE-13, USDA Forest Service, pp 58–65

Hornbeck JW, Kochenderfer JN (2004) A century of lessons about water resources in northeastern forests. In: Ice GG, Stednick JD (eds) A century of forest and wildland watershed lessons. Society of American Foresters, Bethesda, MD, pp 19–31

Hornbeck JW, Pierce RR, Federer CA (1970) Streamflow changes after forest clearing in New England. Water Resour Res 6:1124–1132

Ice GG, Stednick JD (2004) A century of forest and wildland watershed lessons. Society of American Foresters, Bethesda MD, 287 p.

Jackson CR, Sun G, Amatya D, Swank WT, Riedel M, Patric J, Williams T, Vose JM, Tretin C, Aust WM, Beasley RS, Wilson H, Ice GG (2004) Fifty years of forest hydrology in the Southeast. In: Ice GG, Stednick JD (eds) A century of forest and wildland watershed lessons. Society of American Foresters, Bethesda, MD, pp 33–112

Kolka RK, Sebestyen SD, Verry ES, Brooks KN (eds) (2011) Peatland biogeochemistry and watershed hydrology at the Marcell Experimental Forest. CRC Press, Boca Raton, FL

Leaf CF (1975) Watershed management in the Rocky Mountain subalpine zone: the status of our knowledge. Research Paper RM-137, USDA Forest Service

Likens GE, Bormann FH (1995) Biogeochemistry of a forested ecosystem. Springer-Verlag, New York

Likens GE, Bormann FH, Johnson NM, Fisher DW, Pierce RS (1970) Effects of forest cutting and herbicide treatment on nutrient budgets in the Hubbard Brook watershed ecosystem. Ecol Monogr 40:23–47

Lu S-Y (1994) Forest harvesting effects on streamflow and flood frequency in the northern Lake States. PhD dissertation thesis, University of Minnesota, Saint Paul, MN, 213 p

Luoma J (2006) The hidden forest: biography of an ecosystem. Oregon State University Press, Corvallis

National Research Council (NRC) (2008) Hydrologic effects of a changing forest landscape. Washington, DC, The National Academies Press, 180 p.

Neary DG, Gottfried GJ, Beyers JL, Ffolliott PF (2006) Floods and sediment yields from recent wildfires in Arizona. In: Proceedings of the 8th federal interagency sediment conference Las Vegas, NV, 2006 April 3–6 (CD–ROM)

Nichols DS, Boelter DH (1984) Fiber size distribution, bulk density, and ash content of peats in Minnesota, Wisconsin, and Michigan. Soil Sci Soc Am J 48:1320–1328

Nichols DS, Brown JM (1980) Evaporation from a sphagnum moss surface. J Hydrol 48:289–302

Noble DL, Ronco F Jr (1978) Seedfall and establishment of Engelmann spruce and subalpine fire in clearcut opening in Colorado. Research Paper RM-200, USDA Forest Service

Olberding SD, Moore MM (2008) Fort Valley Experimental Forest-a century of research 1908–2008. Conference proceedings. Proceedings RMRS-P-55, U.S. Forest Service

Rich LR, Gottfried GJ (1976) Water yields resulting from treatments on the Workman Creek Experimental Watersheds in Central Arizona. Water Resour Res 12:1053–1060

Rich LR, Thompson JR (1974) Watershed management in Arizona's mixed conifer forests: the status of our knowledge. General Technical Report RM-130, USDA Forest Service

Rothacher J (1970) Increases in water yields following clearcut logging in the Pacific Northwest. Water Resour Res 6:653–658

Schubert GH (1974) Silviculture of southwestern ponderosa pine: the status of our knowledge. Research Paper RM-123, USDA Forest Service

Smith FW (1987) Silvicultural research in coniferous subalpine forests of the Central Rocky Mountains. In: Troendle CA, Kaufmann MR, Hamre RH, Winokur RP (tech coord) Management of subalpine forests: building on 50 years of research. General Technical Report RM-149, USDA Forest Service, pp 15–20

Stednick JD, Troendle CA (2004) Water yield and timber harvesting practices in the subalpine forests of the Central Rocky Mountains. In: Ice GG, Stednick JD (eds) A century of forest and wildland watershed lessons. Society of American Foresters, Maryland, pp 169–186

Stednick JD, Troendle CA, Ice GG (2004) Lessons for watershed research in the future. In: Ice GG, Stednick JD (eds) A century of forest and wildland watershed lessons. Society of American Foresters, Maryland, pp 277–287

Steen HK (1998) Forest Service research: finding answers to conservation's questions. Forest History Society, Durham, NC, 102 p

Swank WT, Crossley DA Jr (eds) (1988) Forest hydrology and ecology at Coweeta. Springer-Verlag, New York

Swank WT, Miner NH (1968) Conversion of hardwood-covered watersheds to white pine reduced water yield. Water Resour Res 4:947–954

Troendle CA, Kaufmann MR (1987) Influence of forests on the hydrology of the subalpine forest. In: Troendle CA, Kaufmann MR, Hamre RH, Winokur RP (tech coord) Management of subalpine forests: building on 50 years of research. General Technical Report RM-149, USDA Forest Service, pp 68–78

Troendle CA, King RM (1985) The effect of timber harvest on the Fool Creek watershed, 30 years later. Water Resour Bull 21:1915–1922

USDA Forest Service (1972) National forests in a quality environment. USDA Forest Service, Washington, DC

USDA Forest Service (2009) Experimental forests and ranges: 100 years of research success stories. General Technical Report, FPL-GTR-182, Gail Wells Communication, pp 8–9

Verma SB, Ullman FG, Shurpali NJ, Clement RJ, Kim J, Billesbach DP (1993) Micrometeorological measurements of methane and energy fluxes in a Minnesota peatland. Suo 43(4–5):285–288

Verry ES (1972) Effect of an aspen clearcutting on water yield and quality in northern Minnesota. In: Csallany SC, McLaughlin TG, Striffler WD (eds) Proceedings of a symposium on "Watersheds in Transition" held at Fort Collins, Colorado, June 19–22,1972, American Water Resources Association, Urbana, IL, pp 276–284

Verry ES (1975) Streamflow chemistry and nutrient yields from upland-peatland watersheds in Minnesota. Ecology 56:1149–1157

Verry ES (1986) Forest harvesting and water: the Lake States experience. Water Resour Bull 22:1039–1047

Verry ES (1997) Hydrological processes of natural, northern forested wetlands. In: Trettin CC et al (eds) Northern forest wetlands: ecology and management. CRC Lewis Publishers, New York, pp 163–188

Verry ES (2000) Water flow in soils and streams: sustaining hydrologic function. In: Verry ES, Hornbeck JW, Dolloff CA (eds) Riparian management in forests of the continental eastern United States. Lewis Publishers, Boca Raton, FL, pp 99–124

Verry ES, Timmons DR (1982) Waterbourne nutrient flow through an upland-peatland watershed in Minnesota. Ecology 63:1456–1467

Verry ES, Lewis JR, Brooks KN (1983) Aspen clearcutting increases snowmelt and stormflow peaks in north central Minnesota. Water Resour Bull 19:59–67

Verry ES, Dolloff CA, Manning ME (2004) Riparian ecotone: a functional definition and delineation for resource assessment. Water Air Soil Pollut Focus 4:67–94

Chapter 15
A History of Watershed Research in Experimental Forests of the Interior Highlands

Daniel A. Marion, Donald Turton and Maria Schleidt

Abstract The history of watershed research in the Interior Highlands can be divided into four periods: Initial Start and Stumble (1930s–1950s), Reestablishment and Renewal (1960–1980), Partnerships and Expansion (1980–1990), and New Scales and Paths (1990–present). While each of these periods was marked by different societal concerns and scientific questions, experimental forests played a central role in accomplishing watershed research during all of these periods. Unlike other regions of the country, there was no dominating theme or inspirational leader to focus watershed research in one particular experimental forest; rather the work shifted between several experimental forests over time. Despite many changes in personnel and research direction, a significant body of knowledge has been developed over the past 70 years that has benefited scientists, forest managers, and the public. Fundamental knowledge has been gained regarding the components of the hydrologic system and how these components are affected by natural disturbances. Timber harvesting impacts on soil and water resources have been quantified and shown to be short-lived. Concerns about acid rain and road erosion also have been addressed and shown to be less severe than initially thought. These findings, coupled with the discovery that, in general, small watersheds responded in similar ways across the Interior Highlands, have been the basis for forest planning across the region. As new research challenges arise, experimental forests will continue to play a critical role in addressing these needs.

Keywords Interior Highlands · Forest hydrology · Paired watershed · Experimental forest · Research history

---

D. A. Marion (✉)
Southern Research Station, USDA Forest Service, P.O. Box 1270, Hot Springs, AR 71902, USA
Phone: 501-623-1880
e-mail: dmarion@fs.fed.us

D. Turton (retired)
Natural Resource Ecology and Management, Oklahoma State University, Stillwater, OK, USA

M. Schleidt
Ouachita National Forest, USDA Forest Service, Mena Ranger District, P.O. Box 1270,
Hot Springs, AR 71902, USA

D. C. Hayes et al. (eds.), *USDA Forest Service Experimental Forests and Ranges*,
DOI 10.1007/978-1-4614-1818-4_15, © Springer New York 2014

15.1 Introduction

The Interior Highlands is a fascinating and diverse region that spans large portions of Arkansas, Oklahoma, and Missouri in the central USA. The forests of this mountainous region have sustained inhabitants for thousands of years, and do so today by providing wood products, recreation, clean water, and economic livelihood. This region presents a unique setting that has challenged scientists in their efforts to understand the region's forest environments and ensure the continued health of these forests.

In our efforts to better understand the forests of the Interior Highlands, watershed studies have played a major role. "Watershed studies," as used here, refers to research on soil and water resources—research that investigates where these resources occur, what are their characteristics, how do their related processes operate, and how are they affected by forestry practices. Watershed studies have been conducted since research first began in this region in the 1930s. To date, over 120 scientific publications have resulted from these studies. Central to the accomplishment of most of this research has been the use of experimental forests.

Experimental forests provide the outdoor laboratories where streams and soils can be examined and measured, and where forests can be manipulated so as to understand how soil and water resources react. Within the Interior Highlands, experimental forests have been created and used not only on US Department of Agriculture Forest Service lands but also on private lands as well. Some have been permanently designated as experimental forests, and are still used as such today, whereas others were assigned this role for a limited period of time and reverted to their previous status when research was completed. Whether permanent or temporary, federal or private, the availability of these experimental forests in the Interior Highlands has been critical in obtaining the significant store of knowledge that we have.

This chapter tells the story of how watershed research evolved in the Interior Highlands and how experimental forests were used to accomplish this work, and summarizes key knowledge we have gained through this effort. The history of watershed research in the Interior Highlands varies from that of other regions in many ways that make this history interesting and unique. Here, societal interactions played a noticeably visible role in affecting which research questions were pursued. Like elsewhere, the research "society" within the Interior Highlands consists of forest managers, the public, and scientists. These groups share many interests about soil and water resources within forests: where it occurs; how much there is and what is its quality; and how soil and water properties and processes are affected by timber harvesting, fire, or recreation. Perhaps more so than elsewhere, the citizens of the Interior Highlands have often expressed their opinions concerning the forests to which they have always been so closely connected. Related interests in watershed resources have caused these groups to interact—sometimes directly, but other times indirectly. In the Interior Highlands, this interaction or dialog has greatly influenced the topics that scientists have investigated and the types of results that have been produced.

In contrast to many other regions of the country, there has been no consistent geographic nexus of watershed research effort within the Interior Highlands; rather work has shifted between several experimental forests over time. No dominating theme or inspirational leader acted to center the work to a particular experimental forest; instead the location changed over time with the shifting interests of the participants and logistical constraints of different sites. Together with uneven funding, vagaries in staffing, and sporadic administrative reorganizations, these circumstances inhibited the development of centralized infrastructure or long-term commitment to a single site.

Along with other influences, the evolving dialog within and between interest groups coupled with the shifting geographic focus produced four distinct periods of watershed research in the Interior Highlands: Initial Start and Stumble (1930s–1950s); Reestablishment and Renewal (1960–1980); Partnerships and Expansion (1980–1990); and New Scales and Paths (1990–present). Each of these four periods is marked by different societal concerns, scientific questions, and locations of research effort. During these periods, the goals of forest managers changed from protecting and resurrecting the forests to "improving" forests, and to restoring ecosystems and maintaining forest integrity, and each of these goals has affected our thinking about watershed resources and research needs. At the same time, public concerns about the effects of forestry practices have shifted from concerns about soil erosion and floods to water pollution and acid rain, to cumulative impacts and to ecosystem restoration.

The remainder of this chapter is organized as follows. First, a brief description is given of the Interior Highlands region and its environmental characteristics. Then, each of the four research periods is examined. When appropriate, a description of the forest conditions at the beginning of the period is given and then the opinions and concerns of forest managers and the public during that time are summarized. How scientists responded to the challenges given them is examined next, and important research results are summarized. Throughout this chapter, our intent is not to exhaustively catalog the studies undertaken and their results; rather we aim to explain the context in which the science occurred, the fundamental role that experimental forests played, and what general lessons have been learned.

15.2 Interior Highlands Environment

The Interior Highlands consists of four distinct subregions: the Ouachita Mountains, the Springfield-Salem Plateau, the Boston Mountains, and the Arkansas River Valley (see Fig. 15.1). The Ouachita Mountains are a series of east–west-trending, parallel ridges composed of alternating sandstone and shale beds that are highly folded and faulted. The Springfield-Salem Plateau consists of flat-lying sedimentary rocks composed primarily of limestone and dolomite, and exhibits karst drainage features with low-relief rolling uplands dissected by entrenched streams with steep valley walls. The Boston Mountains are an east–west-trending range predominantly

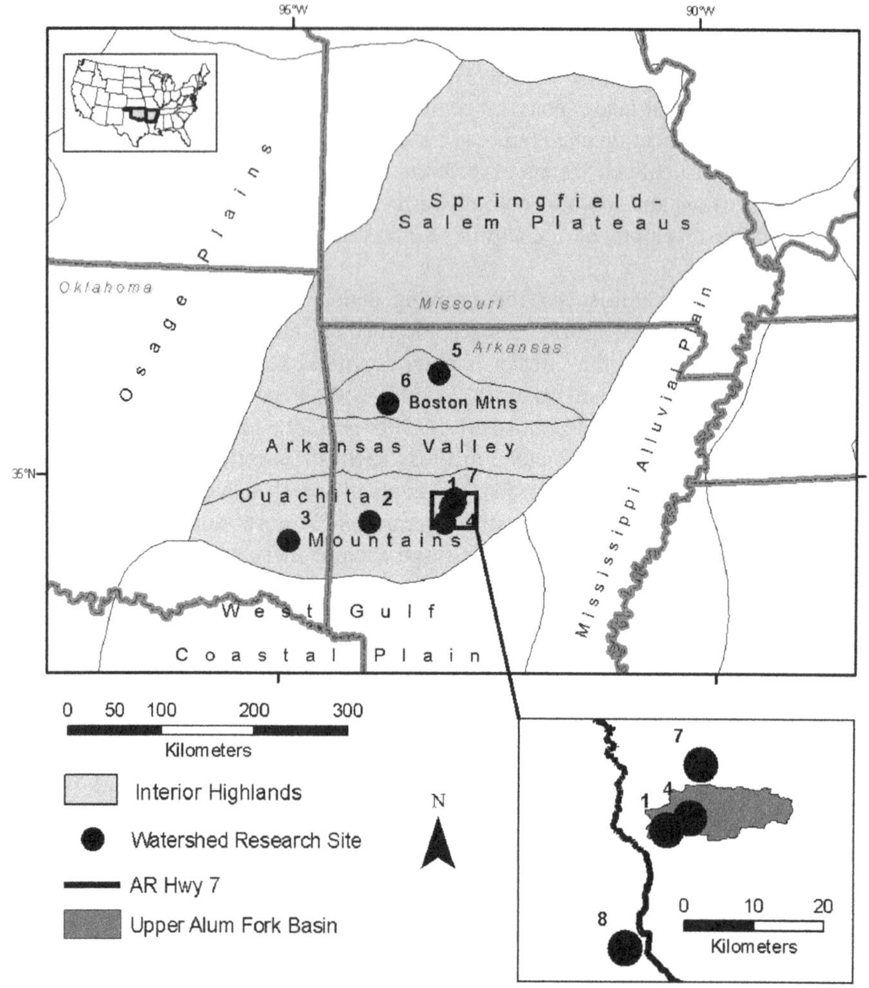

Fig. 15.1 Subregions of Interior Highlands and locations of experimental forests and research areas used for watershed research. Location codes: *1* Alum Creek Experimental Forest, *2* Irons Fork Experimental Forest, *3* Battiest research area, *4* Alum Fork research area, *5* Koen Experimental Forest, *6* Fleming Creek research area, *7* Cedar Mountain research area, and *8* Little Glazypeau research area

underlain by flat-bedded sandstone and shales. The Arkansas River Valley is composed of flat-topped mountains and rolling hills that descend to the Arkansas River, which bisects the subregion.

The Interior Highlands is an old landscape, comprising mostly Paleozoic rocks that were first exposed and began eroding over 300 million years ago. Today, relief within major valleys is around 150 m, slopes angles are typically less than 30%, and mass erosion events are rare. Forest vegetation covers much of the region, and consists of intermixed oak–pine, oak–hickory, and pine forest types with hardwoods increasing in number as one moves northward.

The region possesses a dynamic climate that is very conducive to weathering and erosion. Winters are mild while summers are hot and humid. Precipitation is relatively high (100–140 cm/yr) and evenly distributed throughout the year, with almost all of it occurring as rain. The Interior Highlands is located where dry, cold continental air frequently interacts with warm, moist Gulf air masses, making possible intense thunderstorms, ice storms, tropical storms, or tornados. The combination of this dynamic climate with the highly varied terrain produces a complex set of forest environments across the region.

15.3 Initial Start and Stumble (1930s–1950s)

15.3.1 Land-Use History Prior to 1930

American Indians' occupation of the Interior Highlands dates back at least 10,500–12,000 years (Sabo et al. 1988). Of the impacts that Native Americans had on the Interior Highlands environment, one of the most significant was through their extensive use of fire to clear land and improve game habitat. The combination of natural fires (lightening caused) and Native American burning produced forest conditions in the Ouachita Mountains subregion that were more open than those that typically occur today (Foti and Glenn 1991), though the effect in the Ozarks is less certain (Tucker 1991). By the time Euro-American settlers arrived in the early 1800s, annual burning of forest areas by Native Americans was common (Strausberg and Hough 1997), and was later adopted by the settlers. Still, the Interior Highlands forests remained largely intact and healthy until the late 1800s when increasing population and improved transportation led to the advent of commercial logging in the region (Fig. 15.2).

Commercial logging became firmly established by 1879 in the Interior Highlands with the extension of new rail lines into the region (Strausberg and Hough 1997; Bass 1981). From 1879 until the end of World War II, commercial logging occurred extensively throughout the region. Most logging companies gave no thought to conserving forest resources, leaving cutover areas denuded of vegetation and often severely eroded (Smith 1986). Small farmers added to forest decline by cutting down and selling high-valued trees to supplement their income, repeatedly burning forests to clear land or improve livestock grazing, and then abandoning farm areas to erode once the thin forest soils played out (Strausberg and Hough 1997; Bass 1981). The combination of all of these abusive land-use practices left much of the Interior Highlands' forest in poor condition (Record 1910) and led, in part, to the creation of the National Forests within the region.

Fig. 15.2 Shortleaf pine stand ca. 1931 in Irons Fork Experimental Forest. We speculate that the forest conditions shown here are probably similar to those that existed in typical stands prior to extensive commercial logging in the Ouachita Mountains from 1890 to 1945. The mixed ages, fairly open understory, and lack of dense shrubs suggest frequent, low-intensity burning and natural regeneration. The age of the overstory trees in this photograph predate the beginning of commercial logging in the area

15.3.2 Forest Management Goals and Public Concerns

The first half of the 1900s was a period when growing national concerns for conserving forest soil and water resources were both embraced by some and resisted by others. Within the Interior Highlands, citizens held a great many diverse, and often opposing, attitudes about the best use of public lands and forests, and about the value of the soil and water resources contained within these forests.

National concern for sustaining water supplies from forests was one of the motivations that led to the first laws in the late 1890s establishing the forest reserves that would become the National Forest System (Steen 2005). Recognition of the role that forests play in regulating downstream river flows was the basis of the Weeks Act of 1911 that permitted federal purchase of private lands for national forests in the East. Controversy over the role that forests play in regulating streamflow peaked after the great flood of 1927 in the lower Mississippi River valley, and motivation grew to better understand this role (Douglass and Hoover 1988).

Partly in response to regional forest depletion, the Ouachita and Ozark National Forests were created in 1907–1908. From 1919 to 1941, hundreds of thousands of acres of land, most of it cutover and burned (Bass 1981), were purchased and added to the Ouachita and Ozark National Forests. Forest managers' emphasis on National Forest lands was more on protection and rehabilitation than harvesting timber for its economic value (Strausberg and Hough 1997). By the early 1920s, private forest landowners in the Interior Highlands had begun to change their attitude as well, and began to adopt the idea of managing their lands to sustain timber production (Smith 1986).

Recognizing the damage from past logging and repeated, uncontrolled burning, foresters prescribed methods to rehabilitate forests, regenerate growth, encourage desired species, and restore the forest to a healthy productive state (Smith 1986). Shortleaf pine (*Pinus echinata* P. Mill.) was the desired species, due to its higher economic value, and even-aged management was considered the best way to produce the greatest annual yields (Mattoon 1915).

For generations, small landowners had used the Interior Highlands forest as open lands from which they could freely take timber and over which they could graze their livestock. Initial support for the establishment of the Ouachita and Ozark National Forests was replaced by resentment and vandalism when neighboring landowners realized such practices would now be illegal (Strausberg and Hough 1997). In the generally poor economy of the times, "job fires" were a continuing problem as locals deliberately set the forests ablaze so that they would be hired to put out the fires (Bass 1981). Others advocated that the newly created national forests be made into national parks to better support a nascent tourism industry (Strausberg and Hough 1997).

15.3.3 Interactions and Research Response

In 1921, the Southern Forest Experiment Station was established by the Forest Service in New Orleans, LA, with an area of responsibility that included the Interior Highlands. Initial scouting to identify a suitable location for an experimental forest on the Ouachita National Forest occurred in 1931. However, it was not until 1936, and after repeated urging from the Forest Service Branch of Research in Washington, DC (Munns, personal communications 1932a,1932b,1936), that work began in what would become the first experimental forest in the Interior Highlands—the Irons Fork Experimental Forest, located near Mena, AR.

An ambitious work plan was implemented in 1936 in which research studies were initiated immediately while support infrastructure, such as laboratory, residence, and work buildings, were being constructed onsite. Using Works Progress Administration (WPA) and later Civilian Conservation Corps (CCC) work crews, work commenced in 1936, but it was not until 1940 that the 3,600-ha (9,000-ac) Irons Fork Experimental Forest was officially established.

Though forestry research was the initial impetus behind establishing the Irons Fork Experimental Forest, scientists realized prior to 1935 that watershed research was just as pressing a need (Meginnis 1936). Scientists at the time were just

Fig. 15.3 Structure installed in Irons Fork Experimental Forest to divert streamflow from ephemeral channel to adjacent hillslope where it was spread out and reabsorbed into the soil. The objective was to reduce the amount of and speed at which streamflow from headwater streams moved downstream

beginning to understand how forests and water interact. With the charge to rehabilitate the forests of the Interior Highlands came the realization that no one knew with confidence what effect the proposed forestry practices would have on soil and water processes. Furthermore, basic information was needed on such things as whether existing forest cover was working effectively to control runoff from rainfall, and what roles did surface runoff, shallow seepage, and ground water drainage play in generating flood flows within Interior Highland watersheds.

Research on the experimental forest was conducted from 1936 to mid-1943 and involved a number of studies assessing all aspects of forest watershed science. Plot-scale studies were undertaken to measure the proportion of overland flow from the forest surface within four different condition classes, and to assess the subsequent effect of fire and litter removal on surface runoff and soil erosion. A paired-watershed study was implemented to measure the effects of different forestry practices on streamflow behavior and soil erosion. Another study was devised to evaluate the effect of different engineering structures on reducing downstream peak flows from headwater streams (Fig. 15.3). Road bank erosion and stabilization techniques were assessed in yet another study. To quantify elements of the hydrologic cycle, basic measurements were initiated of precipitation, plant transpiration, soil water content, and water-table depths.

Fig. 15.4 Columbus deep notch weir ca. 1936 built by WPA and CCC labor and used to measure streamflow in Irons Fork Experimental Forest

To accomplish this ambitious research program, crews consisting of either WPA personnel or CCC enrollees were engaged to construct 177 km (110 miles) of trails to access all the study plots, string 18 km (11 miles) of telephone lines, and install over 100 rain gauges, 2 full weather stations, and 2 lysimeters. These men installed rock masonry water control structures, lined stream channels above and below the controls with riprap, and completed a 4.6-m weir used to measure streamflow (Fig. 15.4). The weir alone accounted for 900 bags of concrete. All of this work was conducted from late 1936 to March 1942.

Engaging with the public and management personnel on the Ouachita National Forest was an integral part of the research staff's work. The lead researcher was expected to give talks to local groups, contribute regular articles for the Mena newspaper, and prepare an educational brochure for the public. "Show-me" trips were organized for agencies such as the Arkansas State Planning Board, Water Resources Committee, and Flood Control Commission. Other trips involved Forest Service personnel from the Washington, DC, and regional offices. On April 29, 1941, a Forestry Study Day attended by over 125 individuals was held with the assistance of the Arkansas Extension Service. Individuals attending included the state forester, members of the logging industry, the Soil Conservation Service, and members of the Mena Lions Club.

15.3.4 Research Findings

Despite its energetic start and the clear, pressing need for the information promised by its research, the work at Irons Fork Experimental Forest was never completed. Funding was always a problem. All staff with the exception of the assistant forester was funded from either CCC or WPA funds. As the flow of those funds slowed, the work at the experimental forest was modified or curtailed. In 1939, the CCC program was cut back. By the summer of 1940, the WPA workforce was cut from 42 to 16 men, and in the following year those 16 slots were lost. With the outbreak of World War II (WWII), the situation for Irons Fork went from bad to worse. During the war, both the CCC and WPA programs were ended. The Forest Service staff was reassigned to war duties or, in the case of the project leader, died. Measurements on all watersheds, weather stations, and other field studies ended on July 1, 1943. No other research was conducted on the Irons Fork Experimental Forest after that time. Proposals to restart the research at Irons Fork continued periodically through the late 1950s and early 1960s, but were never implemented. The Irons Fork Experimental Forest was officially disestablished and its lands returned to unrestricted National Forest status in 1969.

While the premature closure of the Irons Fork Experimental Forest precluded completion of the watershed studies, the effort did produce some useful knowledge. Interception data measured at Irons Fork were used by Helvey and Patric (1965) and Helvey (1971) to determine throughfall and stemflow rates for southern hardwoods and conifers, respectively. Practical knowledge was gained concerning how best to measure processes such as plot runoff and streamflow. Initial assumptions about how long it would take to establish relationships between precipitation inputs and streamflow outputs (2 years) were quickly disproved. Later research within the Interior Highlands would benefit from this knowledge.

15.4 Reestablishment and Renewal (1960–1980)

15.4.1 Forest Conditions

The period of WWII and its aftermath saw an interruption in both watershed research and active national forest management. Participation in the war effort and elimination of prewar programs like the CCC and WPA had greatly reduced Forest Service replanting efforts and fire protection within the Interior Highlands. As the postwar economy developed and the workforce returned to the woods, they encountered forests exhibiting reduced productivity and increased fire vulnerability (Strausberg and Hough 1997).

15.4.2 Forest Management Goals and Public Concerns

As the decade of the 1960s began, forest managers saw their primary goal to be increasing forest productivity (Strausberg and Hough 1997). This desire was driven by two expectations: (1) that lumber demand would continue to increase nationally and (2) that new areas would not be converted to forest production. To meet this goal, forest managers sought to reduce forest mortality through suppression of fire, pest, and disease occurrences, and to "improve" stand quality. The most efficient way to improve stand quality was thought to be through the use of even-aged forestry practices that utilized clearcutting, intensive site preparation using herbicides, and replanting with preferred species (shortleaf or loblolly pine; Strausberg and Hough 1997). New harvesting equipment was developed to accomplish this work at reduced cost, but being heavier, this equipment created a greater potential for increased soil erosion (Bass 1981).

The period of 1960–1980 saw a transformation of the public's role in forest management, both nationally and within the Interior Highlands. The period began with passage of the Multiple-Use Sustained-Yield Act of 1960 by Congress, which mandated equal consideration of nontimber resources like soil and water. Over time in this period, the voice of the public grew as concerns about forest soil and water resources were increasingly expressed and argued.

Both the national and regional concern over the role of forest management in reducing floods carried over from the previous period, as evidenced by passage of the Watershed Protection and Flood Prevention Act in 1954. This act mandated that the Forest Service cooperate with state and other federal agencies on flood control (Strausberg and Hough 1997, p. 22). In addition, local communities were very concerned about managing forests to ensure adequate streamflow for recreation and municipal water supplies in the Interior Highlands.

Use of the forest for livestock grazing exemplifies the conflicting attitudes during this period. Livestock owners' desired access to national forest lands for grazing, a common practice in pre-WWII times, but forest managers resisted over concerns about livestock eating or destroying pine seedlings. Hunters also protested that grazing decreased wildlife food sources and spread disease (Strausberg and Hough 1997).

In the 1970s, new federal laws (e.g., the National Environmental Policy Act and what would later become the Clean Water Act) provided still more avenues by which public concerns and values regarding the region's forests could be expressed. Individuals and interest groups increasingly decried the use of clearcutting, which they found aesthetically displeasing and environmentally unsound (Strausberg and Hough 1997). Objections to herbicide use were equally strong (Bass 1981).

15.4.3 Interactions and Research Response

Research scientists in the Forest Service, and later those in universities and in private industry, renewed their efforts to provide answers to the questions and concerns raised by forest managers, users, and the general public. These scientists, like their predecessors, were trained as foresters, and they employed a multidisciplinary approach to their research. Studies that examined how soil and water resources responded to given forestry practices often contained a component evaluating whether the vegetation response to those practices produced the desired productivity improvements. Another feature of the watershed research was a continuing effort to determine basic characteristics (e.g., nutrient amounts within undisturbed soils and streams) and process rates (e.g., precipitation intensity, runoff magnitude, and volume), in addition to how these characteristics and rates changed in response to harvesting. Moreover, these scientists made a fundamental decision: They chose to focus on very small basins that lacked well-developed stream channels (i.e., stream paths had no pronounced incision or features like bars or bedforms) so that the responses they observed could be logically inferred to result from process changes on the surrounding hillslopes and not from in-channel processes such as bank erosion or bed scour.

Recognizing that important environmental differences existed within the Interior Highlands, new experimental forests and watershed research areas were established as representations of intraregional differences. The Henry R. Koen Experimental Forest was established in 1948 within the Springfield-Salem Plateau subregion (see Fig. 15.1). The Koen Experimental Forest is a 4,400-ha area of the Ozark National Forest located near Jasper, AR, in the headwaters of the Buffalo River and covered by a hardwood forest consisting of oaks (*Quercus* spp.), hickories (*Carya* spp.), and white ash (*Fraxinus Americana* L.). In 1959, the Alum Creek Experimental Forest was created within the eastern Ouachita Mountains in the headwaters of the Saline River. Located within the Ouachita National Forest near Jessieville, AR, the Alum Creek Experimental Forest is about 810 ha and is predominantly covered by a mixed pine-hardwood forest type with shortleaf pine, white oak (*Q. alba* L.), and assorted hickories being the dominant species. Late in this period, yet another research area was established in the western Ouachita Mountains on Weyerhaeuser Company lands near Battiest, OK (Fig. 15.1). While not a Forest Service experimental forest, the Battiest research area was used in the same way as the Koen and Alum Creek Experiment Forests. Vegetation cover within the Battiest area was similar to that in the Alum Creek Experimental Forest.

The watershed studies implemented in all three of these research areas were similar in that they used essentially the same experimental design, the same monitoring equipment, and the same type of research sites. The experimental design utilized measurements taken both before and after a change or "treatment" (e.g., clearcutting followed by herbicide application) was imposed on the research sites, and simultaneous measurements taken at both "control" (i.e., undisturbed) sites and "impacted" sites. Instrumentation at each area included precipitation and streamflow gauges, and weather sensors (Fig. 15.5). The type of research sites used were very small watersheds (0.4–3.2 ha in all cases) located at the heads of small, ephemeral streams. Slopes were moderately steep (15–30 %), forest cover was continuous, and no roads were constructed within the watersheds. Each site was assigned a different treatment which was applied across the entire watershed. What differed between research areas was the underlying geology, the forest type, or the specific management treatments used in each study.

Fig. 15.5 Monitoring installation used to measure streamflow amount and collect water samples on Watershed 2 in the Alum Creek Experimental Forest. The H-flume (device with triangular opening in *lower center*) is used to measure the depth of water, which can then be mathematically converted into the streamflow rate (volume per unit time). A Coshocton wheel (device offset from flume outlet) is used to capture 1 % of the streamflow, which is stored in a container for later analysis. Similar installations were used at sites in Koen Experimental Forest and Battiest research area. Photograph was taken prior to the upstream basin being cut and shows typical conditions after 40+ yr of fire suppression: high density of overstory trees and thick shrub and understory growth

15.4.4 Research Findings

Results from the early studies at the Battiest research area and the Koen and Alum Creek Experiment Forests provided important insights into both natural processes and their response to forestry practices. Several of the basic components of the hydrologic cycle in forest ecosystems were documented for the first time in the Interior Highlands. They included streamflow amounts and rate, canopy interception and throughfall, soil water storage, and nutrient concentrations in streamflow (Lawson 1967; Rogerson 1971; Lawson and Hileman 1983). Despite differences in the environment and forestry practices used, these studies showed that small watersheds in the Interior Highlands responded in similar ways. Water yields generally increased

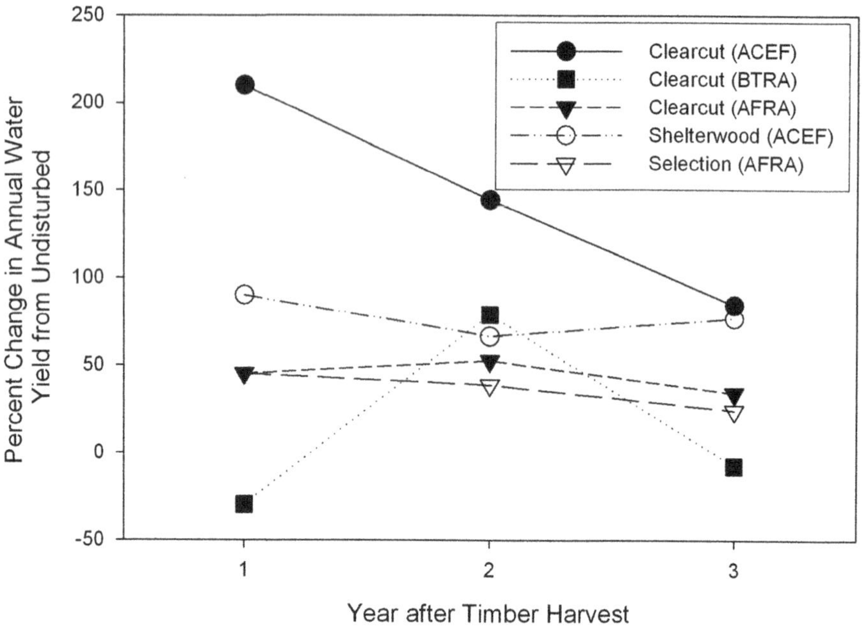

Fig. 15.6 Percent change in annual water yield after timber harvesting within Interior Highland watersheds. Percent change is based on the difference between paired undisturbed watersheds at each location for the same year. Location codes and data sources used: *ACEF* Alum Creek Experimental Forest (Rogerson 1985); *BTRA* Battiest research area (Miller 1984); *AFRA* Alum Fork research area (Miller et al. 1988)

in the 1st year after harvest, but returned to near undisturbed levels by the 3rd year (Fig. 15.6), unless revegetation was suppressed through herbicide applications (e.g., Rogerson 1985). Sediment production also consistently increased in the 1st year following harvest, but returned to undisturbed rates even more quickly than water yield (generally by the 2nd year; Fig. 15.7). Somewhat less consistent was nutrient response. Most nutrient concentrations in streamflow were unchanged by harvesting. The exceptions were nitrogen and phosphorus, which did change, but in variable ways that make the relation to silvicultural methods unclear (Lawson 1985).

15.5 Partnerships and Expansion (1980–1990)

15.5.1 Forest Management Goals and Public Concerns

The 1980s opened with state agencies and the public frequently challenging Forest Service decisions on how best to manage national forests in the Interior Highlands. Nonpoint source (NPS) pollution, particularly from sediment and herbicides, had

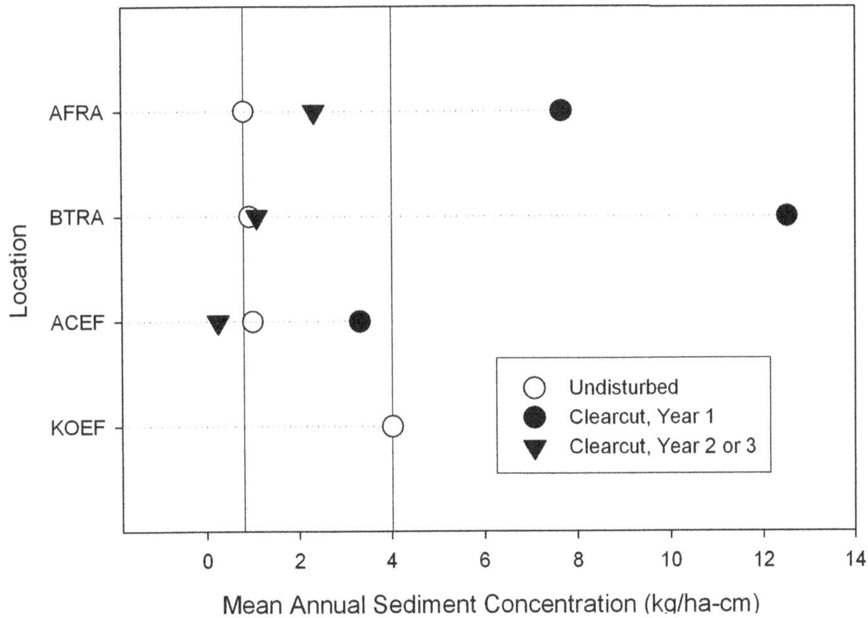

Fig. 15.7 Changes in mean annual sediment concentration after clearcut harvest within Interior Highland watersheds. Only clearcut harvest is shown as it produces the greatest change. *Vertical reference lines* enclose range of observed undisturbed sediment concentrations. Mean annual sediment concentration is computed from the mean annual sediment yield (kg/ha) and mean annual water yield (cm) for all similarly treated watersheds at each location (see Marion and Ursic 1993 for explanation). Same location codes as in Fig. 15.6. (Data sources: ACEF, Rogerson 1985; Lawson 1985; BTRA, Miller 1984; AFRA, Miller et al. 1988)

become a major public concern. Passage of the Clean Water Act in 1985 mandated that states identify their polluted waters and the land uses that were contributing to NPS pollution; yet little information was available about how such pollution was affected by forestry practices. Continued and growing objection to the frequent use of clearcutting and associated herbicide use led to numerous appeals of Forest Service actions through the 1980s (Strausberg and Hough 1997). Another issue of growing concern was acid rain and its effects. Massive fish kills in the northeastern USA had resulted from the acidification of water by dilute sulfuric and nitric acid entering the lakes through rain and snowfall. Both sulfate and nitrate are produced by coal-fired power plants, oil refining, and vehicle engines. At this time, no one knew whether or not acid rain fell within the Interior Highlands, but it was suspected due to its location downwind of major metropolitan and oil-refining regions in Texas, Louisiana, and Oklahoma.

Public dissatisfaction with clearcutting forced Forest Service managers to reexamine uneven-aged forestry practices as an alternative to even-aged methods. Little research was available at the time on how uneven-aged practices affected soils and streams in the Interior Highlands as studies in the prior period had focused on even-aged methods. Private landowners and the forest industry decided to continue their reliance on even-aged methods, but they, along with state forestry agencies, wished to demonstrate that such methods could be used and still meet NPS requirements. Thus, federal, state, and private forest managers needed new information from research.

15.5.2 Interactions and Research Response

This period saw an overall expansion in research effort and important changes in how research was accomplished. Perhaps the most important change was in the marked increase in partnerships between Forest Service, academic, and industry scientists to address mutual research needs in coordinated studies. All parties recognized that working together on joint studies would be more economically and logistically efficient, and increase the likelihood of public acceptance of their research results. This cooperation also would permit the number of sites used for paired-watershed studies to be increased to obtain greater statistical power. Another change was the decision to increase the size of research watersheds from that used in the past. Whereas the previous period focused on basins of 0.4–3.2 ha, new studies would evaluate 4.0–12.1-ha basins to better match with the size of harvest units being used. Yet another important change was in the addition of process-based studies to complement traditional paired-watershed investigations. These process-based studies would measure the amounts and movement of water and its constituents within a given watershed rather than just monitoring what came out at the outlet. And, once again, experimental forests would play a key role in developing the information to answer the questions posed by both the public and forest managers.

To compare the effects of clearcutting (even-aged) and selection harvest (uneven-aged) practices on water quantity and quality, a cooperative watershed study was initiated by the Forest Service, University of Arkansas at Monticello, and Weyerhaeuser Company in 1978. Oklahoma State University joined the effort in 1988. This study utilized nine small watersheds ranging from about 4.0 to 6.1 ha in size (Fig. 15.8). Six of the watersheds were in the Alum Creek Experimental Forest while the other three were on industrial lands a few kilometers to the north near Cedar Mountain (Fig. 15.1). Data were collected for a year prior to harvesting to establish pretreatment behavior, and 5 years afterwards to evaluate the forestry effects.

The Boston Mountains subregion had yet to receive attention from watershed researchers. This knowledge gap was addressed in another research partnership, this time between scientists with the Forest Service and University of Arkansas at Fayetteville. In 1972, a new paired-watershed study was established at the Fleming Creek research area near Brashears, AR. Through an agreement with the Ozark National Forest, four small basins, 6.1–13.4 ha each, were set aside to study the hydrologic response to forestry practices intended to increase pine productivity in a mixed pine–hardwood forest. Though started prior to 1980, the Fleming Creek study fits best with research of the later period because of the size of the watersheds used.

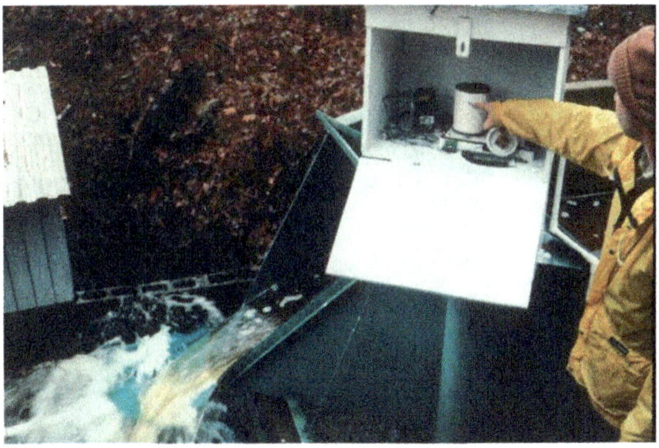

Fig. 15.8 Monitoring installation used to measure streamflow amount and collect water samples on Watershed 11 in the Alum Creek Experimental Forest. Device being pointed to records the water depth on a paper chart while the device immediately to the left records the same data in a digital file

Each basin was instrumented to monitor precipitation, streamflow, water chemistry, and sediment. A meteorological station was installed to measure precipitation, air temperature, and barometric pressure. Clearcutting, thinning, and thinning with herbicides were all tested along with natural regeneration and pine seedling planting. Forestry treatments were applied in 1982 and data collection completed in 1992.

The possible occurrence and effects of acid rain were assessed by another study using the Alum Creek Experimental Forest. The Ouachita and Boston Mountains were both thought to be especially vulnerable to acid rain because the soils and bedrock are naturally acidic and lack acid-buffering minerals such as calcium carbonate. Did acid rain fall in the Interior Highlands? Was there a potential for water and soil acidification in the region? If so, what would be the effect on aquatic life, drinking water supplies, and forest ecosystems? To start answering these questions, a number of small studies were installed and carried out on the Alum Creek Experimental Forest. The Alum Creek Experimental Forest was an ideal place to do this work because information on the geology, soils, vegetation, climate, and hydrology were already available. The small watersheds also allowed investigators to track changes in water chemistry as it entered the watersheds as rainfall, passed through the vegetation and soils, and became streamflow. Work was performed utilizing both undisturbed (control) watersheds as well as the managed watersheds.

Process-based studies were used to better understand the biogeochemical cycle and the relationship of rainfall to streamflow. Past research in the region demonstrated that streamflow occurs very quickly following the onset of rainfall. Past work also indicated that timber harvesting did not increase the streamflows resulting from large storms that produce floods. Since surface runoff outside of streambeds was only observed to occur on highly disturbed areas such as roads and landings, the most likely mechanism by which rainfall could move through watersheds and into streams of the region was by lateral (i.e., subsurface) flow through the soil. In 1989, two stations were installed in an undisturbed watershed on the Alum Creek Experimental Forest to measure subsurface flow and water chemistry. Collectors placed across the slope and at different depths in the soil profile intercepted

Fig. 15.9 Installation for measuring sediment production from a forest road. Equipment is the same as that shown in Fig. 15.5 and 15.8, but has an extended approach section between the flume and the road which collected the bed load sediment eroded from the road. Total sediment was the combined amount deposited in the approach section and the suspended portion that flowed over the flume and was measured using the Coshocton wheel samples

subsurface flow and routed it to flumes where flow rates were measured and water samples for chemical analysis were collected.

Forest roads were recognized as having great potential to erode and deliver sediment to streams. In another process study, Forest Service, University of Arkansas at Monticello, and Weyerhaeuser Company scientists measured sediment production over 17 months in 1982 and 1984 along four road segments in the Bread Creek watershed (Fig. 15.9). The Bread Creek watershed lies adjacent to the Alum Creek Experimental Forest, and would later become part of a larger research area used in the next research period. The measured data were then combined with an inventory of road lengths and conditions throughout the basin to estimate total sediment delivery from roads to the stream system.

15.5.3 Research Findings

The cooperative watershed study showed that 4.0–6.1-ha basins responded much like 0.4–3.2-ha basins. The response patterns were the same: Export of streamflow, sediment, and important nutrients (nitrogen and phosphorus) all increased in the first 1–3 years following harvesting, but then all returned to undisturbed levels as vegetation became reestablished (Figs. 15.6 and 15.7). The study also revealed some new insights into forestry effects. Moderate storms produced higher peak streamflow rates in the postharvest years, but peaks from large storms were unaffected by harvesting. Large storms apparently produce so much water that forest soils become saturated regardless of whether the site was harvested or not (Miller

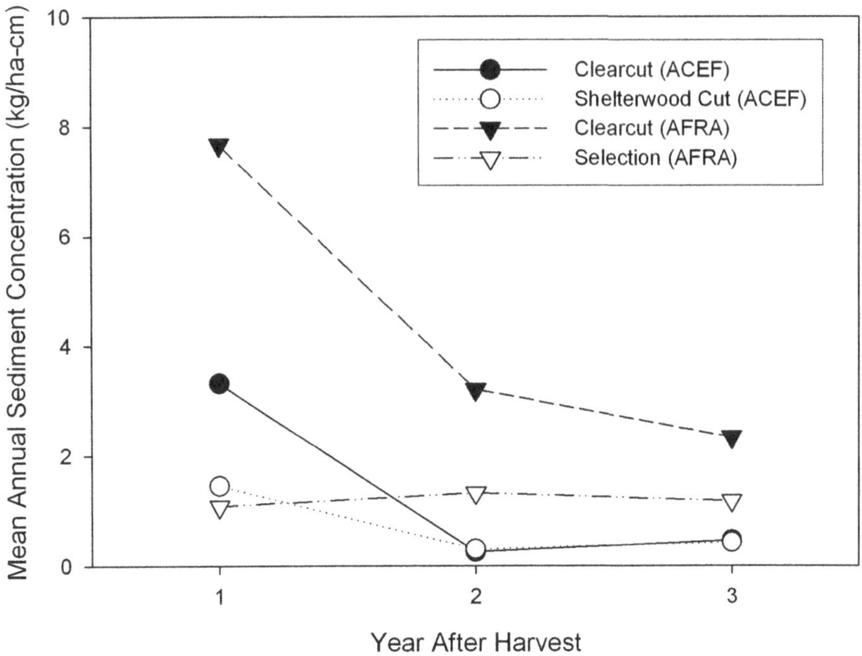

Fig. 15.10 Comparison of mean annual sediment concentrations after harvest for different harvest methods. See Fig. 15.7 for how concentrations were computed, location codes, and data sources

et al. 1988). Another insight was that sediment increases corresponded to the degree of vegetation removal, with clearcuts producing greater increases at a given location than shelterwood or selection cuts (Fig. 15.10). However, it was also clear that none of the forestry practices resulted in sediment increases comparable to those produced by agricultural uses like row crops or pasture. Data from the cooperative watershed study also showed that since nitrogen and phosphorus increases were approximately equal to inputs from the atmosphere, the observed increases reflected the short-term change in vegetation uptake, not actual nutrient depletion of the soils.

Data from studies within the Alum Creek Experimental Forest, Fleming Creek research area, and elsewhere in the Interior Highlands demonstrated that acid rain did occur within the region (Wheeler et al. 2000; Kress et al. 1990; Beasley et al. 1988). However, comparison to nutrient pools measured at the Alum Creek and Fleming Creek sites indicated that excessive leaching of nutrients did not occur. The analyses also showed that nutrient losses following various forestry practices would not deplete soil nutrients or significantly reduce soil productivity.

Subsurface flow research at Alum Creek Experimental Forest revealed much about how water moves through forest soils in the Interior Highlands. This work found that subsurface flow is generated within a few minutes following the onset of rainfall, with flow rates rising and falling in close synchrony with streamflow, and that subsurface flow accounted for all of the flow from the hillslopes.

The rapidity of subsurface water delivery to streams and the correspondence in streamflow chemistry with that of subsurface flow demonstrated the critical role that macropores play in forest hydrology. Macropores are flow conduits within the soil greater than 1 mm in diameter which are formed by old root channels and insect, earthworm, and animal burrows. Macropores form preferential flow paths through which water and its constituent chemicals can travel quickly to streams without being absorbed or transformed by the soil. Studies carried out on the Alum Creek Experimental Forest demonstrated that the potential movement of chemicals like herbicides through macropores to streams was significant under certain conditions (Turton et al. 1995). Thus, the fact that during storms macropores in forest soils can rapidly deliver water and dissolved contaminants to Interior Highland streams is an important consideration for forest managers deciding whether or how to apply fertilizers and herbicides.

Results from the Bread Creek road study were particularly important. This was the first and only study to date that has measured sediment production from forest roads within the Interior Highlands in Arkansas. Prior to this study, road sediment production had been estimated to average almost 390 t/road km/yr for this area, but these estimates were made without the benefit of actual measured data. This study found that road sediment production for the same area was actually 41 t/road km/yr and sediment delivered to the stream system was about 4.5 t/road km/yr (Miller et al. 1985). This study clearly demonstrated that while roads were a potential problem, the magnitude of the problem was far less than was initially thought.

15.6 New Scales and Paths (1990–present)

15.6.1 Forest Management Goals and Public Concerns

The current period of watershed research in the Interior Highlands began with the national public continuing to express frequent dissatisfaction with Forest Service management decisions. Clearcutting was particularly disliked as more people came to value the scenic qualities of their national forests (Robertson 2004). Interest groups increasingly challenged in the courts the tradeoffs inherent in the multiple-use strategy to forest management, using environmental laws like the Endangered Species Act, the National Forest Management Act, and the National Environmental Policy Act. Public objections also led to Congressional inquiries. These factors, plus new thinking among researchers (Franklin 1989), caused the Forest Service to rethink its approach to forest management. From this reassessment came the new management paradigm of "ecosystem management."

Those concerned about national forest management in the Interior Highlands were particularly effective in communicating their opinions through their legislators. US senator David Pryor of Arkansas specifically requested that the Forest Service find ways to eliminate use of clearcutting on the Ouachita National Forest. This request led to the famous "walk in the woods" meeting on the Ouachita National Forest in August 1990 between Senator Pryor, Forest Service Chief Dale Robertson, local Forest Service managers, and research scientists (Robertson 2004). In searching for a compromise acceptable to both the public (represented by Pryor) and the managers, input from scientists regarding alternatives to clearcutting proved significant. An agreement was reached that new harvesting methods would be used on the Ouachita National Forest, and that research would work closely with management to assess how these methods affected other resources like water (Robertson 2004). This initial decision quickly evolved into a 1992 Forest Service-wide directive to do the same throughout the NFS.

Adoption of ecosystem management meant using natural regeneration to restock forests and uneven-aged forestry methods to harvest timber (Guldin 2004). Use of controlled burning would increase, both to replace herbicide use for competition control during forest regeneration and to restore fire as a natural process in the forest. Moreover, the new management paradigm abandoned the past approach of making stand-based decisions for one that encompasses larger areas. It was evident that knowledge gained in the previous periods from small-watershed studies would be insufficient in answering questions about how watersheds would respond at larger spatial scales and to multiple disturbances over time. The effects of this new "landscape" scale of management were largely unknown; thus, a clear need existed for new research.

15.6.2 Interactions and Research Response

The degree of direct interaction between the public, forest managers, and scientists in this period was unprecedented. Through numerous formal and informal meetings among these groups, a program of research work emerged to address the new information need. From the start, new watershed research was recognized as a primary need. A team was formed consisting of scientists from the Forest Service, universities, and private industry, along with hydrologists and soil scientists from the Ouachita National Forest. Bigger scales required bigger research areas. With the cooperation from the Ouachita National Forest, the entire Alum Fork basin upstream of Lake Winona (and including the Alum Creek Experimental Forest) was set aside for use in new research studies (Fig. 15.1). Within this expanded area, three basins became the focus of watershed research: (1) the South Alum Creek watershed, which drained most of the Alum Creek Experimental Forest; (2) North Alum Creek; and (3) Bread Creek. These watersheds, each roughly 600–1,500 ha in area, were selected as good candidates for ecosystem management practices with South Alum serving as a relatively undisturbed situation, Bread Creek as the example of past stand-based forestry, and North Alum as capable of being converted to a historic ecosystem (shortleaf pine-bluestem grass, *Andropogon* spp.) that had disappeared after decades of fire suppression. Furthermore, the Weyerhaeuser Company, an active player in this research, designated one of their nearby watersheds, the Little

Fig. 15.11 Phase III monitoring station on South Alum Creek showing footbridge for high-flow sampling and container for computer-controlled pump sampler (*lower right*). Personnel are measuring streamflow using a sounding weight (*red device* below bridge) and current meter (obscured by vegetation)

Glazypeau Creek basin (Fig. 15.1), to represent a fourth option, industrial forest management. An extensive network of monitoring stations was established within these four watersheds to measure precipitation, air temperature, streamflow, and water temperature (Fig. 15.11). Data collection began in 1996, harvest treatments were applied in 1999–2000, and prescribed burns conducted in 2002. In addition to watershed studies, these four basins also serve as the "core" research area for numerous ongoing studies of geomorphic, aquatic, floral, and faunal studies associated with Phase III of the Ecosystem Management Research Program in Arkansas; thus, the basins are together referred to as the "Phase III Watersheds."

The defining feature of this new research period was the consideration of new and larger scales. Research in the previous periods had focused on the small watershed scale, first at 0.4–3.2 ha then at 4.0–13.4 ha. This new work would evaluate treatments applied to 400 ha (1,000 acres) or more, and assess responses at several scales (e.g., 200 ha, 600 ha, and larger). For the first time in Interior Highlands' research, how responses change as one moves downstream would be examined, providing new insight into how scale affects soil and water processes. At these larger scales, new questions arose. In addition to evaluating streamflow amounts and soil and water quality attributes, new studies were undertaken to assess natural disturbance frequencies (e.g., ice storms and tornados), bed load transport properties, and road conditions.

15.6.3 Research Findings

Findings from this latest research period have only begun to be published, but early results are already having an effect. Whereas results from the previous periods led to developing general guidelines or "best management practices" for limiting negative impacts, new findings from this latest period are allowing more effective application of these practices. One example is a new model for predicting peak flow magnitudes from small watersheds (Marion 2004) which permits more accurate sizing of drainage structures. Another is a study on the effect of aerial fertilizer

Fig. 15.12 Communicating research results to participants in 2006 road erosion modeling workshop. This location was one of the monitoring sites used for the Bread Creek road erosion study during 1982–1984

applications on stream chemistry in an industrial forest which is improving streamside buffer designs (Liechty et al. 2006). Workshops have been offered to more quickly communicate these and other improvements in operational methods and thinking to resource managers and the public (Fig. 15.12).

Results from other studies are expanding our understanding of how soils, streams, and forests coevolved in the Interior Highland forests over time. Bed load transport rates have been quantified for the first time in small forest streams within the Interior Highlands, illustrating a complex relationship between bed load and the channel bed during peak flow events (e.g., Marion and Weirich 2003). Unexpected changes in nutrients have been shown in studies of the long-term effects of pine-bluestem conversion on Ouachita Mountain soils (Liechty et al. 2002). The importance of natural disturbance processes on watersheds and the underappreciated role that forest trees play in soil evolution in the region have been revealed (e.g., Phillips and Marion 2006).

15.7 Summary and Final Thoughts

Over 70 years of watershed research in the Interior Highlands has provided numerous benefits to scientists, forest managers, and concerned citizens. To scientists and others in their related disciplines, this body of research has helped clarify how water, sediment, and nutrients move through forested watersheds. Fundamental knowledge has been gained of the magnitude and flux of interception storage, throughflow, soil–water balance, streamflow, and sediment production throughout the year. The important influence of macropores on subsurface flow and chemical routing through forest soils has been demonstrated, and the important role of natural disturbance events and the primary role that trees play in soil development have both been elucidated.

To forest managers and the public, this research has established the impacts associated with different harvesting methods, and demonstrated that harvesting and site preparation produce only short-term impacts when carefully executed. Knowledge of relative impacts from different forestry practices, coupled with the conclusion that, in general, small watersheds responded in similar ways across the Interior Highlands, have been the basis for forest planning across the region. Research has directly addressed public concern about acid rain, demonstrating that it does occur over the Interior Highlands, but also showing that at present rates, acid rain does not cause excessive nutrients leaching from forested watersheds. Concern about road erosion rates has also been investigated, with research showing that, while certainly deserving of concern, road erosion rates were much less than predicted.

Much has been learned over the past 70 years about the soil and water resources in the forests of the Interior Highlands and how these resources respond to different forestry practices. Scientists have been challenged not only by the varied and dynamic landscape of the region but also by the changing opinions and desires of forest managers and the public. While such challenges are not unique to the Interior Highlands, the degree of advocacy by and interactions between interests groups has been particularly energetic here. In meeting these challenges, scientists have relied heavily on experimental forests as locations for observing how soil and water resources are constructed and how they work, and for manipulating the environment so that their reactions can be assessed. Unlike many other regions, no one experimental forest has served as the primary focus of research activity; rather a number of both permanent and temporary experimental forests have been used. Each has served its purpose to varying degrees. What seems certain is that experimental forests will continue to play a vital role in future watershed studies as scientists continue to build upon and add to this rich legacy of research in the Interior Highlands.

While much has been learned, there remains still more to do. The need for basic scientific investigations into how natural processes occur, operate, and interact will continue, as new understanding inevitably leads to new questions about how the parts work. Such work will logically produce better tools for predicting specific process rates like streamflow and road erosion. However, we think that research into how watershed components and processes interact at larger scales, and predicting their responses given multiple disturbances and condition states over time and space, will grow in importance in the decades to come. This need will be driven both by the desire for sustainable production of forest ecosystem services (e.g., wood products, clean air and water, recreation) and by the desire to restore altered or degraded ecosystems. This research frontier will likely demand new approaches and new concepts to deal with the multiple factors, states, and nonlinear dynamics that watershed systems exhibit. It will need new qualitative and quantitative models to better assess and integrate the critical interrelationships between variables, and increased use of spatial analysis methods and geographic information systems to investigate hydrologic responses at larger scales. However, it will also require continued use of experimental forests to investigate these questions and supply the needed answers.

References

Bass SMW (1981) For the trees: an illustrated history of the Ozark-St. Francis National Forests 1908–1978. U.S. Department of Agriculture, Forest Service, Southern Region, Atlanta, GA, 171 p

Beasley RS, Miller EL, Lawson ER, Stogsdill WR (1988) Contrasting acidic deposition and nutrient pools in an undisturbed forest and a young pine plantation in the Ouachita Mountains. Arkansas Science and Technology Authority. Proposal #86-A-008, 41 p

Douglass JE, Hoover MD (1988) History of Coweeta. In: Swank WT, Crossley DA Jr (eds) Forest hydrology and ecology at Coweeta. Springer, New York, pp 17–31

Foti TL, Glenn SM (1991) The Ouachita Mountain landscape at the time of settlement. In: Henderson D, Hedrick LD (eds) Restoration of old growth forests in the interior highlands of Arkansas and Oklahoma: proceedings of the conference, 1990 September 19–20, Morrilton, AR. Ouachita National Forest, Morrilton, Winrock International, Hot Springs, AR, pp 49–65

Franklin JF (1989) Toward a new forestry. Am For 95 (Nov/Dec):37–44

Guldin JM (2004) Overview of ecosystem management research in the Ouachita and Ozark mountains: phases I–III. In: Guldin JM (tech comp) Ouachita and Ozark Mountains symposium: ecosystem management research. General Technical Report SRS-74, U.S. Department of Agriculture, Forest Service, Southern Research Station, Asheville, NC pp 8–11

Helvey JD (1971) A summary of rainfall interception by certain conifers of North America. In: Biological effects in the hydrological cycle. Proceedings of the third international seminar for hydrology professors, 1971 July 18–30, Lafayette, IN. [No place of publication], [No publisher], pp 103–113

Helvey JD, Patric JH (1965) Canopy and litter interception of rainfall by hardwoods of eastern United States. Water Resour Res 1(2):193–206

Kress MW, Baker R, Ursic SJ (1990) Chemistry response of two forested watersheds to acid atmospheric deposition. J Am Water Resour Assoc 26(5):747–756

Lawson ER (1967) Throughfall and stemflow in a pine-hardwood stand in the Ouachita Mountains of Arkansas. Water Resour Res 3:731–735

Lawson ER (1985) Effects of forest practices on water quality in the Ozark-Ouachita Highlands. In: Blackmon BG (ed) Forestry and water quality: a mid-south symposium, 1985 May 8–9, Little Rock, AR. University of Arkansas, Department of Forest Resources, Monticello, AR, pp 130–140

Lawson ER, Hileman LH (1983) Nutrient distributions in runoff from Ouachita Mountain watersheds. In: Jones EP (ed) Second biennial southern silvicultural research conference, Atlanta, GA, 1982 November 4–5. General Technical Report SE-24, Southeastern Forest Experiment Station, Asheville, NC, pp 477–482

Liechty HO, Nettles J, Wilson S (2006) Impacts of multiple applications of fertilizer on stream chemistry in the Ouachita Mountains. In: Connor KF (ed) Proceedings of the 13th biennial southern silvicultural research conference. General Technical Report SRS-92, U.S. Department of Agriculture, Forest Service, Southern Research Station, Asheville, NC, pp 402–406

Liechty HO, Shelton MG, Luckow KR, Turton DJ (2002) Impacts of shortleaf pine-hardwood forest management on soils in the Ouachita Highlands: a review. South J Appl For 26(1):43–51

Marion DA (2004) An improved flood-frequency model for small watersheds in the upper Ouachita Mountains. In: Guldin JM (tech comp) Ouachita and Ozark Mountains symposium: ecosystem management research. General Technical Report SRS-74, U.S. Department of Agriculture, Forest Service, Southern Research Station, Asheville, NC, pp 207–213

Marion DA, Weirich F (2003) Equal-mobility bed load transport in a small, step-pool channel in the Ouachita Mountains. Geomorphology 55:139–154

Marion DA, Ursic SJ (1993) Sediment production in forests of the Coastal Plain, Piedmont, and Interior Highlands. In: Technical workshop on sediments. Proceedings of the EPA/Forest Service Workshop, 1992 February 3–7, Corvallis, OR, Terrene Institute, Washington, DC, pp 19–27

Meginnis HG (1936) Outline of the proposed work program under a 50-man WPA project to initiate forest influence investigations and to develop an experimental forest in the Ouachita Mountain region. On file at: U.S. Department of Agriculture, Mena Ranger District, Mena, AR

Miller EL (1984) Sediment yield and storm flow response to clearcut harvest and site preparation in the Ouachita Mountains. Water Resour Res 20(4):471–475

Miller EL, Beasley RS, Covert JC (1985) Forest road sediments: production and delivery to streams. In: Blackmon BG (ed) Forestry and water quality: a mid-south symposium, 1985 May 8–9, Little Rock, AR. University of Arkansas, Department of Forest Resources, Monticello, AR, pp 164–176

Miller EL, Beasley RS, Lawson ER (1988) Forest harvest and site preparation effects on stormflow and peakflow of ephemeral streams in the Ouachita Mountains. J Environ Qual 17(2):212–218

Munns (ed) (1932a) [Letter to E.L. Demmon] May 20. On file at: U.S. Department of Agriculture, Oden Ranger District, Oden, AR. File R-SS, Irons Fork, correspondence, 1931–1936

Munns (ed) (1932b) [Letter to E.L. Demmon] June 4. On file at: U.S. Department of Agriculture, Oden Ranger District, Oden, AR. File R-SS, Irons Fork, correspondence, 1931–1936

Munns (ed) (1936) [Letter to E.L. Demmon] June 10. On file at: U.S. Department of Agriculture, Oden Ranger District, Oden, AR. File R-SS, Irons Fork, correspondence, 1931–1936

Phillips JD, Marion DA (2006) Biomechanical effects of trees on soils and regolith: beyond treethrow. Ann Assoc Am Geogr 96(2):233–247

Robertson FD (2004) The history of new perspectives and ecosystem management. In: James M (tech comp) Ouachita and Ozark Mountains symposium: ecosystem management research. General Technical Report SRS-74, U.S. Department of Agriculture, Forest Service, Southern Research Station, Asheville, NC, pp 3–7

Rogerson TL (1971) Hydrologic characteristics of small headwater catchments in the Ouachita Mountains. Research Note SO-117, U.S. Department of Agriculture, Forest Service, Southern Forest Experiment Station, New Orleans 5 p

Rogerson TL (1985) Hydrologic responses to silvicultural practices in pine-hardwood stands in the Ouachita Mountains. In: Proceedings of the Fifth Central Hardwood Forest Conference, 1985 April 15–17, Urbana-Champaign, IL, pp 209–214

Mattoon WR (1915) Shortleaf pine: its economic importance and forest management. Bulletin 308, US Department of Agriculture, Washington, DC, 67 p

Record SJ (1910) The forest resources of Arkansas. Central Printing Co., Little Rock, 38 p

Sabo G III, Early AM, Rose JC et al (1988) Human adaptation in the Ozark and Ouachita Mountains. Research Series 31, Arkansas Archeological Survey, Fayetteville, 296 p

Smith KL (1986) Sawmill: the story of cutting the last great virgin forest east of the Rockies. University of Arkansas Press, Fayetteville, 246 p

Steen HK (2005) The forest service organic act. J For 103(5):234–236

Strausberg S, Hough WA (1997) The Ouachita and Ozark-St. Francis National Forests: a history of the lands and USDA Forest Service tenure. General Technical Report SO-121, USDA Forest Service, Southern Research Station, Asheville, NC, 45 p

Tucker GE (1991). Pre-settlement vegetation of the Ozark National Forest. In: Henderson D, Hedrick LD (eds) Restoration of old growth forests in the interior highlands of Arkansas and Oklahoma: proceedings of the conference, 1990 September 19–20, Morrilton, AR. Ouachita National Forest; Morrilton, Winrock International, Hot Springs, AR, pp 67–75

Turton DJ, Barnes DR, de J Navar J (1995) Old and new water in subsurface flow from a forest soil block. J Environ Qual 24(1):39–146

Wheeler GL, Steele KF, Lawson ER (2000) Water and nutrient movement in small, forested watersheds in the Boston Mountains of Arkansas. For Sci 46(3):335–343

Chapter 16
Research Related to Roads in USDA Experimental Forests

W. J. Elliot, P. J. Edwards and R. B. Foltz

Abstract Forest roads are essential in experimental forests and rangelands (EFRs) to allow researchers and the public access to research sites and for fire suppression, timber extraction, and fuel management. Sediment from roads can adversely impact watershed health. Since the 1930s, the design and management of forest roads has addressed both access issues and watershed health. Road design and management practices developed from research on roads in EFRs in the 1950s are applied throughout the USA and the world. Long-term data sets on watersheds with and without roads have helped us better understand the role of roads in runoff processes. Data collected from roads in EFRs have contributed to the development of hydrology and erosion models used throughout the world. As forest management and utilization practices change, such as gas abstraction, wind energy generation, and off-road vehicle recreation, research will be necessary to address the watershed impacts of roads and other access networks to support those new uses.

Keywords Forest roads · Erosion · Runoff · Watershed health · Forest access

16.1 Introduction

Our nation's forests provide many ecosystem services, including wood products (timber for construction and fuel); recreation (hiking, camping, fishing, and hunting); and gathering of wood, mushrooms, wild flowers, and other forest commodities. The public relies on a widespread road network in order to obtain these services. However, road networks can cause problems such as adverse impacts on water resources (erosion) and habitat fragmentation (Gucinski et al. 2001). Roads in experimental forests have played an important role in our understanding of these

W. J. Elliot (✉) · R. B. Foltz
Rocky Mountain Research Station, USDA Forest Service, 1221 South Main Street, Moscow, ID 83843, USA
Phone: 208-883-2338
e-mail: welliot@fs.fed.us

P. J. Edwards
Northern Research Station, USDA Forest Service, Timber & Watershed Laboratory, Route 219 North – Nursery Bottom, P.O. Box 404, Parsons, WV 26287, USA

D. C. Hayes et al. (eds.), *USDA Forest Service Experimental Forests and Ranges*, DOI 10.1007/978-1-4614-1818-4_16, © Springer New York 2014

Fig. 16.1 Building a road in Deception Creek Experimental Forest in the 1930s. (From photo collection of R.T. Graham)

problems and in identifying management practices to minimize them. From the very beginning of experimental forests, roads have allowed managers and researchers access to research sites, aided in fire suppression, and allowed the public to view ongoing research projects (Fig. 16.1; Wellner and Foiles 1951). Through research that started in experimental forests (Patric 1977), roads are now recognized as a major source of sediment in forested watersheds, with sediment from roads exceeding that from any other source in the absence of wildfire (Elliot 2010). This chapter contains examples of research related to roads in experimental forests: (1) using roads to aid in forest management, (2) identifying roads as a sediment source in forests, (3) predicting the effects of roads on runoff, (4) evaluating the effects of reopening roads that have been closed, and (5) developing predictive models.

Road design in experimental forests often consisted of two categories of roads—climbing roads and contour roads (Hewlett and Douglass 1968). Evidence of this design approach is still apparent in many experimental forests, such as Priest River and Deception Creek in Idaho, although many of the contour roads are now closed and overgrown (Fig. 16.2).

16.2 Background

16.2.1 Using Roads to Aid in Forest Management

Historically, the public needed timber for construction and fuel, and access roads were necessary to extract these products from forests. Between 1920 and 1970, one of the common logging tools in the northern Rocky Mountain Forests was the Idaho jammer, a cable logging machine that was only able to reach logs within

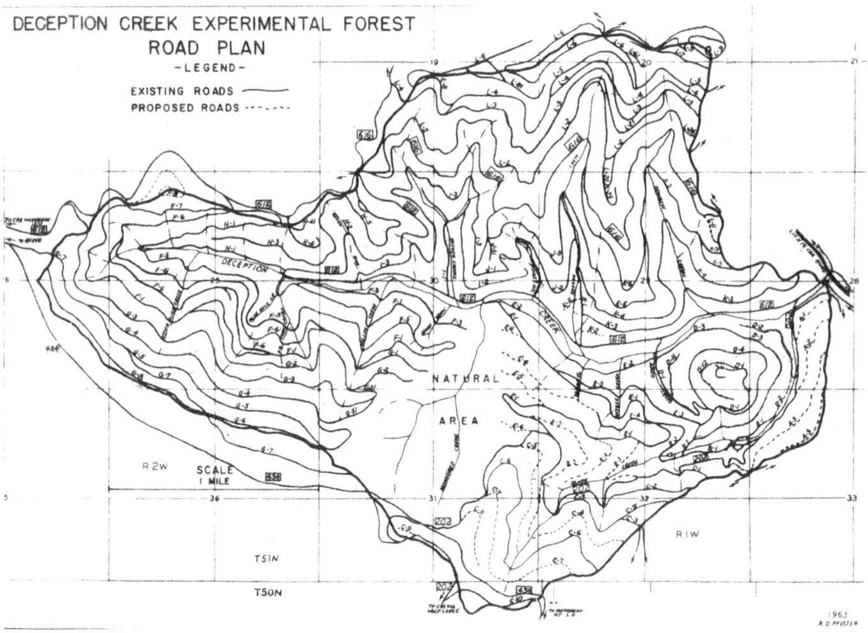

Fig. 16.2 High-density road network within the Deception Creek Experimental Forest drawn in 1963. Several of the proposed roads (*dotted lines*) were subsequently constructed to support forest management research. Note the unroaded research natural area that is still free from roads. (From R.T. Graham)

about 125 m of the road (Fig. 16.3; Stenzel et al. 1985). Yarder logging is necessary on slopes with greater than about 30% steepness, which is typical in many US forests. The Deception Creek Experimental Forest in northern Idaho was a model of a steep forest where the road network was designed so that every tree in the forest was within 250 m of a road (Fig. 16.2). Many of these roads were originally built in the 1930s with the Civilian Conservation Corps program or with funds from timber sales (Wellner and Foiles 1951). Wellner and Foiles (1951) noted that "Development of an intensive utilization road system has been stressed on the Deception Creek Experimental Forest since its establishment." The dense road network (more than 6 km of road in a square kilometer of forest) allowed researchers to carefully select and remove specific trees and evaluate the long-term effects of intensive forest management. The road network was considered a model for the National Forest Systems to follow for forest management (R. Graham, pers. comm., Sept. 2009). Within a decade, however, other managers concluded that not all forests could sustain such a high road density. The climate and geology of many western forests resulted in unstable road segments that led to excessive road failures through mass wasting. Current road densities in this region are about 2.5 km of road in a square kilometer of forest (Elliot 2009).

Modern yarders can reach much further than 250 m, so there is no longer need for such a high road density. In the coming decades, many of the mid-slope roads in

Fig. 16.3 An "Idaho jammer" cable yarder at work in Deception Creek Experimental Forest, December 1935. (From photo collections of D. Ferguson and R.T. Graham)

the Deception Creek Experimental Forest will likely be removed (Fig. 16.2). Watershed researchers are now working with silviculturalists to plan a research project that will evaluate changes in watershed health associated with road removal (R.T. Graham pers. comm., Sept. 2009).

In the 1970s and 1980s, society began to place higher values on water quality and other environmental attributes. Research showed the adverse impacts of roads in watersheds, and the density of roads was considered one measurement of watershed health (Gucinski et al. 2001). In the 1990s, an interagency group carried out an assessment of natural resources in the Interior Columbia Basin (ICB) that included a basin-wide analysis of erosion risk. The assessment revealed that the road density in Deception Creek Experimental Forest far exceeded that of anywhere else in the basin (Quigley et al. 1996). Watershed health was an emerging concept, as was treating watersheds as complex ecosystems and considering multiple ecosystem services along with watershed function (Hascic and Wu 2006; Moryc 2009). The ICB analysis tools were programmed to link road density to watershed health, so it became apparent that the well-intentioned road network created in the 1930s had its drawbacks in the post-1990 era when managing forests for watershed health became as important as managing for timber production.

16.2.2 Identifying Roads as Major Sources of Sediment

As the ecosystem services from forests desired by the public have shifted from timber to other products, in particular clean water, researchers have begun to focus on the impacts of roads on water resources. Experimental forests, particularly those in the eastern USA, played an important role in identifying the role of roads in generating sediment in forested watersheds (Hursh 1935). Much of the research on roads in the eastern USA was carried out in experimental forests or watersheds, whereas road research in the western USA tended to be associated with public or commercial forest operations (Gucinski et al. 2001).

16.3 Road Research on EFRs

16.3.1 Road Research on the Fernow Experimental Forest

Forest products from the steep Appalachian Mountains were critical to the development of the USA. Unplanned and unmanaged trails were constructed throughout these steep forests to meet the needs of a growing US population. The trails were mainly from widespread horse logging in the late nineteenth and early twentieth centuries (Fig. 16.4). The trails resulted in scars on the landscape from soil compaction and erosion (Trimble and Weitzman 1953) and an increase in stream sedimentation (Weitzman 1952a). In the 1930s, forest road studies became an important focus of the research program on the Fernow Experimental Forest in West Virginia. Initially, this research concentrated on skid roads. Because horses were being replaced by mechanical skidding operations (Fig. 16.5) in the 1930s–1950s (Weitzman 1952a), there was a need to study the effects of mechanized skidding and to develop practices to reduce those impacts so that problems of the past erosion and watershed degradation would not be repeated. Given its infancy, initial skid road research on the Fernow was surprisingly sophisticated and broad. Studies focused on quantifying erosion losses from skid roads built to a variety of standards; practices to improve location and construction of skid roads; and time studies and economic analyses of skid road construction, maintenance, and use efficiencies.

These studies showed that preplanning the skid road system was the most important activity needed to reduce the length, cost, and adverse effects of skid roads on water resources. Preplanning skid road location reduced the per area lengths by an average of 37% (and as much as 75%) and reduced the area of skid roads by 40% while maintaining average grades one-third lower than in areas without preplanning (Mitchell and Trimble 1959; Weitzman 1952b). Erosion and water quality impacts also were substantially less from well-planned and well-constructed skid roads. For example, during the first year following skidding, erosion from well-planned roads with grade limits of 10% and water bars installed every 40 m (two chains) was only about half that of losses from skid roads without preplanning, i.e., with no

Fig. 16.4 Horse logging in the Appalachians. (Photo courtesy P. Edwards)

Fig. 16.5 Tractor skidders had much higher capacity than horse logging but caused increased watershed disturbance. (Potlatch Corporation: Historical Photographs © 1999)

location restrictions, grade control, or water control (Weitzman and Trimble 1952; Table 16.1). Roads constructed by loggers with no restrictions or water control requirements also resulted in stream turbidities that far exceeded those from better planned skid roads (Reinhart et al. 1963). Time studies showed the advantages of

Table 16.1 Soil erosion measured from skid roads built to four different standards on the Fernow Experimental Forest. (Weitzman and Trimble 1952)

Skid road characteristics	Length of skid road per 2-ha plot (m)	Erosion per 30 m of skid road (m^3)	Total soil moved per 2-ha plot (m^3)
Skid roads ≤10 % grade except for short distances; water bars installed after logging where needed	160	1.5	7.8
Skid roads ≤20 % grade except for short distances; water bars installed after logging at 40-m (two-chain) intervals	24	1.8	14.6
No restrictions on skid road location or grade (average grade was 28 %); water bars installed after logging but at no closer than 40-m (2-chain) intervals	425	2.0	27.6
No restrictions on skid road location or grade (average grade was 34 %); no water bars or other treatments applied after logging	725	2.6	61.3

maintaining gentle grades (5–15 %) on skid roads—skidding speeds were faster and costs of construction and operation were lower than on flat or steeper roads (Hutnik and Weitzman 1957). Economic analyses also showed that skid roads were far more cost-effective for removing both low- and high-value wood from stands than even low-grade, small truck roads (Trimble et al. 1963).

Results from these early studies formed a body of recommendations for skid road construction that are still applicable today and have been repeated with few alterations numerous times via publications and other technology transfer mechanisms throughout the tenure of research at the Fernow Experimental Forest (e.g., Kochenderfer and Helvey 1989; Patric 1977; Smithson and Phillips 1970; Trimble 1959; Trimble and Weitzman 1953). In brief, these recommendations include (Dunford and Weitzman 1955; Trimble and Weitzman 1953):

- Preplan the skid road system
- Avoid trouble spots
- Keep grades low
- Provide adequate and functioning drainage on the road during use
- Avoid stream crossings if possible and stay as far as possible from streams
- Limit earth disturbance
- Stabilize and maintain skid roads with vegetation, particularly the drainage, once the timber management activity is complete

Many of these recommendations can be found in the current regulations of many state agencies, although their source from research in eastern experimental forests is seldom cited (e.g., California Department of Forestry and Fire Protection and Resource Management 2011; Seyedbagheri 1996; West Virginia Division of Forestry 2005).

Table 16.2 The first set of recommendations (early 1950s) for skid road water bar spacings in the Appalachians developed at the Fernow Experimental Forest. (Weitzman 1952a; Trimble and Weitzman 1953)

Skid road grade (%)	Distance between water bars (m)
2	76
5	40
10	24
15	18
20	14
25	12
30	11
40	9

Fig. 16.6 Water bar constructed on a modern skid trail following guidelines developed in the Fernow Experimental Forest in the 1950s. (Source: http://www.muskingumtrees.com/waterbar.jpg)

Within these studies, early Fernow Experimental Forest researchers acknowledged the important role that water control and drainage played in maintaining usable road conditions and reducing erosion. Consequently, researchers employed erosion data to develop the first set of water bar spacing recommendations for the Appalachians (Table 16.2; Trimble and Weitzman 1953; Weitzman 1952a). These spacing recommendations, along with the recommendations for skid road location, construction, and maintenance practices, formed the foundation for what eventually became the base set of road-related best management practices (BMP) for many eastern states (Fig. 16.6). In fact, West Virginia used these original water bar spacings until a BMP revision in 2005 reduced and simplified the spacing requirements.

From the start of research on the Fernow, scientists recognized the importance of technology transfer and understood target audiences. While some findings were communicated in more research-oriented outlets (e.g., Reinhart et al. 1963; Trimble and Weitzman 1953), many other publications provided descriptive summaries and resulting recommendations for audiences whose interests were in on-the-ground applications. These audiences typically included loggers and foresters whose acceptance and implementation of the recommendations was needed to reduce the adverse impacts of roads on water quality and to improve environmental conditions.

Fig. 16.7 Logging road in the 1930s. Note the lack of an inside ditch and the eroding ruts on the road surface. (Potlatch Corporation: Historical Photographs © 1999)

It is likely that the foresight of the first Fernow researchers who simultaneously studied technical, economic, and environmental outcome-based research resulted in greater and faster acceptance of the recommendations than otherwise would have occurred. They were able to show that implementation of practices to protect soil and water resources were not incompatible with reducing costs and operation times.

As Fernow skid road research developed into a rich body of information, research on haul roads became more dominant (Fig. 16.7). The studies focused on how haul roads could be built more cost-effectively while still controlling erosional losses. Many of the groups and individuals who traditionally used Fernow research (e.g., foresters, consultants, loggers, and private and industrial landowners) expressed a need for methods to develop lower-cost truck road systems. This need translated into the concept of "minimum-standard roads." These are truck roads that are built to the lowest standard necessary to achieve the road use objectives while providing environmental protection (i.e., controlling runoff and erosion) at a reasonable cost (Kochenderfer et al. 1984). Characteristics of a minimum-standard road are:

- The road is constructed from a flagged centerline but has no engineered design or construction staking.
- Bulldozer operators must have experience in forest road construction and use nothing smaller than a 200-kw bulldozer for construction.
- Machine operators are paid by the hour and only one additional, experienced helper is required on the job.
- Right-of-way clearing is done with the bulldozer during road construction by pushing or pulling trees over and off of the road.
- Cutbanks are left vertical unless they are taller than 1.5 m or have a ditchline. In those cases, they are roughly sloped back or benched.
- Culverts are used to handle all live water (streams, springs, seeps, etc.), and where seeps and springs are present on the road system, ditches must be installed.
- Broad-based dips at 60-m spacings and natural grade breaks are used as the primary water control features on the road surface.
- Exposed soil should be seeded as soon as possible and practical to control erosion.

As with the skid trail practices, many of these road practices are still recommended or required by state agencies throughout the USA (e.g., California Department of Forestry and Fire Protection and Resource Management 2011; Seyedbagheri 1996; West Virginia Division of Forestry 2005).

Erosion losses from minimum-standard roads were shown to be comparable or even less than losses from higher-standard roads at many other places in the Appalachians as well as elsewhere throughout the country (Kochenderfer and Helvey 1987). Of course, substantially greater environmental protection and improved usability, especially during wet weather conditions or with heavy traffic, were shown to be possible by graveling roads with large, clean gravel. For example, sediment yields from the road surface were reduced from an average of 20 Mg ha^{-1} to less than 2.5 Mg ha^{-1} by graveling with 75-mm clean limestone gravel 150 mm deep. Clean gravel is preferred on roads because it is has higher bearing strength after compaction than graded aggregates (Kochenderfer and Helvey 1987) and does not include fines that can be washed into streams (Wooten et al. 1999). Unfortunately, graveling a road can double construction costs (Kochenderfer and Helvey 1987), which may be unacceptable, particularly to small private landowners. Consequently, recommendations to gravel only problem areas, such as the bottom of broad-based dips, soft spots, and stream crossings, were developed to provide a compromise that improved erosion protection at acceptable costs (Kochenderfer 1979).

A large body of results concerning erosion losses from skid roads and haul roads has been compiled from the research conducted over the past 60 years by Fernow scientists. However, a major shortcoming is that the results did not include sediment contributions to streams (Kochenderfer and Helvey 1987), which, in many cases, is the problem that most concerns forest managers because of state water quality regulations. Consequently, in 1999, researchers from the Fernow and several university cooperators initiated an 8-year study to quantify sediment delivery to streams, identify the sources of delivered sediment, and identify hillside variables that controlled the delivery in a watershed that underwent road construction. Beatty et al. (2004) describe the study which was recently completed, so detailed results will be forthcoming. In addition to quantifying sediment delivery, a number of adaptive management recommendations for haul road construction are expected from this report, which could greatly reduce sediment delivery in steep Appalachian watersheds.

Roads and their related adverse effects on water resources continue to remain societal concerns, and new challenges related to road design and management continue to develop. In the nation's efforts to become energy independent, there is increasing interest and pressure to extract oil and gas reserves from Federal, state, and private lands (Adams et al. 2011). One result of this goal was the installation of a natural gas well and pipeline on the Fernow Experimental Forest in 2008/2009. Development of renewable energy sources, particularly wind power, is also increasing in the central Appalachians. These new activities require construction of access roads and road-like corridors, including pipelines and electricity transmission lines. Consequently, Fernow scientists are expanding their research to study these new types of disturbances that often are built

Fig. 16.8 Horses deliver logs to a North Carolina landing in 1903. (Source: http://www.unc.edu/~whisnant/appal/Sylfal97.htm)

to much different standards or using different practices or techniques than are traditional on forest roads. Little is known about the impacts of such development, so it is expected that the results of these new studies will be as cutting edge as the initial Fernow skid road research was in the 1950s.

16.3.2 Road Research on the Coweeta Hydrologic Laboratory

The southern Appalachians experienced watershed problems similar to those in the north as southern urban areas required forest resources to support their growth. Instability of roads was soon recognized as an undesirable impact of timber abstraction. Road stabilization research at the Coweeta Hydrologic Laboratory has been part of the research program since the 1930s (Swift 1985, 1988). More recently, Swift (1984a, 1985) carried out a series of studies to evaluate road erosion during construction and the benefits of subsequent mitigation practices.

Swift (1988) reported efforts by Hursh (1935) in the 1930s to minimize road erosion through a number of mulching practices to reduce cut and fill slope erosion risks. Hursh was also credited with some of the first studies in using vegetation to stabilize banks. His approach to stabilization research was followed by both Federal and university researchers in subsequent decades. New research methods and mitigation materials may be evaluated (e.g., Grace et al. 1998), but the basic erosion problems identified in the 1930s have not changed.

Swift (1988) went on to describe that in the 1940s and 1950s, a series of logging demonstrations were developed that required the construction of additional roads and skid trails. Much of the skidding was initially done by horses (Fig. 16.8). Road erosion from these demonstrations became so severe that roads became impassable and had to be closed, cross ditched, and reseeded. Subsequent studies were carried out with roads that were constructed and managed to much higher standards, including incorporating broad-based dips that were first described as a "Coweeeta dip" (Hewlett and Douglass 1968) and later installed in the Fernow Experimental Forest (Kochenderfer et al. 1984).

In 1984, a study at Coweeta on cut and fill slopes found that soil losses were much higher before grass establishment (Swift 1984b). The erosion processes observed were more like mass failures rather than surface erosion. One of the drivers of cut and fill slope erosion was freezing and thawing in the winter months. Another observation from this study was that limiting runoff water from bare fill slopes led to much lower fill slope erosion rates. The study also demonstrated the benefit of adding gravel to the road surface, which resulted in an 80% reduction in erosion. This finding has become a common figure used by watershed managers everywhere to support gravel surfacing as a way to reduce road impacts on watershed health (Swift 1984b).

Swift (1985) expanded his studies to address additional road design and management practices to reduce soil erosion. Swift highlighted the importance of both long-term transportation planning and improved road management practices such as gravel addition, storm water management, and low-impact stream crossings. One of the longer-lasting implications of Swift's work was that when he reported in his findings to the interagency Water Erosion Prediction Project (WEPP) in the late 1980s, the problem of road erosion was recognized to be sufficiently severe that it was specifically identified as requiring special consideration in the development of this technology (Foster and Lane 1987). WEPP currently has the capability for a road-specific submodel, although this option has yet to be developed. Forest Service researchers, however, built on Swift's influence with the WEPP model and have developed templates for and interface to the WEPP model for forest roads that are used by forest managers throughout the world (Elliot 2004; Elliot et al. 2010).

16.4 Impact of Roads

16.4.1 Evaluating the Impacts of Roads and Timber Harvest on Runoff

One of the ecosystem services affected by forest roads is water supply. The quantity, quality, and timing of water from forests are becoming increasingly critical as social demands for water increase (Dissmeyer 2000). Experimental forests play a critical role in providing researchers and managers with long-term data sets, including watershed monitoring data. Such data are critical as scientists and managers seek to understand long-term trends and short-term implications of management practices. However, the interpretation of long-term data may not always lead to the same conclusions. Such was the case when Jones and Grant (1996) studied 34 years of runoff data, including hundreds of runoff events, to determine the effect of roads and timber harvest on peak flows leaving watersheds from the H.J. Andrews Experimental Forest and some nearby gauged watersheds. The authors concluded that "forest harvesting has increased peak discharges by as much as 50% in small basins, and 100% in large basins over the past 50 years. These increases are attributable to

Fig. 16.9 Road in H. J. Andrews Experimental Forest rapidly eroding because of flow diversion. (Grant and Swanson 2007)

changes in both flow routing due to roads, and in water balance due to treatment effects and vegetation succession."

However, not all hydrologists agreed with the conclusions of Jones and Grant (1996). As the watershed data were available to the public, Thomas and Megahan (1998) reached a different conclusion from the same data set. They concluded that the greatest percentage increases occurred on the smallest watersheds and for the smallest floods, and it was difficult to draw any conclusions about the effects of timber management and roads on floods in larger watersheds and with larger flood events.

Jones and Grant (2001) acknowledged the controversial nature of this issue, and analyzed the data set a third time. This time, they concluded that the greatest effects were observed with small flood events. Those events generally occurred during the fall season when untreated forest watersheds generated lower flood flows than watersheds with roads and timber harvest. The return periods for the small events were less than 0.28 years. The important message from this study, as it relates to this chapter, is that long-term data from experimental forests have played and will continue to play an important role in addressing management issues in our forest and rangeland watersheds, whether the forests have roads or not.

16.4.2 Road Location and Sediment Production

More recent work on the H.J. Andrews Experimental Forest road networks has helped to determine the effect of road location on stream interception and sediment generation (Fig. 16.9; Grant and Swanson 2007). Grant and Swanson (2007) found that ridgetop and mid-slope roads tended to generate sediment, whereas toeslope roads tended to collect the sediment generated above. Stafford (2011) found that forest roads at higher elevations tended to generate less sediment in the Kings River Experimental Watersheds than in lower-elevation forests, and the author attributed those differences to more of the precipitation falling as snow in higher elevations.

16.4.3 Evaluating Effects of Reopening Closed Roads

Historically, society believed that wildfire in forests was considered unhealthy, and considerable resources were used for fire suppression. This culture changed in the 1980s when forest managers realized that the lack of fire was leading to a buildup of fuels (dead trees and understory shrubs), resulting in an increased risk of unusually severe wildfire (Agee 1993; Elliot et al. 2010). Reducing these fuel loads while minimizing watershed impacts has become a challenge for public land managers. Fuel management includes selective harvesting and prescribed underburning, requiring access every 10–40 years instead of 50–200 years between harvesting rotations. This means that access roads that had been opened up once in a century for timber harvest could now be opened up once in a decade for fuel management (Elliot et al. 2010). One question this raised was how might runoff and erosion be affected by reopening roads that had been closed.

An ideal site to answer this question was located in the Priest River Experimental Forest where a road was built in 1955 and 1956 and then closed to traffic so that vegetation became established on it. Traffic levels associated with the timber operations could be monitored and road access could be restricted during the study. In 2004, it was opened for a timber sale and had 48 loads of logs and associated traffic pass over it. This resulted in a road condition typical of a high-trafficked forest road. Following the operation, the erodibility of the reopened road was measured with a rainfall simulator (Fig. 16.10) and compared to a nearby road that was constructed in 1976, closed following construction, covered with vegetation, and never reopened (Foltz et al. 2009). Foltz et al. (2009) found that there was little difference in hydraulic conductivity between the newly opened and the closed road, but the newly opened road generated about ten times as much sediment as the closed road. The increase in sediment was attributed to the loss of vegetative cover.

16.4.4 Building Predictive Models for Road Erosion

Rainfall simulation as described in the previous section (Fig. 16.10) has been critical in the development of predictive tools, such as the WEPP model (Laflen et al. 1991; Foltz et al. 2011). Once a model is developed and the erodibility of the soils described in the input files, it is necessary to evaluate the accuracy of model predictions in order to validate the model. Roads in experimental forests can be important sites for such validation. Elliot and Foltz (2001) used data from Swift's (1984a, 1984b) Coweeta studies and Kochenderfer and Helvey's (1987) Fernow studies for such validation. They found that the WEPP:Road tool performed reasonably well in normal years but under-predicted in years with large storm events.

Fig. 16.10 Rainfall simulator (in background under canopy) and plot from previous run (in foreground) measuring changes in forest road soil erodibility when a closed road is reopened for fuel management activities in the Priest River Experimental Forest. (Photo by W. Elliot)

16.5 Future Road Challenges

Because of changing societal needs and changing forest management practices, many roads in experimental forests have been closed as they are no longer needed to reduce their adverse watershed impacts. Most experimental forests allow unregulated public access on the open roads. One of the emerging problems associated with public access is that all-terrain vehicle (ATV) riders may bypass barriers and use closed roads for recreational riding. ATV trails generally generate more sediment than forest roads (Meadows et al. 2008), and forest managers are challenged to develop strategies to minimize ATV impacts to experimental forests and watersheds. Other recreational activities such as fishing, hunting, and berry picking depend on road access. There is scope for research on developing new environmentally neutral road designs to suit these activities rather than relying on past designs that mainly support timber harvest.

Since the mid-1990s, many National Forests have been removing old roads. Data on the immediate and long-term watershed effect of road removal are sparse. Removing roads from experimental forest watersheds with long-term data from the past and a commitment to continue to monitor into the future may provide a better understanding of the short- and long-term costs and benefits of road removal.

Recent Forest Service road network management guidelines state that all roads are assumed to be closed unless declared open (R. Graham, pers. comm., Sept. 2009). This ruling may lead to restricted access to some areas of these forests and, if the roads are not maintained, may lead to delayed response to wildfire, putting long-term vegetation research plots at risk. Experimental forests, however, still have the authority to build or replace roads, whereas the National Forest System does not. This means that there is a greater potential for scientists to carry out road-related research in experimental forests than elsewhere.

16.6 Summary and Implications for the Future

Forest road access is essential to gain many ecosystem services from our nation's forests. Experimental forests have played a critical role in evaluating the impacts of roads on watershed health. In all cases, the ability to access sites with long-term historical records coupled with onsite support for research activities have resulted in experimental forests playing a key role in our ongoing understanding of the interactions between roads and watersheds. Examples were given where changes in forest management and use have meant that new research on roads or related access was needed to support such management. Management guidelines developed from the results of these studies are now prominent in BMPs for roads used throughout the world. The past 80 years of road research in experimental forests have shown that it is important to maintain our network in experimental forests to support future research on the role of roads and on the impacts of the access network on watershed health and other forest attributes.

References

Adams MB, Edwards PJ, Ford WM, Johnson JB, Schuler TM, Thomas-Van Grady M, Wood F (2011) Effects of development of a natural gas well and associated pipeline on the natural and scientific resources of the Fernow Experimental Forest. General Technical Report NRS-78, U.S. Department of Agriculture, Forest Service, Northern Research Station, Newton Square, PA, 28 p

Agee JK (1993) Fire ecology of Pacific Northwest forests. Island, Washington, DC, 493 p

Beatty B, Wang J, Edwards P (2004) Spatial analysis of to-stream sediment delivery in central Appalachian forested watersheds. Presented at the 2004 Council of Forest Engineer "Machines and People, the Interface," 27–30 April. Council on Forest Engineering, Corvalis, OR, 6 p

California Department of Forestry and Fire Protection and Resource Management (2011) California Forest Practice Rules. Sacremento, CA, 355 p

Dissmeyer GE (ed) (2000) Drinking water from forests and grasslands a synthesis of the Scientific Literature. General Technical Report SRS-39,U.S. Department of Agriculture, Forest Service, Southern Research Station, Asheville, NC, USA, 246 p

Dunford EG, Weitzman S (1955) Managing forests to control soil erosion. U.S. Department of Agriculture Yearbook, U.S. Government Printing Office, Washington, D.C., pp 235–242
Elliot WJ (2004) WEPP internet interfaces for forest erosion prediction. J Am Water Resour Assoc 40(2):299–309
Elliot WJ (2009) Watershed analysis for fuel management operations, 15 p. http://forest.moscowfsl.wsu.edu/fswepp/docs/fume/WEPP_FuME.pdf. Accessed Sept 2009
Elliot WJ (2010) Effects of forest biomass use on watershed processes in the western United States. West J Appl For 25(1):12–17
Elliot WJ, Foltz M (2001) Validation of the FSWEPP interfaces for forest roads and disturbances. Presented at the ASAE Annual International Meeting; Jul 30–Aug 1, Sacramento, CA. American Society of Agricultural Engineers, St. Joseph, MI, 13 p
Elliot WJ, Miller IS, Audin L (eds) (2010) Cumulative watershed effects of fuel management in the western United States. General Technical Report RMRS-GTR-231, U.S. Department of Agriculture, Forest Service, Rocky Mountain Research Station, Fort Collins, CO, 300 p
Foltz RB, Copeland NS, Elliot W (2009) Reopening abandoned forest roads in northern Idaho, USA: quantification of runoff, sediment concentration, infiltration, and interrill erosion parameters. J Environ Manag 90:2542–2550
Foltz RB, Elliot W, Wagenbrenner NS (2011) Parent material/soil texture-based parameters compared to using site-specific parameters. Trans Am Soc Agric Biol Eng 54(4):1347–1356
Foster GR, Lane LJ (1987) User requirement USDA-Water Erosion Prediction Project (WEPP) draft 6.3. NSERL Report No. 1, U.S. Department of Agriculture, Agricultural Research Service National Soil Erosion Research Laboratory, W. Lafayette, IN, 43 p
Grace JM, Rummer B, Stokes R, Wilhoit J (1998) Evaluation of erosion control techniques on forest roads. Trans Am Soc Agric Biol Eng 4(2):383–391
Grant G, Swanson F (2007) Effects of forest roads on flow regimes, 1 p. http://www.fsl.orst.edu/wpg/research/roads.htm. Accessed Sept 2009
Gucinski H, Brookes MH, Furniss MJ, Ziemer RR (2001) Forest roads: a synthesis of scientific information. General Technical Report PNW-GTR-509, U.S. Department of Agriculture, Forest Service, Pacific Northwest Research Station, Portland, OR, 120 p
Hascic I, Wu J (2006) Land use and watershed health in the United States. Land Econ 82(2):214–239
Hewlett JD, Douglass JE (1968) Blending forest uses. Research Paper SE-RP-37, U.S. Department of Agriculture, Forest Service, Southeastern Forest Experiment Station, Asheville, NC, 15 p
Hursh CR 1935. Control of exposed soil on roadbanks. TN-12, U.S. Department of Agriculture, Forest Service, Appalachian Forest Experiment Station, Asheville, NC, 4 p
Hutnik RJ, Weitzman S (1957) Gentle-grade roads mean faster skidding. Forest Research Note No. 71, U.S. Department of Agriculture, Forest Service, Northeastern Forest Experiment Station, Upper Darby, PA, 3 p
Jones JA, Grant GE (1996) Peak flow responses to clearcutting and roads in a small and large basins, western Cascades. Water Resour Res 32(4):959–974
Jones JA, Grant GE (2001) Comment on "peak flow responses to clearcutting and roads in a small and large basins, western Cascades, Oregon: a second opinion" by R. B.Thomas and W. F. Megahan. Water Resour Res 37(1):175–178
Kochenderfer JN (1979) Some thoughts on forest roads. West Virginia Tree Farm News, Issue 3. 1 p
Kochenderfer JN, Helvey JD (1987) Using gravel to reduce soil losses from minimum-standard forest roads. J Soil Water Conserv 42:46–50
Kochenderfer JN, Helvey JD (1989) Forest access roads. In: Clark FB, Hutchinson JG (eds) Central Hardwood Notes. North Central Forest Experiment Station, St. Paul, 4 p
Kochenderfer JN, Wendel GW, Smith HC (1984) Cost of and soil loss on "minimum-standard" forest truck roads constructed in the central Appalachians. Research Paper NE-544, U.S. Department of Agriculture, Forest Service, Northeastern Forest Experiment Station, Broomall, PA, 8 p

Laflen JM, Elliot WJ, Simanton JR, Holzhey CS, Kohl KD (1991) WEPP soil erodibility experiments for rangeland and cropland soils. J Soil Water Conserv 46(1):39–44

Meadows D, Foltz R, Geehan N (2008) Effects of all-terrain vehicles on forested lands and grasslands, U.S. Department of Agriculture, Forest Service, Technology and Development Center 0823 1811-SDTDC, San Dimas, CA, 110 p

Mitchell WC, Trimble GR Jr (1959) How much land is needed for the logging transport system. J For 57:10–12

Moryc D (2009) Watershed health is key to clean water supply. J For 107(3):151

Patric JH (1977) Soil erosion and its control in eastern woodlands. North Logger Timber Process. 25:4–5, 22–23, 31, 51

Quigley TM, Haynes RW, Graham RT (tech eds) (1996) Integrated scientific assessment for ecosystem management in the Interior Columbia Basin. General Technical Report PNW-GTR-382. U.S. Department of Agriculture, Forest Service, Pacific Northwest Research Station, Portland, OR, 303 p

Reinhart KG, Eschner A, Trimble GR Jr (1963) Effect on streamflow of four forest practices in the mountains of West Virginia. Research Paper NE-1, U.S. Department of Agriculture, Forest Service, Northeastern Forest Experiment Station, Upper Darby, PA, 79 p

Seyedbagheri KA (1996) Idaho Forestry best management practices: compilation of research on their effectiveness. General Technical Report INT-GTR-339, U.S. Department of Agriculture, Forest Service, Intermountain Research Station, Ogden, UT, 89 p

Smithson P, Phillips RA (1970) Skidroads for logging operations in the Appalachians. In: Proceedings, forest engineering workshop on forest roads. West Virginia University, Morgantown, WV, pp 17–23

Stafford AK (2011) Sediment production and delivery from hillslopes and forest roads in the southern Sierra Nevada, California, Thesis. Colorado State University, Fort Collins, CO, 197 p

Stenzel G, Pearce JK, Walbridge TA Jr (1985) Logging and pulpwood production, 2nd edn. Wiley, New York, 385 p

Swift LW Jr (1984a) Soil losses from roadbed and cut and fill slopes in the southern Appalachian Mountains. South J Appl For 8(4):209–215

Swift LW Jr (1984b) Gravel and grass surfacing reduces soil loss from mountain roads. For Sci 30:657–670

Swift LW Jr (1985) Forest road design to minimize erosion in the southern Appalachians. In: Blackmon BG (ed) Proceedings, forestry and water quality: a mid-south symposium; 1985 May 8–9; Little Rock, AR. University of Arkansas, Department of Forest Resources, Monticello, AR, pp 141–151

Swift LW Jr (1988) Forest access roads: design, maintenance, and soil loss. In: Swank WT, Crossley DA Jr (eds) Forest hydrology and ecology at Coweeta. Ecological studies. Springer, New York, pp 313–324

Thomas RB, Megahan WF (1998) Peak flow responses to clearcutting and roads in small and large basins, western Cascades, Oregon: a second opinion. Water Resour Res 34(12):3393–3403

Trimble GR Jr (1959) Logging roads in Northeastern municipal watersheds. J Am Water Works Assoc 51:407–410

Trimble GR Jr, Weitzman S (1953) Soil erosion on logging roads. Soil Sci Proc 17:152–154

Trimble GR Jr, Gibbs C, Barr CR (1963) Truck roads and small forest products. Northeast Logger 11:16–17, 40, 46–47

Weitzman S (1952a) Mountain logging. South Lumberm 185:199–202

Weitzman S (1952b) Mountain log roads-design and construction. West VA Conserv 16:16–21

Weitzman S, Trimble GR Jr (1952) Skid-road erosion can be reduced. J Soil Water Conserv 7:122–124

Wellner CA, Foiles MW (1951) What to see and where to find it on the Deception Creek Experimental Forest. Miscellaneous Publication No. 2, U.S. Department of Agriculture, Forest Service, Northern Rocky Mountain Forest and Range Experiment Station, Missoula, MT, 75 p

West Virginia Division of Forestry (2005) West Virginia silvicultural best management practices for controlling soil erosion and sedimentation from logging operations. West Virginia Division of Forestry, Charleston, WV, 31 p

Wooten AC, Preer J, Edwards PJ (1999) Geologic and tributary influences on the chemistry of a headwater stream. Research Paper NE-708. Northeastern Research Station. U.S. Department of Agriculture, Forest Service, Radnor, PA, 7 p

Chapter 17
The Role of Experimental Forests and Ranges in the Development of Ecosystem Science and Biogeochemical Cycling Research

James M. Vose, Wayne T. Swank, Mary Beth Adams, Devendra Amatya, John Campbell, Sherri Johnson, Frederick J. Swanson, Randy Kolka, Ariel E. Lugo, Robert Musselman and Charles Rhoades

Abstract Forest Service watershed-based Experimental Forests and Ranges (EFRs) have significantly advanced scientific knowledge on ecosystem structure and function through long-term monitoring and experimental research on hydrologic and biogeochemical cycling processes. Research conducted in the 1940s and 1950s began as "classic" paired watershed studies. The emergence of the concept of ecosystem science in the 1950s and 1960s, the passage of the Clean Air Act and Clean Water Act in the 1970s, the nonpoint source pollution provision enacted in the Federal Water Pollution Control Act, and various other forces led to an increased interest in biogeochemical cycling processes. The ecosystem concept recognized

J. M. Vose (✉)
Department of Forestry and Environmental Resources, College of Natural Resources, North Carolina State University, 5223 Jordan Hall, Box 8008, Raleigh, NC 27695, USA
Phone: 919-513-7367
e-mail: jvose@fs.fed.us

W. T. Swank
Southern Research Station (Emeritus), USDA Forest Service, Coweeta Hydrologic Laboratory, 3160 Coweeta Lab Road, Otto, NC 28763, USA

M. B. Adams
Northern Research Station, USDA Forest Service, Timber & Watershed Laboratory, Route 219 North – Nursery Bottom, P.O. Box 404, Parsons, WV 26287, USA

D. Amatya
Southern Research Station, USDA Forest Service, Santee Experimental Forest, 3734 Highway 402, Cordesville, SC 29434, USA

J. Campbell
Northern Research Station, USDA Forest Service, 271 Mast Road, Durham, NH 03824, USA

S. Johnson
Pacific Northwest Research Station, USDA Forest Service, 3200 SW Jefferson Way, Corvallis, OR 97331, USA

F. J. Swanson
Pacific Northwest Research Station (retired), USDA Forest Service, H.J. Andrews Experimental Forest, P.O. Box 300, Blue River, OR 97413, USA

D. C. Hayes et al. (eds.), *USDA Forest Service Experimental Forests and Ranges*,
DOI 10.1007/978-1-4614-1818-4_17,

that water, nutrient, and carbon cycles were tightly linked, and interdisciplinary approaches that examined the roles of soil, vegetation, and associated biota, as well as the atmospheric environment, were needed to understand these linkages. In addition to providing a basic understanding, several watershed-based EFRs have been at the core of the development and application of watershed ecosystem analysis to ecosystem management, and they continue to provide science to land managers and policy makers. The relevance and usefulness of watershed-based EFRs will only increase in the coming years. Stressors such as climate change and increased climate variability, invasive and noninvasive insects and diseases, and the pressures of population growth and land-use change increase the value of long-term records for detecting resultant changes in ecosystem structure and function.

Keywords Long-term data · Watersheds · Interdisciplinary · Nutrient cycling · Ecosystem management

17.1 Introduction

Forest Service watershed-based Experimental Forests and Ranges (EFRs) have been key for advancing knowledge on ecosystem structure and function through long-term monitoring and experimental research on hydrologic and biogeochemical cycling processes. Indeed, significant knowledge on the linkages among carbon, water, and nutrient cycling has been derived from EFRs whose original and primary mission was to understand the relationship between vegetation and hydrology. In most cases, the initial research conducted in the 1940s and 1950s began as "classic" paired watershed studies (Bosch and Hewlett 1982) where treatment watersheds (e.g., manipulating vegetation, fertilization, herbicide application) were compared to controls using streamflow measurements (amount and timing) as the primary response metric. Physically based water quality measurements (i.e., sediment, temperature, etc.) were often also co-measured to quantify the impacts of forest management activities and to support research efforts on the development of improved management systems that eventually led to the best management practices (BMPs).

R. Kolka
Northern Research Station, USDA Forest Service, 1831 Hwy 169 E,
Grand Rapids, MN 55744, USA

A. E. Lugo
International Institute of Tropical Forestry, USDA Forest Service, 1201 Calle Ceiba,
Jardín Botánica Sur, San Juan, PR 00926-1119, USA

R. Musselman · C. Rhoades
Rocky Mountain Research Station, USDA Forest Servcie, 240 West Prospect Road,
Fort Collins, CO 80526-2098, USA

EFRs have provided powerful empirical tools (i.e., regression models, numerical models, graphical analyses, etc.) that could be used to predict the impacts of forest vegetation changes on water yield and water quality in many areas of the USA. The value of these studies cannot be overstated; however, the watershed ecosystem was treated as a black box with little attention paid to the within watershed structural components and biological processes that regulate hydrologic and biological responses. Understanding hydrologic responses to forest management and natural disturbances remains highly relevant. Greater demand for drinking water and increased frequency of both chronic and acute disturbances continue to increase the value of investments in long-term hydrologic monitoring and watershed manipulation experiments. Many of these long-term watershed EFRs were, and continue to be, valued for more than understanding hydrologic responses.

17.2 The Evolution from Hydrology to Ecosystem Science

The emergence of the concept of ecosystem science in the 1950s and 1960s, the passage of the Clean Air Act and Clean Water Act in the 1970s, the nonpoint source pollution provision enacted in the Federal Water Pollution Control Act, and various other forces led to an increased interest in water quality and the biogeochemical cycling processes that determine how ecosystems cycle carbon and nutrients, and ultimately influence water quality. Many EFRs were uniquely positioned to be major players in ecosystem science and associated biogeochemical cycling research. Small watersheds provided convenient study units for defining ecosystems and testing ecosystem concepts developed by E.P Odum and others in the 1950s and 1960s (Odum 1959; Bormann and Likens 1967; Odum 1969). Furthermore, the strength of watershed-based EFRs came from the long-term hydrologic, climatic, and vegetation records already collected, as well as the capacity to conduct experimental treatments that allowed for testing hypotheses related to the regulatory influences of vegetation on biogeochemical cycling processes. Development of nutrient budgets and fluxes was an approach familiar to Forest Service researchers working at watershed-based EFRs who had been quantifying water fluxes using mass balance and classic paired watershed studies for decades. The transition to biogeochemical cycling required adding chemical analyses to the input and output variables and conducting within watershed process studies to determine linkages among carbon, water, and nutrient cycling processes. As of 2008, 15 EFRs (Table 17.1) are engaged in at least some aspect of both hydrology and biogeochemical cycling research, and their spatial distribution provides a wide coverage of climate, soils, and vegetation gradients (Fig. 17.1) in the continental USA and the islands of Hawaii and Puerto Rico (Adams et al. 2008). Some are relatively new EFRs (e.g., Baltimore, Hawaii), some began incorporating biogeochemical cycling components into their long-term research program within the past several years (e.g., Tenderfoot), while others have a history of using biogeochemical cycling approaches to address hypotheses related to ecosystem structure and function that extends back to the 1950s and 1970s (e.g., Hubbard Brook, Coweeta, Fernow, Marcell, H.J. Andrews, and Calhoun).

Table 17.1 Description of watershed EFRs with long-term monitoring of streamflow, precipitation chemistry, and stream chemistry. (Source: Information developed from a combination of EFR websites, Adams et al. 2008, and input from on-site EFR personnel)

Experimental forest	Location	Major vegetation	Streamflow	Precipitation chemistry	Stream chemistry
Baltimore	Baltimore, MD	Hardwood forest to urban	1999	1999	1999
Coweeta	Western North Carolina	Oak-hickory, cove, northern hardwoods	1934	1972	1972
Fernow	West Virginia	Mixed mesophytic hardwood	1951	1978	1971
Fraser	Colorado	Subalpine forest/alpine tundra	1941	1982	1982
Glacier lakes	Wyoming	Englemann spruce/subalpine fir	1989	1989	1989
Hawaii	Hawaii	Wet and dry tropical forest and grassland	In progress	In progress	In progress
H.J. Andrews	West Cascades, Oregon	Douglas fir/western hemlock	1949	1968	1968
Hubbard Brook	New Hampshire	Northern hardwoods/spruce–fir	1956	1963	1963
Luquillo	Puerto Rico	Evergreen broadleaf tropical forest	1945 (USGS) 1988 (USDA FS)	1988	1986
Marcell	Minnesota	Forested peatlands/upland hardwoods	1961	1978	1967
Sagehen	California	Grassland/shrub mixed conifer	1953 (USGS)	2001	1968 (USGS)
San Dimas	Southern California	Mixed chaparral	1939	1988	1982
Santee	Huber, South Carolina	Loblolly and longleaf pine/mixed pine-hardwood	1967	1976–1979, 2009	1976
Tenderfoot Creek	Central Montana	Subalpine fir	1992	n/a	1992–1993, 2001

USDA FS US Department of Agriculture Forest Service

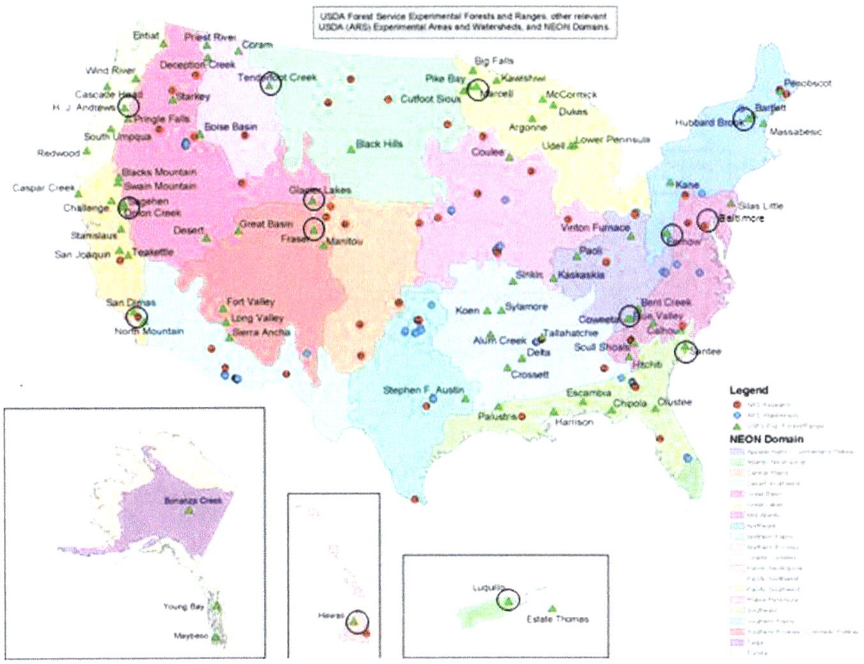

Fig. 17.1 Location of EFRs currently measuring streamflow, atmospheric deposition, and stream chemistry in relation to other monitoring sites in the USA

The ecosystem concept recognized that water, nutrient, and carbon cycles were tightly linked, and interdisciplinary approaches that examined the roles of soil, vegetation, and associated biota, as well as the atmospheric environment, were needed to understand these linkages (Fig. 17.2). The complexity of the issue required non-traditional research approaches; some of the earliest and best examples of interdisciplinary research come from watershed EFRs. For example, starting in the mid-1950s and early 1960s, several EFRs initiated biogeochemical research, usually in collaboration with university partners and with external funding from the National Science Foundation (NSF) and other non-Forest Service-funding sources. The USDA Forest Service, NSF, and NSF International Biological Program (IBP) provided funding for some of the earliest EFR sites to initiate ecosystem studies. Examples include long-term soil nutrient studies at the Calhoun Experimental Forest initiated in the mid-1950s with a sample archive beginning in 1962, biogeochemical cycling research at Hubbard Brook (studies and long-term data collection began in 1963), and the Coweeta Hydrologic Laboratory (IBP-funded studies initiated in 1968; long-term data collection began in 1972). The primary focus of early research at the Calhoun, Hubbard Brook, and Coweeta was on understanding fundamental aspects of nutrient cycling patterns and process (Metz 1952, 1954; Bormann and Likens 1967; Johnson and Swank 1973; Wells and Jorgensen 1975), responses to natural and experimental disturbances, and responses to forest management, and forest growth and development (Wells and Jorgensen 1979; Swank et al. 1981; Swank 1988; Swank

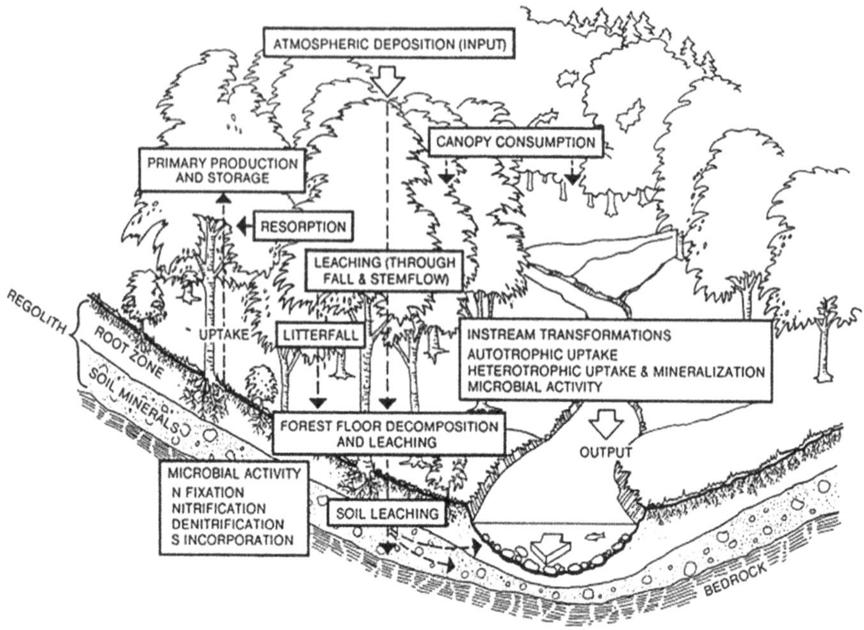

Fig. 17.2 Ecosystem components and processes regulating biogeochemical cycling in forest ecosystems

and Johnson 1994; Swank and Vose 1994; Markewitz et al. 1998). H.J. Andrews (streamflow measurements began in 1953) examined logging impacts on streamflow and water quality in the late 1950s and early 1960s, and expanded into ecosystem studies of biogeochemical cycling in forests and streams in small watershed during IBP in the late 1960s (Sollins et al. 1980; Sollins et al. 1981; Sollins and McCorison 1981;Triska et al. 1984). Research results from these early studies provide some of the best examples of the power of watershed analyses for understanding ecosystem structure and function (Franklin 1989). Building on this history, ecosystem research has continued and expanded at these sites to address contemporary issues using a combination of traditional measurement devices (e.g., weirs) and novel methods and approaches (e.g., stable isotopes, sensor networks, eddy covariance, etc.).

Other EFRs have incorporated biogeochemical cycling research into their research programs and long-term monitoring networks as well. For example, the Fernow Experimental Forest began measuring stream chemistry in 1971 and conducted watershed-scale fertilization studies beginning in 1989 to address issues related to acid deposition and nitrogen saturation (Adams et al. 2000; Edwards et al. 2002; Adams et al. 2006). The Marcell Experimental Forest began measuring stream chemistry in 1967 (Verry 1975) and mercury and organic carbon in the 1990s to determine the influence of forest management and peatlands on water quality and biogeochemical cycles (Kolka et al. 1999). Similarly, precipitation and stream water chemistry measurements at the Santee Experimental Forest (SEF) were initiated in early 1976. As a result, the impacts of prescribed burning on streamflow and water chemistry of the first-order coastal forested watersheds and chemical composition of precipitation at the SEF were examined during the early 1980s (Richter et al. 1982, 1983). Watershed research began in 1988 at the Luquillo Experimental Forest (Scatena 1989), 1 year before Hurricane Hugo passed over the Bisley Experimental Watersheds. This event led to detailed studies of how tropical forests respond to windstorms, an unprecedented record of biogeochemical, structural, compositional, and functional information for tropical forests (Scatena and Lugo 1995; Scatena et al. 1996; Heartsill-Scalley et al. 2007; Heartsill-Scalley et al. 2010; Fig. 8 in Lugo and Heartsill-Scalley, this volume) that is still continuing.

Forest Service EFRs also contribute to a worldwide network of long-term experimental watersheds focused on biogeochemical cycling throughout North America and Europe. Many of these sites were established to examine the impacts of acid deposition on surface water chemistry. In North America, examples include the Walker Branch Watershed in Oak Ridge, TN (established in 1967 by the US Atomic Energy Commission), the Panola Mountain Research Watershed in Georgia (established in1984 by the U.S. Geologic Survey), and The Turkey Lakes Watershed in Canada (established in 1980 by Environment Canada). In Europe, examples include the Plynlimon Watershed in Wales (established in 1968) and the Lake Gardsjon Watershed in Sweden (established in 1978).

Further details of the programmatic research history are available for Coweeta (Douglass and Hoover 1988; Swank et al. 2002), H.J. Andrews (Geier 2007), Calhoun (Metz 1958; Richter and Markewitz 2001), GLEES (Musselman 1994), Hubbard Brook (Likens and Bormann 1995; Groffman et al. 2004); Fernow (Adams et al. in preparation), Fraser (Stottlemyer and Troendle 1987), Marcell (Kolka et al. 2011; Sebestyen and Kolka submitted), and the SEF (Amatya and Trettin 2007a).

17.3 The Importance of Partnerships

A common theme among EFRs actively involved in ecosystem and biogeochemical cycling research is the importance of collaborative partnerships with universities, national programs and other federal agencies, private foundations, and other external funding sources. Several factors contribute to this commonality. Ecosystem science is complex and requires interdisciplinary approaches to understand the interactions among structural and functional components of carbon, nutrient, and water cycling processes. Although many Forest Service scientists have played major roles in ecosystem-based research at EFRs, few EFRs have had the full range of scientific expertise required to address complex and comprehensive ecosystem studies. Hence, large teams of specialists (e.g., hydrologists, soil scientists, plant ecologists, etc.) are often involved to cover the topical diversity. As such, strong collaborative research with universities, other federal and state agencies, and other institutions has developed at many watershed EFR sites (Table 17.2).

The benefits of these partnerships go beyond the value of the science. Forest Service watershed EFRs have played a major role in educating current and future generations of ecosystem and hydrologic scientists, both nationally and internationally.

Table 17.2 Collaborators, participation in national and international Networks, and funding provided from non-USDA Forest Service sources to support overall research at the EFR sites. (Source: Information developed from a combination of EFR websites, Adams et al. 2008, and input from on-site EFR personnel)

Experimental forest	Primary collaborators	National and international networks	Non-USDA FS funding sources
Baltimore	Johns Hopkins, University of North Carolina, Vermont University, University of Maryland, Columbia, Towson University	USGS, NADP, LTER	NSF
Calhoun	Duke University, North Carolina State University, University of Georgia	CZO	NSF, USDA-NRI, Andrew Mellon
Coweeta	University of Georgia, Virginia Tech, Duke, University of North Carolina-Chapel Hill, University of Minnesota, University of Wisconsin, Yale University, N. Carolina State University	NADP, NDDN, CASTnet, LTER, MAB Biosphere Reserves, Ameriflux	NSF, US-EPA, DOD, USDA-NRI
Fernow	West Virginia University, Pennsylvania State University, Virginia Tech.	NADP, CASTNet	US-EPA, NSF, USDA-NRI
Fraser	Colorado State University, University of Colorado,	USGS	US-EPA, USGS, NASA, NOAA
Glacier lakes	Colorado State University, University of Wyoming, Wake Forest University	NADP, CASTNet, Ameriflux, SNOTEL	US-EPA
Hawaii	University of Hawaii		
H.J. Andrews	Oregon State University, NASA	NADP, USGS, HBN, MDN, LTER	NSF
Hubbard Brook	Brown University, Dartmouth, Yale, Cornell, Syracuse, State University of New York, UC-Berkeley, University of New Hampshire, Wellesley College, Cary Institute of Ecosystem Studies, US Geological Survey	NADP, SCAN, MDN, CASTNet, LTER	NSF, DOE, USDA, A.W. Mellon Foundation, EPA, NRCS

Table 17.2 (continued)

Experimental forest	Primary collaborators	National and international networks	Non-USDA FS funding sources
Luquillo	University of Puerto Rico, U. New Mexico, UC-Berkeley, U. Pennsylvania, University of New Hampshire, Utah State, U. Connecticut, State University of New York-Syracuse, University of Georgia	NADP, USGS, LTER, CZO, MAB Biosphere Reserves, Smithsonian Biodiversity Plots	NSF
Marcell	University of Minnesota, University of Toronto, University of Wisconsin, Michigan Tech, Michigan State, University of Nebraska, North Carolina State, Gustavus Adolphus College	NADP, MDN	US-EPA, NASA, DOE, NSF, NSERC
Sagehen	UC-Berkeley	NADP	
San Dimas	UC-Riverside, University of Georgia, University of Iowa, Cal Poly-Pomona, Cal State-Long Beach,	NADP	
Santee	College of Charleston, Clemson, University of South Carolina, Virginia Tech, University of New Hampshire	USGS	US-EPA, South Carolina EPD
Tenderfoot	Montana State University, University of Montana	USGS, SNOTEL, AmeriFlux	

CZO Critical Zone Observatory, *DOD* Department of Defense, *DOE* Department of Energy, *EPD* Environmental Protection Division, *MAB* Man and the Biosphere Programme, *MDN* Mercury Deposition Network, *NADP* National Atmospheric Deposition Program, *NASA* National Aeronautics and Space Administration, *NDDN* National Dry Deposition Network, *NRCS* Natural Resources Conservation Service, *USDA* United States Department of Agriculture-National Research Initiative, *USDA FS* US Department of Agriculture Forest Service, *US-EPA* United States Environmental Protection Agency

For example, nearly 1,000 students have received graduate degrees from work conducted at Coweeta (~270), Hubbard Brook (~190), Fernow (~50), Fraser (~50), H.J. Andrews (~240), Marcell (~30), Luquillo (~100), and the Calhoun and Santee (~20). National monitoring programs such as the National Atmospheric Deposition Program (NADP), the Mercury Deposition Network (MDN), National Dry Deposition Network (NDDN, CASTnet), and other federal agencies such as the US Geological Survey (USGS) have collaborated or provided data on precipitation chemistry, water quality, and streamflow at several of the sites (Table 17.2). For example, USGS and National Park researcher Bob Stottlemyer added complementary biogeochemical research to ongoing Forest Service hydrologic studies at the Fraser Experimental Forest and supported this partnership for two decades.

Success in leveraging funding in addition to that provided by the USDA Forest Service has also been an important key to the success of many of these programs. Most notably, the NSF recognized that long-term approaches were required for understanding ecosystem processes, and the Coweeta Hydrologic Laboratory and the H.J. Andrews Experimental Forests were among the original eight NSF Long-Term Ecological Research (LTER) Sites established in 1980; they were followed by Bonanza Creek in 1987, Hubbard Brook in 1988, Luquillo in 1988, and Baltimore in 1997. Current and future relevance of the long-term data records is evidenced by involvement of several EFRs in the formation of two emerging national networks: (1) National Ecological Observation Network (NEON) and (2) Urban Long-Term Research Area (ULTRA). The LTER program is not the only additional funding source that helps sustain EFR research; however, most of these other funding sources are typically shorter term. Other sources of funding include the Environmental Protection Agency, the NSF, private foundations, various state agencies, USDA National Research Initiative, NASA, Department of Defense, Electric Power Research Institute, National Council for Air and Stream Improvement, Department of Energy, and various others. Another recent example of a partnership involving the SEF is the Turkey Creek Watershed initiative, a multiagency eco-hydrological research collaboration to address the critical issues of sustainable water management for the low-gradient coastal landscape (Amatya and Trettin 2007b).

17.4 Applying Ecosystem Science to Forest Management

Long-term partnerships with USDA Forest Service managers, resource specialists, and decision makers ensure that research conducted at the EFRs is responsive to the resource management challenges on National Forests. EFR scientists and staff provide a ready outlet for science delivery to land managers, and most EFRs contribute frequently to technical tours and educational field trips for the public. Demonstration areas and self-guided tours are also active outlets for sharing research findings.

Several watershed-based EFRs have been at the core of the development and application of watershed ecosystem analysis to ecosystem management (Kessler et al. 1992). For example, researchers at Hubbard Brook (Likens 1989; Hornbeck and Swank 1992), Coweeta (Swank and Johnson 1994; Meyer and Swank

1996), H.J. Andrews (Franklin et al. 1981), and Luquillo (Lugo and Scatena 1995; Lugo et al. 1999) were leaders in providing the conceptual basis for ecosystem management. The combination of an ecosystem-based research approach with watershed scale experimental treatments that included forest management practices (e.g., logging, road construction, etc.) provided data and real-world examples to test the often "fuzzy" concepts of ecosystem management (Christensen et al. 1996). For example, research at Hubbard Brook and Coweeta provided watershed ecosystem analysis methods to evaluate effects of harvesting practices, acidic deposition, and past land use (Hornbeck and Swank 1992). Studies at Marcell and Coweeta have been critical to developing BMPs for forestry in the midwest (Verry 1976; Verry et al. 1983) and southeast regions of the USA (Phillips et al. 2000; Riedel et al. 2007). Results from research at EFRs have had a significant impact on many forest management and environmental policy issues in the USA. For example, watershed research and studies of old-growth forest ecosystems at H.J. Andrews contributed substantially to the development and early implementation of the Northwest Forest Plan (Cissel et al. 1994).

A few EFRs expanded their research beyond the EFR boundary to demonstrate and test the application of ecosystem management concepts in partnership with natural resource managers. For example, Coweeta initiated the Wine Spring Creek Ecosystem Management Project (WSC) with the objective of using ecosystem-based concepts, principles, and technology to achieve desired resource conditions (Swank and Van Lear 1992). Participants in the project included an interdisciplinary team of over 55 scientists and managers in five research units in the Southern Research Station; the National Forest Systems and seven universities; state agencies; environmental conservation groups; and the public. A consensus-building process comprising of workshops attended by all stakeholders was conducted over an 18-month period. From this consensus-fielding process, 35 desired future resource conditions were identified for the project area. Management prescriptions were applied to achieve desired future conditions, and then monitored for response, followed by application of adaptive management if needed. Findings from individual studies proved useful in making management decisions and an EMERGY-based environmental systems assessment to integrate and quantify the balance of ecological, economic, and social demands placed on land resources (Tilley and Swank 2003). Watershed ecosystem principles were at the core of the research–management interface since water transports materials within and from the forested landscape, and water cycles are tightly linked with carbon and nutrient cycling processes. Other examples include research synthesis products that have been developed based on work in northeastern (Hubbard Brook) and midwestern (Marcell) experimental forests to inform forest managers of ecosystem impacts of management (Verry et al. 2000). More recent work has integrated research results from Marcell to help understand the effects of land use and fragmentation on stream systems and biologic communities in midwestern landscapes (Verry 2004). At the Luquillo Experimental Forest, watershed research led to a new design for water extraction, which prevented the damming of the river and allows for water extraction without affecting the two-way migration of critical stream fauna (March et al. 2003). Like many EFRs, the ecosystem studies at the Fraser Experimental Forest were designed to quantify the effects of vegetation manipulation associated with forest management activities on soil nutrient cycling and stream water nutrient export (Reuss et al. 1997; Stottlemyer and Troendle 1999). Extensive overstory mortality caused by mountain pine beetle outbreak throughout much of the interior west has taken Fraser's watershed studies in a new direction. Historic studies now will allow comparison of the response of paired basins with varying management history and stand structure (e.g., regenerating vs. old-growth forest) to a natural forest disturbance. For example, in the years immediately following beetle infestation resulting in the loss of 40 % of total overstory basal area, nitrate concentrations increased in old-growth basins compared to the pre-infestation stream water record (Rhoades et al. 2008).

17.5 Challenges and Opportunities

EFRs have been critical for addressing fundamental questions related to ecosystem structure and function and for applying ecosystem concepts to the management of forest and range ecosystems. This national network of long-term climate, hydrologic, biogeochemical, vegetation, and land-use records that address fundamental resource management and ecosystem concepts provides a unique niche that facilitates collaboration with academic institutions and other federal and state agencies. Despite their important historic contributions and their current role in development of continental-scale ecological networks aimed at addressing climate and land-use change, EFRs face significant challenges maintaining long-term data collection and research facilities management. These challenges are consistent with those discussed in recent reviews by Ice and Stednick (2004) and Stednick et al. (2004) who compiled lessons learned from watershed research throughout the country, with most information derived from Forest Service EFRs. Most EFRs struggle with the substantial (and growing) fixed costs associated with the collection, analyses, and data management (QA/QC, storage, and access systems) required for monitoring precipitation volume and chemistry, streamflow volume and chemistry, ground water levels, and all other associated measurements (e.g., climate, soils, vegetation, etc.) required to understand ecosystem structure and function.

The value of long-term ecosystem measurements is substantial and the return on investment will continue to grow. As an example, stream NO_3 concentration has been measured for more than 30 years after clearcutting a hardwood forest at Coweeta (Fig. 17.3). The three time trends noted on the graph represent patterns observed over a ~5-year period. If the monitoring had been stopped with the assumption that the response trajectory observed over the previous 5-year period was going to continue into the future, the assumption (and associated interpretations of physical and biological factors regulating the response) would have been wrong in every case. While this example is based on measurements from a manipulated watershed, long-term baseline measurements from reference watersheds are also critical for detecting responses to forcings such as climate change, vegetation development, atmospheric deposition, insects and disease, and whatever else the future holds. These data also provide important reference data for evaluating restoration

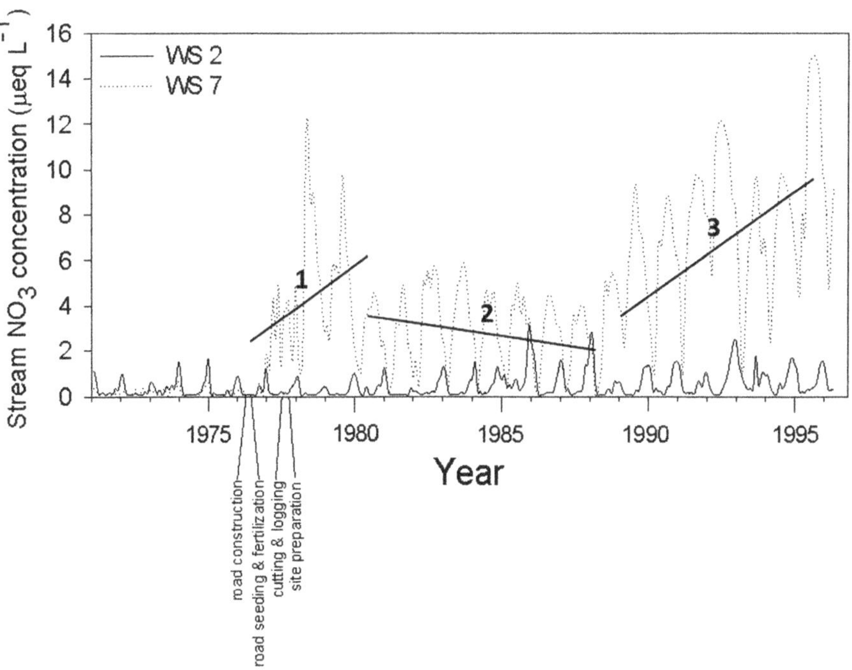

Fig. 17.3 Nitrate nitrogen concentration in stream water from a reference (*solid line, WS2*) and clearcut watershed (*dashed line, WS7*) at Coweeta. Three distinct tends in stream NO_3 were observed: *1*=an increase in response to cutting and road bank fertilization, *2*=a decrease in response to vegetation regrowth and N uptake, and *3*=an increase in response to unknown factors (e.g., black locust mortality, enhanced decomposition) that are currently being examined.

success on degraded watersheds. Second, the relevancy of EFRs will require looking beyond the EFR boundary. For example, many contemporary issues are global (i.e., climate change, invasive) or regional (e.g., land-use change, drought) in scale, yet EFRs have been traditionally inward-looking with the majority of research activities focused on the site. To continue to be relevant, EFRs will be challenged to scale site-based research to larger spatial scales and expand experimental and observational approaches beyond EFR boundaries, while at the same time, maintaining the in-depth, long-term research within the EFR. Expanding beyond the watershed boundary will require that the human dimension of ecosystem science be directly addressed—humans are both a part of ecosystems and depend upon them for the services they provide. Despite the importance of the human dimension, few EFRs have the scientific expertise or experience in integrating social and ecological sciences. Finally, it has become increasingly difficult to conduct manipulative experiments at EFRs due to the challenges associated with meeting National Environmental Policy Act (NEPA) requirements. Novel experiments are at the very core of testing complex ecosystem hypotheses, yet the "experimental" component of the EFR has been considerably restricted in recent years. Indeed, historical whole watershed manipulations such as preventing regrowth after cutting, species conversions, grazing, herbicides, acidification, long-term prescribed burning have yielded (and continue to yield) considerable insight into ecosystem processes. The ability of EFRs to efficiently conduct manipulations required to address contemporary issues will be a critical determinant of their value in the twenty-first century.

Despite these challenges, the relevance and usefulness of watershed-based EFRs will increase in the coming years. Stressors such as climate change and increased climate variability, invasive and noninvasive insects and diseases, and the pressures of population growth, and land-use change increase the value of long-term records for detecting resultant changes in ecosystem structure and function. Much of these long-term records are high quality, and improvements in networking and accessibility via electronic data bases such as HydroDB and ClimDB (http://www.fsl.orst.edu/climdb/harvest.htm) make them available to the greater scientific community.

17.6 Conclusions

Forest Service EFRs have played an important role in the development of ecosystem science. Early approaches focused heavily on biogeochemical and hydrologic cycling processes as key metrics for testing ecosystem hypotheses, but have expanded to include understanding linkages between climate change and carbon cycling as well. EFRs will continue to play an important role in ecosystem science and will be critical for measuring and predicting impacts of an altered atmospheric environment and other forest health threats in the future. Long-term data and experiments are available from watershed-based EFRs spanning across wide geographic, vegetation, and climate gradients to test and develop models required to predict ecosystem responses to contemporary and future forcing variables such as climate change, invasive species, and other pressures associated with increased human population growth and growing demand for ecosystem services.

References

Adams MB, Burger JA, Jenkins AB, Zelazny L (2000) Impact of harvesting and atmospheric pollution and nutrient depletion of eastern U.S. hardwood forests. For Ecol Manage 138(1):301–319

Adams MB, DeWalle DR, Hom JL (eds) (2006) The Fernow watershed acidification study. Springer-Verlag, New York, 279 p

Adams MB, Loughry L, Plaugher L (comp.) (2008) Experimental Forests and Ranges of the USDA Forest Service. General Technical Report NE-321 Revised, U.S. Department of Agriculture, Forest Service, Northeastern Research Station, Newtown Square, PA, 178 p, [CD ROM]

Amatya DM, Trettin CC (2007a) Development of watershed hydrologic research at Santee Experimental Forest, Coastal South Carolina. In: Furniss M, Clifton C, Ronenberg K (eds) Advancing the fundamental sciences: proceedings of the Forest Service National Earth Sciences Conference, San Diego, CA, October 18–22, 2004, PNW-GTR 689, US Department of Agriculture, Forest Service Pacific Northwest Station, Portland, OR

Amatya DM, Trettin CC (2007b) An eco-hydrological project on Turkey Creek Watershed, South Carolina, U.S.A. In: Meire P et al (eds) Integrated water management: practical experiences and case studies, NATO science series IV: earth and environmental sciences. Springer, Netherlands, pp 115–126

Bormann FH, Likens GE (1967) Nutrient cycling. Science 155:424–429

Bosch JM, Hewlett JD (1982) A review of catchment experiments to determine the effect of vegetation changes on water yield and evapotranspiration. J Hydro 55:3–23

Christensen NL, Bartuska AM, Brown JH, Carpenter S, D'Antonio C, Francis R, Franklin JF, MacMahon JA, Noss RF, Parsons DJ, Peterson CH, Turner MG, Woodmansee RG (1996) The report of the ecological society of America committee on the scientific basis for ecosystem management. Ecol Appl 6:665–691

Cissel JH, Swanson FJ, McKee WA, Burditt AL (1994) Using the past to plan the future in the Pacific Northwest. J For 92(8):30–31, 46

Douglass JE, Hoover MD (1988) History of Coweeta. In: Swank WT, Crossley DA Jr (eds) Forest hydrology and ecology at Coweeta. Ecological studies, vol 66. Springer-Verlag, New York, pp 17–31

Edwards PJ, Kochenderfer JN, Coble DW, Adams MB (2002) Soil leachate responses during 10 years of induced whole-watershed acidification. Water Air Soil Pollut 140:99–118

Franklin JF (1989) Importance and justification of long-term studies in ecology. In: Likens GE (ed) Long-term studies in ecology: approaches and alternatives. Springer-Verlag, New York, pp 3–19

Franklin JF, Cromack K Jr, Denison W, McKee A, Maser C, Sedell J, Swanson F, Juday G (1981) General Technical Report PNW-GTR-118, U.S. Department of Agriculture, Forest Service, Pacific Northwest Research Station, Portland, OR, 48 p

Geier MG (2007) Necessary work: discovering old forests new outlooks, and community on the HJ Andrews Experimental Forest, 1948–2000. USDA, Forest Service, Pacific Northwest Research Station. General Technical Report PNW-GTR687, 357 p

Groffman PM, Driscoll CT, Likens GE, Fahey TJ, Holmes RT, Eagar C, Aber JD (2004) Nor gloom of night: a new conceptual model for the Hubbard Brook ecosystem study. BioScience 54:139–148

Heartsill-Scalley T, Scatena FN, Estrada C, McDowell WH, Lugo AE (2007) Disturbance and long-term patterns of rainfall and throughfall nutrient fluxes in a subtropical wet forest in Puerto Rico. Journal of Hydrology 333:472–485

Heartsill-Scalley T, Scatena FN, Lugo AE, Moya S, Estrada-Ruiz C (2010) Changes in structure, composition and nutrients during 15 year of hurricane-induced succession in a subtropical wet forest in Puerto Rico. Biotropica 42:455–463

Hornbeck JW, Swank WT (1992) Watershed ecosystem analysis as a basis for multiple-use management of eastern forests. Ecol Appl 2(3):238–247

Ice G, Stednick JD (eds) (2004) A century of forest and wildland lessons. Society of American Foresters, Bethesda, 287 p

Johnson PL, Swank WT (1973) Studies on cation budgets in the southern Appalachians on four experimental watersheds with contrasting vegetation. Ecology 54:70–80

Kessler WB, Salwasser H, Cartwright CW Jr, Caplan JA (1992) New perspectives for sustainable natural resources management. Ecol Appl 6:738–740

Kolka RK, Grigal DF, Verry ES, Nater EA (1999) Mercury and organic carbon relationships in streams draining forested upland/peatland watersheds. J Environ Qual 28:766–775

Kolka RK, Sebestyen SD, Verry ES, Brooks KN (eds) (2011) Peatland biogeochemistry and watershed hydrology at the Marcell experimental forest. CRC Press, Boca Raton

Likens GE (ed) (1989) Long-term studies in ecology: approaches and alternatives. Sringer-Verlag, New York

Likens GE, Bormann FH (1995) Biogeochemistry of a forested ecosystem, 2nd edn. Springer-Verlag New York Inc., New York, 159 p

Lugo AE, Scatena FN (1995) Ecosystem-level properties of the Luquillo experimental forest with emphasis on the tabonuco forest. In: Lugo AE, Lowe C (eds) Tropical forests: management and ecology. Springer, New York, pp 59–108

Lugo AE, Baron JS, Frost TP, Cundy TW, Dittberner P (1999) Ecosystem processes and functioning. In: Szaro RC, Johnson NC, Sexton WT, Malk AJ (eds) Ecological stewardship: a common reference for ecosystem management. Elsevier Science, Oxford, pp 219–254

March JG, Benstead JP, Pringle CM, Scatena FN (2003) Damming tropical island streams: problems, solutions, and alternatives. Bio Sci 53:1069–1078

Markewitz D, Richter DD, Allen HL, Urrego JB (1998) Three decades of observed soil acidification at the Calhoun experimental forest: has acid rain made a difference? Soil Sci Soc Am J 62:1428–1439

Metz LJ (1952) Weight and nitrogen and calcium content of the annual litter fall of forests in the South Carolina Piedmont. Soil Sci Soc Am Proc 16:38–41

Metz LJ (1954) Forest floor in the Piedmont region of South Carolina. Soil Sci Soc Am Proc 18:335–338

Metz LJ (1958) The Calhoun experimental forest. Southeastern Forest Experiment Station, Asheville, p 24

Meyer JL, Swank WT (1996) Ecosystem management challenges ecologists. Ecol Appl 6(3):738–740

Musselman RC (technical editor) (1994) The glacier lakes ecosystem experiments site. General Technical Report RM-249. U.S. Department of Agriculture, Forest Service, Rocky Mountain Forest and Range Experiment Station, Fort Collins, CO, p 96. http://www.fs.fed.us/rm/pubs_rm/rm_gtr249.html

Odum EP (1959) Fundamentals of ecology, 2nd edn. Saunders, Philadelphia

Odum EP (1969) The strategy of ecosystem development. Science 164:262–270

Philips MJ, Swift LW Jr, Blinn CR (2000) Best management practices for riparian areas. In: Verry ES, Hornbeck JW, Dolloff CA (eds) Riparian management of forests of continental eastern. U.S. Lewis Publishers, Boca Raton, pp 272–285

Reuss JO, Stottlemyer R, Troendle CA (1997) Effect of clearcutting on nutrient fluxes in a subalpine forest at Fraser, Colorado. Hydrol Earth Syst Sci 1:333–344

Rhoades C, Elder K, Hubbard R, Dixon M (2008) Streamwater chemistry and nutrient export during five years of bark beetle infestation of subalpine watersheds at the Fraser experimental forest, Eos Trans. AGU, 89(53), Fall Meeting Suppl, AbstractH13C-0929

Richter DD, Ralston CW, Harms WR (1982) Prescribed fire: effects on water quality and forest nutrient cycling. Science 215:661–663

Richter DD, Ralston CW, Harms WR (1983) Chemical composition and spatial variation of bulk precipitation at a coastal plain watershed in South Carolina. Water Resour Res 19(1):134–140

Richter DD, Markewitz D (2001) Understanding soil change: sustainability of soils over millennia, centuries, and decades. Cambridge University Press, UK

Riedel MS, Swift LW Jr, Vose JM, Clinton BD (2007) Forest road erosion research at the Coweeta hydrologic laboratory. In: Furniss M, Clifton C, Ronnenberg K (eds) Advancing the fundamental sciences, pp 197–205. Proceedings of the Forest Service National Earth Science Conference, 18–22 October 2004, San Diego, CA PNW-GTR-689, USDA Forest Service, Pacific Northwest Research Station, Portland, OR

Scatena FN (1989) An introduction to the physiography and history of the Bisley experimental watersheds in the Luquillo Mountains of Puerto Rico. General Technical Report SO-72, USDA Forest service, Southern Forest Experiment Station, New Orleans

Scatena FN, Lugo AE (1995) Geomorphology, disturbance, and the soil and vegetation of two subtropical wet steepland watersheds of Puerto Rico. Geomorphology 13:199–213

Scatena FN, Moya S, Estrada C, Chinea JD (1996) The first five years in the reorganization of aboveground biomass and nutrient use following Hurricane Hugo in the Bisley Experimental Watersheds, Luquillo Experimental Forest, Puerto Rico. Biotropica 28:424–440

Sebestyen S, Kolka R (submitted) Experimental forest and range synthesis: Marcell experimental forest. General Technical Report WO-XX, USDA Forest Service, Washington Office, Washington DC

Sollins P, McCorison FM (1981) Nitrogen and carbon solution chemistry of an old growth coniferous forest watershed before and after cutting. Water Resour Res 17(5):1409–1418

Sollins P, Grier C, McCorison FM, Cromack K Jr, Fogel R, Fredriksen RL (1980) The internal element cycles of an old-growth Douglas-fir ecosystem in western Oregon. Ecol Monogr 50(3):261–285

Sollins P, Cromack K Jr, McCorison FM, Waring RH, Harr RD (1981) Changes in nitrogen cycling at an old-growth Douglas-fir site after disturbance. J Environ Qual 10(1):37–42

Stednick JD, Troendle CA, Ice GG (2004) Lessons for watershed research in the future. In: Ice GG, Stednick JD (eds) A century of forest and wildland watershed lessons. Society of American Foresters, Bethesda, pp 277–287

Stottlemyer R, Troendle CA (1987) Trends in streamwater chemistry and input-output balances, Fraser experimental forest, Colorado RP-RM-275. USDA Forest Service, Rocky Mountain Forest and Range Experiment Station, Fort Collins

Stottlemyer R, Troendle CA (1999) Effect of subalpine canopy removal on snowpack, soil solution, and nutrient export, Fraser Experimental Forest, CO. Hydrol Process 13:2287–2299

Swank WT (1988) Stream chemistry responses to disturbance. In: Swank WT, Crossley DA Jr (eds) Forest hydrology and ecology at Coweeta. Ecological Studies, vol 66. Springer-Verlag, New York, 339–357

Swank WT, Johnson CE (1994) Small catchment research in the evaluation and development of forest management practices. In: Moldan B, Cerny J (eds) Biogeochemistry of small catchments: a tool for environmental research. Scientific Committee on Problems of the Environment (SCOPE) 51. Wiley, Chichester, pp 383–408

Swank WT, Van Lear DH (1992) Multiple—use management: ecosystem perspectives of multiple—use management. Ecol Appl 2(3):219–220

Swank WT, Vose JM (1994) Long-term hydrologic and stream chemistry responses of southern Appalachian catchments following conversion from mixed hardwoods to white pine. In: Landolt R (ed) HydrologiekleinerEinzugsgebiete: Gedenkschrift Hans M. Keller. BeitragezurHydrologie der Schweiz 35.Bern, Schweizerische: SchweizerischeGesellschaft fur Hydrologie und Limnologie, pp 164–172

Swank WT, Waide JB, Crossley DA Jr, Todd RL (1981) Insect defoliation enhances nitrate export from forest ecosytems. Oecologia 51:297–299

Swank WT, Meyer JL, Crossley DA Jr (2002) Long-term ecological research: Coweeta history and perspectives Chapter 8. In: Barrett GW, Barrett TL (eds) Holistic science—the evolution of the Georgia Institute of Ecology (1940–2000). Taylor & Francis, New York, pp 143–163

Tilley DR, Swank WT (2003) EMERGY-based environmental systems assessment of a multipurpose temperate mixed-forest watershed of the southern Appalachian Mountains, USA. J Environ Manage 69:213–227

Triska FJ, Sedell JR, Cromack K Jr (1984) Nitrogen budget for a small coniferous forest stream. Ecol Monogr 54(1):119–140

Verry ES (1975) Streamflow chemistry and nutrient yields from upland-peatland watersheds in Minnesota. Ecology 56:1149–1157

Verry ES (1976) Estimating water yield differences between hardwood and pine forests. Research paper NC-128, USDA Forest Service, St. Paul, MN, 12 p

Verry ES (2004) Land fragmentation and impacts to streams and fish in the central and upper Midwest. In: Ice GG, Stednick JD (eds) Lessons for watershed research in the future: a century of forest and wildland watershed lessons. Society of American Foresters, Bethesda, pp 129–154

Verry ES, Lewis JR, Brooks KN (1983) Aspen clearcutting increases snowmelt and storm flow peaks in north central Minnesota. Water Resour Bull 19(1):59–67

Verry ES, Hornbeck JW, Todd AH (2000) Watershed research and management in the Lake States and northeastern United States. In: Ffolliott PF, Baker MB, Edminster CB, Dillon MC, Mora KL (eds) Land stewardship in the 21st century: the contributions of watershed management. Conference proceedings, Tuscon, AZ, 13–16 March 2000, USDA Forest Service, Fort Collins, CO, pp 81–92

Wells CG, Jorgensen JR (1975) Nutrient cycling in loblolly pine plantations. In: Bernier B, Winget CH (ed) Forest soils and forest land management, Fourth North American forest soils conference. Laval University Press, Quebec, pp 137–158

Wells CG, Jorgensen JR (1979) Effect of intensive harvesting on nutrient supply and sustained productivity. In: Leaf A (ed) Impact of intensive harvesting on forest nutrient cycling. State University of New York at Syracuse, USA, pp 212–230

Chapter 18
Evolution of Soil, Ecosystem, and Critical Zone Research at the USDA FS Calhoun Experimental Forest

Daniel deB. Richter, Allan R. Bacon, Sharon A. Billings, Dan Binkley, Marilyn Buford, Mac A. Callaham, Amy E. Curry, Ryan L. Fimmen, A. Stuart Grandy, Paul R. Heine, Michael Hofmockel, Jason A. Jackson, Elizabeth LeMaster, Jianwei Li, Daniel Markewitz, Megan L. Mobley, Mary W. Morrison, Michael S. Strickland, Thomas Waldrop and Carol G. Wells

The past is never dead. It's not even past.
(Faulkner, Requiem for a Nun, 1951)

Abstract The US Department of Agriculture (USDA) Forest Service Calhoun Experimental Forest was organized in 1947 on the southern Piedmont to engage in research that today is called restoration ecology, to improve soils, forests, and watersheds in a region that had been severely degraded by nearly 150 years farming. Today, this 2,050-ha research forest is managed by the Sumter National Forest and Southern Research Station. In the early 1960s, the Calhoun Experimental Forest was closed as a base of scientific operations making way for a new laboratory

D. deB. Richter (✉) · A. R. Bacon · P. R. Heine · M. Hofmockel · M. L. Mobley
Nicholas School of the Environment, Duke University, Durham, NC 27708, USA
Phone: 919 613-8031
e-mail: drichter@duke.edu

S. A. Billings · J. Li
University of Kansas, Lawrence, KS, USA

D. Binkley
Colorado State University, Fort Collins, CO, USA

M. Buford
USDA Forest Service National Headquarters, Washington, DC, USA

M. A. Callaham
Southern Research Station, USDA Forest Service, 320 Green Street, Athens, GA 30602-2044, USA

J. A. Jackson
Southern Research Station, USDA Forest Service, P.O. Box 12254, Research Triangle Park, Durham, NC 27709, USA

T. Waldrop
Southern Research Station, USDA Forest Service, 233 Lehotsky Hall, Clemson University, Clemson, SC 29634, USA

D. C. Hayes et al. (eds.), *USDA Forest Service Experimental Forests and Ranges*,
DOI 10.1007/978-1-4614-1818-4_18, © Springer New York 2014

in Research Triangle Park, NC. Many papers were written during the Calhoun's 15 years of existence, papers that document how land-use history creates a complex of environmental forcings that are hard to unwind. One Calhoun field experiment remains active, however, and over nearly six decades has become a model for the study of soil and ecosystem change on timescales of decades. The experiment contributes greatly to our understanding of the effects of acid atmospheric deposition on soils, forests, and waters and of decadel changes in carbon and nutrient cycling in soils and forests. Perhaps the long-term experiment's major contribution is its clear demonstration that soils are highly dynamic systems on timescales of decades and that this dynamism involves both surface and deep subsoils. The ongoing experiment's success is attributed to relatively simple experimental design, ample plot replication, rigorous (but not too arduous) protocol for resampling and archiving, and to its ability to address changing scientific and management priorities that are important to society and the environment. In the last decade, the experiment has become a platform for research and education that explore basic and applied science. As this manuscript goes to press, the Calhoun Experimental Forest has been designated to become one of the National Science Foundation's national Critical Zone (CZ) Observatories, a development that will allow researchers to, return to the questions that originated the Calhoun Experimental Forest in the first place: how and why severely disturbed landscapes evolve through time.

C. G. Wells
Southern Research Station, USDA Forest Service, 3041 E. Cornwallis Road,
Research Triangle Park, Durham, NC 27709, USA

A. E. Curry
Millersville University of Pennsylvania, Millersville, PA, USA

R. L. Fimmen
Geosyntec Consultants, Columbus, OH, USA

A. S. Grandy
University of New Hampshire, Durham, NH, USA

M. Hofmockel
Iowa State University, Ames, IO, USA

E. LeMaster
Francis Marion and Sumter National Forests, USDA Forest Service, 20 Work Center Road,
Whitmire, SC 29178, USA

M. L. Mobley
University of Wyoming, Laramie, WY, USA

D. Markewitz
University of Georgia, Athens, GA, USA

M. W. Morrison
Francis Marion and Sumter National Forests, USDA Forest Service, 4931 Broad River Rd.,
Columbia, SC 29212, USA

M. S. Strickland
Virginia Polytechnic and State University, Blacksburg, VA USA

Keywords Pedology · Ecosystem ecology · Forest soil · Forest dynamics · Long-term soil experiment · Sample archive · Nutrient cycling · Soil change · Acidification · Nutrient depletion · Nutrient resupply · Carbon sequestration · Carbon cycling · Acid deposition · Iron–Carbon redox cycling · ^{10}Be · Earth's critical zone · Biogeochemistry · Granite · Mineral weathering

18.1 The Calhoun Experimental Forest Rooted in Southern Environmental History

Louis J. Metz (1958a) began his remarkable booklet that described research on the John C. Calhoun Experimental Forest, "The Calhoun Experimental Forest...was established in 1947 for work on Piedmont forest, soil, and water problems. Located in the Sumter National Forest, near Union, South Carolina, the forest was chosen because it represented poorest Piedmont conditions."

The "poorest Piedmont conditions" were an aftermath of farming mainly for cotton, tobacco, corn, and wheat that geographer Stanley Trimble (2008) estimated eroded 17 cm of soil between 1750 and 1950 from the Piedmont from Virginia to Alabama. The Fisk University sociologist Charles S. Johnson (Johnson et al. 1935) described the 1930s' Piedmont and the rest of the cotton-producing South as:

> a miserable panorama of unpainted shacks, rain-gullied fields, straggling fences, and rattle-trap Fords, dirt, poverty, disease, drudgery, and monotony that stretches for a thousand miles across the cotton belt.

The US Department of Agriculture (USDA) Forest Service ambitiously organized the Calhoun Experimental Forest with high-quality scientists including Metz, Marvin Hoover, Jim Douglass, Otis Copeland, and Carol Wells, all of whom were based in rural Union County, SC, to work "principally at soil improvement" thereby promoting forests and stabilizing watersheds. They explicitly stated that if they succeeded in their Calhoun efforts, they could succeed anywhere in the South where soils and landscapes had been so adversely transformed by agriculture.

Much is written about the history of the South, but mainly about the region's remarkable human history, its political and cultural strife, tragedy, even its mystique (Irish 1952; Roland 1982; Tindall 1964). Simkins (1947) even declared, "History, not geography, made the solid South." However, a history of the South can also be written about the strong human-historical interactions with the region's dynamically changing environment (i.e., the region's dynamic geography). According to Cowdrey (1996), this environmental history was only beginning to be explored, but since that time has experienced a flourish of activity (Sutter and Manganiello 2009). Consider that in the early eighteenth century, a colonial agriculture was extensively being initiated in the southern Coastal Plain and Piedmont, thus beginning the human–environment interactions that can motivate new discussions in "Southern Studies." Profitable cotton markets, an amenable climate, and the native fertility of soils spurred clearing and conversion of southern forests and the formation of an enormous agricultural economy. In South Carolina, as across the entire South,

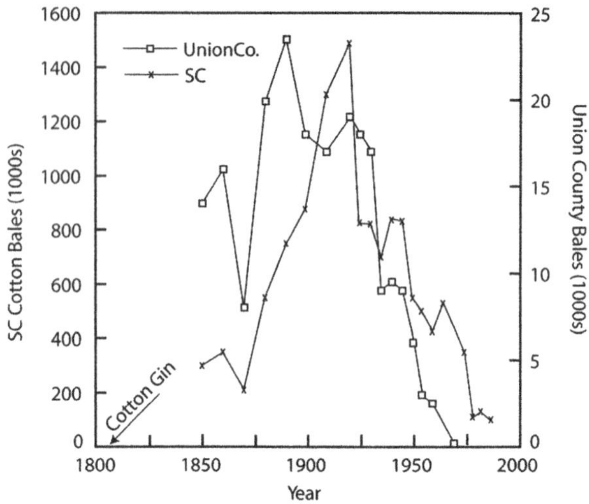

Fig. 18.1 Nineteenth-century and twentieth-century patterns of cotton production in Union County where the Calhoun Experimental Forest is located and South Carolina. (US Department of Commerce Agricultural Census data as cited in Richter and Markewitz 2001). In recent years, cotton has returned to South Carolina, not in the Piedmont but on intensive forms on the upper Coastal Plain

cotton production grew rapidly through the nineteenth century, interrupted only by the Civil War (Gray 1933), finally peaking between 1910 and 1920 (Fig. 18.1).

Cotton's expansion until the early twentieth century was followed by a rapid collapse (Fig. 18.1) which we attribute to forces that derive from changing human–environment relations. After the Civil War, Piedmont plantations and farms were increasingly worked by tenant farmers and sharecroppers. The new economic system brought land degradation and extensive erosion and poverty to many of its farmers. Tenant farmers and sharecroppers cultivated fields to grow cotton in farming systems that required hand labor throughout hot and humid summers. By 1920, southern farmers found themselves fighting a losing battle against the devastating boll weevil in fields that were severely eroded (Gesen 2011). In retrospect, one is struck by how and why farmers held onto the land so long. Piedmont farmers abandoned the land bound for the industrial southern and northern cities or for more promising agricultural regions across the nation While the experience of all tenant farmers and sharecroppers was difficult that of black farming families is particularly moving (Scott 1919; Rosengarten 1974).

Much of this story today seems like ancient history. Viewed from interstate highways, it is easily seen that the Piedmont is no longer much cultivated and the regeneration of trees appears most impressive. Southern pines and southern oaks, kudzu, and Bermuda grass now blanket most of Charles Johnson's "unpainted shacks, rain-gullied fields, straggling fences," superficially obscuring agriculture's severe effects on the soil and land. To know the South, it is therefore necessary to travel into the rural landscape, hiking through old-field forests, learning to read the land (Leopold 1949; Wessel 1997) like an ecologist, forester, geomorphologist, archeologist, and soil scientist, and learning to talk with, and ask questions of, rural residents. Such a tour of the southern Piedmont will readily turn up rain-gullied fields under the blanket of green and quickly lead us to a conclusion that not only is the history of the South about people *and* the environment, but that the passage of time has hidden more than healed cultivation's severe impacts on the land. A southern history is all about the transformation of soils and ecosystems, and if we are content to view the secondary green blanket as a restoration, we have much more to learn from these deep changes in the land. To rephrase Simkins (1947), history and geography have made the dynamic South.

18.2 Ecological "Restoration" at the Calhoun Experimental Forest

In the 1930s, the Sumter National Forest was assembled from tax-delinquent and abandoned farmland as well as from land owned by both willing and unwilling sellers. Land was typically purchased for a few dollars an acre. New insights are found in a thesis by Curry (2010) about the cultural history of the severely eroded and deteriorated land that became the Sumter National Forest and about the New Deal's programs to improve the Sumter's land and the lives of Piedmont farmers. Many informative files about the federal land purchases that created the National Forest are stored in land-purchase files at the Sumter National Forest's Tyger River and Enoree District Forest Offices.

In 1947, the Calhoun Experimental Forest was created within the Sumter National Forest (Dunford 1947; Metz 1958a). The Calhoun's goals were ambitious: to stimulate recovery of soils and forests on the severely eroded southern Piedmont. The Calhoun laboratories and offices were headquartered in the area today occupied by the Fairforest Firing Range. Metz (1958a) remarked, "We want to find the cheapest, quickest, most effective ways of speeding tree growth, increasing plant nutrients, and improving soil structure so that the land stores water for plant use."

The scientists were impressed with, and even appalled by the ecological condition of the region, and were highly motivated to improve forests, soils, and watersheds (Fig. 18.2). Leno Della-Bianca meticulously photographed the Calhoun and Sumter with many hundreds of Kodachrome slides. Marvin Hoover (1950), who would become one of the nation's leading hydrologists, began one of his Calhoun papers, "Nowhere in the country have hydrological processes in the soil been altered by past land use to a greater extent than in the South Carolina Piedmont." The Calhoun scientists published widely of an array of topics, but most focused on the serious environmental effects that had arisen from land degradation of the Southern Piedmont.

Perhaps more than anything, the early research at the Calhoun Experimental Forest demonstrated that the Piedmont's land-use history had created a complex of ecological problems that would prove difficult to unwind (Staff 1951). Not only were soils often severely eroded and gullied but such conditions greatly increased surface and subsurface runoff, creating enormous problems with sedimentation and jeopardizing soil-water storage and accentuating plant-water stress and plant disease (Hoover 1950, 1952a, b, 1954; Metz 1958a). Papers were written to gain a better understanding of forest O horizons (Hoover and Lunt 1952; Metz 1954), which greatly increase soil stability even when present as a relatively light blanket of leaves and needles. Advanced soil-moisture measurements were tested and adapted to Calhoun conditions (Olson and Hoover 1957) including the neutron probe deployed to

Fig. 18.2 **a** and **b** Two scenes of the eroded land impacted by cotton on the Sumter National Forest and the Calhoun Experimental Forest. Metz (1958a) stated that the location of the Calhoun Experimental Forest was selected as it represented "poorest Piedmont conditions." (USDA FS photographs from the 1950s)

5-meter depth (Hewlett and Metz 1960). Time series data demonstrated how contrasting vegetation types utilized soil moisture and altered soil temperature, and often these measurements were made to great soil depths (Douglass 1960, 1966; Greene 1953; Hoover 1950; Hoover et al. 1953; Metz and Douglass 1959; Patric et al. 1965), often in collaboration with Coweeta scientists.

These studies began a series of Calhoun studies that conceive of soil as potentially deep systems, much deeper than the superficial *solum* (Bacon et al. 2012; Markewitz et al. 1998; Richter and Markewitz 1995a, 2001; Richter et al. 2011; Richter and Yaalon 2012). Four small watersheds collected 10–15 years of precipitation and runoff data from contrasting land uses including exploitive farming and restored forest complete with gully control and terracing (Douglass 1972; Hewlett and Metz 1960). Precipitation's interception by tree canopies and litter layers was measured to ensure that all major forest water fluxes were quantified (Hoover 1953; Metz 1958b).

With the recognition that formerly cultivated Piedmont soils were associated with littleleaf disease and that the disease affected both shortleaf and loblolly pine (*Pinus echinata* and *Pinus taeda*), Calhoun scientists determined that severity of soil erosion was highly correlated with a disease complex associated with the pathogen *Phytophthora cinnamomi,* low nitrogen availability, and poor soil internal drainage (Copeland 1949a, b, 1954; Copeland and McAlpine 1955). Several studies examined root mortality of diseased trees (Copeland 1952). And by using rain shelters to impose drought stress on mature shortleaf pine trees, an alternative hypothesis was rejected that drought stress promoted littleleaf disease (Copeland 1955). Littleleaf hazard rating systems for the South were developed based on soil characteristics, one of which is still in use today (Campbell and Copeland 1954; Mistretta 1998). The studies of littleleaf spurred interest in shortleaf pine (Haney 1955, 1957), a species that in the twentieth and twenty-first centuries is having major trouble with its regeneration.

There was also great interest in how low soil fertility and high soil acidity might limit reforestation on eroded soils (Copeland and McAlpine 1955; Metz 1958a). Calcium and nitrogen were of greatest concern, and the beginnings of nutrient cycles were estimated. Fluxes of major nutrients were measured in contrasting forest stands including forest litterfall, tree foliage, wood, and soils (Metz 1952a, b; Metz and Wells 1965; Wells 1965; Wells and Metz 1963).

For reasons that today are not entirely clear, in the early 1960s, after about 15 years of work, the Piedmont Research Center and the Calhoun Experimental Forest were closed. According to Douglass (1972), research at the Calhoun Experimental Forest was terminated as part of a relocation of its scientists to Coweeta Hydrologic Laboratory and to a new, modern Forestry Sciences Laboratory in the Research Triangle Park, NC. Although research on the Calhoun Experimental Forest was no longer funded by the USDA Forest Service, the Calhoun continues to appear on national lists of Experimental Forests and Ranges and to maintain a director, currently USDA Forest Service scientist, Dr. Thomas Waldrop. Calhoun Experimental Forest records, files, data, and many hundreds of high-quality Kodachrome slides that document research during the 1940s and 1950s (e.g., Figs. 18.2 and 18.3) were moved for storage to the Coweeta Hydrologic Laboratory. Some files, mainly office

Fig. 18.3 The pine spacing study designed and installed by Lou Metz. Planted in the winter of 1956–1957 on old cotton fields with four tree spacings and four experimental blocks (i.e., 16 permanent plots), the study is a model of research continuity. Thanks to the career-long work of Carol Wells, the field study grew into what is today Calhoun's Long-term Soil-Ecosystem Experiment. (USDA FS photo, Calhoun Experimental Forest)

communications and miscellaneous correspondence, remain behind in the Sumter's Tyger River and the Enoree District Forest Offices in Union and Whitmire, SC. Daniel deB. Richter (lead author of this chapter, currently at Duke University) inherited files (and the Calhoun sample archive) from the laboratory of Dr. Carol Wells who had served on Richter's PhD committee from 1977 to 1980.

18.3 The Calhoun Experimental Forest's Versatile Long-Term Field Experiment

After the closure of the Calhoun's Piedmont Research Center, only one of the original experiments remains active today. This experiment has endured entirely due to Carol Wells' career-long interest in forest soils, becoming today a >50-year field experiment that is synonymous in many circles with "the Calhoun Experimental Forest." The experiment was one of the last installed by Lou Metz and was a meticulously designed, large-scale, loblolly pine spacing study (Fig. 18.3). Today, known formally as the Calhoun Long-Term Soil-Ecosystem Experiment (LTSE), the experimental plots reside on broad, gently sloping interfluves that were previously variably eroded cotton fields. The pine seedlings were planted in 16 plots in a randomized complete block design with four blocks of different soil conditions and four seedling densities (at 6 × 6-, 8 × 8-, 10 × 10-, and 12 × 12-ft spacings). Soon after planting, the experiment caught the attention of a young soil scientist from the University of Wisconsin, Carol Wells, hired by Metz to add soil and analytical chemistry expertise to the Calhoun's staff. In 1962, Wells began a time series of periodic samples of the ecosystem, aboveground and belowground, using a rigorous resampling protocol that continues to this day. The soil-sampling protocols involve sampling plots about once every 5 years by compositing 20 within-plot samples (each 2-cm dia) from each of four soil depths within 0–60 cm. After air-drying, samples are passed through a 2-mm screen, and become part of an extensive sample archive, housed at Duke University's Nicholas School of the Environment.

What Metz and Wells set in motion was very special, as their field experiment, resampling, and archiving protocols have created one of the world's finest studies of soil and ecosystem biogeochemical change *over the life of a forest*. The study's success is attributable to its relatively simple experimental design, its resampling and archiving protocols that are rigorous but not overly arduous, its relatively deep resampling of the root zone, and its ability to address a series of scientific and management issues that are critically important to society and the environment.

The great pedologist Walter Kubiena (1970) once suggested that soils and finely made watches have much in common, in that to understand how a soil or a watch works, one must observe and study the system as a whole. Like Kubiëna, we credit Metz and Wells for being critical of a too reductionist approach to soil science, in which studies reduce the soil to its parts and take little advantage that can come from observing and experimenting with the whole soil and the plants it supports. To paraphrase Kubiëna's words, Metz and Wells created a field study that offers scientists the potential to study the relations of the soil's component parts and the functioning of the whole system.

In the beginning, the Calhoun's LTSE was known as "the Calhoun spacing study" because it was established to learn how tree density affects water relations and the growth of individual trees and stands of loblolly pine. Metz (1958a) indicated that the study was "to be used in the future for litter and soil-moisture studies." Wells was interested in what the study might explain about soil fertility, tree and stand nutrition, and how a growing forest altered soils physically and chemically.

From its inception, the Calhoun spacing study generated regional interest as indicated by historic files. Other spacing studies in North Carolina and around the South were directly modeled after the Calhoun study. A loblolly pine spacing study based directly on the Calhoun study was even planted in Maui, HI (DeBell et al. 1989). The Calhoun study was used by Wells in the IBP or the International Biological Program (Coleman 2010).

As the pine trees grew in the Calhoun spacing study, their periodically measured diameters and heights were converted to yield data that were used in PhD dissertations, regional stand-growth models, and forest industry databases. The data were used to quantify relationships among tree density, mortality, volumetric dimensions of tree and stand biomass, and ecosystem productivity (Balmer et al. 1975; Harms and Lloyd 1981; Buford 1983, 1991; DeBell et al. 1989; Hafley et al. 1982; Hafley and Buford 1985). Buford (1983, 1991) developed and evaluated several sophisticated growth models from the observed density-dependent trajectories of growth. DeBell et al. (1989) compared the trajectories of growth and mortality at the Calhoun with those of the identically planted loblolly spacing study in Maui, HI. This latter comparison allowed DeBell et al. (1989) to develop ideas about stockability, a novel concept that suggested new ways that stand density–tree size relationships are related to environment as well as genotype.

During the second half of the twentieth century, the South grew to become the largest industrial wood-producing region in the world. The harvested wood was mainly pine and much of it originated from old-field stands not dissimilar to the old-field forests of the Calhoun spacing experiment. A number of scientists were more than a little interested in and concerned about how old-field soils, so seriously altered by land-use history, would support the new intensive pine management. To increase understanding about the nutrition of these new forests, Wells devoted his research to the science of ecosystem nutrient cycling. Using the Calhoun's field experiment (i.e., the Calhoun spacing study) and a network of pine stands across the Piedmont, Wells was among the first to estimate pine forest nutrient uptake from soils; nutrient recycling in canopy litterfall, throughfall, and retranslocation; and overall soil-nutrient demands of intensive pine management (Wells and Jorgensen 1975, 1979).

By the end of Wells' career in the late 1980s, the Calhoun soil-ecosystem experiment, then 30 years since planting, was providing indications that soil nutrients were being substantially altered by pine forest development. In a paper evaluating effects of intensive harvesting on soil nutrient supply and sustained productivity, Wells and Jorgensen (1979) used the Calhoun field experiment (then nearly 20 years since planting) to illustrate several important points about soil fertility:

> In the South after a conversion of old fields or low-quality hardwood stands to pine, the mineral soil is the primary source of the nutrients retained in trees and accumulated in forest floor during the first 20 years of stand development when nutrient requirements are large. In a *Pinus taeda* plantation on a field converted from agriculture (sic, the long-term Calhoun field experiment), N content of the surface 60 cm of mineral soil decreased from 2392 kg/ha at age 5 to 2010 kg/ha at age 15 (Wells and Jorgensen 1975). Extractable P, K, Ca, and Mg also declined. Mineral soil changes appear to be small between the ages of 20 to 40 in *Pinus taeda* stands because during this period the forest floor provides nutrients at an increasing rate with stand maturity. It is possible, however, that the mineral soil has been depleted of readily available elements by the demands of the growing stand.

According to Stone (1979), Wells' research (Wells and Jorgensen 1979) was some of the most advanced work of its kind. In addition, at the moment of Wells' retirement in the late 1980s, the Calhoun field experiment was poised like few others to directly address two of the most vexing and highest-stakes environmental issues of our time: that of acid deposition effects on soils and forests and that of forest carbon sequestration and cycling. Wells was also well aware that the field experiment was able to quantify ecological processes that are basic to ecosystem science, for example, the rate at which biologically available nutrients are released by mineral weathering (Carol Wells' personal communication with Daniel Richter, late 1980s).

The remainder of this chapter is organized into four sections, two of which involve how the long-running Calhoun soil-ecosystem experiment has been used since the late 1980s to address the internationally significant environmental issues of: the effects of acid atmospheric deposition on soils, water, and forests (Sect. 18.1.4), and the rates and processes with which forests cycle and sequester carbon (Sect. 18.1.5). Subsequently, we describe the many ways the Calhoun experiment is exploring basic science questions about ecosystems and ecosystem development (Sect. 18.1.6). The chapter concludes with a brief evaluation about this evolutionary research trajectory that has been shaped both by information needs of society and by the gradually increasing duration of the Calhoun field experiment itself (Sect. 18.1.7).

18.4 What are the Effects of Acid Atmospheric Deposition on Soils and Forests?

Rain and snow across the eastern USA and throughout northern and central Europe had turned acidic as a result of post-World War II emissions of SO_2 from uncontrolled coal burning and NO_x emissions from internal combustion engines. During the 1970s and 1980s, acidification of lakes, streams, soils, forests, and even architectural structures grew to become major concerns among scientists, public policy analysts, and the public. The environmental and financial stakes over acid deposition effects were high. While the processes and rates by which atmospheric acid deposition acidified soils and natural waters were widely studied (Binkley et al. 1989a; Johnson and Lindberg 1992; Likens et al. 2002), key issues remained unresolved. Uncertainties were exacerbated by poor understanding of a number of basic ecosystem processes, including soil mineral weathering reactions that ultimately buffer acid inputs to soils. While much was known about soil acidity via measurements of soil and water pH, aluminum solubility and reaction chemistry, short-term acidification experiments, and computer models, long-term direct observations of acidification in forest and wildland soils were notably absent. The situation would probably have been criticized by the whole-system advocate Kubiena (1970), as the acid rain issue revealed that soil and ecosystem sciences had been far too uniformly reductionist. *In other words, scientists knew a lot about how soil pH, acidity, and chemistry varies across space (at a moment in time), but we had little understanding and few direct observations of the rates at which individual soils were acidifying though time.*

The Calhoun field experiment, however, with its repeated soil samplings, archiving, and chemical analyses demonstrated significant acidification throughout the upper 60 cm of the rooting zone, a trend that was probably initiated soon after the last liming of cotton in 1955 (Fig. 18.4). Even by 1962, however, the soil's base saturation remained >70 % throughout the upper 60 cm of mineral soil. To grow cotton, the acidic Ultisol had been transformed into a "cultural Alfisol" by liming, an arable soil with relatively high exchangeable base saturation (Fig. 18.4). With the rapid growth of the forest, however, in the absence of continued liming and with acid deposition, the soil rapidly re-acidified. Binkley et al. (1989b) compared acidity in archived samples from 1962 and 1982, and estimated that decreasing pH was mainly due to reductions in exchangeable base cations, and that the soil's acid-neutralizing capacity was decreasing at about 1.3 kmol ha^{-1} y^{-1}. The growing trees were trading protons for much-needed nutrient cations of Ca, Mg, and K. Subsequent studies of Calhoun's soil acidification combined observations of soil chemical change with nutrient-cycling information and concluded that mineral weathering release of calcium approached zero during the first three decades of forest development (Bacon 2014; Markewitz 1995; Richter et al. 1994). Markewitz et al. (1998) analyzed soil-solution chemistry down to 6-m depth and estimated that up to 40 % of the acidification in the upper 60 cm of soil was attributable to acid atmospheric deposition. Below 60 cm, however, sulfate adsorption effectively reduced cation leaching. Acid deposition was thus acidifying surficial layers of soil

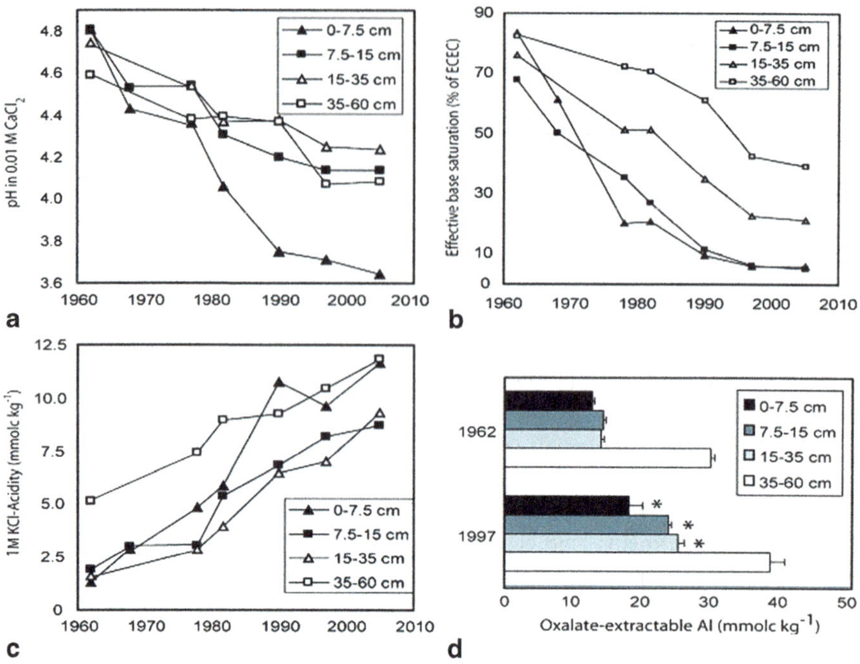

Fig. 18.4 The development of the Calhoun pine forest has been accompanied by substantial soil acidification, here indexed by soil **a** pH in 0.01 M $CaCl_2$, **b** effective base saturation, **c** KCl-exchangeable acidity, and **d** oxalate-extractable Al

by translocating exchangeable base cations more deeply into the root zone and soil profile (from O, A, E, and upper B horizons into middle and lower B horizons). In this sulfate-adsorbing system, acid deposition did not leach base cations completely out of the ecosystem, but rather moved them lower in the soil profile (Markewitz et al. 1998). These observations of soil acidification were featured in national assessments of acid deposition and air pollution (Richter 1991; Richter and Markewitz 1995b) and results from Markewitz et al. (1998) were widely reported in the popular press.

Stepping back from the technical details of soil acidification, these environmental issues could only have been addressed with an exceptional long-term field study supported by a long-term sample archive. Worldwide, fewer than 40 long-term field experiments have sample archives that are as old as those assembled in the Calhoun experiment. Metz and Wells can be credited with creating a unique and especially valuable forest-soil experiment, given that nearly all of these 40 long-term experiments test changes in agricultural soils. The Calhoun's sample archive is today as valuable as the long-term field plots themselves, and the entire program of study will only increase in value in years to come. This latter point is amply demonstrated by the Calhoun's growing contributions to scientific understanding of forest carbon sequestration and cycling (Fimmen 2004; Fimmen et al. 2008a, b; Galik et al. 2009;

Gaudinski et al. 2001; Grandy et al. 2009; Harrison et al. 1995; Mobley 2011; Mobley and Richter 2010; Mobley et al. 2013; Richter 2007; Richter et al. 1999, 2001; Richter and Mobley 2009; Richter and Markewitz 1996; Richter et al. 1995, 1999, 2006b, 2007a, b; Smith et al. 1997; Strickland 2009; Strickland et al. 2010).

18.5 Forest Carbon Sequestration and Cycling: Where has all the Carbon Gone?

National and international demand for quantitative information about soil and whole ecosystem gains and losses of carbon are drawing upon research sites such as that at the Calhoun for carbon cycling information (e.g., Galik et al. 2009; Richter et al. 1999). The Calhoun study is particularly well situated to address both soil and ecosystem carbon losses and sequestration: (1) The long-term Calhoun field experiment is part of a highly eroded landscape, and a major and contentious question concerns whether soil erosion is an atmospheric CO_2 source or sink (Billings et al. 2010; Van Oost et al. 2007). (2) The long-term field study quantitatively estimates carbon changes in the whole ecosystem aboveground *and* belowground over nearly five decades (Billings and Richter 2006; Mobley 2011; Mobley et al. 2013; Richter and Markewitz 1996; Richter et al. 1999). The Calhoun experiment's repeated soil-carbon sample archive and inventory to a soil depth of 60 cm makes exploring soil-carbon dynamics especially fruitful, as only about 10% of the >300 of the world's soil-carbon change studies have sampled soil deeper than 30 cm (Post and Kwon 2000; Richter and Mobley 2009; West and Post 2002).

Although cultivation might conventionally be considered to affect only losses of soil carbon, erosion recently has been proposed to be an important carbon sink due to transport and burial of eroded soil particles that contain soil organic carbon. A modeling exercise using data from across the Calhoun Experimental Forest was developed to provide a range of potential carbon effects associated with soil erosion (Billings et al. 2010). Modeling the influence of erosion associated with 150 years of agriculture on the strength of net soil-carbon sources or sinks suggests that erosion has been a substantial driver of carbon dynamics. Depending on the fate of eroded carbon—whether it was oxidized or retained as soil organic matter—the modeling suggests that up to ~9 kg of soil carbon m^{-2} was accrued, or up to ~3 kg carbon m^{-2} was released as CO_2, as a result of erosional forcings.

In 1995, the Calhoun's long-term field experiment and long-term time series of carbon data were selected to be the forest data used by several dozen carbon modelers at the International Soil Organic Matter Network (SOMNET) workshop. The Calhoun carbon data complemented data sets from the famous Broadbalk Wheat and the Park Grass experiments that represented cultivated and grassland systems, respectively. Outcomes of this carbon modeling were summarized in Smith et al. (1997), which, among other conclusions, emphasized the need for more concerted efforts at modeling soil-carbon change over decades' timescales and the need for many more long-term soil experiments such as the Broadbalk, Park Grass, and the Calhoun.

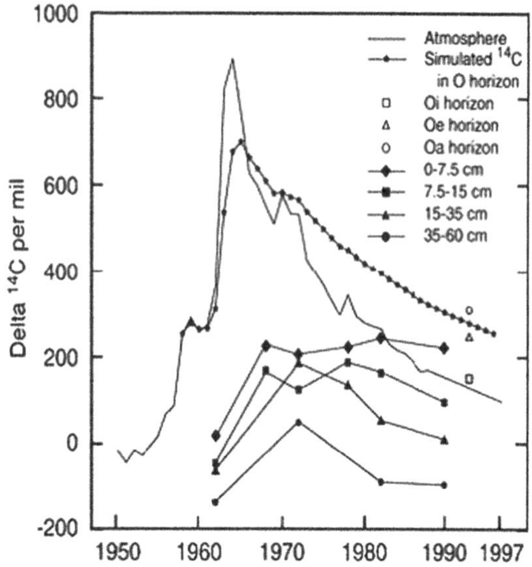

Fig. 18.5 Time trends of ^{14}C in atmospheric CO_2 (1950–1997), three layers of 1992 forest floor (Oi, Oe, and Oa), and mineral soil (in 1962, 1968, 1972, 1977, 1982, and 1990) at the Calhoun Experimental Forest. Model-simulated changes in ^{14}C in O horizons (1957–1996) are estimated from a decomposition model of Wells and Jorgensen (1979). Estimates of litterfall inputs over the four decades were set to coincide with the pattern of ^{14}C in atmospheric CO_2

In the soils of the Calhoun's long-term field experiment, long-practiced cultivation prior to reforestation had greatly depleted organic carbon from the sandy A horizons. Under the secondary forest, enormous amounts of forest organic carbon have been added to the soil, and the historic effects of cultivation are slowly receding from the soil's memory (Targulian and Goryachkin 2004). Yet, the drastic increases in soil-carbon inputs have not resulted in much soil sequestration due to active decomposition and plant and microbial demands for nutrients. Ongoing research focuses on these processes, which is reflected in both the quantity and composition of the soil's changing organic matter.

Overall, the aggrading forest in the Calhoun experiment has been a strong sink for atmospheric carbon from planting to the late 1990s. For example, the forest as a whole has sequestered over 16 kg C m^{-2} in its first four decades, a rate in excess of 400 g C m^{-2} y^{-1} (Galik et al. 2009; Harrison et al. 1995; Richter et al. 1999). Nearly all of this carbon accumulation is in tree biomass, aboveground and belowground, and secondarily in the soil, i.e., the forest floor and mineral soil. Remarkably, surface mineral soils have sequestered minimal carbon over five decades of forest development, and the experiment demonstrates how rapidly the decomposer system in the mineral soil oxidizes organic carbon inputs. This rapid rate of decomposition is also clear in the cycling of ^{14}C (radiocarbon) through the forest and soil (Fig. 18.5). Forest carbon inputs to the soil profile are labeled with ^{14}C because the pine forest was planted in the winter of 1956–1957 and has grown entirely within the era of aboveground thermonuclear bomb testing. By 1964, atmospheric CO_2 had nearly doubled its content of ^{14}C (Fig. 18.5) and while forest photosynthates were within several years distributed throughout the upper 60 cm of soil, the radiocarbon rapidly passed through mineral soils deeper than 15 cm (Fig. 18.5).

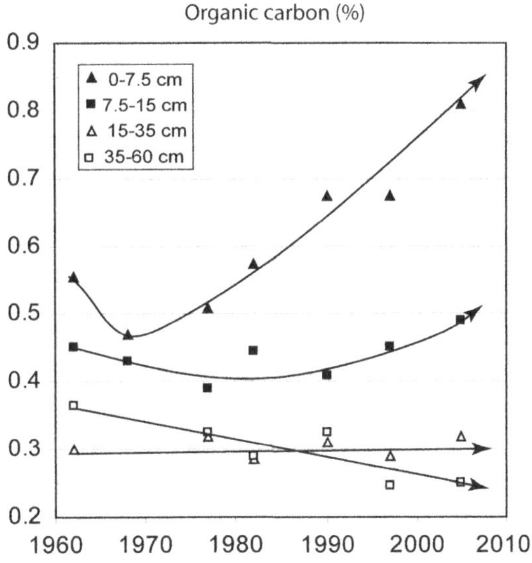

Fig. 18.6 Mineral-soil carbon in permanent plots at the Calhoun Experimental Forest. The randomized complete block repeated analysis indicates significant increases over time in the 0–7.5-cm layer and decreases in the 35–60-cm layer

Thus, in spite of the substantial carbon sink of the whole aggrading forest during the period of active forest growth, there has been little carbon sequestration in the mineral soil. This result has varied somewhat with soil depth, however. In the four layers of mineral soil sampled, the most surficial 7.5 cm has significantly increased in its carbon concentration and content as an A horizon has redeveloped under the forest after long-practiced cultivation of cotton (Fig. 18.6). The mean content of carbon sequestered in the surficial 0–7.5-cm layer is about 2.9 Mg ha^{-1} from 1962 to 2005, a statistically significant increase. In contrast, the deepest depth sampled, 35–60 cm, significantly lost about 4.2 Mg C ha^{-1} during the same period counterbalancing the modest gains observed in surficial layers (Fig. 18.6). Our hypothesis for this loss of soil carbon at that depth is that organic matter's decomposition was promoted by forest growth due to the forest's greatly increased transpiration and the forest's demand for nutrients. Especially from the deepest depth sampled, 35–60 cm, water uptake by trees is much larger than that by cotton and this has hypothetically promoted oxidative conditions required for significant mineralization to proceed. Ongoing research is exploring this hypothesis and that of forest-priming effects on soil organic carbon decomposition (Mobley 2011).

18.6 The Calhoun Long-Term Field Experiment and Basic Soil and Ecosystem Science

Over the years, Calhoun investigators have evaluated how decades of forest growth have altered soil chemistry and used the soil-change data to study difficult-to-measure ecosystem processes involving chemical elements not only of carbon (see

Section 18.5), but of nitrogen, phosphorus, potassium, calcium, magnesium, and trace elements boron, copper, iron, manganese, and zinc. Such processes include not only acidification and organic carbon sequestration but also nitrogen fixation, atmospheric nitrogen deposition, turnover of organic and mineral fractions of soil phosphorus, dynamics of Fe and Al oxy-hydroxides, and mineral weathering reactions involving calcium, magnesium, potassium, and trace elements. Promising preliminary studies have been conducted on the mobility of atmospheric lead derived from leaded gasoline and on the legacies of arsenic historically applied to cotton as pesticides to combat the boll weevil. We have also begun studies of the effects of land-use history on soil biology, soil ecology, and fungal biodiversity. And not only have human-affected changes in Calhoun soils and ecosystems been studied but also the natural pedogenic system has been a focus of research as well. Recently, we proposed a new hypothesis for crustal weathering involving redox reactions initiated in rhizospheres (Fimmen et al. 2008b; Richter et al. 2007b). In 2010, soils were sampled and inventoried for cosmogenic beryllium (specifically, the radionuclide ^{10}Be) to estimate soil residence time on the Calhoun's biogeomorphically most stable interfluves, which we hypothesize to be much older than generally appreciated (Bacon 2010; Bacon et al. 2012). This is an active period of research at the Calhoun Experimental Forest.

18.6.1 Soil and Ecosystem Nitrogen

Soil nitrogen in the Calhoun's long-term experiment has been a subject of intense study. Four decades of forest growth has resulted in the transfer of about 30 % of the total nitrogen in the upper 60 cm of mineral soil into new forest biomass and forest floor, via soil organic matter mineralization and subsequent tree uptake of bioavailable nitrogen. The forest grew itself into a state of acute nitrogen deficiency, with low LAI, (< 2.5 m2m-2) and foliar N Concentrations in upper crowns of 1.06%. This progressive nitrogen limitation (Johnson 2006) is attributed to the rapid rate of tree uptake of nitrogen combined with the extremely low rates of nitrogen mineralization in the pine forest floor during the first four decades of forest development. The repeated samplings of soil indicated that the source of soil N came initially from the most surficial layers of mineral soil but that over the life of the forest, the 35–60-cm layer of soil was the source of a very large fraction of the N accumulated in biomass and forest floor (Billings and Richter 2006; Richter et al. 2000). Forest development also imparted pronounced depth dependence in ^{15}N in soil organic matter over several decades, due to the transfer of > 800 kg N ha^{-1} into aggrading plant biomass and forest floor. The deepest layer sampled, 35–60-cm soils, experienced the greatest change in ^{15}N, reaching a maximum of 9.1 ‰, during the 7.6 ‰ shift over four decades (Billings and Richter 2006). A modest whole-ecosystem accretion of nitrogen (in plants, forest floor, plus soil) suggests that the most likely source of the additional nitrogen in the ecosystem is from incorporation of atmospheric nitrogen deposition rather than N_2 fixation (Richter et al. 2000).

18.6.2 Soil and Ecosystem Potassium, Calcium, and Magnesium

Four decades of forest growth in the old-field soils of the Calhoun field experiment have significantly depleted soil-exchangeable cations. Mineral weathering release rates of the three principle nutrient cations, calcium, magnesium, and potassium were evaluated from estimates of mineral-soil depletions, nutrient accumulations in biomass and forest floor, atmospheric inputs, and soil leaching removals. The rates of weathering resupply differed greatly among the three cations. For calcium, accumulations in tree biomass and forest floor plus that lost to leaching were comparable to observed depletions of soil-exchangeable calcium. Thus, for calcium, mineral weathering and deep-root uptake did not buffer soil's exchangeable pools even on timescales of decades. For potassium, however, the patterns could hardly have contrasted more. Removals of soil potassium by tree uptake plus leaching losses exceeded observed depletions of soil-exchangeable potassium by nearly 20 times. Bioavailable potassium was readily replenished by non-exchangeable mineral sources, patterns also observed in laboratory and greenhouse experiments and attributable to relatively large bioavailable potassium-containing minerals in the Calhoun soils (Markewitz and Richter 2000). The behavior of magnesium shared attributes of the contrasting patterns of calcium and potassium. While forest growth significantly depleted soil-exchangeable magnesium, depletions were not to the full extent of magnesium contained in biomass and forest floor and that lost to leaching (Richter et al. 1994; Richter and Markewitz 2001). Additional studies need to identify the minerals involved in supplying bioavailable fractions of potassium and magnesium and to make progress on soil water movement and leaching (Gnau 1992).

18.6.3 Soil and Ecosystem Phosphorus

Forest growth also drew heavily upon fractions of mineral-soil phosphorus to supply the >80 kg P ha^{-1} contained in biomass and forest floor in three decades. While agricultural fertilization had built up phosphorus in organic compounds and that associated with calcium, iron, and aluminum phosphates, surprisingly, the most labile forms of bioavailable phosphorus, that indexed by exchange resins, $NaHCO_3$, and Mehlich-III extractants, remained relatively high throughout this period of forest development (Richter et al. 2006a). Decreases in soil phosphorus were statistically significant and most substantial in slowly cycling organic and inorganic phosphorus associated with iron and aluminum oxides and especially from calcium compounds, and it was these latter fractions that accounted for nearly all of the phosphorus contained in biomass and O horizons. These dynamics in soil phosphorus are attributed to the strong phosphorus sink strength of the aggrading forest (at 2.9 kg ha^{-1} year^{-1} over 28 years), the legacies of fertilization, and the ability of relatively slowly cycling phosphorus to supply substantial amounts of bioavailable phosphorus over decades' timescales. Forest development had clearly restructured the chemistry of soil phosphorus in only a few decades (Richter et al. 2006a).

18.6.4 Soil and Ecosystem Trace Elements

The range of responses among the ecosystem's trace elements provides more evidence that soil systems are highly dynamic over timescales of decades (Li 2009; Li et al. 2008). Soil-extractable boron and manganese were significantly depleted by four decades of forest growth with depletions comparable to accumulations in biomass and forest floor. Thus, uptake of boron and manganese by trees greatly outpaced resupplies from atmospheric deposition, mineral weathering, and deep-root uptake. The changes in boron and manganese contrasted with those of soil-extractable zinc and copper which changed relatively little during forest growth, indicating that zinc and copper resupplies (viatmospheric deposition, mineral weathering, and deep-root uptake) kept pace with accumulations by the aggrading forest (Li et al. 2008). Lastly, forest iron cycling was qualitatively different from that of all other macronutrients and micronutrients. Forest floor accumulations of iron dwarfed iron inputs in litterfall and canopy throughfall. Moreover, soil oxalate-extractable iron, taken to represent iron in short-range-order forms of oxides, *increased* about tenfold more than tree-biomass accumulations. We hypothesize that the forest's large soil inputs of organic matter over four decades combined with the ecosystem's profound acidification have altered the surface chemistry of the soil's relatively large crystalline pools of iron oxy-hydroxides, in effect increasing more reactive forms of short-range-order components (Li et al. 2008). These changes in iron oxide surfaces may have altered the bioavailability and retention of other nutrients, contaminants, and organic matter, components which intimately interact with oxides.

18.6.5 How do Land-Use Histories Alter Soil Ecology, Organic Matter Chemistry, and Soil Heterogeneity?

The Calhoun's LTSE provides special insight into soil biogeochemical changes that accompany ecosystem development and land-use history. To augment this ecosystem study and explore processes not well captured in the soil archive-based study, we have investigated how soil ecology, soil heterogeneity, and organic matter chemistry were altered by past land uses and their developmental trajectories. A network of ecosystems (on biogeomorphically stable interfluves and similar soil series to those found in the Calhoun's LTSE) was established with the main difference among sites being land-use history. The different histories include remnant hardwood stands never cultivated or fertilized serving as reference sites and three other ecosystems being historically cultivated for cotton and then continuing to be cultivated for corn and wheat, converted to grasslands for hay and pastures, and converted to secondary pine forests which at the time of sampling were about 50 years in age. Three replicates of each of the four ecosystems were located across the Calhoun Experimental Forest, the Sumter National Forest, and local private farmlands.

To examine how soil macrofauna were influenced by land-use histories (Callaham et al. 2006), soils of the four ecosystems were collected quarterly. Hardwood stands that were never cultivated or fertilized supported by far the most taxonomically diverse communities, followed by pine stands, pastures, and cultivated fields in order of decreasing diversity. For earthworms, scarab and carabid beetles, diplopods, chilopods, gastropods, and Diptera, there were long-term effects of soil disturbance and plant composition, with less diverse invertebrate communities typically having a few, often nonnative, disturbance-tolerant taxa in the most disturbed sites (Callaham et al. 2006).

To determine how microbial diversity and activity were land-use dependent, Strickland et al. (2010) determined that land use and soil-extractable phosphorus (an index of past fertilization) were both strongly associated with the mineralization of ^{13}C-labeled glucose applied in the field across seasons. Measures of microbial community size, activity, and composition all appeared to be relatively poor predictors of mineralization. In a related study, Strickland et al. (2009) used contrasting litters and demonstrated that the concept of "litter quality" is not simply determined by the chemical characteristics of the plant detritus but is largely determined by the history of the microbial communities responsible for the initial stages of decomposition.

To evaluate how land uses had affected fungal communities, Jackson et al. (2005) not only quantified high species diversity of fungi in O and A horizon soils of hardwoods, cultivated fields, grasslands, and in old-field pines but also observed deep taxonomic shifts in the makeup of these communities that were associated with land-use change (Jackson 2010). At the species level, all land uses were distinct, while phylogenetic analysis at higher taxonomic levels revealed shifts from a broad range of ruderal, disturbance-tolerant, lower fungi in cultivated and grassland soils to the dominance of several lineages of ectomycorrhizal fungi associated with the forest's return on the landscape. This pattern was also evident in molecular surveys of soil microeukaryotes, as well as assays of fungal propagules. Jackson's (2010) work highlights the very powerful effects of land-use change on soil microbial diversity.

To begin to examine how land use has altered the chemistry of soil organic matter, soil samples were collected from the four ecosystems and key attributes of the chemical structure of organic matter determined (Grandy et al. 2009). Structure of organic matter was evaluated using pyrolysis–gas chromatography/mass spectroscopy. The dominant chemical components of soil organic matter included common pyrolysis products of lignin (e.g., 4-vinylguaiacol and 4-acetylguaiacol), polysaccharides (e.g., levoglucosenone and furfural), nitrogen-bearing compounds (e.g., pyridine and pyrrole), as well as a variety of lipids and compounds of unknown origin. In these analyses, there were no detectable effects of land-use history on soil organic matter chemistry expressed as either individual moieties or chemical groups. Across all soils, the nitrogen content of organic matter was strongly correlated with soil biological processes including enzyme activities and fungal/bacterial ratios. Soil texture strongly predicted both the abundance of lignin derivatives and nitrogen-containing compounds. The study showed the potential for strong relationships between soil biological processes and soil organic matter chemistry and for the overriding influence of edaphic soil properties such as soil texture on these relationships.

Lastly, to examine effects of land-use history on soil spatial heterogeneity of chemical elements and soil properties, Li et al. (2010) sampled surficial soils (0–7.5-cm mineral soils) using a spatially explicit design within three 0.09-ha plots in three replicates of the never-cultivated hardwoods, cultivated agricultural fields, and old-field pine forests about 50 years in age. Results indicated that land-use history altered soil properties' central tendencies and their spatial heterogeneities; within-plot variations were generally much higher in hardwood and pine forest soils than in cultivated soils; for soil carbon and major and trace elements, trend surface analysis, correlograms, and interpolation maps indicated pronounced spatial patterns were evident in surface soils under hardwood and pine forests and much less so in cultivated soils. Relative to soils that have never been cultivated, spatial heterogeneity is greatly reduced in many soil properties by plowing, fertilization, and other practices associated with agricultural crop production, but pine forest growth on previously cultivated soils reestablishes at least some of the heterogeneity of soil properties within a few decades. Overall, within-plot variances were high for most properties especially of the forested soils and emphasize that researchers should pay careful attention to soil-sampling designs and sample sizes, given the variability of soil properties they are studying. Li et al. (2010) demonstrate clearly that optimal sample sizes are a function of land-use history.

18.6.6 The Discovery of a New Biologically Driven Mineral Weathering Reaction?

Detailed investigations of biologically driven acidification and weathering have for several decades been a theme of research at the Calhoun (Richter and Markewitz 1995a, b, 2001; Richter et al. 1994; Markewitz and Richter 1998). Research has recently focused on rhizosphere effects on pedogenesis in general (Richter et al. 2007b) and on rhizogenic iron–carbon redox cycling (Fimmen et al. 2008b) as an underexplored ecosystem process with many implications (Fig. 18.7).

Since the early days of research at the Calhoun Experimental Forest, internal soil drainage and aeration were understood to be key characteristics of upland Piedmont soils (Copeland and McAlpine 1955; Hoover 1950). In many upland soils with inhibited drainage, precipitation events periodically inhibit oxygen diffusivity, especially in subsoil B horizons, which allows the oxidation of organic reductants to mobilize manganese and iron via reductive dissolution (Fimmen et al. 2008b; Richter et al. 2007b). Rhizospheres are microsites in subsoils with concentrations of organic reductants and where reduction is most pronounced. Without oxygen, manganese and iron are reduced and mobilized from near-root microsites only to oxidatively precipitate in surrounding, more oxidizing soil environments (Fig. 18.7). Although upland Calhoun soil profiles in most conditions are aerobic, these periodic suboxic and anoxic events are most pronounced in rhizosphere microsites that create redoximorphic features that are strongly depth dependent (Fig. 18.7). Because oxidation and reduction reactions of redox-active metals are accompanied by significant

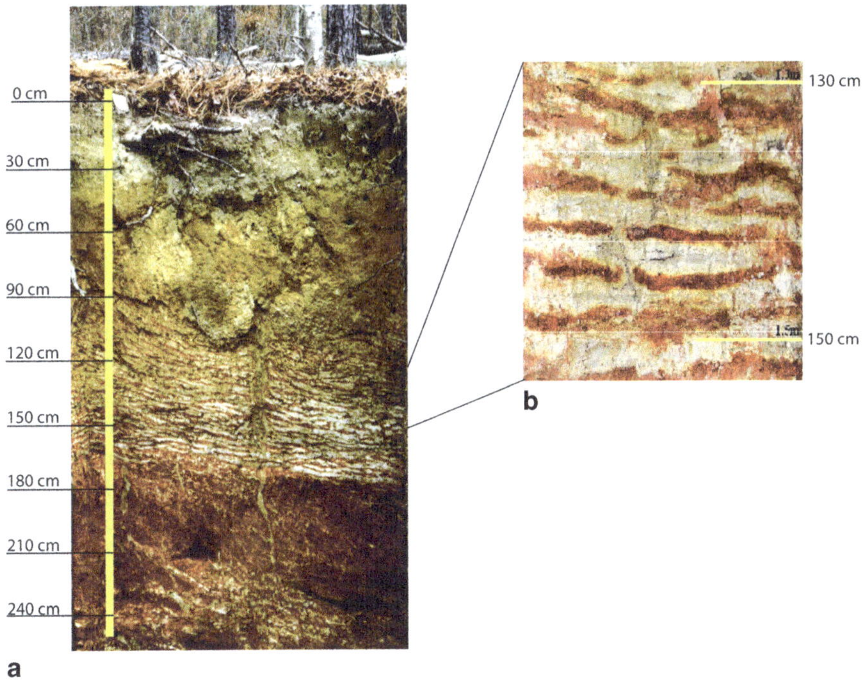

Fig. 18.7 Calhoun soil profile illustrating major soil horizons, 0–240-cm depth. The inset photo from 130 to 150 cm illustrates prominent redoximorphic microsites of the lower B horizons

transfers of electrons and protons, we hypothesize that these iron–carbon redox reactions are part of a potent weathering process, complementing the more well-studied acid-promoted and ligand-promoted reactions as biogeochemical processes that drive crustal weathering (Fimmen et al. 2008b; Richter et al. 2007b). Moreover, we hypothesize that rhizogenic iron–carbon redox cycling is significant due to the deep and extensive rooting and mottling of many upland subsoils across a wide range of plant communities, lithologies, and soil-moisture and temperature regimes.

18.6.7 Just How Old are Piedmont Soils and Land Surfaces?

On the granite–gneiss bedrock of the southern Piedmont, soil development on biogeomorphically stable surfaces is not dissimilar to some of the advanced weathering-stage soils in the humid lowland tropics. Soil systems of the Calhoun Experimental Forest are extremely acidic, all weatherable primary minerals have been exhausted for many meters, and soil profiles have accumulated to fantastic depths, i.e., to several tens of meters, all over unweathered bedrock (Bacon 2010; Bacon et al. 2012; Richter and Markewitz 1995a, 2001). Mass-balance calculations using total elemental contents referenced to insoluble elements such as zirconium or titanium indicate that nearly all calcium-bearing and sodium-bearing minerals are depleted for 5 m to greater than 10 m in depth (Oh and Richter 2005; Rasmussen et al. 2011).

To estimate residence times of such soils and interfluves, geomorphologists inventory meteoric ^{10}Be in soil profiles. The minimum age of a land surface can be estimated from the quotient of a soil's ^{10}Be inventory and the estimated rate of atmospheric deposition of ^{10}Be, accounting for radioactive decay and erosional losses. In the past, the approach assumes immobility of ^{10}Be, i.e., that atmospheric deposition is accumulated in the soil profile with no solution leaching. Beryllium solubility, however, is strongly controlled by pH, and in chemical systems with pH less than 5, beryllium dehydroxylates and reacts as a divalent cation, thereby becoming subject to cation-exchange reactions like those affecting the mobile cations calcium and magnesium. Mass-balance estimates of total beryllium (^{9}Be) in this Calhoun soil profile indicate that at least 60 % of the beryllium in unweathered rock has been lost by leaching. By coupling this mass balance of total beryllium with the meteoric ^{10}Be inventory, we have redefined pedogenic time on the Piedmont and estimate that soils on Piedmont interfluves have resided at Earth's surface for much if not all of the Quaternary. (Bacon 2010; Bacon et al. 2012).

18.7 Conclusions: The Future of the Calhoun LTSE and Critical Zone Observatory

In May of 2007, 8 of the 16 sampling plots of the Calhoun's LTSE were clearcut harvested. A local logger was contracted to clear all trees from the plots planted in the winter of 1956–1957 and in surrounding buffer zones and to do so in an operational manner. In the winter of 2008–2009, the cutover plots were replanted with pine seedlings by Duke University students, and seedling survival as of the summer of 2011 is more than 95 %. A comprehensive sampling of fresh logging slash was initiated on the eight cutover plots, and this organic material has been resampled in 2010, 3 years after logging.

Looking back, the long-term Calhoun experiment has quantified pine forest growth effects on a bare old-field soil; looking forward, the Calhoun experiment will quantify pine forest growth effects on a cutover forest soil. The dynamics of the variable amounts of logging slash among the eight cutover plots provide great experimental interest to the renewed long-term study. The long-term experiment will be testing differences in soil and ecosystem biogeochemistry between a second-rotation pine plantation with that of forests planted in 1956–1957 that are transitioning to hardwoods.

We know remarkably little about how soils change over decades and centuries, not only because soils are so spatially variable, hard to observe, and highly complex but also because so few long-term field experiments have been set out to directly observe changes in soils over time. In the last century, while long-term measurements of weather, floods, water quality, human health, wildlife, earthquakes, and air pollution have all become indispensable for guiding decision-making in natural resource and environmental management, long-term observations of soils have lagged far behind (Richter et al. 2007a). Humanity is rapidly changing the Earth's soil on local to global scales (Richter 2007; Richter et al. 2011), and long-term soil experiments such as those which Metz and Wells set out on the Calhoun Experimental Forest are becoming increasingly valuable (Richter et al. 2007a; Richter and Yaalon 2012).

The continuity of the long-term Calhoun experiment has allowed researchers to learn much about the functioning of soils and whole ecosystems, and about the importance of environmental history to the environmental present and future. The more we have studied the Calhoun ecosystems, the more we appreciate the importance of historical impacts that still affect many components and processes of contemporary Calhoun soils and ecosystems. There is still much to learn at the Calhoun's LTSE, just as there was when the Calhoun Experimental Forest was first established. One difference is that in 2010, the region's soils, so long cultivated for cotton, have now been greatly altered by old-field forest regeneration over more than half a century of forest development. Both historic cultivation and reforestation have now transformed Calhoun soils like many others across the South, and nearly all soils of the region are now thoroughly influenced by human culture.

After Wells' retirement in the late 1980s, the field experiment has been continued by Daniel Richter in a Duke University–USDA FS collaboration with key USDA Forest Service support from Mac Callaham, Tom Waldrop, Mary Morrison, and Beth LeMaster. Many others have participated along the way. Since the 1980s, researchers have competed successfully for financial support from a wide range of research organizations, sustaining the Calhoun's long-term research entirely from short-term competitive research grants. The USDA's competitive grants programs supported Calhoun research on acid deposition effects in the late 1980s and early 1990s. After passage of the 1990 Clean Air Act diminished the research priority of acid deposition, forest carbon cycling became a national research priority, and from the mid-1990s to the present, Calhoun research has been supported to document both the forest's accumulation rate of carbon and the biogeochemical details of the forest carbon cycle. The carbon research has been funded by several competitive research programs of the National Science Foundation and the USDA as well.

After 2001 and the publication of *Understanding Soil Change,* the Calhoun field experiment became a platform that investigated a range of basic biology, ecosystem, and biogeochemistry questions. This research has been supported by NSF's programs in the Biology and Geosciences Directorates, the Andrew W. Mellon Foundation, the Trent Foundation, the Forest History Society, and Duke University. This support has encouraged the evolution of the Calhoun field experiment toward one that couples soil and ecosystem processes, quantifies biogeochemical changes in ecosystems aboveground *and* belowground on timescales of decades, documents soil biological community responses to ecosystem development, and examines the importance of environmental history in forests and soils.

Calhoun researchers are convinced that LTSEs are critical to understanding and managing ecosystem sustainability. We have therefore established a Web site-driven inventory of the world's LTSEs to initiate a global network of LTSE experiments and scientists. The LTSE inventory now numbers about 250 studies on all continents, and the project encourages scientists to write cross-LTSE review papers and initiate new cross-LTSE research (Richter and Yaalon 2012).

Calhoun researchers also welcome the news received in June 2013 that the research site will become one of the nation's nine Critical Zone (CZ) Observatories.

The concept of the CZ is related to the ecologist's "ecosystem" (Pickett and Cadenasso 2002), and Calhoun researchers view the CZ as an "expanded ecosystem," ranging as it does from the atmosphere and uppermost plant canopy boundary layers down through the root zone and soil to the deepest penetration of groundwater (Amundson et al. 2007; Brantley et al. 2006; Lin 2010). While the lower bounds of the ecosystem are often marked by the depth of rooting, the subsurface CZ extends throughout the full biogeochemical zone of Earth's crustal weathering. The CZ is an "expanded ecosystem" due also to its scientific breadth that marshals participants from biology, ecology, geology, pedology, hydrology, biogeochemistry, climatology, geomorphology, sedimentology, geochemistry, geophysics, engineering, materials research and even social science, environmental history, and perhaps someday environmental humanities. The US National Research Council (NRC 2001) described the integrated study of Earth's CZ as one of the most compelling research areas in the Earth sciences in the twenty-first century, and Calhoun researchers anticipate addressing questions about how natural and human forcings affect CZ dynamics and evolution, specifically to improve understanding and management of the Earth's CZ in the face of land-use change and land degradation, and how human-forced CZs alter human well-being.

References

Amundson R, Richter DD, Humphreys GS, Jobbágy EG, Gaillardet J (2007) Coupling between biota and earth materials in the critical zone. Elements 3:327–332

Bacon AR (2014) Pedo-genesis and anthropedo-genesis on the Southern Piedmont. PhD dissertation, Duke University, Durham, NC

Bacon AR, Richter DD, Bierman PR, Rood DH (2012) Coupling meteoric ^{10}Be inventories with pedogenic losses of ^{9}Be to improve estimates of soil residence time on an ancient North American interfluve. Geology 40:847–850

Balmer WE, Owens EG, Jorgensen JR (1975) Effects of spacing on loblolly pine growth 15 years after Planting. USDA Forest Servivce Southeastern Forest Experiment Station, Asheville, NC, Research Note SE-211

Billings SA, Richter DD (2006) Changes in stable isotopic signatures of soil nitrogen and carbon during forty years of forest development. Oecologia 148:325–333. doi:10.1007/s00442-006-0366-7

Billings SA, Buddemeier RW, Richter DD, Van Oost K, Bohling G (2010) A simple method for estimating the influence of eroding soil profiles on atmospheric CO_2. Glob Biogeochem Cycles 24 GB2001. doi:10.1029/2009GB003560

Binkley D, Driscoll CT, Schoeneberger P (1989a) Acid deposition and forest soils: context and case studies of the Southeastern United States. Ecosystem studies 72. Springer, New York

Binkley D, Valentine D, Wells C, Valentine U (1989b) An empirical analysis of the factors contributing to 20-yr decrease in soil pH in an old-field plantation of loblolly pine. Biogeochemistry 8:39–54

Brantley SL, White TS, White AF, Sparks D, Richter D, Pregitzer K, Derry L, Chorover J, April R, Anderson S, Amundson R (2006) Frontiers in exploration of the critical zone: report of a workshop sponsored by the National Science Foundation, p 30

Buford MA (1983) Probability distributions as models for mortality in loblolly pine (*Pinus taeda* L.) plantations. Ph. D. dissertation, North Carolina State University, Raleigh, NC

Buford MA (1991) Performance of four yield models for predicting stand dynamics of a 30-year-old loblolly pine (*Pinus taeda* L.) spacing study. For Ecol Manag 46:23–38

Callaham M, Richter DD Jr, Coleman DC, Hofmockel M (2006) Long term land-use effects on soil invertebrate communities in southern Appalachian Piedmont soils. Eur J Soil Biol 42:S150–S156

Campbell WA, Copeland OL (1954) Littleleaf disease of shortleaf and loblolly pines. Circular 940. U.S. Department of Agriculture, Washington, DC

Coleman DC (2010) Big ecology. University of California Press, Berkeley

Copeland OL (1949a) Relationship of soils to littleleaf disease in pine. For Farmer 8:6

Copeland OL (1949b) Some relations between soils to the littleleaf disease of pine. J For 47:566–568

Copeland OL (1952) Root mortality of shortleaf and loblolly pine in relation to soils and littleleaf disease. J For 50:21–25

Copeland OL (1954) Estimating the littleleaf hazard in South Carolina Piedmont shortleaf pine stands based on site index. USDA Forest Service Southeastern Forest Experiment Station Research Note 57, Asheville, NC

Copeland OL (1955) The effects of an artificially induced drought on shortleaf pine. J For 53:262–264

Copeland OL, McAlpine RG (1955) Interrelationships of littleleaf, site index, soil, and ground cover in Piedmont shortleaf pine stands. Ecology 36:635–641

Cowdrey AE (1996) This land, this South. University of Kentucky Press, Lexington

Curry AE (2010) Emancipation and erosion: tenant farming, share-cropping and soil erosion on the South Carolina Piedmont, 1860–1925. Thesis, Millersville University of Pennsylvania

DeBell DS, Harms WR, Whitesell CG (1989) Stockability: a major factor in productivity differences between *Pinus taeda* plantations in Hawaii and the Southeastern United States. For Sci 35:708–719

Douglass JE (1960) Soil moisture distribution between trees in a thinned loblolly pine plantation. J For 58:221–222

Douglass JE (1966) Effects of species and arrangement of forests on evapotranspiration. In: International symposium on forest hydrology. Pergamon Press, New York, pp 451–461

Douglass JE (1972) Annotated bibliography of publications on watershed management by the Southeastern Forest Experiment Station, 1928–1970. USDA Forest Service Research Paper SE-93, Asheville, NC

Dunford EG (1947) Research in the central Piedmont. For Farmer 6(4):8

Faulkner W (1951) Requiem for a nun. Random House, New York

Fimmen RL (2004) Organic geochemistry of the South Carolina Piedmont: decomposition, mineral associations, and ferrolysis. Ph. D. Dissertation, Duke University, Durham, NC

Fimmen RL, Richter DD, Vasudevan D (2008a) Determination of DON speciation in soil solution: peptide hydrolysis and florescent amine analysis. J Environ Sci 20:1273–1280

Fimmen RL, Richter DD, Vasudevan D, Williams MA, West LT (2008b) Rhizogenic Fe-C redox cycling: a hypothetical biogeochemical mechanism that drives crustal weathering in upland soils. Biogeochemistry 87:127–141. doi:10.1007/s10533-007-9172-5

Galik CS, Mobley ML, Richter DD (2009) A virtual "field test" of forest management carbon offset protocols: the influence of accounting. Mitig Adapt Strateg Global Change 14:677–690

Gaudinski JB, Trumbore SE, Davidson EA, Cook AC, Markewitz D, Richter DD (2001) The age of fine-root carbon in three forests of the eastern United States measured by radiocarbon. Oecologia 129:420–429

Giesen JC (2011) Boll weevil blues. University of Chicago Press, IL

Gnau CB (1992) Modeling the hydrologic cycle during 25 years of forest development. Master's project, Duke University, Durham, NC

Grandy AS, Strickland MS, Lauber CL, Bradford MA, Fierer N (2009) The influence of microbial communities, management, and soil texture on soil organic matter chemistry. Geoderma 150:278–285

Gray LC (1933) History of agriculture in the Southern United States to 1860. The Carnegie Institution of Washington, Washington, DC

Greene GE (1953) Soil temperatures in the South Carolina Piedmont. USDA Forest Service Southeastern Forest Experiment Station Research Paper 29, Asheville, NC

Hafley WL, Buford MA (1985) A bivariate model for growth and yield prediction. For Sci 31:237–247

Hafley WL, Smith WD, Buford MA (1982) A new yield prediction model for unthinned loblolly pine in plantations. South Forest Research Center Technical Report 1, North Carolina State University, Raleigh, NC

Haney GP (1955) Shortleaf pine bibliography. USDA Forest Service Southeastern Forest Experiment Station Research Paper 48, Asheville, NC

Haney GP (1957) Seed production of shortleaf pine in the Piedmont. USDA Forest Service Southeastern Forest Experiment Station Research Note 113, Asheville, NC

Harms WR, Lloyd FT (1981) Stand structure and yield relationships in a 20-year old loblolly pine spacing study. Southern J Appd For 5:162-166

Harrison K, Post W, Richter DD (1995) Recovery of soil carbon by reforestation of old-field soils. Global Biogeochem Cycles 9:449–454

Hewlett JD, Metz LJ (1960) Watershed management research in the southeast. J For 58:269–271

Hoover MD (1950) Hydrologic characteristics of South Carolina Piedmont forest soils. Soil Sci Soc Am Proc 14:353–358

Hoover MD (1952a) Influence of plant cover on soil moisture in the Piedmont. Southern agricultural workers proceedings 49:172

Hoover MD (1952b) Water and timber management. J Soil Water Conser 7:75–78

Hoover MD (1953) Interception of rainfall in a young loblolly pine plantation. USDA Forest Service Southeastern Forest Experiment Station Research Paper 21, Asheville, NC

Hoover MD (1954) Soil sampling for pore space and percolation. USDA Forest Service Southeastern Forest Experiment Station Research Paper 42, Asheville, NC

Hoover MD, Lunt HA (1952) A key for the classification of forest humus types. Soil Sci Soc Am Proc 16:368–370

Hoover MD, Olson DF, Greene GE (1953) Soil moisture under a young loblolly pine plantation. Soil Sci Soc Am Proc 17:147–150

Irish MD (1952) Recent political thought in the South. Am Polit Sci Rev 46:121–141

Jackson JA (2010) Fungal community change in response to land use: a molecular approach. PhD. Dissertation, Duke University, Durham, NC

Jackson JA, Vilgalys R, Richter DD (2005) Using ITS clone libraries to assess soil fungal diversity in the South Carolina Piedmont. Soil Ecology Society, 10th biennial conference, Argonne, IL

Johnson DW (2006) Progressive N limitation in forests: review and implications for long-term responses to elevated CO_2. Ecology 87:64–75

Johnson DW, Lindberg SE (1992) Atmospheric deposition and forest nutrient cycling: a synthesis of the integrated forest study. Springer-Verlag, NY

Johnson CS, Embree ER, Alexander WW (1935) The collapse of cotton tenancy. University of North Carolina Press, Chapel Hill

Kubiena WL (1970) Micromorphological features of soil geography. Rutgers University Press, New Brunswick

Leopold A (1949) A Sand County Almanac. Oxford University Press, UK

Li JW (2009) Effects of land-use history on soil macro- and trace elements in the Southern Piedmont of North America. PhD Dissertation, Duke University, Durham, NC

Li JW, Richter DD, Heine P (2008) Four-decade responses of soil trace elements to an aggrading old-field forest: B, Mn, Zn, Cu, and Fe. Ecology 89:2911–2923

Li JW, Richter DD, Mendoza A, Heine P (2010) Effects of land-use history on soil spatial heterogeneity of macro- and trace elements in the Southern Piedmont USA. Geoderma 156:60–73

Likens GE, Driscoll CT, Buso DC, Mitchell MJ, Lovett GM, Bailey SW, Siccama TG, Reiners WA, Alewell C (2002) The biogeochemistry of sulfur at Hubbard Brook. Biogeochemistry 60:235–316

Lin H (2010) Earth's critical zone and hydropedology: concepts, characteristics, and advances. Earth Syst Sci 14:25–45

Markewitz D (1995) Soil acidification, soil potassium availability, and biogeochemistry of aluminum and silicon in a 34-year-old Loblolly Pine (*Pinus taeda* L.) ecosystem in the Calhoun Experimental Forest, South Carolina. Ph.D. Dissertation, Duke University, Durham, NC, USA

Markewitz D, Richter DD (1998) The bio- in Al and Si geochemistry. Biogeochemistry 43:1–17

Markewitz D, Richter DD (2000) Long-term soil potassium availability from a Kanhapludult to an aggrading loblolly pine ecosystem. For Ecol Manage 130:109–129

Markewitz D, Richter DD, Allen HL, Urrego JB (1998) Three decades of observed soil acidification at the Calhoun Experimental Forest: has acid rain made a difference? Soil Sci Soc Am J 62:1428–1439

Metz LJ (1952a) Weight and nitrogen and calcium content of the annual litter fall of forests in the South Carolina Piedmont. Soil Sci Soc Am J Proc 16:38–41

Metz LJ (1952b) Calcium content of hardwood litter four times that from pine: nitrogen double. USDA Southeastern Forest Experiment Station Research Notes 14, Asheville, NC

Metz LJ (1954) Forest floor in the Piedmont region of South Carolina. Soil Sci Soc Am J Proc 18:335–338

Metz LJ (1958a) The Calhoun Experimental Forest. USDA Forest Service Southeastern Forest Experiment Station, Asheville, NC, 24 p

Metz LJ (1958b) Moisture held in pine litter. J For 56:36

Metz LJ, Douglass JE (1959) Soil moisture depletion under several Piedmont cover types. USDA Forest Service Technical Bull. No. 1207

Metz LJ, Wells CG (1965) Weight and nutrient content of the aboveground parts of some loblolly pines. USDA Southeastern Forest Experiment Station Research Paper SE-17, Asheville, NC

Mistretta PA (1998) Littleleaf disease. Forest insect and disease leaflet 20. USDA Forest Service, Pineville, LA

Mobley ML (2011) An ecosystem approach to dead plant carbon over 50 years of old-field forest. PhD dissertation, Duke University, Durham, NC

Mobley ML, Richter DD (2010) Soil carbon change during fifty years of old-field forest development. Presented to American Geophysical Union, San Francisco, CA

Mobley ML, Richter DD, Heine PR (2013) Accumulation and decay of coarse woody detritus in a humid subtropical secondary pine forest. Can J For Res 43:109–118

National Research Council (2001) Basic research opportunities in earth science. National Academies Press, Washington, DC

Oh NH, Richter DD (2005) Elemental translocation and loss from three highly weathered soil-bedrock profiles in the Southeastern United States. Geoderma 126:5–25

Olson DF, Hoover MD (1957) Methods of soil moisture determination under field conditions. USDA Forest Service Southeastern Forest Experiment Station Paper 38, Asheville, NC

Patric JH, Douglass JE, Hewlett JD (1965) Soil water absorption by mountain and Piedmont forests. Soil Sci Soc Am J Proc 29:303–308

Pickett STA, Cadenasso ML (2002) The ecosystem as a multidimensional concept: meaning, model, and metaphor. Ecosystems 5:1–10

Post WM, Kwon KC (2000) Soil carbon sequestration and land-use change: processes and potential. Glob Change Biol 6:317–327

Rasmussen C, Brantley S, Richter DD, Blum A, Dixon J, White AF (2011) Strong climatic and tectonic control on plagioclase weathering in granitic terrain. Earth Planet Sci Lett 301:521–530
Richter DD (1991) Effects of acidic deposition on soils. In: Irving PM (ed) Terrestrial, material, and visibility effects of acidic deposition, vol III, pp 16/73–78. State of Science and Technology, US National Acid Precipitation Assessment Program, Washington, DC
Richter DD (2007) Humanity's transformation of earth's soil: Pedology's new frontier. Soil Sci 127:957–967
Richter DD, Markewitz D (1995a) How deep is soil? BioScience 45:600–609
Richter DD, Markewitz D (1995b) Atmospheric deposition and soil resources of the southern pine forest. In: Medlarz S, Mickler R (eds) Air pollutants and southern pine forests, ecological studies series. Springer-Verlag, New York, pp 315–336
Richter DD, Markewitz D (1996) Soil carbon dynamics during the growth of an old-field loblolly pine forest at the Calhoun Experimental Forest, USA. In: Powlson DS, Smith P, Smith JU (eds) NATO ASI Series 38. Springer-Verlag, Berlin, pp 397–407
Richter DD, Markewitz D (2001) Understanding soil change. Cambridge University Press, UK
Richter DD, Mobley M (2009) Monitoring the earth's critical zone. Science 326:1067–1068
Richter DD, Yaalon D (2012) 'The changing model of soil' revisited. Soil Sci Soc Am J 76:766–778
Richter DD, Markewitz D, Wells CG, Allen HL, April R, Heine P, Urrego B (1994) Soil chemical change during three decades in an old-field loblolly pine (*Pinus taeda* L.) ecosystem. Ecology 75:1463–1473
Richter DD, Markewitz D, Wells CG, Allen HL, Dunscombe JK, Harrison K, Heine PR, Stuanes A, Urrego B, Bonani G (1995) Carbon cycling in a loblolly pine forest: implications for the missing carbon sink and for the concept of soil. In: McFee WW, Kelly JM (eds) Carbon forms and functions in forest soils. Soil Science Society of America, Madison, pp 233–252
Richter DD, Markewitz D, Trumbore SA, Wells CG (1999) Rapid accumulation and turnover of soil carbon in a re-establishing forest. Nature 400:56–58
Richter DD, Markewitz D, Heine PR, Jin V, Raikes J, Tian K, Wells CG (2000) Legacies of agriculture and forest regrowth in the nitrogen of old-field soils. For Ecol Manage 138:233–248
Richter DD, Li JW, Markewitz D, Raikes J, Allen HL (2006a) Bioavailability of slowly cycling soil phosphorus: major restructuring of soil-P fractions over four decades in an aggrading forest. Oecologia 150:259–271. doi:10.1007/s00442-006-0510-4
Richter DD, Callaham MA, Markewitz D, Vilgalys R (2006b) Understanding ecosystem change at the Calhoun Experimental Forest's long-term soil experiment. In: Irland LC, Camp AE, Brissette JC, Donohew ZR (eds) Long-term Silvicultural and Ecological Studies, pp 169–175. Global Institute of Sustainable Forestry Research Paper 5, Yale University, New Haven
Richter DD, Hofmockel M, Callaham MA, Powlson DS, Smith P (2007a) Long-term soil experiments: keys to managing earth's rapidly changing ecosystems. Soil Sci Soc Am J 71:266–279
Richter DD, Oh NH, Fimmen R, Jackson JA (2007b) The rhizosphere and soil formation. In: Cardon Z (ed) The rhizosphere: an ecological perspective. Elsevier, New York, pp 179–200
Richter DD, Bacon AR, Mobley ML, Richardson CJ, Andrews SS, West L, Wills S, Billings S, Cambardella CA, Cavallaro N, DeMeester JE, Franzluebbers AJ, Grandy AS, Grunwald S, Gruver J, Hartshorn AS, Janzen H, Kramer MG, Ladha JK, Lajtha K, Liles GC, Markewitz D, Megonigal PJ, Mermut AR, Rasmussen C, Robinson DA, Smith P, Stiles CA, Tate RL III, Thompson A, Tugel AJ, van Es H, Yaalon D, Zobeck TM (2011) Human-soil relations are changing rapidly: proposals from SSSA's new cross-divisional working group on soil change. Soil Sci Soc Am J 75:2079–2084
Roland CP (1982) The ever-vanishing South. J South Hist 48:3–20
Rosengarten T (1974) All God's dangers, the life of Nate Shaw. A. A. Knopf, New York
Scott EJ (1919) Letters of Negro migrants of 1916–1918. J Negro Hist 4:290–340
Simkins FB (1947) The everlasting South. J South Hist 13:307–322
Smith P, Powlson DS, Smith JU, Elliott ET (1997) Evaluation and comparison of soil organic matter models using datasets from seven long-term experiments. Geoderma 81:1–225

Staff (of the Calhoun Experimental Forest) (1951) Eroded Piedmont old fields are difficult sites. USDA Southeastern Forest Experiment Station, Research News No. 13, Asheville, NC

Stone EL (1979) Nutrient removals by intensive harvest—some research gaps and opportunities. In: Leaf A (ed) Impact of intensive harvesting on forest nutrient cycling. State University of New York at Syracuse, USA, pp 366–386

Strickland MS (2009) Ecosystem carbon cycling: relationships to soil microbial community structure. PhD Dissertation, University of Georgia, Athens, GA, USA

Strickland MS, Osburn E, Lauber C, Fierer N, Bradford MA (2009) Litter quality is in the eye of the beholder: initial decomposition rates as a function of inoculum characteristics. Funct Ecol 23:627–636. doi:10.1111/j.1365-2435.2008.01515.x

Strickland MS, Callaham MA, Davies CA, Lauber CL, Ramirez K, Richter DD, Fierer N, Bradford MA (2010) Rates of in situ carbon mineralization in relation to land-use, microbial community and edaphic characteristics. Soil Biol Biochem 42:260–269

Targulian VO, Goryachkin SV (2004) Soil memory: types of record, carriers, hierarchy and diversity. Revista Mex de Cienc Geológicas 21:1–8

Tindall GB (1964) The benighted South: origins of a modern image. Va Q Rev 40:281–294

Trimble SW (2008) Man-induced soil erosion on the southern Piedmont—1700–1970. Soil and Water Conservation Society, Ankeny

Van Oost K, Quine TA, Govers G, De Gryze S, Six J, Harden JW, Ritchie JC, McCarty GW, Heckrath G, Kosmas C, Giralde JV, Marques da Silva JR, Merckx R (2007) The impact of agricultural soil erosion on the global carbon cycle. Science 318:626–629

Wells CG (1965) Nutrient relationships between soils and needles of loblolly pine (*Pinus taeda*). Soil Sci Soc Am J 29:621–624

Wells CG, Jorgensen JR (1975) Nutrient cycling in loblolly pine plantations. In: Bernier B, Winget CH (eds) Forest soils and forest land management, fourth North American forest soils conference. Laval University Press, Quebec, pp 137–158

Wells CG, Jorgensen JR (1979) Effect of intensive harvesting on nutrient supply and sustained productivity. In Leaf A (ed) Impact of intensive harvesting on forest nutrient cycling. State University of New York, Syracuse, pp 212–230

Wells CG, Metz LJ (1963) Variation in nutrient content of loblolly pine needles with season, age, soil, and position on the crown. Soil Sci Soc Am J 27:90–93

Wessel T (1997) Reading the forested landscape. Countryman Press, Woodstock

West TO, Post WM (2002) Soil organic carbon sequestration rates by tillage and crop rotation: a global data analysis. Soil Sci Soc Am J 66:1930–1946

Chapter 19
Research in the Luquillo Experimental Forest Has Advanced Understanding of Tropical Forests and Resolved Management Issues

Ariel E. Lugo and Tamara Heartsill Scalley

Abstract Long-term research on the response of wet forests in the Luquillo Experimental Forest (LEF) to natural and anthropogenic disturbances yielded information useful for the management of these forests and to a better understanding of the functioning of tropical forests and how species composition changes under different disturbance regimes. We summarize studies on basal area removal, response to ionizing radiation, and the effects of hurricanes and landslides on forested watersheds. We also review studies on forested stream biota following hurricane, drought, and flooding events. This chapter also evaluates reforestation of degraded lands and recovery of forests after abandonment of paved roads. All the studies combined cover the major land cover changes that take place throughout the tropics and which require attention to conserve tropical biodiversity. These changes range from limited extractions of resources from forests to deforestation and conversion to pastures. When tropical forests are converted to pastures, more intensive management actions are needed to restore lands, including planting of introduced species capable of growing on degraded lands. Results from the LEF have demonstrated the high resistance and resilience of tropical forests and the success of plantings in the restoration of forest cover on degraded lands. In both streams and forests, species composition shifts from native to introduced species when anthropogenic disturbance regimes become prevalent over the natural disturbance regimes.

Keywords Disturbance · Management · Recovery · Resilience · Succession · Anthropogenic · Species · Streams · Tropical

19.1 Introduction

Tropical watersheds are subject to many types of human modification ranging from removal of individual organisms to deforestation and stream channel modification. These systems are also subjected to a wide range of natural disturbances that in

A. E. Lugo (✉) · T. Heartsill Scalley
International Institute of Tropical Forestry, USDA Forest Service, 1201 Calle Ceiba, Jardín Botánico Sur, San Juan, PR 00926-1119, USA
Phone: 787-766-5335
e-mail: alugo@fs.fed.us

D. C. Hayes et al. (eds.), *USDA Forest Service Experimental Forests and Ranges*,
DOI 10.1007/978-1-4614-1818-4_19, © Springer New York 2014

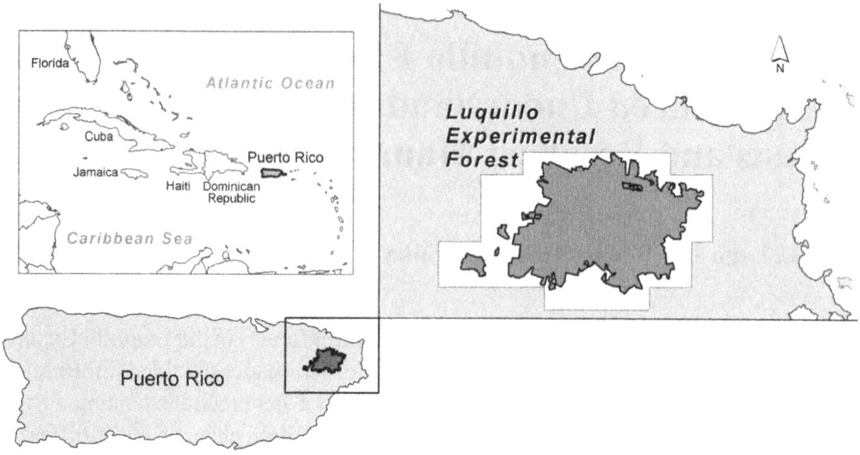

Fig. 19.1 The Luquillo Experimental Forest in northeast Puerto Rico

the Caribbean include hurricanes, landslides, floods, and drought. The net result of the synergy between anthropogenic activity and natural disturbances is a landscape with a mosaic of land covers that range from bare lands to mature forest stands, and ecosystems in different stages of succession following different types of disturbance events. The high density of species that characterize tropical ecosystems and the poor scientific understanding of tropical forest watersheds exacerbate managing this complexity of land covers and ecosystem states.

In this chapter, we review the results of studies on disturbance of ecology that led to better understanding of tropical forest watersheds in the wet Caribbean and to their management and restoration. Specifically, we focus on research that manipulated forest basal area; exposed forest stands to ionizing radiation; experimented with tree cutting and reforestation with introduced species; and studied hurricane, landslide, drought, and flood effects on forests and streams. After summarizing the research, we discuss the societal implications of the research and how these studies have led to long-term research programs and networking with other research sites.

Our main focus is on the Luquillo Experimental Forest (LEF) located in the Luquillo Mountains in eastern Puerto Rico (Fig. 19.1). Within the LEF, we review mostly research in watersheds within the subtropical wet forest life zone (*sensu* Holdridge 1967) and forest stands dominated by *Dacryodes excelsa,* known as tabonuco forest. However, where appropriate, we present research from the subtropical moist forest at lower elevation from the tabonuco forest or forests above the cloud condensation point, which occurs at 600 m in the LEF. Tabonuco forests occur from about 200 to about 600 m elevation.

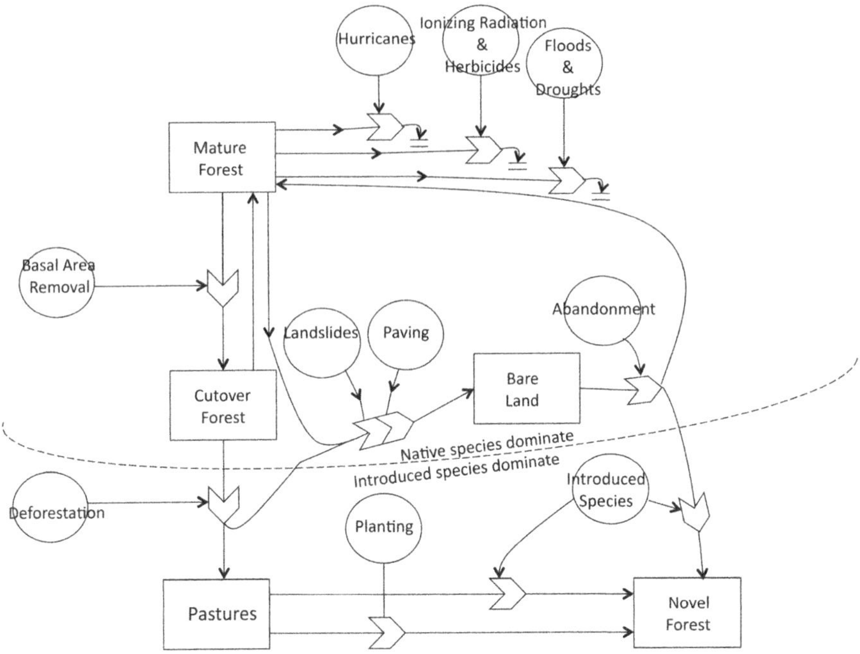

Fig. 19.2 Diagram of land cover change and successional pathways of forests in relation to anthropogenic and natural disturbances in the Luquillo Mountains of Puerto Rico. *Circles* represent external forces that cause change in the land cover and biota. The text contains a discussion of each of these disturbances or human actions. *Boxes* represent a land cover including forested, pastures, or bare lands. The *large arrow* symbol illustrates the interaction between an action or disturbance and a flux or change in land cover or forest state. For hurricanes, ionizing radiation, herbicides, floods, and droughts, the effect is a stress on the system and does not involve a change in cover or successional pathway. *Lines with arrows* illustrate the successional pathways and their direction. Native species dominate those states *above the dotted line,* and introduced species dominate those *below the dotted line.* This system is driven by climate and solar energy, but those forces are not shown. The text contains more discussion of this diagram

19.2 Land Cover Changes and Disturbances in the Study Area

To place the studies that we review in context, we developed a heuristic diagram of land cover change and disturbance effects in the Luquillo Mountains (Fig. 19.2). The diagram shows three distinct land covers (forested, bare land, and pastures), three forest states (mature, cutover, and novel), four natural disturbances (hurricanes, floods, drought, and landslides), three experimental disturbances (removal of basal area, ionizing radiation, and planting), two anthropogenic interventions (deforestation and land abandonment), and a variety of successional pathways and responses to disturbances powered by sunlight (not shown).

Land covers change from forested to pasture or bare land depending on the intensity of the anthropogenic or natural disturbance, but all land covers can be reversed towards forest maturity through natural succession or through intervention by planting. Species composition can change if plantings involve introduced species or if conditions through succession favor introduced species. Understanding these changes has been one of the main objectives of long-term research in the LEF.

The processes of land cover and ecosystem state changes and responses to disturbances depicted on Fig. 19.2, with the exception of the experimental treatments, occur in most tropical landscapes within the hurricane belt. Outside the hurricane belt, other natural disturbances such as fires become predominant. Within the Luquillo Mountains, the Forest Service faced a management challenge when it tried to restore deforested lands with temperate zone-based approaches (Wadsworth 1995). Research was needed to develop land rehabilitation strategies, particularly species selection, and to understand the importance of biodiversity to ecological functions. The results of the studies that follow, although conducted independently over a period of about 60 years, provide insights into the management of complex tropical landscapes by shedding light into the response of tropical forests and streams to both anthropogenic and natural disturbances in terms of changes in structure, species composition, and rates of ecological processes.

19.3 Hurricanes and Other Stressors Shape Forest Structure and Functioning

Schimper (1903), one of the early ecophysiologists to write about tropical rain forests, noted the importance of abundant rainfall to the delimitation of rain forests. However, while abundant rainfall is vital for sustaining forest growth and its quantity correlates with species richness and epiphyte abundance (Gentry 1982), when a forest is exposed to too much rainfall, water becomes a stress to plants and adaptations are required for their survival. This stress is true in parts of the LEF, where research has revealed a long list of plant attributes for coping with high rainfall (Table 19.1). In general, forest complexity measured as the product of basal area, tree density, maximum tree height, number of tree species, and the constant 10^3 (all expressed per 0.1 ha), increases with rainfall (Holdridge 1967). Paradoxically, the wettest elfin forests in the LEF are shorter and have lower rates of primary productivity than tabonuco forests that receive less rainfall (Weaver and Murphy 1990).

While rainfall can influence the physiological behavior of plants and animals at the LEF (Lugo 1986; Lugo and Scatena 1995), it is the influence of wind that appears to be the dominant factor in the shaping of forest structure and tree species density (Lugo 2008). After Hurricane Hugo affected the Caribbean in 1989, Brokaw et al. (2004) found support for the role of hurricanes in shaping forest canopy

Table 19.1 Effects of water on, and forest response to, too much water in wet and rain forests. (Lugo 1986)

Effects
Tree growth, seed germination, explosive seedling growth, leaf fall, flowering, and fruiting are all synchronized to slight changes in rainfall, which was documented in the Luquillo Experimental Forest
Bromeliads and other epiphytic organisms store water within their leaves, and large and diverse populations of animals utilize these as habitats and for reproduction
Tap roots, abundance of deciduous species, and tree growth rings, which are normal responses to moisture seasonality, are not usual features in the forests of the Luquillo Experimental Forest
Response to too much water
Epiphytic coverage of surfaces increases with increasing moisture, which in turn, contributes to an even distribution of throughfall by temporarily storing water and reducing its impact on other surfaces
Epiphytes also absorb nutrients from incoming waters and this contributes to a reduction in the loss of minerals to downstream ecosystems
Anatomical and morphological characteristics of plants at high elevations and low saturation deficits contribute to the increase in transpiration rates. For example, number and size of stomata increase with altitude
Where saturation deficits are high, anatomical and morphological characteristics of plants reduce water loss
Palms develop massive adventitious roots laden with lenticels that contribute to root gas exchange in anaerobic soils
Surface and adventitious roots increase dramatically with increasing water logging of soils
Trees maintain epiphyte-laden old leaves for long time periods in spite of the low Photosynthesis over respiration ratio of these leaves. It appears that their role in mineral cycling and nutrient conservation has more selective advantages than their role as net organic matter producers
Forests have extensive root mats that are essentially mineral-tight
Plants flower for longer periods in the wetter sites and depend on insects and birds for pollination

structure. In Puerto Rico, the LEF has the richest flora in the island because the mountain gradient encompasses an annual rainfall range from about 2,000 to more than 5,000 mm. Nevertheless, at the scale of a hectare, all island forests, regardless of rainfall, have a similar tree species density of about 60–70 species (Lugo 2005). This is in spite of an insular rainfall gradient from about 800 to more than 5,000 mm. The similarity in species density is not an insular effect because similar observations have been made elsewhere (insular or continental) where hurricanes are the dominant natural disturbance (Lugo 2008). The point to be made is that hurricane winds, persistent trade winds, and high rainfall are physical factors that strongly influence the biota in terms of its species composition, community structure, and ecosystem functioning. These are the primary natural factors to be considered when interpreting long-term phenomena in Caribbean forests. As we will see below, anthropogenic factors also exert strong influences on forests and they require consideration, particularly outside the LEF. The LEF is an insular standard for the maturity of its vegetation and the reduced effects of anthropogenic disturbances on its ecosystems.

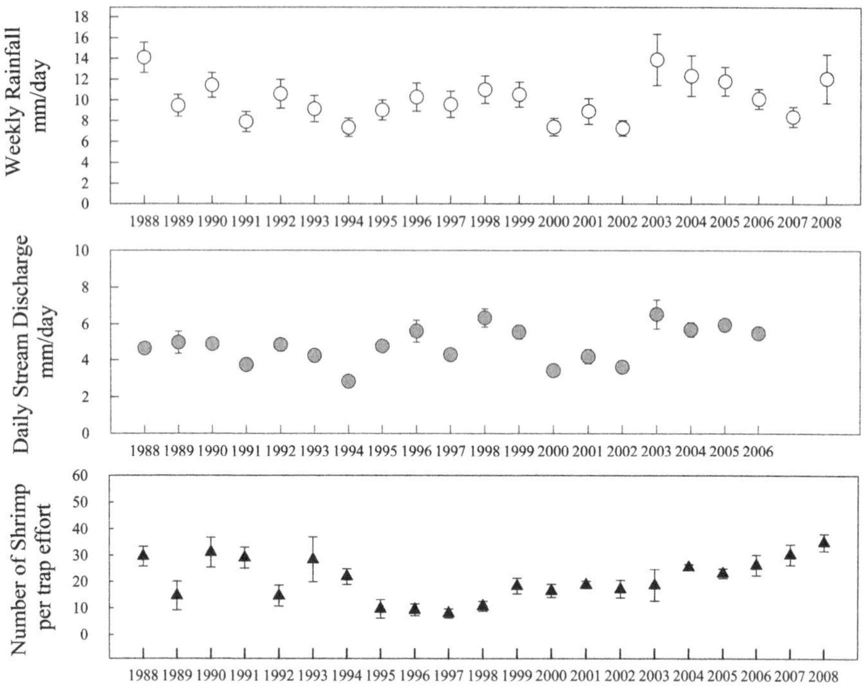

Fig. 19.3 Mean weekly rainfall and daily stream discharge in the Bisley Experimental Watersheds, and number of shrimp per trapping effort in Prieta Stream, Luquillo Experimental Forest, Puerto Rico. (Data sets are available at http://luq.lternet.edu/data/luqdata)

19.4 Droughts and Floods Have Greater Short-Term Effects on Shrimp Populations than Hurricanes

Stream ecosystems in the LEF are tightly coupled with the environmental conditions of their surrounding forest landscape. Even though there is relatively high mean daily rainfall in the LEF (Heartsill Scalley et al. 2007), there is relatively little water storage capacity in the river systems, which have high gradient streams that drain into the ocean within a very short distance traveled to the coastal plain. There is a tight link between rainfall, headwater stream discharges, and shrimp abundances in these streams (Fig. 19.3). Both rainfall events with record-breaking intensity and prolonged drought periods have been observed in the LEF (Table 19.2). These decreased rainfall events affect the stream environment by dramatically decreasing water flow levels, pool depths, and altering stream substrata. Consequently, these drought effects alter short-term shrimp population dynamics more than past hurricane events (Covich et al. 2006; Table 19.3). Extremely low flows persist during prolonged drought and reduce physical habitat availability for stream fauna, as well as affect other aspects such as migration, dispersal, and flow-related fauna communication among prey and their predators (Crowl and Covich 1994; Covich

Table 19.2 Weeks with extreme rainfall events in the past 20 years, 1988–2008 in the Luquillo Experimental Forest, Puerto Rico

Number of weeks				
Year	>40 mm/d	50–60 mm/d	>60 mm/d	Local drought
1988	1	0	0	1
1989	0	0	0	3
1990	0	0	0	0
1991	1	0	0	1
1992	0	1	0	3
1993	0	1	0	2
1994	0	0	0	3
1995	0	0	0	4
1996	1	1	0	3
1997	1	0	0	5
1998	2	0	0	3
1999	0	0	0	3
2000	0	0	0	2
2001	1	0	0	7
2002	0	0	0	2
2003	0	0	2	2
2004	2	0	1	2
2005	1	0	0	2
2006	0	0	0	1
2007	0	0	0	3
2008	2	0	2	0
Total	*12*	*3*	*5*	*52*

Local drought events are defined as 5 % of the times in the 20-year record with low or no throughfall, equivalent to 0.5 mm per day (mm/d)

et al. 2003). Stream channel structure and heterogeneity are altered during droughts, as riffles become constrained and dry out, causing pools to become disconnected, which in turn leads to a decrease in downstream dispersal of larvae; upstream migration of postlarvae; and downstream flow of leaves, branches, wood, and other organic matter. Organic matter accumulation and low or no flow downstream reduce pool volume and can further alter the quality of the physical habitat by decreasing dissolved oxygen. From the perspective of the stream physical habitat, hurricanes can produce a combination of cumulative effects that are also produced by other disturbances. For example, some are similar to those that occur during droughts, as pools can become disconnected due to debris dams and organic matter accumulation, while other effects are similar to those of floods, such as washout and scouring. The aquatic fauna of these streams have life history strategies that can sustain their populations following the change in physical conditions after the passing of hurricanes, and their populations have been observed to recover rapidly after such events (Covich et al. 1991). However, the cumulative effects of more frequent and intense hurricanes, coupled with more drought effects, could greatly alter these populations. Moreover, anthropogenic activities such as harvesting of larger species for human

Table 19.3 Observed effects of droughts, floods, and hurricanes on shrimp stream habitat in the Luquillo Experimental Forest, Puerto Rico

←Less resilience		More resilience→
Drought	Flood	Hurricane
Habitat loss: riffles dry out, shallow headwater pools disappear, and stream channels narrow	Temporary habitat loss: high water flows wash out individuals from headwater reaches	High water flows occur and debris dams are formed throughout the stream elevation gradient
Physical habitat fragmentation	Scouring and washout of physical habitat	Debris dams decrease washout of individuals
	Temporary habitat loss: high overland water flows can increase landslides and increase silt levels enough to decrease quality of habitat in headwater reaches	Debris dams retain water and organic matter
	Access to quality habitat and food sources decreased (silted surfaces affect periphyton and algal growth)	Debris dams slowly release water and organic matter during following months
Increased predation: lower pool water levels and fragmentation increases encounters with predators	Increased predation in new temporary habitat (lower elevation pools)	
Decrease of water oxygen levels from lack of stream flow		Retained particles are slowly released, an alternate food source to particulates of green and brown litter fall at various stages of decomposition
Loss of dissolved organic matter from lack of stream flow and habitat fragmentation		Movement towards headwater reaches from mid-elevation sites
Possible increase in anoxic conditions in decreased and fragmented pool riffle habitat		Restructuring of stream pools and riffles, but no net loss or fragmentation occurs

We use temporary habitat loss because immediate upstream migration can occur

consumption and dam construction can diminish populations when combined with increased natural disturbances such as more frequent hurricanes and droughts.

Most of the tropical island stream fauna in the LEF, including fish, shrimp, and snails, have to migrate between the stream freshwater habitats to the saltwater estuarine and coastal habitats to complete their life cycles. Aquatic insects disperse by flying as adults and freshwater crabs complete their life cycle in headwater streams and do not migrate to coastal waters. The freshwater shrimp that dominate the streams of the LEF spend their lives as adults in the headwater streams and rivers where they feed, grow slowly, and reproduce (Cross et al. 2008). The gravid, egg-carrying, female shrimp then release larvae into the flowing water. Larvae drift with the current downstream all the way to estuarine waters where they will remain in this drifting, planktonic stage until they develop into postlarval juveniles (Covich and McDowell 1996; Benstead et al. 2000; Scatena 2001; March et al. 2003). As post larvae, they migrate upstream where they can occupy pools that serve as spatial refugia from fish predators once they climb steep waterfalls that are barriers to predatory fishes (Covich et al. 2009; Kikkert et al. 2009). This migratory life cycle is classified as amphidromous, and it is also observed in other native aquatic species such as gobiid fishes and neritina snails (March et al. 2003; Blanco and Scatena 2006; Covich 2006).

During a 15-year span that included two hurricane events (Hurricane Hugo in 1989 and Georges in 1998) and one prolonged, island-wide drought in 1994, Covich et al. (2006) measured shrimp abundances along the elevation gradient of Quebrada Prieta, a stream that flows into the Espiritu Santo, one of the major rivers of the LEF. They found that hurricanes and storm flows had no persistent measurable effect on pre- and post-event shrimp abundances when compared with the drought, which decreased shrimp abundance in headwater streams. During low flows, there was a loss of chemical cues (Crowl and Covich 1994) and increased predation (Covich et al. 1996). Droughts primarily alter physical habitat and spatial refugia from predatory fishes and these changes affect populations of aquatic species over several years. In contrast, flood events have different effects on particular species and these effects relatively short term.

Flood events displace more of the shredder than the filter feeder shrimp guilds. Increased vulnerability to washout during high-flow events of shredders is related to their behavior, as the primary species of shredders swim into in the upper open pool areas compared to the main filter feeder species which stay close to the bottom of pools and among crevices in the stream banks and between boulders. However, during extreme flooding events when stream substrata are scoured, both of these guilds can be displaced from the higher-elevation stream reaches to the mid- and lower-elevation stream sections where deeper pools and crevices in the banks can serve as refugia (Covich et al. 1996).

Following high-wind events, leaves and woody branches removed from riparian trees form organic debris dams. During prolonged droughts, these dams can contribute to loss of connectivity among pools and lower water quality by decreasing available oxygen levels in pools with accumulated and decomposing organic matter. In general, the organic debris dams ameliorate the effects of high flows during hurricanes and other tropical storms. Hurricanes can produce sustained floods in headwater reaches and drainage-wide high-flow events, which can be buffered by debris dams resulting from high-wind events. In the wider channels downstream, larger flows tend to wash out debris dams and deposit wood along riverbanks. During hurricane events, dams maintain habitat heterogeneity, which sustain shrimp abundances in headwater streams (Covich et al. 2006).

Stream ecosystems in headwaters are tightly coupled with rainfall and inputs from forested riparian areas that serve as an energy source for aquatic fauna (Crowl et al. 2006). High rainfall and almost no water storage capacity in the river systems means that the role of riparian vegetation in providing structure and connectivity is crucial to sustaining the stream ecosystem and the services it provides. Coupled rainfall, stream discharge, and shrimp abundance (Fig. 19.3) indicate that watershed management needs to consider the intrinsic forest and stream connections, which means not disrupting the migration of native fauna and minimizing the alteration of riparian and wetland zone vegetation. In one of the main rivers whose headwaters are in the LEF, the Mameyes River, an innovation in water resources management allows water flow to remain unobstructed while it is able to provide water for local consumption. Backed by research that spans over two decades, the river contains a unique water extraction system. This low-impact system allows for continuous fauna migrations without interrupting water flow and stream ecosystem connectivity as low-head dam barriers do. This within-channel withdrawal system is located near the center of the channel nestled within its substrate and away from the bulk of the water flow (Scatena 2001; March et al. 2003). The placement of this system resulted in minimum river channel and upstream habitat modification, in addition to a lack of interruption of larval migration and unobstructed base flow levels that also address local water consumption needs.

19.5 Basal Area Reduction Experiments and Monitoring Uncover Controls of Tree Growth

Since 1943, 420 plots of 0.1 ha were established in the tabonuco forest of the LEF to monitor tree growth in support of the preparation of a land management plan and to study tree growth in complex tropical forests. These plots were dispersed on the landscape ranging in elevation from 200 to 640 m. Thousands of trees with diameter at breast height (dbh)>9.1 cm were identified to species, tagged, and measured. Plots were remeasured in 1947, 1952, 1958, 1965, 1976, and 1982. Additional tree growth plots were established in other mature forest stands including tabonuco stands. In some of these plots, 50% of the basal area was experimentally removed to ascertain growth responses. These additional plots were also remeasured periodically over the next several decades.

Reduction of basal area led to increased growth of remaining trees (Fig. 19.4). Closer analysis of growth data revealed the importance of canopy position to tree growth. Trees with dominant canopies (all canopy exposed to light) had the faster growth followed by those with codominant canopies (canopy receives light from the top), intermediate canopies (canopy below the main canopy), and suppressed trees whose canopy is shaded most of the time. In one species, the growth of trees with dominant canopies was four times higher than the growth of trees with suppressed canopies (Parresol 1995). Tree growth also varied with species and topographic position (ridge, slope, and valley), with some species doing better in one or another position (Parresol 1995). Moreover, Wadsworth et al. (2010) identified crop trees from the data set and was able to identify which trees were growing at rates twice as fast as the mean and the conditions associated with the fast growth of these crop trees.

Tree growth information and knowledge of what factors influenced growth rates were used by Wadsworth to develop a scheme for liberating crop trees from

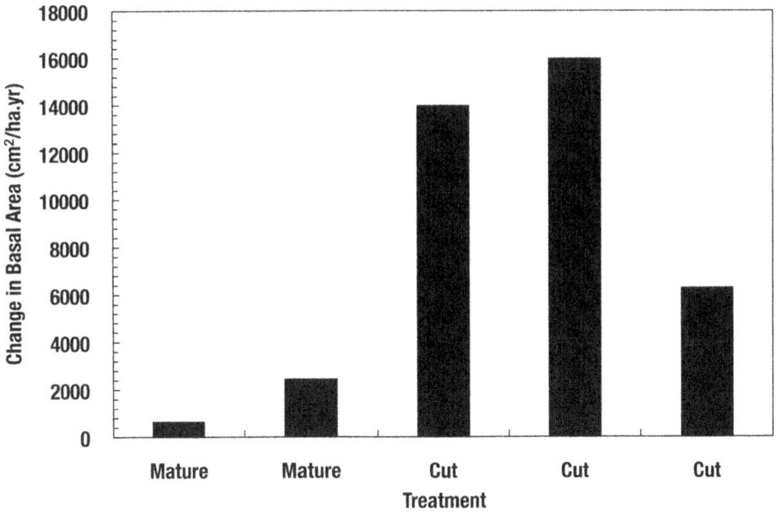

Fig. 19.4 Basal area increase of trees in mature tabonuco forests when compared with trees in tabonuco forests with 50% basal area reduction (cut). (Data are summarized in Brown et al. 1983)

competition, thus achieving the maximum growth possible (Wodsworth 1997). This scheme was successfully tested in Puerto Rico and exported to Brazil (Wadsworth and Zweede 2006) and Costa Rica (Hutchinson and Wadsworth 2006). In Brazil, the system proved economically feasible as the increased productivity of wood could pay the entire cost of the liberation. Also, increased growth rate reduced by 25% the time in the wait for the next harvest.

19.6 Ionizing Radiation and Other Disturbances Uncover the Resilience of Tropical Forests

In January 19, 1965, about 2 ha of the LEF was irradiated for 92.8 days with a 10,000-curie cesium (Cs^{137}) ionizing radiation source (Odum and Drewry 1970). The impetus for the study was to learn about how tropical forests might respond to ionizing radiation, should nuclear devices be used to excavate a second Panama Canal between the Pacific and Atlantic Oceans. The study area, which formed a canopy gap as a result of tree mortality, was studied for 23 years (Taylor et al. 1995).

The level of ionizing radiation applied to the tabonuco forest in the LEF was high and initially the forest exhibited a high resistance both at the tree population and forest level. For example, Odum et al. (1970a) reported live trees that had received up to 100,000 R (roentgens) and many with normal appearance in spite of being exposed to 12,000 R. For a year, it was difficult to establish a clear zone of tree mortality around the ionizing radiation source. However, with time, the trees within 40 m of the ionizing radiation source all died and a large gap was created (Taylor et al. 1995). Tree growth declined in several species that survived the ionizing radiation,

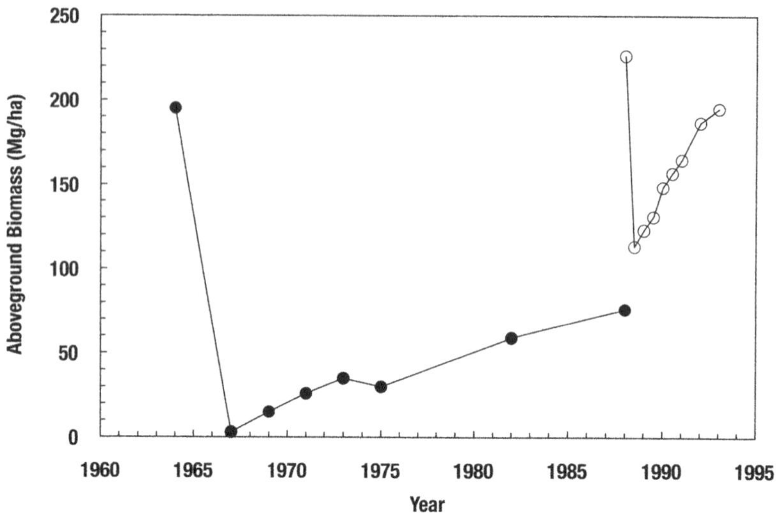

Fig. 19.5 Aboveground biomass after gamma irradiation of a tabonuco forest stand at El Verde (*solid circles*) and after a hurricane passed over a tabonuco forest at Bisley (*open circles*), Luquillo Experimental Forest. Data are from Taylor et al. (1995) and Scatena et al. (1996). The high value in each data set corresponds to the pre-disturbance of aboveground biomass. The reduction after the hurricane was instantaneous while that after irradiation was progressive over several years

but after the event they returned to normal rates of growth, while other species remained at slower rates of growth or increased (Murphy 1970). Successional tree species like *Cecropia schreberiana* accelerated tree growth after ionizing radiation, while primary species like *D. excelsa* reduced growth.

Although many scientists were seeking to document dramatic ionizing radiation effects after the radiation source was shut down, they were mostly unsuccessful in the short term (Odum et al. 1970a). However, long-term data and comparisons of forest response to ionizing radiation with responses to an experimental cut, addition of herbicides, and hurricanes yielded numerous insights on forest resilience and allow us to position ionizing radiation effects in perspective relative to the other disturbances. We focus first on a comparison between ionizing radiation and hurricane disturbance effects on biomass accumulation and productivity, and then compare various disturbances in relation to species composition responses.

Fig. 19.5 shows the aboveground biomass loss and accumulation in two tabonuco forest stands in the LEF, one exposed to ionizing radiation and the other to a category-3 hurricane. The reduction in biomass due to ionizing radiation was much greater than the reduction of biomass due to the hurricane. The hurricane reduced forest stand biomass by 50 %, while ionizing radiation reduced biomass by almost 100 %. This comparison requires some explanation to properly understand how each of the two disturbances affects the forest. The loss of biomass due to ionizing radiation has two caveats. First, the loss, while very high, is limited to a small area receiving high radiation dosages. A meter away from the limit of the mortality zone shows no apparent ionizing radiation effects on trees. Second, the loss of biomass

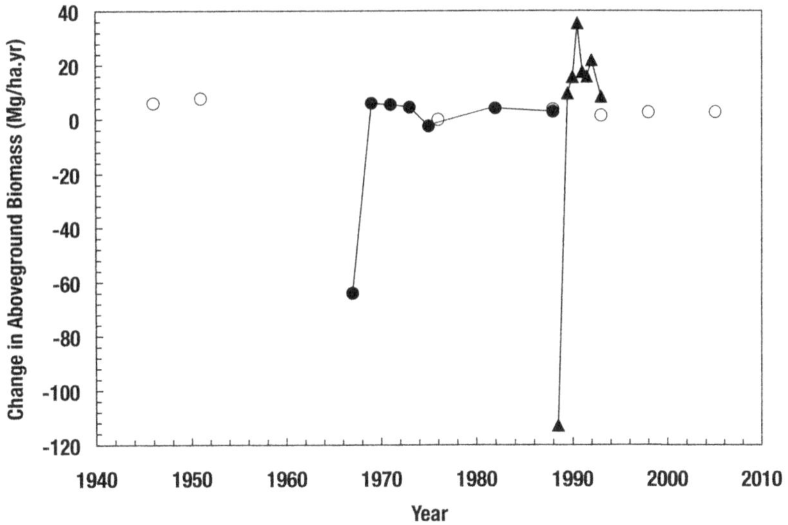

Fig. 19.6 Changes in aboveground biomass of tabonuco forests in different sectors of the Luquillo Experimental Forest. Positive values correspond to growth and negative values to losses to mortality. The long *open circle* data series corresponds to El Verde 3, a mature tabonuco plot (see Lugo et al. in this volume and Drew et al. 2009 for data source and analysis of the behavior of this plot over 62 years). *Solid circles* correspond to the tabonuco plot that was irradiated at El Verde and the *solid triangles* correspond to a tabonuco plot in Bisley (Scatena et al. 1996)

associated with ionizing radiation occurs over a period of years, which is a slow response. Thus, anionizing radiation disturbance delivers a high-intensity effect over a small area (square meters) and over a relatively long time (years). In contrast, the effect of the hurricane disturbance is brief (hours) and over a larger area of landscape (hectares). These different modes of action have contrasting ecological effects in terms of the coupled ecological processes such as wood decomposition, mineral cycling, and carbon dynamics (Lugo and Scatena 1996).

Figure 19.5 also shows that the accumulation of biomass after the disturbance effect follows different slopes for ionizing radiation and hurricanes. This is better illustrated in Fig. 19.6 where rates of loss and accumulation are shown. Biomass accumulation after a hurricane is much faster than biomass accumulation after ionizing radiation. Yet, the rates of biomass accumulation after ionizing radiation are in the same order of magnitude as those measured in a nearby mature tabonuco stand not subject to ionizing radiation. Thus, it appears that the slower recovery of biomass after the ionizing radiation experiment is due mostly to the lower quantity of live biomass that remained after the event reached its peak effects. Of course, the level of residual live biomass was a function of the intensity of the ionizing radiation source. After a hurricane, more live biomass remains standing, and rates of accumulation reach very high values, leading the stand to pre-hurricane biomass in less than 5 years or so for a hurricane such as the one depicted in Fig. 19.7. Another factor that might influence the rate of biomass accumulation is the physiological state of irradiated trees, which could have exhibited reduced photosynthetic capacity as a result of the stress.

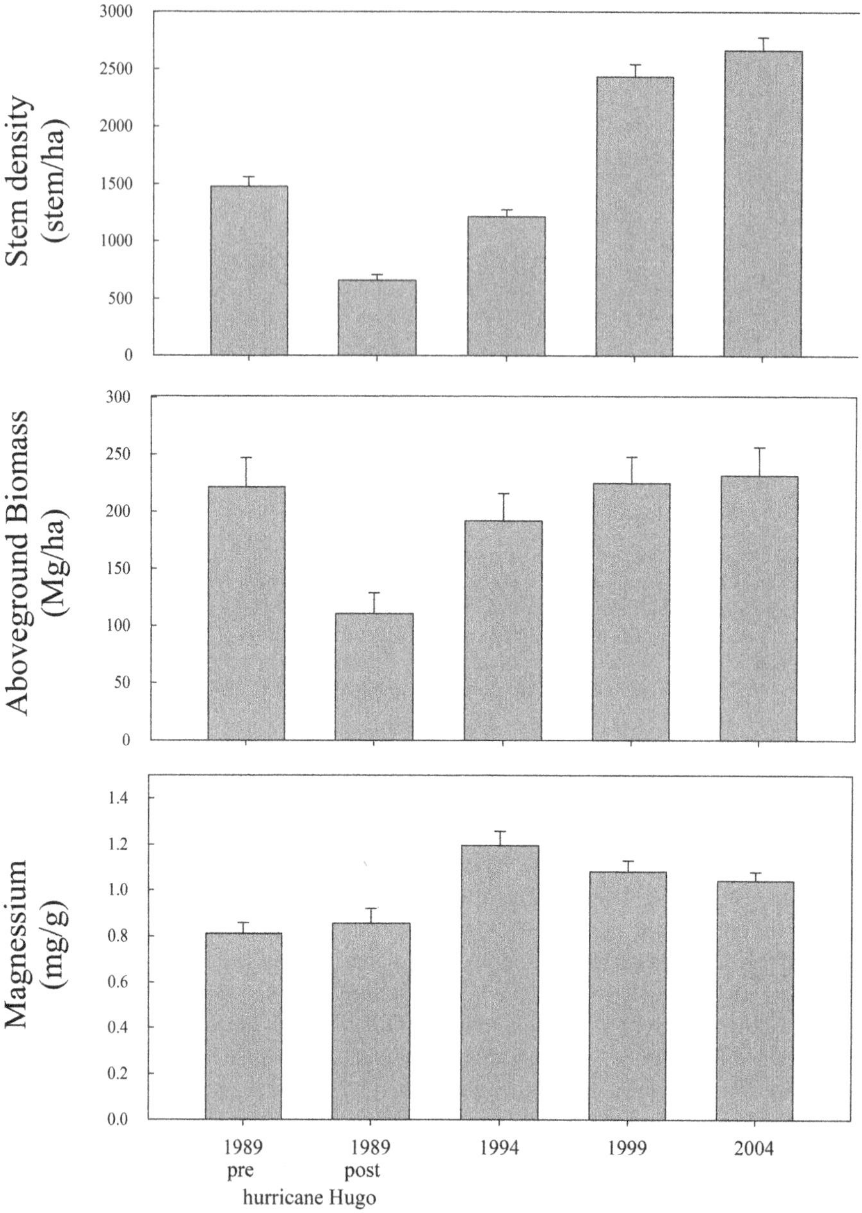

Fig. 19.7 Stem density, aboveground biomass, and magnesium concentration in aboveground biomass in the Bisley Experimental Watersheds, Luquillo Experimental Forest, Puerto Rico

Forest response to disturbances also includes changes in species composition, and at the LEF it has been possible to compare species responses to ionizing radiation, herbicides, hurricanes, and cutting (data, pictures, and discussion in Brown et al. 1983; Odum et al. 1970a; Dowler and Tschirley 1970; Smith 1970). Among the many observations, three stand out from a long-term perspective:

1. In all instances, the regeneration after the disturbances (ionizing radiation, hurricanes, defoliants, or cutting) was composed of native species. Introduced species were not able to regenerate after any of these disturbances.
2. The diversity of plant life-forms increased during regeneration but was particularly notable after irradiation. Normally, the arboreal life-form maintains dominance during regeneration after hurricanes, application of defoliants, and cutting. Seedlings, saplings, and vines dominate the understory. After ionizing radiation, radiation killed most of the original plants including seeds, and the ensuing vegetation was conspicuously herbaceous dominated by graminoids, herbs, vines, root sprouts, ferns, and some palms. These life-forms gave way to seedlings and saplings, but seedlings of primary forest species had not reappeared on the site after 25 years of succession.
3. The disappearance of primary forest species in the succession after irradiation underscores a point about the speed of succession after the various disturbances. Hurricanes and cutting were both followed by vigorous growth of seedlings and saplings, both from the seed and seedling banks of the forest. The same was true of defoliation, although herbicide treatment of soil at high dosages arrested grass regeneration for about a year; the soil was not sterilized for a long period. In terms of initial regeneration, the ranking of disturbed sites in terms of species diversity was highest in cut forest, followed by irradiated forest, and herbicided forest last. Irradiation eliminated the seed and seedling banks and thus the speed of succession was slowed and skewed to those native species capable of reaching and establishing at the site. Even today, 44 years since the ionizing radiation, the species composition at the radiation site has not returned to what it was before, while the cut, herbicided, and hurricane-affected sites recovered relatively quickly, i.e., within 20 years (Heartsill Scalley et al. 2010).

19.7 Forest Recovery from Landslides and Road Abandonment Is Rapid

The ecosystem dynamics of primary succession from bare land in the LEF are rapid, initial biomass accumulation is followed by increasing vegetation cover and finally structure approaches that of forest stands. Consistently, in both landslides and abandoned roads, after 60 years, basal area values are similar to those found in adjacent forests. The observed recovery, or reclaiming of bare land areas by vegetation in the LEF, is also scale dependent because it is affected by the size of the landslide and heterogeneity, as adjacent forest type and its conditions were the most important factors influencing the direction and level of recovery.

Primary succession in the exposed surfaces of landslides and abandoned roads are part of the forest mosaic in the LEF. During a 50-year record (1936–1988), landslides comprised <1% of LEF area, while active roads were estimated at close to 0.5% of the LEF (Guariguata 1990; Karecha 1997). Even though these road and landslide disturbances occupy a small area within of the forest, they are conspicuous features of the landscape, particularly after heavy rains. Landslides and roads are part of the natural and anthropogenic disturbance regime of this forest. Landslides expose weathered rocks and clays in the lower elevations, and saturated silts at the higher elevations of the LEF. Abandoned asphalt roads of a lane and a half width are surrounded by contiguous forest and must first accumulate soils in order to begin any forest recovery processes.

Heyne (2000) found that a strong influence on recovery of soil physical characteristics and nutrients in these abandoned roads was exerted by adjacent forest conditions. This follows the documented high heterogeneity of soils in the LEF, where nutrient pools are influenced by forest dynamics at small spatial scales (Silver et al. 1996). Abandoned paved roads, which ranged in age since abandonment from 4 to 60 years, had litter and soil characteristics similar to their adjacent forest areas after 11 years (Fig. 19.8). Net nitrogen mineralization (grams per gram of dry soil) and percent soil organic matter were not different between abandoned roads and adjacent forest, while soil pH required more than 60 years of abandonment for recovery to forest conditions. Physical accumulation of soil above the pavement was the delaying factor to the full recovery of soil properties. Some sites had litter accumulation levels similar to those observed in the adjacent forest at 11 years of abandonment, but in one site litter recovered after only 4 years (Fig. 19.8). Litter mass on newly deposited soils above roads contributes to changes in environmental conditions by minimizing soil temperatures and providing a physical structure for retaining seeds, creating microhabitat for soil fauna, and minimizing erosive soil loss due to rainfall (Heyne 2000). In contrast, soil characteristics of landslides that were <1 year old were similar to forest sites only in their bulk density. Landslide soils had lower nutrient concentrations and organic carbon, but similar total phosphorus concentrations to those in forest soils (Guariguata 1990).

Ferns, seedlings, and woody vines dominate understory vegetation developing on abandoned roads (Heyne 2000). Ferns dominate landslides after 30 years of succession following an initial dominance by herbs and grasses, whereas in abandoned roads of similar age, tree densities and number of species reached the same values as those observed in adjacent forests. Canopy structure was similar between abandoned roads and adjacent forests after 40 years of succession. There are different constraints on succession after these two types of disturbances, as seen in how their species composition changes through succession. On landslides, there was an abundance of *C. schreberiana,* a small-seeded, light-demanding pioneer tree species, but this species was not common in the abandoned road sites. Initial differences in species composition result in different successional paths to maturity depending on adjacent forest conditions and scale of disturbance. Small landslides are quickly colonized and progress rapidly towards tree cover because soil disturbance is slight. Larger slides are more heterogeneous and include zones where the soil profile is exported, exposing rock or saprolite and setting succession back to primary suc-

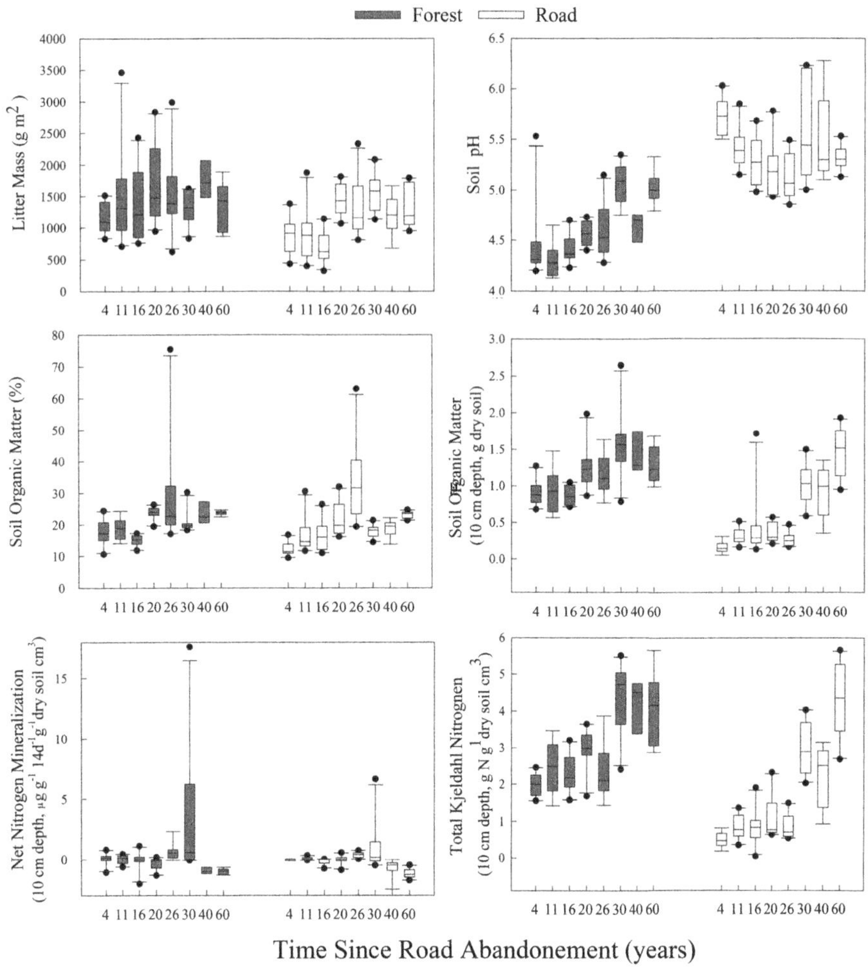

Fig. 19.8 Changes in various soil parameters in chronosequences of abandoned roads and adjacent forests of similar age in the Luquillo Experimental Forest, Puerto Rico. *Boxplots* represent distribution of data points (Heyne 2000), each box encompasses the 25th through 75th percentile, and the *horizontal lines* mark the median and 10th and 90th percentiles

cession. In older landslides and roads abandoned for more than 30 years, the palm *Prestoea montana* reached similar or greater abundances than those in adjacent forest areas (Guariguata 1990; Heyne 2000). However, the vegetation on these newly exposed terrains (landslides, abandoned roads, and roadside filled areas) remained different in species composition from that of nearby forest areas even after 60 years. Floristic composition began to approximate, but did not attain, mature forest species composition at any of the measured sites (Guariguata 1990; Walker et al. 1996; Olander et al. 1998; Heyne 2000).

19.8 Tree Planting on Abandoned Pastures Results in Long-Term Carbon Sinks

A common land management problem in the tropics is overcoming arrested succession on deforested and degraded lands that are invaded by grasses. These grasslands or abandoned pastures are hostile environments for trees, and remain as pastures for decades because tree regeneration without active management does not occur (Parrotta 1999). This problem of land restoration was resolved in the LEF through extensive experimental plantings of native and introduced tree species (Marrero 1950). One such planting involving nine native and four introduced tree species was evaluated for its carbon sequestration outcome 55–61 years after the initial plantation (Silver et al. 2004).

The forest was planted on degraded pasturelands in the subtropical moist forest life zone of Holdridge (1967). Over the next 61 years, it had accumulated 75 tree species or 62 more than were planted, with 60% of the importance value accounted by native tree species (Table 19.4). The growth of trees was not uniform over the 61-year period as trees increased in basal area faster as the forest matured (Fig. 19.9). The forest rapidly accumulated carbon both aboveground and belowground, with greater carbon storage belowground (Table 19.4). Fine root biomass was in the same order of magnitude as measured in mature native forests (Silver et al. 2004).

Planting trees on degraded pasturelands not only resulted in a species-diverse forest but it also transformed the distribution and accumulation of carbon on the site. Isotope work showed how forest-derived carbon accumulates steadily over the soil profile, while pasture carbon declines (Fig. 19.10). However the increase in forest-derived soil carbon was faster than the decrease, causing a net carbon sink of about 33 Mg/ha over the 61-year period. Moreover, the accumulation of belowground fine root and aboveground biomass, as well as litterfall rates, was all high and contributed to the overall carbon sink of the planted forest (Table 19.4).

19.9 Introduced Species Cannot Dominate Native Forests but Form Novel Forests on Degraded Lands

We noted in the previous section that native species were the only ones regenerating after hurricanes, clearcuts, irradiation, and application of defoliants to tabonuco forests. We also pointed out that the vegetation of the LEF is among the most pristine in Puerto Rico and contains the largest area of primary forest in the island (Lugo 1994). A reason why introduced species have so little success might be the dominance of native species and the undisturbed condition of these closed-canopy forests within the LEF. However, there are many introduced and established populations of plants and animals in this experimental forest. They help explain why the balance between introduced and native species favors the native species.

Along highways and recreation areas where human activity is concentrated, one finds many introduced species, mostly because they were planted or abandoned and fed (in the case of stray dogs and cats). As we will see subsequently, introduced plants,

Table 19.4 Long-term outcomes in ecosystem attributes of a planted forest on degraded pasturelands, Luquillo Experimental Forest

Ecosystem attribute	Outcome
Accumulated number of tree species	75 in 4.64 ha
Importance value of planted species	40%
Importance value of introduced species	5%
Aboveground carbon	80 Mg/ha
Soil carbon to 60 cm	102 Mg/ha
Fine root biomass	2.5 Mg/ha
Dead fine root biomass	2.3 Mg/ha
Live fine root biomass	0.1 Mg/ha
Rate of increase in forest soil carbon	0.9 Mg/ha.yr
Rate of loss of pasture soil carbon	0.4 Mg/ha.yr
Net rate of soil carbon accumulation	0.5 Mg/ha.yr
Net aboveground biomass accumulation	1.4 Mg/ha.yr
Fine root biomass accumulation	0.09 Mg/ha.yr
Litterfall	10–12 Mg/ha.yr

The forest was initially planted in the mid- to late 1930s with 13 tree species (nine native and four introduced). The outcome after 61 years is from Silver et al. (2004). Trees with diameter at breast height > 9.1 cm were measured. Importance value is the sum of relative density, relative basal area, and frequency by species based on 116 plots and expressed in percent

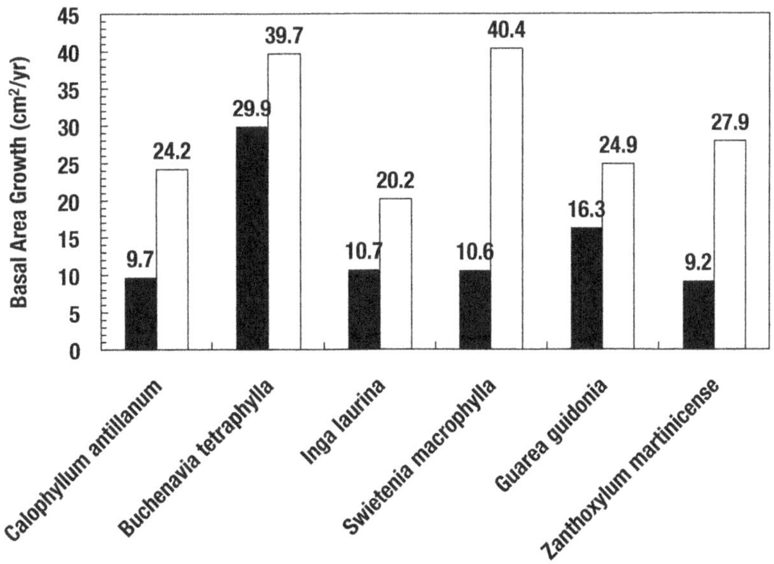

Fig. 19.9 Tree growth in a planted forest on degraded pasturelands, Luquillo Experimental Forest. The *solid bar* corresponds to the time interval of 1937–1959, and the *open bars* to the interval of 1959–1998. Data correspond to trees with diameter at breast height > 9.1 cm. The increased rates during the second time interval were all significant at $p < .05$ (Silver et al. 2004). *Swietenia* is an introduced species while the others are native tree species

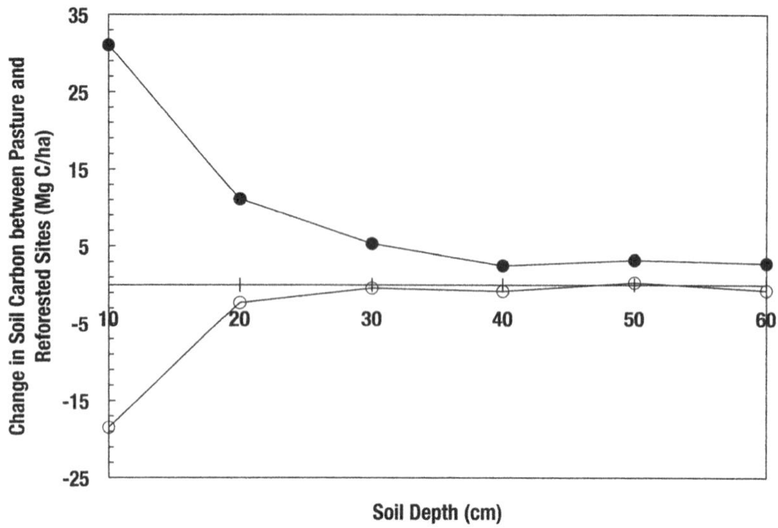

Fig. 19.10 Change in soil carbon over a period of 61 years in a planted forest on degraded pasturelands, Luquillo Experimental Forest. The data are based on stable isotope determinations (^{13}C and ^{12}C) whereby the soil carbon attributed to forests (C3, *solid points*) and grasses (C4, *open points*) is tracked through a 60 cm deep soil profile. (Details in Silver et al. 2004)

even if planted seldom, invade closed canopy native forests in the LEF. Some introduced animals like rats, mongooses, toads, mosquitoes, and bees do invade and establish populations in pristine forest stands at LEF, but they do not attain high abundances. In fact, introduced rats have been present in the forest for centuries and now appear naturalized at about 40/ha (Odum et al. 1970b). Odum et al. (1970b) assessed the situation thusly (p E-14): "There may be a pattern to finding the abundant, dominant, species of mans' farms and houses present in the rain forest as a scarce specialist there…."

In the restored forests on degraded pasturelands (above), introduced trees survived and reproduced after 62 years of succession, but they lost importance relative to native species over the long term. Thompson et al. (2007) examined the long-term dynamics of introduced species that are common in tabonuco forest recovering from past human activity. When people inhabited these sites, they planted trees for food, shade, or crops and many of these species are commonly seen in these recovering forests. The study was conducted in a 16-ha plot where all trees from >1 cm have been tagged, identified, and measured. A hurricane affected the plot 1 year before the first inventory and also between the second and third inventory, which allowed the assessment of the influence of a natural disturbance on the dynamic of introduced species in closed canopy tabonuco forests. The following are the findings of Thompson et al. (2007) who analyzed population trends for 12 introduced tree species in the plot:

- Most introduced species were planted prior to 1932.
- As a group, introduced species constituted a small component of the stand, i.e., <7 individuals/ha and <1.6 m^2/ha (representing <1 % of the stands basal area).

- Introduced species had lower mortality rates than native ones, and similar growth rates.
- The populations of four introduced species changed little over time.
- The populations of six introduced species declined between the first and second census, but increased after the hurricane.
- The population of one introduced species increased somewhat on all censuses and the population of another introduced species declined on all censuses.
- The populations of two introduced species increased under a closed canopy in areas that had a history of past anthropogenic disturbance.
- Most introduced populations have not migrated from where they were initially planted.

In summary, it appears that recurrent natural disturbances open the forest canopy and allow some regeneration of introduced species. However, the population density of these introduced species is low and as soon as the canopy closes, many individuals die, and the populations of introduced species do not increase, remaining at low densities, as do introduced animal species in the LEF.

Odum (1970) said it best (p I-275):

> When exotic organisms are introduced into a rain forest, the niche they can occupy is small and their role is that of a specialist; but, when the forest system is eliminated, the exotic may become the dominant since its programs of control are absent or left behind. Widespread destruction of native forest systems creating disturbance ecosystems allows invasions and upsets, which unharnessed may become epidemic, affecting desired trees, agricultural productions, etc. Some compromise with diversity may reduce the problem of disturbance from weed species. As discussed in connection with viruses and the *Aedes aegyptii* in the forest… diverse native forest may protect exotics as minor components and reservoirs for enormous multiplication potential. Irradiation and human disturbance of the El Verde forest seemed to develop mosquitoes and viruses. Straight control by killing is difficult, since to destroy the last reservoir or the undesired species would require disturbance of the natural forest reservoir thus increasing the situation favorable to the undesired species. The patterns for these organisms are like that of forest man who also has potential for dominance after the forest is cut but who is a protected minor component of the climax forest.

Earlier we discussed the carbon sink function of a planted forest on degraded pastureland in the LEF (Silver et al. 2004). The planting included 4 introduced tree species out of 13 that were planted and, within 61 years, the forest had 75 tree species. Of those species, introduced ones comprised 10.9% of the importance value (5% accounted by the planted species and the rest by one that was not planted), but two of the introduced species were in the top ten species and one ranked fourth in importance with a 5.9%. This species (*Syzygium jambos*) arrived to the plots by natural dispersal.

The species composition of this planted forest is new to Puerto Rico, as it includes introduced species and proportions of species that have not occurred in the island before. Hobbs et al. (2006) and Lugo and Helmer (2004) termed these new communities novel forests. It appears that these novel forests are a natural response to anthropogenic effects as the novelty of the species combinations is more dramatic as site degradation increases. The extremes occur outside of the LEF, where monocultures of introduced species colonize for discrete periods of time abandoned and degraded agricultural lands (Lugo 2004). After several decades, what appeared as a monoculture diversifies into new combinations of native and introduced species, much as was documented in the LEF by Silver et al. (2004).

19.10 Conclusions, Social Implications, and Research Networking

Research in the Luquillo Experimental Forest illustrates many examples of anthropogenic and natural disturbances and recovery events leading to successional pathways with different speeds and different species outcomes (Fig. 19.2), even if the end states are forests with similar structural characteristics. When the disturbances lead to arrested successions, the forest recovery process can be restarted by planting trees on abandoned pasture lands, which eventually not only leads to a net carbon sink aboveground and belowground but also provides habitat that facilitates colonization by native trees species that otherwise could not have so quickly established and dominated in abandoned pastures. Forest biomass and structure can recover quickly after hurricanes and localized disturbances such as forest road construction particularly when there is adjacent forest that can provide seed sources and organic matter inputs. The resilience of tabonuco forest to various types of disturbances is evidenced by the recovery of biomass and structure, but species composition does not recover in the same way. The apparent trade-off to this recovery is that the recovered forest produces new combination of species not observed previously. These new combinations of tree species are part of the response to disturbance, and part of the new successional trajectories created. Detailed observations on tree canopy dynamics and natural variation in growth rates of forest trees led to the application of the basal area reduction experiments that resulted in improvements of wood tree production in the tropics. This approach maintains only a few trees maturing concurrently for harvest, and this allows for management and harvest of other forest products to be possible.

In the LEF, the continuous and long-term monitoring of forest ecosystem components and processes, climate, and hydrology has been the basis for developing practical approaches to deal with watershed management needs at local and regional scales, and these also have potential for wider application. The impetus for much of this research was the dire social situation in Puerto Rico during the first half of the twentieth century. Overpopulation, an agrarian society, and dependency on fuelwood led to dramatic deforestation and land degradation, with the resulting environmental and social problems that develop when forest cover is lost on a moist tropical climate. The suite of disturbances afflicting forests, and depicted in Fig. 19.2, led to the long-term studies summarized here. The development of an experimental infrastructure in the LEF in turn led to an impressive array of collaborative work worldwide (Table 19.5). These collaborations plus the basic forestry research conducted over the past 100 years at the site contribute to improving understanding of these complex forests and to management actions designed to sustain forest productivity and resilience under constantly changing environmental conditions.

Table 19.5 Selected collaborative research and inter-site projects conducted at the Luquillo Experimental Forest (LEF) as part of collaborative research among experimental forests and ranges and experimental watersheds and with other research networks. (Information compiled by E. Meléndez Colóm, available at http://luq.lternet.edu/research/CrossSiteStudies and http://luq.lternet.edu/research/LTERProjectsList2)

Activity	Description
Luquillo Critical Zone Observatory, LCZO	The overarching focus of the LCZO is how critical zone processes and water balances differ in landscapes with contrasting bedrock but similar climatic and environmental histories. Sampling sites and a unified data management system will allow critical zone processes to be contrasted by bedrock, landscape position, depth, forest type, and location. https://criticalzone.org/luquillo/
Water in a changing environment	Research on impacts of climate change and variability on water supply from forested watersheds. Caspar Creek, Coweeta, Fraser, Fernow, H.J. Andrews, Hubbard Brook, LEF, Marcell, Santee, and San Dimas and other experimental watersheds
Experimental Forests and Ranges, International Cooperation Program (ICP) Level II	The EFR as a network established 18 sites to conduct the ICP level II climate and atmospheric monitoring. Each site will monitor climate parameters, atmospheric deposition, and ozone
Vegetation net productivity responses to precipitation variability	Collaboration among USDA ARS, USDA Southwest Watershed Research Center, University of Arizona, University of Technology Sydney, USDA-FS Northern Research Station, USDA Northwest Watershed Research Center, USDA-FS Pacific Southwest Research Station, and USDA-FS International Institute of Tropical Forestry
Water, Energy, and Biogeochemical Budgets, WEBB	This project compares the energy, water, and chemical budgets between forested basins in the Luquillo Experimental Forest and agriculturally developed basins. http://pr.water.usgs.gov/public/webb/
National Atmospheric Deposition Program, NADP	Precipitation chemistry data from El Verde, site PR20, at the LEF since 1985. http://nadp.sws.uiuc.edu/sites/siteinfo.asp?id=PR20&net=NTN
Relationship between nutrient inputs, faunal diversity, and abundance	Dominica, Saba, Costa Rica, Brazil, LEF
Long-term Intersite Decomposition Experiment Team, LIDET	The LIDET experiment is designed to test the effects of substrate quality and macroclimate on long-term decomposition and nutrient release dynamics of fine litter. 27 LTER and international sites http://knb.ecoinformatics.org/knb/metacat/nceas.328.26/knb
Tropical Montane Cloud Forest Network	Volunteer network of sites worldwide, United Nations Environment Programme, UNEP
Tropical nutrient limitation studies	LEF, Guánica, La Selva, Monte Verde, Barro Colorado Island, Tapajós, 64 sites total
Lotic Intersite Nitrogen Experiment, LINX	10 LTER and other sites, http://www.biol.vt.edu/faculty/webster/linx/

Table 19.5 (continued)

Activity	Description
Luquillo Forest Dynamics Plot, LFDP	The LFDP, previously known as the Hurricane Recovery Plot and the Luquillo long-term ecological research grid, is a 16-ha forest plot located near El Verde Field Station. Information from the LFDP contributes to the efforts of the Center for Tropical Forest Science (CTFS, Smithsonian Institution) network of large tropical forest plots in order to improve our understanding of tropical forests, to elucidate tree life histories, species interactions, and population changes in order to determine the forest response to environmental changes and disturbance. LEF + 17 tropical sites
Carbon, nitrogen, and phosphorus dynamics in tropical forest ecosystems	LEF, Tapajós, Brazil
Dissimilatory nitrate reduction in humid ecosystems	LEF, Bonanza Creek Experimental Forest, La Selva
Comparative study of terrestrial and aquatic decomposition rates	LEF, Coweeta Hydrologic Laboratory
Earthworms and soil processes in tropical ecosystems	LEF, Xishuangbanna, China
World Wide Aquatic Leaf Decomposition Experiment, WW-DECOEX	LEF + 10 tropical sites
Comparisons of hydrology, nutrient cycling, and canopy dynamics following severe storm damage	LEF, Hubbard Brook Experimental Forest, Taiwan
Comparison of bromeliad phytotelmata in tabonuco and elfin forests	LEF, Dominica
Network analysis of food webs	LEF and six LTER sites
UNESCO Help Program	LEF, 12 watersheds in the USA, including H. J. Andrews Experimental Forest and others worldwide
Landscape fragmentation and forest fuel accumulation: effects of fragment size, age, and climate	LEF, Bonanza Creek Experimental Forest, Idaho
Comparison of aquatic insect emergence	LEF, La Selva, Costa Rica

Acknowledgments This study was done in collaboration with the University of Puerto Rico. We thank Alan P. Covich and three anonymous reviewers for the review of this manuscript. Mildred Alayón edited the final draft of the paper and Gisel Reyes provided bibliographic support.

References

Benstead JP, March JG, Pringle CM (2000) Estuarine larval development and upstream postlarval migration of freshwater shrimps in two tropical rivers, Puerto Rico. Biotropica 32(3):545–548

Blanco JF, Scatena FN (2006) Hierarchical contribution of river-ocean connectivity, water chemistry, hydraulics, and substrate to the distribution of diadromous snails in Puerto Rican streams. J N Am Benthol Soc 25(1):82–98

Brokaw N, Fraver S, Grear JS, Thompson J, Zimmerman JK, Waide RB, Everham-III EM, Hubbell SP, Foster RB (2004) Disturbance and canopy structure in two tropical forests. In Losos E, Leigh EG Jr (eds) Tropical forest diversity and dynamism: results from a long-term tropical forest network. University of Chicago Press, Chicago, pp 177–194

Brown S, Lugo AE, Silander S, Liegel L (1983) Research history and opportunities in the Luquillo Experimental Forest. USDA Forest Service, Southern Forest Experiment Station, General Technical Report SO-44, New Orleans, LA

Covich AP (2006) Dispersal-limited biodiversity of tropical insular streams. Polish J Ecol 54:523–547

Covich AP, McDowell WH (1996) The stream community. In: Reagan DP, Waide RB (eds) The food web of a tropical rain forest, Chap. 13. University of Chicago, Chicago, pp 434–459

Covich AP, Crowl TA, Johnson SL, Varza D, Certain D (1991) Post-Hurricane Hugo increases in atyid shrimp abundances in a Puerto Rican montane streams. Biotropica 23:448–454

Covich AP, Crowl TA, Johnson SL, Pyron M (1996) Distribution and abundance of tropical freshwater shrimp along a stream corridor: response to disturbance. Biotropica 28:484–492

Covich AP, Crowl TA, Scatena FN (2003) Effects of extreme low flows on freshwater shrimps in a perennial tropical stream. Freshw Biol 48:1199–1206

Covich AP, Crowl TA, Heartsill Scalley T (2006) Effects of drought and hurricane disturbances on headwater distributions of palaemonid river shrimp (*Macrobrachium* spp.) in the Luquillo Mountains, Puerto Rico. J N Am Benthol Soc 25:99–107

Covich AP, Crowl TA, Hein CL, Townsend MJ, McDowell WH (2009) Importance of geomorphic barriers to predator-prey interactions in river networks. Freshw Biol 54:450–465

Cross WF, Covich AP, Crowl TA, Benstead JP, Ramirez A (2008) Secondary production, longevity, and consumption rates of freshwater shrimps in two tropical streams with contrasting geomorphology and foodweb structure. Freshw Biol 53:2504–2519

Crowl TA, Covich AP (1994) Responses of a freshwater shrimp to chemical and tactile stimuli from a large decapod predator. J N Am Benthol Soc 13:291–298

Crowl TA, Welsh V, Heartsill Scalley T, Covich AP (2006) Effects of different types of conditioning on rates of leaf-litter shredding by *Xiphocaris elongata*, a Neotropical freshwater shrimp. J N Am Benthol Soc 25(1):198–208

Dowler CC, Tschirley FH (1970) Effects of herbicides on a Puerto Rican rain forest. In: Odum HT, Pigeon RF (eds) A tropical rain forest. National Technical Information Service

Drew AP, Boley JD, Zhao Y, Johnston MH, Wadsworth FH (2009) Sixty-two years of change in subtropical wet forest structure and composition at El Verde, Puerto Rico. Interciencia 34(1):34–40

Gentry AH (1982) Patterns of neotropical plant species diversity. Evol Biol 15:1–83

Guariguata MR (1990) Landslide disturbance and forest regeneration in the upper Luquillo Mountains of Puerto Rico. J Ecol 78:814–832

Heartsill Scalley T, Scatena FN, Estrada C, McDowell WH, Lugo AE (2007) Disturbance and long-term patterns of rainfall and throughfall nutrient fluxes in a subtropical wet forest in Puerto Rico. J Hydrol 333:472–485

Heartsill Scalley T, Scatena FN, Lugo AE, Moya S, Estrada C (2010) Changes in structure, composition and nutrients during 15 year of hurricane-induced succession in a subtropical wet forest in Puerto Rico. Biotropica. doi:10.1111/j.1744-7429.2009.00609.x

Heyne CM (2000) Soil and vegetation recovery on abandoned paved roads in a humid tropical rain forest, Puerto Rico. Department of Biological Sciences. Las Vegas, USA, University of Nevada. 153 p

Hobbs RJ, Arico S, Aronson J, Baron JS, Bridgewater P, Cramer VA, Epstein PR, Ewel JJ, Klink CA, Lugo AE, Norton D, Ojima D, Richardson DM, Sanderson EW, Valladares F, Vilà M, Zamora R, Zobel M. (2006) Novel ecosystems: theoretical and management aspects of the new ecological world order. Glob Ecol Biogeogr 15:1–7

Holdridge LR (1967) Life zone ecology. Tropical Science Center, San José
Hutchinson ID, Wadsworth FH (2006) Efectos de la liberación en un bosque secundario de Costa Rica. Recs. Nats. y Amb. 46:155–160
Kharecha P (1997) Energy evaluation of the effects of human activities on the Luquillo Experimental Forest, Puerto Rico. M.S. thesis in Center for Environmental Policy and Environmental Engineering Sciences. University of Florida, Gainesville, 39 p
Kikkert DA, Crowl TA, Covich AP (2009) Upstream migration of amphidromous shrimp in the Luquillo Experimental Forest, Puerto Rico: temporal patterns and environmental cues. J N Am Benthol Soc 28:233–246
Lugo AE (1986) Water and the ecosystems of the Luquillo Experimental Forest. USDA Forest Service, Southern Forest Experiment Station General Technical Report SO-63, New Orleans, LA
Lugo AE (1994) Preservation of primary forests in the Luquillo Mountains, Puerto Rico. Conserv Biol 8:1122–1131
Lugo AE (2004) The outcome of alien tree invasions in Puerto Rico. Front Ecol Environ 2:265–273
Lugo AE (2005) Los bosques. In: Joglar RL (ed) Biodiversidad de Puerto Rico. Vertebrados terrestres y ecosistemas. Editorial del Instituto de Cultura Puertorriqueña, San Juan, pp 395–548
Lugo AE (2008) Visible and invisible effects of hurricanes on forest ecosystems: an international review. Austral Ecol 33:368–398
Lugo AE, Scatena FN (1995) Ecosystem-level properties of the Luquillo Experimental Forest with emphasis on the tabonuco forest. In: Lugo AE, Lowe C (eds) Tropical forests: management and ecology. Springer, New York, pp 59–108
Lugo AE, Scatena FN (1996) Background and catastrophic tree mortality in tropical moist, wet, and rain forests. Biotropica 28:585–599
Lugo AE, Helmer E (2004) Emerging forests on abandoned land: Puerto Rico's new forests. For Ecol Manag 190(2):145–161
March JG, Benstead JP, Pringle CM, Scatena FN (2003) Damming tropical island streams: problems, solutions, and alternatives. Bioscience 53(11):1069–1078
Marrero J (1950) La reforestación de tierras degradadas de Puerto Rico. Caribb For 11:16–24
Murphy PG (1970) Tree growth at El Verde and the effects of ionizing radiation. In: Odum HT, Pigeon RF (eds) A tropical rain forest: a study of irradiation and ecology at El Verde, Puerto Rico. National Technical Information Service, Springfield, pp D141–D171
Odum HT (1970) Summary: an emerging view of the ecological systems at El Verde. In: Odum HT, Pigeon RF (eds) A tropical rain forest. National Technical Information Service, Springfield, pp I191–I289
Odum HT, Drewry G (1970) The cesium source at El Verde. In: Odum HT, Pigeon RF (eds) A tropical rain forest. National Technical Information Service, Springfield, pp C23–C36
Odum HT, Murphy P, Drewry G, McCormick F, Schinham C, Morales E, McIntyre J (1970a) Effects of gamma radiation on the forest at El Verde. In: Odum HT, Pigeon RF (eds) A tropical rain forest. National Technical Information Service, Springfield, pp D3–D75
Odum HT, Drewry G, McMahan EA (1970b) Introduction to section E. In: Odum HT, Pigeon RF (eds) A tropical rain forest. National Technical Information Service, Springfield, pp E3–E15
Olander L, Scatena FN, Silver WL (1998) Impacts of disturbance initiated by road construction in a subtropical cloud forest in the Luquillo Experimental Forest, Puerto Rico. For Ecol Manag 109:33–49
Parresol BR (1995) Basal area growth for 15 tropical tree species in Puerto Rico. For Ecol Manag 73:211–219
Parrotta JA (1999) Productivity, nutrient cycling, and succession in single- and mixed-species plantations of *Casuarina equisetifolia*, *Eucalyptus robusta*, and *Leucaena leucocephala* in Puerto Rico. For Ecol Manag 124:45–77
Scatena FN (2001) Ecological rhythms and the management of humid tropical forests: examples from the Caribbean National Forest, Puerto Rico. For Ecol Manag 55(28):1–12
Scatena FN, Moya S, Estrada C, Chinea JD (1996) The first five years in the reorganization of aboveground biomass and nutrient use following Hurricane Hugo in the Bisley Experimental Watersheds, Luquillo Experimental Forest, Puerto Rico. Biotropica 28:424–440

Schimper AFW (1903) Plant-geography upon a physiological basis. Clarendon, Oxford
Silver WL, Scatena FN, Johnson AH, Siccama TG, Watt F (1996) At what temporal scales does disturbance affect below ground nutrient pools? Biotropica 28(Special Issue 4a):441–458
Silver WL, Kueppers LM, Lugo AE, Ostertag R, Matzek V (2004) Carbon sequestration and plant community dynamics following reforestation of tropical pasture. Ecol Appl 14:1115–1127
Smith RF (1970) The vegetation structure of a Puerto Rican rain forest before and after short-term gamma irradiation. In: Odum HT, Pigeon RF (eds) A tropical rain forest. National Technical Information Service, Springfield, pp D103–D140
Taylor CM, Silander S, Waide RB, Pfeiffer WJ (1995) Recovery of a tropical forest after gamma irradiation: a 23-year chronicle. In: Lugo AE, Lowe C (eds) Tropical forests: management and ecology. Springer, New York, pp 258–285
Thompson J, Lugo AE, Thomlinson J (2007) Land use history, hurricane disturbance, and introduced species survival in a subtropical wet forest in Puerto Rico. Plant Ecol 192:289–301
Wadsworth FH (1995) A forest research institution in the West Indies: the first 50 years. In: Lugo AE, Lowe C (eds) Tropical forests: management and ecology. Springer, New York, pp 33–56
Wadsworth FH (1997) Forest production for tropical America. USDA Forest Service, Washington, DC. (Agriculture Handbook 710)
Wadsworth FH, Zweede JC (2006) Liberation: acceptable production of tropical timber. For Ecol Manag 233:45–51
Wadsworth FH, Bryan B, Figueroa-Colón J (2010) Cutover tropical forest productivity potential merits assessment, Puerto Rico. Bois For Trop 305:33–41
Walker LR, Zarin DJ, Fetcher N, Myster RW, Johnson AH (1996) Ecosystem development and plant succession on landslides in the Caribbean. Biotropica 28:566–576
Weaver PL, Murphy PG (1990) Forest structure and productivity in Puerto Rico's Luquillo Mountains. Biotropica 22:69–82

Chapter 20
The Discovery of Acid Rain at the Hubbard Brook Experimental Forest: A Story of Collaboration and Long-term Research

Gene E. Likens and Scott W. Bailey

Abstract The 3,519-ha Hubbard Brook Experimental Forest (HBEF) was established in 1955 as the primary hydrological research facility in the northeastern USA. In 1963, FH Bormann, GE Likens, NM Johnson, and RS Pierce initiated the Hubbard Brook Ecosystem Study (HBES) to assess mass balance water and chemical budgets using gauged watersheds. From the study's inception, rain and snow inputs to the HBEF were unusually acid. Using back trajectories for air masses, HBES long-term data showed clearly that sulfate deposition at HBEF was strongly related to SO_2 emissions hundreds or thousands of kilometers distant. Other research showed that acid rain started in eastern North America in the 1950s. Reductions in emissions since 1970, primarily of SO_2 due to federal regulations, caused ~60% decline in acidity at HBEF since 1963. It required 18 years of continuous measurement to fit a significant linear regression to these data, showing the value of long-term measurements. HBEF data showed calcium depletion as a major impact of acid deposition. Other results showed slowed forest growth. In 1999, wollastonite (a calcium silicate mineral) was added experimentally to an entire watershed in an amount roughly equivalent to the amount estimated to have leached in the previous 50 years. Early results suggest positive survival and growth responses in sugar maple. The long-term data from the HBES suggest that changes in federal regulations to reduce emissions have reduced sulfate in both precipitation and stream water, demonstrating a positive link between high quality long-term research and public policy.

Keywords Hubbard Brook Ecosystem Study · Acid rain · Calcium depletion · Clean Air Act · Sulfate deposition · Long-term measurements

---

S. W. Bailey (✉)
Northern Research Station, USDA Forest Service, 234 Mirror Lake Road,
North Woodstock, NH 03262, USA
Phone: 603-726-8902
e-mail: swbailey@fs.fed.us

G. E. Likens
Cary Institute of Ecosystem Studies, Millbrook, NY, USA

Department of Ecology and Environmental Biology, University of Connecticut, Storrs, CT, USA

D. C. Hayes et al. (eds.), *USDA Forest Service Experimental Forests and Ranges*,
DOI 10.1007/978-1-4614-1818-4_20, © Springer New York 2014

20.1 Introduction

The Hubbard Brook valley (43°56′N, 71°45′W) is an area of 3,519 ha drained by Hubbard Brook, a fifth-order river. This valley is mostly (90%) comprised of the Hubbard Brook Experimental Forest (HBEF) and has Mirror Lake and its watershed near its base. The HBEF and Mirror Lake are remarkable, long-term research sites in the White Mountains of New Hampshire. The HBEF was established in 1955 to be the primary hydrological research facility in the northeastern USA. The valley and its river, and at one time Mirror Lake, were named or misspelled, for an early pioneer, William Hobart, in this region (Likens 1972; Cogbill 1989 unpublished.).

It is appropriate that research in forest hydrology found such a significant home in the White Mountain National Forest as the Weeks Act of 1911, which first authorized federal purchase of forestlands in the East, was inspired by questions about the effects of forest harvesting watersheds in this region. Beginning in the 1860s, and peaking in 1907 (Belcher 1980), the forests of the White Mountains were mostly liquidated in a series of large clearcuts executed by timber barons and land speculators. Poor logging practices, including use of splash dams and riparian areas to transport logs, and extensive railroad development, which often promoted slash fires, led to extensive erosion and sedimentation. So, while scenic protection may have been the initial impetus for the outcry to protect forests in the White Mountains, it was not until Congress was faced with testimony from downstream mill-owners about the effects of floods and droughts on mill productivity that Congress authorized federal action, in an act sponsored and named for Representative John Weeks, a native of the White Mountains (see Hornbeck 2001).

Questions about the influence of forests and forest management on water resources, flooding, and sedimentation were first studied scientifically in western Europe in the nineteenth century, although such questions were raised in ancient times as revealed by the writings of Roman philosopher Pliny the Elder (McGuire and Likens 2011). A breakthrough in the study of forest effects on hydrology was made with the development of the paired-watershed approach, in which the effects of vegetation and management on a treated watershed were compared against an untreated reference watershed. This approach was pioneered by USDA Forest Service (USDA FS) research at Wagon Wheel Gap, Colorado in 1909 (Bates and Henry 1928). It was subsequently applied to various regions of differing climate and vegetation with the establishment of the Coweeta Hydrologic Laboratory in the southern Appalachians in 1936, the Fernow Experimental Forest in the central Appalachians in 1950, and the HBEF in the northern Appalachians in 1955.

The HBEF was established by the USDA FS White Pine-Hardwood Research Center, located in Laconia, NH to study the question, "Is good timber management also good watershed management?" (Hornbeck 2001). Beginning in 1955, gauged watersheds were established in the headwaters of Hubbard Brook. While paired-watershed manipulation experiments were to be the primary approach, the first decade was devoted to building infrastructure, including locating watersheds, building rain gauges and weirs, and initiating basic characterization studies of the site including soil, geologic, and topographic surveys. Plans were developed by the

USDA FS to initiate studies of flood flows, water quality, and water supply; basic studies also were initiated on components of the hydrologic cycle including snow accumulation, soil frost, and canopy interception and stemflow (Hornbeck 2001).

20.2 The Hubbard Brook Ecosystem Study

In 1963, the Hubbard Brook Ecosystem Study (HBES) was initiated in the Hubbard Brook Valley by F. H. Bormann, G. E. Likens, N. M. Johnson, and R. S. Pierce. The approach was to initiate comprehensive ecosystem and biogeochemical studies, using the gauged watersheds in the headwaters of the HBEF, and focus on water and chemical budgets. Initially, the concept of the HBES was based on a simple, medical metaphor: Can the chemistry of stream water be used to diagnose the extremely complicated functions ("health") of a forest watershed ecosystem and associated aquatic ecosystems, as a physician would diagnose the health of a human patient by analyzing the chemistry of the patient's blood and urine? To do this, careful measurements and quantitative budgets (mass balances) of all inputs and outputs of water and chemicals for watershed ecosystems needed to be determined (Bormann and Likens 1967). These watershed mass balances turned out to be a powerful analytical tool for addressing questions about how large and complicated watersheds of the valley are structured and function, and how serendipity frequently led HBEF scientists into pioneering efforts including the first nonmedical use of an atomic absorption spectrophotometer for critical measurements of base cations in precipitation, stream, and lake water (see Lindenmayer and Likens 2010, for a fuller account of this story).

Bormann (plant ecologist), Likens (aquatic ecologist), and Johnson (geologist) were faculty members at Dartmouth College, and Pierce (soil scientist) was a USDA FS project leader at HBEF at the initiation of the HBES. Originally, Bormann and Likens "...felt that slower growth [of the project] would be more manageable and would allow for substantial interaction among all senior investigators to ensure proper coordination and development of the overall study" (from the Preface in Likens et al. 1977). Pierce was an exceptional facilitator of collaboration and cooperation in the early days of the HBES. Early collaborators in the HBES included Donald Fisher of the USGS, Robert H. Whittaker of Cornell University, Margaret B. Davis of the University of Minnesota, John S. Eaton of Dartmouth College, James W. Hornbeck and C. Anthony Federer of the USDA FS, Richard T. Holmes of Dartmouth College, Frank W. Sturges of Shepherd's College, and a number of graduate students and postdoctoral associates. Nevertheless, in the 1980s, the HBES grew in size, complexity, and number of collaborators. Indeed, the HBES was a magnet for researchers wanting to work on terrestrial and associated aquatic ecosystems, all contributing to the wealth of long-term data (Likens 2004).

The HBES has been continuously funded by the National Science Foundation (NSF) since 1963, a Long-Term Ecological Research (LTER) site since 1988, a Long-Term Research in Environmental Biology (LTREB) site since 1993, a Clean

Air Status and Trends Network (CASTNet) site for monitoring precipitation chemistry since 1989, an Environmental Protection Agency (EPA)-funded National Dry Deposition Network (NDDN) site since 1988 (now part of CASTNet since 1990), National Acid Deposition Program / National Trends Network (NADP/NTN) site since 1978, a US Geological Survey (USGS) site for hydrologic studies on Mirror Lake since 1978, and a site for characterizing groundwater flow and chemical transport in fractured rock since 1990. Currently, some 50 scientists do research in the Hubbard Brook Valley.

For additional information about the HBEF and the HBES, see Likens (1995), Bormann and Likens (1979), Likens (1985), and www.hubbardbrook.org.

20.3 The Mirror Lake Ecosystem Study

Mirror Lake, located at the base of the Hubbard Brook Valley, has been a large and important component of the HBES since the early 1960s with many publications and two books produced (Likens 1985; Winter and Likens 2009). In 1978, Thomas C. Winter of the USGS joined in the long-term studies of Mirror Lake and added his extensive experience and talents relative to groundwater to these studies. James LaBaugh and Donald Rosenberry of USGS also joined in these studies of the hydrology of Mirror Lake.

20.4 Acid Rain

We will use the discovery of acid rain in the Hubbard Brook Valley as a case study to illustrate the collaboration among the founders and other scientists of the HBES, as well as the value of long-term monitoring.

It was obvious from the first sample of rain collected at the HBEF on 24 July 1963, as the HBES began its attempt to measure all of the inputs to and outputs from forest watershed ecosystems, that the rain was very acid (the pH of this rain event was 3.70). The first complete water year of the HBES (1964–1965) had a volume-weighted average pH of 4.12. The precipitation has been consistently acid since then, with many individual rainstorms at pH levels less than 4. The lowest pH measured in a rainstorm at the HBEF was 2.85! The annual average pH of precipitation in northeastern North America is between 4 and 5.4 (Fig. 20.1).

Although precipitation collected at the HBEF was unusually acid, the cause was unclear. It was not known whether the acidity was some unusual feature of the White Mountains in New Hampshire and the relatively base-poor geology characterizing the area. Two serendipitous events helped to clarify why the precipitation was so acid, and to determine that it was a widespread phenomenon. In 1969, Likens moved from a faculty position at Dartmouth College to one at Cornell University.

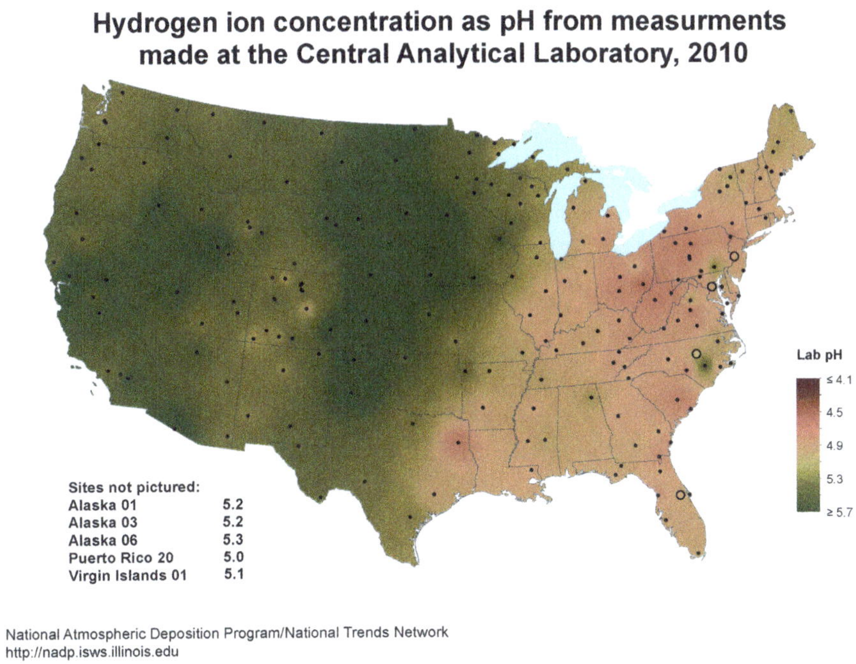

Fig. 20.1 Annual, mean pH of precipitation across the USA in July 2010. (Source: National Acid Deposition Program)

As part of his early research at Cornell University, he established precipitation collection sites around Cayuga Lake and Seneca Lake in the Finger Lakes region of New York State and discovered that the chemistry of rain and snow in this part of upstate New York was essentially identical to that measured in the White Mountains of New Hampshire (Likens 1989). This was the first clue that acid rain was a regional, not local phenomenon in North America (Likens and Bormann 1974; Cogbill and Likens 1974). In 1969, Likens travelled to Sweden on a NATO Senior Fellowship and met Prof. Svante Odén during a visit to the University of Uppsala. He learned that Odén had been studying the chemistry of precipitation in Scandinavia; Odén had found that the precipitation was quite acid, and had related this abnormal acidity to emissions of sulfur dioxide from the more urbanized and industrialized countries of Europe (Likens 1989; Odén 1968). The first paper on acid rain in North America (titled "Acid Rain") was published in 1972 (Likens et al. 1972). We were unaware at the time that Robert Angus Smith had referred to acid rain 100 years earlier (Smith 1872, p 444 and 555) in his studies of air pollution in urbanized areas of Great Britain.

Acid rain is formed when fossil fuels, primarily coal and oil, are combusted emitting sulfur dioxide and nitrogen oxides to the atmosphere. In the atmosphere, some of these gases are converted to acids, which fall back to the surface of the Earth along with acidifying particles. Acid rain is the popular term for the atmo-

Fig. 20.2 Bulk precipitation collector used at Hubbard Brook Experimental Forest. (Photo by G. E. Likens)

spheric deposition of various forms of acidic substances, i.e., abnormally acidic rain, snow, sleet and hail, cloud water, fog and rime ice, and the dry deposition of acidifying particles and gases (Likens et al. 1972). Air pollution leading to acid rain can cause many serious environmental problems, including degradation of aquatic and terrestrial ecosystems, corrosion of structures and monuments exposed to air pollution and acid rain, and impacts on human health, often amounting to large environmental and economic costs (see below; Weathers et al. 2007; Likens 2010).

In 1963, many important questions needed to be answered to make progress in understanding this emerging environmental problem and to guide policy in dealing with it. First, a collector for clean, uncontaminated samples of precipitation, both rain and snow, had to be developed. Our attempts resulted in a collector for rainwater that we called a bulk collector; it consisted of a simple funnel connected to a reservoir with the tubing in a loop between the funnel and the reservoir. This loop, when partially full of water, served as a vapor barrier to prevent evaporation (Fig. 20.2). More sophisticated kinds of collectors, such as the Aerochem metrics wet/dry collector, also have been used in the HBES (Likens 2013).

Sulfur and nitrogen oxide pollutants can be transported long distances in the atmosphere by prevailing winds, but the winds bringing precipitation events to the HBEF are much more complicated than just west to east flow. Because inventories

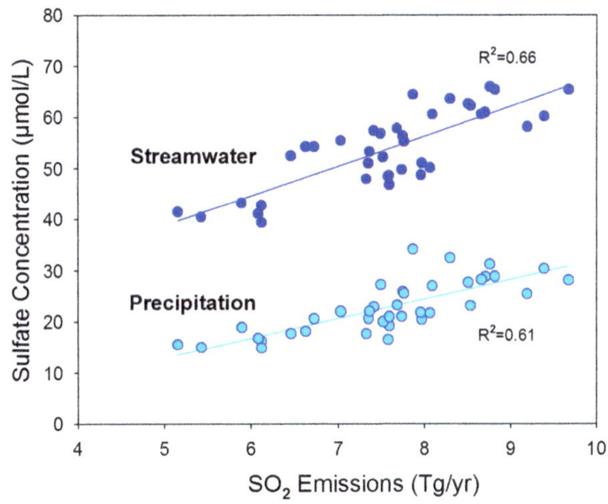

Fig. 20.3 Relation between SO_2 emissions from the 24-h source area (southeastern Canada, Maine, New Hampshire, Vermont, Connecticut, Rhode Island, Massachusetts, New York, New Jersey, Pennsylvania, Michigan, Ohio, Delaware, Maryland, West Virginia, Virginia) for the HBEF and sulfate concentrations in precipitation and stream water. (Modified and updated from Likens et al. 2002, 2005; also Likens 2010)

of emissions are done on a state-by-state basis by the EPA, it was necessary to develop procedures for following storm tracks, determining how long it took a parcel of air to come to the HBEF from upwind locations, and what the prevailing storm tracks were in order to match these tracks with the source areas for emissions of sulfur and nitrogen oxides. Using the high-split 4-model for calculating back trajectories for air masses (Draxler and Hess 1998), we could determine the back trajectories for air masses that were providing precipitation at the HBEF and then analyze these back trajectory patterns by cluster analysis (Stunder 1996). Using this approach, we were able to develop strong relationships between the emissions of sulfur dioxide from the relevant source area for HBEF with measured sulfate concentrations in precipitation and stream water at the HBEF (Likens et al. 2002, 2005). These relations were extremely important because in the early days of the acid rain debate (sometimes called the "acid rain wars" of the 1980s; Likens 2010), there was enormous controversy about whether the emissions from the midwestern USA were actually the source of the unusual acidity in precipitation in the northeastern USA. The long-term data from the HBES and the relationships observed (Fig. 20.3) showed clearly that sulfate deposition at the HBEF was strongly related to SO_2 emissions hundreds or thousands of kilometers distant. We developed a similar relationship for NO_X emissions and nitrate concentrations in precipitation, but the emission–deposition relationship was more difficult because there had been smaller reductions in the emissions of NO_X since 1970 and the NO_X pollutants are more reactive in the atmosphere (Butler et al. 2003).

We had to answer other fundamental questions, such as was the acidity abnormal, and how long had the precipitation been abnormally acidic, i.e., what was the acidity of precipitation prior to the Industrial Revolution? To answer such questions, Likens and colleagues went to some of the most remote regions of the world—the southern tip of Chile, the southern tip of Africa, an island (Amsterdam Island) in the middle of the Indian Ocean, a remote site in Australia, etc.—to measure the precipi-

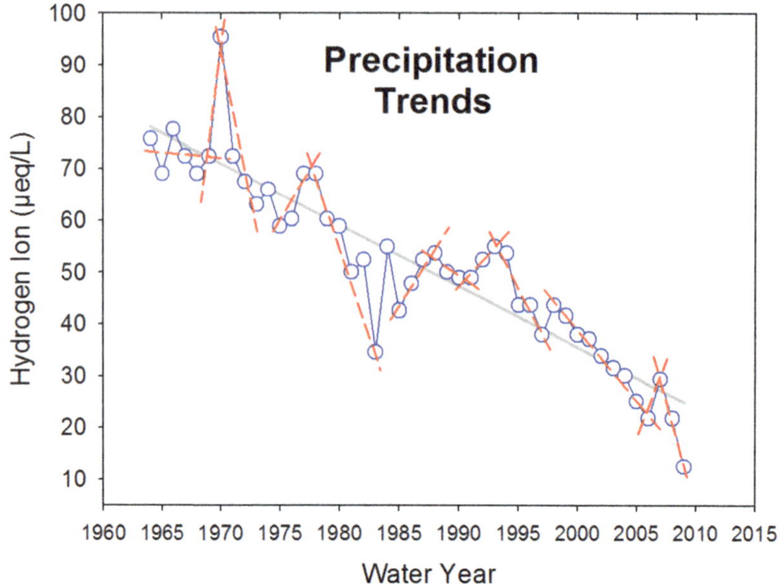

Fig. 20.4 Long-term record of annual precipitation acidity (hydrogen ion in μeq/L) at HBEF. (Updated and redrawn from Likens 1989)

tation chemistry in those areas, which were relatively unimpacted by anthropogenic activities. From these studies, a background pH of about 5.1 for mineral acids was determined (Likens et al. 1987). This reference was used for judging how much the acidity of precipitation had increased because of the pollution generated primarily by the combustion of fossil fuels. Analyses of historical data for the eastern USA suggested that acid rain as a phenomenon started in eastern North America around the mid-1950s (Cogbill and Likens 1974, 1976; Cogbill et al. 1984; Likens and Butler 1979; Butler et al. 1984). Reductions in emissions since 1970, primarily of sulfur dioxide since 1970 because of federal regulations, have resulted in a decline in acidity of about 60% in the 47 years since 1963, in precipitation measured at HBEF. The long-term record (Fig. 20.4) is quite interesting in this regard—it shows that the volume-weighted, annual values of hydrogen ion since 1963 are declining, although there is much variability from year to year. Indeed, there are periods up to 9 years long showing decreasing or increasing trends, but these short-term trends do not characterize the long-term trend. It required 18 years of continuous measurement to fit a significant linear regression to these data showing that the acidity of precipitation at the HBEF had actually decreased (Likens 1989). This is a sobering example of the value of long-term measurements (see Lindenmayer and Likens 2010). Shorter-term measurements can be quite misleading regarding the overall long-term trend. It is important to note in Fig. 20.4 that the acidity of rainfall is still some 2–4 times higher than it should be, if the air were not polluted. In 1963, sulfuric acid contributed about 65% of the acidity and nitric acid about 25%. Because of changes in emissions with time, these two strong acids are currently approaching

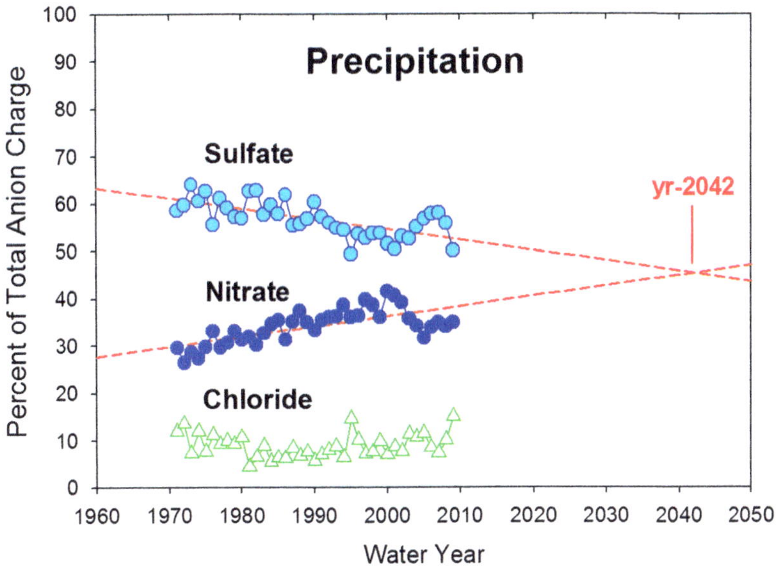

Fig. 20.5 Long-term percent of total sulfate, nitrate, and chloride acids in precipitation at the HBEF. The regression lines for sulfate and nitrate are projected to cross in about 2042. (Updated and redrawn from Likens and Fallon 1998)

similar values of contribution to the total acidity, and projections suggest that nitric acid might become the dominant acid in precipitation in the next decade or two (Fig. 20.5). This result is extremely important because the ecological impact of nitric acid on forest and aquatic ecosystems is different from sulfuric acid.

The early 1980s were a time of intense policy discussions related to the acid rain problem. President Jimmy Carter's administration had begun a bilateral assessment in 1978 of air pollution across the US/Canada border. Technical working groups working under the US/Canada Memorandum of Intent on Transboundary Air Pollution issued reports in early 1983. This report, however, expressed quite divergent views from the two nations (Oreskes and Conway 2010, pp. 74–76).

In 1983, Likens led a small delegation of scientists to the White House to brief President Ronald Reagan and his cabinet on the issues of acid rain. This briefing had been arranged by William Ruckelshaus, administrator of the EPA. These scientists had a full hour with the President and the Cabinet to present a case for a reduction in sulfur dioxide emissions that they believed would lead to amelioration of acid rain impacts. The cabinet and President Ronald Reagan listened intently for an hour, but in January 1984, David Stockman, the director of the Office of Management and Budget, announced that it would cost too much to deal with acid rain and decided to postpone action on the issue, to ignore the advice, and instead to continue to study the problem of acid rain for a decade in a program called the National Acid Precipitation Assessment Program (NAPAP; see Sun 1984; Oreskes and Conway 2010, pp. 93–94). The stated intention was that after 10 years of study, policymakers would know all of the relevant scientific answers to questions related to acid rain and could make intelligent decisions toward solving the problem (see Likens 2010). The USA spent about US $570 million on NAPAP during the 1980s, rather than making a policy decision to deal with the problem. Unfortunately, although much research was done, a formal, thorough assessment *per se* was not done, and the 1990 Clean Air Act Amendments, signed into law prior to the issue of NAPAP's final report, called for reductions in SO_2 emissions that were almost exactly what had been proposed in the meeting with the President and the Cabinet in 1983.

20.5 The Calcium Response to Acid Rain

The HBES pioneered the development of the watershed ecosystem as an experimental unit to study element fluxes. Applying the technique to nutrient budgets led to the discovery of calcium depletion as a major impact of acid deposition (Likens et al. 1996). Calcium along with magnesium, potassium, and sodium are ecologically and biogeochemically important base cations in the HBEF. Early work at HBEF followed the then accepted geochemical model that the difference between stream water export and precipitation input could be used as an estimate of base cation denudation, equivalent to mineral weathering (Bormann and Likens 1967; Likens et al. 1967; Johnson et al. 1968; Likens 2013). As the ecosystem approach was developed, this view expanded to account for net changes in long-term storage of nutrients in forest biomass (Likens et al. 1977; Likens 2013). It was generally accepted that annual changes in storage within available nutrient pools in forest soils are negligible as the available soil pool is large relative to other nutrient pools and annual fluxes. Furthermore, soils develop on the timescales of centuries to millennia, so changes at the annual to decadal scale were assumed to be very small or negligible.

However, Likens et al. (1996) challenged this assumption by independently estimating the mineral weathering flux, thereby solving the watershed mass balance equation to estimate change in the available calcium soil pool. This approach suggested that a substantial reduction in the soil available calcium pool had occurred since the onset of acid deposition, with the peak depletion occurring in the early 1970s. Depletion of soil calcium was confirmed at the nearby Cone Pond watershed, based on a modified mass balance approach using an independent estimate of weathering flux gained by study of strontium isotopes (Bailey et al. 1996). It was further confirmed by Bailey et al. (2003) who used a mass balance for Na and mineral stoichiometry to place error limits on the HBEF mineral weathering estimate. In subsequent years, other experimental approaches corroborated this important finding, including studies of mass balance and repeated soil sampling in an experimentally acidified watershed in Maine (Fernandez et al. 2003, 2010) and by direct measurement via soil sampling repeated on a multi-decadal timescale (Bailey et al. 2005). The degradation of terrestrial ecosystems through the accelerated loss of base cations has now been found to occur in several base-poor regions of the eastern USA and Canada (e.g., Bailey et al. 2005; Long et al. 2009; Warby et al. 2009; Watmough and Dillon 2001, 2004). Calcium decline has also been observed widely in surface freshwaters (e.g., Buso et al. 2009; Jeziorski et al. 2008). In Mirror Lake, calcium has declined by about 20 % since 1980 (Buso et al. 2009). Accordingly, the average pH of Mirror Lake only increased from 6.36 (1967–1981) to 6.5 (1981–2000). Prior to the Industrial Revolution, the average pH of Mirror Lake was estimated to be about 7.0–7.5 (Buso et al. 2009). The acidity of Mirror Lake is strongly buffered by base cations in groundwater inputs (about 63 % of annual acid-neutralizing capacity influx to the lake comes from groundwater (Buso et al. 2009). The loss of base cations from soils and surface waters makes ecosystems even more sensitive (less buffering available) to the continued inputs of acid rain.

There have been major changes in calcium pools and fluxes, e.g., in both precipitation and stream water at the HBEF since the beginning of the long-term studies. Calcium and magnesium concentrations declined markedly in precipitation from 1963 to about 1972, and then more slowly since (Fig. 20.6). The potential causes for the steep decline are a decrease in emissions of particles containing calcium or magnesium from power plants and from cement production, changes in dust from agricultural activities, and paving or sealing of gravel roads (again, in terms of dust added to the atmosphere). The most compelling reason for the decline in particulate emissions is the large change in industrial processes, fuel combustion, and solid waste incineration that occurred between 1970 and 1987, primarily as a result of the 1970 Clean Air Act Amendments, which were focused on reducing particle emissions from smokestacks (Table 20.1).

There also have been sizeable declines in calcium (~50 %) and magnesium (~30 %) concentrations in stream water, but they were less marked than in precipitation (~75 and ~80 %, respectively) (Fig. 20.6). The export of these base cations in stream water is balanced by the mobile anions, sulfate, and nitrate (correlation coefficient of 0.94; Fig. 20.7), which originate primarily from the atmosphere in the HBEF.

An interesting relationship has developed in the long-term record of stream water chemistry in response to inputs of atmospheric pollutants to watershed ecosystems of the HBEF. Starting with estimates for pre-Industrial Revolution values, increasing atmospheric deposition of sulfate and nitrate has increased the concentrations of the mobile anions, sulfate, and nitrate, acidifying stream water. With the reduction in emissions to the atmosphere because of federal rulemaking (Likens 2004, 2010), there has been a concurrent decline of sulfate since about 1970 and more recently a decline in nitrate concentrations in stream water until the present time, leading to acidification recovery (Likens et al. 1996, 2002; Likens and LaBaugh 2009, p. 314; Judd et al. 2011; Likens and Buso 2012). For the first time in these long-term studies at the HBEF, stream water now has an acid-neutralizing capacity (ANC) of 0 or slightly positive. Thus, the ANC and acidic waters in the headwater streams at the HBEF and in Mirror Lake, now, are recovering slowly toward pre-Industrial Revolution values (e.g., Likens et al. 1996; Buso et al. 2009; Likens and Buso 2012).

Geochemical process for the neutralization of acid inputs in precipitation to the ecosystem was first proposed in an influential paper by Johnson et al. (1981). They proposed that neutralization is a two-step process whereby inputs of hydrogen ion from the atmosphere are first neutralized by the dissolution of reactive alumina in

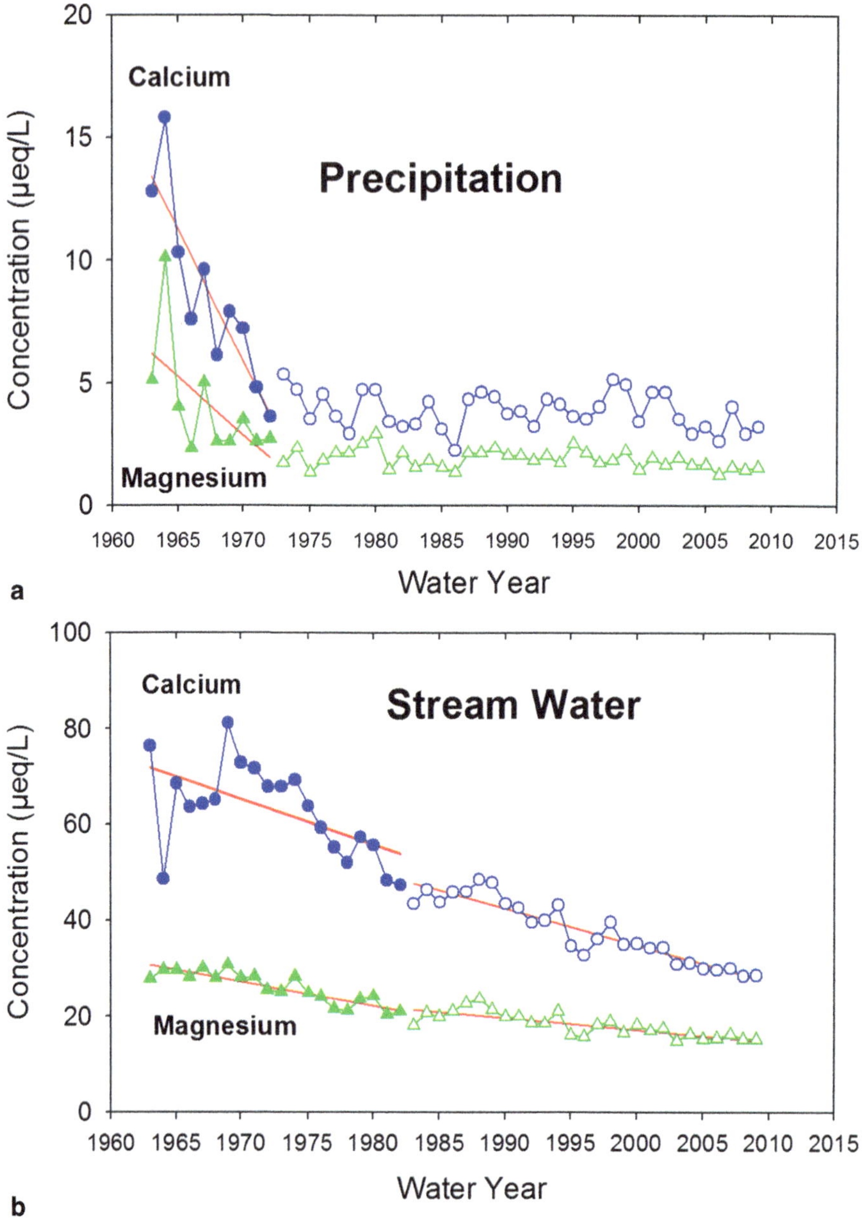

Fig. 20.6 Long-term declines in annual, volume-weighted calcium and magnesium concentrations in precipitation (**a**) and stream water (**b**) for Watershed 6 of the HBEF. (Updated and redrawn from Likens et al. 1998)

Table 20.1 Estimates of particulate matter emissions for the United States by source category. (metric tons × 10^6/year; data from EPA 1989)

Source	1970	1987	% Change
Industrial processes	10.5	2.5	−76
Fuel combustion	4.6	1.8	−61
Solid waste incineration	1.1	0.3	−73
Miscellaneous[a]	1.1	1.0	−9
Transportation	1.2	1.4	+17
Total	18.5	7.0	−62

[a]e.g. Forest fires

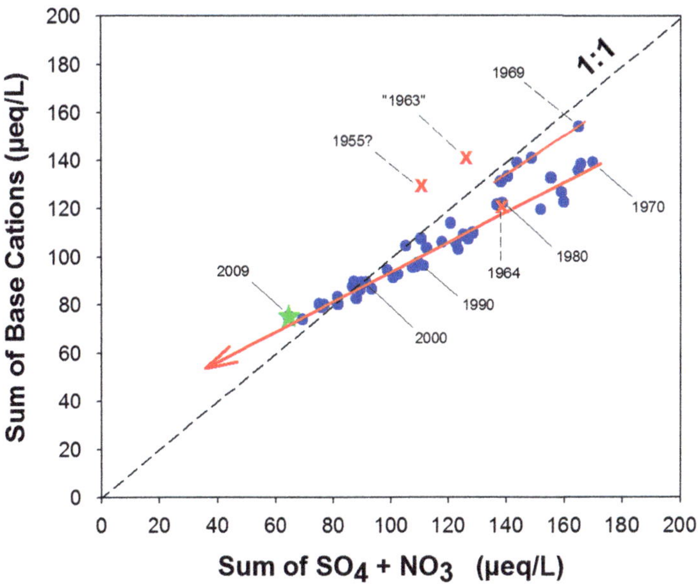

Fig. 20.7 Relationship between sum of base cations (calcium, magnesium, potassium, sodium) export in stream water to export of mobile anions, sulfate plus nitrate, for Watershed 6 of the HBEF (updated from Likens et al. 1998). PIR is estimated pre-Industrial Revolution values. (Modified and updated from Likens et al. 1998)

the soil zone, followed by neutralization of both hydrogen ion and aluminum acidity by chemical weathering of primary silicate minerals. They initiated longitudinal measurement of stream water chemistry from headwaters to third-order tributaries at the HBEF to confirm this hypothesis. Not only was this paper important regarding neutralization of acid rain, but it showed the importance of this process in generating toxic, dissolved aluminum within the rooting zone of forest soils and in stream water.

A number of soil characteristics conspire to make HBEF, and much of the surrounding northern Appalachian region, sensitive to the impacts of acid deposition. The ability of an ecosystem to buffer the inputs of acids rests largely in the soil. The rate at which minerals in the soil decompose largely determines this buffering capacity since mineral weathering is the primary process in forests that neutralizes acidity. In turn, the rate of mineral weathering, which might be expressed as the amount of mineral weathered per unit area of the landscape, per unit time is largely governed by three factors—the types of minerals that are present, the distribution of soil particle sizes, and the depth of soils and unconsolidated geologic deposits. Upland forests of the northern Appalachians are characterized by silicate minerals that decompose slowly, by relatively coarse soils with high sand and low clay content, and correspondingly small surface area, which limits weathering reactions, and by shallow depths to bedrock (Bailey 2000).

As sulfuric and nitric acid pass through the forest and soils of the ecosystem, after precipitation lands on the canopy or directly on the forest floor, calcium, magnesium, and aluminum are exchanged with the hydrogen ion in the precipitation; subsequently, calcium, magnesium, and aluminum in dissolved form exit the system in stream water, along with the mobile anions, sulfate, and nitrate. As a result, very large quantities of base cations, particularly calcium, are leached from the system by acid rain. It was estimated that some 840 kg of calcium/ha were depleted from soil pools of the HBEF during 1940–1995 in this way (Likens et al. 1996).

20.6 How did the Forest Respond to Acid Rain?

Starting in 1965, measurements in Watershed 6 (W6) showed that the forest rapidly increased in aboveground biomass until 1982, then leveled off (Likens et al. 1994; Fahey et al. 2005; Siccama et al. 2007). In fact, the most recent measurements, in 2007, showed that the forest is now declining in biomass (Battles et al. 2014; Lindenmayer and Likens 2010, p. 123). Long-term measurements of belowground carbon are difficult to make quantitatively at the watershed ecosystem scale, but current estimates suggest that there are no significant trends in this value. Soil carbon is a very large pool in the ecosystem at the HBEF, and annual changes are expected to be small, thus difficult to quantify (Huntington et al. 1988, 1989; S. P Hamburg, personal communication). Probably, it is fair to state that the forest in W6 is no longer accumulating biomass, but is now emitting carbon to the atmosphere rather than sequestering it (Likens and Franklin 2009; Lindenmayer and Likens 2010).

What could be the explanation for this remarkable change in the HBEF? It might be a natural pattern; a response to climate change (e.g., Likens 2000; Campbell et al. 2007); an effect of air pollution, a direct effect of acid rain, an increased impact of ozone (e.g., Ollinger et al. 1997); or maybe disease was the primary causal factor. There has been significant beech bark scale disease (e.g., Lovett et al. 2006) causing great mortality of American beech (*Fagus grandifolia*) trees at the HBEF and elsewhere in eastern North America; perhaps, it is the nutrient limitation—nitrogen usually is the element thought to be most limiting to growth in northern hardwood forests like these (e.g., LeBauer and Treseder 2008); possibly because of the major depletion, calcium has now become the limiting nutrient. In an attempt to answer this complicated question, a watershed scale experiment was initiated in 1999 whereby wollastonite, a mineral composed of calcium silicate, was added experimentally to an entire watershed (W1) of the HBEF. The amount of calcium added in this experimental manipulation (1,189 kg calcium/ha) was roughly equivalent to the amount leached from the ecosystem by acid rain in the previous 50 years or so. It was thought that the main reason for the decline in forest biomass was due to elevated mortality of trees, primarily sugar maple (*Acer saccharum*), and the early results of the experiment are encouraging (e.g., Juice et al. 2006; Halman et al. 2008). The seedlings, germinants, and canopy of sugar maple are responding to the addition of the wollastonite to W1, in comparison with sugar maple in adjacent, untreated areas. Likewise, red spruce (*Picea rubens*) showed decreased winter injury in the treated area in comparison to the reference area (Hawley et al. 2006). It will be important to learn how this experimental manipulation plays out during the next 50 years or so from both a scientific and a management point of view.

Horsley et al. (2002) reviewed several decades of research on sugar maple decline disease and found that imbalances in base cations, including low Ca and Mg supply relative to available Al and Mn, were a common factor in decline episodes reported across the range of sugar maple in the second half of the twentieth century. Bailey et al. (2004) proposed soil thresholds for levels of available Ca and Mg that protect sugar maple from health impacts due to secondary stressors, based on study of a widespread decline episode that occurred in northern Pennsylvania in the 1990s. In a retrospective soil sampling study in this same area of Pennsylvania, Bailey et al. (2005) showed that forest soils at all sites, where archived soil samples were available, were above these thresholds in 1967, but had fallen below by 1997. This change is consistent with the expansion of sugar maple decline in the 1980s and 1990s in this region. Bailey et al. (2005) suggested that the region where soils are in a nutrient range supportive of healthy sugar maple had declined due to air pollution-induced soil acidification. Hallett et al. (2006) showed that even though severe decline was not present across a larger region, including northern New England during this time, there was measureable deterioration of sugar maple crown condition associated with sites having low Ca and Mg supply. Long et al. (2009) further showed that these impacts extend to sugar maple growth as well as health. Across the region, including Pennsylvania, New York, and northern New England, sugar maple grew at similar rates on all sites through the late 1960s, whereas since that time, growth has declined on sites with lower supply of Ca and Mg from the soil.

20.7 Policy Implications of Long-Term Acid Rain Research

Based on data from the EPA, US emissions of SO_2 have declined from ~28 Tg/year in 1970 to about 12 Tg/year in 2009 (57% decrease), and of NO_X from ~24 Tg/year in 1970 to ~16 Tg/year in 2009 (33% decrease). Currently, some 38% of total US sulfur dioxide emissions are generated in 12 midwestern US states (IA, IL, IN, KY,

MI, MN, MO, OH, PA, TN, WI, WV) in the production of electrical power, primarily in the Ohio and Tennessee Valleys (T. J. Butler, personal communication). Some 70 % of US emissions of SO_2 are generated by electrical utilities (EPA).

Development of policy for reducing the impact of acid rain has spanned across three US presidents, one pope, and about 27 years from the time that acid rain was discovered at the HBEF. Unfortunately, the problem is still not solved politically, and arguably, it is worse in terms of impact than was thought 20 years ago when the 1990 Clean Air Act Amendments were enacted because of the increased sensitivity of the buffer-depleted system (Likens 2010). Did taxpayers get their money's worth from the large costs of implementing the Clean Air Act of 1990? The long-term data from the HBES would suggest an emphatic positive answer! With reductions in sulfur dioxide emissions as a result of expensive federal regulation, there has been a significant corresponding reduction in sulfate in both precipitation and stream water (Fig. 20.3; Likens 2010). The US EPA has estimated that the rules relating to these air pollutants, including acid rain, have resulted in benefits, including health benefits, of between US$ 101 and 119 billion per year at costs of US$ 8–8.8 billion per year over the 10 years preceding 2003. It normally takes a very long time from the discovery of an environmental problem to political action (Likens 1992, Likens 2010), but special scientific resources like the HBEF and the scientific research that is possible in such sites can help to speed and guide this process.

20.8 Some Major Findings from the Long-Term HBES

There have been several discoveries from the long-term HBES, including:

- There is net retention (precipitation inputs > stream water outputs) of H^+, N, Cl, and P and net losses (stream water outputs > precipitation inputs) of Ca, Mg, Na, K, S, Si, and Al for this northern hardwood forest ecosystem (see Likens 2013; Likens 2004).
- Following deforestation disturbance, loss of biological regulation occurs and stream water outputs increase; with time, gradual recovery of biological regulation of outputs develops (e.g., Likens et al. 1970; Bormann and Likens 1979; McGuire and Likens 2011). Deforestation disturbance caused a reduction in evapotranspiration with a shift from evaporative water loss to runoff of liquid water, and to increased stormflow. Forest disturbance also affected microbial activity, affecting nutrient output (e.g., Likens et al. 1970; Bormann and Likens 1979), by accelerating decomposition and nitrification with the production of H^+ and NO_3^- losses in stream water. An increase in the erodibility of the devegetated system occurred with time (Likens 2004).
- Calcium and other plant nutrients have been markedly leached from the soils of the HBEF by acid deposition (Likens et al. 1996, 1998; Likens 2004).
- Accelerated loss of nutrients, including calcium, has reduced forest growth (e.g., Likens et al. 1994; Fahey et al. 2005; Siccama et al. 2007).

Acknowledgments We thank S. Horsley, D. Jeffries, and S. Stout for thoughtful comments and suggestions on the manuscript. Financial support for long-term research and monitoring at the HBEF was provided by the NSF (including the Long-Term Research in Environmental Biology and Long-Term Ecological Research programs) and The Andrew W. Mellon Foundation. The HBEF is operated and maintained by the Northern Research Station in Newtown Square, PA. D. Buso assisted with the illustrations, and P. Likens helped with manuscript preparation.

References

Bailey SW (2000) Ecological setting of the northern forest—geologic and edaphic factors influencing susceptibilty to environmental change. In: Birdsey R, Hom J, Mickler R (eds) Responses of northern US forests to environmental change, vol 139 (Ecological studies series). Springer-Verlag, New York. pp 27–49

Bailey SW, Hornbeck JW, Driscoll CT, Gaudette HE (1996) Calcium inputs and transport in a base-poor forest ecosystem as interpreted by Sr isotopes. Water Resour Res 3(3):707–719

Bailey SW, Buso DC, Likens GE (2003) Implications of sodium mass balance for interpreting the calcium cycle of a forested ecosystem. Ecology 3(2):471–484

Bailey SW, Mayer B, Mitchell MJ (2004) Evidence for influence of mineral weathering on stream water sulphate in Vermont and New Hampshire (USA). Hydrol Process 18:1639–1653

Bailey SW, Horsley SB, Long RP (2005) Thirty years of change in forest soils of the Allegheny Plateau, Pennsylvania. Soil Sci Soc Am J 69:681–690

Bates CG, Henry AJ (1928) Forest and streamflow experiment at Wagon Wheel Gap, Colorado. U.S. Weather Bureau monthly water review, suppl 30, 79 p

Battles JJ, Fahey TJ, Driscoll CT, Blum JD, Johnson CE (2014) Restoring soil calcium reverses forest decline. Environmental Science & Technology Letters 1(1):15–19

Belcher CF (1980) Logging railroads of the White Mountains. Appalachian Mountain Club, Boston, 242 p

Bormann FH, Likens GE (1967) Nutrient cycling. Science 3(3761):424–429

Bormann FH, Likens GE (1979) Pattern and process in a forested ecosystem. Springer-Verlag, New York, 253 p

Buso DC, Likens GE, LaBaugh JW, Bade D (2009) Nutrient dynamics. In: Winter TC, Likens GE (eds) Mirror Lake: interactions among air, land and water. University of California Press, California, pp 69–203

Butler TJ, Cogbill CV, Likens GE (1984) Effect of climatology on precipitation chemistry. Bull Am Meteorol Soc 3(6):639–640

Butler TJ, Likens GE, Vermeylen FM, Stunder BJB (2003) The relation between NO_X emissions and precipitation NO_3^- in the eastern USA. Atmos Environ 3(15):2093–2104

Campbell JL, Driscoll CT, Eagar C, Likens GE, Siccama TG, Johnson CE, Fahey TJ, Hamburg SP, Holmes RT, Bailey AS, Buso DC (2007) Long-term trends from ecosystem research at the Hubbard Brook Experimental Forest. General Technical Report NRS-17, U.S. Department of Agriculture, Forest Service, Northern Research Station, Newtown Square, PA, 41 p

Cogbill CV (1989) Hubbard Brook revisited: land use history of the Hubbard Brook Valley, New Hampshire (unpublished manuscript)

Cogbill CV, Likens GE (1974) Acid precipitation in the northeastern United States. Water Resour Res 3(6):1133–1137

Cogbill CV, Likens GE (1976) Reply to comment on 'acid precipitation in the northeastern United States'. Water Resour Res 3(3):571

Cogbill CV, Likens GE, Butler TJ (1984) Uncertainties in historical aspects of acid precipitation: getting it straight. Atmos Environ 3(10):2261–2270

Draxler RR, Hess GD (1998) An overview of the HYSPLIT 4 modelling system for trajectories, dispersion and deposition. Aust Meteorol Mag 47:295–308

Fahey TJ, Siccama TG, Driscoll CT, Likens GE, Campbell J, Johnson CE, Aber JD, Cole JJ, Fisk MC, Groffman PM, Holmes RT, Schwarz PA, Yanai RD (2005) The biogeochemistry of carbon at Hubbard Brook. Biogeochemistry 3(1):109–176

Fernandez IJ, Rustad LE, Norton SA, Kahl JS, Cosby BJ (2003) Experimental acidification causes soil base-cation depletion at the Bear Brook Watershed in Maine. Soil Sci Soc Am J 67:1909–1919

Fernandez IJ, Adams MB, SanClements MD, Norton SA (2010) Comparing decadal responses of whole-watershed manipulations at the Bear Brook and Fernow experiments. Environ Monit Assess 171:149–161

Hallett RA, Bailey SW, Horsleyand SB, Long RP (2006) Influence of nutrition and stress on sugar maple at a regional scale. Can J For Res 36:2235–2246

Halman JM, Schaberg PG, Hawley GJ, Eagar C (2008) Calcium addition at the Hubbard Brook Experimental Forest increases sugar storage, antioxidant activity and cold tolerance in native red spruce (*Picea rubens*). Tree Physiol 3(6):855–862

Hawley GJ, Schaberg PG, Eagar C, Borer CH (2006) Calcium addition at the Hubbard Brook Experimental Forest reduced winter injury to red spruce in a high-injury year. Can J For Res 3(10):2544–2549

Hornbeck JW (2001) Events leading to establishment of the Hubbard Brook Experimental Forest. http://hubbardbrook.org/pdfs/hbef_history.pdf accessed 29 April 2014.

Horsley SB, Long RP, Bailey SW, Hallett RA, Wargo PM (2002) Health of eastern North American sugar maple forests and factors affecting decline. North J Appl For 19:34–44

Huntington TG, Ryan DF, Hamburg SP (1988) Estimating soil nitrogen and carbon pools in a northern hardwood forest ecosystem. Soil Sci Soc Am J 52:1162–1167

Huntington TG, Johnson CE, Johnson AH, Siccama TG, Ryan DF (1989) Carbon, organic matter and bulk density relationships in a forested spodosol. Soil Sci 3(5):380–386

Jeziorski A, Yan ND, Paterson AM et al (2008) The widespread threat of calcium decline in fresh waters. Science 322:1374–1377

Johnson NM, Likens GE, Bormann FH, Pierce RS (1968) Rate of chemical weathering of silicate minerals in New Hampshire. Geochim Cosmochim Acta 32:531–545

Johnson NM, Driscoll CT, Eaton JS, Likens GE, McDowell WH (1981) "Acid rain," dissolved aluminum and chemical weathering at the Hubbard Brook Experimental Forest, New Hampshire. Geochim Cosmochim Acta 3(9):1421–1437

Judd KE, Likens GE, Buso DC, Bailey AS (2011) Minimal response in watershed nitrate export to severe soil frost raises questions about nutrient dynamics in the Hubbard Brook Experimental Forest. Biogeochemistry 106:443–459. doi:10.1007/s10533-010-9524-4

Juice SM, Fahey TJ, Siccama TG, Driscoll CT, Denny EG, Eagar C, Cleavitt NL, Minocha R, Richardson AD (2006) Response of sugar maple to calcium addition to northern hardwood forest. Ecology 3(5):1267–1280

LeBauer DS, Treseder KK (2008) Nitrogen limitation of net primary productivity in terrestrial ecosystems is globally distributed. Ecology 3(2):371–379

Likens GE (1972) Mirror Lake: its past, present and future? Appalachia 3(2):23–41

Likens GE (ed) (1985) An ecosystem approach to aquatic ecology: Mirror Lake and its environment. Springer-Verlag, New York, 516 p

Likens GE (1989) Some aspects of air pollution on terrestrial ecosystems and prospects for the future. Ambio 3(3):172–178

Likens GE (1992) The ecosystem approach: its use and abuse. Excellence in ecology, vol 3. Ecology Institute, Oldendorf, 167 p

Likens GE (2000) A long-term record of ice cover for Mirror Lake, New Hampshire: effects of global warming? Verh Intern Ver Limnol 3(5):2765–2769

Likens GE (2004) Some perspectives on long-term biogeochemical research from the Hubbard Brook Ecosystem Study. Ecology 3(9):2355–2362

Likens GE (2010) The role of science in decision making: does evidence-based science drive environmental policy? Front Ecol Environ 8(6):e1–e8. doi:10.1890/090132

Likens GE, Bormann FH (1974) Acid rain: a serious regional environmental problem. Science 3(4142):1176–1179

Likens GE (2013). Biogeochemistry of a forested ecosystem. Springer Science & Business. New York.

Likens GE, Buso DC (2012) Dilution and the elusive baseline. Environ Sci Tech 46(8):4382–4387, DOI: 10.1021/es3000189

Likens GE, Butler TJ (1979) Trends in acid precipitation in the eastern U.S. In: MacCracken M (coord) The multistate atmospheric power production pollution study—MAP3S. Progress Report for FY1977 and FY1978. DOE/EV–0040, pp 104–108

Likens GE, Franklin JF (2009) Ecosystem thinking in the northern forest—and beyond. BioScience 3(6):511–513

Likens GE, LaBaugh JW (2009) Mirror Lake: past, present and future. In: Winter TC, Likens GE (eds) Mirror Lake: interactions among air, land and water. University of California Press, California, pp 301–328

Likens GE, Bormann FH, Johnson NM, Pierce RS (1967) The calcium, magnesium, potassium, and sodium budgets for a small forested ecosystem. Ecology 3(5):772–785

Likens GE, Bormann FH, Johnson NM, Fisher DW, Pierce RS (1970) Effects of forest cutting and herbicide treatment on nutrient budgets in the Hubbard Brook watershed-ecosystem. Ecol Monogr 3(1):23–47

Likens GE, Bormann FH, Johnson NM (1972) Acid rain. Environment 3(2):33–40

Likens GE, Bormann FH, Pierce RS, Eaton JS, Johnson NM (1977) Biogeochemistry of a forested ecosystem. Springer-Verlag, New York, 146 p

Likens GE, Keene WC, Miller JM, Galloway JN (1987) Chemistry of precipitation from a remote, terrestrial site in Australia. J Geophys Res 92(D11):13299–13314

Likens GE, Driscoll CT, Buso DC, Siccama TG, Johnson CE, Ryan DF, Lovett GM, Fahey T, Reiners WA (1994) The biogeochemistry of potassium at Hubbard Brook. Biogeochemistry 25:61–125

Likens GE, Driscoll CT, Buso DC (1996) Long-term effects of acid rain: response and recovery of a forest ecosystem. Science 272:244–246

Likens GE, Driscoll CT, Buso DC, Siccama TG, Johnson CE, Lovett GM, Fahey TJ, Reiners WA, Ryan DF, Martin CW, Bailey SW (1998) The biogeochemistry of calcium at Hubbard Brook. Biogeochemistry 3(2):89–173

Likens GE, Driscoll CT, Buso DC, Mitchell MJ, Lovett GM, Bailey SW, Siccama TG, Reiners WA, Alewell C (2002) The biogeochemistry of sulfur at Hubbard Brook. Biogeochemistry 3(3):235–316

Likens GE, Buso DC, Butler TJ (2005) Long-term relationships between SO_2 and NO_X emissions and SO_4^{2-} and NO_3^- concentration in bulk deposition at the Hubbard Brook Experimental Forest, New Hampshire. J Environ Monit 3(10):964–968

Likens, GE and Fallon Lambert K. (1998) The importance of long-term data in addressing regional environmental issues. Northeastern Naturalist 5(2):127–136

Lindenmayer DB, Likens GE (2010) Effective ecological monitoring. CSIRO Publishing and Earthscan, Collingwood, Victoria 3066 Australia. 170 p

Long RP, Horsley SB, Hallett RA, Bailey SW (2009) Sugar maple growth in relation to nutrition and stress in the northeastern United States. Ecol Appl 19:1454–1466

Lovett GM, Canham CD, Arthur MA, Weathers KC, Fitzhugh RD (2006) Forest ecosystem responses to exotic pests and pathogens in eastern North America. BioScience 56:395–405

McGuire KJ, Likens GE (2011) Chapter 1: Historical roots of forest hydrology and biogeochemistry. In: Levia DF, Carlyle-Moses D, Tanaka T (eds) Forest hydrology and biogeochemistry: synthesis of past research and future directions (Ecological studies series No 216). Springer-Verlag, Heidelberg, pp 3–26

Odèn S (1968) The acidification of air and precipitation and its consequences on the natural environment. Bulletin 1, Swedish National Science Research Council, Ecology Committee

Ollinger SV, Aber JD, Reich PB (1997) Simulating ozone effects on forest productivity. Ecol Appl 7:1237–1251

Oreskes N, Conway EM (2010) Merchants of doubt. Bloomsbury Press, New York.

Siccama TG, Fahey TJ, Johnson CE, Sherry T, Denny EG, Girdler EB, Likens GE, Schwarz P (2007) Population and biomass dynamics of trees in a northern hardwood forest at Hubbard Brook. Can J For Res 37:737–749

Smith RA (1872) Air and rain. Beginnings of a chemical climatology. Longmans Green, London

Stunder BJB (1996) An assessment ofthe quality of forecast. J Appl Meteorol 35:1319–1331

Sun M (1984) Acid rain allegedly suppressed. Science 225:1374

Warby RAF, Johnson CE, Driscoll CT (2009) Continuing acidification of organic soils across the northeastern USA: 1984–2001. Soil Sci Soc Am J 73:274–284

Watmough SA, Dillon PJ (2001) Base cation losses from a coniferous catchment in central Ontario, Canada. Water Air Soil Pollut Focus 1:507–524

Watmough SA, Dillon PJ (2004) Major element fluxes from a coniferous catchment in central Ontario, 1983–1999. Biogeochemistry 67:369–398

Weathers KC, Likens GE, Butler TJ (2007) Acid rain. In: Rom WN (ed) Environmental and occupational medicine, 4th edn. Lippincott Williams & Wilkins, Philadelphia, pp 1507–1520

Winter TC, Likens GE (eds) (2009) Mirror Lake: interactions among air, land and water. University of California Press, California, 361 p

Part VI
Research Trajectories in Fire

Chapter 21
One-Hundred Years of Wildfire Research: A Legacy of the Priest River, Deception Creek, and Boise Basin Experimental Forests of Idaho

Russell T. Graham, Theresa B. Jain, Kathy L. Graham, Robert Denner and Colin Hardy

Abstract The 1910 fires, which burned more than 1.3 million ha of northern Rocky Mountain forests, provided a mission and management objectives for the newly created Forest Service. By 1911, the Priest River Experimental Station (Forest-PREF) was established in northern Idaho to help meet the needs of the Forest Service. Harry T. Gisborne, whose work was centered at PREF, proved to be one, if not *the* most influential and far-seeing fire researcher in the history of the Forest Service. Examples of his contributions include the fire danger rating system, fuel moisture sticks, short- and long-term specialized fire-weather forecasting, and the beginnings of predicting fire behavior. After Gisborne's death in 1949, Jack Barrows, one of Gisborne's assistants, led the fire program and introduced high-tech approaches to fire research. Barrows was instrumental in creating the state-of-the-art Fire Sciences Laboratory in Missoula, Montana. The McSweeney–McNary Act (1928) laid the groundwork for a nationwide system of forest experiment stations and experimental forests, and in 1933 Deception Creek (DCEF) and Boise Basin Experimental Forests (BBEF) were established. DCEF was located in a productive mixed conifer forest in northern Idaho. Fire was integral to studies conducted at DCEF on harvesting, regenerating, and tending western white pine stands. Research at BBEF in southern Idaho emphasized timber production within interior ponderosa pine forests and prescribed fire was studied as a means of preparing seedbeds and minimizing grass and shrub competition to trees. Similar to other dry forests of the West, wildfires were aggressively controlled at BBEF, causing portions of it to be overrun with seedlings and saplings, which created dense forests. As such, BBEF was well suited for investigating ways of restoring ponderosa pine forests. After nearly 100 years of fire research, we still strive to effectively manage forests in the face of ever-growing threats of urbanization and unwanted wildfires. Building on

---

R. T. Graham (✉) · T. B. Jain · K. L. Graham · R. Denner
Rocky Mountain Research Station, USDA Forest Service, 1221 South Main, Moscow, ID 83843, USA
Phone: 208.883.2325
e-mail: rtgraham@fs.fed.us

C. Hardy
Rocky Mountain Research Station, USDA Forest Service, 5775 US West Highway 10, Missoula, MT 59808, USA

D. C. Hayes et al. (eds.), *USDA Forest Service Experimental Forests and Ranges*,
DOI 10.1007/978-1-4614-1818-4_21, © Springer New York 2014

the legacy of research accomplished on the Idaho experimental forests and the basic understanding of fire and its effects the early researchers developed, these forests are now more valuable than ever.

Keywords Fire research · Fire danger rating system · Harry T. Gisborne · Priest River Experimental Forest · Ponderosa pine · Western white pine

21.1 Introduction

When Raphael Zon, head of the Forest Service's Office of Silvics, recommended the establishment of experiment stations in 1906, tree cultivation and planting were identified as key information needs by the newly created Forest Service (Schmaltz 1980). By 1910, the need for information regarding forest fires and predicting fire danger became a priority for the newly formed research organization. Of course, large wildfires had occurred in the northwestern USA before the Forest Service was established. The 182,000-ha Yaquina fire of 1846, the 130,000-ha Nestucca fire of 1853 in Oregon, and other large fires, as noted by John Lieberg, burned in the Priest River (Idaho) and Bitterroot (Montana) Forest Reserves prior to 1897 (Cohen 1978; Pyne 2001). The wildfires that burned in the northern Rocky Mountains during the summer of 1910, however, were of such magnitude and intensity that they shaped the American fire landscape more than any other fire of the twentieth century (Pyne 2001, 2010).

During the late 1800s and early 1900s throughout eastern Washington, northern Idaho, and western Montana forest fires were common, especially fires used to clear timber for railroad right-of-ways. Abnormally low amounts of precipitation fell during 1909 and 1910, and temperatures for the month of April were the highest on record. By August, more than 1,700 fires were burning within the nearly 16 million ha of District One[1] of the Forest Service. Pushed by dry winds blowing from the southwest on August 20 and 21, the fires burned more than 1.3 million ha of northern Idaho and western Montana forests (Pyne 2001). The fires killed at least 85 people and destroyed billions of cubic meters of highly valued western white pine (*Pinus monticola*)-dominated forests. Elers Koch, the supervisor of the Lolo National Forest in western Montana, described the 1910 fires as a "complete defeat for the newly organized Forest Service force" because of the damage they caused (Baker et al. 1993).

The devastation from the fires was extensive. The smoke drifted as far away as Saskatchewan, Canada, Denver, Colorado, and Watertown, New York. The Savenac Nursery near Haugen, Montana, the largest forest nursery in the USA, was decimated. The towns of Taft, Saltese, and De Borgia in Montana were partially destroyed as was a large portion of Wallace, Idaho (Pyne 2001). District One became the focal

[1] In 1911, District One of the Forest Service included Montana, northeastern Washington, northern Idaho, northwestern South Dakota, northern Michigan, northern Minnesota, and southwestern North Dakota. The District office was located at the Hammond Block, Missoula, MT.

point because of the impact of the fires and its Forest Service personnel became the "experts" on fire. Nevertheless, it was apparent there was a need to have better information, equipment, and manpower. According to William Greeley, District One Forester, "Congress and the Forest Service now realize that fire protection was the number one job of the Forest Service. We knew this before, but the 1910 experience burned it in terms of sweat, labor, and human life. Protection was it—we must lick the fire problem" (Spencer 1956).

The Priest River Forest Reserve near the Canadian border in northern Idaho was one of the several forest reserves that President Cleveland established across the western USA in 1897. The Reserve was 261,071 ha and all of the forest types that grew in the western part of District One were represented. John Leiberg, a dendrologist assigned by the General Land Office to review the Reserve in 1897 described it as a "magnificent forest…of western white pine and tamarack (larch, *Larix occidentalis*)" (Graham 2004). Leiberg's description no doubt influenced Raphael Zon's choice to establish the region's experiment station within the Reserve. The Reserve had areas suitable for reforestation, access to transcontinental railroads, lands subject to withdrawal for exclusive use by the Forest Service, and suitable building sites (Graham 2004). Other areas within the District were considered for an experiment station, but 290 ha near the Benton Ranger Station, which by 1911 was located on the Kaniksu National Forest (formerly the Priest River Forest Reserve), was set aside for the Station (Fig. 21.1).

21.2 Priest River Experiment Station

In August of 1911, Raphael Zon, along with Robert Y. Stuart, Assistant District Forester- Silviculture and F. I. Rockwel, Director of Silvics, from the District Office, visited the Benton Ranger Station in the Priest River Valley and brought along the basic supplies needed to establish the Priest River Experiment Station. By September 1, 1911, Zon, Donald H. Brewster, the first director of the Station (1911–1917), and another ten men set up camp and started building the facilities (Fig. 21.2). They completed the preliminary work by the end of October and the remainder of work was left to Brewster and Douglas MacDonald (cook) who subsequently had been appointed Forest Guard (Wellner 1976; Fig. 21.3).The meteorological instruments were installed on September 4, 1911, and have been continuously recording at Priest River since then.

By the summer of 1912, the Priest River Experiment Station was firmly established and the chief of the Forest Service, the district forester, and other prominent foresters visited and approved of the area (Fig. 21.4). Julius A. Larsen[2] from the Flathead National Forest in western Montana and his family joined Brewster at the Station in 1913 and initiated studies relating weather observations to forest changes

[2] In 1912, J. A. Larsen and some 218 of the first 300 graduates (holding M.F. degrees or certificates) of Yale were employed or had been employed by the Forest Service (Hoar et al. 1981).

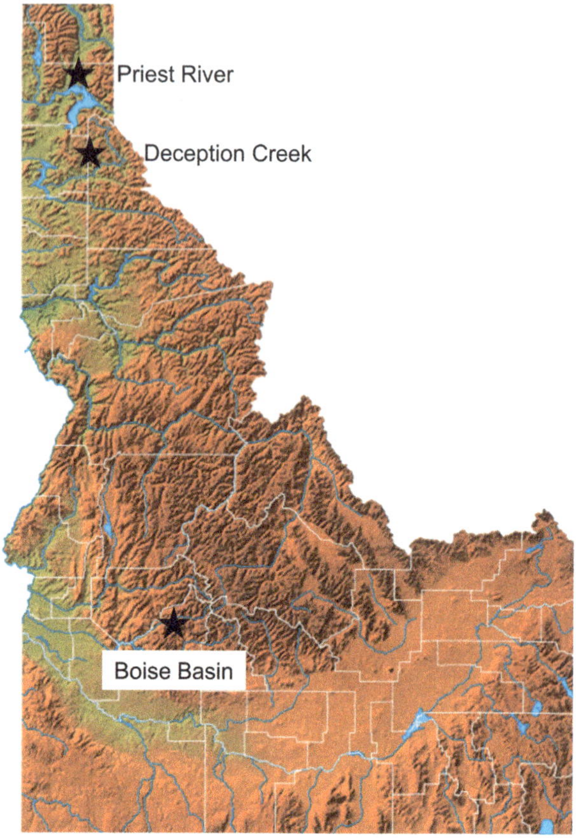

Fig. 21.1 The Priest River Experimental Station (Forest) was established in 1911, Boise Basin Branch Station (Experimental Forest) in 1933, and Deception Creek Experimental Forest in 1933. These Idaho Experimental Forests are administered by the Rocky Mountain Research Station headquartered in Fort Collins, CO

Fig. 21.2 Part of the construction party in the fall of 1911 consisted of (from *left*): Howard Simpson, Raphael Zon, W. W. Morris, Ed Brown, Donald Brewster, John Kirk, "dad" Crosby, and F. I. Rockwell

Fig. 21.3 The first living quarters at the Priest River Experiment Station in 1911. Ronald Mac Donald, the cook's son, stands in front of the tent

Fig. 21.4 The Priest River Experiment Station was inspected in 1912 by (from *left*): William Greeley District (Regional) Forester, E. B. Tanner, David Mason, Ferdinand Silcox, James Girard, M. H. Wolff, Henry Graves (Chief of the Forest Service), and Mallory Stickney

(e.g., tree growth, disease occurrence) occurring on the Experiment Station's forest (Graham 2004). By 1915, the Forest Service Experiment Stations were fully operational throughout the USA and Chief Henry Graves established the Branch of Research administered by Earle H. Clapp. During this time, the Investigative Committee within District One directed research activities at the Priest River Experiment Station. At their 1916 meeting, recognizing the proximity of the Station to the area burned by the 1910 fires, the Committee requested studies be initiated to determine ways to detect and control forest fires. They emphasized the work should identify factors that affect fire spread and rate of spread as influenced by weather and site conditions (Graham 2004).

21.2.1 Genesis of Fire Research

Based on the Investigative Committee's recommendations, Clapp advised that forest fire research be started at Priest River. Larsen had already linked the Station's weather records to duff and surface soil moisture concentrations, thereby making him the logical choice for starting studies of fire hazard and liability (Graham 2004; Hardy and Hardy 2007). In response, Larsen immediately set out to relate Priest River's weather data with fire records of the Kaniksu National Forest that adjoined the Experiment Station. He divided the Kaniksu into climatic units, determined their meteorological conditions, described their topography, acquired fire spread rates that occurred in each unit, and described the fuels the fires burned. From this work, Larsen published *Wind and Its Relation to Forest Fires, Sunshine and Air Temperature In Relation To Forest Fires,* and *Relative Humidity of the Atmosphere and Its Relation to the Fire Problem* in 1921. Larsen's work was the beginning of predicting dangerous conditions for the ignition and spread of wildfires or the earliest formulation of "fire-danger rating" (Graham 2004; Hardy and Hardy 2007, Larsen 1921a, b, c; Wellner 1976).

When Clapp provided the impetus for the beginning of fire research in 1916, he went on to say "that anyone who successfully worked out solutions would receive the highest type of recognition, both within and outside the Forest Service and the men who were the leaders of fire research would become the most important forest researchers in the country." During World War I, Congress drastically cut funds to the Station and Larsen was temporally transferred to District One to work with W. C. Lowdermilk. Together they studied ways of using fire to dispose of logging slash. In addition to conducting fire studies, Larsen published and kept the ecological, growth and yield, conifer seed, silvicultural, and many other studies alive at Priest River during these war years. Larsen went on to be an excellent forest scientist and later Dean at Iowa State University; however, he would not reach the prominence in fire research that Clapp described. In 1921 after the War, the Station received a substantial increase in funds and Clapp directed Robert H. Weidman, Station Director to transfer Harry T. Gisborne from the Whitman National Forest in Oregon to the Station. It would be Gisborne's sole responsibility to concentrate on fire research. Harry Gisborne would go on to exceed Clapp's expectations (Graham 2004).

21.2.2 Fire Danger

Upon his arrival in District One, Gisborne found the relative fire hazard descriptions for the western parts of the District (northern Idaho, northeastern Washington, and western Montana) to be in utter chaos. Forest officers had employed a variety of creative tactics to describe their fire hazards in order to acquire extra firefighting funds and forces (Hardy and Hardy 2007). Gisborne inherited the data and work of Larsen and Lowdermilk which he used as a starting point, but Gisborne aimed to develop a simple set of numbers, or a "common language," that fire managers could use to

Fig. 21.5 Along with Matt Dunlap from the Forest Service Forest Products Laboratory in Madison, WI, Harry Gisborne developed a duff hygrometer (*top*), which measured the moisture content hence inflammability of the dead leaves, twigs, etc. and an anemohygrograph (*bottom*), which was intended to measure fine fuel moisture, duff moisture, and wind speed

communicate to anybody, whether a woodsworker, a settler, a ranger, or an administrator of the fire hazard for a given forest. Before the end of his first fire season at the Station (1922), Gisborne established fire weather stations throughout the Kaniksu, Clearwater, and Nezperce National Forests of northern Idaho (Graham 2004).

Gisborne quickly recognized the inadequacies of available instrumentation for determining fuel (i.e., twigs, down logs, etc.) and weather characteristics used to predict forest fire hazard. Drawing on Larsen's work, Gisborne began refining the role fuel moisture plays in fire danger. He evaluated the moisture content of duff, twigs, and down logs in relation to conditions such as air temperature and relative humidity and looked for ways to measure these fuels and weather conditions. Since the right instruments did not exist for this research in 1923, he worked with Matt Dunlap from the Forest Service Forest Products Laboratory in Madison, Wisconsin, to develop a duff hygrometer that measured the moisture content, hence inflammability, of the dead leaves, twigs, and other forest floor organic materials and an anemohygrograph intended to measure fine fuel moisture, duff moisture, and wind speed (Fig. 21.5). Both of these instruments, however, were either too expensive or difficult to calibrate, and by 1940 both were discontinued (Graham 2004).

Fig. 21.6 An anemometer that George Jemison calibrated while riding on a hood of a car as his wife drove. Note each gage was numbered

Fire research had an annual budget of US$ 5,000 in 1927, and even though the McSweeney–McNary Act of 1928 augmented research funds, none were designated for Gisborne. Major W. Evan Kelley,[3] who became District One Forester in 1929, was specifically assigned to Missoula to solve the fire suppression problem. Major Kelley enthusiastically supported fire research, and in 1931 the work of Gisborne and his associates took on both regional and national importance. With the support that Major Kelley provided, Gisborne was able to hire George Jemison, a University of Idaho forestry graduate, as his first full-time professional assistant (Graham 2004).

Jemison had been so impressed with Gisborne's lectures at the University of Idaho, he applied for a summer job working for Gisborne in 1930. After graduation and receiving his appointment, one of Jemison's first duties was to identify and collect fuel and weather data at the fuel inflammability stations at Priest River. Another task of Jemison's was to calibrate the inexpensive anemometers (wind speed gauges) that Gisborne had a local machinist build (Fig. 21.6). Each instrument varied in craftsmanship and had to be calibrated manually (circa 1932). Jemison mounted each gauge on a car fender, and as his wife drove at various speeds, he lay on the hood and counted the revolutions, thus calibrating each instrument (Graham 2004).

While on sabbatical in 1923, J. A. Larsen visited several European experiment stations and reported to Gisborne that in Denmark they were using wood blocks as a criterion of atmospheric humidity (Hardy and Hardy 2007). Gisborne took this idea further and had Jemison test various twigs, wood blocks, and dowels to find a material and configuration that behaved uniformly with humidity changes. They settled on 1.3-cm ponderosa pine (*Pinus ponderosa*) dowels and ultimately assembled sets of four dowels, trimmed to weigh exactly 100 g when oven-dried. These sticks became the standard for measuring fine fuel moisture concentrations

[3] During WW I, Kelley went overseas with the 10th Forestry Engineers where he commanded all sawmilling, logging, and road construction operations in France. He retained his military rank after the war, hence the title of Major.

Fig. 21.7 The control weather station and George Jemison atop the 46-m western larch tree where weather instruments were maintained at the Priest River Experimental Forest in 1932

and were an essential part of all fire weather stations. Not only was the research for developing fuel sticks conducted at Priest River, but from 1948 to 1952 the manufacture and distribution of fuel sticks for all fire protective agencies west of the Mississippi River was done at Priest River (Wellner 1976). To this day, the 100-g fuel sticks that Gisborne and Jemison developed are the standard manual method for measuring the moisture concentration of small and dead woody fuels (Fischer and Hardy 1976). In the fall of 1937, Jemison transferred to the Forest Service Appalachian Research Station (now Southern Station) and subsequently established a fire research program there. Jemison came back to the Northern Rocky Mountain Station as Director in 1950 and finished his Forest Service career as Deputy Chief of Research in 1969 (Fig. 21.7).

With these and other trials, Gisborne and his associates were able to assemble a set of relatively inexpensive instruments that could be used at fire weather stations (Hardy and Hardy 2007). The basic set included scales for weighing fuel moisture sticks, rain gauges, four-cup anemometers for measuring wind speed, thermometers, and a visibility meter. By 1934, Gisborne had helped establish more than 50 fire weather stations stretching from Nevada and Utah in the Southwest to Yellowstone and Glacier National Parks in the northern Rocky Mountains. By this time with Gisborne's urging, the US Weather Bureau had also established the Fire Weather Warning Service network and Congress had appropriated money for forest fire weather forecasting. The first daily reporting of local fire weather was telegraphed from the Priest River Experimental Forest (PREF)[4] to the Spokane, Washington Weather Bureau in 1927. Since then, there has been regular broadcasting of fire weather forecasts and special warnings by radio and television stations throughout the USA (Graham 2004).

Gisborne investigated general weather conditions, and he also studied the relationship between lightning and fires, and how to predict when lightning storms were approaching (Gisborne 1931). He examined the effect of lightning on soils, rocks, forest cover, and ways to control lightning (Gisborne 1933). At Priest River, he strung a wire from a ridge top to the office so he could measure the amount of static electricity in the air. During lightning storms, he would often sit in the office and discharge electricity from his fingertips (Hardy 1983). He predicted the effects of lightning strikes and investigated ways of altering forest characteristics to control them. By obtaining more than 1,300 storm reports, he began to determine storm patterns and discovered that most storms were not single and well defined, but tended to be numerous and intermixed. Using sunspot forecasts, he investigated ways to forecast 10-day fire weather and used long-term precipitation records in an attempt to predict fire season rainfall. He hoped this information, combined with fuel moisture data and weather reports from the west coast, would provide an early warning system as to the potential fire danger within the Inland Northwest. He discovered, however, that this approach poorly predicted fuel inflammability and overall fire danger (Graham 2004; Hardy 1983).

In his quest to develop a common language for predicting fire danger, Gisborne sought ways to integrate what he defined as the three main drivers of fire danger. Using a Kodak exposure meter that combined light, exposure time, and lens opening into a single set of values, Gisborne substituted these values with fuel moisture, wind velocity, and relative humidity thus providing the first fire danger meter (Hardy and Hardy 2007). Because the relationships of these three components were not linear, the meter had to be modified. He needed to integrate these three variables, but he also wanted to get potential users to accept the meter. Gisborne brought experienced firefighters from administration and research together at the PREF and asked each to draw a set of curves expressing how they thought these three factors should fit together. Gisborne organized this expert knowledge into a composite set of curves. He packaged this information into a pocket-sized slide rule, which expressed the relative fire danger within a range of 1–6. The fire danger meter, model one was used in the 1932 fire season, and by 1942 a sixth model was developed that incorporated a burning index and also adjusted the fire danger for both human- and lightning-caused ignition risk (Hardy and Hardy 2007; Fig. 21.8).

At the 1958 national meeting of the American Meteorological Society, a National Fire Danger Rating System (NFDRS) was proposed. A joint committee com-

[4] In 1922, the headquarters of the Priest River Experiment Station was moved to Missoula, MT, and in 1925 the Station was renamed the Northern Rocky Mountain Forest and Range Experiment Station with Robert H. Weidman as Director. In 1930, the Priest River Experimental Forest was recognized and continued to be the center of research for Gisborne and many others.

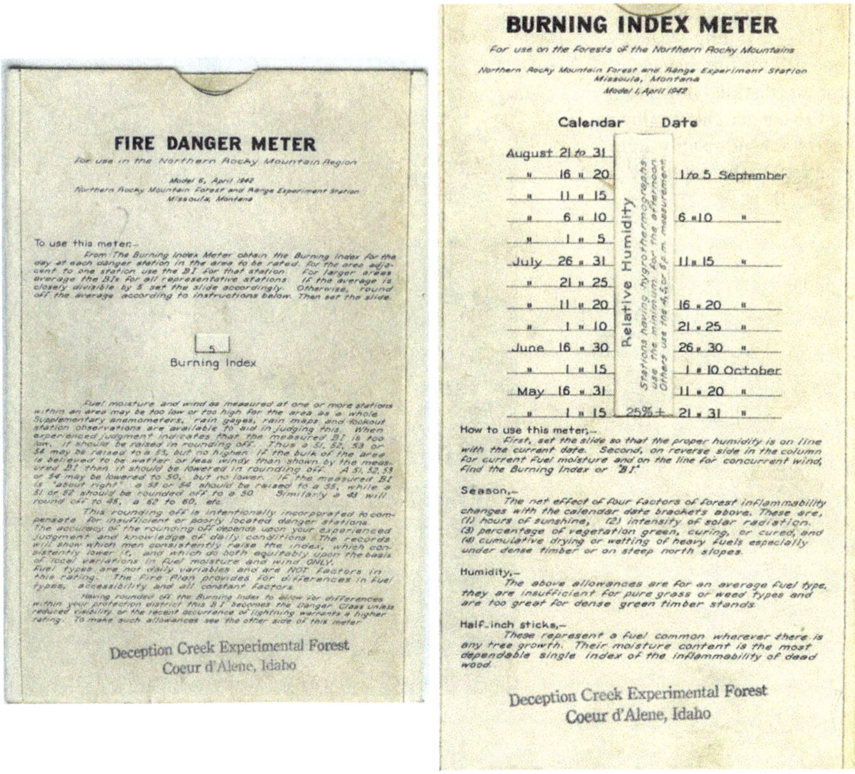

Fig. 21.8 Using expert knowledge, Harry Gisborne produced the first fire danger meter (model 6 shown). By 1942, model 6 incorporated a burning index along with fuel moisture, wind velocity, relative humidity, and visibility into fire danger

prised of both fire management and research personnel determined that a national system was feasible and a team was formed to implement the program. By 1961, a system was developed and it was tested in 1962 (Hardy and Hardy 2007), but the system lacked quantitative rigor and was very subjective necessitating further work. In 1968, a NFDRS Forest Service Research Work Unit was established in Fort Collins, CO. By extending Gisborne's work and the research from the facilities and experiments conducted at Missoula and other fire laboratories, the NFDRS began being used throughout the USA in 1972 (Deeming et al. 1972). The rating system informed fire management decisions, but did not readily communicate fire danger to the public as Gisborne stressed such a system should. To address this shortcoming, in 1974 the metrics (e.g., energy release component, ignition component) within the NFDRS were distilled to five classes—low, moderate, high, very high, and extreme which have been widely displayed in conjunction with Smokey Bear on signs (Helfman et al. 1975, Hardy and Hardy 2007; Fig. 21.9).

Fig. 21.9 The National Fire Danger Rating System developed for fire management was distilled down to five fire danger classes (low to extreme) and readily communicated to the public by Smokey Bear

21.2.3 *Fire Research Expansion*

Lloyd G. Hornby was hired by the Northern Rocky Mountain Forest and Range Experiment Station in 1931 to launch the first Research, Development, and Application program within the Forest Service. The program was housed at Priest River with Gisborne so Hornby could apply Gisborne's research in a form that could be used to make fire control planning decisions. Hornby, trained both as an engineer and forester, had spent 15 years working within National Forest Systems in positions ranging from a smoke chaser to forest supervisor. He developed fuel classifications that described fire spread rates and their resistance to control from low to extreme (four levels) and mapped the fuel types (classifications) for District One. He also developed methods for mapping the "seen-area" from lookouts that influenced the probability of a fire being detected. Using his engineering background and the new field of operations research, he produced eight principles of fire control planning relevant to District One, but also to fire control planning throughout the USA (Gisborne 1939). Hornby's research and development contributions and collaboration with Gisborne were cut short in 1937 by his heart attack and death on the Toboggan Creek Fire in the Clearwater National Forest of Idaho (Graham 2004).

Gisborne described the elements of fire danger, but he also worked on determining how these elements influenced fire behavior. In the early 1930s, again drawing on the work of J. A. Larsen, Gisborne began investigating the effect terrain, aspect, and elevation had on fire danger and fire behavior (Larsen 1922). In the fall of 1934, G. Lloyd Hayes, who came to Priest River in the summer as an Assistant Silvicultural Technician, was assigned to Gisborne's staff as a Junior Forester. Hayes's first assignment under Gisborne was to conduct the altitude and aspect study. In 1941, using weather observations and fuel conditions occurring on north- and south-facing slopes over a range of elevations (700–1,675 m), Hayes described how fire danger varied with aspect and altitude. In 1942, he added diurnal changes to his fire danger findings and firmly established the occurrence of a thermal belt in mountainous terrain. Within a thermal belt (e.g., at 900–1,220 m at Priest River), burning conditions change less from daytime to nighttime than they do in either the valley bottoms or on the mountaintops (Gisborne 1948). Hayes (1942) described the thermal belt as the "altitude of most dangerous fire behavior."

Gisborne and his assistants Jemison and Hayes evaluated fire behavior by observing burning wildfires as close to the fire line as possible. As the fires approached, they would measure the slope angle and aspect of where the fire was burning and characterize the size, kind, abundance, and arrangement of fuels it was burning. These observations of fire behavior were supplemented with studies Gisborne established at Priest River determining the inflammability and heat retention of different sized fuels with varying moisture levels. These studies provided the fundamentals of fire behavior that are still used today (Graham 2004).

The McSweeney–McNary Act (1928) laid the groundwork for a nationwide system of forest experiment stations and experimental forests. Funds from the New Deal programs in the 1930s such as the National Industrial Recovery Act (NIRA), the Emergency Work Corps (EWC), and the Economic Recovery Act (ERA) helped foster experimental forest establishment. In 1933, the Investigative Council of Region One (formerly known as District One) recommended the establishment of four additional experimental forests within the Region. One of the forests approved by the Washington Office was the Deception Creek Experimental Forest (DCEF; Fig. 21.1; Graham 2004).

21.3 Deception Creek Experimental Forest

In 1933, DCEF (1,315 ha) was established in the heart of one of the most productive forest areas in the Rocky Mountains (Jain and Graham 1996; Fig. 21.10).The activities at Priest River had been supervised directly by Northern Rocky Mountain Station Director Lyle Watts,[5] but when he left the responsibility largely passed to Gisborne's Fire Research Division. With the influx of money and manpower from the New Deal, fire research activities at Priest River were flourishing, leaving no room

[5] Lyle Watts became director of the Northern Rocky Mountain Forest and Range Experiment Station in 1931 and left in 1936 to become Regional Forester in Milwaukee, Wisconsin, and later Chief of the Forest Service.

Fig. 21.10 The Deception Creek Experimental Forest, established in a western white pine forest, had areas that once contained trees more than 60 m tall. Such stands still exist today (2012), however most pines were killed by blister rust. (Fig. 21.11)

for a strong silvicultural program. So within the Northern Rocky Mountain Forest and Range Experiment Station, Priest River became the center for fire research and Deception Creek, some 22 miles east of Coeur d'Alene, Idaho, and 88 miles from Priest River, became the center for silvicultural studies with considerable rivalry between the two divisions (Wellner 1976).

Kenneth P. Davis,[6] a former ranger on the Gallatin Forest in Montana, first worked at PREF during the summer of 1932 while he studied at the University of Michigan. After earning a master's degree, he returned to Priest River in 1933 and was sent by Director Watts to the newly established DCEF to serve as its first superintendent. In 1935, he was assigned the responsibility for studying the silvics and silviculture of western white pine and its associates. At DCEF, Ken Davis, Charles

[6] In 1937, Ken Davis became chief of the Silvics Division of the Northern Rocky Mountain Forest and Range Experiment Station until 1940 when he became an assistant to I. T. Haig in Washington, D.C. In the late 1930s, he took educational leave to work on a Ph.D. at the University of Michigan. Davis later became Dean of the Forestry School at the University of Montana, then a professor of forestry first at the University of Michigan and later at Yale University.

Fig. 21.11 White pine blister rust, an imported disease from Eurasia in 1910 attacks and most often kills western white pine throughout the western United States

Wellner, and others conducted major studies on harvesting, site preparation, planting, and growing western white pine (Graham 2004).

21.4 Boise Basin Experimental Forest

The McSweeney–McNary Act also paved the way for the Intermountain Forest and Range Experiment Station in Utah and adjoining states. The station's headquarters were established on July 1, 1930, in Ogden, UT, thus providing strong ties to the Region 4 headquarters. As it did for the Northern Rocky Mountain Station, the New Deal also facilitated the expansion of the Intermountain Station with the Boise Branch Station facilities at Idaho City, Idaho, being built by the Civilian Conservation Corps (CCC) in 1933 (Fig. 21.1). This Branch Station included 1433 ha within the Boise National Forest, which ultimately became part of the Boise Basin Experimental Forest (BBEF; Sloan and Steele 1996).

The BBEF was established in highly productive ponderosa pine (*Pinus ponderosa*) forests near Idaho City, Idaho. In the late 1800s, the city was a major mining center and used large amounts of wood from the surrounding forests. Because of the harvesting and mining activities, all ages of ponderosa pines from seedlings to mature yellow-barked trees were included in the forest when it was established.

Research on BBEF emphasized timber production within interior ponderosa pine forests and prescribed fire was studied as a means of preparing seedbeds and minimizing grass and shrub competition to trees. Different methods of selecting mature trees for harvest were also studied along with how cutting impacted the remaining trees, regeneration, and competing vegetation. These silvicultural studies produced methods for producing timber and protecting the pines from insects (bark beetles), animals (e.g., porcupines), and fire (Sloan and Steele 1996).

21.5 Fire Control

Research funding was drastically reduced in the 1940s due to World War II. Annual appropriations for the entire Forest Service research program averaged US$ 105,000 or about half the record high of the 1930s (Jemison 1950). The war also caused staff reductions and in 1942, more than 30% of regular Forest Service personnel entered military service. In 1944, only 13 technical staff kept the research programs functioning at the Northern Rocky Mountain Forest and Range Experiment Station, the lowest number since 1930, with the majority of their work conducted at the PREF and DCEF. Similarly, the Intermountain Station[7] suffered severe cuts in funding and the Boise Basin Branch Station was closed for the duration of World War II (Klade 2006).

By 1945, some return to normalcy occurred within both experiment stations and new studies were started. One of the most notable changes occurring after the war was the increase in the number of women working in research; however, they were mainly clerks and stenographers. Within the Northern Rocky Mountain Station, the Division of Silviculture and the Division of Forest Protection were reestablished and Gisborne, now the head of Forest Protection, was able to enlarge his staff (Graham 2004).

Jack Barrows joined Gisborne's staff in 1946. Barrows exhibited many of the work and research traits of Gisborne and he was able to incorporate many wartime technologies into fire research (Hardy 1983). In the 1930s, Barrows conducted fire control and behavior workshops for the National Park Service and later became their Chief of Fire Control Training. During this time, Gisborne and Barrows became close friends and Gisborne asked Barrows to continue the fire control planning research started by Lloyd Hornby. But after less than a month of refining Hornby's work, Barrows was detailed to lead a new aerial bombing project. Even with this assignment, he was able to compile the backlog of National Forest fire reports that had been recorded on 23,000 punch cards. This work improved fire control planning by incorporating new fire behavior knowledge and new equipment and techniques that became available for firefighting (Graham 2004).

[7] On January 1, 1954, the Northern Rocky Mountain Forest and Range Experiment Station merged with the Intermountain Forest and Range Experiment Station with its headquarters in Ogden, UT.

Barrows's energy, organizational skills, and military and political connections allowed him to introduce "high-tech" methods into wildland fire research and management (Klade 2006). Because Barrows was a Lieutenant Colonel during World War II and prepared aerial bombing strategies and tactics, he was a natural choice for leading the aerial bombing project. Gisborne had tried such methods in the 1930s when he dropped water in barrels, iron cans, and in 100-gallon tanks from airplanes. These methods proved to be ineffective in making an impact on even the smallest fire. Gisborne turned the entire project over to Barrows. Even though dropping water from airplanes to extinguish wildfires showed promise and received abundant publicity, the program was terminated in 1948. It was not until 1954 after California tested dropping cascading water on fires and new retardant technology was developed, that aerial application of water and retardant became a vital part of fire research and control efforts throughout the world (Klade 2006).

Though the aerial bombing program was terminated, Gisborne and Barrows started another high-tech project by testing cloud seeding as a way to control lightning. Gisborne used his connections to collaborate with Irving Langmuir and Vincent J. Schaefer of the General Electric Company. In 1948 at Priest River, Barrows and Gisborne, along with Schaefer, devised a strategy for seeding clouds and in the summer of 1949 they rigged a C-47 aircraft to do so. Bob Johnson, founder of the Johnson Flying Service in Missoula, Montana flew a C-47 to 26,000 ft with a dry-ice hopper manned by Gisborne and Barrows in the rear. Apparently, the oxygen tubes that Gisborne and Barrows were using as they chopped dry ice became disconnected requiring a panicky and blue-faced Gisborne to tell Johnson to descend rapidly. The C-47 proved unsuitable for cloud seeding, but Barrows was able to acquire a B-29 from Fairchild Air Force Base in Spokane, WA. Although this plane made several test runs near Priest River, no clouds appeared that were suitable for a proper experiment (Hardy 1983; Klade 2006).

On August 5, 1949, the Mann Gulch fire on the Helena National Forest in Montana trapped 12 smokejumpers and 4 other firefighters. Eleven men were burned to death by the fire and two others died the next day from their injuries. Because of his heart condition, Gisborne's activities were limited, but his interest was piqued by the abnormal behavior of this fire. Gisborne was eager to see the effects of the fire and was asked to investigate the fire's behavior. On November 9, 1949, accompanied by Robert Jansson, the Ranger of the Canyon Ferry District and a survivor of the fire, Gisborne visited the fire site. Because of Gisborne's physical condition, the half-hour hike turned into a 2-h trip. Knowing that Gisborne was showing signs of distress, Jansson convinced the stubborn Gisborne to stop, so that they could return to the gulch the next day to evaluate what they had found. Although excited about the potential of a new theory on fire behavior, Gisborne reluctantly agreed. But, about 800 m from their truck, Gisborne suddenly had a fatal heart attack. In August 1999, on the 50th anniversary of the Mann Gulch fire, Forest Service Chief Mike Dombeck and Montana Governor Marc Racicot recognized Gisborne as the 14th victim of the fire (Graham 2004; Maclean 1992; Rothermel 1993).

Lightning detection, cloud seeding, and related work by Barrows and Gisborne, along with Irving Langmuir and Vincent Schaefer of General Electric provided the genesis of Project Skyfire (Gisborne 1931). At Priest River in 1952, its first project was training lookout personnel to track lightning storms and make cloud surveys. Barrows included many cooperators (e.g., Boeing, universities, US Weather Bureau, and Park Service) and in 1953 Project Skyfire became a formal research program within the Northern Rocky Mountain Forest and Range Experiment Station. Meteorologist Don Fuquay was hired to gather basic information on the occurrence, behavior, and control of lightning-caused forest fires (Barrows et al. 1957). This work led to lightning detectors being placed on mountaintops throughout the Rocky Mountains, which ultimately became part of a network of remote automated weather stations (RAWS) recording lightning conditions and feeding the data into the National Interagency Fire Center in Boise, Idaho (Klade 2006).

Both Barrows and Gisborne were dedicated fire researchers and had the skills and work ethic to meet the ambitious goals they set. However, Gisborne demanded perfection from himself, his subordinates, and cooperators and often alienated both coworkers and cooperators. Barrows also appreciated excellent work, but he tempered it with diplomacy. Barrows was able to heal divisions Gisborne created with the Weather Bureau in Project Skyfire and brought them in as an important fire research partner. Gisborne openly criticized the Forest Service's budget for fire research, which in turn complicated the budget process rather than helped. In contrast, Barrows worked indirectly with key individuals within the Forest Service and political circles, which allowed him to acquire the million dollars to build the Fire Sciences Laboratory at Missoula, Montana. It was a need that Gisborne recognized as early as 1936 noting that his field experiments lacked the precision and controlled conditions he needed for estimating fire danger (Graham 2004; Hardy 1983; Klade 2006).

The Fire Sciences Laboratory opened in 1960 and as chief administrator, Barrows hired physicists, mathematicians, engineers, and technicians. Barrows found such talent at the Idaho National Engineering Laboratory (INEL) where a nuclear-powered aircraft program was being closed. From this program, Barrows hired Hal Anderson, Stan Hirsch, and Dick Rothermel. These men proved to be leaders and innovators in producing information on how to detect wildfires, as well as providing an understanding about how they developed and spread (Klade 2006).

21.6 Prescribed Fire and Mechanical Fuel Treatments

White pine blister rust (*Cronartium ribicola*) was introduced into western North America in 1910. White pines in the Puget Sound area of Washington became infected by 1913, and by 1923 the disease was found in several locations in Idaho (Geils et al. 2010). During this time, western white pine was by far the most valuable tree species growing in the northern Rocky Mountains and it was being attacked and killed by the disease (Fig. 21.11). Blister rust requires two hosts to complete its life cycle, a white pine and a *Ribes* (current) bush. Hand-pulling of *Ribes* was tested in

Fig. 21.12 In the 1920s through 1960s, intense controlled burns were the preferred method of disposing of logging slash and the inferior tree species left after clearcutting in western white pine forests

1923 by the Office of Blister Rust Control at PREF as a method for controlling the spread and impact of the disease. During this test, 15 men pulled 53,555 bushes on 690 ha of the Benton Creek drainage at PREF, but it proved to be unsuccessful in stopping the spread of the disease (Wellner 1976).

At DCEF and the adjoining Coeur d'Alene National Forest, the first work that Davis and Wellner undertook was to investigate ways of ensuring that western white pine regenerated after harvest. They also studied ways to create conditions that minimized the blister rust hazard. In the 1930s, it was believed the spores that traveled from *Ribes* to infect white pines lost their effectiveness in approximately 550 m (Spaulding 1922). Thus, approaches were tested that would provide such a buffer between western white pines and *Ribes* bushes. Large clearcuts were created (e.g., hundreds of hectares in size) and the inferior species such as western hemlock (*Tsuga heterophylla*) and grand fir (*Abies grandis*) were felled and/or sometimes poisoned (Foiles 1950). The resulting fuels (often large amounts) were intensely burned with the heat generated by the fire stimulated the sprouting of *Ribes* (Fig. 21.12). These plants would subsequently be hand-pulled or sprayed with a herbicide, thereby protecting the western white pine plantation from blister rust. It was discovered that *Ribes* spores were viable over distances greater than 550 m and no matter how many "workings" an area received, *Ribes* could not be eradicated. Hutchinson and Winters (1942) described *Ribes* control like "bailing an ocean with a teacup." Even though *Ribes* control was futile in protecting western white pines, tree improvement programs started in 1949 by Richard Bingham, a research scientist at the Moscow, Idaho Forest Sciences Laboratory, were very effective in producing rust resistant trees suitable for planting (Bingham 1983). Using silvicultural methods such as, but not limited to, planting rust resistant stock, tree pruning, and mass selection, the future in 2012, not without some problems, is very bright for western white pine (Geils et al. 2010; Graham et al. 1994; Schwandt et al. 2010).

Control burning research in the 1930s and 1940s at Deception Creek and the surrounding Coeur d'Alene National Forest was aided by Gisborne's inflammability and fuel moisture studies. Controlled burns developed the fundamental understanding of what is now called prescribed fire. These studies investigated different tools to ignite fires including drip torches, backpack propane torches, and truck-mounted flame throwers. Season of burn, time of day the burn occurred, and onsite weather variables vital for having a successful burn were tested. Most often, a successful burn was one that stayed within the fire line perimeter and severely burned the woody material and left the forest floor clean (Fig. 21.12).

At PREF in 1952, studies were conducted to identify variables besides fuel moisture and weather that could be used to plan and execute prescribed fires. The site of old F-127 Civilian Conservation Corp Camp along the Priest River was used for the slash burning experiments conducted by George Fahnestock from the Station and Dave Olson from the University of Idaho. Most of the experiments were completed by 1957 when Fahnestock transferred to the Southern Station, but he returned in 1960 to complete the study by burning 5-year-old slash (Fahnestock 1953, 1960; Wellner 1976). Similar to how Larsen's work influenced Gisborne, Fahnestock's work provided the foundation for Jim Brown of the Fire Sciences Laboratory to develop slash inventory methods and other allied information on slash and its consumption by fires (Brown 1974; Brown et al. 2003).

Fire danger and fire control research decreased at PREF when the Fire Sciences Laboratory in Missoula opened. However, PREF along with the BBEF and DCEF became integral in studying the effects of prescribed fire in the moist and dry forests. At BBEF, as in most ponderosa pine forests of the western USA, wildfires were aggressively suppressed. As a result, both ponderosa pine and Douglas fir (*Pseudotsuga menziesii*) regeneration proliferated over much of the forest and dense multi-canopied forests prevailed (Fig. 21.13). At Boise Basin, mechanical and fire methods are being studied as ways to restore the character of these forests. In addition, because tree densities increased in the dry forests due to fire exclusion, the forest floor accumulated layers of needles and bark slough. The amount of these materials would have been minimal if the native fire regimes would have continued. These uncharacteristically deep layers are most noticeable around large yellow-barked trees (Fig. 21.14). Fine roots can accumulate in these layers and, if they are destroyed through fire or mechanical means, the tree can be stressed and often succumbs to bark beetles. Studies are being conducted on ways to reduce these layers by increasing decomposition through mechanically mixing the surface layers and burning the organic layers in snow wells (Fig. 21.14). The results of these

Fig. 21.13 Prior to masticating, many small trees occurred in this ponderosa pine stand (*top*). After masticating with a track-mounted machine, the majority of the small trees were removed leaving the larger trees (*bottom*)

tests have shown that up to two snow well burns or surface mixings may be required depending on the layers' depth before the majority of the fine roots will occur in the mineral layers. It has also been shown that the best time to disturb the surface organic layers is when the lower duff layers (e.g., humus and fermentation layer) have a moisture concentration near or exceeding 100% and their temperatures are below 4.4 °C. After such duff layers have been reduced, a prescribed fire can more readily be used without unduly stressing the residual trees (Graham et al. 2007, 2010; Fig. 21.15).

In the moist forests at Priest River and Deception Creek, treatments are being tested that reduce the fuels, but leave a high forest canopy (Jain et al. 2004, 2008). Such conditions are often valued for wildlife and provide a sense of security and/or place to people. Treatments are being tested both in the wildland urban interface and matrix lands. The tests have shown that using a mechanical masticator that leaves large chunks will create material and conditions that favor wood decomposi-

Fig. 21.14 Because of fire exclusion, uncharacteristically deep layers of needles and bark slough can accumulate beneath large yellow-barked ponderosa pine trees (*top*). These layers can contain abundant fine roots and if the layers are destroyed mechanically or by fire, these valuable trees can be stressed and succumb to bark beetles. By judicially applying fire when the lower organic layers are moist and cool, the fine roots when they start growing (when the soil warms) will remain in the mineral soil (*bottom*)

tion (Fig. 21.16). Even though fine fuels are increased immediately after treatment, within 3 years the fire hazard is reduced as the material readily decomposes. Similar results were found in masticating ponderosa pine fuels in the BBEF as long as summer monsoon rains occasionally wetted the forest floor (Graham et al. 2010; Fig. 21.13).

There is a great deal of research on how to manipulate moist forests for the purpose of timber production. Priest River and Deception Creek provided valuable in-

Fig. 21.15 It may take multiple snow well burns in ponderosa pine forests where fire has been excluded before the root architecture is such that prescribed fire can be broadcast through the forest safely (Fig. 21.14)

formation on growth, yield, site preparation, planting, disease resistance, and other information applicable for growing timber crops (Bingham 1983; Haig 1932; Haig et al. 1941). Using this foundation, both forests are being used to test how fuel treatments (e.g., mechanical, fire) can be designed and implemented over landscapes to affect wildfire behavior and burn severity if a fire was to occur (Jain et al. 2008). FARSITE and FlamMap, two fire models developed at the Fire Sciences Laboratory, have been used to project how fuel treatments may alter fire behavior (Finney and Andrews 1998; Finney 2006). These analyses have shown that the fuel treatments would not stop a fire nor necessarily reduce its ultimate size. However, the fuel treatments, no larger than 1.6 ha in size would disrupt the progression of a fire and offer suppression opportunities. Also within fuel treatments, predicted flame lengths were less than 30 cm compared to more than 3.0 m in untreated areas. After the simulations, it was noted that the heterogeneous forest landscapes created by fuel treatments would leave a mixture of green, brown, and black forest conditions distributed across the forest compared to all black conditions left after the simulated fire in the untreated landscape. Such mixed burn severities present in the treated forest would offer greater opportunities for a forest to recover compared to forests that were completely blackened after a forest fire (Jain et al. 2008).

Fig. 21.16 Decomposition of fuels can be maintained and enhanced by mastication if pieces left are of sufficient size as to not wet and dry readily and are in close contact to the forest floor. With the increase in fine fuels, the fire hazard can be exacerbated in the short term from mastication but within 2–3 years it can be minimal

21.7 Post-Wildfire Treatments

Information about the impact of post-wildfire forest treatments (e.g., salvage logging) on vegetation, soil, and water is needed throughout the western USA (Peterson et al. 2009). Most often data are not available on forest structure and forest floor conditions before a wildfire burns. Replicates of a wildfire, in conjunction with well-documented post-wildfire treatments, are also hard to come by. Therefore, at both PREF and BBEF sediment catchments were installed at the mouths of eight small watersheds (e.g., 4–6 ha; Fig. 21.17). Vegetation (mixed conifer forest) was burned on two watersheds at each forest and for one, the burning was followed by a salvage treatment (Fig. 21.18). The salvage operation was conducted to leave conditions that offered the greatest opportunity for forest recovery and, depending on how severely the watershed was burned, the number of trees remaining ranged from a few to many. In addition to these wildfire treatments, two watersheds at each forest remained undisturbed and in two watersheds the fuels were treated. The results of these studies are still forthcoming, but will show how wildfire, wildfire followed by salvage logging, and fuel treatments affect the soil, water, and vegetation in the moist and dry forests (Elliot et al. 2006).

Fig. 21.17 Sediment catchments were established at the mouths of eight small watersheds at the Priest River Experimental Forest to test the effects of prescribed wildfire (shown) and wildfire followed by logging on soil, vegetation, and water properties

21.8 Conclusion

After nearly 100 years of fire research, we still strive to effectively manage forests in the face of ever-growing threats of urbanization and unwanted wildfires. Building on the legacy of research accomplished on the Idaho experimental forests and the basic understanding of fire and its effects that the early researchers developed, these forests are more valuable now than ever. They are outdoor laboratories where observational and manipulative studies can occur. Fire and fire-related studies have been a part of the Idaho experimental forests since in 1912 when J. A. Larsen began quantifying duff moisture and related this information to fire occurrence. The drive and perfection demanded by Harry T. Gisborne set the standard for fire scientists that many would aspire to, but few would achieve (Graham 2004; Hardy 1983; Klade 2006; Maclean 1992). The tone and direction of fire research he started at Priest River in 1922 is still relevant for meeting today's challenges (Hardy and Hardy 2007).

The experimental forests of Idaho produce short- and long-term studies and data applicable for understanding and managing dry and moist forests. In particular, they

Fig. 21.18 Prescribed wildfire (pictured at PREF) and wildfire followed by salvage logging are being evaluated at both the Priest River and Boise Basin Experimental Forests for their impacts on soil, water, and vegetation. In addition, fuel treatments and undisturbed watersheds are included in the studies at both forests

can help inform how fire can be used to manage North American forests. All three forests have replicates of interdisciplinary studies investigating ways stands to landscapes can be treated to interrupt and decrease the burn severity of a wildfire if one was to occur (Elliot et al. 2006; Jain et al. 2008). Fuel treatments such as mastication, grapple piling, and prescribed fire are being studied singly and in combination as to how they can emulate the effects of low and mixed severity fires. Because of fire exclusion, especially on BBEF, these Idaho experimental forests provide abundant research opportunities to study innovative methods of forest restoration. No one knows exactly what issues or informational needs will arise in the next 100 years. By ensuring that experimental forests provide a wide variety of forest structures and compositions they will be high value assets to researchers, managers, and the citizens of the USA for addressing future information needs.

Acknowledgments We would like to thank all of the individuals over the years who have been diligently collecting and protecting the historical records pertaining to experimental forests. Also, we would like to thank the anonymous reviewers who added greatly and also thank Jonathan Sandquist of the Rocky Mountain Research Station for his excellent contributions.

References

Baker RD, Burt L, Maxwell RS, Treat VH, Dethloff HC (1993) National Forests of the northern region: living legacy. Intaglio, College Station

Barrows JS et al (1957) Project skyfire. In: Final report of the advisory committee on weather control, vol 2. Unnumbered reprint. United States Department of Agriculture, Forest Service, Intermountain Forest and Range Experiment Station, Ogden, UT, pp 105–125

Bingham RT (1983) Blister rust resistant western white pine for the inland empire: the story of the first 25 years of the research and development program. General Technical Report INT-GTR-146, U.S. Department of Agriculture, Forest Service, Intermountain Forest and Range Experiment Station, Ogden, UT

Brown JK (1974) Handbook for inventorying downed woody material. General Technical Report INT-GTR-16, United States Department of Agriculture, Forest Service, Intermountain Forest and Range Experiment Station, Ogden, UT

Brown JK, Reinhardt ED, Kramer KA (2003) Coarse woody debris: managing benefits and fire hazard in the recovering forest. General Technical Report RMRS-GTR-105, U.S. Department of Agriculture, Forest Service, Rocky Mountain Research Station, Ogden, UT

Cohen S (1978) The Big Burn: the Northwest's forest fire of 1910. Pictorial Histories Publ Co, Missoula

Deeming JE, Lancaster JW, Fosberg MA, Furman WR, Schroeder MJT (1972) Fire-danger rating system. Research Paper RM-RP-84, U.S. Department of Agriculture, Forest Service, Rocky Mountain Forest and Range Experiment Station, Fort Collins, CO

Elliot WJ, Miller IS, Glaza BD (2006) Using WEPP technology to predict erosion and runoff following wildfire. 2006 ASABE annual international meeting, Portland, OR. ASABE Paper No. 068011, Sponsored by ASABE, Portland Convention Center: American Society of Agricultural and Biological Engineers

Fahnestock GR (1953) Inflammability of the current year's logging slash. Research Note INT-RN-120, U.S. Department of Agriculture, Forest Service, Intermountain Forest and Range Experiment Station, Ogden, UT

Fahnestock GR (1960) Logging slash flammability. Research Paper INT-RP-58, U.S. Department of Agriculture, Forest Service, Intermountain Forest and Range Experiment Station, Ogden, UT

Finney MA (2006) An overview of FlamMap fire modeling capabilities. In: Andrews PL, Butler BW (comps) Fuels management-how to measure success: conference proceedings; Portland, OR. General Technical Report INT-GTR-213WWW. U.S. Department of Agriculture, Forest Service, Rocky Mountain Research Station, Fort Collins, CO, pp 213–220

Finney MA, Andrews PL (1998) FARSITE: fire area simulator—a model for fire growth simulation. Fire Manage Note 59(2):13–15

Fischer WC, Hardy CE (1976) Fire-weather observer's handbook. Agriculture handbook 494, U.S. Department of Agriculture, Washington, DC

Foiles MW (1950) Recommendations for poisoning western hemlock. Research Note NRM-RN-77, Northern Rocky Mountain Research Station, U.S. Department of Agriculture, Forest Service, Missoula, MT

Geils BW, Hummer K, Hunt RS (2010) White pines, *Ribes*, and blister rust: a review and synthesis. For Pathol 40:147–185

Gisborne HT (1931) Lightning on the lookouts. Am For 37(9):515–518, 574–575

Gisborne HT (1933) Lightning and forest fires. Pulp Pap Can 34(6):327–329

Gisborne HT (1939) Hornby's principles of fire control planning. J For 4(1):292–296

Gisborne HT (1948) Fundamentals of fire behavior. Fire Control Note 9:13–24

Graham KL (2004) History of the Priest River Experiment Station. General Technical Report RMRS-GTR-129, U.S. Department of Agriculture, Forest Service, Rocky Mountain Research Station, Fort Collins, CO

Graham RT, Tonn JR, Jain TB (1994) Managing western white pine plantations for multiple resource objectives. In: Baumgartner DM, Lotan JE, Tonn JR (eds) Proceedings interior cedar-hemlock-white pine forests: ecology and management; Spokane, WA. Department of Natural Resource Sciences, Washington State University, Pullman, WA, pp 357–362

Graham RT, Jain TB, Sandquist J (2007) Free selection: a silvicultural option. In: Powers R (ed) Restoring fire-adapted forested ecosystems. 2005 national silviculture workshop, Lake Tahoe, CA. General Technical Report PSW-GTR-203, U.S. Department of Agriculture, Forest Service, Pacific Southwest Research Station, Albany, CA, pp 121–156

Graham RT, Jain TB, Mathews SJ (2010) Chapter 3: Fuel management in forests of the Inland West. In: Elliot WJ, Miller IS, Audin L (eds) Cumulative watershed effects of fuel management in the western United States. General Technical Report RMRS-GTR-231, pp 19–68

Haig IT (1932) Second-growth yield, stand, and volume tables for western white pine type. Technical Bulletin No. 323, U.S. Department of Agriculture, Washington, DC

Haig IT, Davis KP, Weidman RH (1941) Natural regeneration in the western white pine type. Technical Bulletin No. 767, U.S. Department of Agriculture, Washington, DC

Hardy CE (1983) The Gisborne era of forest fire research legacy of a pioneer. FS-367, U.S. Department of Agriculture, Forest Service, Washington, DC

Hardy CC, Hardy CE (2007) Fire danger rating in the United States of America: an evolution since 1916. Int J Wildland Fire 16:217–231

Hayes GL (1942) Differences in fire danger with altitude, aspect, and time of day. J For 40(4):318–323

Helfman RS, Deeming JE, Straub RJ, Furman RW (1975) User's guide to AFFIRMS: time-share computerized processing for fire-rating. Department of Agriculture, Forest Service, Rocky Mountain Forest and Range Experiment Station, Fort Collins, CO

Hoar CA, Noyes DK, Larsen JA, Jeffers DS, Newins HS, Watzek CH (1981) "Dear professor Chapman": letters from Yale forestry graduates, 1910–1912. J For Hist 25(4):197–209

Hutchison SB, Winters RK (1942) Northern Idaho forest resources and industries. Miscellaneous Publication 508, U.S. Department of Agriculture, Washington, DC

Jain TB, Graham RT (1996) Deception Creek Experimental Forest. In: Schmidt WC, Friede JL (comps) Experimental forests, ranges, and, watersheds in the Northern Rocky Mountains—a compendium of outdoor laboratories in Utah, Idaho, and Montana; General Technical Report INT-GTR-334, U. S. Department of Agriculture, Forest Service, Intermountain Research Station, pp 45–50

Jain TB, Graham RT, Morgan P (2004) Western white pine growth relative to forest openings. Can J For Res 34:2187–2197

Jain TB, Graham RT, Sandquist J, Butler M, Brockus K, Frigard D, Cobb D, Han H, Halbrook J, Denner R, Evans J (2008) Restoration of Northern Rocky Mountain moist forests: integrating fuel treatments from the site to the landscape. In: Deal R (ed) Integrated restoration efforts for harvested forest ecosystems; Ketchikan, AK. General Technical Report PNW-GTR-733, U.S. Department of Agriculture, Forest Service, Pacific Northwest Research Station, Portland, OR, pp 147–172

Jemison G (1950) Forty years of forest research in the Northern Rocky Mountain region: fortieth annual report. U.S. Department of Agriculture, Forest Service, Northern Rocky Mountain Forest and Range Experiment Station, Missoula, MT

Klade RJ (2006) Building a research legacy—the intermountain station 1911–1997. General Technical Report RMRS-GTR-184, U.S. Department of Agriculture, Forest Service, Rocky Mountain Research Station, Fort Collins, CO

Larsen JA (1921a) Relative humidity of the atmosphere and its relation to the fire problem. Appl For Note 6, U.S. Department of Agriculture, Forest Service, Northern Rocky Mountain Forest and Range Experiment Station

Larsen JA (1921b) Sunshine and air temperature in relation to forest fires. Appl For Note 9, U.S. Department of Agriculture, Forest Service, Northern Rocky Mountain Forest and Range Experiment Station, Missoula, MT

Larsen JA (1921c) Wind and its relation to forest fires. Appl For Note 7, U.S. Department of Agriculture, Forest Service, Northern Rocky Mountain Forest and Range Experiment Station
Larsen JA (1922) Weather records at lookout stations in northern Idaho. J For 20(3):215–219
Maclean N (1992) Young men and fire. University of Chicago Press, Chicago
Peterson DL, Agee JK, Aplet GH, Dykstra DP, Graham RT, Lehmkuhl JF, Pilloid DS, Potts DF, Powers RF, Stuart JD (2009) Effects of timber harvest following wildfire in western North America: issues and scientific evidence. General Technical Report PNW-GTR-776, U.S. Department of Agriculture, Forest Service, Pacific Northwest Research Station, Portland, OR
Pyne SJ (2001) Year of fires: the story of the great fires of 1910. Penguin Books, New York
Pyne SJ (2010) America's fires: a historical context for policy and practice (Rev edn). Forest History Society, Durham
Rothermel RC (1993) Mann Gulch fire: a race that couldn't be won. General Technical Report INT-299, United States Department of Agriculture, Forest Service, Intermountain Forest and Range Experiment Station, Ogden, UT
Schmaltz NJ (1980) Forest researcher: Raphael Zon. J For Hist 24(1):24–39
Schwandt JW, Lockman JB, Kliejunas JT, Muir JA (2010) Current issues and management strategies for western white pines in the western United States and Canada. For Pathol 40:226–250
Sloan J, Steele R (1996) Boise Basin Experimental Forest. In: Schmidt WC, Friede JL (comps) Experimental forests, ranges, and, watersheds in the Northern Rocky Mountains—a compendium of outdoor laboratories in Utah, Idaho, and Montana. General Technical Report INT-GTR-334, U. S. Department of Agriculture, Forest Service, Intermountain Research Station, pp 39–44
Spaulding P (1922) Investigations of the white pine blister rust. Bulletin No. 957, U.S. Department of Agriculture, Washington, DC
Spencer BG (1956) The big blowup. Caxton Printers, Ltd, Caldwell
Wellner CA (1976) Frontiers of forestry research—Priest River Experimental Forest 1911–1976. U.S. Department of Agriculture, Forest Service, Intermountain Forest and Range Experiment Station, Ogden

Chapter 22
The Fire Research Program at the Silas Little Experimental Forest, New Lisbon, New Jersey

Kenneth L. Clark, Nicholas Skowronski and Michael Gallagher

Abstract In this chapter, we document the development and current research efforts of the fire research program at the Silas Little Experimental Forest of the Northern Research Station, USDA Forest Service, in the Pinelands of southern New Jersey. The 450,000-ha (1.1 million-acre) Pinelands National Reserve contains some of the most challenging fuel types for wildland fire managers in the eastern USA. These highly flammable forests occur adjacent to extensive wildland–urban interface and major transportation corridors. We first briefly discuss the ecological setting of upland forests in the Pinelands, highlighting how fire, past industrialization, and other disturbances have shaped the composition and structure of these forests. We then document the establishment of the Experimental Forest and the fire research program. We focus on the career of Dr. Silas Little, a silviculturist with the Northeastern Forest Experiment Station, Upper Darby, PA. Beginning in the late 1930s, his research on prescribed burning practices and silviculture in the Pinelands resulted in cost-effective methods to reduce wildfire risk while promoting the regeneration of commercially important timber species. We discuss how many of the prescribed burning practices developed during these research efforts are now used operationally by the New Jersey Forest Fire Service (NJFFS) and

---

Silas Little, Jr., received a B.S. degree from Massachusetts State College, and M.F. and Ph.D. degrees from Yale University. In 1937, Dr. Little was assigned to the Forest and worked there until his retirement in 1979, much of that time serving as the research project leader. He was instrumental in conducting early research on forest regeneration and silviculture following intense land-use change (charcoaling for iron forges, mills, glassworks, and towns); thus, the use of prescribed fire to reduce wildfire hazard was designed simultaneously to promote tree growth in the New Jersey Pinelands. Dr. Little also conducted primarily silvicultural, fire, and forestry-related research in Maryland and Pennsylvania. He produced more than 100 publications, many of which focused on the Pinelands of New Jersey.

K. L. Clark (✉) · M. Gallagher
Silas Little Experimental Forest, Northern Research Station, USDA Forest Service,
501 Four Mile Road, New Lisbon, NJ 08064, USA
Phone: 609-894-0325
e-mail: kennethclark@fs.fed.us

N. Skowronski
Northern Research Station, USDA Forest Service, 180 Canfield Street,
Morgantown, WV 26505, USA

D. C. Hayes et al. (eds.), *USDA Forest Service Experimental Forests and Ranges*,
DOI 10.1007/978-1-4614-1818-4_22, © Springer New York 2014

federal wildland fire managers in the Pinelands. Finally, we highlight current and future research efforts at the Experimental Forest.

Keywords Prescribed fire · Wildfire · Forest management · Pinelands National Reserve · LiDAR · Hazardous fuels management

22.1 Forests of the New Jersey Pinelands

The potential for destructive wildfires in the pine-dominated forests of southern New Jersey has been dramatically illustrated numerous times. On average, 10% or more (>40,000 ha) of the predominantly pine- and oak-dominated forests could burn in a single year, and did at approximately 20-year intervals before the adoption of modern fire-suppression activities. The most notable recent major event occurred on the weekend of April 20–21, 1963, when wildfires burned 74,000 ha of forest, destroyed or damaged 186 homes and 197 buildings, and were responsible for seven deaths. More recently, a wildfire burned 7,786 ha in Ocean County in 1995, and during the late spring of 2007, 6,298 ha were burned near Warren Grove and the Garden State Parkway (Fig. 22.1, Table 22.1).

It is believed that park-like stands of pines with an open understory prevailed in the original upland forests as a result of frequent fires set by Native Americans, mostly during the fall and winter (Wacker 1979). The present-day forests have been shaped by previous land uses and extensive fire history, especially the occurrence of large, hot wildfires (Forman and Boerner 1981; Little 1979b). Most of the upland forest stands in the Pinelands have regenerated naturally following the cessation of logging and charcoaling activities in the late 1800s. As in the past, fire continues to be the major factor affecting stand age, species composition, and biomass and structure of these forests.

Wildfires burn primarily in upland forests in the Pinelands. Dominant tree species include pitch pine (*Pinus rigida*), shortleaf pine (*P. echinata*), and mixed oaks, primarily black (*Quercus velutina*), white (*Q. alba*), and chestnut (*Q. prinus*) oak. Understory vegetation is dominated by shrubs, including huckleberries (*Gaylussacia baccata*), blueberries (*Vaccinium* spp.), and scrub oaks (*Q. ilicifolia*, *Q. marilandica*). Pine canopies are burned far more frequently than oaks because most fires in the Pinelands occur during the spring when oaks are leafless (Fig. 22.1). During wildfires, pines may have only their foliage killed, or with greater fire intensity, apical buds and branches, yet individuals can survive and resprout new canopies from "epicormic buds" that occur in cambial tissue on stems and branches (Little and Somes 1951, 1956). This capacity to resprout from stems and branches following fire is retained into maturity in pitch and shortleaf pines, and contrasts with many other pine species. Following the most intense wildfires, regrowth of shoots from stump sprouts frequently occurs, but growth is retarded by the age of the stumps from which they originate, and by competition among large numbers of sprouts (Little and Somes 1951, 1964). Most of the fire damage to oaks occurs near the base, damaging or kill-

Fig. 22.1 A wildfire burns pitch pine (*Pinus rigida* Mill.) and scrub oaks in the Pine Plains in March 2006. (Photo credit Dr. Nicholas Skowronski)

Table 22.1 Recent major wildfires in the New Jersey Pine Barrens. (Data are from the New Jersey Forest Fire Service)

Date	Ha	County	Township
August 3, 2007	983	Burlington	Washington
May 15–19, 2007	6,300	Burlington, Ocean	Stafford, Bass River
August 15, 2002	1,100	Ocean	Plumstead
June 2, 2002	517	Ocean	Berkeley
June 10, 2001	600	Ocean	Little Egg Harbor
April 30, 1999	4,773	Burlington	Bass River
July 29, 1997	800	Atlantic	Hammonton
April 4, 1995	7,786	Burlington	Woodland
June 13, 1992	2,140	Ocean	Lacey
May 23, 1992	400	Burlington	Woodland
May 3, 1992	3,131	Ocean, Burlington	Lacy, Woodland

ing the cambium. When all of the cambium is killed and the aboveground portion of the stem dies, sprouts may initiate from buds located just belowground. Following severe wildfires, a dense understory of oaks can form from sprouts, along with scattered pine sprouts and seedlings.

Four major upland forest types can be recognized: Pine Plains, pine–scrub oak, pine–oak, and oak–pine (Forman 1979; Lathrop and Kaplan 2004; McCormick and Jones 1973; Skowronski et al. 2007). In the Pine Plains, the low stature of pitch pine and scrub oaks is a result of repeated severe wildfires (Little 1945, 1946, 1973, 1979b; Lutz 1934; Fig. 22.1). Severe fires at fairly frequent intervals (<20 years) have eliminated species that do not bear seed at an early age, such as shortleaf pine and canopy oaks (Little 1946, 1979b). Frequent wildfires have also favored a race of pitch pines that is relatively slow growing, develops a mature, often crooked form relatively early, and has serotinous cones. Nearly all individuals have crooked stems in "mature" stands because they have originated from stump sprouts following the last severe wildfire.

Pitch pine–scrub oak stands, which are similar in composition to those in the Pine Plains, arise from slightly less frequent and/or less intense wildfires. Pines are typically slow growing because many stems likely started as sprouts following a severe wildfire, and many individuals have lived through one or more large fires that damaged their crowns. Pine–oak stands are dominated by a mix of oaks and pines. In contrast to Pine Plains and pitch pine–scrub oak stands, the frequency and intensity of wildfires are lower, and the importance of oaks in the canopy is greater. Oak–pine stands are dominated by oaks with scattered pitch and shortleaf pines that survived the last and, often, earlier fires. This composition apparently results from severe fires at intervals of approximately 30–65 years, and certainly at longer intervals than in the Pine Plains and pitch pine–scrub oak stands.

Among mature upland stands of approximately the same age, canopy stature and biomass are typically greatest in oak–pine stands and least in pine–scrub oak stands (Forest Inventory and Analysis data at www.FIA.gov; Skowronski et al. 2007). In contrast, these forest types can be equally productive in terms of net CO_2 exchange in the absence of disturbance or drought (Clark et al. 2010b), and pine-dominated stands typically have higher gross primary productivity on an annual basis because pine foliage is displayed year-round. For example, Clark et al. (2010b) reported that net CO_2 uptake occurred in a pine–scrub oak stand throughout the winter months when air temperature during the previous night did not drop below 0 °C. Because foliage is displayed year-round, annual ecosystem respiration also can be higher in pine-dominated stands than in oak-dominated stands.

If fires are excluded and no other disturbances such as harvesting or insect infestations occur, hardwoods will gradually replace pitch and shortleaf pines and eventually dominate the stand. The succession from pines to hardwoods is due to two main factors: Hardwood seeds, being larger, can become established in the relatively thick litter and organic matter layers that accumulate in unburned stands, and hardwoods can survive and grow under lower light conditions than pines (Little 1973, 1979a). Fire history also affects the structure and composition of the understory (Buell and Cantlon 1953). For example, frequent light fires tend to reduce the shrub cover and favor herbaceous plants, especially along roads or under open stands.

22.2 Historic Use of Forest Resources in the Pinelands

Industrial development beginning in the late 1600s had a major impact on upland forests in the Pinelands (Pearce 2000; Wacker 1979). Harvesting of forest for timber, pitch, and charcoal to operate bog iron furnaces, forges, and glassworks occurred until the mid-1800s. In addition, a number of pulp and paper mills existed in the Pinelands until the late 1800s. These industries produced numerous goods for the large markets in Philadelphia and New York City. Following extensive harvesting for high-quality timber, second-growth and third-growth stands were harvested by axe for charcoal and pulpwood on an approximately 30-year rotation (Mounier

1997; Pearce 2000). "Colliers" or charcoal tenders would stack logs and then bury them with sand and humus for charring using oxygen-depleted fires. Covered log stacks typically measured 4–11 m in diameter, and were 2–3 m tall. The center was formed into a chimney or archway to ignite the logs and to control combustion. Separate holes in the charcoal "pits" were used to control the oxygen supply to the charring wood. Logs were charred for several days, depending on the size of the pile, the initial moisture content and condition of the wood, and weather conditions. Many of these charcoal "pits" are still visible throughout the Pinelands today.

Two industrial developments in the vicinity of the Experimental Forest are notable: the Mount Misery mill and the Lebanon glass works (Beck 1961). Mount Misery, located ca. 10 km from the forest, was one of the centers of the timber and charcoaling industries in the northern portion of the Pinelands in the 1800s. First settled in the early 1700s, Mount Misery once contained more than 100 homes, a store, and a hotel. It is also notable because it was the location of one of the early prescribed fire trials described below. The Lebanon glass works, located ca. 5 km from the Experimental Forest, employed more than 200 people and had 20 homes on-site at its peak. Sand in the vicinity contained few impurities, and the glassworks produced artistic as well as production pieces from 1851 to 1867, when it was finally abandoned because of local charcoal depletion (Beck 1961).

The Pinelands' population peaked around 1859 (Pearce 2000; Wacker 1979). The discovery of new sources of iron ore in Pennsylvania and the depletion of bog iron beds in the Pinelands led to the decline of the iron industry in the mid-1800s. By this time, industrial-scale charcoaling was also waning. Transportation of coal from Pennsylvania to the major urban centers by rail and local exhaustion of forest resources were major factors in the demise of the charcoal industry in the Pinelands. After the iron, charcoal, and glass industries collapsed, people gradually moved to other areas.

22.3 A Degraded Forest and the Beginnings of Silas Little Experimental Forest

Following the abandonment of industrial uses of forests in the Pinelands, most stands had been harvested repeatedly for timber and charcoal. Wildfires were apparently unchecked through the late 1800s and early 1900s. For example, one wildfire burned 50,600 ha in 1894, approximately 41,000 ha were burned in 1915, approximately 400,000 ha were burned in 1923, and eight wildfires burned 70,000 ha in May 1930. As an example closer to the site of the Experimental Forest, one scarlet oak (*Q. coccinea*) that was harvested to create a fire break in the late 1930s showed wildfire damage in 1878, 1884, 1889, 1900, and 1902. The essentially unregulated harvesting activities since the 1700s and repeated wildfires had resulted in poorly stocked stands of pitch and shortleaf pine with a dense understory of oaks that had originated from burned stumps. Many upland stands had entered into essentially a positive feedback loop, producing stands of wildfire-damaged pines and oaks that had little to no commercial value, and were prone to repeated wildfires. In a report to the State Geologist of New Jersey in 1899, Gifford Pinchot noted that wildfires in southern New Jersey needed to be controlled, and that without forest fire protection, forests could not be managed successfully for public benefit. As these forests regenerated and burned, a new interest in fire management and silviculture arose in the Pinelands.

By the late 1920s, state foresters considered the large contiguous stands of oak- and pine-dominated forests in the Pinelands to be highly degraded and in need of rehabilitation. State policy at the time was to remove the hardwoods in favor of planted conifers. It was also concluded that any rehabilitation of oak–pine stands that had originated from sprouts following wildfire had to be based on intensified fire protection and aggressive stand improvement techniques, including thinning and restocking by seed and/or seedlings.

It is within this context that the Silas Little Experimental Forest began (Adams et al. 2004). The fire research program dates back to 1927, when forest yield plots were established and weather measurements recorded at Camp Ockanickon. In 1933, the Northeastern Experiment Station and the Board of Conservation and Development, State of New Jersey, signed a cooperative agreement for a new location in Burlington County in response to increased recreation pressures at Camp Ockanickon. The State placed at the disposal of the Station 239 ha (591 acres) "for the purpose of conducting studies, experiments, and demonstrations in silvics and silviculture… to solve forest problems of the region typified by conditions in southern New Jersey. These may include experiments in obtaining natural reproduction of the forest after cutting, in thinning to stimulate growth, and in artificial reforestation; also, more fundamental studies of the factors which affect tree growth" (http://www.nrs.fs.fed.us/locations/nj/silas-little). Over the next few years, the USDA Forest Service (USDA FS) constructed an office, part of the current garage and a fire tower, and moved a bunkhouse to the site from Medford Lakes as part of a Works Progress Administration project. The Experimental Forest was initially known as the Lebanon Experimental Forest because of its location in the Lebanon State Forest.

In 1937, Dr. Silas Little, Jr., was assigned to the Forest. In addition to research on prescribed fire to reduce wildfire risk and silviculture of tree species in the Pinelands, major themes of his research were the ecology and silviculture of Atlantic white cedar (Little 1950, 1959; Little and Somes 1965) and tree improvement and the production of a pitch/loblolly hybrid (Garrett 1981).

22.4 Prescribed Fire in the Pinelands

When the fire research program got under way, limited use of prescribed fire was already in place in the Pinelands. The Lenni-Lenape Indians were the first to introduce prescribed burning to New Jersey's forests, using fire to facilitate travel, improve hunting, drive away insects, and also increase the availability of browse, acorns, and berries (Wacker 1979). Starting in the late 1600s, settlers in the

Pinelands cleared and burned forest and agricultural residue, two centuries before industrial forest clearing and burning (Pearce 2000; Wacker 1979). Beginning in the early 1900s, cranberry and blueberry growers protected their property by using prescribed fire to remove heavy accumulations of forest fuels from around their fields and buildings.

The New Jersey Forest Fire Service (NJFFS) first used prescribed burning practices in 1928 to protect state forestlands by burning along roadside safety strips (Little 1979b; Section Forest Fire Wardens of Division B 2006). Prescription burning along roads involved using low-intensity fires, mostly in the winter and early spring. Protection strips along roads were normally between 10 and 70 m wide. Following the large and destructive wildfires in 1930, state agencies expanded this practice to include larger blocks of woodland. There was also a precedent for silvicultural management of upland forests in the Pinelands, following ideas regarding regeneration of sprout oak stands. For example, the Civilian Conservation Corps undertook numerous projects in the state forests between 1933 and 1940. One of the most common treatments was thinning in predominantly oak stands on upland sites. Thinnings varied from very light, involving only the removal of trees that would die soon, to very heavy. Thinning treatments were usually justified on the assumption that growth of the residual stems would be stimulated. Thus, when the fire research program began at the Experimental Forest, the use of prescribed fire and silvicultural management techniques were accepted in the Pinelands, and it is likely that little public sentiment existed against these management practices. In fact, because of the long history of destructive wildfires, burning of hazardous fuels was viewed as an essential survival strategy for residents of the Pinelands (McPhee 1968).

22.4.1 Early Research Efforts on the Use of Prescribed Fire

In 1935, the State Forester of New Jersey proposed a large study of the effects of fire and the Experiment Station agreed to cooperate. A 30-year effort of active fire suppression by the State had not solved the problem of protecting the oak–pine lands against destructive wildfires. Two objectives of the 1935 agreement were to evaluate damage from wildfires and to compare losses from wildfires with the much lighter losses expected to occur from prescribed fires set under favorable conditions. The scope of this research included measurements of biomass and productivity of the sites, measuring changes in water-holding capacity of litter and soil and monitoring effects on wildlife.

Out of this agreement, the first studies of prescribed burning were initiated in 1936. An experimental area consisting of 16.2-ha blocks, each bounded by disked firebreaks, was installed at the Experimental Forest. Prescribed burns were to be conducted at intervals of 1, 2, 3, 4, 5, 10, and 15 years, with an unburned control. All blocks except the control were first burned in March 1937 (Little and Moore 1945; Fig. 22.2). All trees and tree reproduction in transects through the center of each block were recorded. Shrubs and litter were also sampled in subplots in each block. The New Jersey Agricultural Experiment Station investigated the effects of

Fig. 22.2 Prescribed burning near the Lebanon Experimental Forest (now the Silas Little Experimental Forest) in 1937. Note the leafless deciduous trees, indicating that this burn was conducted in late winter or early spring. (Photo credit Dr. Silas Little)

prescribed burning on soils, and soil samples were collected from preburning and postburning blocks and analyzed over the years. A second 14-block experimental area was installed and monitored near Mount Misery in 1940 (Little and Moore 1945).

Prescribed burning of large plots and blocks in interior forest in the Pinelands was more involved than burning plots along roads and required knowledge of forest fuels, fire behavior, suppression techniques, and local weather conditions. An orderly progression of sections in each block was burned sequentially to reduce the accumulation of fine and 10-h fuels on the forest floor and to reduce ladder fuels consisting of shrubs, saplings, and lower branches. Prescribed fires were typically conducted during the winter and early spring (note leafless canopies in Fig. 22.2). Horrace Somes, Jr., a recently retired Fire Warden for Division B of NJFFS, recalls that when he was a child, his father (Horrace Somes, Sr.) and Silas Little would conduct some prescribed fires at night to better control fire behavior.

Although prescribed burns were conducted primarily to reduce the wildfire hazard, prescribed burning also tended to create stands of seedling or seedling sprout origin, which permitted trees to develop without the deformities created by severe wildfires. Thus, these research efforts served a dual purpose. In oak–pine stands, use of prescribed fires favored the regeneration of pines, which had a higher commercial value than oaks at that time. One of the early papers published by Silas Little concludes that prescribed burning in the form of low-intensity winter fires in the oak–pine stands of New Jersey's pine region lessens the danger of severe spring fires and may also be used advantageously for preparing the seedbed for pine reproduction and the control of hardwoods (Little 1946). Further research indicated that frequent light fires under an overstory of pines or oaks cause significant mortality of acorns and seedlings and provide seedbeds unfavorable for most hardwood seedlings. Frequent prescribed fires did not damage the current stand appreciably

Table 22.2 Years and total area burned using prescribed fire in Lebanon State Forest (now Brendan Byrne State Forest) over a 9-year period from 1949 to 1957

Year	Area burned, ha	% Reburns
1949–1950	1,305	0
1950–1951	1,137	36
1951–1952	686	39
1952–1953	611	67
1953–1954	709	75
1954–1955	1,128	100
1955–1956	926	86
1956–1957	997	97

Percentages denote percentage of burns that were conducted in areas that had been previously treated with prescribed fire

but did provide conditions that favor pine seedlings to become established (Little and Moore 1949; Little et al. 1948).

By around 1940, active wildfire suppression by the NJFFS with better equipment had considerably reduced the average size of wildfires in oak–pine forests in the Pinelands (Forman and Boerner 1981; Section Forest Fire Wardens Division B 2006). Analyses by Forman and Boerner (1981) indicated that the average area burned annually dropped sharply from about 20,000 ha (50,000 acres) during 1906–1939 to 8,000 ha after 1940. Similarly, the frequency of severe wildfires decreased, and after 1940, oak- and pine-dominated stands burned at an average of 65-year intervals, compared with 20-year intervals earlier in the nineteenth and twentieth centuries.

However, large wildfires still occurred, especially during the late spring. One of the most interesting early accounts of the effectiveness of prescribed burning in reducing wildfire risk occurred on April 26, 1946. A wildfire had ignited to the north of the Experimental Forest near the railroad tracks in New Lisbon and burned towards the south. Firefighters from NJFFS attempted numerous times to backfire along roads moving from north to south to suppress the crown fire but failed because it was a class-five fire day, and weather conditions were causing erratic fire behavior and spotting from wind-driven embers. By 5:00 to 5:15 pm, a large head fire approached and then burned into a prescribed fire block just to the north of the Experimental Forest (similar to the treatment block in Fig. 22.2). Reduced fuel loading in the treatment block caused the head fire to lose intensity, and crews were able to suppress the spot fires that burned into the block.

In 1948, the practice of prescribed burning as a management tool for silvicultural purposes and fuels reductions on state and private lands was introduced to the public. Prescribed fires were used operationally in large forest blocks starting in 1949, and hectares burned per year in Lebanon State Forest are shown in Table 22.2. An attempt to conduct prescribed fires at various intervals is apparent, as indicated by the increasing number of burns conducted in previously burned areas. These areas are consistent with current burning practices in Brendan Byrne State Forest (previously Lebanon State Forest). For example, the NJFFS burned 501 ha (1,236 acres) in this state forest in 2007–2008.

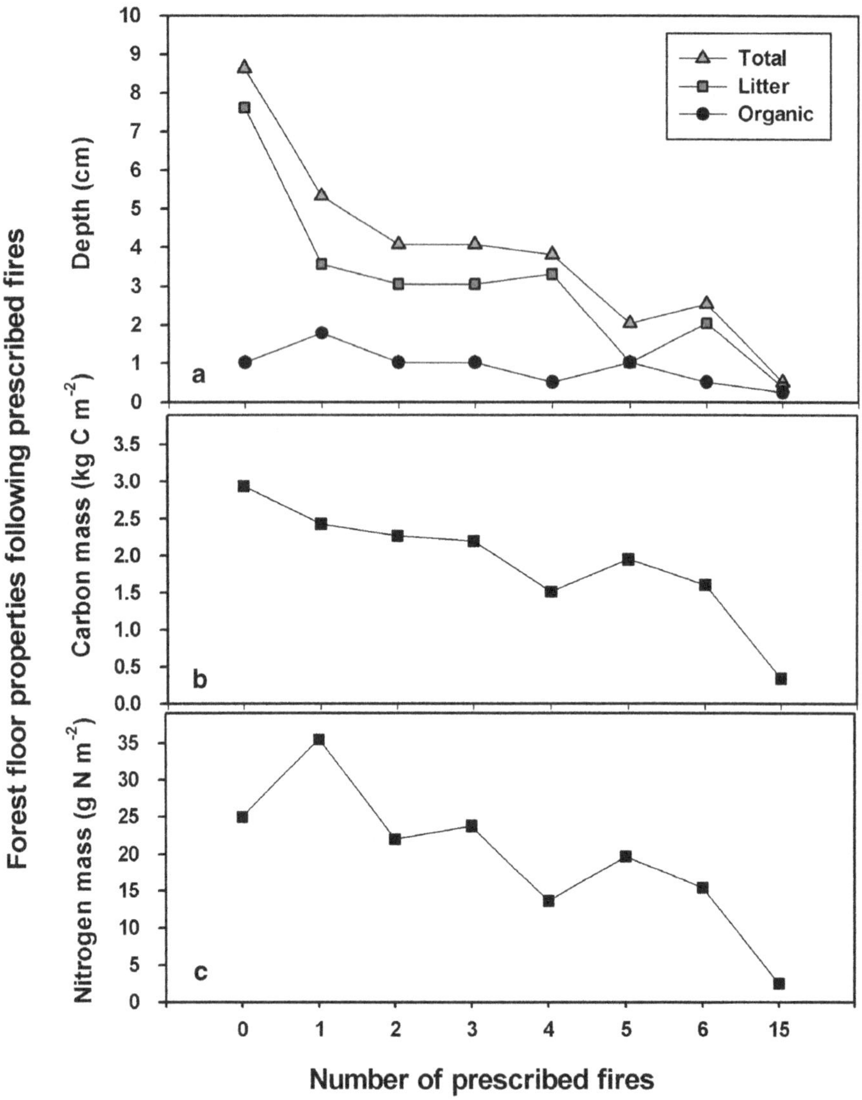

Fig. 22.3 Forest floor properties as a function of the number of prescribed fires conducted starting in 1937 at the Lebanon Experimental Forest. Variables are (**a**) litter layer (L horizon) depth, organic matter layer (O horizon) depth, and total depth; (**b**) total carbon content of the forest floor; and (**c**) total nitrogen content of the forest floor. (Data are plotted from tables in Burns 1952)

Further results of the study initiated in 1937 were published in the 1940s and 1950s. Repeated prescribed fires reduced the depth, mass, and nitrogen (N) content of the litter layer (L horizon) but had relatively little effect on the depth of the humus layer (O horizon; Burns 1952; Fig. 22.3). In terms of soil chemistry, repeated, low-intensity winter burns increased exchangeable calcium and pH but had only

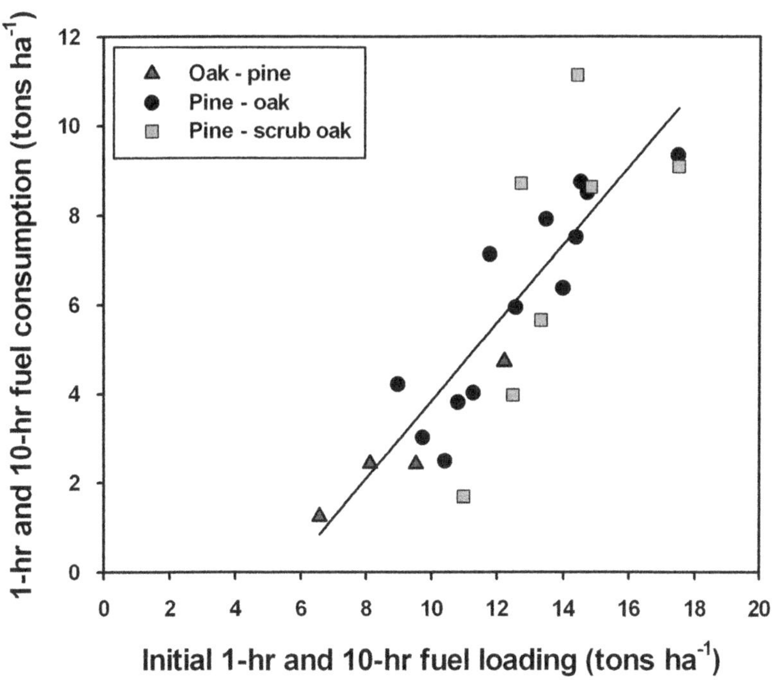

Fig. 22.4 The relationship between initial 1-h + 10-h fuel loading and amount of fuel consumed by forest type during 24 prescribed fires conducted in the Pinelands National Reserve in 2004–2009. Prescribed fires conducted in the Pine Plains were included in the pine–scrub oak category. Consumption =0.898 (fuel loading)—5.319, r^2=0.722, P<0.001. (Redrawn from Clark et al. 2010b)

minor effects on nitrogen and phosphorus (P) contents, which overall were very low in the sandy soils at the Experimental Forest. Thus, low-intensity prescribed burning conducted in the late winter and early fall was effective at reducing 1- and 10-h fuels in the litter layer but had only limited impact on the organic horizon and surface soil layers. These results are consistent with current research results, which also indicate minimal effects of prescribed burning on organic matter layers, soil, and soil nutrients in Pineland ecosystems (Clark et al. 2009; Neill et al. 2007; Fig. 22.4).

Further research efforts addressed the effects of a series of prescribed fires and then the suspension of fire management activities on hardwood regeneration (Little 1973). Low-intensity winter fires used for a short period from 1954 to 1962 had relatively little effect in checking the development of hardwood understory after 18 years (1953 to 1971). In fact, the increase in the number of stems of primarily black and scarlet oak, but also white, post, and chestnut oaks, was greater in plots burned between 1954 and 1962 than in unburned plots. Results suggested that if succession were to be checked by fire alone, periodic winter burns should be followed by one or more summer burns, which are more effective at killing hardwood rootstocks. It was also concluded that periodic winter fires would be far more effective in reducing fuel loading and oak sprouts than a short period of use of prescribed burns followed by fire exclusion.

Dr. Little's research results and field observations were condensed into several "rules of thumb" for conducting prescribed fires in the Pinelands. In interior forest blocks that were used for firebreaks, prescribed burns were conducted at 4–6-year intervals. Less strategic blocks were burned at longer intervals, and roadside safety strips were burned at 1–2-year intervals. Two aspects of the research conducted by Dr. Little were unique. First, responses of different ecosystem components to varying frequencies of prescribed fire, as described above, had rarely been studied until then. Second, the arrangement of burn block locations and return intervals was designed to form a landscape-scale pattern of firebreaks. The early burn maps indicate a strategic orientation of the burn blocks, and in the earliest published results it is clear that Dr. Little had taken a landscape-scale perspective on the use of prescribed fires to protect against severe wildfires in the Pinelands. These landscape-scale firebreaks were oriented in a north–south axis because many of the severe wildfires burned from west to east on strong westerly winds in spring (Fig. 22.5).

Dr. Little first considered a checkerboard pattern of prescribed burns at the landscape scale but apparently realized early on that this was impractical to accomplish with 15–20 "safe" burning days occurring per year across more than 450,000 ha of forest. However, the landscape-scale distribution of rivers, wetland forest, and cranberry bogs provided an excellent opportunity to break up the largest extents of upland forest. For example, obvious fuel breaks occur in what is now Wharton State Forest between the Batsto and Mullica rivers and in two other north–south-oriented wildfire protection areas shown in Fig. 22.5. This effort was likely assisted by aerial reconnaissance; one of the first complete set of aerial photographs of the Pinelands was taken in 1937. Many of these original images and maps are available in the library at the Silas Little Experimental Forest today.

22.4.2 Prescribed Burning in the Pinelands today

Today, most of the practices initiated in the 1930s are still in use. The NJFFS is authorized to conduct prescribed burning by the authority of New Jersey Statutes, Title 13:9-2, General powers of department, and as specified in the NJ Air Pollution Control Code (Title 7, Subchapter 27). The prescribed burning season is limited to the period between October 1 and March 31, and the NJFFS burns an average of 4,000–6,000 ha of public lands and 2,000 ha of private lands annually. In addition to reducing hazardous fuel accumulations, prescribed burns provide a foundation for safer, more effective fire suppression and protection operations during wildfires. The second goal of Dr. Little's research, to provide commercially important timber, was never fully realized. The creation of the Pinelands National Reserve by the US Congress in 1978, expanded tourism on the coast, and the demise of local timber and pulp mills and the rising cost of hauling logs from the Pinelands essentially eliminated the timber industry from southern New Jersey.

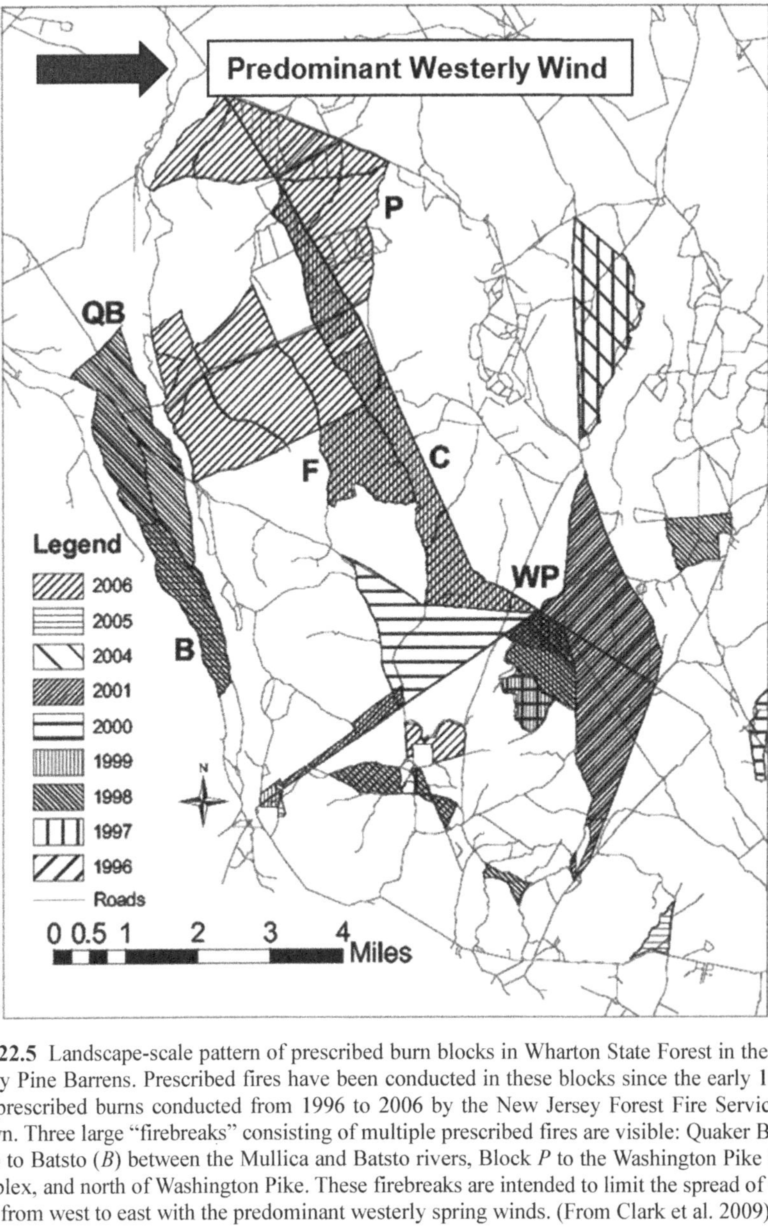

Fig. 22.5 Landscape-scale pattern of prescribed burn blocks in Wharton State Forest in the New Jersey Pine Barrens. Prescribed fires have been conducted in these blocks since the early 1950s, and prescribed burns conducted from 1996 to 2006 by the New Jersey Forest Fire Service are shown. Three large "firebreaks" consisting of multiple prescribed fires are visible: Quaker Bridge (*QB*) to Batsto (*B*) between the Mullica and Batsto rivers, Block *P* to the Washington Pike (*WP*) complex, and north of Washington Pike. These firebreaks are intended to limit the spread of wildfires from west to east with the predominant westerly spring winds. (From Clark et al. 2009)

Burn blocks in the Pinelands are arranged as they have been since the 1940s and 1950s. For example, three large fuel breaks are indicated on the map of prescribed burn blocks in Wharton State Forest (Fig. 22.5). Prescribed burning today follows many of the traditional practices, with some mechanized improvements. A written plan is first developed detailing the proposed burn, and once the plan is approved by the New Jersey Department of Environmental Protection, perimeter and internal firing lines are located and cleared using a tractor and fire plow unit, or occasionally a gyrotrack. The fire plow cuts a path through the forest floor down to the sand and gravel, usually between 30 and 60 cm deep. The lines are plowed in the fall after leaf senescence but before the ground freezes. As in the past, only about 15–20 optimum burning days occur during the burning season. These sometimes require coordination with the National Weather Service for prescribed fire "spot forecasts." Fire managers in the Pinelands also can now utilize real-time fire weather data provided by the USDA FS and the State Climatologist of New Jersey (available at http://climate.rutgers.edu/usfs/monitoring.php). On the correct day, burning crews first gather for a coordination meeting to summarize the burning plans. Crews then ignite the control and interior lines in a systematic and progressive manner to assure that the entire area is burned, utilizing drip torches to ignite the lines. All communications are by portable radio and cell phones, and aircraft or helicopter observations are used for the more difficult burns with heavy fuels. The NJFFS has recently acquired an aerial ignition device, which was first used successfully in spring 2009 to burn large prescribed burn blocks in the Pine Plains that had limited road access.

Prescribed burning in a given area is normally repeated on an interval of 4–8 years. Hazard reduction blocks and safety strips require burning at annual or biennial intervals. Other less strategic blocks are burned at longer intervals. The reduction of 1- and 10-h fuels during prescribed fires is largely a function of the initial fuel loading, with loading explaining 72% of the variability in consumption when forest types are considered together (Fig. 22.4). Typically, only a portion of the litter layer is burned during prescribed fires, leaving the humus layer intact, consistent with results from the original study initiated in 1937 (e.g., Burns 1952; Fig. 22.3). The reduction of ladder fuels from 1- to 4-m height due to repeated prescribed fires over a 15-year period is shown in Fig. 22.6 (see Skowronski et al. 2007, 2011 for a description of Light Detection and Ranging, LiDAR, sampling in the Pinelands; Clark et al. 2009). These data generated from light detection and ranging (LiDAR) indicate that the 5–8-year burn interval is effective at reducing ladder fuels to below 20% cover, while the landscape-level mean is ca. 39% (Fig. 22.6).

Most wildfires continue to occur in Pine Plains and pitch pine–scrub oak forest types, as in the past. These stands are concentrated towards the eastern margin of the Pinelands, where, unfortunately, land-use pressures have resulted in a proliferation of residential subdivision and developments adjacent to these forests with heavy fuel loads. The "wildland–urban interface" or "WUI" is the term used to describe the placement of residential communities within forested areas. Forest fires burning into unprotected developments can take a large toll on lives and property. Prescribed burning to reduce forest fuels, coupled with other fire protection measures, can help provide an effective level of fire protection for homes and businesses in the WUI today.

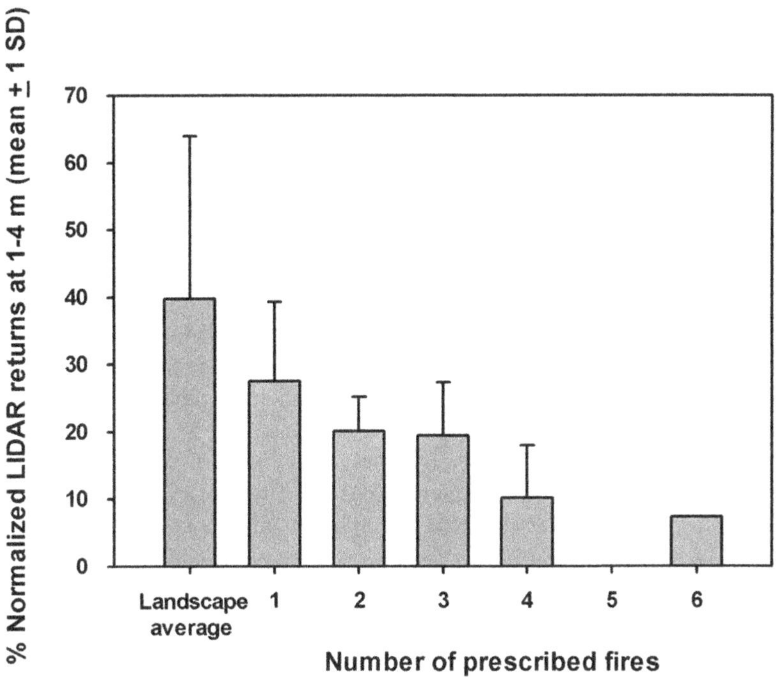

Fig. 22.6 Density of "ladder fuels" from 1- to 4-m height as a function of the number of prescribed fires conducted over a 15-year period (1991–2006) in Wharton State Forest. (Data are derived from profiling LiDAR, and were analyzed to estimate cover in 1-m height classes and then averaged from 1 to 4 m. From Clark et al. 2009)

22.5 Current and Future Research Efforts at Silas Little Experimental Forest

Following the retirement of Silas Little in 1979, the Experimental Forest was leased to Rutgers University. The Silas Little Experimental Forest was then reinstated with National Fire Plan Funding in 2003 to conduct research on fire weather and fire danger models specific to the Pinelands. Since then, substantial progress has been made in monitoring and delivering fire weather information to wildland fire managers (real-time fire weather data for the Pinelands available at http://climate.rutgers.edu-usfs-monitoring.php), measuring hazardous fuel loads (Clark et al. 2009; Skowronski et al. 2007, 2011; Wright et al. 2007), and assessing fire danger in the Pinelands. Current and future research efforts directed specifically towards the prescribed burning program encompass three major topics: (1) quantification and management of smoke from prescribed fires (Clark et al. 2010a), (2) quantifying trade-offs between hazardous fuels management and carbon (C) sequestration by forests in the Pinelands (Clark et al. 2009, 2010b), and (3) determining trade-offs between WUI protection and land-use change. These efforts are summarized in the following paragraphs.

In addition to the fire research program, scientists at the Silas Little Experimental Forest are also active in a number of the problem areas assigned to Northern Research Station Group 6, specifically an understanding of carbon and water exchanges between the land and atmosphere, and development of quantitative methods for ecosystem science. We have operated a number of eddy covariance towers in the dominant upland forest types since 2005 to quantify carbon and energy exchanges between these forests and the atmosphere (Clark et al. 2010b). Tower research sites are included in the USDA FS climate tower network and the Ameriflux network (http://public.ornl.gov/ameriflux/), and research efforts are consistent with efforts of the North American Carbon Program. In conjunction with forest inventory plots nested at 0.1, 1, and 9 km^2 scales around each flux tower, this monitoring network is capable of closing the landscape-scale carbon budget, i.e., accounting for all major exchanges of carbon between the land and atmosphere. The combination of disturbances that have recently occurred in the Pinelands, including fire, repeated invasive insect defoliation, and severe drought, has made the Silas Little Experimental Forest an ideal site for studying ecosystem processes during disturbance and recovery (e.g., Amiro et al. 2010). For example, all three flux tower sites were defoliated by gypsy moth in 2007, providing us the opportunity to quantify the impacts of invasive insects on disturbance and subsequent recovery of carbon, water, and nutrient cycling (Clark et al. 2010b; Schafer et al. 2010). Research efforts are consistent with Northern Research Station science themes—Managing with Disturbance, Sustaining Forests, and Providing Clean Air and Water.

Researchers at the Experimental Forest also have pioneered the use of LiDAR technology, which uses laser measurements to accurately quantify canopy and understory structure. We have utilized this technology to map hazardous fuels and evaluate the effectiveness of fuel reduction treatments conducted by the NJFFS and federal fire managers (Skowronski et al. 2007, 2011; Clark et al. 2009; Fig. 22.6). LiDAR technology can also be used to accurately project plot-based biometric measurements across the landscape for forest carbon inventories. Current research in this area at the Silas Little Experimental Forest is focused on improving the estimation of leaf area at the ecosystem and landscape scales for physiological studies and modeling purposes, and to estimate variations in canopy structure and fuel loading due to management and disturbance regimes. These efforts have been recently funded by the Joint Fire Sciences Program.

Management of smoke from prescribed burning is now a critical issue. Prescribed fires are subject to federal and state air pollution laws, and can affect air quality, highway traffic, and properties adjacent to urban areas that sometimes exceed state and federal standards for criterion air pollutants. Recent changes in the federal air pollution standards require reduced emissions of particulate matter, PM 2.5, to conform to hourly ambient air concentrations of 35 $\mu g\ m^{-3}$ rather than 65 $\mu g\ m^{-3}$, as well as reduced gaseous emissions. All adjacent smoke-sensitive areas must now be identified in the burning plan. Wind direction, wind speed, and smoke dispersal are some of the atmospheric characteristics that must be considered before conducting a burn. Firing techniques also affect smoke emissions, with backfires producing considerably lower emissions than other firing techniques.

We are currently using E-BAM Beta particle attenuators (Met One Instruments, Grants Pass, OR) to measure ambient PM 2.5 concentrations in the air at Silas Little Experimental Forest (Clark et al. 2010a). Mean seasonal ambient air concentration of PM 2.5 are highest in the summer, averaging 8.5 ± 8.0 μg m^{-3}, whereas burning season (January–May) PM 2.5 concentrations averaged 4.0 ± 5.0 μg m^{-3} in 2008. Seasonal patterns are consistent with previously measured PM 2.5 concentrations in ambient air at the margins of the Pinelands. E-BAM collectors were also used to measure ambient air concentrations of PM 2.5 during a series of prescribed fires near the Cedar Bridge Fire Tower in March 2008. Highest concentrations were measured in and near prescribed fires at the end of March 2008, peaking at 4,155 μg PM 2.5 m^{-3}. Following this peak, which was associated with flaming combustion within 1 m of the instrument, PM 2.5 concentrations dropped to below 35 μg m^{-3} within 12 h.

Using the combustion data presented in Fig. 22.4, published emission factors for particulate emissions during prescribed fires from the Environmental Protection Agency (WebFIRE at http://www.epa.gov/ttn/chief/efpac/index.html, Battye and Battye 2002) and a database of prescribed fire locations and sizes, we calculated emissions from prescribed fires conducted by the NJFFS from 2002 to 2009. Average fuel consumption for 24 prescribed fires conducted from 2004 to 2008 was 606 ± 319 g m^{-2}, and average emission of PM 2.5 was estimated at 8.2 ± 3.8 g m^{-2} during prescribed fires. Annual emissions from prescribed fires conducted by the NJFFS for 2002–2008 averaged 365 t PM 2.5 $year^{-1}$ and ranged from 256 to 611 t $year^{-1}$. For comparison, PM 2.5 emission estimated to have occurred during the Warren Grove wildfire (6,280 ha or 15,500 acres) on May 15–19, 2007, was 727 t (range of 535–913 t), approximately twice the annual emissions from prescribed fires. When considered in the context of overall total metric tons PM 2.5 emitted annually in New Jersey in 2002 (29,103 t $year^{-1}$), prescribed fire in New Jersey added an estimated 1.3 %, with a range of 0.9–2.1 % of total PM 2.5 emitted to the atmosphere across the state (Clark et al. 2010a).

Current research efforts involve conducting “fireflux” experiments by monitoring prescribed burns using an array of flux towers to measure turbulence and meteorological variables at multiple heights through the canopy in and around the burn block. These data will be used to estimate parameters for predictive smoke dispersion models, specifically CalPUFF, BLUESKY, and others used by the Eastern Area Modeling Consortium (http://www.ncrs.fs.fed.us/eamc/). This effort is part of a Joint Fire Science Program project to better understand and predict smoke dispersion from low-intensity fires. We also are continuing our research efforts with the USDA FS Fire and Environmental Research Applications Team, which is using consumption data from prescribed fires in the Pinelands to further develop models of forest floor and shrub consumption during prescribed fires (CONSUME 3.0).

A second set of currently unresolved questions involves the trade-offs between prescribed fires and wildfires in the context of carbon dioxide (CO_2) release to the atmosphere, and how both of these disturbances impact long-term rates of carbon sequestration by upland forests in the Pinelands. Our current research indicates that 24 prescribed fires conducted by the NJFFS from 2004 to 2009, where we collected preburn and postburn fuel loading data, released an average of 312 ± 138 g C m^{-2} to the atmosphere (Fig. 22.4). In contrast to these relatively minor losses due to prescribed fires, we estimated that the Warren Grove wildfire in May 2007 released approximately 874 ± 370 g C m^{-2} (Clark et al. 2010a). Following the initial combustion losses during prescribed burns, further CO_2 emissions from the forest floor were not enhanced appreciably over preburn levels during "fireflux" experiments conducted in 2006 and 2008. For example, the prescribed fire conducted at the Cedar Bridge flux tower in March 2008 resulted in an average combustion loss of 447 g C m^{-2} and killed aboveground stems of nearly all understory shrubs and scrub oaks. Following the prescribed fire, resprouting of pines and understory vegetation resulted in the rapid recovery of leaf area (Clark et al. 2010b, 2012). Excluding combustion losses, net CO_2 exchange for 2008 at the Cedar Bridge stand totaled 48 g C m^{-2} year^{-1}, a small net uptake of carbon. In 2009, leaf area had nearly recovered to pre-prescribed burn values, and daytime net CO_2 exchange rates are very similar to preburn levels, resulting in an annual uptake of 169 g C m^{-2} year^{-1}. Net CO_2 exchange had completely recovered by 2010, and averaged 194 g C m^{-2} year^{-1} from 2010 to 2012. For comparison, mature undisturbed stands sequester 100–200 g C m^{-2} year^{-1} (Amiro et al. 2010; Clark et al. 2010b; Pan et al. 2006), suggesting that burned stands begin to sequester C rapidly following prescribed fires.

The protection of homes and property in and around the margins of the Pinelands is an ongoing problem. Early on, Silas Little advocated 60-m-wide protection strips around structures and housing developments in the Pinelands. Currently, the NJFFS suggests using 30-m protection areas ("Do you have what it takes? 100 ft of defensible space!"; http://www.state.nj.us/dep/parksandforests/fire/aboutus.html) and has devoted considerable effort to educate homeowners as to proper fuel management techniques. However, the amount of impervious surfaces and lawns or other landscaping around homes impacts peak flows in streams during storms, and where these plantings require N and P fertilizers, they potentially degrade water quality (e.g., Dow 2007). One of us (NS) is working with the Center for Remote Sensing and Spatial Analysis at Rutgers University to quantify trade-offs between fuel reduction in the WUI and impacts on surface waters.

Future research efforts using the tools and monitoring network developed at the Silas Little Experimental Forest will involve the interactions of invasive insects, fire, and climate change. Although timber harvesting now occurs only on a very limited basis in the Pinelands National Reserve, disturbances from insect infestations, intense storms, and wildfire may increase with climate change. One new threat that will likely have a large impact on pine-dominated stands in the Pinelands is the northern migration of southern pine beetle, recently detected in southern New Jersey (http://www.state.nj.us/dep/parksandforests/forest/njfs_spb.html). Gypsy moth defoliation has already resulted in substantial oak mortality and an increase in coarse fuels on the forest floor. Mortality due to other insects has the potential to increase wildfire intensity, which may be further accelerated by a changing climate.

References

Adams MB, Loughry L, Plaugher L (Comps) (2004) Experimental Forests and Ranges of the USDA Forest Service. General Technical Report NE-321, USDA Forest Service, Northeastern Research Station, Newtown Square, 178 p

Amiro BD, Barr AG, Barr JG, Black TA, Bracho R, Brown M, Chen J, Clark KL, Davis KJ, Desai AR, Dore S, Engel V, Fuentes JD, Goulden ML, Kolb TE, Lavigne MB, Law BE, Margolis HA, Martin T, McCaughey JH, Montes-Helu M, Noormets A, Randerson JT, Starr G, Xiao J (2010) Ecosystem carbon dioxide fluxes after disturbance in forests of North America. J Geophys Res 115:G00K02. doi:10.1029/2010JG001390

Battye W, Battye R (2002) Development of emissions inventory methods for wildland fire. 'Final Report Contract 68-D-98-046, US Environmental Protection Agency.' (Research Triangle Park, North Carolina)

Beck HC (1961) Forgotten towns of Southern New Jersey. Rutgers University Press, New Brunswick, 278 p

Buell MF, Cantlon JE (1953) Effects of prescribed burning on ground cover in the New Jersey pine region. Ecology 34:520–528

Burns PY (1952) Effect of fire on forest soils in the Pine Barren region of New Jersey. Yale Sch For Bull 57:1–50

Clark KL, Skowronski N, Hom J, Duveneck M, Pan Y, Van Tuyl S, Cole J, Patterson M, Maurer S (2009) Decision support tools to improve the effectiveness of hazardous fuel reduction treatments. Int J Wildland Fire 18:268–277

Clark KL, Skowronski N, Gallagher M, Heilman W, Hom J (2010a) Fuel consumption and particulate emissions during fires in the New Jersey Pinelands. In: Proceedings of the 3rd Fire Behavior and Fuels Conference, October 25–29, 2010, Spokane, Washington, USA. International Association of Wildland Fire, Birmingham, Alabama. http://www.nrs.fs.fed.us/pubs/38885

Clark KL, Skowronski N, Gallagher M, Renninger H, Schäfer K (2012) Effects of Invasive Insects and Fire on Forest Energy Exchange and Evapotranspiration in the New Jersey Pinelands. Agricultural and Forest Meteorology 166-167:50-61.

Clark KL, Skowronski N, Hom J (2010b) Invasive insects impact forest carbon dynamics. Glob Chang Biol 16:88–101

Dow CL (2007) Assessing regional land-use/cover influences on New Jersey Pinelands streamflow through hydrograph analysis. Hydrol Process 21:185–197

Forman RTT (ed) (1979) Pine Barrens: ecosystem and landscape. Academic Press, New York, 601 p

Forman RTT, Boerner RE (1981) Fire frequency and the pine barrens of New Jersey. Bull Torrey Bot Club 108:34–50

Garrett PW (1981) The Northeast pitch x loblolly hybrid program. In: Gusies RP, Kang HC (eds) Research needs in tree breeding. Proceedings of the 15th North American Quantitative Forestry Genetics Group Workshop. Coeur d'Alene, Idaho. NC State University, College of Forest Resources, Raleigh, pp 71–79

Lathrop R, Kaplan MB (2004) New Jersey land use/land cover update: 2000–2001. New Jersey Department of Environmental Protection, Trenton, 35 p

Little S (1945) Influence of fuel types on fire danger. J For 43:744–749

Little S (1946) The effects of forest fires on the stand history of New Jersey's pine region. Forest Management Paper No. 2, USDA Forest Service, Northeastern Forest Experiment Station, Upper Darby, 43 p

Little S (1950) Ecology and silviculture of white cedar and associated hardwoods in southern New Jersey. Yale Sch For Bull 56:103

Little S (1959) Silvicultural characteristics of Atlantic white-cedar (*Chamaecyparis thyoides*). Station Paper 118, USDA Forest Service, Northeastern Forest Experiment Station, Upper Darby, PA, 16 p

Little S (1973) Eighteen-year changes in the composition of a stand of *Pinus echinata* and *P. rigida* in southern New Jersey. Bull Torrey Bot Club 100(2):94–102

Little EL (1979a) Checklist of United States trees (native and naturalized). Agriculture Handbook 541. USDA Forest Service, Washington, DC

Little S (1979b) Fire and plant succession in the New Jersey pine barrens. In: Foreman RTT (ed) Pine Barrens: ecosystem and landscape. Academic Press, New York, pp 297–314

Little S, Moore EB (1945) Controlled burning in south Jersey'0146s oak-pine stands. J For 43:499–506

Little S, Moore EB (1949) The ecological role of prescribed burns in the pine-oak forests of southern New Jersey. Ecology 30:223–233

Little S, Somes HA (1951) Age, origin, and crown injuries affect growth of South Jersey pines. Research Note 8, USDA Forest Service, Northeastern Forest Experiment Station, Upper Darby, 4 p

Little S, Somes HA (1956) Buds enable pitch and shortleaf pines to recover from injury. Research Paper 81, USDA Forest Service, Northeastern Forest Experiment Station, Upper Darby, 14 p

Little S, Somes HA (1964) Releasing pitch pine sprouts from old stools ineffective. J For 62:23–26

Little S, Somes HA (1965) Atlantic white-cedar being eliminated by excessive animal damage in south Jersey. Research Note NE-33, USDA Forest Service, Northeastern Forest Experiment Station, Delaware, 3 p

Little S, Allen JP, Moore EB (1948) Controlled burning as a dual-purpose tool of forest management in New Jersey's pine region. J For 46:810–819

Lutz HJ (1934) Ecological relations in the pitch pine plains of southern New Jersey. Yale Univ Sch For Bull 38:1–80

McCormick J, Jones L (1973) The Pine Barrens: vegetation geography. Research report number 3, New Jersey State Museum, Trenton, NJ, 76 p

McPhee J (1968) The Pine Barrens. Farrar, Straus and Giroux, New York, 157 p

Mounier RA (1997) Black and dirty work: archaeology amidst the relict charcoal kilns of Southern New Jersey. Bulletin 52, Archaeological Society of New Jersey, 113 p

Neill C, Patterson WA III, Crary DW Jr (2007) Responses of soil carbon, nitrogen and cations to the frequency and seasonality of prescribed burning in a Cape Cod oak-pine forest. For Ecol Manage 250:234–243

Pan Y, Birdsey R, Hom J, McCullough K, Clark K (2006) Improved estimates of net primary productivity from MODIS satellite data at regional and local scales. Ecol Appl 16:125–132

Pearce JE (2000) Heart of the pines: ghostly voices of the Pine Barrens. Batsto Citizens Committee, Batsto, 904 p

Schäfer KVR, Clark KL, Skowronski N, Hamerlynck EP (2010) Impact of insect defoliation on forest carbon balance as assessed with a canopy assimilation model. Glob Chang Biol 16:546–560

Section Forest Fire wardens of Division B (2006) New Jersey Forest Fire Service. Arcadia Publishing, Mount Pleasant 127 p

Skowronski N, Clark K, Nelson R, Hom J, Patterson M (2007) Remotely sensed measurements of forest structure and fuel loads in the Pinelands of New Jersey. Remote Sens Environ 108:123–129

Skowronski NS, Clark KL, Duveneck M, Hom J (2011) Three-dimensional canopy fuel loading predicted using upward and downward sensing LiDAR systems. Remote Sens Environ 115:703–714

Wacker PO (1979) Human exploitation of the New Jersey Pine Barrens before 1900. In: Foreman RTT (ed) Pine Barrens: ecosystem and landscape. Academic Press, New York, pp 3–23

Wright CS, Ottmar RD, Vihnanek RE (2007) Stereo photo series for quantifying natural fuels. Volume VIII: Hardwood, pitch pine, and red spruce/balsam fire types in the Northeastern United States. PMS 840. Boise, ID: National Wildfire Coordinating Group, National Interagency Fire Center 91 p

Part VII
Intersite and Network Trajectories

Chapter 23
The Key Roles of Four Experimental Forests in the LTSP International Research Program

Robert F. Powers, David H. Alban, Robert Denner, John D. Elioff, Gary O. Fiddler, Deborah Page-Dumroese, Felix Ponder, Allan E. Tiarks, Peter E. Avers, Richard G. Cline and Nelson S. Loftus

Abstract Four Experimental Forests were pivotal in piloting the long-term soil productivity (LTSP) cooperative research program—one of the most successful and extensive collaborative science efforts yet undertaken by the USDA Forest Service. Launched on the Palustris, Challenge, Marcell, and Priest River Experimental Forests, LTSP traces to a seminal discussion during a field tour in central Louisiana in 1986. P. E. Avers, National Soils Program Leader in Washington DC, described to D. H. Alban and R. F. Powers a problem arising from the National Forest Management Act of 1976 (NFMA). That conversation sparked an idea that quickly caught fire. This chapter documents how LTSP came to be and why four Experimental Forests were central to its success. It began with a ripple effect of the NFMA.

Keywords Challenge Experimental Forest · Marcell Experimental Forest · Palustris Experimental Forest · Priest River Experimental Forest · Long-term research · Sustainable productivity

---

Robert F. Powers and Felix Ponder Jr. are deceased.

R. G. Cline (✉)
703 Kingsley Rd., S.W., Vienna, VA 22180, USA
e-mail: rgcline@verizon.net

R. F. Powers
Pacific Southwest Research Station, USDA Forest Service, 3644 Avtech Parkway, Redding, CA 96002, USA

D. H. Alban
Northern Research Station, USDA Forest Service, 1831 Hwy 169 East, Grand Rapids, MN 55744, USA

R. Denner · D. Page-Dumroese.
Rocky Mountain Research Station, USDA Forest Service, 1221 South Main Street, Moscow, ID 83843, USA
e-mail: bdenner@fs.fed.us

D. Page-Dumroese
e-mail: ddumroese@fs.fed.us

J. D. Elioff
Northern Research Station, USDA Forest Service, 1831 Hwy 169 East, Grand Rapids, MN 55744, USA

D. C. Hayes et al. (eds.), *USDA Forest Service Experimental Forests and Ranges*,
DOI 10.1007/978-1-4614-1818-4_23, © Springer New York 2014

23.1 Background

23.1.1 The Challenge of NFMA

In 1989, the annual meeting of the American Society of Agronomy (ASA) was held in New Orleans. One of ASA's tri-societies is the Soil Science Society of America, and its S-7 Division then was called "Forest and Range Soils." That Division's meeting included a field trip to tour sites of forest soils of interest in the region. Seated on a tour bus behind Forest Service researchers Robert Powers and David Alban was Peter Avers, National Soils Program Leader for the administrative arm of the Forest Service. At one point, Peter leaned forward and remarked that his national program needed help from Forest Service Research.

The problem that Avers described concerned language in National Forest Management Act (NFMA) that echoed earlier mandates in the Multiple-Use Sustained-Yield Act of 1960 and the Natural Resources Planning Act of 1974. Namely, that public resources should be managed *without impairment of the productivity of the land* (USDA Forest Service 1993). The essence of NFMA was that the Forest Service should conduct research, monitoring, and assessment to ensure sustained yield in perpetuity while protecting all resource values. This statement of public land ethic is remarkable in that it precedes by more than a decade the Dutch Soil Protection Act of 1987 and Australia's National Forest Policy Statement of 1992 (Nambiar 1996; Powers et al. 1998).

Provisions in NFMA called for the creation of a national committee to counsel and advise the Secretary of Agriculture during the development of National Forest planning regulations, and the Secretary asked the National Academy of Sciences for help. The eight-member Committee of Scientists, formed in 1977 and chaired by A. W. Cooper of North Carolina State University, forged a framework for implementing NFMA. Recommendations were finalized in 1979 and incorporated into the 1985 Code of Federal Regulations for Forest Planning. A notable

G. O. Fiddler
Pacific Southwest Research Station, USDA Forest Service, Region 5, 3644 Avtech Parkway, Redding, CA 96002, USA
e-mail: gfiddler@fs.fed.us

F. Ponder
Northern Research Station, USDA Forest Service, c/o Mrs. Mevelene Ponder, 3501 Woodland Drive, Columbia, MO 65202, USA

A. E. Tiarks
Southern Research Station, USDA Forest Service, 825352 McPherson Ave, Council Bluffs, IA 51503-5990, USA
e-mail: aetiarks@att.net

P. E. Avers
Avers Soil Surveys Inc., 1015 Cromwell Cove, Snellville, GA 30078, USA

N. S. Loftus
Washington Office, USDA Forest Service, 207 Horizon Dr., State College, PA 16801, USA
e-mail: bnloftus@yahoo.com

section required the Forest Service to monitor the effects of forest management prescriptions, including "significant changes in land productivity" (USDA Forest Service 1985)—a requirement preceding by more than a decade the "Criteria and Indicators of Sustainable Forestry" of the Montreal Process (Canadian Forest Service 1995) and the environmental surge toward "green certification" (Anonymous 1995). Thus, it is landmark. It also was a tall order because it demanded of Forest Service managers a level of resource knowledge absent in any single individual.

23.1.2 Defining the Problem

The Forest Service knew that unambiguous definitions were central to carrying out its monitoring charge. At the fore was a clear and objective definition of "land productivity." In the broadest sense, land productivity might mean the capacity of a site to produce a sustainable flow of a myriad of forest and range values. However, many values are difficult to assess because they are subjective. Accordingly, and with guidance from the US Office of General Council (OGC), a working definition emerged. *Land productivity* was interpreted as the carrying capacity of a site for sustaining growth of native vegetation. In turn, *carrying capacity* was seen as the average periodic dry matter production when the site was fully stocked, and analogous to net primary productivity in a mature forest community. Finally, *significant change* was defined as a reduced level of carrying capacity induced by management that could be detected within practicable levels of operational monitoring (USDA Forest Service 1987). Given the vagaries of climatic flux in dry matter production and realistic operational limitations, discussions with Land Management Planning and OGC led to a decision that productivity declines of 15 % or greater would have to occur to be measurable under operational conditions. And this became the working threshold for "significant decline" (USDA Forest Service 1987).

23.1.3 Monitoring What?

The question facing the USDA Forest Service was how to monitor possible productivity change. Measuring vegetation directly seemed logical but beset with problems. What if vegetation was not at a stocking level that reflected stability? What if current productivity reflected unknown conditions in the past, such as unusual storm damage or pest outbreaks? Departures of any given forest stand from a productivity baseline required both a solid knowledge of the baseline and the passing of enough time to detect a substantive change. Taking the lead in this, the Watershed and Air Management division of National Forest Systems (NFS) adopted a monitoring philosophy based on the following principles:

1. Management practices create soil disturbances.
2. Soil disturbances affect soil and site processes.
3. Soil and site processes control fundamental forest productivity.

Soil and site processes clearly govern fundamental productivity within the limits of climate and biology. But monitoring soil and site processes directly simply was not feasible, nor was our knowledge of baselines and processes perfect. The questions confronting soil scientists were how to determine baseline "soil productivity," and how to monitor management-caused changes to it. This is the concept since recognized as "soil quality" (Doran et al. 1994; Karlen et al. 1997; National Research Council 1993). Broad definitions of soil quality existed, but they were too qualitative and conceptual to be useful. Consequently, the Watershed and Air Management division proposed that a national template be developed. The template would list measurable soil properties linked to potential productivity. These properties would be monitored and thresholds set to indicate significant risks to productivity. Based largely on professional judgment, threshold standards were set regionally to alert managers of serious, possibly irreversible problems stemming from management practices.

23.2 Birth of a Plan

23.2.1 Tackling the Problem

Standards based on professional judgment are a good first step. But subjective standards are controversial and readily challenged. To be convincing, standards needed to be anchored in objective science. Powers, following a recent research sabbatical to New Zealand, had been particularly impressed by a small-scale experiment aimed at solving a similar problem (Murphy et al. 2004). Recognizing that the New Zealand experiment was additive and not factorial, and that it had several drawbacks, Powers saw the NFS predicament as a unique opportunity for a new research thrust along the lines of the New Zealand experiment, and over several months, he, Avers, and Alban discussed this further. Avers had scheduled a field review of the California Region's soils and watershed programs in spring, 1987, and Powers invited key scientists from the several Research Stations to attend. Forged at this meeting was the nucleus for a possible cooperative program. Returning home, scientists discussed the concept for a national program with their Station administrators. Directors were supportive of the grassroots concept, but advised that coordination with the Washington Office (headquarters in Washington DC) was essential in launching a program of this scope. Subsequently, administrative leaders in Forest Management and Soil staffs, both in Research and Development (R&D) and the NFS, were briefed and a cadre of Avers, R. G. Cline, R. O. Fitzgerald, and N. S. Loftus Jr. from the Washington Office, plus Alban, Powers, R. E. Miller, A. E. Tiarks, and C. G. Wells from the North Central, Pacific Southwest, Pacific Northwest, Southern, and Southeastern Research Stations, respectively, assembled in St. Louis, MO, in 1988 to discuss an approach.

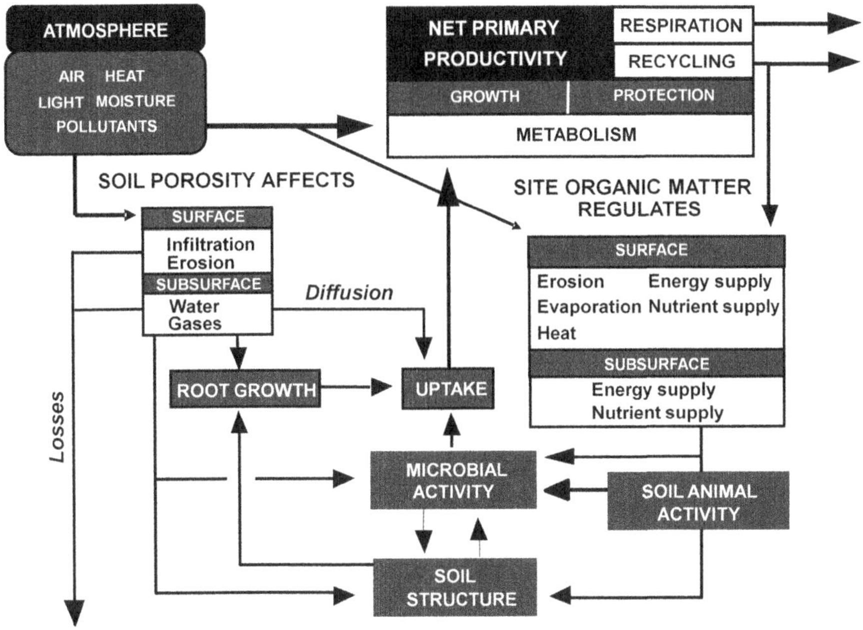

Fig. 23.1 Conceptual model of how soil porosity and site organic matter regulate net primary productivity within the framework of environmental controls. (from Powers et al. 1990)

The St. Louis meeting had several aims. We needed to clarify issues stemming from NFMA and to reach consensus on how to address the issues in a way that would further both the management and research arms of the Forest Service. We agreed that the scientific basis for soil quality monitoring was shaky. Some of what helped shape professional judgment was merely opinion and much of the remainder rested on anecdotal evidence. We agreed that another round of expert opinion might help in the short run. But what truly was needed was an objective and robust series of original experiments to provide the scientific foundation for sound policy. Thus, a second aim was to develop consensus on, and commitment for, an entirely new cooperative research approach.

Apportioning literature review among the researchers, several weeks were spent examining published evidence purporting to show productivity decline caused by management impacts on soil. We concluded that most claims could be attributed to other factors. However, two soil characteristics with promising linkages to lowered productive capacity were losses in soil organic matter and soil porosity—properties governing fundamental site processes that regulate net primary productivity within a climatic and genetic framework (Fig. 23.1).

Based on this conceptual model, we drafted an experimental design to test the following null hypotheses (Powers 2006):

1. Pulse changes in site organic matter and (or) soil porosity do not affect the sustained productive potential of a site to capture carbon and produce phytomass.

2. If productivity changes do occur, they are universal.
3. Such changes are irreversible.
4. Plant diversity has no impact on the productive potential of a site.

23.2.2 Presenting the Concept

Our conclusions on purported productivity declines and our call for a new research approach were showcased at the 7th North American Forest Soils Conference held in Vancouver, British Columbia, in 1988. The paper received broad international attention and technical review, and it was published 2 years later in the conference proceedings (Powers et al. 1990).

Immediately following the 1988 conference, we prepared a plan for a nationally coordinated study to determine the long-term effects of soil disturbance on fundamental productivity. The Washington Office sent the plan for technical review to all Forest Service research silviculturists and soil scientists and their counterparts in Regional Offices, as well as individuals in other forestry research institutions both domestically and abroad. As such, it stands as the most widely reviewed research study plan in the history of the USDA Forest Service. Review comments were assembled in the Washington Office and final adjustments to the plan were made there by the authors. In 1989, the plan was presented to the Deputy Chiefs for R&D and the NFS and approved as a national study of long-term soil productivity (LTSP; Powers et al. 1989). Installation cost was to be covered from excess timber sale receipts administered from the Washington Office through the Forest Service Regional Offices.

23.3 The Hard Part Begins

23.3.1 Core Treatments

The LTSP study centered on commercial forest types on National Forests across the nation. Requirements were that sites must support young-mature forests typifying those under active timber management. Soils typical of each forest type and spanning a range of potential productivities would be identified, thus ensuring broad latitude in soil properties. Once sites were selected, treatments were to follow a standard template of three levels of site organic matter removal (commercial bole only, whole tree, and whole tree plus forest floor) crossed with three levels of soil compaction (none, moderate, and severe). Each treatment plot was to encompass at least 1 acre (0.4 ha) with an appropriate buffer that separated treatment plots from surrounding timber.

Plots were to be harvested by the most practicable means and regenerated with species characterizing the previous forest. Treatment plots were then to be split, one half receiving regular vegetation control (trees developing without competing vegetation), the other half allowed to regenerate to natural understory vegetation. In this simplicity, there is elegance, and a factorial matrix of 18 possible treatment combinations was created. Response was to be measured as dry matter production per unit area over time (net primary productivity). Split plot treatments afforded two measures of productivity: timber production free of competing vegetation and ecosystem production that included both trees and emerging understory. This twist allowed the testing of all hypotheses stated previously.

23.3.2 Just Enough Organization to Avoid Chaos

Participation in LTSP was voluntary, requiring only an agreement between a principle investigator from the Research Station and the chief soil scientist and silviculturist in the participating Forest Service Region. National oversight was provided by four members of the Washington Office representing timber and soils interests—two from R&D and two from the NFS. Technical guidance was provided by a national committee consisting of principal investigators from Research Stations and the chief regional soil scientists from participating Forest Service administrative Regions. The Technical Committee was chartered to have at least one formal meeting annually led by an appointed chair, with the meeting venue shifting about the country. Regional guidance was secured through steering committees consisting of the research principal investigators from the participating Experiment Stations, plus the lead soil scientist and silviculturist for each participating Forest Service administrative region (initially regions 4, 5, 8, and 9). The Regional Committees identified the soil and forest types for focusing the LTSP experiment, encouraged cooperation among National Forests and Ranger Districts, and met at least annually to review progress, set priorities, and find solutions to emerging problems. Financial backing was secure. With this tidy level of organization, all that remained was to choose sites and implement the study.

23.3.3 Overcoming Inertia

The problem was how to begin. Rounds of discussion were held among the Technical Committee on how to choose sites and apply treatments. All agreed that the forest cover types and field sites would be identified by the Regional Steering Committees to span a gradient of soils and that actual approach in installing treatments would be at the discretion of the principal investigators. Individual ingenuity was freed to achieve an end product, rather than a means of getting there. Since there was no historical guideline for any of this, all of us agreed that the pilot installations would be on Experimental Forests under the control of the Southern, Pacific Southwest, North Central, and Intermountain Research Stations. Experimental Forests had the advantage of being under the jurisdiction of the Research Stations and treatments could be applied under categorical exclusion, thereby avoiding administrative

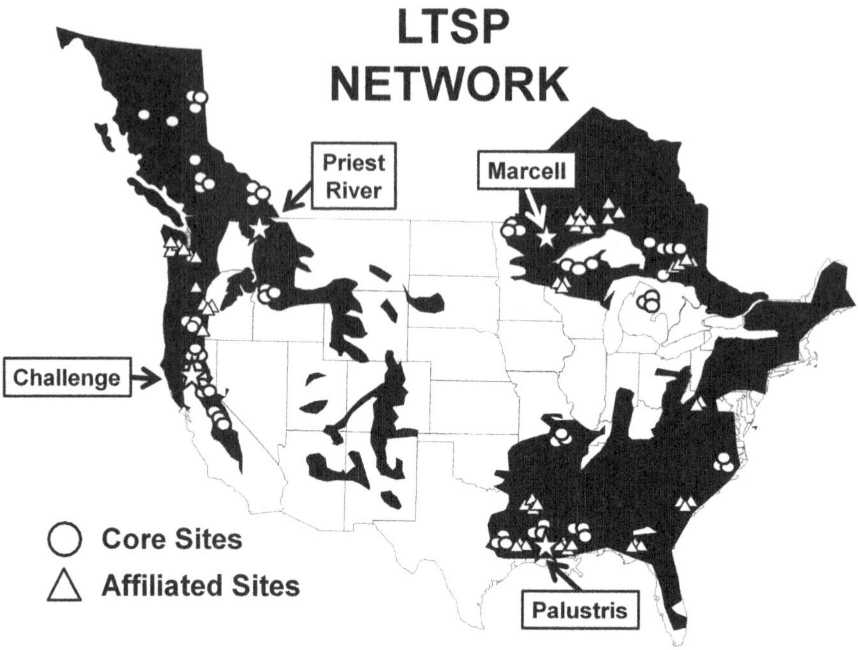

Fig. 23.2 Distribution of North American core and affiliate LTSP installations. Shading shows forest area capable of mean annual increments of 20 ft^3 $acre^{-1}$ of merchantable wood. The first four installations all were on USDA Forest Service Experimental Forests as indicated by stars

delays. They also had historically high visibility (making them useful for demonstration) and a close rapport with Ranger District and National Forest personnel. Finally, most Experimental Forests had lodging facilities that would reduce the cost of travel. Thus, they had tangible, practical advantages for piloting the program.

We discussed how best to get started, but we soon reached the point of diminishing returns on "what ifs." To break the gridlock, A. E. Tiarks boldly volunteered to be the guinea pig and establish the first installation on the Palustris Experimental Forest in Louisiana. The pioneering experiment was installed in 1990 under a memorandum of understanding among the Station Directors of the Southern and Southeastern Forest Experiment Stations (now combined into the Southern Research Station) and the Regional Forester of the Southern Region, USDA Forest Service. Challenge Experimental Forest in California was the second replication, and the Marcell in Minnesota and Priest River in Idaho soon followed. The Canadian government quickly became a partner, first through the British Columbia Ministry of Forests and then the Canadian Forest Service, each securing their own sources of funding. Affiliate installations also were recognized on public and private forest lands in Canada and the USA. In 2012, Mainland China's National Academy of Science also became an active partner following reciprocal visits. Today, more than 100 LTSP installations of core and affiliate sites have been established to create the world's largest network of studies aimed at understanding how management affects the fundamental productivity of forest land (Fig. 23.2). Most core installations are

Table 23.1 General characteristics of the first four Experimental Forests at the time of LTSP establishment. (Forest cover types follow Eyre (1980))

Experimental Forest (and State)	Palustris (Louisiana)	Challenge (California)	Marcell (Minnesota)	Priest River (Idaho)
Location	Kisatchie National Forest, Lat. 31.02 N, Long. 92.38 W	Plumas National Forest, Lat. 39.50 N, Long. 121.22 W	Chippewa National Forest, Lat. 47.30 N, Long. 90.30 W	Idaho Panhandle National Forest, Lat. 48° 21′ N, Long. 116° 41′ W
Elevation (feet)	150	2,765	1,411	2,040
Annual ppt. (inches)	46	67	30	33
Soil parent material	Coastal Plain sediments	Metamorphosed basalt	Des Moines lobe till	Lacustrine sediment overlain with volcanic ash
Soil family	Fine-loamy, siliceous, thermic Plinthic Paleudults	Mesic Typic Palexerults	Arenic Eutroboralf	Frigid, Andic Fragiudalf
Forest cover type	Loblolly pine (SAF Type 81)	Mixed-conifer (SAF Type 243)	Aspen (SAF Type 16)	Mixed-conifer (SAF Type 215)
Stand age (years)	37	108	70	90
Preharvest biomass (tons/acre)				
Trees	40.95	211.46	94.12	71.43
Understory	2.28	0.26	1.24	0.08
Forest floor	0.00	27.18	44.16	10.72
Year of establishment	1990	1991	1992	1992

on national and state forests, but expansion would have been doubtful had we not pioneered LTSP on four key Experimental Forests whose characteristics are shown in Table 23.1. Our trial and error experiences in establishing the experiment on these four forests provide insights into the best approaches for large-scale manipulative studies and demonstrate what is possible through a grassroots effort of determined scientists and administrators.

23.4 The First Installations

23.4.1 Palustris Experimental Forest

Managers from both the Southern Region and the Kisatchie National Forest agreed with Tiarks that the southern LTSP effort should center on the widespread loblolly pine (*Pinus taeda*) forest type and that the first site should be on Louisiana's Palustris Experimental Forest. Palustris was established in 1935 and draws its name from the virgin forests of *P. palustris* (longleaf pine) that once covered a vast area of the South. Climate is subtropical and typical of the Southern Coastal Plain, with rainfall spaced fairly evenly throughout the year. Like other Experimental Forests, the Palustris had long-term records, high visibility, and a good degree of protection. Furthermore, the Palustris had completed the National Environmental Policy Act (NEPA) public process for comment for harvesting. Because the installation required a uniform stand and soils of broad extent, site selection narrowed to one suitable stand recommended by Eugene Shoulders, a scientist familiar with its history. It was a good place to begin.

23.4.1.1 The Study Site

Shoulders recalled that the existing stand of loblolly pine had been established as a direct seeding demonstration area in 1953. Previously, the site supported old-growth longleaf pine, harvested in the 1930s with poor subsequent regeneration. It was burned regularly as prescribed in range management studies, so that there was little woody understory and negligible forest floor. Assisted by the National Forest soil scientist, a detailed soil survey was made in each 0.5- and 1-acre plot to ensure uniformity. The upper 12 in. of soil was sampled for bulk density and chemical properties at nine locations per plot. Diameters of all living pines were measured and three trees in each plot were selected for destructive biomass measurements. All the preharvest soil and vegetation samples had been collected by early fall 1989.

23.4.1.2 Administrative Hurdles

Discussions with the silviculture, sales, and purchasing personnel in the Kisatchie Forest Supervisor's office determined that the best logging approach was through two timber sales. The first sale of the surrounding buffer trees was straightforward, and in 30 days LTSP plots were isolated for further treatment (Fig. 23.3).

The second sale was tailored to plots with no compaction. By locating an access alley between the understory control and no understory control halves of the plots, logging equipment could lift trees from the uncompacted plots without disturbing the soil (Fig. 23.4). As the plots were logged, tops and limbs were removed or left on the plots depending on treatment. Close cooperation between sales administration and the logger ensured that the work was completed in late 1989 according to specifications. Remaining treatments were applied and the site was planted in February 1990.

23.4.1.3 Achieving Compaction

First attempts of soil compaction during the summer before harvest clearly showed the futility of compacting dry soils that have the property of "hard when dry." Thus, compaction began in early winter when logging was completed and soils were moist.

Fig. 23.3 Aerial view of the first LTSP site on the Palustris Experimental Forest showing logged buffers. This isolated individual plots for applying LTSP treatments. Note that the 1-acre units are divided, allowing equipment access

Fig. 23.4 Full-tree suspension of trees by a loader to avoid machine traffic and possible soil compaction at the Palustris Experimental Forest installation

Fig. 23.5 Tractor-drawn Wobble Wheel used in our first soil compaction effort. Note ballast for adjusting the Wobble Wheel load

A pneumatic tired compactor called a "Wobble Wheel" (Fig. 23.5) was adjusted to two different axle loads to achieve intermediate and severe levels of soil compaction. The axle loads were estimated from published literature and confirmed on small test areas in the buffer. The compactor, towed by a small crawler tractor, was chosen because its axles flexed to the uneven terrain, it allowed various loadings, and it was narrow enough to fit between tree stumps. It was towed across each plot three times in one direction and then three more passes perpendicularly. On plots where logging residues remained, limbs and tops were pulled aside to form a path during compaction and then replaced. Because the site had been burned regularly, there was no woody understory. The forest floor was thin enough that it did not interfere with the compaction which was completed before any forest floor removal.

The compaction treatment produced bulk densities averaging 1.47, 1.54, and 1.60 Mg m^{-3} in the surface 4 in. for the uncompacted, moderate, and severe treatments, respectively (Tiarks et al. 1991). Where logging slash was retained, materials were distributed evenly by hand. Where the thin forest floor was removed, hand equipment was used to mow and rake the grasses and herbs present. Materials were moved manually and stacked off the plot. Once the treatments were completed in February 1990, containerized loblolly seedlings were planted at 8 × 8-foot spacing. A weather station was erected on the site to collect temperature, rainfall, and several other parameters.

23.4.1.4 Findings

One striking early finding was that removing more than the main bole reduced new seedling growth substantially. This was attributed to the increased nutrient mass removed in tree foliage and understory and underscores the significance of surface residues to the nutrient cycle in humid climates and impoverished soils. These early results led both the National Forest and private land managers to limit whole tree harvesting whenever possible. None of the treatments reduced the soil organic carbon concentration within the top 180 cm after a full decade (Tiarks et al. 2004).

However, the porosity of the soil did increase as the soil carbon concentration increased, showing the importance of maintaining organic carbon within the soil profile in respect to soil physical properties.

As this was the first LTSP site, successful installation demonstrated to the cooperating scientists and National Forest managers that the study could be implemented as proposed to meet the criteria of the study plan. It also served to demonstrate the field design for LTSP to prospective cooperators in extending the experiment to other sites in the South. Visits to the Palustris pilot site by personnel from the forests in Mississippi and Texas overcame doubts about the potential returns from the long-term study.

23.4.2 Challenge Experimental Forest

We identified the Sierra Nevada mixed-conifer forest type as having the highest priority for soil disturbance research. Characterized by *Abies concolor, Calocedrus decurrens, Pinus lambertiana, P. ponderosa,* and *Pseudotsuga menziesii* in the north, this forest type is second only to *Sequoia sempervirens* in potential productivity in California, and ranks first in forest management issues. The first of our 12 replicate California installations would be at the Challenge Experimental Forest.

Located on the Plumas National Forest, Challenge was chartered in 1942 to tackle management questions concerning young-growth forests of the Sierra Nevada's west slope. Forests were even-aged, century-old stands that regenerated following clearcutting, railroad logging, and slash fires. Challenge was a prime candidate because it was managed by the Pacific Southwest Research Station's (PSW) unit with lead responsibility for LTSP and it typified low-elevation mixed-conifer forests of superior productivity, anchoring the upper end of the site quality spectrum. Challenge also is well-roaded, accessible year-round, and a convenient drive from our research offices in Redding. It attracts many tours that showcase silvicultural research.

23.4.2.1 Approach

PSW held an advantage. The California Region's Silvicultural Development Unit (SDU) was administratively assigned to PSW to assist with field operations and technology transfer. SDU personnel were experienced in inventory methods, timber sale contracts, and inspection of contracted operations. Temporary plots of 1 and 0.5 acres were established and sampled to describe spatial variability in edaphic and stand conditions, and those with similar spatial variation were accepted.

Experience at the Palustris Experimental Forest had little application here because Challenge vegetation and forest floor masses were greater by half an order of magnitude (Table 23.1). Therefore, we improvised new methods to meet our needs. Plot buffers needed to be as wide as the height of the bordering forest to minimize

Fig. 23.6 A half-acre treatment plot exposed by logging the surrounding buffer. Dominant tree heights approach 160 ft

edge effect, and these wide buffers more than doubled the area to be harvested (35 acres).Harvesting contracts for the buffer and main plots were advertised separately with the buffer contract awarded first. Special clauses required hand-felling and skidding to a landing without trafficking the treatment plots. In all, 1.125 million board feet were harvested in about 30 days, leaving treatment plots isolated for further work (Fig. 23.6).

We developed diameter-based biomass equations by sampling more than three dozen trees spanning the range of sizes and species (including understory hardwoods). Understory biomass and forest floor mass were measured by sampling randomized subplots of fixed area, and results appear in Table 23.1.

Harvesting operations were more complicated for the treatment plots. Large tree sizes, gentle slopes, and our ban on ground traffic posed a challenge that the logger met with ingenuity and improvisation. Trees were removed by full suspension, using a truck-mounted 60-ft tower and sufficient cable drums to serve as a live skyline yarder. Tail blocks were hung in large trees at the back of each plot and placed at least 90 ft high for adequate lift (Fig. 23.7). A practical advantage of having a single logger deal both with the buffers and the treatment plots was that potential tail block trees could be marked and left standing in the buffer to provide lift if no suitable trees were present at the back of the treatment plot. In all, 432 thousand board feet were removed from treatment plots in about 30 days.

Whole trees were removed by full suspension if plots were not to be compacted (Fig. 23.8). Slash was retained by delimbing the felled trees and removing boles. For compacted plots, slash was stored in the buffer until it could be returned following compaction. Through a separate service contract, hand crews raked forest floors onto tarps which were carried to the nearest buffer. A ten-person raking crew could complete 1 acre a day.

Compared with harvesting procedures, soil compaction operations were simple. The thick forest floor prevented appreciable compaction, so it had to be removed before compaction, then replaced and distributed uniformly. Compaction was applied

Fig. 23.7 To get adequate deflection for full suspension, tail blocks were set at least 90 ft above the ground

Fig. 23.8 Whole-tree removal was accomplished either by a shovel loader operating from the buffer as shown or by cable suspension. Both methods avoid vehicular traffic on the plots

with a Dynapac vibrating roller weighing 32,000 £ (Fig. 23.9). We established a test strip in one of the buffers to determine the number of roller passes needed to reach target bulk densities. Compaction proceeded in perpendicular passes on each plot at the rate of 1–2 acres per day, depending on the target density level. Forest floors and logging slash were respread at the rate of 2–3 acres per day through a special service contract.

Fig. 23.9 Achieving uniform soil compaction at Challenge through perpendicular passes of a vibrating roller

We planted species reflecting the composition of the dominant conifer overstory at the time of harvest. Given California's summer drought, we knew that survival hinged on first-year root growth. Accordingly, each planting spot was drilled to a depth of about 12 in. using a gasoline-powered auger. Bare root seedlings were planted at 8 × 8-foot spacing in an alternating arrangement of species (*Abies concolor, Pinus ponderosa, P. lambertiana, Pseudotsuga menziesii*). Survival is greater for *Pinus ponderosa* than for other species, so to ensure full stocking, each planting spot had two augered planting holes. One hole received the sequential species noted above, the other a *Pinus ponderosa.*

23.4.2.2 Subsequent Operations

Trees and emerging vegetation were left to develop for 3 years. Then, half of each 1-acre plot received glyphosate applied as a directed spray and *Pinus ponderosa,* the "insurance" species, was rogued if the other conifer from the mix of four seemed apt to survive. Otherwise, the pine was left. Plots have been remeasured every 5 years.

23.4.2.3 Important Findings

Challenge was not meant as a stand-alone experiment, but simply the first of a dozen California installations spanning the spectrum of soil types and site qualities in the mixed-conifer forest. Yet, several "firsts" emerged from this trial. Our pretreatment harvests at Challenge provided the first detailed assessment of mass and nutrient content of any California forest ecosystem (Powers 2002). Of the 6,807 lbs nitrogen contained per acre in vegetation and soil at Challenge, less than 8 % was in standing vegetation and only a small fraction of this (0.7 %) was in the understory. Although the forest floor mass was 12 % of the standing forest mass, it contained 43 % of aboveground nitrogen, illustrating its significance as a major nutrient reservoir and one that is affected readily by disturbances. We found that forest floor removal has a greater effect on reducing soil nitrogen availability than does whole-tree harvesting (Craigg 2006). Yet, productivity was not affected over the first decade on such fertile California soils (Powers et al. 2005), and we conclude that lesser biomass removals as in commercial thinning or whole-tree harvesting are not apt to affect the productive potential of a site as fertile as this.

Challenge showed us other positive benefits of the forest floor beyond its significance as a nitrogen reservoir. Average summer soil temperatures in the fine root zone were 8 °F cooler if a forest floor was present than if it was absent, and available soil moisture was extended 3 months longer—both critical factors affecting seedling survival and performance in a Mediterranean climate. Where soil cover was present, soil compaction had little effect on soil erosion rate measured over the first 3 years. But removing the forest floor accelerated soil erosion two- to eightfold (Powers 2002).

To our knowledge, the experiment at Challenge stands as the first rigorous testing of the modern recording soil penetrometer in North American forestry. Measurements taken in midsummer of the third year at Challenge showed that strengths in the fine root zone averaged 2 MPa or less where the forest floor was present, but a full MPa greater where it was absent, indicating how moisture evaporation from exposed soil increased the difficulty for root penetration (Powers and Fiddler 1997). Compaction increased soil strength by an average of 1.5 MPa during the growing season and conifer growth was reduced severely (Gomez et al. 2002a; Powers and Fiddler 1997). Experience gained at Challenge and expanded to other sites led us to recommend the recording penetrometer as an important tool for operational soil quality monitoring (Powers et al. 1998). Carbon research at Challenge and subsequent LTSP sites established that isotopic ratios of ^{13}C to ^{12}C were sensitive indicators of cumulative water stress from soil compaction (Gomez et al. 2002b).

The Challenge LTSP experiment was the first to demonstrate that competing vegetation can mask the fundamental effect of soil compaction. Because compaction can reduce the density and abundance of returning vegetation, auger-planted trees growing in compacted soils may find less competition for soil resources (Powers and Fiddler 1997). By eliminating competing vegetation, the detrimental effect of soil compaction on tree growth was apparent. Subsequent studies revealed that depending on soil texture and climate, moderate soil compaction may reduce, promote, or have little effect on tree growth because of changes in pore size and water retention (Gomez et al. 2002a; Powers et al. 2005)—possibly the first report of this in forestry, and one of international significance.

23.4.3 Marcell Experimental Forest

Established as a formal research site in 1962, the 898-acre Marcell Experimental Forest focused on both small-plot and watershed-scale research in several forest types of fire origin, including *Pinus banksiana* and *P. resinosa,* and mixed stands of *Picea mariana, Larix laricina, Betula* spp., and *Populus* spp. Granite and gneiss

Fig. 23.10 Litter removal "party" at Marcell Experimental Forest. *Left* to *right*: Peter Roussopoulos, David Lothner, and Donald Boelter, research administrators at the then North Central Research Station

underlay glacial tills of clay, sand, and gravel. Upland soils are sandy to clayey, and lowland soils are humic, including peatlands. Marcell is dominated by a continental climate with moist, warm summers and cold, dry winters with abundant sunlight. The site was adjacent to the first NADP site in the nation which had been operating since 1978. And, of course, the Experimental Forest had both the forest type (aspen) and soil type (moderate moisture holding capacity) we sought for beginning the study.

23.4.3.1 Approach

Because the study would require substantial commitment of personnel and capital for an extended period, we presented several seminars describing the proposal to all interested parties. Helpful in selling the program in the Lake States was that it was a nationwide effort and that all parties were interested in answering the basic questions of the impacts of soil changes on site productivity. We selected a unit of 70-year-old mixed hardwoods dominated equally by *Populus tremuloides* and *P. grandidentata*. Field treatments began in February 1991.

Our aspen site on the Marcell had several characteristics that simplified treatment operations. To minimize unwanted soil compaction, logging was done in winter when the soil was protected by 12 in. of frost capped with 17 in. of snow. Trees were leafless during harvest, so we were restricted to whole-tree removal and whole-tree plus forest floor removal. Reforestation was simplified because of the natural suckering of aspen, and it regenerates to such a high density that understory vegetation control was not considered. Spring was approaching, providing us a narrow window for installing treatments.

To consolidate support, we hosted a postharvest "litter removal party" on the Experimental Forest and our guests included several Station administrators (Fig. 23.10). Snow fell the night before litter raking began. It melted by midday, but the litter was left damp and heavy enough to tax office-softened muscles. Other attendees included representatives from the Forest Service Washington office, the Eastern Region office in Milwaukee, and others from the North Central Forest Experiment Station and all the forests on which future plots were to be established. The hard physical work and the evening discussions created enthusiasm for the study and helped establish "buy-in" and favor future cooperation and funding.

Soil compaction was accomplished using the "Wobble Wheel" method used at Palustris Experimental Forest, but breakdowns were frequent. The challenge faced at Marcell was keeping the Wobble Wheel working on our plots and out of the repair shop. Two unharvested plots were retained, and these are remeasured on the same schedule as our treated plots. Treatments were completed and the experiment was established on May 29, 1992.

The Marcell Experimental Forest was the pilot for techniques we applied over the next 3 years to replicated plots on other forests. At Marcell, we learned that compaction with a Wobble Wheel as used at Palustris was not very effective, and we shifted to a front-end loader at all other installations. Our Marcell experiment afforded good estimates of the effort needed to remove the forest floor from the other sites which previous to the Marcell experience was nearly impossible to estimate.

23.4.3.2 Important Findings

Our Marcell LTSP site was not meant as a stand-alone experiment, but merely the pilot in an extensive series throughout the aspen forest of the Lake States. Still, some results are notable. While compaction at Marcell significantly increased soil strength and soil bulk density, essentially no recovery was found in the first 5 years following treatment (Stone and Elioff 1998). By 15 years, some recovery had occurred, but soil strength and bulk density were still considerably greater than pretreatment levels. Total vegetation biomass 15 years after harvest showed that both forest floor removal and soil compaction had major effects, supporting those observed earlier (Alban et al. 1994). Plots with compaction or forest floor removal averaged 31 % less vegetation biomass than plots without these treatments.

23.4.4 Priest River Experimental Forest

Located on a western spur of Idaho's Selkirk Mountains, Priest River was established for research in 1911, placing it among the earliest Experimental Forests in the nation. Encompassing 2,400 acres, Priest River describes a cool, temperate forest with a modified maritime climate. Annual precipitation averages 32 in., falling mainly as winter snow. Soils include volcanic ash deposits over granitics and metamorphic Precambrian rocks of the Belt Series. We were presented a list of three candidate sites on the Experimental Forest that might meet our needs. All were even-aged, mixed stands of *Pinus monticola, Larix occidentalis,* and *Pseudotsuga menziesii* of harvest age. After inspecting these stands, we enlisted the aid of the

Priest Lake Ranger District Supervisory Forester and the Idaho Panhandle National Forest Soil Scientist. This group (District, Forest, Experimental Forest, and Research Scientist) reached consensus on an initial site after visiting each stand and conducting initial soil surveys.

23.4.4.1 Administrative Hurdles

We found that the experimental design and criteria for establishing plots with no compaction and removing slash and surface organic matter would be difficult to administer under current timber sale standards. The Supervisory Forester had to author a set of special provisions for the sale contract to ensure rigid adherence to the overall LTSP study design. This took considerable time at the Ranger District, but provisions were set soon enough to advertise the sale and begin logging in fall 1991.

23.4.4.2 Approach

The Priest River Experimental Forest site was ideal from a soils perspective. It is a level area adjoining the Priest River, has a deep (30 cm) volcanic ash-cap over lacustrine sediments, and had few coarse fragments in the surface soil. Because the soils were fairly uniform, we concerned ourselves with ensuring that the aboveground components were also fairly uniform. During the summer of 1991, temporary 1-acre plot boundaries were established in the areas that had similar tree diameter distributions and understories. Soil samples were collected on a 66-ft (20 m) grid to a depth of 12 in. (30 cm). Understory vegetation was characterized by type and samples were collected for total biomass. *Armillaria* spp. root disease also was a concern, so all plots were surveyed preharvest, and one treatment plot had all stumps pulled to see if we could rid that plot of *Armillaria* infection. Forest floor samples were collected at each soil sampling point and analyzed for dry mass and nutrient content. Five downed wood transects were completed in each treatment plot.

In early fall 1991, after all the preharvest soil and vegetation samples had been collected, work began to clear a landing/staging area and to improve the approach of the haul road to the county road. All buffers were harvested first and trees were felled directionally to avoid damaging a study plot. Buffers were harvested before study plots and slash was piled outside the plot boundaries where appropriate. Because harvesting equipment was excluded from plots, we established a skid trail down the center of each plot so that a grappler could reach and extract trees. Logging was halted when fall rains and snow made completion impossible without damage to the study plots. Logging resumed the following spring and the sale was completed.

The next hurdle following harvest was applying LTSP treatments. With some adjustments, surface organic matter removal was straightforward where compaction was intended. Our plan was to use a bulldozer to mechanically pile slash and surface organic matter off the plot, but we found that pushing materials with a dozer

Fig. 23.11 Both research and national forest personnel monitor logging operations. Close cooperation in monitoring is critical to success

blade gouged the soil and caused tracks to spin and mix the surface mineral soil. Thus, the operator was instructed to back drag the blade (traveling in reverse with the blade down) and this achieved the desired objectives (Fig. 23.11). The plot was then compacted and, if required, organic matter replaced with a grapple arm. On plots with organic matter removal but no compaction, we used the grapple arm to remove what could easily be reached and we hand-raked the rest. Materials were carried from the plot on tarps and then returned following compaction. Compaction level was monitored after each equipment pass so that we could achieve the levels of compaction necessary for the study. Stumps made uniform compaction difficult and the bulldozer had to drive closely on all sides to compact the soil.

Plot treatments were completed in late April 1992 and tree planting began in May. Researchers placed the trees on an 8 × 8-foot grid and seedlings were dibble planted through the organic matter. No "scalps" were created and in some cases, trees were planted directly into decayed (class V) logs. *Pinus monticola* and *Pseudotsuga menziesii* were double planted in pure blocks in each quarter acre. After 2 years, the inferior tree was rogued from the double plantings.

23.4.4.3 Subsequent Treatments

Naturally occurring understory vegetation was controlled with hexazinone on half of each treatment plot in the first year and double-planted trees were rogued in the third year by removing the least vigorous tree. Plots have been remeasured at regular 5-year intervals.

23.4.4.4 Important Findings

The Priest River installation was a single replicate and not meant to stand alone, but one clear finding was how readily the volcanic ash-cap soil could be compacted.

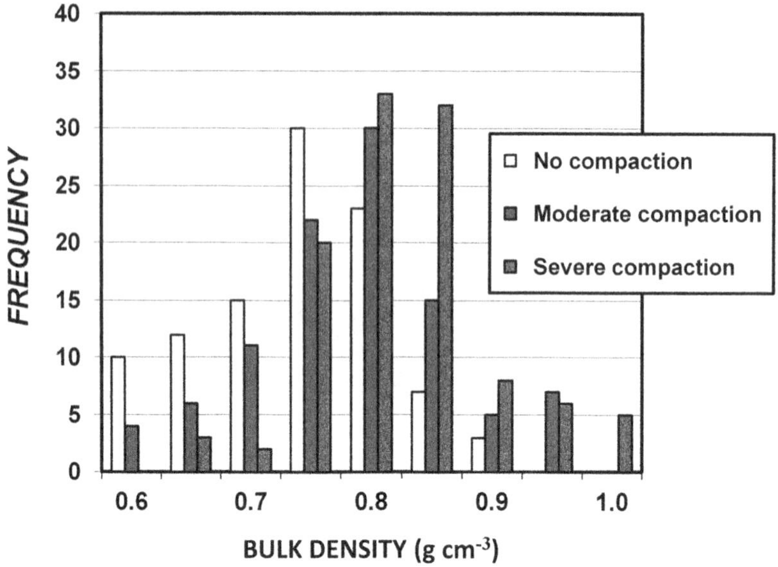

Fig. 23.12 Change in distribution of ash-cap soil bulk density in each compaction treatment at the Priest River Experimental Forest

Fewer than five passes were needed to achieve moderate compaction on this fine-textured soil, and fewer than ten passes achieved severe compaction. The ash-cap soil has a very low bulk density (0.6–0.7 Mg m^{-3}), and bulk density increased ~60% above preharvest conditions on severely compacted plots (Page-Dumroese et al. 2006). Work detailed the first intensive information on how the distribution of compaction (measured by bulk density cores) changes after deliberate compaction (Fig. 23.12), and established the soil-monitoring value of a recording soil penetrometer in ash-cap soils (Page-Dumroese et al. 2006; Johnson et al. 2007). Additionally, the LTSP installation on ash-cap soils was key to our finding of relatively minor short-term impacts on tree growth from compaction (Fleming et al. 2006) but large impacts on ectomycorrhizal colonization (Amaranthus et al. 1996). Data from this site also launched a northwest-wide workshop and proceedings on the properties and management of ash-cap soils (Page-Dumroese et al. 2007). Information from this installation has helped define the characteristics of ash-cap soils in revising regional soil quality standards.

The Priest River LTSP installation also was the first in a global series of studies that used a standard substrate (*Pinus taeda* and *Populus tremuloides*) in the mineral soil to evaluate compaction and organic matter removal impacts on belowground decomposition processes (Jurgensen et al. 2006). Current literature on litterbag studies suggested that clearcutting slowed decomposition on the soil surface and that unharvested stands had much faster surface decomposition rates. However, our work showed that decomposition accelerated in the mineral soil in the harvested plots, particularly the compacted plots, because—unlike soil surface conditions—

soil moisture was usually higher. In the uncut stand, mineral soil temperatures usually were cooler, thereby slowing decomposition rates (Collins 2009).

23.5 Lessons Learned

Our collective experience in establishing a large-scale field experiment using four Experimental Forests as trial horses led to agreement on points that should be helpful to future scientists contemplating manipulative experiments on federal lands and perhaps beyond. We group them into three categories.

23.5.1 Administrative and Philosophical Issues

- Experimental Forests have a major advantage over other federal lands because they are dedicated to research and demonstration, which may streamline the NEPA process.
- Success for a national network requires both vertical and horizontal buy-in from research scientists and the local Ranger Districts, vertically through research and administrative arms of the USDA Forest Service in Washington.
- "Grassroots" studies can expand to national and international scope if the concept is appealing, there is careful planning, and there is commitment from a cadre of individuals treated as peers.
- Innovation, rather than rigid conformity, is a critical element because it sparks creativity.
- To endure, experiments must have attributes robust enough to address concerns important now and in the future.

23.5.2 Technical Installation Issues

- Research must be involved directly with developing contracts for accomplishing the various tasks as well as overseeing all on-the-ground operations.
- Pretreatment sampling of prospective treatment plots is important. Watch for microrelief subtleties suggesting localized poor drainage once the existing forest is harvested and the transpiration pump is gone.
- To provide adequate buffers and staging area, the candidate area for an LTSP installation needs a total area twice to three times that needed for treatment plots.
- Buffers should surround a treatment plot. The minimum practical width is 20 ft. Buffers should be logged before the treatment plots.
- Daily inspection of contracted field work is critical.
- Submerchantable stems should be felled prior to felling large trees in a treatment plot. Felling large trees atop smaller trees cushions the impact, reducing unwanted compaction and ground damage during yarding operations.

- Directionally fell all large trees toward the nearest plot side by premarking the direction of fall and jacking trees as necessary to ensure this.
- To achieve the desired compaction levels, test strips must be established at each new site because soil types differ.
- Soil compaction cannot be achieved satisfactorily if logging slash or forest floors are present.
- Soil compaction can be followed with soil penetrometer measurements to quickly assess the uniformity of compaction.
- Slash and forest floor should be scalped at each planting spot to enhance seedling survival. Otherwise, air pockets may be created in the planting hole, leading to seedling mortality in a droughty climate.
- Herbicides are the most effective means for controlling unwanted vegetation in trees-only plots. Hand grubbing or mowing will be ineffective and will require yearly maintenance.

23.5.3 Issues Involving Subsequent Measurements

- If scheduled measurements cannot be accomplished with research personnel, they can be handled efficiently with service contracts, provided that (1) quality control standards are built into the contract; (2) there is continual Forest Service inspection; and (3) there is rapid turnaround in quality control of collected data.
- Use multiple-year contracts for data collection. This allows the same contractor to become familiar with standards and apply these same standards evenly to all sites.
- Contracts must include remeasurement penalties for inaccurate data. This illustrates the need for timely quality control.

23.6 Summary: The Pivotal Role of Four Experimental Forests

The Palustris, Challenge, Marcell, and Priest River Experimental Forests were pilots for LTSP. Unique treatments were applied in strange combinations, and we had no precedent for how to install them efficiently. Experimental Forests gave us the freedom and flexibility needed for innovation and daily adjustments, and offered unique advantages. Each Experimental Forest came with a charted history of treatment and forest development. Further, categorical exclusions for research of limited extent reduced the time and cost spent in the NEPA process. Unless we recognize the importance of this independence, the value of Experimental Forests will be lessened. Each forest had sufficient infrastructure (good roads and in several cases lodging, office, storage, and laboratory facilities) to expedite the work. And in all cases, Experimental Forests afforded administrative stability and exceptional protection from intrusion. They also have high visibility and draw many visitors each year from scientific, professional, and lay sectors of society. They are showcases for both past and ongoing research.

We proved that a grassroots effort by a small cadre of motivated individuals could expand to a major network if you believe strongly enough in the cause. The LTSP effort stands as an extraordinary example of collaboration between research and management arms of the USDA Forest Service at all levels. Our collective experiences gained on these four Experimental Forests set a research trajectory that has bridged affiliation boundaries and even political borders, and our efforts have influenced other programs. Coordination for the National Fire and Fire Surrogate Study (Weatherspoon 2000) was patterned on the successful organizational structure of the LTSP network. Internationally, the Center for International Forestry Research coordinated a sustainable productivity research network in short-rotation tropical plantations patterned on LTSP and a similar program in Australia. Core treatments involved several intensities of organic matter treatment (Tiarks et al. 1998). In common with LTSP, each site had a complete set of treatments that could be analyzed independently of other sites. And based on the experience of LTSP, workshops were held every 1–2 years to coordinate the research. Proceedings of these workshops were published by CIFOR, the latest in 2008 (Nambiar 2008).

Several major questions remain: (1) Will soil properties return to pretreatment levels? (2) How will soil compaction and organic matter removal, applied as a pulse disturbance, effect total biomass yield at the end of a rotation? (3) How do we maintain the essentials of a long-term experiment in the face of an economic crisis of unprecedented proportion?

Acknowledgments We are grateful to P. S. Aune (USDA FS, retired), C. B. Goudey (USDA FS, retired), R. E. Miller (USDA FS Pacific Northwest Research Station, retired), D. A. Scott (USDA FS Southern Research Station), and to R. O. Fleming (Canadian Forest Service) for reviews of early drafts and for their helpful comments. We also owe an immense debt to our colleagues on the ground and in Forest Service administrative offices who expedited installations and measurements and who collectively made this study possible. At several points in the manuscript, trade names are mentioned. The purpose is for accuracy and does not necessarily indicate endorsement by the USDA Forest Service.

References

Alban DH, Host GB, Eliof JD, Shadis D (1994) Soil and vegetation response to soil compaction and forest floor removal after aspen harvesting. Research Paper NC-315, USDA Forest Service, North Central Research Station, St. Paul, MN, 8 p

Amaranthus MP, Page-Dumroese DS, Harvey A, Cazares E, Bednar LF (1996) Soil compaction and organic matter affect conifer seedling nonmycorrhizal and ectomycorrhizal root tip abundance and diversity. PNW-RP-494, USDA Forest Service, Pacific Northwest Research Station, Portland, OR, 12 p

Anonymous (1995) The certified forest: what makes it green? J For 93:1–41

Canadian Forest Service (1995) Criteria and indicators for the conservation and sustainable management of temperate and boreal forests. The Montreal process. Canadian Forest Service, Natural Resources Canada, Quebec, 17 p

Collins A (2009) Impact of soil compaction and organic matter removal on wood decomposition rates in the mineral soil. MSc Thesis. Michigan Technological University, Houghton

Craigg TL (2006) Evaluation of methods used to assess changes in forest soil quality. MSc Thesis, University of California, Davis

Doran JW, Coleman DC, Bezdicek DF, Stewart BA (eds) (1994) Defining soil quality for a sustainable environment. SSSA Special Publication No. 35, Soil Sci Soc Am, Madison, WI

Eyre FH (1980) Forest cover types of the United States and Canada. Society of American Foresters, Washington, DC, 148 p

Fleming RL, Powers RF, Foster NW, Kranabetter JM, Scot DA, Ponder F Jr, Berch S, Chapman WK, Kabzems RD, Ludovici KH, Morris DM, Page-Dumroese DS, Sanborn PT, Sanchez FG, Stone DM, Tiarks AE (2006) Effects of organic matter removal, soil compaction and vegetation control on 5-year seedling performance: a regional comparison of long-term productivity sites. Can J For Res 36:529–550

Gomez A, Powers RF, Singer MJ, Horwath WR (2002a) Soil compaction effects on growth of young ponderosa pine following litter removal in California's Sierra Nevada. Soil Sci Soc Am J 66:1334–1343

Gomez GA, Singer MJ, Powers RF (2002b) Soil compaction effects on water status of ponderosa pine assessed through $^{13}C/^{12}C$ composition. Tree Physiol 22:459–467

Johnson LR, Page-Dumroese DS, Han H-S (2007) Effects of machine traffic on the physical properties of ash-cap soils. In: Page-Dumroese DS, Miller RE, Mital J, McDaniel P, Miller D (tech eds) Volcanic ash-derived forest soils of the Inland Northwest: properties and implications for management and restoration. RMRS-P-44, USDA Forest Service, Rocky Mountain Research Station, Fort Collins, pp 69–83

Jurgensen M, Reed D, Page-Dumroese D, Laks P, Collins A, Mroz G, Degorski M (2006) Wood strength loss as a measure of decomposition in northern forest mineral soil. Eur J Soil Biol 42:23–31

Karlen DL, Mausbach MJ, Doran JW, Cline RG, Harris RF, Schuman GE (1997) Soil quality: a concept, definition, and framework for evaluation. Soil Sci Soc Am J 61:4–10

Murphy G, Firth JG, Skinner MF (2004) Long term impacts of forest harvesting related soil disturbance on log product yield sand economic potential in a New Zealand forest. Silva Fenn 38:279–289

Nambiar EKS (1996) Sustained productivity of forests is a continuing challenge to soil science. Soil Sci Soc Am J 60:1629–1642

Nambiar EKS (2008) Site management and productivity in tropical plantation forests. Proceedings of multiple workshops in Piracicaba (Brazil) 22–26 November 2004 and Bogor (Indonesia) 6–9 November 2006. Center for International Forestry Research, Bogor, Indonesia

National Research Council (1993) Soil and water quality: an agenda for agriculture. National Academy Press, Washington, DC, 542 p

Page-Dumroese DS, Jurgensen MF, Tiarks AE, Ponder F Jr, Sanchez FG, Fleming RO, Kranabetter JM, Powers RF, Stone DM, Elioff JD, Scott DA (2006) Soil physical property changes at the North American Long-Term Soil Productivity sites: 1 and 5 years after compaction. Can J For Res 36:551–564

Page-Dumroese DS, Miller RE, Mital J, McDaniel P, Miller D (eds) (2007) Volcanic ash-derived forest soils of the inland northwest: properties and implications for management and restoration. RMRS-P-44, USDA Forest Service, Rocky Mountain Research Station Fort Collins, CO, 220 p

Powers RF (2002) Effects of soil disturbance on fundamental, sustainable productivity of managed forests. In: Verner J (comp.) The kings river sustainable ecosystem project. USDA Forest Service General Technical Report PSW-GTR-183, USDA Forest Service Pacific Southwest Research Station, Albany, CA, pp 63–82

Powers RF (2006) Long-term soil productivity: Genesis of the concept and principles behind the program. Can J For Res 36:519–528

Powers RF, Fiddler GO (1997) The North American long-term soil productivity study: progress through the first 5 years. In: Cooper SL (comp.) Proceeding of 18th annual forest vegetation conference, University of California Cooperative Extension, Redding, CA, pp 88–102

Powers RF, Alban DH, Ruark GA, Tiarks AE, Goudey CB, Ragus JF, Russell WE (1989) Study plan for evaluating timber management impacts on long-term site productivity: a research and national forest system cooperative study. Unpublished study on file with the USDA Forest Service, Washington, DC, 33 p

Powers RF, Alban DH, Miller RE, Tiarks AE, Wells CG, Avers PE, Cline RG, Fitzgerald RO, Loftus NS Jr (1990) Sustaining site productivity in North American forests: problems and prospects. In: Gessel SP, Lacate DS, Weetman GF, Powers RF (eds) Sustained productivity of forest soils. Proceedings of the 7th North American forest soils conference, Faculty of Forestry, University of British Columbia, Vancouver, BC, pp 49–79

Powers RF, Tiarks AE, Boyle JR (1998) Assessing soil quality: practicable standards for sustainable forest productivity in the United States. In: Adams MB, Ramakrishna K, Davidson EA (eds) Criteria and indicators of soil quality for sustainable forest productivity. Special Publication 53, Soil Sci Soc Am, Madison, WI, pp 53–80

Powers RF, Scott DA, Sanchez FG, Voldseth RA, Page-Dumroese D, Elioff JD, Stone DM (2005) The North American long-term soil productivity experiment: findings from the first decade of research. For Ecol Manage 220:31–50

Stone DM, Elioff JD (1998) Soil properties and aspen development five years after compaction and forest floor removal. Can J Soil Sci 78:51–58

Tiarks AE, Kimble MS, Elliott-Smith ML (1991) The first location of a national, long-term forest productivity study: methods of compaction and residue removal. In: Coleman SS, Neary DG (comps.) Proceedings of the 6th Biennial southern silvicultural research conference, vol 1. General Technical Report 70, USDA Forest Service, Southeastern Forest Experiment Station, Asheville, NC, pp 431–442

Tiarks A, Nambiar EKS, Cossalter C (1998) Site management and productivity in tropical forestry plantations. CIFOR Occasional paper 16, Center for International Tropical Reseach, Bogor, Indonesia, 11 p

Tiarks A, Elliott-Smith M, Stagg R (2004) Effects of logging residue management on the growth and nutrient distribution of a *Pinus taeda* plantation in central Louisiana, USA. In: Nambiar EKS, Tanger J, Tiarks A, Toma T (eds) Site management and productivity in tropical plantation forests. Proceedings of workshops in Congo July 2001 and China February 2003.Center for International Forestry Research, Bogor, Indonesia, pp 151–162

USDA Forest Service (1985) Code of Federal Regulations. Evaluation and monitoring. In: Title 36, part 219: Planning. Subpart A: National forest system land resource management planning. Federal Register, 36 CFR 219.12. USDA forest Service, Washington, DC pp 45–71

USDA Forest Service (1987) Soil quality monitoring. In: Soil Management Handboook 2509.18, Chap 2 (October). USDA Forest Service, Washington, DC

USDA Forest Service (1993) The principle laws relating to Forest Service activities. U.S. Government Printing Office, Washington, DC

Weatherspoon CP (2000) A proposed long-term national study of the consequences of fire and fire surrogate treatments. In: Proceedings Joint Fire Science Conference, Boise, ID, June 15–17, 1999. University of Idaho Press, Moscow, ID, pp 117–126

Chapter 24
Networked Science Among Experimental Forests and Ranges: Past Experience and a Vision for the Future

Douglas F. Ryan and Frederick J. Swanson

Abstract Over its long history, the collection of now more than 80 Forest Service Experimental Forests and Ranges (EFRs) has produced studies involving several EFRs and also participated in other monitoring and research networks. This legacy of networked science involving EFRs has included participation in environmental monitoring networks, cross-site experiments and syntheses, and multiagency research communities, such as National Science Foundation's Long-Term Ecological Research (LTER) Network. Participation in these networks has produced significant benefits, but has been limited to a small number of EFR sites and Forest Service scientists. A culture of idea and data sharing is a necessary, but not sufficient, ingredient for fostering a highly-functional, well-integrated EFR Network. Additional requirements include sustained leadership and technical, financial, and administrative support for network scientific activities. Some of the EFRs have high quality records of climate, hydrology, vegetation, and biogeochemistry spanning many decades, providing a rare opportunity to examine effects of environmental change on vital ecological services. Several pioneering syntheses on topics such as climate, streamflow, and streamwater chemistry have come from EFR and allied environmental research networks, and demonstrate the potential contributions of these networks for fruitful scientific work of importance to society.

Keywords Research network · Environmental monitoring network · Research syntheses · Cross-site experiments · Data sharing

24.1 Introduction

For most of their century-long history, Forest Service Experimental Forests and Ranges (EFRs) were selected and operated as stand-alone units to represent a broad array of forest and grassland types and conditions. In the first half century of EFRs,

D. F. Ryan (✉)
Pacific Northwest Research Station, USDA Forest Service, 3625 93rd Ave. SW, Olympia, WA 98512, USA
Phone: 360-753-7652
e-mail: dfryan@fs.fed.us

F. J. Swanson
Pacific Northwest Research Station, USDA Forest Service, 3200 Jefferson Way, Corvallis, OR 97331, USA

D. C. Hayes et al. (eds.), *USDA Forest Service Experimental Forests and Ranges*, DOI 10.1007/978-1-4614-1818-4_24, © Springer New York 2014

their scientists interacted primarily through informal or scientist-to-scientist ties. More formal linkages among sites developed after 1960. Some EFRs joined environmental monitoring networks; a number of cross-site experiments and science syntheses were conducted; and some sites participated in multiagency groups, such as the National Science Foundation's (NSF) Long-Term Ecological Research (LTER) Network, founded in 1980, with substantial Forest Service engagement in 6 of the 26 LTER programs. The recognized value of long-term studies and records of observations at EFRs, the emergence of broad-scale science questions related to environmental change, the inception of new research and monitoring networks, and the success of the LTER Network have prompted Forest Service leadership to promote the transition of EFRs from a collection of individual sites to a functional research network. This chapter focuses on the most recent era within EFRs' long history (see chapter by Shapiro in this volume) in which the Forest Service joined other agencies to push formation of continent-spanning research and monitoring networks (Coleman 2010). In many cases, these new networks have capitalized on the much longer history of EFR work and the EFR system has benefited from the expanded research scope and the new scientific talent that came from participating in networks.

We have examined the experience of EFRs in the constellation of environmental monitoring and science networks that they have participated in over the past several decades, in cross-site science syntheses, and in developing a data-sharing system to find what lessons may be useful for further forming EFRs into a research network. We also examined agency efforts to foster scientific interactions among scientists at multiple sites that would be needed to design and implement science experiments and analyses across EFRs. Finally, drawing on lessons learned from EFRs' experience doing networked science, we discuss what future directions would move EFRs toward achieving the goal of forming an integrated research network that can contribute to addressing large-scale environmental issues.

24.2 Efforts to Integrate Science Among EFRs

The concept of EFRs as a research network received major impetus in the Forest Service in 2004 when the deputy chief for research and development (R&D) proclaimed the formation of an EFR Network as a national goal for R&D. Several national-level efforts toward that goal followed, including publication of a national summary of EFR sites (Adams et al. 2008), establishment of a national EFR Web site (http://www.fs.fed.us/research/efr/) and business plan, formation of a national-level committee on EFR research, and formulation of an information management plan for EFRs. Important first steps have also been taken toward building a community among EFR scientists, including holding two national meetings for EFR lead scientists, and more frequent monthly meeting of an EFR Working Group with scientist-representatives from each research station to discuss issues that affect large segments of the EFR research community. Other activities, primarily led by site scientists taking the initiative to work together, have also built a variety of scientific linkages among EFRs (Adams et al. 2010). Many of these bottom-up initiatives predate the current national-level networking activities and have cross-site connections that could make important contributions toward the goal of a functional research network. We review what has been accomplished by existing networking approaches among EFRs to develop a vision of how to make EFRs into a more functional research network that both capitalizes on already existing connections and adds or strengthens needed network components that are currently either weak or do not yet exist.

24.2.1 *Why a Network?*

The purpose of forming EFRs into a more functional research network is to enhance their collective scientific capacity to address pressing, regional- to continental-scale environmental issues, such as predicting environmental effects on forests and grasslands of widespread influences, such as global climate change, altered atmospheric chemistry, and growth at the urban–wildland interface. Through better coordination and less duplication of efforts, a network could also make better use of the long-term investment in scientific research that the Forest Service has made at these sites. An information-sharing infrastructure to facilitate access to and synthesis of cross-site data is a necessary foundation for a research network.

24.3 Network Approaches and the Ties They Produce Among EFRs

Applying the term "network" to EFRs is complicated, because over time "networks" and activities to form them associated with EFRs have taken many forms. Scientists at EFRs have a long history of collaborating in formal and informal ways with many partners, including scientists at other EFRs, at universities, and in other agencies, and with land managers and policymakers (Swanson et al. 2010). EFR programs and scientists have participated in other networks; some of them interagency, such as the LTER Network and the National Atmospheric Deposition Program (NADP), and some international, such as UNESCO's Man and the Biosphere (MAB) Programme (see Table 24.1). Understanding the nested sets of networks within which EFRs operate helps to give perspective on what kinds of linkages different network approaches foster among sites and what additional actions may be needed to make EFRs a functioning research network. We broadly categorized networks in which EFRs have participated to illustrate the kinds of relationships each has created among its members, including participating EFRs.

Table 24.1 A selective list of EFR sites that engage in cross-site activities and some networks in which they participate

EFRs	Networks									
	LTER[a]	MAB[b]	NEON[c]	NADP[c]	Ameriflux[c]	LTSP[d]	LINX[d]	LIDET[d]	Clim/HydroDB[e]	Stream chemistry syntheses[f]
H.J. Andrews	x	x	x				x	x	x	x
Hubbard Brook	x	x		x			x	x	x	x
Coweeta	x	x					x	x	x	x
Bonanza Creek	x						x	x	x	x
Luquillo	x	x					x	x	x	
Baltimore	x								x	
Marcell				x		x			x	x
Fernow									x	x
Fraser		x							x	x
San Dimas		x								x
Tenderfoot									x	x
Santee									x	x
Wind River			x		x					
Olympic Experimental State Forest									x	
Caspar Creek									x	
Hawai'i				x						
GLEES				x	x					
Bartlett			x		x					

LTER NFS Long-Term Ecological Research Network, *MAB* UNESCO Man and the Biosphere Programme, *NEON* NSF National Ecological Observatory Network, *NADP* National Atmospheric Deposition Program, *LIDET* Long-term Inter-site Decomposition Experiment Team, *ClimHydroDB* Cross-site meteorological and hydrological data-sharing web portal, *LINX* Lotic Intersite Nitrogen eXperiment (http://www.faculty.biol.vt.edu/webster/linx//), *Stream Chemistry Syntheses* science syntheses of long-term stream chemistry monitoring at EFRs, *LTSP* Long-Term Soil Productivity study (http://forest.moscowfsl.wsu.edu/smp/ltsp/index.html), *Ameriflux* network of long-term atmospheric flux monitoring sites

[a] Integrated Research Network

[b] Network in name

[c] Environmental monitoring network

[d] Cross-site experimentation network

[e] Cross-site data sharing

[f] Cross-site science syntheses

24.3.1 Networks-in-Name

Networks-in-name provide recognition, but little or no significant network functionality for their members. For example, nine EFRs have been named as sites in the MAB Programme's World Network of Biosphere Reserves (http://www.unesco.org/new/en/natural-sciences/environment/ecological-sciences/man-and-biosphere-programme/), recognizing them as members of an international group of significantly intact ecosystems set aside for study, with some opportunities for experimentation. Membership indicates that these sites have high intrinsic value for understanding natural and managed ecosystems, and has produced occasional scientist-to-scientist contacts. However, MAB's programmatic and funding support for scientific exchange among its member sites, including participating EFRs, has been very limited.

24.3.2 Environmental Monitoring Networks

EFRs have participated for many years in a variety of environmental monitoring networks, primarily as sites for collecting observational data. Data gathered through these networks usually follow standard protocols, and data management, analysis, interpretation, dissemination, and publication are usually done by organizations that produce relatively little interaction among the participating sites, including EFRs. Monitoring networks with EFR members address physical parameters, such as the US Geological Survey (USGS) stream gauging system (National Water Information System, NWIS) and NADP, an interagency network measuring wet and dry atmospheric deposition, as well as biological components of ecosystems, such as the National Audubon Society's Christmas Bird Count and the USA National Phenology Network. In some cases, participation requires dedicated effort on the part of Forest Service personnel (e.g., to maintain and operate an NADP collection site), or little effort, when data are collected by partners (e.g., bird counts or phenology observations). These networks provide data that have been used in inter-EFR studies, but participation in most has not led to cross-site interactions among EFRs. A potential exception may be the NSF's National Ecological Observatories Network (NEON). Three EFRs have been selected as candidate "core sites" for intensive environmental monitoring that will be highly coordinated. Additional observational sites that may potentially include other EFRs are planned along selected environmental and/or land-use gradients, or as part of multisite, manipulative experiments. The number and strength of linkages that NEON will actually create among EFRs, however, will depend on details of implementation that have not yet been decided (see http://neoninc.org/).

24.3.3 Experimentation Networks

EFRs have been connected by scientists who have collaborated on cross-site experimentation and observational programs. The focus has usually been one-time or fixed-term studies, although some individual experiments may run for a long time, in some cases for decades. Some experiments are designed and contained within a subset of the EFR system and other studies have been organized by scientists partially or entirely outside the EFR system. In the Pacific Northwest, for example, university and Forest Service scientists initiated a three-site experiment examining root decomposition along the environmental gradient from moist-marine to dry-continental climate represented by the Cascade Head—H.J. Andrews—Pringle Falls EFs and a forest canopy gap manipulation experiment replicated at the Wind River and Andrews EFs. Other experiments have had wider participation, such as the extensive Long-Term Inter-site Decomposition Experiment Team (LIDET) study of fine litter decomposition, which spans 28 sites, including 5 EFRs, extending from the North Slope of Alaska to the Caribbean and Central America (see http://andrewsforest.oregonstate.edu/research/intersite/lidet.htm). Such cross-site experiments have engaged scientists in experimental design, data collection, analysis protocols, and interpretation and dissemination of results. Study outcomes can provide important insights into how ecological processes vary over environmental gradients at large geographic scales. However, most have focused on one-time experiments, and made relatively little provision for carryover of data or relationships to future cross-site experiments.

24.3.4 Integrated, Long-Term Research Networks

Integrated, long-term research networks are composed of study sites, collaborating research programs, and a social network of scientists from across sites in which participants are committed to sharing ideas and cooperating on multiple, long- and short-term projects. Linking science capacity among sites makes it possible to address network-scale questions spanning the extent of the network. Both LTER and EFRs are attempting to reach a high level of research network functionality, even though both were initially set up as a collection of sites rather than as a functioning network. Research networks combine aspects of environmental monitoring, experimentation, and synthesis. Critical ingredients for network science include a culture that encourages and rewards network science, support for network projects, management of collaborations that build and sustain a strong and open community, and infrastructure for cross-site investigations, especially information-sharing systems.

Interactions among EFRs have been fostered in some cases by their participation in research networks that have originated outside of the Forest Service. Participation of three EFRs in the NSF's International Biosphere Program (IBP) in the 1970s produced only modest cross-site activity because IBP's emphasis was primarily on individual sites. Membership of six EFR sites in IBP's successor, the LTER Network, which began in 1980, has linked LTER member EFRs to one another as well as to the non-EFR sites within LTER. Beginning in the 1990s, NSF pushed LTER sites to function as a research network. LTER Network activities expanded to include several network-wide synthesis efforts, the beginnings of a network-wide information management system, and development of an LTER Network Office tasked with supporting network activities. The Forest Service EFR sites within LTER have shared leading roles in many LTER networking activities. At H.J. Andrews EF, for example, the Forest Service and its academic partner Oregon State University were especially active in building a dynamic capability to share climate and hydrological information across the LTER Network. Forest Service involvement in the LTER Network has linked the six EFRs that are formal members of LTER, and has in various ways also linked other non-LTER EFRs. The Forest Service–LTER partnership provides several examples of how participation of multiple institutions in a research network can provide mutual benefits.

24.4 EFR Network Accomplishments: Science Syntheses and Information-Sharing Infrastructure

Focusing on areas in which the EFR Network has produced significant accomplishments, we examine examples of cross-site science syntheses and EFR Network information-sharing infrastructure. Both of these activities have been the focus of considerable effort and are related because a primary purpose of information sharing has been to facilitate synthesis work. Although neither of these activities has reached the full potential that we envision could be achieved by an integrated EFR Network, they do illustrate some of the benefits that could be gained by more fully integrating EFRs into a network. Lessons learned in pursuing these activities indicate networking approaches that have worked as well as some remaining obstacles that will need to be overcome to reach the goal of an integrated network.

24.4.1 Cross-Site Science Syntheses

Science syntheses have knitted together existing evidence from multiple EFRs in *thought experiments* that are analogous, in some ways, to cross-site experiments based on new treatments and observations. Science syntheses draw together the weight of evidence from past studies or directly from analyses of long-term data to seek answers to new science, land management, or policy-related questions. Syntheses among EFRs have a long history (e.g., Lull and Sopper 1966), and some have provided the science basis for major land management and policy decisions (Lugo et al. 2006). For example, the Northwest Forest Plan (USDA/USDI 1994), which set management guidelines for 9.8 million ha (24 million acres) of Federal land in the Pacific Northwest, drew upon syntheses of findings from several EFRs.

Earlier syntheses were usually one-time efforts that did not facilitate follow-up studies. Multisite syntheses were also often meta-analyses of published studies with the drawback that analytical methods and assumptions often varied among studies. Initial attempts to apply common analytical approaches across all sites and data sets required onerous commitments of time and effort to find, assemble, and interpret data that were often managed differently at each EFR. Overcoming these practical issues was a formidable obstacle to analytically rigorous multi-EFR syntheses. However, the demand for science syntheses has increased as the scope and scale of environmental issues have expanded. As a result, more recent synthesis efforts have become better coordinated and have increasingly been supported by information management infrastructure designed to make data more readily available to answer new synthetic questions using the same analytical frame for all participating sites.

For example, a recent synthesis of carbon dynamics under the USDA Global Climate Change Program (Birdsey 2009) combines ground-based measurements of carbon stocks, forest growth, and climate from several EFRs and high-resolution measurements of carbon exchange between terrestrial ecosystems and the atmosphere at AmeriFlux sites (http://public.ornl.gov/ameriflux/) with wide-area land-use information from remote sensing and forest inventory data. Participating investigators at the Bartlett EF in New Hampshire, Marcell EF in Minnesota, Fraser EF in Colorado, and Glacial Lakes Ecosystem Experiments Site (GLEES) in Wyoming are contributing to this effort linking large-scale monitoring to local-scale land management decisions. This project combines environmental monitoring (AmeriFlux and Forest Inventory and Analysis data) with cross-site experimentation (vegetation plots and experimental methods) and synthesizes across sites using models. Information gathered was assembled into a database that will facilitate future syntheses.

Two capabilities will need to be added for EFRs to function as an integrated research network: the ability to seamlessly link long-term research data with associated explanatory documentation (metadata) and the formation of a strong social network among scientists from across EFR sites to facilitate network science. In the following sections, we discuss what progress has been made and what lessons have been learned by recent efforts to create data-sharing infrastructure among EFRs and what work has been done and what remains to be accomplished in developing a viable cross-site scientific community.

24.4.2 Cross-Site Data-Sharing Infrastructure

The LTER Network began data sharing among EFRs in the 1990s, when it asked all participating LTER sites, including the EFR members, to make their long-term site meteorological data available on the Web. A cross-site synthesis of these data from all LTER sites was undertaken to examine trends for the period 1980–1990 (Greenland et al. 1997). That initial synthesis encountered several obstacles to gathering and interpreting information, even though the relevant data had been nominally made available on the Web for each individual LTER site. Data were often incompatible across sites for reasons such as nonuniform data formats, units of measure, or sampling frequency, and/or inconsistencies of metadata, which explain data collection, formatting, analysis protocols, and site characteristics. To address these shortcomings, information managers at the H.J. Andrews EF, with NSF funding, led the design of a Web-based, cross-site, data-sharing system named "ClimDB" to assemble long-term site climate data using an approach called "Web harvesting" (Henshaw et al. 1998). Content and format were guided by consensus developed at meetings of lead scientists and information managers from across the LTER Network. This group agreed that local control of data was essential while still providing both a cross-site repository of consistent data and metadata and a single portal that serves users by making the data easily accessible with search, display, and download capabilities. Participation in ClimDB was open to any long-term study site, including EFRs, regardless of whether they were an LTER member.

Forest Service R&D was interested in building a cross-site data-sharing capacity for long-term EFR streamflow data based on the Web harvesting approach used in ClimDB. An initial effort to synthesize streamflow data from six EFR sites (Post and Jones 2001) had encountered difficulties locating and interpreting data from multiple sites similar to those encountered in the initial LTER climate synthesis. Gathering streamflow data from six EFRs took almost 3 years to complete, although these archived data were nominally publicly available. To address these shortcomings, in 2000 the Forest Service funded "HydroDB", a Web harvester to make EFR cross-site hydrologic data and metadata available on the web. Modeled after ClimDB, the content and format of HydroDB were designed by consensus among lead scientists and information managers from several interested EFRs. To get broad participation, Forest Service R&D provided funds to individual EFRs to offset costs to enter data and metadata in HydroDB, and NSF's LTER program funded similar efforts at LTER sites to encourage them to join. Subsequently, HydroDB was merged with ClimDB to form a unified web portal at the H.J. Andrews EF named "ClimHydroDB" and operation of the combined central Web harvester and data warehouse was migrated to the LTER Network Office in 2010. As of this writing, ClimHydroDB participation has expanded to include a total of 45 sites that share long-term data including EFRs, LTERs, International LTER (ILTER), and other sites (see http://www.fsl.orst.edu/climhy/hydrodb/).

The Web harvester was an early approach within a whole class of possible future solutions that can and will facilitate data discovery and integration in the EFR Network. Although that larger class of solutions is appropriate when looking to information sharing for the future, a number of accomplishments of the Web harvester approach and lessons learned from its implementation give some indications of the potential benefits and obstacles for future information-sharing efforts for the EFR Network.

24.4.2.1 Accomplishments of the Web Harvester

Practical Benefits of the Web Harvester Individual EFRs have realized a number of "value added" benefits from participating in the Web harvester. EFRs' long-term data are more likely to be included in cross-site syntheses, providing opportunities for sites to demonstrate that their data are being used to address important, large-scale scientific or natural resource questions. For example, a new member of the EFR Network, the Olympic Experimental State Forest (OESF), was able to participate in a national-scale synthesis on climate change (Jones et al. 2012) in only its 3rd year as an EFR Network member because ClimHydroDB had made it possible for OESF to share its long-term site data. As a result, in a remarkably short time, Washington Department of Natural Resources, which administers OESF, was able to show significant progress toward one of its goals for joining the EFR Network: participation of OESF in cross-network science. Prior to joining the Web harvester, many EFRs did not provide convenient access to local long-term environmental data. The Web harvester offers templates for organizing long-term data and assembling appropriate metadata, as well as means to update and perform quality control checks on the data. Additionally, the Web harvester includes a tool kit that allowed scientists to easily search, download, and graphically compare both local and cross-site data. ClimHydroDB benefited participating sites by giving them a persistent Web presence from which site scientists and the public could access daily, monthly, and annual aggregations of site data. Leaders of several EFRs that joined the Web harvester commented that they had planned to eventually build local capabilities analogous to the Web harvester, but were spared the effort by joining. Federal agencies participating in ClimHydroDB were also complying with the E-Government Act of 2002 (http://www.archives.gov/about/laws/egov-act-section-207.html) that directs them to make federally funded data freely available to the public.

Dr. Robert Waide has used the Web harvester to teach a graduate/undergraduate course at University of New Mexico called "Ecosystem Dynamics." "We used Clim/Hydro DB in a lab exercise in which we asked students to obtain data on precipitation for a given period from any three LTER sites and graph the data together," said Waide. "What seems to be a straightforward task is more complicated than they realize, and few if any of them complete the task after an hour of work. We then tell them to try ClimHydroDB to complete the same task, and most are able to do so in a few minutes. It's a very good example of the value of this kind of data tool. I would think that this lesson could be applied in many educational settings."

24.4.2.2 Benefits to the EFR Network of ClimHydroDB

The large number and variety of intensive study sites that participate in ClimHydroDB provides a rich diversity of ecosystems from which to draw data for synthesis studies. Most participating research sites are locations of multiple ecological investigations, so their value for examining ecosystem processes and environmental trends is much greater than if only a small number of environmental variables were measured. As sites and data themes are added to the Web harvester, the number of relationships among environmental variables and locations that are potentially available for analysis expands geometrically.

The long-term data and metadata stored on ClimHydroDB are tapped by several groups of users. On average, 30 users per day are served by this publicly accessible site. Users have self-identified their purposes as 40 % for research, 35 % for education, and 25 % for exploring or participant testing (Suzanne Remillard, Oregon State University, pers. comm. 2010).

Although education was not the primary purpose for developing ClimHydroDB, the site has received substantial educational use. Among the 35 % of users who self-identified their use as educational, 90 % were students. For example, students in the NSF's Research Experience for Undergraduates (REU) program at the H.J. Andrews EF have used ClimHydroDB in their individual research projects (Don Henshaw, USDA Forest Service, pers. comm. 2010), and university faculty have drawn on ClimHydroDB for content in undergraduate/graduate courses (see Text Box).

ClimHydroDB has formed or strengthened linkages among EFRs and with other environmental networks. Participation of 20 EFR sites in ClimHydroDB has expanded interactions of EFRs with the LTER Network considerably beyond the 6 EFRs that are formally LTER members. The Consortium of Universities for the Advancement of Hydrologic Science, Inc. (CUAHSI) is a research network that is planning to link ClimHydroDB to its Hydrologic Information System (CUAHSI-HIS), a Web-based research library of hydrologic data sources to facilitate large-scale syntheses on water resources (see http://www.cuahsi.org/).

Joint operation of ClimHydroDB with LTER sites has also brought new technical capabilities that give EFRs access to additional environmental networks. For example, the information manager at Georgia Coastal Ecosystems (GCE) LTER developed a computer application that provides data mining of the USGS NWIS real-time and historical streamflow data and transmission of these data into ClimHydroDB. Twelve participating sites, with a total of 65 USGS real-time gauges, have adopted the GCE application to automatically include their streamflow data in ClimHydroDB. When USGS recently changed the data format for NWIS, GCE adapted the application to the new format, which restored operations for all sites with little duplication of effort or disruption of data flow. Recently, GCE extended these capabilities to include weather data from the National Oceanic and Atmosphere Administration's National Climate Data Center sites.

Synthesis efforts that used data and metadata from ClimHydroDB have contributed to scientific advance. A comparative analysis of streamflow at six EFRs representing different roles of snow and mixes of deciduous versus evergreen forest cover revealed varying types and strengths of vegetation control on streamflow (Post et al. 1998), and led to the development of a new research field called ecohydrology. ClimHydroDB was identified in subsequent studies (e.g., Jones and Post 2004; Jones et al. 2012) as a tool that made possible development of this new cross-disciplinary research field by greatly expanding and facilitating access to long-term hydrology and climate data for synthetic analyses.

The Web harvester has also stimulated support from Forest Service R&D to incorporate an additional theme of long-term stream chemistry data from EFRs to address new, policy-relevant synthesis questions. Ten EFRs are contributing water quality data to answer a series of questions about how stream nutrients respond to forest harvest and other disturbances across North America (Johnson et al. 2009). Data and metadata being assembled are incorporated into an information-sharing Web portal called "StreamChemDB" that is currently being developed http://web.fsl.orst.edu/streamchem/. The forest products industry has supported this effort, which has the potential to provide critical data needed by state and federal agencies formulating water quality standards for forestry operations. A critical next step is to develop a "VegDB" for vegetation data for small, experimental watersheds where long-term climate, streamflow, and biogeochemistry data exist. This would make possible analysis of a suite of questions concerning effects of changing climate, atmospheric chemistry, and vegetation growth on ecological services, such as water quantity and quality, carbon sequestration, wood production, and biodiversity.

24.4.2.3 Elements Contributing to the Success of the Web harvester

Successful development and use of ClimHydroDB is the result of a combination of technical capacity, the needs of the science community, and the existence of incipient EFR and LTER Networks that could rally several dozen sites to participate (Henshaw et al. 2006). Given below are the critical factors in this development.

Scientific and Practical Interest A key point for persuading Forest Service and LTER leadership to support developing ClimHydroDB was the selection of themes with demonstrated utility for answering important synthesis questions of value to basic scientists, to land managers, and to policy makers.

Scientific Leadership Site-level science leaders gave their time and effort freely to the development of a consensus for the design of the Web harvester and made individual site personnel, time, data, and resources available to support site participation. Leadership by research administrators in both the Forest Service and the NSF was crucial in securing funding.

Technical Leadership Information managers from several sites, led by H.J. Andrews EF, brought computing and electronic networking capability to the project that could achieve user-friendly connectivity and search capabilities that were required by its scientific goals.

Incentives for Site Participation Incentives have encouraged voluntary participation of EFRs and LTER sites in ClimHydroDB. The Forest Service provided small grants for EFR site participation. Grants to individual sites to initially join HydroDB totaled about US$ 200,000 (approximately US$ 10,000 per site), which exceeded the initial software development cost of the central Web harvester of about

US$ 100,000. The LTER program likewise offered modest financial incentives to LTER sites to participate in ClimDB and later in HydroDB. These small grants to individual sites were an effective way to quickly expand voluntary participation.

Bottom-Up and Top-Down Development Approaches Both the Forest Service and LTER combined bottom-up and top-down approaches to developing ClimHydroDB. The scientific content and technical design were largely the result of bottom-up initiatives that drew upon the personal and professional interests of scientists and information managers from across EFR and LTER sites. Site personnel willingly worked together, developing consensus on the formats and content to which all sites conform to participate in the Web harvester. The Web harvester has benefited from the collective creativity of cross-site cooperation. Top-down leadership primarily consisted of concurrence on overall direction and provision of financial support. Experience developing ClimHydroDB illustrated that an appropriate mixture of both bottom-up and top-down leadership could achieve a collaborative network among a heterogeneous group of independent study sites. A contributing factor in achieving multiparty cooperation was the long experience of most partners working together toward a common goal of sustaining long-term ecological studies.

Interagency Cooperation NSF/LTER and the Forest Service took leadership of different phases and functions in ClimHydroDB development. For example, NSF/LTER funded initial development of ClimDB and the cost of LTER sites to join, while the Forest Service funded the initial development of HydroDB and the cost of EFRs to join. More recently, the Forest Service has funded initial development of StreamChemDB, while the LTER Network Office has assumed the operation of the central Web harvester function of ClimHydroDB. Contributions of both agencies have been largely complementary. Together they brought participants, funding, and technical capabilities to the enterprise more easily than either could have assembled on its own.

24.5 Discussion

Each of the various network approaches contains ingredients that could contribute toward the goal of an integrated EFR network (Fig. 24.1). Networks in name provide visibility and recognition. Environmental monitoring networks collect standardized observational data that can be used in cross-site analyses. Cross-site experimentation fosters cooperation among scientists from multiple sites and creates new observations and manipulations designed to address scientific questions at large scales. Integrated research networks have the potential to facilitate a richer suite of interactions and information sharing among scientists that crosses site boundaries. These networking approaches are not mutually exclusive and could be linked synergistically; for example, scientific cooperation within an integrated research network could encourage more active cross-site experimentation, cross-site syntheses, and other cross-site collaboration.

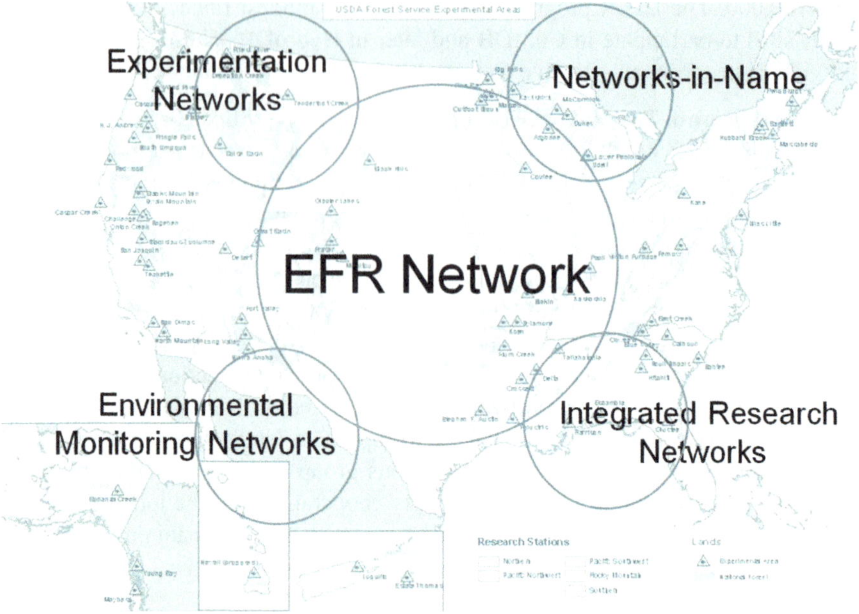

Fig. 24.1 An Experimental Forest and Range (*EFR*) Network links Forest Service long-term, dedicated research sites into an integrated research network and includes interactions with numerous other networks in which EFR sites participate, including networks in name, environmental monitoring networks, experimentation networks, and externally centered integrated research networks.

However, for achieving an integrated EFR Network with limited resources, the key question is what network approaches will most effectively address the factors that limit the capability of an EFR Network to respond to emerging, large-scale environmental issues? EFRs already have considerable experience and capability in achieving public recognition, participating in environmental monitoring, and performing cross-site experimentation. Advances in these areas might add increments to an EFR Network's capabilities, for example, if future NEON development increased environmental monitoring capacity of a subset of EFRs. However, efforts in these areas, as they have been done in the past, have created linkages among only the limited subset of EFRs that participate. An integrated EFR Network that can respond to emerging large-scale environmental issues requires more wide-ranging cooperation and communication among EFRs on scientific matters than these approaches have provided in the past.

Interactions among EFRs within networks centered outside the Forest Service have clearly benefited the EFR Network by developing models of how to integrate across research sites, and, in cases such as LTER, by working as partners facilitating integrated research among EFRs that are LTER members. However, it would be unrealistic to expect networks centered outside the Forest Service, even integrated research networks, to contribute more than peripherally to the goal of integrating EFRs into a functioning research network. Forest Service R&D is the only organization likely to have a strong enough stake in integrating EFRs into a functioning research network to provide the sustained leadership and support that will be needed.

Forming EFRs into an integrated research network requires more than just data sharing. Probably the greatest challenge will be developing a viable community of EFR scientists to foster sharing of ideas and cooperation in designing and carrying out multiple long- and short-term cross-EFR research projects. Despite some recent activities building cross-site cooperation, enthusiasm and participation of EFRs in Network activities has been highly variable. Building a culture of cross-site sharing among scientists within a Network will take persistence, practice, and leadership to overcome attitudes and habits developed in the past when sites were quite independent.

Scientists working at EFRs must play a central role in developing cross-site science, but these network-wide social and scientific interactions are likely to grow only if the tasks required are rewarding for EFR scientists. Recognition of scientists' achievements in the EFR Network and funding for cross-site experimentation, synthetic analysis, and publications will be critical to motivate and enable scientists to reach this goal. In addition, Network scientific activities would also be encouraged and enhanced if a support staff were dedicated to facilitating and coordinating Network interactions among EFR scientists. The LTER Network Office provides an example of the kinds of support that network staff can provide (see http://lno.lter). Although the specifics of support that would suit the EFR Network would likely differ from the LTER Network Office, the general kinds of support functions an EFR Network would need are probably analogous to the LTER Network. For example, the LTER Network Office provides communications among network member sites, including a network-wide, online newsletter, and arranges venues for cross-site interactions among scientists and technical personnel, including facilitating cross-site working groups on several topics, holding all-scientists' meetings every 3 years, and coordinating and operating the Network's data-sharing infrastructure. The LTER Network Office is a work in progress that has evolved over time to serve the LTER Network's needs. An EFR support staff would likewise adapt as needs of the EFR Network develop and mature, and could probably benefit from the experience of the LTER Network Office.

A compelling research agenda could mobilize EFRs as a research network, and the EFR system has a unique opportunity to set such as an agenda with its long-term record of climate, streamflow, vegetation, and biogeochemistry observations from several dozen experimental watershed sites across the Nation. These records could be used to frame and explore a suite of questions concerning effects of changing climate, atmospheric chemistry, and vegetation on ecosystem services, such as water quantity and quality, carbon sequestration, wood production, and biodiversity. Research questions would include, for example, distinguishing changes in ecosystem services (e.g., streamflows) that can be attributed to changing climate from those with other causes, for example, from vegetative succession related to land use and natural disturbances. Jones et al. (2012) suggested an example of such a new approach to EFRs, when they proposed that rather than only viewing traditional experimental watersheds as decades-long study sites of forest hydrology and forest management effects, these installations should be reconceptualized as "headwater ecosystem" research sites with high potential for addressing emerging issues such as global climate change and ecosystem services. This work would be greatly facilitated through use of information-sharing infrastructure for access to data about streamflow (HydroDB), climate records (ClimDB), biogeochemistry (StreamChemDB), and changing vegetation (VegDB). In such a project, it would be important to present data in terms that are relevant to a suite of ecosystem services. Site vegetation data, for example, would need to include characteristics such as leaf area to be relevant to ecosystem services such as carbon stocks and sequestration rates, and streamflows. Simply building information infrastructure, however, will not be enough: Achieving an integrated research network will require funds to sustain activities such as ongoing entry of site monitoring data into these data bases, for the continued operation, maintenance, and upgrades of these information-sharing systems and for active use of data they contain in cross-site synthetic analyses.

The experience developing ClimHydroDB showed that technology was not the limiting factor in making data and metadata widely accessible. Rather, it was bringing groups of scientists and information managers from across multiple sites to consensus on how to design and implement an approach. That lesson applies to other parts of building an integrated research network as well. Engaging groups of scientists from multiple sites to focus and agree on common solutions to priority research questions takes leadership, negotiation, and, above all, time. Forming EFRs into a research network that can respond in timely ways to pressing issues requires developing agreement in advance and at a broad enough scope to be ready to respond quickly and effectively to emerging issues and funding opportunities. A research network that starts from scratch on each new issue will have great difficulty responding to pressing issues expeditiously. Developing a network and a culture that is prepared to address emerging issues requires support and organization to engage scientists from across sites in network design and implementation. Although social networking technologies, Web meetings, and video conferences may reduce the logistics of consensus building, gaining agreements among diverse individuals will probably continue to require hard work. Support staff could facilitate this necessary work but cannot be expected to provide the foresight and leadership to anticipate national-scale issues.

National-level leadership has an important strategic role to play in forging a research network. National R&D leaders need to make building an EFR research network a high priority, for example, by providing rewards and incentives, and also by guiding development of network capabilities to ensure that the EFR Network can anticipate national-scale issues. Another role of national R&D leaders will be to mobilize the EFR Network by providing the resources required to perform research that will address large-scale environmental issues as they arise.

24.6 A Vision for the Future

A fully functional and integrated EFR Network will have the capability to effectively address scientific questions concerning the implications that are relevant to the policy and management of emerging, large-scale environmental issues on the nation's forest and grassland ecosystems. The Network will accomplish this goal by building working relationships among scientists doing research within and across sites throughout the EFR Network. The foundation of the Network will be a culture of cooperation that encourages, facilitates, and rewards EFR scientists working together. EFR scientists will have a strong community that fosters mutual trust and facilitates interactions among EFR sites and their programs of research. For example, the EFR Network will enhance cross-site interactions by sponsoring network-wide events similar to those of the LTER Network. The EFR Network will hold EFR all-scientists meetings every 3 years to build community by serving as a forum for collaboration on topics of major interest to scientists, providing a seedbed for multisite research projects. Gatherings of EFR scientists would also include, when appropriate, non-Forest Service cooperating scientists and EFR Network science users, including representatives of the policy and management communities.

EFR sites and their research programs will have the scientific capacity to be a critical mass capable of participating at a high level in major EFR Network science topics. Not all EFR sites can have full capability on all topics; so subsets of sites will have developed the capacity to address critical questions, so that together they can address national-scale questions related to vegetation dynamics, biogeochemistry, or hydrology, even though not all sites will have all of these capabilities.

The Network will provide selected services that enhance science at all sites and their science programs throughout the Network. Common protocols will be established for environmental data collection and documentation. Information-sharing capacity will be developed that enables scientists to build upon science done across the entire network and is readily shared with the larger scientific community and the public. EFRs will participate in research networks managed by other institutions, such as LTER, NEON, and NADP, that can extend the reach and inferences of findings from EFRs and share lessons learned about networking processes. Dedicated administration and leadership will be in place to support network science with an appropriate mix of top-down and bottom-up approaches. Funding and resources for network science will be sufficient to provide a reliable and merit-based process to sustain a significant level of cross-site science to address major, large-scale issues for forests and grasslands.

References

Adams MB, Loughry L, Plaugher L (2008) Experimental Forests and Ranges of the USDA Forest Service. General Technical Report NE-321 Revised, Department of Agriculture, Forest Service, Northeastern Research Station, Newtown Square, PA, 178 p. [CD-ROM]. http://nrs.fs.fed.us/pubs/gtr/gtr_ne321R.pdf. Accessed 29 April 2014

Adams MB, McNeel J, Rodriguez-Franco C (2010) Meeting current and future conservation challenges through the synthesis of long-term silvicultural and range management research. General Technical Report WO-84, Department of Agriculture, Forest Service, Washington, DC, 82 p

Birdsey R (2009) Linking landscape-scale carbon monitoring with forest management. Final Report for NASA Project CARBON/04-0225-0191

Coleman DC (2010) Big ecology: the emergence of ecosystem science. University of California Press, Berkeley, p 236

Greenland D, Kittel T, Hayden BP, Schimel DS (1997) A climatic analysis of long-term ecological research sites. http://intranet2.lternet.edu/sites/intranet2.lternet.edu/files/documents/Scientific%20Reports/Climate%20and%20Hydrology%20Database%20Projects/CLIMDES.pdf. Accessed 29 April 2014

Henshaw DL, Stubbs M, Benson BJ, Baker K, Blodgett D, Porter JH (1998) Climate database project: a strategy for improving information access across research sites. In: Michener WK, Porter JH, Stafford SG (eds) Data and information management in the ecological sciences: a resource guide. LTER Network Office, University of New Mexico, Albuquerque, pp 123–127

Henshaw DL, Sheldon WM, Remillard SM, Kotwica K (2006) CLIMDB/HYDRODB: a web harvester and data warehouse approach to building a cross-site climate and hydrology database. In: Proceedings of the 7th International Conference on Hydroscience and Engineering (ICHE-2006); Philadelphia, PA. Drexel University, College of Engineering, Philadelphia, PA (Not paged). http://hdl.handle.net/1860/1434. Accessed 29 April 2014

Johnson SL, Adams MB, Amatya DM, Bailey SW, Rhoades CC, Jones JB, Knoepp JD, McCaughey W, McDowell WH, McGuire KJ, Sebestyen SD, Wohlgemuth PM (2009) Stream nutrient responses to forest harvest and disturbance: lessons from long-term research at USDA FS experimental forests. (Abstract). Seventh North American Forest Ecology Workshop, Logan, UT

Jones JA, Post DA (2004) Seasonal and successional streamflow response to forest cutting and regrowth in the northwest and eastern United States. Water Resour Res 40:W05203. doi:10.1029/2003WR002952. http://www.agu.org/pubs/crossref/2004/2003WR002952.shtml. Accessed 28 April 2014

Jones JA, Creed IF, Hatcher KL, Warren RJ, Adams MB, Benson MH, Boose E, Brown WA, Campbell JL, Covich A, Clow DW, Dahm CN, Elder K, Ford CR, Grimm NB, Henshaw DL, Larson KL, Miles ES, Miles KM, Sebestyen SD, Spargo AT, Stone AB, Vose JM, Williams MW (2012) Ecosystem processes and human influences regulate streamflow response to climate change at long-term ecological research sites. BioScience 62(4):390–404. http://andrewsforest.oregonstate.edu/pubs/pdf/pub4679.pdf. Accessed 29 April 2014

Lugo AE, Swanson FJ, González OR, Adams MB, Palik B, Thill RE, Brockway DG, Kern C, Woodsmith R, Musselman R (2006) Long-term ecological research at the USDA Forest Service's Experimental Forests and Ranges. BioScience 56(1):39–48. http://andrewsforest.oregonstate.edu/pubs/pdf/pub3843.pdf. Accessed 29 April 2014

Lull HW, Sopper WE (1966) Factors that influence streamflow in the northeast. Water Resour Res 2:371–379

Post DA, Jones JA (2001) Hydrologic regimes of forested, mountainous headwater basins in New Hampshire, North Carolina, Oregon and Puerto Rico. Adv Water Resour 24:1195–1210

Post DA, Grant GE, Jones JA (1998) New developments in ecological hydrology expand research opportunities. EOS 79:517–526

Swanson FJ, Eubanks S, Adams MB, Brissette JC, DeMuth C (2010) Guide to effective research-management collaboration at long-term environmental research sites. General Technical Report PNW-GTR-821, Department of Agriculture, Forest Service, Pacific Northwest Research Station, Portland, OR, 12 p

USDA/USDI (US Department of Agriculture, US Department of Interior) (1994) Record of decision for amendments to Forest Service and Bureau of Land Management planning documents within the range of the northern spotted owl (place of publication unknown), 74 p

Part VIII
Emerging Human Dimension Research Trajectories

Chapter 25
Social Science Research at Experimental Forests and Ranges

Susan Charnley and Lee K. Cerveny

Abstract For a century, US Department of Agriculture Forest Service experimental forests and ranges (EFRs) have served as a resource for scientists conducting long-term research relating to forestry and range management and ecosystem science. Social science research has not comprised a significant portion of the research endeavor at EFRs to date, despite their past history of occupation and their current human uses. The EFR network presents a rich, though largely untapped opportunity for social scientists to engage in long-term, comparative, and interdisciplinary research related to human-natural resources interactions. This chapter explores the potential for social science research at EFRs. We synthesize the human dimensions research that has been pursued there to date by social scientists and others. This research falls into six areas: human uses, prehistorical and historical studies, economics, human dynamics at the wildland–urban interface, human values relating to forests, rangelands, and their management, and interdisciplinary studies of socio-ecological systems. Discussions with EFR scientists and site administrators revealed the potential for several types of future social science research. However, lack of awareness, limited budgets and networking, and the historic predominance of biophysical scientists who administer and conduct research at EFRs appear to have inhibited the development of social science research there. Nevertheless, we see signs that it is on the rise, and expect it to increase in the future. We suggest ways of encouraging social science research at EFRs, and describe its potential contributions.

Keywords USDA Forest Service · Social science research · Experimental forests and ranges · Human dimensions of public lands management

---

S. Charnley (✉)
Pacific Northwest Research Station, USDA Forest Service,
620 SW Main St., Suite 400, Portland, OR 97205, USA
Phone: (503) 808-2051
e-mail: scharnley@fs.fed.us

L. K. Cerveny
Pacific Northwest Research Station, USDA Forest Service,
400 N. 34th St., Seattle, WA 98103, USA

D. C. Hayes et al. (eds.), *USDA Forest Service Experimental Forests and Ranges*,
DOI 10.1007/978-1-4614-1818-4_25, © Springer New York 2014

25.1 Introduction

Research conducted at experimental forests and ranges (EFRs) over time has often been shaped by societal issues relevant to particular times and places. Unlike the long-term trajectory of biophysical research that has occurred at EFRs since their establishment, most social science research that has occurred there has been spotty and more recent. This is surprising, given the extent of human interactions with EFRs, and the fact that 35% of them are located within 50 miles of an urban area (Cerveny and Charnley, this volume). Existing summaries of research conducted at EFRs make little mention of social science. One recent overview of 79 EFRs lists for each site the reasons they were established, the key research topics pursued, major research accomplishments, and the long-term databases they maintain (Adams et al. 2008). Economics research is mentioned as having been conducted at 8 of the 79 sites; there is otherwise no mention that social science research has figured prominently at any sites. One exception is the Baltimore Ecosystem Study, an urban long-term ecological research site.[1] There, social science has been an integral component of the research endeavor, which has focused on measuring interactions between the social, physical, and ecological components of the urban ecosystem (Grove, this volume). Another recent Forest Service publication highlighting *100 Years of Research Success Stories* at EFRs (Wells 2009) makes no mention of social science research.

This chapter argues that EFRs hold a great, but as yet largely untapped, potential for social science research. We synthesize the social science and other human dimensions research that has been conducted at or in association with EFRs across the USA over the past century in order to take stock of the range of studies that have occurred and their contributions. We also discuss future directions for social science research at EFRs identified by scientists and site administrators who work there. We then explore why more social science research has not occurred at these sites, and suggest some ways forward that could help social scientists take advantage of the many research opportunities EFRs offer as the network heads into its next century.

25.2 Background: Social Science Research and Federal Lands

Social science is the field of scientific knowledge and scholarship related to the study of humans, including individuals, groups, societies, and systems (Kuper and Kuper 1996). The social sciences include a broad range of disciplines (geography, economics, sociology, demography, anthropology, political science, psychology, and history, among others). Social scientists have long been exploring the dynamics of human–natural resources interactions related to federal forests and rangelands. A brief review of this history provides context for understanding the nature of social science research that has occurred around EFRs, and perhaps, why more has not.

[1] Long-term ecological research sites are places where scientists conduct research on ecological processes at broad spatial scales and over long time periods. The network was established in 1980 by the National Science Foundation. There are currently 26 long-term ecological research sites representing a wide range of ecosystem types, environmental conditions, and human characteristics (http://www.lternet.edu/).

25.2.1 Forestry and Federal Lands

Much of the social science research relating to forestry and federal lands in the USA has focused on forest communities and how economic conditions, social organization, well-being, and social change in these communities have been linked to the timber industry and timber management and production on federal lands (Field et al. 2005; Sturtevant and Donoghue 2008). Because commercial timber production was the management focus on federal forestlands from the 1940s through the 1980s, this made sense. However, the 1970s and 1980s saw increasing emphasis on federal lands management for multiple uses. As a result, one research area that developed in the 1980s pertained to public values relating to forests, and how to incorporate them into forest management practices and policies (Shindler et al. 2003). The 1970s and 1980s also saw a transition from top-down, government-controlled management of federal forestlands to a more participatory, collaborative, community-based approach (Steel and Weber 2003). Consequently, another social science research area that emerged focused on institutions for forest management.

In the 1990s, timber production from federal forestlands dropped dramatically, and social scientists investigated the effects on forest communities, and their resilience and capacity to adapt to change. They also researched the potential for federal lands to continue to provide social and economic benefits to local communities, be it through recreation and tourism, amenity migration, or new jobs associated with ecosystem management and forest restoration. Ecosystem management became the paradigm for forest and rangeland management on federal lands, and with this change came an emphasis among social scientists on encouraging forest managers to view people as a part of, rather than as separate from, forest ecosystems (Cordell and Bergstrom 1999). This more holistic way of thinking about people as a part of ecosystems opened the door for integrated, interdisciplinary research studies. It also increased interest in researching the effects of human activities on forest ecosystems, and called for studies to help forest managers understand the social context in which they were managing federal lands. With this change, the potential of EFRs to serve as sites for social science research grew.

Economics research relating to forestry on federal lands dates back to the 1920s, though the majority of forestry activity occurred on private lands up until World War II. Early economics research focused on the economics of forest resource use, such as timber utilization, lumber prices, and lumber exports (Flora 2003). The economics of forest management was another topic of interest, including the costs of fire prevention and control, and the economics of replanting versus natural regeneration following a harvest. In the 1930s, research about how to optimize costs associated with timber production (through different silvicultural systems) and lumber production (e.g., log transport, milling) emerged. Research also looked at the basis for sustained yield management on federal lands and its consequences for private timberland owners. By the mid-1900s, economics research relating to federal forestry had evolved to focus on timber supply questions—what kind of trees, from where, and how to increase supply in response to demand. Production questions continued to be important for providing insight into how to make forestry pay in the face of a changing market and natural resource base. Economics research on timber supply and production economics continued for the next several decades.

The multiple use mandate for federal forestlands codified in the Multiple-Use and Sustained-Yield Act of 1960 caused non-commodity economics research to emerge, for example, analyzing the economic trade-offs of establishing wilderness areas in one location versus another. Research on non-timber objectives and constraints grew in the 1980s, as did the need to estimate the economic impacts of restricting federal harvest as part of habitat conservation strategies. Ecosystem management in the 1990s caused this trend to continue. Examples are the economics of managing for salmon and riparian habitat, how to incorporate nonmarket values into forest management and planning, design of hazardous fuels reduction strategies, and the economics of forest carbon sequestration.

Other notable topics in economics research relating to federal forestry over the past few decades include marketing economics research (market structure, market statistics, price formation, regional market dynamics, regional resource and market projections, and timber trade interactions); the competitiveness of federal timber sale procedures and the influence of federal timber sale programs and harvest levels on stumpage prices for various land owners; and the local and regional impacts of federal land management activities on surrounding local communities and residents. Regional economic assessments were also undertaken (Flora 2003).

As recreation uses of public lands became more popular in the 1960s and 1970s, recreation research also emerged. Throughout the 1970s and 1980s, recreation researchers in the Forest Service developed a worldwide reputation as leaders in the generation of scientific knowledge (Cerveny and Ryan 2008). This research was conducted in close collaboration with university researchers and managers working on the ground. Early research in the 1960s focused on estimating recreation use, the impacts of recreation use on developed sites, and identifying recreation site attributes. In the 1970s, emphasis shifted to trying to understand more about recreation visitors and their experiences. In the 1990s, recreation research emphasized visitor connections to nature, places, and resources. More recently, recreation research has helped forest managers integrate recreation data into large-scale planning efforts. Today, major research themes include: understanding recreation and leisure experiences; developing tools for establishing visitor limits to designated areas, such as wilderness and parks; studies to understand and manage recreation conflict; exploring wilderness values, experiences, and site attributes/preferences; understanding the social acceptability of recreation activities; conducting large-scale assessments (national and regional surveys) of recreation participation and preferences; and understanding how different cultural and ethnic groups relate to natural places and outdoor recreation (Cerveny and Ryan 2008).

Against this backdrop of research relating to federal forests and rangelands and their management, the social science research that has occurred to date at EFRs, and areas for potential future inquiry, can be better understood. Many of the topics described earlier are represented in the kinds of social science research that has occurred at EFRs. This research largely focuses on forests and forestry; little social science research relating to rangelands and ranching has occurred to date.

25.2.2 Ranching and Federal Lands

Most ranchers in the west are economically dependent on a combination of private base lands that they own and grazing allotments or leases on federal lands to which they have access for grazing (Sheridan 2007). Thus, social science research relating to rangelands often touches on federal lands grazing issues. A key focus of social science research on ranching has been the family ranch as an enterprise, and its adaptive strategies—both environmental and economic (Sheridan 2007). The cultural importance of ranching as a way of life has been another area of inquiry. The relationship between ranchers and the government vis-à-vis range management is a third area of focus (e.g., Starrs 1998). Over the past two decades, conflict over the environmental impacts of grazing on federal lands has been another topic of study (e.g., Sayre 2002). And, as private ranchlands have increased in value as real estate, social science research has examined how ranchers can maintain their land and lifestyles given the changing political economy of ranching. Finally, social science research has explored how working ranches and federal lands grazing can be maintained in ways that are environmentally sound and socially and economically sustainable (Sheridan 2007).

25.3 Methods

We collected information about the social science research that has been conducted at and around 80 EFRs in the Forest Service network during summer 2009 using three main methods.[2] First, we conducted a review of the published and gray literature resulting from social science research at or associated with EFRs. Although we tried to obtain and review as many of these documents as possible, many were old or not readily available and could not be obtained for purposes of this study. Other social science research did not, or has not yet, resulted in publications. Thus, our overview of social science research at EFRs is not comprehensive, but portrays the range of social science topics pursued there and potential areas for future inquiry. Second, we contacted EFR coordinators and scientists using e-mail correspondence and telephone interviews to request information about the history of social science research associated with each site, and possible future studies, among other things. A total of 80 individuals across 58 EFRs were contacted during this data gathering phase. Third, we carried out a formal, online survey that was administered to Forest Service personnel affiliated with each site using SurveyMonkey. Its purpose was to systematically verify our findings about social science research at EFRs. A total of 68 surveys were completed, representing an 85% response rate for all EFRs in the Forest Service network.

[2] The official number of EFRs is reported at 81 (Wells 2009). However, the total number of EFRs changed during the course of this study. It began at 79 (Adams et al. 2008), but then two new EFRs were added to the network (Olympic and Heen Latinee), and one was removed (Young Bay). Thus, our final sample size was 80.

25.4 Social Science Research at EFRs

More social science and other human dimensions research has occurred at or in association with EFRs than is evident at first glance, and several studies are currently ongoing or have been proposed.[3] In some cases, this research has been one component of a larger study focusing on biophysical research questions. In other cases, it has been conducted on the national forest or rangeland, or in the larger region, where the EFR occurs and has included it, but has not exclusively focused on it. Several studies have explicitly focused on the EFR itself. We group the social science and human dimensions research that has occurred at or in association with EFRs into six topic areas.

Humans have long been a part of the ecosystems that are currently managed as EFRs, and people continue to use EFRs in ways that can influence ecosystem processes and functions there. Understanding the nature of human uses and how to manage them has emerged as one area of research at EFRs. Another related area of research consists of prehistorical and historical studies of areas now being managed as EFRs, with an emphasis on their environmental and social histories. Third, economics research has been carried out at several sites. A fourth area of inquiry concerns human–EFR interactions at the wildland–urban interface. As mentioned earlier, one EFR is urban (Baltimore), and many (35%) lie within 50 miles of a metropolitan area (Cerveny and Charnley, this volume). The remaining EFRs are rural, but rapid population growth over the past two decades in rural counties containing federal lands has been well documented, especially for the western states, and is predicted to continue (Charnley et al. 2008; Stein et al. 2007). This area of social science research looks at the effects of nearby development and changing demographic patterns on EFRs and the broader landscapes of which they are a part. A fifth research topic focuses on human values relating to forest and rangeland ecosystems, and stakeholder preferences for how these ecosystems should be managed. Finally, a number of integrated, interdisciplinary social and ecological research studies have occurred or are underway, especially at EFRs that are long-term ecological research sites (LTERs). We provide an overview of these studies subsequently and highlight examples from each category to give a sense of the diversity and potential for social science research at EFRs, and what it can contribute to our understanding of forest and rangeland ecosystems and their management.

25.4.1 Human Uses

Much of the research on human uses at EFRs has been conducted by biophysical scientists rather than social scientists because the main object of study has been the environmental effects of human activities, rather than the people engaged in these

[3] Some research at EFRs has had a human dimension, but has not been undertaken by social scientists.

activities, with humans treated as a disturbance factor. This is particularly true for earlier studies, and studies having experimental designs. We include these studies in our review because they nevertheless represent human dimensions of the research at EFRs. They also represent the types of studies that could be undertaken using an interdisciplinary approach that would include social scientists in order to provide insight into the motivations and behaviors of forest and range users.

25.4.1.1 Human Uses Research by Nonsocial Scientists

One reason the EFR network was originally established was to study the effects of grazing and silvicultural practices on forest and rangeland ecosystems, and to use this understanding to inform management (Shapiro 2008). Thus, one type of research relating to human uses at EFRs focuses on the environmental effects of people's natural resource use and management practices. For example, in 1912 the Great Basin Experimental Range (UT) was established to conduct studies of the impacts of grazing on plants and soils (Shapiro 2008). Research on rangeland ecology and management has continued there since that time, contributing to the development of the discipline of range management (Adams et al. 2008). The Fort Valley Experimental Forest (AZ) is another place where research on grazing and its impacts has been ongoing for a century (Pearson et al. 2008).

Research on the effects of forestry practices date to 1909, when a study on the Wind River (OR) focused on the impacts of silvicultural practices in Douglas Fir forests (Wind River was later established as an experimental forest in 1932; Adams et al. 2008). This research continued into the 1960s. The Bent Creek Experimental Forest is a place where long-term silvicultural research in hardwood forests has been conducted since the 1920s or so (Adams et al. 2008).

Much less common have been studies of non-timber forest products harvesting at EFRs. We are aware of only one such study (Haywood et al. 1998), which took place at the Palustris Experimental Forest (LA). These authors investigated the effects of pine straw harvesting on longleaf pine (*Pinus palustris*) productivity, straw yields, needle fall patterns, soil conditions, and herbaceous plants using an experimental research design. Pine straw is a commercially valuable non-timber forest product that is used for mulch in farming, horticulture, and landscaping. The results point to best management recommendations for harvesting pine straw in ways that minimize negative impacts to soils and plants in the forest understory, and to timber stand volume production.

Recreation research began at EFRs in the 1950s. EFRs have served as sites for conducting experimental research relating to recreation and its impacts, as well as sites where ongoing recreation activities have been studied. Numerous recreation activities are currently permitted on EFRs, with 96% reporting some kind of recreation use (Cerveny and Charnley, this volume). Much of this research examines how recreation uses affect wildlife and forest and range conditions.

The earliest recreation study at an EFR that we are aware of took place on the San Joaquin Experimental Range (CA) in 1955–1956 (Philpott et al. 1958). This research was prompted by unregulated hunting and associated problems that had caused many private landowners to close their properties to hunting by others. The purpose of the study was to develop and test a permit-registration card that could be used to regulate the number and timing of hunters hunting on private lands. A method of using access permits to regulate sport hunting was tested on the San Joaquin by conducting two quail hunts, using an experimental design.

Another recreation research experiment was carried out at Hubbard Brook Experimental Forest (NH) between 1978 and 1980. There, an experimental trail system was used to undertake controlled hiking at three different levels of intensity in order to assess the effects of trampling on ground flora (Kuss and Hall 1991). The trails were closed to hiking in 1981 and the recovery of the disturbed area studied for the next 5 years to observe changes in soil compaction and plant species composition, density, and abundance.

Since the 1990s, the leading topic of inquiry among scientists studying recreation and its impacts at EFRs has been motorized use (e.g., off-highway vehicles, OHVs and all terrain vehicles, ATVs), hunting, and how they affect wildlife. We found that research on these topics had occurred, was ongoing, or was proposed at and around at least five different EFRs, including Long Valley (AZ), Manitou (CO), Palustris (LA), Black Hills (SD), and Starkey (OR). At four additional EFR sites, scientists interviewed expressed a desire for future research on hunting and OHV/ATV use because of concerns about the effects of these activities on natural resources and on study plots.

Perhaps the best example of this type of study comes from the Starkey Experimental Forest and Range (OR), where research on hunter–wildlife interactions has been ongoing since 1991 (Johnson et al. 2004; Naylor et al. 2009; Wisdom et al. 2004). There, a 10,117-ha (25,000-acre) ungulate-proof enclosure was constructed in 1987 that makes it possible to study interactions between humans and ungulates using controlled research experiments. Between 1991 and 2000, scientists conducted a 10-year experimental study of the effects of hunting pressure (measured by hunter density and associated motorized vehicle traffic) on elk and mule deer, comparing rifle and archery hunting. Another study at Starkey evaluated the impacts of off-road recreation—including ATV use, horseback riding, mountain biking, and hiking—on elk and mule deer. The results of these studies can be used to inform hunting regulations and management guidelines for off-road recreation in national forests throughout the West where recreation, elk, and mule deer cooccur.

25.4.1.2 Human Uses Research by Social Scientists

Human uses research at EFRs by social scientists is, for the most part, more recent and sparse. On the San Joaquin Experimental Range (CA), one study examined the effects of grazing on cultural resources. Classes of anthropology students from a local community college conducted surveys of cultural resources over a 17-year period (1978–1995; Beck 1995). The survey data were used to investigate the effects of cattle foraging, water resource development, corrals, access roads, fences and gates, feeding, salt and watering stations, cattle trails, and wallowing on cultural resources. The study found that most of these factors either had no negative impacts or their impacts were difficult to separate from those of other factors like human and rodent activity. Guidelines for monitoring and managing the impacts of grazing on cultural resources were developed that are broadly applicable to rangelands elsewhere.

Research on hunting at EFRs has not been limited to recreational hunting. At the Bonanza Creek Experimental Forest in Alaska where subsistence hunting occurs, studies are currently being implemented to investigate the social and economic aspects of moose hunting by residents of local communities. Moose are one of the most important subsistence resources in the area.

Elsewhere, studies are underway that examine the effects of ongoing recreation activities occurring at EFRs. At Bent Creek Experimental Forest (NC), social scientists are involved in a study that measures recreation use, visitor preferences and satisfaction, and knowledge about the experimental forest (Payne 2009). At Manitou Experimental Forest, there are plans for a future study to assess the impacts of OHV use on natural resources. Research at both sites addresses conflicts over recreation and its effects on scientific experiments in an effort to manage recreation and scientific uses in ways that are compatible, using collaborative approaches.

25.4.2 Prehistorical and Historical Studies

Not only do EFRs contain valuable natural resources but they often contain important cultural resources as well. Our survey of EFR contacts found that historical, archaeological, and cultural heritage sites occur at many EFRs (Table 25.1). More than half (56%) contain historic structures, and many have archaeological, cultural, or sacred sites. Examples include historic structures installed by the Civilian Conservation Corps during the 1930s and 1940s, sites that remain from early European settlement, archaeological resources such as petroglyphs, tool-making sites, and pre-Columbian dwellings, and cultural heritage sites that are sacred places or sites where important events in human history occurred. The presence of these sites invites research on the human prehistory and history of EFRs.

The only archaeological study we are aware of at an EFR (mentioned in the preceding section) comes from the San Joaquin Experimental Range (CA; Beck 1995). As mentioned under Human Uses, between 1978 and 1995, students in the anthropology department at Fresno City College conducted surveys of cultural resources from the Indian and early ranching and mining periods. The students learned archaeological survey methods, and the Range gained valuable information about its cultural resources. This information included an inventory and description of the sites and their archaeological, ethnographic, and historic backgrounds, and maps of their geographic locations.

Studies of prehistory and history have examined how indigenous peoples used and managed natural resources on and around EFRs in the past. One area of inquiry

Table 25.1 Cultural resources at EFRs

Type of site	No. of EFRs	Percentage of EFRs
Historical sites	38	55.9%
Archaeological sites	20	29.4%
Cultural heritage sites	17	25.0%
Spiritual or sacred sites	11	16.2%
Do not know	14	20.6%

concerns Indian burning practices and the use of fire for forest management in historic and prehistoric times (Hines et al. 2000 for Vinton Furnace, OH; Natcher et al. 2007 for Bonanza Creek, AK). Understanding how burning practices enhanced culturally important plant and animal species in the past can inform forest management today to favor these species and support the cultural practices that depend on them. Information about past fire use can also help with contemporary forest restoration efforts.

A second area of inquiry is illustrated by a study that reconstructs the use of plant and animal resources by indigenous peoples (Zobel 2002). Drawing on anthropological records from the early 1900s, archaeological data, and secondary sources, this researcher developed a list of species he believed to have been used by a subgroup of the Tillamook Tribe called the Salmon River people. The Salmon River people occupied the area that is now the Cascade Head Experimental Forest (OR) in prehistoric and historic times. He checked this species list against scientific data about the modern presence and distribution of plants and animals in the Cascade Head area. The long history of research at the experimental forest and the neighboring Cascade Head Scenic Research Area produced detailed ecological data about the region that made it possible to cross-reference sources. Zobel found that at least 68 plant species and 56 animal species from Cascade Head had been used by the Salmon River people, and documented at least 308 different uses of these species. Such information provides insight into how indigenous peoples who occupied EFRs lived in the past, and how these ecosystems have changed since European settlement.

The earliest historical study at an EFR comes from the Bent Creek Experimental Forest (NC), and contains a detailed reconstruction of the land-use history of individual ownership tracts that now makeup the Forest (Nesbitt 1941). Since the 1990s, a number of additional studies that examine how historic settlement patterns and land-use activities shaped present forest conditions at experimental forests have been conducted. One such study (Garcia-Montiel and Scatena 1994) reconstructs the land-use history of the Luquillo Experimental Forest in Puerto Rico since the time of European settlement, assessing how the present structure and composition of the forest reflects past human activities there. These activities included agroforestry, selective logging, charcoal production, and timber management. A second study (Gragson and Bolstad 2006) documents the nature and extent of human land use, past and present, at the Coweeta LTER site in Georgia (of which the Coweeta Experimental Forest is a part) to identify how it has influenced the structure and function of terrestrial and aquatic communities in southern Appalachia, and implications for future change. A third set of studies documents the history of individual EFRs and research endeavors there, but does not attempt to link this history to forest and range conditions. Examples are Douglass and Hoover (1988) for Coweeta (NC) and Geier (2007) for H.J. Andrews (OR).

Another study looked at how past forest management practices on an EFR shaped the social and economic history of the region in which it is located. The Crossett Lumber Company was the first industrial-scale lumber company in southern Arkansas, and was one of the few in the south to try to continue logging operations on its lands after all of the primary forest had been cut in the 1930s (Darling and Bragg 2008; Reynolds 1980). The Crossett Experimental Forest was established on land given to the Forest Service's Southern Research Station by the Company in 1934 to research forest management in second growth stands (Adams et al. 2008), about which little was known at the time (Reynolds 1980). Historical research on logging and milling operations in primary loblolly and shortleaf pine forests on Crossett Lumber Company lands describes how these activities affected southern Arkansas' lumber industry in the late nineteenth and early twentieth centuries.

25.4.3 Economics

Economics research represents the longest-standing social science research at EFRs, dating to the 1930s. This is not surprising, given the broader history of social science research relating to federal lands described in Sect. 25.2. From the 1930s to the 1950s, economics research took place at the Desert Experimental Range (UT) to evaluate the economic impacts of alternative grazing strategies (Hutchings 1946, 1958, 1966; Hutchings and Stewart 1953). The economics of grazing was also studied at Palustris Experimental Forest (LA) from the 1960s to the 1980s, according to our survey results. Studies to assess the economics of different silvicultural treatments in the context of forest management for timber production have also occurred for decades, having been carried out at Penobscot (ME), Big Falls (MN), Cascade Head (OR), Blacks Mountain (CA), Palustris (LA), Hitchiti (GA), Argonne (WI), and Escambia (AL; Adams et al. 2008, survey results). Economic surveys have also been conducted at Silas Little (NJ).

More recent economics studies at EFRs have focused on recreation. For example, research at the Starkey Experimental Forest and Range (OR) examined hunters' willingness to pay for improvements in the quality of their elk hunting experience (Fried et al. 1995). Current studies at Bent Creek Experimental Forest (NC) are investigating the economic impacts of recreation activities on the local economy, and the broader economic benefits of recreation use of the Forest.

Rarely has comparative social science research of any kind been conducted across more than one EFR. An exception to this is a study of forest–economy relations at six LTERs, four of which are EFRs (Courant et al. 1997). The purpose of this study was to develop a framework for describing and assessing the multiple relationships between forest ecosystems and regional economies, and to improve understanding of the variables that influence these relationships. The framework for evaluating forest ecosystem–economy relations developed at these case-study sites can be applied elsewhere to help managers evaluate what the economic impacts of forest management decisions will be.

25.4.4 Human Dynamics at the Wildland–Urban Interface

A common trend around many national forests and rangelands is population growth and increased development driven by an influx of urban residents, retirees, second homeowners, and telecommuters who wish to live closer to natural areas and the benefits they provide (Charnley et al. 2008). According to US Census data, in the year 2000 about one-fifth (20.9%) of the US population lived within 50 miles of an EFR (Arthaud n. d.). On average, population density within 50 miles of an EFR was higher in 2000 (110 persons/square miles) than the US average (78 persons/square miles). Population growth within 50 miles of the five EFRs that saw the most growth ranged from 43.8 to 67.8% between 1990 and 2000 (Arthaud n. d.). Although these data do not necessarily reflect population dynamics in the wildland–urban interface directly surrounding EFRs, they do shed light on demographic conditions and trends near EFRs, and the potential for population growth and development along their borders.

Few studies have been conducted around EFRs that focus on human dynamics at the wildland–urban interface specifically, although two have been proposed, both in the south. At the Coweeta LTER in the southern Appalachians, a study is commencing that examines how key ecosystem processes, biodiversity, and water quantity and quality will be affected by changes in land use caused by exurbanization in urban, peri-urban, and wildland areas. The study will also explore how these changes in land use will interact with climate change to produce social and environmental impacts in rural areas.[4] The second study is the Southern Forest Futures Project, which will forecast future changes in forest conditions and the social and economic benefits associated with forests, in response to changing demographic trends and values.[5] This study is regional in nature, and its findings will be useful for informing forest management and science programs on EFRs in the south.

25.4.5 Human Values Relating to Forests, Rangelands, and their Management

Studies of people's values relating to forests, rangelands, and their management were initiated in association with EFRs in the 1970s and 1980s. These early studies

[4] (http://coweeta.uga.edu/ecology/renewal/cwt_vi_proposal_2_1_2008).

[5] (http://www.srs.fs.usda.gov/futures/process/draftplan/socialeconomic/).

were undertaken by scientists associated with experimental forests to examine people's aesthetic preferences regarding alternative silvicultural treatments (e.g., Benson 1980, Benson and Ullrich 1981 for Coram, MT; McGee 1970 for Bent Creek, NC). They were apparently motivated by emerging public concern over the visual impacts of clearcutting, a harvest method that was used in experimental treatments at these forests. Rather than bringing people on site to see and respond to different treatment types, however, photographs of treatments were shown to people living in the surrounding area. At least one such study included photos taken of treated stands in an experimental forest (Pings and Hollenhorst 1993 for Kane, PA). The study findings are relevant to silvicultural research taking place at experimental forests because they provide information about the social acceptability of different harvest practices, an important consideration in determining which practices to implement more widely.

Once the emphasis of federal forest management had shifted from sustained-yield timber production to ecosystem management, research examining people's attitudes toward different silvicultural practices focused on treatments associated with the goals of ecosystem management, and their scenic preferences (e.g., Herrick and Rudis 1994 for Alum Creek, AR; Shindler and Mallon 2006 for H.J. Andrews, OR). The H.J. Andrews study focused on whether forest management practices such as timber harvesting and prescribed burning that try to mimic past natural disturbance regimes are socially acceptable. Residents of the watershed that contains the H.J. Andrews and of nearby cities were surveyed to find out their knowledge about and perceptions of disturbance-based management, and their level of support for it. The survey findings will help managers communicate with and engage the public in developing ecosystem management strategies, and overcome barriers to implementing disturbance-based management regimes.

Human values relating to forests, rangelands, and their management also reflect people's perceptions of their relationship to the natural world. At Oregon's H.J. Andrews Experimental Forest, scientists, creative writers, and environmental philosophers have been invited to contemplate this relationship using a humanistic mode of inquiry as part of the Long-Term Ecological Reflections program that started in 2003. These scientists, creative writers, and environmental philosophers spend time at the H.J. Andrews, visit research plots and other special places on the forest, interact, and reflect upon how ecosystems change over time and the evolving relationship between humans and the natural world (Swanson et al. 2008; Thompson 2008). They are encouraged to communicate their insights through journal entries that are eventually published as essays, poems, and books. This work contributes to our understanding of forest ecosystems with a perspective from the humanities that complements the natural science perspective.

Another kind of human values study that has been carried out at EFRs solicits input from members of the public on specific forest and range management issues. One project at Deception Creek (ID) asked stakeholders their opinions of forest road closures, and used the findings in the environmental impact statement associated with transportation management planning there. Another study at Cascade Head (OR) asked members of the public how they valued the Salmon River Estuary, which is one of two estuaries located on Forest Service lands in the 48 contiguous United States (Adams et al. 2008; Greene 2009). Beginning in 1979, the Salmon River estuary was the focus of restoration efforts, and information gathered from the public was used in planning restoration projects there. It has since been the site of important research on the role of estuaries in the life cycle of anadromous fish.

25.4.6 Integrated Interdisciplinary Social–Ecological Research

EFRs offer unique opportunities to engage in interdisciplinary research that integrates the biophysical and social sciences in addressing common research questions. Six EFRs participate in the LTER network: Baltimore (MD), Bonanza Creek (AK), Hubbard Brook (NH), H.J. Andrews (OR), Coweeta (NC), and Luquillo (Puerto Rico). Because these sites receive funding from the National Science Foundation, which strongly encourages human dimensions research and interdisciplinary research, the best examples of integrated social–ecological studies come from these sites.

The most prolific site in terms of interdisciplinary research has been the Baltimore Ecosystem Study (MD), where numerous studies pertaining to human ecology in urban ecosystems have been published over the past decade (see Grove, this volume). Around Bonanza Creek (AK), the emphasis has been on studying adaptation and resilience in arctic socio-ecological systems in response to global social and environmental change (e.g., Chapin et al. 2006). Studies at Bonanza Creek have also examined human–fire interactions (e.g., Chapin et al. 2003). New research being proposed will focus on “moose–human systems,” investigating place-based feedback between ecological disturbance–moose interactions, human uses of moose, and human responses to change.

Since the mid-1990s, interdisciplinary research at Coweeta has included studies of the social, economic, and environmental factors driving land cover change in the southern Appalachians region, and the consequences of these changes for water quality and quantity, terrestrial and aquatic biodiversity, and regional carbon cycles (Swank et al. 2001). The National Science Foundation intentionally augmented funding for Coweeta in order to strengthen the social science component of research there so that it would be more interdisciplinary, and to build a regional research program (Swanson 2010).

The Luquillo Experimental Forest has been the site of integrated research about how natural and anthropogenic disturbances interact to influence forest characteristics (Foster et al. 1999; Garcia-Montiel and Scatena 1994; Grau et al. 2003; Thompson et al. 2002; Uriarte et al. 2009). Research questions include: (1) How have social and economic forces affected patterns of land use and land-use change? (2) How has land-use change led to land cover change? (3) How do anthropogenic changes in land cover interact with natural disturbances to influence forest composition, diversity, and structure? Economic globalization caused a shift from agriculture to manufacturing in Puerto Rico beginning in the late 1940s. This shift caused people to abandon their farmlands and migrate to urban centers. Farm abandonment, in turn, led to forest recovery in places where agriculture had been marginal. Scientists reconstructed the land-use histories of different forested areas by examining artifacts, historic documents, photographs, and forest surveys, and interviewing local residents to understand how these legacies have shaped current forest conditions. They found that processes of forest recovery varied, depending on the land-use legacy of a particular area. These forests are also subject to natural disturbances. Hurricanes are the leading natural disturbance agent in Puerto Rico. Thus, researchers also investigated how hurricanes interact with human disturbance to influence forest characteristics. Their findings demonstrate that forest dynamics cannot be understood without considering how both natural and anthropogenic forces influence ecological processes.

Interdisciplinary social–ecological research has yet to materialize in a substantive way at Hubbard Brook or H.J. Andrews. A number of social science research projects have been conducted around these sites, however. Topics range from the role of natural scientists in policy advocacy (Lach et al. 2003) to public perceptions of visibility around forests in light of air pollution (Kimball et al. 1990). According to the EFR representatives we contacted, more interdisciplinary research is likely to emerge at these sites in the near future.

25.5 Future Social Science Research at EFRs

We asked the EFR coordinators and scientists we interviewed and surveyed what kinds of social science research, if any, they thought would be pertinent to undertake at their sites. Respondents expressed particular interest in research pertaining to the people who use EFRs, and the nature of visitation to EFR sites. For example, what characterizes the typical EFR visitor? What activities are they involved in? What prompts them to come to the EFR as opposed to another site? What value do they gain from visiting the site? Are they aware of the science mission of the EFR, and does that play a role in their visit? How do visitors value ongoing research activities? Some EFR administrators were also eager to gain more insight about visitation rates over time and the seasonality of visits.

People interviewed and surveyed for this study also expressed a high level of interest in better understanding the implications of population and land-use change around EFRs. Some EFR administrators identified a need for studies of shifting population trends at the wildland–urban interface and its implications for land use, as well as studies of land-use policy. For example, public land exchanges resulting in the privatization of public land, along with the sale of private industrial forestlands to real estate developers near national forests, are causing changes in the number and makeup of local residents, and an increase in structures around EFR boundaries. In addition, several EFR scientists mentioned a need for research to understand changing human values toward forests and rangelands that accompany the migration of urban and amenity-driven residents to rural areas around EFRs.

Respondents were also interested in understanding new residents' attitudes about the value of science, and their perceptions of EFRs. Scientists worried that the arrival of new residents could threaten the research mission of EFRs.

EFR scientists and administrators identified ecosystem services as a desirable area for future economics research. EFRs could be used to help economists and other scientists understand how forests are valued for the variety of ecosystem services they provide and to develop methods, tools, and approaches for valuing ecosystems.

Several contacts also mentioned research on human values as being desirable. They expressed interest in continued research on attitudes toward silvicultural treatments. The EFR network provides a unique opportunity to bring people on site to respond to forest treatments, generating information about the social acceptability of different forest management practices. Information could also be gathered on site about people's perceptions of postfire conditions, road characteristics, wildlife distribution, recreation density, and other features. Such research could take place at different sites within one EFR, or through comparative studies at multiple EFRs in the network. EFRs could also be used to learn more about how people connect with the natural world by exposing them to different kinds of forests and natural places. Such studies could explore the process by which places become special and develop meaning for people, and whether specific forest features lead to the development of special place attachments and connections. Another area of interest mentioned by study participants was public trust and how it affects perceptions of and relations with the Forest Service and research at EFRs. Again, understanding public perceptions of EFRs and their science missions was a priority for administrators and scientists we communicated with.

25.6 Overcoming Barriers to Social Science Research at EFRs

EFRs hold enormous potential for contributing to our understanding of human–natural resources interactions on and around federal forests and rangelands, past, present, and future. As described in this chapter, a broad array of social science topics have been investigated at EFRs, ranging from the economics of grazing, public perceptions of silvicultural treatments, the effects of recreation activities on wildlife, and how past land use has shaped present forest ecosystems, to human ecology in cities. Nevertheless, the volume of social science research that has been conducted at EFRs is quite small compared with that of biophysical research, as this book attests. Moreover, the majority of social science studies have been undertaken as discrete, terminal studies rather than as long-term research trajectories. And, rarely has comparative work been undertaken between sites. Why have not social scientists taken advantage of the unique opportunities to conduct research at EFRs, and why have not EFR personnel more actively recruited social scientists to work there?

Part of the explanation may lie in the history of social science research relating to federal lands, described earlier. The fact that social scientists (other than economists) working in this area have typically focused on communities (place based, occupational, or interest based) or individuals (e.g., forest workers and their families, family ranches) as the unit of analysis may explain why they have overlooked EFRs as places to conduct research. In contrast, the long history of economics research associated with forestry on federal lands likely explains why economics is the only social science research mentioned in published summaries of research at EFRs. Moreover, because some EFRs are remote and others receive little public use, there may simply be no pressing social science research questions that need addressing there. And, studies of the socioeconomic impacts of forest and range management activities typically consider national forests and grasslands as a whole as their unit of analysis, not individual EFRs.

Part of the explanation may also lie in the way that EFRs have historically been used and viewed by the scientists and administrators who work there. Social conditions on and around EFRs are not the same today as they were 100 years ago. For much of the twentieth century, the main uses of national forests and grasslands were extractive; management decisions were for the most part made internally without the extensive public participation and collaborative stakeholder processes common today; and human intrusions into EFRs were resisted (Swanson 2010). Population and development pressures were also generally not as high as they have been in recent decades. In this context, social science research, apart from timber production and utilization economics, may have seemed less relevant.

When we asked EFR scientists and administrators why more social science had not occurred at EFRs, they offered several explanations. A widely shared view was that limited budgets and staffing were a barrier. Forest Service Research and Development (R&D) has always been dominated by biophysical scientists. Many research laboratories lack social scientists on site. With few social scientists on staff and limited research budgets, biophysical research projects have taken precedence. Biophysical scientists, who have dominated both the research and administrative aspects of these sites, may also be unaware of what social science has to offer. Perhaps as a result, they have not made an effort to encourage social science research at EFRs, though many with whom we spoke were open to working with social scientists. The fact that the most extensive social science research at EFRs has occurred at the LTERs—where NSF funding and mandates drive research—is telling in this regard.

Even at sites where there is a strong interest in incorporating social science, biophysical scientists sometimes do not know whom to engage because of a lack of people having social science expertise nearby, or lack of awareness of existing social science networks. Research networks—both social and biophysical—can be highly influential in shaping who does what kinds of research where. Social scientists may also be unaware that EFRs exist and of the research opportunities they present. An inquiry we conducted of Forest Service R&D research social scientists and economists (there are roughly 50) found that only one of those currently on staff had conducted research at an EFR.

Finally, biophysical and social scientists alike may be unclear on how to link their work in interdisciplinary research projects. Although there has been emphasis recently within Forest Service R&D and research funding organizations on integrated research and coupled human–natural systems, the practice of working across scientific disciplines can pose challenges.

Highlighting the broad range of social science research topics associated with EFRs to make social and biophysical scientists aware of the untapped potential these sites offer for social science and interdisciplinary research is one important step toward overcoming some of these barriers. We hope that with this chapter we have taken this step. Another step is to make scientists and site administrators aware of the contributions social science research can make to the body of knowledge produced at EFRs so that they will support and allocate funding for it. These contributions are many.

EFRs are places where biophysical scientists test and experiment with alternative forest and range management techniques related to fuels reduction, timber harvesting, grazing, watershed restoration, and carbon sequestration, to name a few. Especially on public lands, natural resource management goals are socially defined; and implementing effective management approaches to achieving these goals will depend on whether they are economically feasible and socially acceptable (Shindler et al. 2002). Thus, including social and economic components in these kinds of studies can help managers understand what values people want federal lands to be managed for, and help research results be successfully applied.

EFRs are also places where people currently engage in a variety of human uses, and where people have lived and used natural resources in the past. Forest and range ecosystems have been shaped by their human histories, and they continue to be shaped by people today. EFRs are places where the ongoing dynamics between social and natural processes can be studied to better understand how they interact to influence forest and rangeland conditions, and socioeconomic conditions on and around them.

Biophysical scientists may be concerned about the ecological effects of human uses at EFRs, and the impacts of these uses on scientific research there. Social science can help document the nature, extent, and motivations behind different forest and range uses and their importance to people. This information can be used to develop solutions to management issues, and to promote sustainable human uses of EFRs. It can also generate information on how the public views and values the science mission of EFRs to ensure the relevance of the research conducted there.

Social science research can also improve understanding of the social environment surrounding EFRs, how it is changing, and the implications of this change for EFRs. This can help EFRs develop effective education and outreach programs, establish positive collaborative relationships with local communities, engage citizens in their science, address management issues, and garner support for the science mission at their sites.

Another step is to make social scientists aware of the advantages of undertaking research at EFRs. These sites provide an opportunity to undertake long-term, comparative, and experimental research. They also have biophysical data sets that can be used. They offer a community of researchers with whom to interact and pursue interdisciplinary studies. And many have existing research facilities and infrastructure. All of these features make them attractive for social scientists who are seeking places to carry out long-term, comparative, and interdisciplinary research in particular.

A number of lessons can be learned from scientists involved in the LTER network about how to include more social science at EFRs. These scientists have discussed how to enhance social science research at LTERs and how to promote interdisciplinary social–ecological research there (see Redman et al. 2004). Suggestions include collaboratively building a conceptual framework for integrated research, focusing research on the interactions between the social and ecological components of ecosystems, developing a multi-scaled approach to research questions, and identifying tools and models that both natural and social scientists can utilize. One excellent example from the LTER network is the Maps and Locals Project. This comparative research project is being developed across several LTER sites to study patterns of landscape change over time, the anthropogenic and non-anthropogenic drivers of ecosystem change, and the implications of this change for human livelihoods. Holding meetings (local to national) of all EFR scientists and inviting social scientists to participate, as occurs among the LTER scientists, would be one way to promote communication, networking, and relationship building that could result in future studies. Social scientists could also be involved in the strategic planning process associated with EFRs in which future science questions are developed.

25.7 Conclusions

We hope that by describing the broad range of social science research that has been conducted on and around EFRs, the potential for social science there, and what social science research has to offer, this chapter will encourage social and biophysical scientists alike to reach out to one another and promote more social science research at EFRs. We are optimistic that the next 100 years at EFRs will include a substantial social science research component. Several signs point in this direction. The need and desire for social science research at EFRs is recognized by a number of scientists and administrators associated with these sites. Some funding organizations that support EFR research, including the Forest Service and the National Science Foundation, are increasingly calling for interdisciplinary research that addresses both the natural and human dimensions of forest and rangeland ecosystems and their management. One of the newest EFRs in the network—Heen Latinee, or "River Watcher" in Tlingit (AK)—was created in part because it is accessible to Juneau where there is an opportunity for Forest Service scientists to work together across disciplines and develop integrated studies, and to involve the community in outreach and education efforts (Kruger 2009). Social science will thus be incorporated at the outset, providing a model for future EFR sites. A century of research at EFRs has yielded valuable insights about the workings of forest and range ecosystems. We hope that the next century will do the same, with a greater contribution from the social sciences.

Acknowledgments We are greatly indebted to Daniel Hunter, our research assistant on this project, for his excellent work in gathering the primary data on which this chapter is based. We also deeply appreciate the time and assistance given to us by the many research scientists, officers in charge, and EFR coordinators who participated in the study. We thank Anne Hoover, Richard Haynes, and Pat Winter for their consultation in developing the chapter. Finally, we thank Marla Emery, Fred Swanson, and the book editors for their very helpful reviews, which significantly improved it.

References

Adams MB, Loughry LH, Plaugher LL (2008) Experimental forests and ranges of the USDA Forest Service, Revised. General Technical Report NE-321, US Department of Agriculture, Forest Service, Northeastern Research Station, Newtown Square, PA

Arthaud G (n. d.) Spatial arrangement of the network of USDA Forest Service Experimental Forests and Ranges. Unpublished white paper, USDA Forest Service, Research and Development, Washington, DC, p 11

Beck A (1995) San Joaquin experimental range archaeological reconnaissance report with a study of the impact of cattle grazing on cultural resources at the San Joaquin experimental range, vol 1. Submitted to the USDA Forest Service, Pacific Southwest Research Station, Fresno City College, Fresno, CA, pp i–78

Benson RE (1980) Esthetic impacts of harvesting and residue management: public preference ratings. Environmental consequences of timber harvesting in Rocky Mountain coniferous forests: symposium proceedings. General Technical Report INT-90, Missoula, MT, 11–13 September 1979, pp 433–440. US Department of Agriculture, Forest Service, Intermountain Forest and Range Experiment Station, Ogden, UT

Benson RE, Ullrich JR (1981) Visual impacts of forest management activities: findings on public preferences. Research Paper INT-262, US Department of Agriculture, Forest Service, Intermountain Forest and Range Experiment Station, Ogden, UT, p 14

Cerveny LK, Ryan CM (2008) Agency capacity for recreation science and management: the case of the U.S. Forest Service. General Technical Report-PNW-GTR-757, US Department of Agriculture, Forest Service, Pacific Northwest Research Station, Portland, OR

Chapin FS, Rupp TS, Starfield AM, DeWilde L, Zavaleta ES, Fresco N, Henkelman J, McGuire AD (2003) Planning for resilience: modeling change in human-fire interactions in the Alaskan boreal forest. Front Ecol Environ 1(5):255–261

Chapin FS, Hoel M, Carpenter SR, Lubchenco J, Walker B, Callaghan TV, Folke C, Levin SA, GoranMaler K, Nilsson C, Barrett S, Berkes F, Crepin A, Danell K, Rosswall T, Starrett D, Xepapadeas A, Zimov SA (2006) Building resilience and adaptation to manage arctic change. Ambio 35(4):198–202

Charnley S, McLain RJ, Donoghue EM (2008) Forest management policy, amenity migration, and community well-being in the American West: reflections from the Northwest Forest Plan. Hum Ecol 36:743–761

Cordell HC, Bergstrom JC (1999) Integrating social sciences with ecosystem management. Sagamore Publishing, Champaign

Courant PN, Niemi E, Whitelaw W (1997) The ecosystem-economy relationship: insights from six forested LTER sites, 90 p

Darling OH, Bragg DC (2008) The early mills, railroads, and logging camps of the Crossett Lumber Company. Ark Hist Q LXVII(2):107–140

Douglass JE, Hoover MD (1988) History of Coweeta. In: Swank WT, Crossley DA Jr (eds) Forest hydrology and ecology at Coweeta. Ecological Studies, vol 66. Springer Verlag, New York, pp 17–31

Field DR, Luloff AE, Krannich RS (2005) Rural sociology and natural resources: building on Firey. In: Lee RG, Field DR (eds) Communities and forests, where people meet the land. Oregon State University Press, Corvallis, pp 31–51

Flora DF (2003) Forest economics research at the Pacific Northwest Research Station, to 2000. PNW-GTR-562, US Department of Agriculture, Forest Service, Pacific Northwest Research Station, Portland, OR

Foster DR, Fluet M, Boose ER (1999) Human or natural disturbance: landscape-scale dynamics of the tropical forests of Puerto Rico. Ecol Appl 9(2):555–572

Fried BM, Adams RM, Berrens RP, Bergland O (1995) Willingness to pay for a change in elk hunting quality. Wildl Soc Bull 23(4):680–686

Garcia-Montiel DC, Scatena FN (1994) The effect of human activity on the structure and composition of a tropical forest in Puerto Rico. For Ecol Manage 63:57–78

Geier MG (2007) Necessary work: discovering old forests, new outlooks, and community on the H.J. Andrews Experimental Forest: 1948–2000. PNW-GTR-687, US Department of Agriculture, Forest Service, Pacific Northwest Research Station, Portland, OR

Gragson TL, Bolstad PV (2006) Land use legacies and the future of southern Appalachia. Soc Nat Resour 19(2):175–190

Grau HR, Aide TM, Zimmerman JK, Thomlinson JR, Helmer E, Zou X (2003) The ecological consequences of socioeconomic and land-use changes in post agriculture Puerto Rico. Bioscience 53(12):1159–1168

Greene S (2009) Personal communication, September 2009. President, Sitka Center for Art and Ecology, Otis, OR

Haywood JD, Tiarks AE, Elliott-Smith ML, Pearson HA (1998) Response of direct seeded *Pinus Palustris* and herbaceous vegetation to fertilization, burning, and pine straw harvesting. Biomass Bioenerg 14(2):157–167

Herrick TA, Rudis VA (1994) Visitor preference for forest scenery in the Ouachita National Forest. Proceedings of the symposium on ecosystem management research in the Ouachita Mountains: pretreatment conditions and preliminary findings, Hot Springs, AR, 26–27 October 1993. General Technical Report SO-112, US Department of Agriculture, Forest Service, Southern Forest Experiment Station, pp 212–222

Hines EB, Salick J, Sutherland EK, Hutchinson TF, Sutherland S (2000) Ethnoecology of fire: an experimental approach in the Ohio Valley. Workshop on fire, people, and the central hardwoods landscape. General Technical Report NE-274, US Department of Agriculture, Forest Service, Northeastern Research Station, Richmond, KY, p 126

Hutchings SS (1946) Drive the water to the sheep. Natl Wool Grow 36(4):10–11, 48

Hutchings SS (1958) Hauling water to sheep on western ranges. Leaflet 423, US Department of Agriculture, Forest Service, Washington, DC, p 8

Hutchings SS (1966) Grazing management of salt-desert shrub ranges in the western United States. IX International Grassland Congress Proceedings, pp 1619–1625

Hutchings SS, Stewart G (1953) Increasing forage yields and sheep production on Intermountain winter ranges. USDA Circular 925, US Department of Agriculture, Forest Service, Washington, DC, p 63

Johnson BK, Ager AA, Noyes JH, Cimon N (2004) Elk and mule deer responses to variation in hunting pressure. Transactions of the 69th North American Wildlife and Natural Resources Conference, Wildlife Management Institute, Spokane, WA, pp 625–640

Kimball KD, Hammond JE, Doucette JE, Allen G, Ferrin R (1990) Visibility in the White Mountain National Forest: the potential role of air pollutants and the public's perception. AMC Technical Report 90-2, p 20

Kruger L (2009) Personal communication, September 2009. Research Social Scientist, US Department of Agriculture, Forest Service, Juneau Forestry Sciences Lab, Juneau, AK

Kuper A, Kuper J (1996) The social science encyclopedia. Taylor & Francis, London

Kuss FR, Hall CN (1991) Ground flora trampling studies: five years after closure. Environ Manage 15(5):715–727

Lach D, List P, Steel B, Schindler B (2003) Advocacy and credibility of ecological scientists in resource decisionmaking: a regional study. Bioscience 53(2):170–178
McGee CE (1970) Clearcutting and aesthetics in the southern Appalachians. J For 68(9):540–544
Natcher DC, Calef M, Huntington O, Trainor S, Huntington HP, DeWilde L, Rupp S, Chapin FS (2007) Factors contributing to the cultural and spatial variability of landscape burning by native peoples of interior Alaska. Ecol Soc 12(1):7
Naylor LM, Wisdom MJ, Anthony RG (2009) Behavioral responses of North American Elk to recreational activity. J Wildl Manage 73(3):328–338
Nesbitt WA (1941) History of early settlement and land use on the Bent Creek Experimental Forest, Buncombe County, N.C., 47 p. Appalachian Forest Experiment Station, Asheville, NC
Payne C (2009) Tracking recreation at Bent Creek. Compass Perspect Tools Benefit South For Resour South Res Station 13:8–9
Pearson HA, Clary WP, Moore MM, Sieg CH (2008) Range Management Research, Fort Valley Experimental Forest. In: Olberding SD, Moore MM (eds) Proceedings of the Fort Valley Experimental Forest—A Century of Research 1908–2008 Conference, Proceedings RMRS-P-53CD, US Department of Agriculture, Forest Service, Rocky Mountain Research Station, Fort Collins, CO, 7–9 August 2008, pp 68–85
Philpott GW, Howard WE, Graham CA (1958) A method of managing sportsmen on rangeland. J Range Manage 11(6):290–292
Pings P, Hollenhorst S (1993) Managing eastern hardwood forests for visual quality. Proceedings of the 1993 Northeastern Recreation Research Symposium, Saratoga Springs, NY, 18–20 April 1993. General Technical Report NE-185, US Department of Agriculture, Forest Service, Northeastern Forest Experiment Station
Redman CL, Grove JM, Kuby LH (2004) Integrating social science into the long-term ecological research (LTER) network: social dimensions of ecological change and ecological dimensions of social change. Ecosystems 7:161–171
Reynolds RR (1980) The Crossett story: the beginning of forestry in southern Arkansas and northern Louisiana. General Technical Report SO-32, US Department of Agriculture, Forest Service, Southern Forest Experiment Station, 50 p
Sayre NF (2002) Ranching, endangered species, and urbanization in the southwest: species of capital. The University of Arizona Press, Tucson
Shapiro A (2008) A grand experiment: early Forest Service research. Forest Service Today, 30 May
Sheridan TE (2007) Embattled ranchers, endangered species, and urban sprawl: the political ecology of the New American West. Annu Rev Anthropol 36:121–154
Shindler B, Mallon A (2006) Public acceptance of disturbance-based forest management: a study of the Blue River landscape strategy in Oregon's Central Cascades Adaptive Management Area Corvallis, Oregon State University, OR
Shindler BA, Brunson M, Stankey GH (2002) Social acceptability of forest conditions and management practices: a problem analysis. General Technical Report PNW-GTR-537, US Department of Agriculture, Forest Service, Pacific Northwest Research Station, Portland, OR
Shindler BA, Beckley TM, Finley MC (2003) Seeking sustainable forests in North America. In: Shindler BA, Beckley TM, Finley MC (eds) Two paths toward sustainable forests: public values in Canada and the United States. Oregon State University Press, Corvallis, pp 4–17
Starrs PF (1998) Let the cowboy ride: cattle ranching in the American West. The Johns Hopkins University, Baltimore
Steel BS, Weber EP (2003) Ecosystem management and public opinion in the United States. In: Shindler BA, Beckley TM, Finley MC (eds) Two paths toward sustainable forests: public values in Canada and the United States. Oregon State University Press, Corvallis, pp 76–92
Stein SM, Alig RJ, White EM, Comas SJ, Carr M (2007) National forests on the edge: development pressures on America's national forests and grasslands. General Technical Report PNW-GTR-728, US Department of Agriculture, Forest Service, Pacific Northwest Research Station, Portland, OR, p 26

Sturtevant VE, Donoghue EM (2008) Community and forest connections: continuity and change. In: Sturtevant VE, Donoghue EM (eds) Forest community connections. Resources for the Future Press, Washington, DC, pp 3–26

Swank WT, Meyer JL, Crossley DA (2001) Long-term ecological research: Coweeta history and perspectives. In: Barrett GW, Barrett TL (eds) Holistic science: the evolution of the Georgia Institute of Ecology (1940–2000). Sheridan Books, Ann Arbor, pp 143–163

Swanson FJ (2010) Personal communication, August 2010. Research Geologist, US Department of Agriculture, Forest Service, Corvallis Forestry Sciences Lab, Corvallis, OR

Swanson FJ, Goodrich C, Moore KD (2008) Bridging boundaries: scientists, creative writers, and the long view of the forest. Front Ecol Environ 6(9):499–504

Thompson J (2008) Long-term ecological reflections: writers, philosophers, and scientists meet in the forest. Science Findings, US Department of Agriculture, Forest Service, Pacific Northwest Research Station (105), pp 1–5

Thompson J, Brokaw N, Zimmerman JK, Waide RB, Everham EM, Lodge DJ, Taylor CM, Garcia-Montiel D, Fluet M (2002) Land use history, environment, and tree composition in a tropical forest. Ecol Appl 12(5):1344–1363

Uriarte M, Canham CD, Thompson J, Zimmerman JK, Murphy L, Sabat AM, Fetcher N, Haines BL (2009) Natural disturbance and human land use as determinants of tropical forest dynamics: results from a forest simulator. Ecol Monogr 79(3):423–443

Wells G (2009) Experimental forests and ranges: 100 years of success stories. General Technical Report FPL-GTR-182, US Department of Agriculture, Forest Service, Washington, DC, 29 p

Wisdom MJ, Ager AA, Preisler HK, Cimon NJ, Johnson BK (2004) Effects of off-road recreation on mule deer and elk. Transactions of the 69th North American Wildlife and Natural Resources Conference, Wildlife Management Institute, Spokane, WA, pp 531–550

Zobel DB (2002) Ecosystem use by indigenous people in an Oregon coastal landscape. Northwest Sci 76(4):304–314

Chapter 26
Public Use and Citizen Outreach at Experimental Forests and Ranges

Lee K. Cerveny and Susan Charnley

Abstract Experimental Forests and Ranges (EFRs) in the United States Department of Agriculture (USDA) Forest Service historically have served as a resource for scientists conducting longitudinal and comparative research on aspects of forestry and range management. In addition to being sites of scientific investigation, EFRs are widely used by people for activities ranging from resource extraction and subsistence harvesting to nature appreciation and recreation. They also are used for education, training, technology transfer, and citizen science. Using a mixed-methods approach, including surveys, conversations with EFR contacts, literature review, and web searches, we explored the variety of ways that humans interact with EFRs. We received data on 68 EFRs in the network and found evidence of human use at every site. In this chapter, we present data on EFR visitor patterns as well as unsanctioned uses. The most frequent visitor activities were "walking, hiking, and running," hunting, and wildlife viewing. Several EFR sites contain extensive trail networks or are located within or adjacent to public lands with special designations that encourage recreation use. Variations in use existed between rural and urban EFRs. We also found that Forest Service scientists engaged with communities and stakeholders through a wide range of public outreach and education activities. Continued efforts toward stakeholder collaboration may yield future opportunities to expand the science mission at EFRs, involve citizens in data gathering and monitoring, and forge strong ties with nearby communities.

Keywords Experimental forests · USDA Forest Service · Human use · Public outreach

L. K. Cerveny (✉)
Pacific Northwest Research Station, USDA Forest Service,
400 N. 34th St., Seattle, WA 98103, USA
Phone: (206) 732-7832
e-mail: lcerveny@fs.fed.us

S. Charnley
Pacific Northwest Research Station, USDA Forest Service,
620 SW Main St., Suite 400, 97205 Portland, OR, USA

D. C. Hayes et al. (eds.), *USDA Forest Service Experimental Forests and Ranges*,
DOI 10.1007/978-1-4614-1818-4_26, © Springer New York 2014

26.1 Introduction

The United States Department of Agriculture (USDA) Forest Service Experimental Forests and Ranges (EFRs) have been host to a century of long-term scientific research related to forestry, range sciences, and resource management. The integrated network of EFRs presents a unique resource to scientists for conducting experimental research and long-term studies in a dedicated area where the research mission is predominant. While EFRs are settings where research can be conducted with minimal risk of disturbance, they should not be viewed as isolated outdoor laboratories devoid of human influence. People have lived on, around, and within EFRs for centuries. Humans have long been a part of these ecosystems and have influenced present conditions there. EFRs have strong ties to the past and often are sites of historical or cultural significance. They are also sites used by neighboring residents and nonlocal visitors for recreation, hunting, fishing, and gathering and for connecting to sacred or special places. In addition, EFRs serve as natural classrooms for students, educators, scientists, and resource professionals. These places have meaning for the people who visit them and the scientists who work there. They also are places where people go to learn and explore science. Thus, EFRs are not simply a valued resource for the scientific community; they also play a meaningful role in the communities that surround them.

In this chapter, we dispel the myth of the EFR as an isolated entity by exploring the range of human uses and interactions at EFRs, and the implications of these activities for the scientific mission of EFRs. We present data from a systematic survey of site contacts who described the nature and frequency of human uses and their effects on facilities and resources. We delve into variations in use between urban and rural EFRs and discuss the implications of these differences. We also talk about challenges associated with human interactions at these sites and identify unsanctioned uses observed by EFR contacts (e.g., vandalism, theft), and the potential effects of these activities upon the research mission. Finally, we address ways that EFR scientists have involved citizens and students in science, education, and outreach. The examples of public engagement we highlight suggest many ways that EFRs can build upon their role in local communities and be valued as a resource for learning and scientific exchange.

26.2 Methods

We are not aware of any previous attempts to summarize human dimensions information relating to the EFR network as a whole. Therefore, in the summer of 2009, we collected information about the human dimensions of EFRs for each of the 80 sites in the USDA Forest Service network.[1] We relied upon both primary and secondary sources and employed multiple methods to gain an understanding about the variety of human uses.

[1] The official number of EFRs is reported at 81 (Wells 2009). However, the total number of EFRs changed during the course of this study. It began at 79 (Adams et al. 2008), but then two new EFRs were added to the network (Olympic and HeenLatinee), and one was removed (Young Bay). Thus, our final sample size was 80.

Phase I involved gathering background information about each EFR, including existing facilities, geographic features, and current research programs, as well as policies and regulations related to public use. Secondary data were collected using published sources that describe and summarize features of each EFR, such as location, primary ecosystems represented, key research topics, on-site facilities, and collaborator relationships (Adams et al. 2008). This information was augmented with e-mail and telephone correspondence with EFR coordinators and scientists. Typically, a Forest Service scientist at each research station serves as the "Officer in Charge" of the EFR, and these became our primary contacts. Data were entered into a database and organized according to a set of predetermined parameters: basic information about the EFR, human activities and public uses, social science research, and public outreach and education efforts. Initial lists of key EFR contacts were obtained from the lead EFR coordinator at each Forest Service research station. In some cases, we contacted one scientist who was responsible for multiple EFRs, while in other cases we sought multiple contacts for information on a single EFR. Follow-up correspondence helped to clarify and augment information about public use and outreach. We also created a map of EFR locations and their proximity to urban areas. Proximity to EFRs was determined using Geographic Information Systems (GIS) shapefile analysis of EFR coordinates obtained from the Forest Service website. Minimum driving distances from EFRs to nearest urban areas were calculated using a standard Internet map program.

Phase II comprised a formal, online survey of primary contacts for each EFR in the Forest Service network. Many of the phase I participants also served as respondents (primary contacts) for phase II surveys. Other survey respondents were identified by phase I participants as being primary contacts for a particular EFR. An invitation to participate in the survey was emailed to 63 individuals designated as the primary contacts for the 80 EFRs in the Forest Service network. (A few respondents completed multiple surveys because they were the primary contact for more than one site.) We requested only one survey response per EFR. After initial contact, recipients received two follow-up reminders. The instrument was made available for 2 weeks from August 7 to 21, 2009 and a third week from September 16 to 23, 2009. Of the 80 EFRs in the Forest Service network, we received survey data from 68 EFRs, representing an 85% response rate. Survey data were entered into a spreadsheet and analyzed.

26.3 Human Uses of Experimental Forests and Ranges

The conduct of scientific research is the primary objective of the EFR network. Most sites are not equipped with visitor services or facilities other than those available to scientists and technicians involved in data collection. Nevertheless, people do come to EFRs to engage in recreation and other activities they seek on public lands.

26.3.1 Common Public Uses of EFRs

Survey respondents were asked about the types of human uses and activities that occur in the EFR they administer. From a list of common outdoor activities, respondents selected those that take place in their EFR. Table 26.1 shows the number of EFRs at which different human uses occur. The three most common uses were walking/hiking/running, hunting, and wildlife viewing, each of which took place on nearly 90% of EFRs in our sample. Other common activities found at more than 75% of EFRs were vehicle driving, scenic viewing, and dog walking. Timber harvesting was also mentioned on 78% of EFRs, though some of this harvesting was conducted as part of the site's actual research mission. Less common human activities were motor boating, mining, grazing, commercial recreation, and outdoor events. Motorized activities, such as snowmobiling, off-highway vehicle (OHV), or all-terrain vehicle (ATV) use, driving vehicles, and motorboating were prevalent on EFRs to varying degrees.

We also asked respondents to explain whether activities "occurred frequently" or "occurred occasionally." Hunting was the activity that was most commonly said to occur frequently (72% of EFRs). Respondents identified only two other activities—vehicle driving and walking/hiking/running—that occurred frequently, in more than half of EFRs. There were a few activities respondents said they knew little about, including cultural, religious, and spiritual uses and geocaching. Other human uses of EFRs mentioned by survey respondents that we did not include on the survey were oil and gas development, trapping/baiting, bog-stomping (driving oversized vehicles through wetlands), and skateboarding.

These data suggest that EFRs are used in a variety of ways, with recreation being most typical, although more consumptive activities such as hunting, timber harvesting, firewood gathering, and special forest products harvest are also common. The gathering of special forest products was frequently discussed in conversations with site administrators, who described a wide variety of products being harvested, including edibles and herbals such as berries, nuts, seeds, fruits, mushrooms, ginseng, bittersweet, and ferns, as well as craft products like birch bark, pinecones, and balsam bows.

A comparison of these findings with the National Visitor Use Monitoring (NVUM) Program summary data from 2007 suggests that visitor activity at EFRs may vary slightly from overall visitation to national forests (USDA Forest Service 2008). Direct comparisons cannot be made with NVUM data because of differences in methodologies and units of analysis. NVUM relies on visitor intercept surveys and reports the percentage of visitors to national forests engaged in a particular activity. Our study was based on a survey of EFR scientists and site administrators and reflects the percentage of EFRs where these activities occur. Still, many of the activity categories we used are similar to NVUM categories, and rudimentary observations can be made.

According to the 2007 NVUM national report, the five most common activities taking place in national forests were: viewing nature, viewing wildlife, hiking/walking, relaxing, and driving for pleasure (USDA Forest Service 2008).

Table 26.1 Human activities that occur on EFRs by frequency (N=68)

Activity	Occurs on EFR (%)	Occurs frequently (%)	Occurs occasionally (%)	Does not occur (%)	Do not know (%)
Walking/hiking/running	89.7	54.4	35.3	8.8	1.5
Hunting	89.6	71.6	17.9	9.0	1.5
Wildlife viewing	88.2	47.1	41.2	7.4	4.4
Vehicle driving	83.6	53.7	29.9	11.9	4.5
Timber harvest	77.9	17.6	60.3	22.1	0.0
Scenic viewing/sight-seeing	76.5	32.4	44.1	17.6	5.9
Dog-walking	74.6	28.4	46.3	14.9	10.4
Mountain biking	70.6	17.6	52.9	16.2	13.2
Picnicking	70.6	19.1	51.5	19.1	10.3
Horseback riding	61.8	23.5	38.2	27.9	10.3
OHV/ATV use	58.8	30.9	27.9	33.8	7.4
Fishing	55.9	32.4	23.5	39.7	4.4
Firewood/Christmas tree cutting	51.5	10.3	41.2	47.1	1.5
Camping	50.0	19.1	30.9	41.2	8.8
Special forest products gathering	44.8	10.4	34.3	46.3	9.0
Target shooting	44.8	13.4	31.3	47.8	7.5
Snow-machining	38.8	25.4	13.4	55.2	6.0
Skiing	37.3	13.4	23.9	58.2	4.5
Cultural/spiritual/religious use	27.3	3.0	24.2	37.9	34.8
Geocaching/orienteering	26.9	9.0	17.9	28.4	44.8
Kayaking/canoeing/rafting	26.5	8.8	17.6	72.1	1.5
Organized events (festivals, races)	23.5	2.9	20.6	66.2	10.3
Commercially guided recreation	23.5	8.8	14.7	67.6	8.8
Grazing	21.2	12.1	9.1	72.7	6.1
Mining	14.9	1.5	13.4	83.6	1.5
Motorboating	11.8	7.4	4.4	85.3	2.9

These activities were quite consistent with those found in our study. (In our survey, we asked about "vehicle driving," with no reference to pleasure.) High on our list of visitor activities in EFRs was hunting (89%), yet this activity did not appear to be prominent for visitors to national forests, where it was the ninth (of 28) activity (9.1% of visitors participated). Gathering of special forest products also was a common activity on EFRs (45%), but was not noted as a highly popular activity in the NVUM study (4.4% of visitors participated). There is a similar pattern with horseback riding, which occurred in 62% of EFR sites and was at the bottom of the NVUM list (1.3% of visitors participated.) Again, direct comparisons cannot be made, but these data suggest that EFRs may be serving a niche for particular visitor types. EFRs also may represent an opportunity for researchers to evaluate the NVUM program and the appropriateness of NVUM data at the site level. A study to understand visitor motivations and activities, as proposed in Chapter 25, could illuminate what features of EFRs draw visitors (Chap. 25, Social science research at Experimental Forests and Ranges).

26.3.2 Urban and Rural Differences

To understand the nature and extent of human uses, it is helpful to know something about where EFRs are located in relation to human communities. Table 26.2 shows that more than one third of EFRs (36%) are located within 80 km (50 miles) of an urban area, and just over 15% are located within 40 km (25 miles) of an urban area.[2] Conversations revealed that a few EFRs receive considerable use from nearby urban residents, though no visitation data exist to measure use trends. One example is Bent Creek, located 21 km (13 miles) from Asheville, NC, USA. According to scientists interviewed in phase I, Bent Creek's vast trail network is often used by Asheville residents and workers during the lunch hour and after work. Recent visitor studies indicate that 87% of all visitors to Bent Creek come from within a 40 km (25 mile) radius (Bowker and Zarnoch 2008). In contrast, 25% of EFRs are fairly remote, located more than 160 km (100 miles) from an urban area, where the high-frequency users are more likely to be rural residents or adjacent property owners. Figure 26.1 shows a map of EFRs and the relationship of these sites to nearby urban centers.

Table 26.3 shows a breakdown of EFRs by urban (within 80 km of an urban area) and rural (greater than 80 km of an urban area). Preliminary conversations with EFR contacts revealed that urban-proximate EFRs may face different pressures from nearby development. We grouped the EFRs into rural and urban categories and then noted the proportion of EFRs in each category that hosted the visitor activities listed in our survey. For many of the activities, there are minimal differences in the proportion of uses at urban and rural sites. However, there are a few notable activities that are greater at either rural or urban EFRs. For example, dog walking, biking, skiing, firewood and tree cutting, and target shooting were more common in urban EFRs. Activities more commonly occurring in the rural EFRs were fishing, camping, special forest products harvesting, timber harvesting, and cultural, religious, or spiritual uses. It appears that EFRs in more rural locales serve a more prominent role in subsistence for users. These data suggest that site managers of urban EFRs may have different needs than their rural counterparts.

[2] According to the US Census Bureau, an urban area (UA) "consists of contiguous, densely settled census block groups and census blocks that meet minimum population density requirements, along with adjacent densely settled census blocks that together encompass a population of at least 50,000 people" (Bureau of the Census 2000).

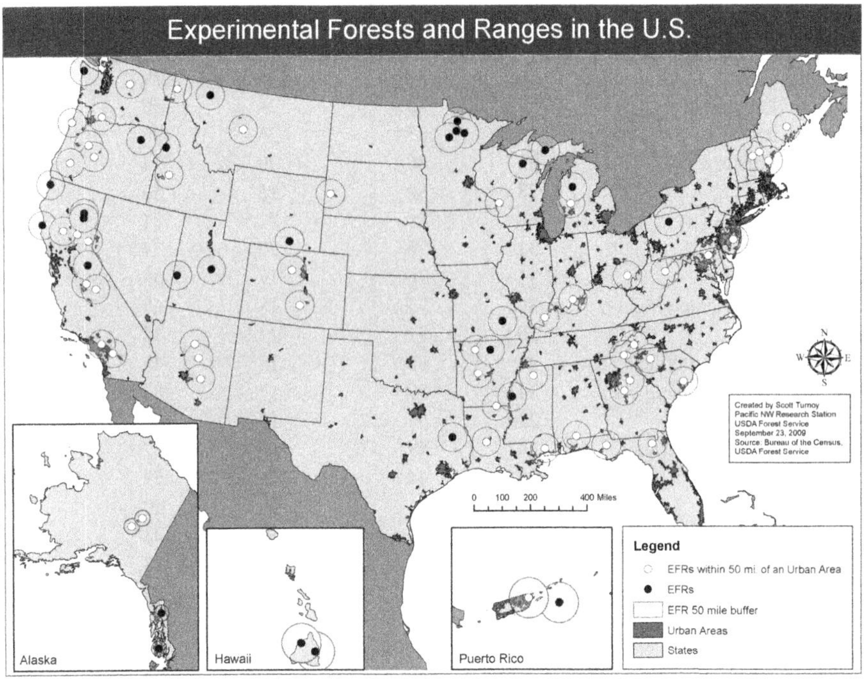

Fig. 26.1 Experimental Forests and Ranges in the USA

Table 26.2 Distance of EFRs from urban areas ($N=79$)

Miles from urban area (50,000 or more residents)	Number of sites	Percent of total EFR sites (%)
0–25 miles (0–40 km)	12	15.2
26–50 miles (41–80 km)	16	20.3
51–75 miles (81–121 km)	18	22.8
76–100 miles (122–161 km)	13	16.5
101–150 miles (162–241 km)	15	19.0
151–200 miles (242–322 km)	3	3.8
Greater than 200 miles (322 km)	2	2.5

26.3.3 Unsanctioned Visitor Activities

Conversations with scientists and site administrators revealed that some human activities occur at EFRs that are illegal, unauthorized, or deemed undesirable by scientists or site administrators. Our survey sought to identify the extent of what we are collectively labeling as "unsanctioned activities." Table 26.4 shows that vandalism and trash dumping were reported on more than three fourths of EFRs, and at least half of the EFRs listed incidents of partying, illegal timber harvesting, poaching of

Table 26.3 Breakdown of resource activities by rural (N=41) and urban (N=27) EFRs

Activities and uses	No. rural	Pct. rural (%)	No. urban	Pct. urban (%)
Hunting	36	88	24	89
Wildlife viewing	34	83	24	89
Dog-walking	34	83	26	96
Timber harvesting	33	80	20	74
Fishing	33	80	15	56
Snowmobiling	33	80	22	81
OHV/ATV use	31	76	20	74
Camping	31	76	17	63
Geo-caching, orienteering	30	73	18	67
Picnicking	29	71	21	78
Non-timber forest products	27	66	13	48
Walking, hiking, running	24	59	17	63
Driving	22	54	12	44
Cultural, religious, spiritual	20	49	10	37
Firewood and tree-cutting	18	44	16	59
Biking	17	41	16	59
Kayaking, canoeing, rafting	16	39	9	33
Races and events	12	29	4	15
Commercial guides	11	27	6	22
Grazing	10	24	4	15
Horseback riding	10	24	7	26
Scenic viewing	9	22	7	26
Motorized boating	8	20	8	30
Skiing	7	17	10	37
Mining	6	15	4	15
Target shooting	3	7	5	19

Table 26.4 Unsanctioned activities at EFRs (N=68)

Activity	Occurs on EFR (%)	Occurs frequently (%)	Occurs occasionally (%)	Does not occur (%)	Do not know (%)
Trash dumping	75.0	13.2	61.8	20.6	4.4
Vandalism	72.1	1.5	70.6	22.1	5.9
Partying	66.2	7.4	58.8	13.2	20.6
Illegal harvesting/poaching	61.8	4.4	57.4	16.2	22.1
Theft	52.9	1.5	51.5	27.9	19.1
Arson/fire rings	50.0	4.4	45.6	38.2	11.8
Narcotics sales/production	23.9	3.0	20.9	28.4	47.8

plants and animals, arson, and theft. Few of these activities occurred frequently, according to EFR contacts. Other activities perceived to be undesirable by survey respondents included: paint-balling, unpermitted firewood harvest, illegal grazing, and unauthorized races and events (bicycle races, ATV or motorbike rallies).

Table 26.5 Breakdown of unsanctioned activities by rural and urban EFRs

Activity	Rural (N=41)	Pct. Rural (%)	Urban (N=27)	Pct. Urban (%)
Vandalism	30	73	19	70
Trash dumping	29	71	22	81
Partying	26	63	18	67
Poaching	25	61	17	63
Theft	20	49	16	59
Arson, fire rings	19	46	15	56
Narcotics	5	12	10	37

In Table 26.5, we compared these activities among rural and urban EFRs and found notable differences. Narcotics activity (production, exchange) appears far more common in the urban-proximate EFRs. With the exception of vandalism, which is reportedly higher in rural EFRs, the proportion of EFRs experiencing unsanctioned activities is greater in the urban-proximate EFRs, particularly for trash dumping, theft, and arson.

Unsanctioned activities present management challenges and often require mitigation, administration, and repair. The cost of dealing with these challenges for agency budgets and personnel was widely noted. As one EFR contact explained, "All illegal activities are in conflict with our objectives…There is a loss of time associated with clean-up from vandalism, theft, and trash dumping/partying." Another EFR respondent concurred "…they take time away from research activities to do enforcement, cleanup, etc." With research budgets stretched thin, the cost of mitigating vandalism and replacing damaged equipment was perceived to be a challenge to site administrators.

26.4 Factors Influencing Visitor Use

Human interactions varied considerably among the EFRs in the sample. Conversations with site coordinators during phase I and qualitative responses from the survey in phase II allowed us to learn more about factors that may shape visitor use at EFRs. Variations may be based on proximity to urban areas with a large population base, the presence of unique topographic features, or natural amenities, such as lakes and rivers, or an abundance of special forest products. Public visitation also may depend on overall accessibility to the site (roads) and the EFR's own policy of public use.

26.4.1 Urban Proximity

The proximity of some EFRs to urban areas means that they serve as potential sites for recreation and consumptive uses for a larger number of people. Suburban development as well as demographic change and population growth in the wildland–urban interface also affect the way neighboring EFRs are used and valued. One EFR contact explained:

> Real estate land trusts—former timber companies—are buying up valuable forest land, clearcutting it, and then selling off the land to wealthy second-home developers, impacting both the environmental and socio-demographic conditions of nearby resource-dependent communities.

EFRs located near these areas of forest fragmentation and development may face unique pressures resulting from increased human use. Another EFR contact in the southern region observed:

> Several areas near the [named] Experimental Forest are experiencing considerable growth, driven by Hurricane Katrina. [People are] moving away from the waterfronts, moving to higher ground, [due to the] costs of rebuilding near the beach, and costs of insuring near the beach, to name a few.

Natural occurrences and socioeconomic drivers encourage regional shifts in population with implications for EFRs and their management. New studies underway at the Bent Creek Experimental Forest (NC) are exploring trends in population growth, demography, urbanization, and their effects on forests in the wildland–urban interface (Payne 2009). In this way, the EFR can serve as a living laboratory for understanding broader patterns of land-use change.

26.4.2 Natural Amenities

Proximity of EFRs to high-interest natural amenities and high-use recreation destinations may affect visitation to EFRs. Some EFRs are situated in areas with prime scenic views, unique topographical features, or offer a one-of-a-kind recreation experience that visitors cannot achieve on nearby public lands. In some cases, the proximity of EFRs to specially designated public lands such as national recreation areas, wild and scenic rivers, and scenic byways means that visitors to these special areas may be entering the EFR. As one EFR respondent described:

> With its adjacency to the [National Park] at its north edge and the…[Named] National Forest to the south, the…[Named] experimental forest is ideally situated to receive broad visitation. Visitors from across the United States and around the world use the interpretive trail and picnic area.

The location of an EFR in or near public lands with high visitor use has implications for human activity in the EFR, particularly in areas with open access. One site manager mentioned that the proximity of national forest campgrounds to the EFR site meant that campground visitors tended to use the EFR as a recreation playground and sometimes for overflow camping. Conversely, the absence of nearby natural amenities of keen interest to visitors also affects use of EFRs. As one respondent explained:

> EFRs vary in opportunities for public uses. [Named] Experimental Forest is limited in this way because it does not have features that tend to attract recreational use and does not have archaeological or cultural landmarks.

The quality or abundance of natural resources available in some EFRs relative to those in surrounding areas can provide an incentive for visitors. One respondent indicated that hunting was less crowded (and more desirable) on the EFR where he works, than on the nearby national forest. At one EFR in the study, the abundance of a particular edible plant was perceived to be greater than on surrounding public lands, primarily because the product had not been previously overharvested. In addition, the prohibition of motor vehicles on this EFR made it less accessible to commercial harvesters. In other instances, better quality recreation experiences were perceived to exist in the EFRs than on nearby public lands. The EFR may be the only convenient site where a certain activity can occur. Finally, crowding or use restrictions on neighboring public lands can also lead to a spillover effect, displacing activities to EFRs that did not previously exist.

26.4.3 EFR Policies Toward Public Access

EFRs exist in a variety of land ownership configurations (federal, state, private), and public access may depend on land management agency rules as well as the specific research mission, or an individual science team's responses to human use. Enabling legislation that designates each EFR stipulates the site's research mission and may offer explicit rules about public access. (Due to schedule and budget constraints of this study, we were not able to track down or analyze enabling documents or specific policies.) According to conversations with EFR contacts, only two EFRs in the Forest Service network are completely closed to the general public. Sixteen other EFRs implement special limitations on human use, such as closing off sensitive areas, using seasonal closures, or restricting certain activities.

EFRs have employed a variety of strategies to restrict access. In one EFR, public access was limited because it was located entirely on private lands, although private hunting clubs with leases were allowed to camp and hunt there seasonally. Some EFRs have restricted certain uses, such as overnight camping, motorboats, or firewood cutting. Some activities like hunting, beekeeping, and plant gathering are monitored by issuance of permits, though there is evidence of illegal poaching. Sometimes efforts to curb behavior are more passive, such as signage explaining the experiments going on and encouraging people to stay away from particularly sensitive areas. Some EFRs are posted, gated, or marked with signs that preclude entry by motorized vehicle, though foot traffic is permitted. Some site administrators explained that their efforts to restrict access often go ignored, and attempts to control or curtail use can be ineffective. Primary contacts reported examples where EFR visitors have knocked down gates or removed fencing to gain access to the sites.

In other EFRs, public use is actively encouraged, desired, monitored, or is a focus of scientific study. Many EFRs contain developed trail networks for biking, snow machining, hiking, or skiing that go through the EFR. As one site administrator described, "A major trail passes through the experimental forest and this is used often by groups for hikes, bird-watching, biking, cross-country skiing, and occasionally for illegal ORV use." The Bent Creek Experimental Forest (NC), for example, is a showcase for world-class mountain biking, and the trail network attracts bikers passing through the heart of the forest. Other EFRs have camping areas used by youth camps, forest trail work groups, and families. In some EFRs, there are private in-holdings, such as cabins or vacation homes that are accessible by road, though these may be gated. (One EFR has an 80-lot subdivision located within its boundaries.) Other examples of public facilities include boat ramps, picnic areas, beaches, and historic fire towers.

In many instances, visitors arrive to EFRs entirely unaware of the site's designated status, the location of its boundary lines, or the presence of experimental research. The lack of signage identifying an EFR as an experimental study area means that visitors often continue their recreation activity as if they were on other public lands. As one respondent observed, "To my knowledge, no attempt has ever been made to inform the local public about the existence of the [EFR], or to develop visitor attractions like interpretive trails or signage. " Visitors using trails may therefore be unaware that they are crossing a boundary and entering an EFR. One EFR scientist explained:

> The [named] Experimental Forest is included within the boundaries of the [named] National Forest with little if any signage designating when you cross from one to the other. Recreational activities permitted on the [named] National Forest are permitted on the [named] Experimental Forest because there is no one stationed on the [EFR] to oversee recreational activities.

While signage may improve public awareness and visitor behavior, some site administrators expressed hesitancy to erect signs in their study areas for fear of drawing attention to the facility or the studies underway, or of attracting more people to the area. Operating in a low-profile fashion appears to be part of the organizational history of some EFRs in the Forest Service network.

26.5 Human Uses and the Science Mission

We asked survey respondents whether any human activities taking place at EFRs conflicted with research goals. Vandalism and theft, in addition to being a nuisance, did sometimes result in damage to important research equipment, facilities, and instruments. As one EFR contact explained, "People have damaged research sites in the past by pulling up flags, posts, and seedlings." As a result of vandalism, theft, or other destructive activity, site administrators in some EFRs erected fencing and gates to protect equipment, or to fence off roads to reduce access. Another EFR contact told this story about an EFR where organized hunts were taking place:

> We just opened the gate on weekends, but we began to have some vandalism problems. For that reason, we finally ended the hunts and closed the experimental forest to public access. There was some theft: a portable generator was stolen, a trailer was stolen. It wasn't chronic. Those thefts happened over about five years, but we decided the last time it happened we didn't want it to happen anymore.

Another EFR contact talked about a bonfire at a campsite that got out of control and destroyed sensitive research equipment. As a result, the road to the campsite was closed. In these instances, scientists found ways to continue their research, although there were repercussions for public visitors, who were now restricted from these areas. Other scientists mentioned that the destruction and theft of signs and plot markers that identified the names of treatment processes were harmful to their research. Hunting, which was widely prevalent in EFRs, did not appear to affect the research underway. However, in one case, hunting interfered with annual bird surveys. Another scientist mentioned that illegal hunting affected the study animals in his big-game research project. In one EFR, concerns about stray bullets reaching containers of explosive materials led to safety precautions.

Recreation activities also had an effect on research goals in some areas. One scientist mentioned that recreationists build rock dams for swimming, and that these makeshift pools may interfere with long-term studies of aquatic species. Elsewhere, mountain bikers were responsible for creating new trails off of the existing system, causing bikers to drift into study areas. Neighboring landowners' actions also led to some undesirable human uses. One EFR contact mentioned that a landowner removed fencing along the EFR boundary, opening access to the EFR by the public, particularly for OHVs. In another EFR, ATV users were responsible for running over young trees and saplings involved in an experiment. Concern over activities such as partying, shooting, poaching, and narcotics production was also mentioned in the context of safety for on-site staff. Scientists noted that they simply did not always feel a sense of security on the EFR knowing that these activities were taking place in their midst.

Perceived human conflicts with the scientific mission pushed one site administrator to seek closure of the EFR for certain activities, and temporal or spatial restrictions on others:

> We would like to have a general closure of our experimental range, except for approved activities. We have been unsuccessfully trying to realize this for many years now. We need to limit activities to known times and places to avoid interference with research projects.

While some site administrators and scientists sought to close sites or restrict use, others welcomed the public and sought ways to open roads and facilities to the local population. This difference raises philosophical and practical questions about the extent to which the public should be entitled or invited to access EFRs, and how public use can occur without impeding the advancement of science. It also points to the fact that, although almost all EFRs experience human use, the majority of research there does not include people as part of the ecosystems under study, but rather focuses on understanding natural processes and functions uninfluenced by human disturbance.

Human interactions with EFRs raise a number of social science research questions ripe for future investigation. The fact that most EFRs are open to the public has implications for the integrity of these sites as research units, the sustainability of the natural resources there, and for potential social science research topics.

Rather than view human "encroachment" as a liability, the presence of visitors on EFRs may suggest potential opportunities for fresh areas of inquiry. For example, there is potential to conduct experimental work on recreation behavior, to explore visitor preferences for certain natural, social, and built environmental conditions, to engage in longitudinal studies of visitor use and resource effects (which is aligned with the original purpose of EFRs), and to launch comparative studies of recreation and human use patterns among several EFRs in the network. Moreover, the presence of visitors also creates opportunities for public engagement and building awareness of scientific endeavors.

26.6 EFRs and a "Sense of Place"

The term "sense of place" refers to an individual or group's identification with a particular place based on previous interactions with that area (Kruger and Williams 2007). People can develop strong ties to a place associated with the experiences and events that have occurred and the memories and associations they have about it. When applied to natural resource management, the concept generally refers to connections the public makes with natural areas through repeated recreation use, multigenerational family visits, cultural and historical meanings, emotional ties, and other connections. For example, recreation visitors to Bent Creek expressed strong attachment to the trail system which appeared to play an important role in the Asheville community.

Our conversations with site coordinators and scientists revealed that EFRs are not only meaningful to area residents and visitors, but EFRs can evoke strong place connections for scientists. Although we did not collect specific data on this question, we learned through many hours of discussion with scientists that EFRs can become treasured places that have special meaning for scientists. Many scientists have conducted work in these same plots and forests for decades and have developed strong attachment to these places. They talked about bringing scholars, students, and family members out to the EFRs to share their science legacy and present unique features. It was not uncommon for scientists to refer to these places as "my forests," or to feel a sense of protectiveness about potential disturbances to their special EFR, both natural and human. Nor was it unusual for the scientists to state concerns about policies or regulations that affect the status of their EFR as a long-term research site.

This connection between scientists and EFRs is important to recognize, particularly in exercises of strategic planning and in determining policy and regulation related to human use. EFR administrators and scientists often express hesitancy about "opening up" their sites to visitors. Responses to the presence of people engaged in outdoor activities may reflect a combination of concerns about maintaining the integrity of scientific activities, protecting scientific resources from damage, and ensuring safety of the research team, as well as preserving the scientist's unique "sense of place."

26.7 Citizen and Community Engagement with EFRs

An important phase of the research process is disseminating findings to the public and to resource managers who may implement them. EFRs have been an important part of that process of science exchange and technology transfer for the past 100 years—serving as a gathering place for scientists, students, resource professionals, and to some extent, citizens. In fact, some EFRs, such as Desert Experimental Forest (UT) and Escambia Experimental Forest (AL), were established with an explicit technology transfer component, for example, developing effective and economical timber production and grazing practices for adoption by ranchers and private landowners (Wells 2009; Hoyle 2009). Others, like Hubbard Brook (NH), were established as outdoor laboratories for ongoing study and involvement with university students and scientists. Today, Hubbard Brook has an extensive system of on-site classrooms, conference facilities, and laboratories.

EFRs use a variety of approaches for involving stakeholders, communities, other scientists, students, and citizens in their scientific endeavors, some of which go well beyond the communication of science findings. Each EFR is different in terms of how it interacts with the public. In some cases, community and stakeholder involvement is minimal or nonexistent. Other EFRs have multiple strategies for involving partners and individuals in the scientific process. At a few EFRs, collaborative stewardship is fundamental to the way the unit operates. The nature and extent of these interactions vary, depending on proximity to communities, universities, and overall accessibility. In addition, institutional factors affect the nature of interactions, such as the establishment of long-term ecological research (LTER) units at EFRs, a designation that encourages a high degree of public engagement and education. In this section, we explore the variety of ways EFRs engage with the public. Our discussion is based on information from websites, conversations with EFR scientists and administrators, and survey questions about outreach activities. We also feature case studies of EFRs that embody a unique approach to citizen outreach.

26.7.1 Public Outreach

Table 26.6 presents survey results on citizen and community outreach activities at EFRs and their frequency. According to our survey, outdoor education programs and field trips are the most frequent ways that EFR scientists connect with the public, with 82.4 % of EFRs offering some kind of educational program or field experience. Many examples were provided where visiting scientists, students, stakeholders, and others participated in a guided field trip at the EFR, or engaged in some sort of educational program, such as a professional training (short course), seminar or lecture, or classroom experience. One EFR scientist reported conducting 50 tours per year with educational groups, including high schools, college students, and visiting scientists. Although most of these tours were for special study groups, at least

Table 26.6 Outreach activities at EFRs (N=68)

Activity	Number	Pct. (%)
Outdoor education/field trips	56	82.4
Technology transfer to local stakeholders	46	67.6
Interpretative trails/tours	44	64.7
Outreach to local communities	32	47.1
Work with volunteers and partners for forest stewardship	26	38.2
Collaborative decision making about EFR research	21	30.9
Citizens involved in monitoring/data collection	15	22.1
Other	5	7.4

one EFR noted the potential for field trips that include the local community, as this EFR administrator explained:

> We occasionally have workshops and field trips where research results are presented or for recreational/educational purposes (e.g., bird or wildflower walks). Both local stakeholders and members of the local community are invited. For example, we recently had a centennial celebration where various field trips were offered in the morning and presentations on past and current research were presented, as well as visions for the future.

The communication of science findings through technology transfer to various groups is an important part of the field trip and classroom experience at many EFRs (67.6 %). These groups include private landowners, ranchers, timber corporations, resource managers, and education professionals. Technology transfer takes many forms and is associated with the use of trails, forest tours, and training sessions. As one EFR contact described:

> We have published a self-guided tour of some of the research plots; we take all kinds of groups, including private forest landowners, on guided tours of forest stewardship. We have offered annual training for more than 30 years to a very broad geographic and disciplinary range of professional land and resource managers.

Another EFR sponsors programs in tree-climbing and environmental leadership in cooperation with partners. When EFR programs reach out to local schools, the connections that form can foster community ties that extend over many generations.

One example of an EFR that is reaching out to surrounding communities through education is the Sagehen Experimental Forest (CA), which engages school children, regional university students, and community members in various education and outreach programs related to watershed restoration and hydrologic systems. The Sagehen Creek Field Station had been used by scientists from University of California (UC) Berkeley for studying hydrology, stream fauna, wildlife, climate change, and the biology of wetlands since the 1950s. In 2005, the Sagehen was formally designated an EFR and citizen science was promoted as a way to engage the local community in scientific discovery. A number of education programs have been undertaken. Sixteen hectares (40 acres) were committed to the local school district for science programs, and a program funded by the National Science Foundation partners UC Berkeley with local elementary schools. Another program engages non-native English-speaking teens in outdoor education, environmental literacy, and community leadership. The Sagehen is host to an environmental camp in partnership with a local children's museum that provides outdoor learning for preschool children. A summer speaker series engages the public with science. EFR staff members collaborate with the Truckee River Watershed Council on watershed restoration projects. And, the Sagehen website offers links to a Fish-Cam, the news blogs, and audio podcasts about ongoing research.

It is also common for EFRs to develop and maintain interpretive trails and tours (64.7%). Trails may be oriented to the public and include descriptions of the natural environment. Trails also may feature explanations of the long-term research taking place at the site. At Fernow Experimental Forest (WV), a 1.6-km (1 mile) trail was built on a contour featuring different grading standards and demonstration plots dating back to the 1940s, marked by interpretive signs. Some EFRs have developed partnerships with local organizations, including Boy Scouts, city agencies, and civic organizations to maintain their trails. One EFR administrator described the collaborative effort involved in creating a trail with disability access:

> We have a handicapped accessible nature trail that was established in collaboration with several local civic organizations that included the Jobs Training Partnership Act, Future Foresters of America, Army Reserves and [community] chamber of commerce.

A site manager explained how the network of trails allows them to offer field trips to a variety of stakeholders:

> We have a number of signed nature and demonstration trails. We do a lot of field trips and outdoor education for college students, practitioners of forestry, woodland owners, researchers, and educators. The Boy Scouts are occasionally active on the EF, assisting with trail construction or other physical labor. …We occasionally have tours in cooperation with the neighboring museum for the public and general interest articles in the paper.

Trails were seen as an effective way to introduce groups to science and to the EFR. However, the cost of maintaining the trails can be prohibitive. As one EFR administrator offered, "The [EFR] had an extensive set of walking trails with interpretive signage when the forest was being actively managed. Most trails are hard to find and signage is mostly gone."

Just under half of EFR contacts (47.4%) identified outreach to local communities as something that they regularly do. Some EFR administrators described close-knit relations and communications with communities. As these two EFR contacts stated:

> [EFR] is highly valued by the local community for its research on forest practices and is seen as a compatible use with the rest of the state forest it is part of. We have an extensive trail system leading to 21 gauging sites. These are used primarily by mountain bikers, but have also provided access to vandals. We adapt our treatments to the concerns of adjacent landowners including private landowners and a state park.

> We have hosted local communities at anniversary events; we hosted a public meeting to plan a research response to a windstorm that affected the experimental forest as well as many acres of public and private land throughout the region at a time when salvaging wind-thrown trees was very controversial.

However, not every EFR is located near a formal community. For remote sites, these kinds of partnerships and programs would be difficult to orchestrate.

Working with partners on stewardship projects are other ways that EFRs conduct outreach (38.2 %). Many EFRs had worked with volunteers and partners on projects ranging from bird counts, bridge construction, and trail maintenance to restoration.

26.7.2 Collaborative Decision Making

Collaborative decision making is an important part of the operation of some EFRs (30.9 %). Site managers felt there was much to gain by establishing collaborative planning processes for EFRs in general or to address specific issues. One EFR, the Coulee State Experimental Forest (WI), employed a participatory planning process with clients and stakeholders during their strategic planning efforts to meet the goals of sustainable forestry. The management plan that resulted focused on long-term partnerships with regional communities and highlighted the site's recreation opportunities (Wisconsin Department of Natural Resources 2010).

The collaborative planning process used at the Bent Creek Experimental Forest (NC) to address mountain biking represents one example of how EFRs are in some places linked to the social landscape of the broader region, and can benefit from community partnerships and collaborative planning. Established in 1925, the Bent Creek Experimental Forest in Asheville, North Carolina, has been the site of ongoing research in upland hardwood ecosystem management and studies of wildlife responses to silvicultural treatments. Just 15 min from downtown Asheville, the Bent Creek Experimental Forest is also a highly accessible area for outdoor recreation visitors. Each year, visitors arrive to hike, ride horses, camp, swim, canoe, and fish in the Lake Powhatan Recreation Area. The site has also become a premier destination for mountain-bikers nationwide. Miles of single-track bike trails were developed on roads once used by foresters for silvicultural experiments. When the Pisgah National Forest circulated the 2003 Bent Creek Complex Environmental Assessment, hundreds of community members gave feedback about the proposal to introduce new study plots near the bike trails, citing a history of repeated use. Bent Creek scientists collaborated with forest officials and representatives of bike associations to establish a dedicated trail system in the EFR. They produced a trail map, with proceeds directed at trail maintenance. Volunteers from local bike shops repair damaged trails at the EFR and instruct customers to be mindful of the research mission. Collaborative planning meant that the public now has access to large areas of the Bent Creek Experimental Forest, while researchers are able to plan and conduct experiments in areas they know face minimal chances of human disturbance.

26.7.3 Monitoring and Data Collection

Some EFRs (22.1%) reported public involvement in monitoring or data collection. For example, at the Coweeta Experimental Forest (GA), students from local schools were involved in collecting salamanders and assessing the health of existing streams. Another project at Coweeta involved students collecting climate change data and building a long-term database. We did not find evidence of participatory monitoring or citizen science that focused on social science.

One innovative example of citizen science is at Manitou Experimental Forest (CO), where an ecological monitoring system is being developed to evaluate the effects of vegetation conditions on the severity of fire. The Manitou Experimental Forest contains within its boundaries numerous private holdings and residential subdivisions that were severely impacted by the Hayman Fire in 2002 (Adams et al. 2008). In collaboration with the Coalition for the Upper South Platte and the Colorado Forest Restoration Institute, Manitou scientists contribute their expertise to the Woodland Park Healthy Forest Initiative. This research explores pre- and postfire monitoring protocols that integrate ecological, economic, and social indicators to bridge communication gaps between forest managers and community stakeholders. Working with residents of nearby Woodland Park, researchers are creating a demonstration area designed for ecological monitoring systems that evaluate the effects of vegetation on fire severity and study the effects of fire on forest regeneration. Economists are studying the potential for wood product extraction, biomass conversion, and job creation associated with forest vegetation in pre- and postfire environments. A major goal of the project is to develop monitoring tools that can be implemented by local volunteers. Citizen scientists contribute to data collection and monitoring in ways that inform future adaptive management decisions.

26.7.4 Improving Engagement

A wide range of outreach activities takes place at EFRs, and most combine multiple strategies to engage with stakeholders and the public. We suggest that the majority of EFRs are very involved in science exchange with other scientists, students, resource professionals, and members of the public through formal events such as field trips, courses, and events. Less common are efforts to directly engage local citizens in stewardship, collaborative planning, and environmental monitoring. There is potential for working with local residents and indigenous groups to incorporate local knowledge and indigenous environmental knowledge into the scientific process at EFRs; and, there are places where this is being proposed (Bonanza Creek, AK). Citizen science represents a potential pathway for engagement with communities, investing in the health and sustainability of the sites, their resources and facilities, and for furthering the scientific mission critical to the EFR system. Interest in expanding engagement with communities and citizens among EFR coordinators is apparent, as one EFR contact explained, "I would like to see more interchange between science and partners, stakeholders, especially on the special role of an experimental forest as a location to conduct research. I think most of our neighbors and visitors like and use the area for recreation."

Finally, EFRs also represent opportunities to share knowledge and build ties between scientists and resource managers. In a study of recreation managers in the Forest Service, the gap between agency research and management was often noted as a barrier in the exchange of science information. Managers in the study preferred direct, personal interaction between scientists and managers as a way to learn about available scientific tools and relevant results (Cerveny and Ryan 2008). Field trips, on-site trainings, and short courses are some examples of how science can be disseminated. In addition, EFRs offer opportunities for collaborative problem solving with resource managers, scientists, and stakeholders working together to identify problems, determine research approaches to address these problems, conduct research, and engage in monitoring. A new guide to effective research management collaboration at long-term research sites shows how EFRs can bring managers and scientists together in various ways, with researchers and managers playing both unique and shared roles.

26.8 Conclusions

EFRs are far from being fenced, isolated outdoor laboratories visited by a select group of scientists with waterproof notebooks and sturdy shoes. Rather, EFRs represent places with a long history of human use, and many continue to be vibrant areas of human activity. By acknowledging and assessing use, site administrators and scientists can plan strategically how to allow visitation that is safe, appropriate, and sustainable, and that does not impair scientific research underway. Thus far, high visitor volumes have not been cited as a problem in EFRs, but more accurate monitoring may be warranted in some areas that receive increasing use over time. The presence of visitors comes with some negative externalities, such as vandalism, litter, and theft. Dealing with these problems can consume scarce resources and take money away from the core science mission. Understanding and planning for these human uses can help to minimize risks. The presence of visitors offers opportunities for scientists to share their research mission and build public support for science. EFR visitors and stakeholders may serve as stewards of both science and the resource itself, offering new ways to build connections to the surrounding community.

Our investigation revealed a variety of forms of outreach underway. Examples from around the Forest Service network demonstrate the potential of EFRs to be vibrant outdoor venues for science communication and exchange, education, interpretation, and stewardship with a diversity of audiences. Neighboring landowners, local schools, scout troops, environmental organizations, and recreation groups all represent potential partners for collaborative stewardship of these facilities.

The sites also suggest opportunities to promote collaboration between scientists and resource managers in dealing with shared resource challenges or developing common research agendas. The EFR network is a living legacy not just for the community of scientists engaged in research on forest and rangeland science but also for the community of citizens and practitioners engaged with the thrill of science and discovery.

Acknowledgments We are greatly indebted to Daniel Hunter, our research assistant on this project, for his excellent work in gathering the data on which this chapter is based. We also deeply appreciate the time and assistance given to us by the many research scientists and EFR coordinators who participated in the study. We thank Anne Hoover and Pat Winter for their consultation in developing the chapter. We also thank Fred Swanson for reading and commenting on an earlier version of this material. Finally, we appreciate the anonymous reviewers, who provided helpful feedback and whose suggestions helped us make important changes.

References

Adams MB, Loughry LH, Plaugher LL (2008) Experimental Forests and Ranges of the USDA Forest Service, Revised. General Technical Report NE-321, U.S. Department of Agriculture, Forest Service, Northeastern Research Station, Newtown Square, PA, 178 p

Bowker M, Zarnoch B (2008) Personal communication

Bureau of the Census (2000) Federal Register, March 15, 2002. (vol 67, Number 51), pp 11663–11670. From the Federal Register Online via GPO Access, http://www.access.gpo.gov, DOCID: fr15mr02-43, Department of Commerce. Accessed 14 May 2009

Cerveny LK, Ryan CM (2008) Agency capacity for recreation science and management: the case of the U.S. Forest Service. General Technical Report-PNW-GTR-757, U.S. Department of Agriculture, Forest Service, Pacific Northwest Research Station, Portland, OR, 78 p

Hoyle Z (2009) The Escambia. Compass: perspectives and tools to benefit southern forest resources from the Southern Research Station 13:26–30

Kruger LE, Williams DR (2007) Place and place-based planning. In: Kruger LE, Mazza R, Lawrence K (eds) Proceedings: national workshop on recreation research and management. General Technical Report PNW-GTR-698, U.S. Department of Agriculture, Forest Service, Pacific Northwest Research Station, Portland, OR, pp 83–88

Payne C (2009) Tracking recreation at Bent Creek. Compass. Perspectives and tools to benefit southern forest resources from the Southern Research Station 13:8–9. http://ddr.nal.usda.gov/bitstream/10113/32225/1/IND44222663.pdf. Accessed 14 May 2009

U.S. Department of Agriculture, Forest Service (2008) National visitor use monitoring program. National Summary Report, 2007. http://www.fs.fed.us/recreation/programs/nvum/nvum_national_summary_fy2007.pdf. Accessed 14 May 2009

Wells G (2009) Experimental Forests and Ranges: 100 years of success stories. USDA Forest Service General Technical Report FPL-GTR-182, 29 p

Wisconsin Department of Natural Resources (2010) About the Coulee Experimental State Forest Planning Process. http://www.dnr.state.wi.us/master_planning/coulee/process.htm. Accessed 5 March 2011

Chapter 27
Expanding the Vision of the Experimental Forest and Range Network to Urban Areas

J. Morgan Grove

Abstract After 100 years, the USDA Forest Service has emerging opportunities to expand the Experimental Forest and Range (EFR) network to urban areas. The purpose of this expansion would be to broaden the types of ecosystems studied, interdisciplinary approaches used, and relevance to society of the EFR network through long-term and large-scale social–ecological projects in urban areas. The goals of these urban long-term research areas (ULTRAs) are to create scientifically rigorous knowledge of our urban regions and provide information needed to address current and future social–ecological problems.

The first of the four sections of this chapter explains the importance of understanding cities. Motivations for understanding cities are practical, scientific, and fundamental to the Forest Service's mission. The extension of the EFR network can play a critical role in understanding the social and ecological dynamics of cities and serve the majority of Americans directly, where they live. In the second section, I identify key traits of ULTRA projects: What features they share with existing EFR sites and how they differ? A key feature of the EFR network is its focus on applied research. An example from the Baltimore Field Station illustrates how ULTRAs can catalyze and inform partnerships between research and decision making. In the third section, I detail the current status of the ULTRA effort. I conclude with a look at how the ULTRA program can continue its contribution to science and society.

Keywords Interdisciplinary · LTER · Social–ecological · Sustainability · ULTRA

27.1 Introduction

One hundred years since the establishment of the US Department of Agriculture, Forest Service, opportunities are emerging for the agency to extend the Experimental Forest and Range (EFR) network to urban areas. The expansion would be designed to broaden the types of ecosystems studied, increase the variety of

---

J. M. Grove (✉)
Northern Research Station, USDA Forest Service, 5523 Research Park Drive, Suite 350, Baltimore, MD 21228, USA
Phone: 803-238-4328
e-mail: mgrove@fs.fed.us

D. C. Hayes et al. (eds.), *USDA Forest Service Experimental Forests and Ranges*,
DOI 10.1007/978-1-4614-1818-4_27,

interdisciplinary approaches used, and enhance the EFR network's relevance to society by incorporating long-term and large-scale social–ecological projects into urban areas. These urban long-term research areas (ULTRAs) are intended to generate scientifically rigorous knowledge of our urban regions and provide information needed for the identification and solution of current and future social–ecological problems (Rains, this book).

ULTRAs exist in several forms. One type comprises the urban field stations initiated by the USDA Forest Service and its partners in New York City, Philadelphia, Baltimore, Chicago, and Los Angeles. The Baltimore Field Station is also a key member of the Baltimore Ecosystem Study (BES), which is a long-term ecological research (LTER) project funded by the National Science Foundation (NSF). Another type of ULTRA consists of sites funded jointly by the Forest Service and NSF and selected through a competitive process. The current form of these projects is called ULTRA-Ex. Twenty-one ULTRA-Ex projects are currently funded. Despite differences in types, ULTRAs share many traits in terms of goals, requirements, approaches, and relevance to the EFR network. A primary aim of this chapter is to discuss these traits.

In the first section of this chapter, I explain why it is important to understand cities. Understanding of the social and ecological dynamics of cities is needed for reasons that are practical, scientific, and fundamental to the Forest Service's mission. The extension of the EFR network can enhance knowledge about urban areas and directly serve the majority of Americans where they live: in urban areas. In the second section, I identify key traits of ULTRA projects: the features they share with existing EFR sites and the ways they differ from them. A key trait of the EFR network is its focus on applied research. I use an example from the Baltimore Field Station, which participates in the BES urban LTER project, to illustrate how ULTRAs can catalyze and inform partnerships between research and decision making. While ULTRAs draw substantially from the experiences and lessons learned from EFR sites, I note lessons from ULTRA projects that are likely to contribute to new avenues of research and application on EFR sites. The third section describes the current status of the ULTRA effort. I conclude with a sense of future directions for the ULTRA program.

27.2 The Need for Understanding Cities

27.2.1 Practical Motivations

27.2.1.1 Urbanization of the Planet

There are several motivations for expanding the EFR network to urban areas that are practical, scientific, and central to the Forest Service's mission. From a practical perspective, it is essential to recognize that *we are an urban planet* (Fig. 27.1). In 1800, about 3 % of the world's human population lived in urban areas. By 1900, this propor-

Fig. 27.1 NASA global composite of nighttime lights. Urbanization, indicated by the density of nighttime lights, is now visible from space. (http://visibleearth.nasa.gov/)

tion had risen to approximately 14 % and in 2008 exceeded 50 %. Nearly every week, 1.3 million additional people arrive in the world's cities for a total of about 70 million a year (Brand 2006; Chan 2007). The urban population on a global basis is projected by the United Nations to climb to 61 % by 2030 and eventually reach a dynamic equilibrium of approximately 80 % urban to 20 % rural dwellers that will persist for the foreseeable future (Brand 2006; Johnson 2006). This increase in the share of the urban population from 3 % to the projected 80 % is a massive change in the social–ecological dynamics of the planet (Brand 2009; Seto et al. 2010).

The spatial extent of urban areas is growing as well. In industrialized nations, land is being converted from wild and agricultural uses to urban and suburban settlement at a faster rate than the growth in urban population. Cities are no longer compact (Pickett et al. 2001); they sprawl in fractal or spider-like configurations (Batty 2011; Makse et al. 1995) and increasingly intermingle with rural and wild lands. Even in many rapidly growing metropolitan areas, suburban zones are growing faster than other zones (Katz and Bradley 1999). The resulting new forms of urban development include edge cities (Garreau 1991) and a wildland–urban interface in which housing is interspersed in forests, shrublands, and desert habitats.

Accompanying this spatial change is a change in constituencies and perspectives among the world's population. Although these habitats were formerly dominated by agriculturists, foresters, and conservationists, they are now increasingly dominated by people possessing resources from urban systems, drawing upon urban experiences and expressing urban habits. An important consequence of these trends in urban growth is that cities have become *the dominant global human habitat* of this century in terms of geography, constituency, experience, and influence. This reality has important consequences for social and ecological systems at global, regional, and local scales, as well as for natural resource organizations attempting to integrate ecological function with human desires, behaviors, and quality of life across all of these scales.

27.2.1.2 Cities' Vulnerability to Climate Change

Urbanization creates both ecological vulnerabilities and efficiencies. For instance, coastal areas, where many of the world's largest cities occur, are home to a wealth of natural resources that are rich with diverse species, habitat types, and productive potential. They are also vulnerable to land conversion, changes in hydrologic flows, outflows of waste, and sea level rise (Grimm et al. 2008). In the USA, 10 of the 15 most populous cities are located in coastal counties (National Oceanic and Atmospheric Administration 2004) and 23 of the 25 most densely populated counties are in coastal areas. These areas have already experienced ecological disruptions (Couzin 2008).

27.2.1.3 Cities' Critical Role in Climate Change: Mitigation and Adaptation

Although ecological vulnerabilities are significantly associated with urban areas, urbanization also fosters ecological efficiencies. The ecological footprint of a city, i.e., the land area required to support its inhabitants, is quite large (Folke et al. 1997; Grimm et al. 2008; Johnson 2006). Cities consume enormous amounts of natural resources, while the assimilation of their wastes—from sewage to the gases that cause global warming—also are distributed over large areas. For example, London occupies 170,000 ha and has an ecological footprint of 21 million ha—125 times its size (Toepfer 2005). In Baltic cities, the area needed from forest, agriculture, and marine ecosystems corresponds to approximately 200 times the area of the cities themselves (Folke et al. 1997).

Ecological footprint analysis can be misleading, however, for numerous reasons (Deutsch et al. 2000). It ignores the more important question of efficiency, defined here as persons-to-area: How much land area (occupied area and footprint area) is needed to support a certain number of people? From this perspective, it becomes clear that urbanization is critical to delivering a more ecologically sustainable and resource-efficient world because the per-person environmental impact of city dwellers is generally lower than that of people in the countryside (Brand 2006; Grimm et al. 2008; Johnson 2006). For instance, the average New York City resident generates about 29% of the carbon dioxide emissions of the average American. By attracting 900,000 more residents to New York City by 2030, New York City can actually save 15.6 million t of carbon dioxide a year relative to the emissions of a more dispersed population (Chan 2007). The effects of urbanization on ecological efficiency may mean that social–ecological pressures on natural systems can be dramatically reduced in terms of resources used, wastes produced, and land occupied. Cities thus have the potential to provide essential solutions of mitigation and adaptation to the long-term social–ecological viability of the planet given the current population trends for this century.

Current global demographic trends are paralleled by changing conceptions of cities and urbanization. In very broad historical terms, we have begun a new paradigm for cities. Since the 1880s, a great deal of focus has centered on the "Sanitary

City," with concern for policies, plans, and practices that promoted public health (Melosi 2000). While retaining the fundamental concern for the Sanitary City, we have begun to envelope the Sanitary City paradigm with a concern for the "Sustainable City," which places urbanization in a social–ecological context at local, regional, and global scales (Grove 2009).

Urban ecology and long-term, place-based approaches have a significant role to play in advancing our concepts of sustainable cities (Grove et al. 2013; Pickett et al. 2013). Urban ecology already has an important applied dimension as an approach used in urban planning, especially in Europe (Singh et al. 2013). Carried out in city and regional agencies, the approach combines ecological information with planning methodologies (Hough 1984; Pickett and Cadenasso 2007; Pickett et al. 2004; Schaaf et al. 1995; Spirn 1984; Thompson and Steiner 1997).

Major investments in urban ecology theory, data, and practices are required to meet the needs of cities and urbanizing areas. Cities face increasingly complex and uncertain challenges. Many of these complexities are associated with changes in climate, demography, economy, and energy at multiple scales. Because of these complex, interrelated changes, concepts such as resilience, vulnerability, and ecosystem services may be particularly useful for both addressing current issues and preparing for future scenarios requiring long-term, and frequently capital-intensive, change.

Cities have already begun to address these challenges and opportunities in their policies, plans, and management. For example, on June 5, 2005, mayors from around the globe took the historic step of signing the Urban Environmental Accords—Green City Declaration with the intent of building ecologically sustainable, economically dynamic, and socially equitable futures for urban citizens. The Accords covered seven environmental categories to enable sustainable urban living and improve the quality of life for urban dwellers: (1) energy, (2) waste reduction, (3) urban design, (4) urban nature, (5) transportation, (6) environmental health, and (7) water (www.urbanaccords.org). International associations such as *ICLEI-Local Governments for Sustainability* (http://www.iclei.org/) are developing and sharing resources to deal with these issues. The ability to address these seven categories will require numerous, interrelated strategies and scientific domains. New York City's "plaNYC for a Greener, Greater New York" (City of New York 2007), for example, includes 127 different but interrelated strategies for making the city more sustainable, dynamic, and equitable.

27.2.2 Scientific Motivations

Existing EFRs and LTERs have broadened our scientific understanding of ecological systems in rural and wildland areas. ULTRAs would complement the existing EFR network by providing a more complete and comparative understanding of social–ecological systems in urban areas through the enhancement of existing theories, methods, and data previously developed in rural and wildland areas as well as the development of novel theories, methods, and data for built environments.

For instance, the existing urban LTER sites—BES and Central Arizona-Phoenix (CAP)—demonstrate the high level of scientific diversity and productivity of urban long-term social–ecological research platforms. BES and CAP study regions that span their central cities and the surrounding urban, suburban, and rural counties that constitute the Metropolitan Statistical Area for the Baltimore and Phoenix regions, respectively. Both sites include numerous investigators from academic, governmental, and nongovernmental organizations as well as biophysical, social, and technical sciences. All told, more than 80 principal investigators currently work on BES and CAP projects.

The diversity of organizations and expertise enables BES and CAP to address a wide range of social–ecological topics, such as links between urbanization and changes in hydrologic and atmospheric cycles; long-term trends in environmental justice and environmental amenities, disamenities, and ecosystem services; urbanization and changes in plant and animal communities; and long-term drivers of and responses to land-use change. Through these and other investigations over their first 10 years, BES and CAP produced more than 470 journal articles and 120 book chapters, and included more than 250 undergraduate and 240 graduate students, and 30 postdoctoral fellows. Results from BES and CAP research have led to changes in regional policies, the development of new management strategies such as urban tree canopy (UTC) goals (Grove et al. 2013; Pickett et al. 2007), and support for sustainability programs at local and state levels.

27.2.3 Caring for the Land and Serving People: Fulfilling the Forest Service Mission through ULTRAs

The ULTRAs play a fundamental role in the Forest Service's mission to "Care for the Land and Serve the People." The Forest Service was established to fulfill this mission in 1906 and the Experimental Forests and Ranges program was initiated soon after in 1908 (see Shapiro, Chapter 1, this book). At that time, the nation's population was predominantly rural. After this early phase of the Forest Service's history, the nation's population reached an equal proportion of rural to urban in the 1920s and an equilibrium of 20% rural to 80% urban that persists today and is likely to continue into the future. During this time, however, the location of EFR sites has not followed the changing demographics of the nation. There are no Experimental Forest or Range sites located in urban areas, and the two LTER sites—Baltimore and Phoenix—were established only in 1997 (Fig. 27.2).

The Forest Service recognizes that "Caring for the Land, Serving the People" also means "all lands and all people." The ULTRAs represent a foundation for the Forest Service to fulfill its mission by developing scientific information and practices that can serve the majority of Americans directly, where they live. A critical feature of this initiative is to address the fact that public land ownership is not extensive in urban areas. For example, the division between public and private ownership in the City of Philadelphia is 33% public and 67% private (Fig. 27.3). Of those

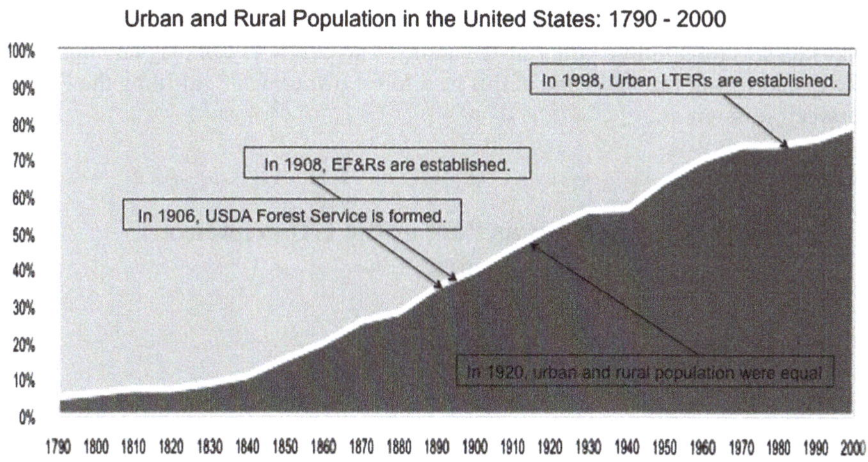

Fig. 27.2 Changing proportion of urban and rural populations in the USA (1790–2000) and the establishment of long-term research sites. The establishment of long-term research sites focused on urban areas has lagged significantly behind the growing proportion of urban population in the USA. Rural population corresponds to *light gray* area, urban population to *dark gray* area

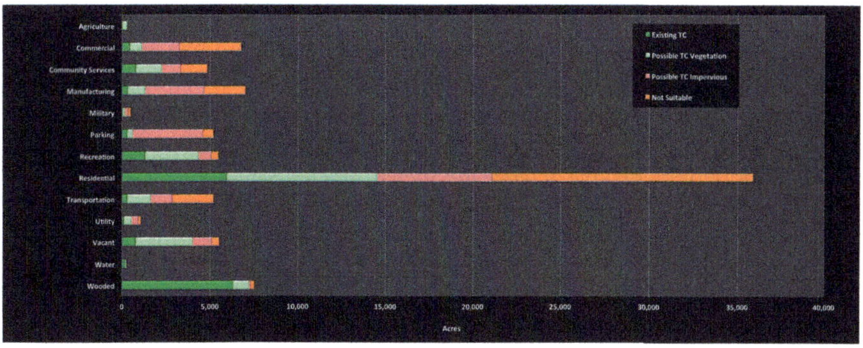

Fig. 27.3 Distribution of existing and possible urban tree canopy (UTC) in the City of Philadelphia by land ownership type. The distribution of existing and possible UTC is dominated by residential areas. (O'Neil-Dunne and Grove 2011)

private lands, 85 % are residential lands, with 459,524 individual parcels. Thus, the new "forest owner" in urban areas is in many cases a residential homeowner.

This social–ecological fact in urban areas parallels much of the Forest Service's experience and organizational culture east of the Mississippi River, where nearly 90 % of all lands are in private ownership. But this situation is very different from the Forest Service's experience west of the Mississippi, where only 10 % of all lands are in private ownership, and the landscape is dominated by extensive federal ownership. Regardless of region, however, private ownership prevails in urban areas. The role of this new forest owner in the stewardship of urban ecologies will be particularly important to understand and assist in order to mitigate and adapt to climate change and promote urban sustainability. The ULTRAs can play an important role in addressing the challenge of this new forest owner while fulfilling the Forest Service's mission.

27.3 The Ultra Program as Part of the Experimental Forest and Range Network

27.3.1 A Vision for ULTRA Projects

The USDA Forest Service has worked to develop different permutations of the ULTRA concept since the first urban LTER projects—BES and CAP—were funded in 1997. A key partner has been NSF, and this partnership has built upon the Forest Service and NSF's considerable interest in the long-term, dynamic interactions among people and natural ecosystems. As the LTER network has developed since 1980, the Forest Service has collaborated in supporting seven LTERs—six EFR sites and BES. Furthermore, BES is recognized as a "cooperating" member of the EFR network.

The momentum for a concerted, program-level focus on urban and urbanizing areas has been growing. Recent strategic planning by the LTER community has highlighted the need for greater integration of the social and ecological sciences across the LTER network, as evidenced in its decadal plan and the strategic research initiative entitled *Integrative Science for Society and the Environment* (Collins et al. 2007, 2011). This planning, in combination with the success of the urban LTERs and of the Dynamics of Coupled Natural and Human Systems Program (also co-funded and coordinated by NSF and the Forest Service), has led Forest Service and NSF leaders to jointly explore possibilities for the development of an additional program of large-scale ULTRA projects, described later in this chapter.

27.3.2 Traits Common to the ULTRAs and the Existing EFR Network

ULTRA projects share many traits with existing EFR and LTER sites. For instance, the EFR and LTER networks have similar structural characteristics. Because of the need to study ecological patterns and processes that cannot be understood with short-term funding cycles or other existing funding opportunities, the EFR and LTER networks provide long-term agency support and oversight while supporting research initiatives by individual scientists, teams, and sites. Research in the EFR and LTER sites has occurred over the long term, blending sustained, long-term studies with short-term initiatives focused on contemporary issues. Many studies outlasted an individual scientist's career. Research spans spatial, temporal, and hierarchical scales, and addresses multiple stressors (Grove et al. 2013).

The EFR and LTER networks include four types of environmental research—long-term monitoring, experiments, comparisons, and modeling (Carpenter 1998; Grove et al. 2013)—and an emphasis on education. To support analyses of long-term data and comparisons among sites, the EFR and LTER networks emphasize standardized methods for data collection and information management, including long-term data storage, curation, access, and discovery (see Ryan and Swanson, Chap. 24, in this book). Additionally, LTER projects include partnerships among the arts, humanities, and sciences (Swanson et al. 2008; www.ecologicalreflections.com).

A distinguishing characteristic of the EFRs has been a concern for the connection between research and application. The EFR Program has had an enduring philosophy of addressing land management problems at local and regional scales where they occur. EFRs are living laboratories or "seedbeds of discovery" that serve as demonstration, education, and training sites for cooperators and stakeholders. Similarly, the involvement of decision makers and educators is critical to the success of individual ULTRAs. Each ULTRA needs to engage decision makers and educators in the research-application cycle of (1) identifying questions, (2) collecting and analyzing data, (3) interpreting results, (4) disseminating and applying findings, and (5) identifying new questions.

EFRs are located in a wide variety of eco-regions and in varying states of ecological health, including sites that might be considered the "worst of the worst": deforested, overgrazed, or degraded. Thus, studies of restoration or reclamation can be an important focus of study. EFRs range in size from 47 ha (Kawishiwi Experimental Forest in Minnesota) to 22,500 ha (Desert Experimental Range in Utah). EFRs have not been limited to Forest Service lands and have included state and privately owned properties. Likewise, ULTRAs facilitate comparisons among urban regions in different climates and with different ecological, cultural, and economic histories. They include areas in need of restoration or reclamation and diverse land ownerships.

A broad range of topics have been studied in the EFR network. EFRs have examined the long-term role of climate change and land-use change on carbon sequestration, water yield, biodiversity, ecosystem productivity, and other ecosystem goods and services. Management and restoration are important considerations. Long-term studies of silviculture, hydrology, fire ecology, and other aspects of vegetation change have been essential to EFR sites. Most of these basic and applied topics are relevant ULTRA topics as well. To consider these topics in an urban context, however, the ULTRAs will have to address two important challenges. The first challenge is how to understand these topics in a coupled social–ecological system framework where people live. The second challenge is the development of novel methods for understanding these topics and their social–ecological dynamics over extensive areas, the long-term and multiple social and ecological scales.

27.3.3 Differences Between the ULTRAs and the Existing EFR Network

The existing EFRs and the ULTRAs overlap in many ways but also differ significantly. First, the ULTRAs add certain traits related to the types of sites examined, the conceptual framework and data employed, and the topics studied. Second, ULTRA projects are different from EFRs in that they focus on lands dominated by human influence *and* human settlement. As part of a network, ULTRA projects stimulate comparisons among urban areas in different climates and with different ecological, cultural, and economic histories. Third, ULTRAs require site-specific conceptual frameworks that incorporate methods and data from the geophysical, biological, social, and engineering sciences.

The most widely accepted template for social–ecological research is the press-pulse dynamics (PPD), which has been adopted by the LTER network and some of its international partners (Collins et al. 2007, 2011). It is useful for advancing social–ecological research and promoting multisite comparisons. The PPD was developed over 3 years by members from the ecological and social science communities in the USA to promote long-term social–ecological research. The PPD (Fig. 27.4) incorporates methods and data from the geophysical, biological, social, and engineering sciences but is not a theory or a model in and of itself. The PPD focuses on (1) "press and pulse events" that may drive social–ecological systems and (2) the linkages between social and ecological templates in terms of changes in the quantity and quality of ecosystem services. The PPD adds to the traditional topics of existing LTER—i.e., the biophysical template (structure and function) and regulating and provisioning ecosystem services shown in Fig. 27.4—by including topics such as cultural and supporting ecosystem services, the social template, social pulse and press drivers, and the relationships between these topics.

The intention of the PPD is to provide a generalizable, scalar, mechanistic, and hypothesis-driven framework to promote social–ecological research within existing LTER projects, the development of new long-term socio-ecological research, and comparisons among existing and new projects and networks (Grove et al. 2013). The PPD can be used to focus a long-term social–ecological research agenda through the identification of and connections among six strategic research questions (Collins et al. 2011):

1. How do long-term press disturbances and short-term pulse disturbances interact to alter ecosystem structure and function (H1)?
2. How can biotic structure, including built structure, be both a cause and consequence of ecological fluxes of energy and matter (H2)?
3. How do altered ecosystem dynamics affect ecosystem services (H3)?
4. How do changes in vital ecosystem services alter human outcomes (H4)?
5. How do changes in human perceptions and outcomes affect human behaviors and institutions (H5)?
6. Which human actions influence the frequency, magnitude, or form of press and pulse disturbance regimes across ecosystems and what determines these actions (H6)?

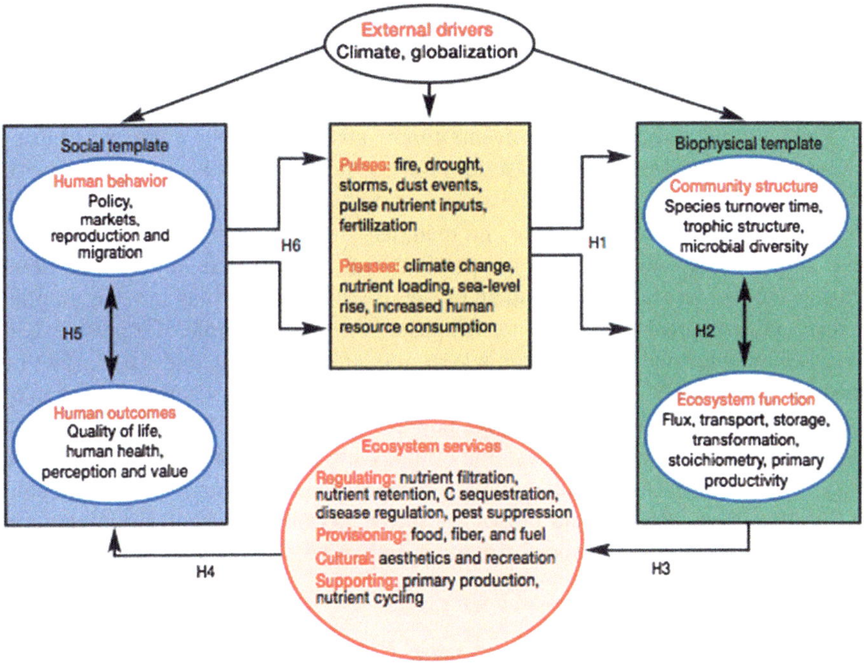

Fig. 27.4 Press–pulse dynamics (PPD) framework. The PPD framework provides a basis for long-term, integrated, social–ecological research. The *right-hand side* represents the domain of traditional ecological research; the *left-hand side*, traditional social research associated with environmental change. The two are linked by pulse and press events influenced or caused by human behavior and by *ecosystem services*, *top* and *bottom*, respectively (Collins et al. 2011). Individual items shown in the diagram are illustrative and not exhaustive

Because the PPD framework focuses on press and pulse types of disturbance, it is important to define the term. Disturbance is a technical term when used in social–ecological research. It was originally used to refer to events with sharp onset, short duration, and the ability to affect the physical structure of an ecological system. The term was provocative when first introduced because such events disrupted some aspects of ecological systems yet generated positive results for some other features of ecological systems. For example, disturbance often created opportunities for disadvantaged species to persist or enter an ecosystem, or provided locations in which resource conversion rates increased, facilitating access by suppressed or newly establishing organisms. Disturbance as originally introduced was distinct from ecological stress, which was often a longer-lasting event that directly affected the function or metabolism of a system. Both disturbance and stress can be unified as perturbations, and this term reminds researchers that the effects on any specific system component or entire system may be positive, negative, or neutral. Since its introduction in the mid-1980s, the concept of disturbance has been refined, and has led to a new consideration of ecological events in general. Events are now considered to be complex occurrences characterized by an onset, a duration in time, and potentially a later decline. Ecologists recognize that the complexities of onset, duration, and demise of events will result in different effects. A short flood may not kill many plant species on a floodplain, while an unusually long flood may cause mortality and may even remove sensitive species from the system.

The complexity of ecological events can be abstracted in the contrast of pulses and presses. Although this contrast does not consider all possible combinations of sharpness of onset, duration, or existence or rate of decline (Pickett and Cadenasso 2009), it focuses attention on two end members of that rich array of events: those that are transient and those that are persistent, at least for a relatively long time. Pulse events have sharp attack and quick demise, though they may have substantial effects on ecological systems. An earthquake is a good example of a pulse event. Press events alter the conditions in a system over a long time, and may in fact be more akin to stresses. A shift in a climate regime from wet to dry and the injection of a new level of resource supply through pollution are examples of biological pulses. Social pulses and presses are also important. New investment in a neighborhood may be a pulse. A shift in demographic composition in a district of a city would illustrate a press. Presses and pulses are raw materials for advancing integration between bioecological and social structures and processes in human ecosystems.

27.3.4 An Illustration of Dynamic Feedbacks Between Science and Decision Making: From Urban Hydrologic Drought to an UTC Goal

The ULTRAs can lead to a dynamic coupling between scientists and decision makers (Fig. 27.5). An example from BES illustrates this opportunity. The Baltimore region is characterized by ecologically functional watersheds and stream valleys that have contributed to Baltimore's economic and cultural history. An early test of the BES project was to apply and demonstrate the utility of forested watershed studies from the Coweeta (North Carolina), H.J. Andrews (Oregon), and Hubbard Brook Experimental Forests (New Hampshire)/LTERs (Bormann and Likens 1979) to an urban watershed system. One of the initial questions that BES asked, using a watershed approach, was: "Do riparian zones, thought to be an important sink for nitrogen in many non-urban watersheds, provide a similar function in urban and suburban watersheds?"

Somewhat surprisingly, BES analyses found that riparian areas had the potential to be sources—rather than sinks—of nitrogen in urban and suburban watersheds. This finding could be explained by the observation that hydrologic changes in urban watersheds, particularly incision of stream channels and reductions in infiltration in uplands due to stormwater infrastructure, led to lower groundwater tables in riparian zones. The resulting "hydrologic drought" created aerobic conditions in urban riparian soils which decreased denitrification, an anaerobic microbial process that converts reactive nitrogen into nitrogen gases and removes it from the terrestrial system (Groffman and Crawford 2003; Groffman et al. 2002, 2003).

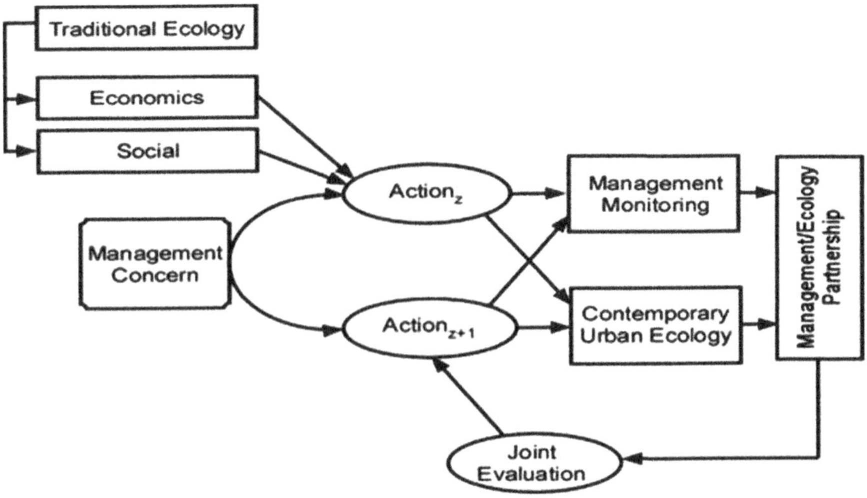

Fig. 27.5 An abstracted cycle of interaction between research and management. The cycle begins with the separate disciplines of ecology, economics, and social sciences interacting with a management or policy concern. In the past, ecology has neglected the urban realm as a subject of study, leaving other disciplines to interpret how ecological understanding would apply to an urban setting. A management or policy action (*Action$_z$*) results. Management monitors the results of the action to determine whether the motivating concern was satisfied. Contemporary urban ecology, which integrates with economics and social sciences, is now available for scientists to conduct research that recognizes the meshing of natural processes with management and policy actions. Combining this broad, human ecosystem and landscape perspective with the concerns of managers can generate a partnership to enhance the evaluation of management actions. New or alternative management actions can result. (*Actions$_{z+1}$*; Pickett et al. 2007)

Based upon these urban riparian results, the Chesapeake Bay Program reassessed its goals for riparian forest restoration in urban areas. Given that riparian zones in deeply incised urban channels were not likely to be functionally important for nitrate attenuation in urban watersheds, the program focused instead on establishing broader UTC goals for entire urban areas (Fig. 27.6). Increases in canopy cover across the city were expected to have important hydrologic and nutrient cycling benefits to the bay (Raciti et al. 2006).

This science-decision making cycle is dynamic and iterative. The UTC example has already progressed through seven cycles (Grove et al. 2013). After establishment of the City of Baltimore's UTC goal, analyses of the relationship between property regimes and UTC found that an "All Lands, All People" approach would be critical for achieving the city's goal (Action$_{z+2}$). Private lands under the control of households are a critical component to reaching any vegetation management goal in the city. Total existing canopy cover is 20%, with 90% of that cover located on private lands. Likewise, about 85% of the unplanted land area where potential planting could occur in the future is on private land, as compared to less than 15% on public rights of way (Galvin et al. 2006).

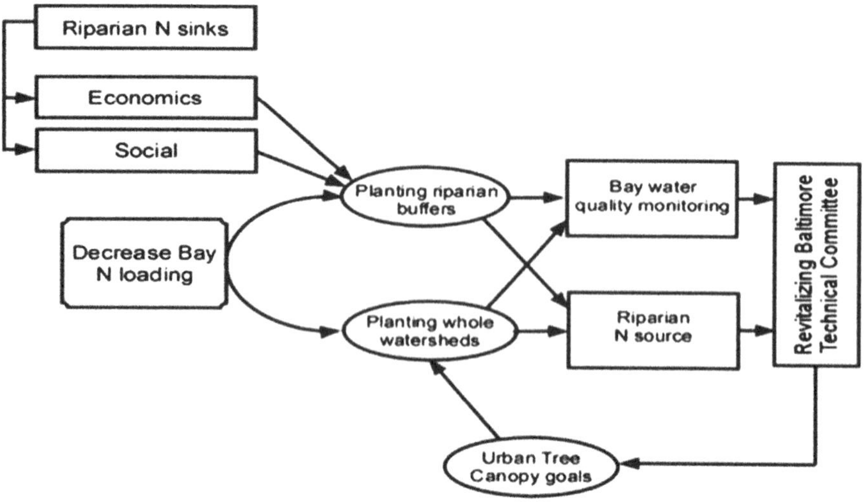

Fig. 27.6 An example of the management–research interaction in Baltimore City Watersheds. Traditional ecological information indicated that riparian zones are nitrate sinks. The management concern was to decrease nitrate loading into the Chesapeake Bay. In an effort to achieve that goal, planting trees in riparian zones was proposed. Management monitoring indicated that progress toward decreasing bay nitrate loadings was slow. Results from BES research suggested that stream channel incision in urban areas has resulted in riparian zones functioning as nitrate sources rather than sinks. In partnership with managers and policy makers in Baltimore City and the Maryland Department of Natural Resources, BES reevaluated strategies to mitigate nitrate loading, leading to a decision to increase tree canopy throughout the entire Chesapeake Bay watershed. Baltimore City adopted an urban tree canopy goal, recognizing both the stormwater mitigation and other ecological services such canopy would provide. (Pickett et al. 2007)

The importance of residential households to achieving Baltimore's UTC goal led to research addressing the relationships among households, their lifestyle behaviors, and their ecologies (Boone et al. 2009; Grove et al. 2006; Troy et al. 2007; Zhou et al. 2009). A critical finding from this research was that lifestyle factors such as family size, life stage, and ethnicity may be weakly correlated with social–economic status, but these lifestyle factors also play a critical role in determining how households manage their properties. These findings suggested the need for novel marketing campaigns that promoted UTC efforts while differentiating the marketing approach by type of neighborhood (Action_{z+3}). The need to "market" differently to different neighborhoods led to the need to understand existing and potential gaps in stewardship networks (Romolini 2013; Romolini and Grove 2010; Svendsen and Campbell 2008)—both functional and spatial dimensions of the network as a mechanism to communicate and organize local, private stewardship (Action_{z+4}).

Benefits from ULTRA projects are not limited to an ULTRA site. The findings and methods developed in Baltimore through these successive science-decision making cycles have had widespread utility in other urban areas. For instance, the tools developed in Baltimore to assess and evaluate existing and possible UTC have been disseminated through existing Forest Service networks and applied to more than 70 urban areas in the USA and Canada.

27.3.5 The ULTRAs' Potential Contributions to the Existing EFR Program

The ULTRAs can contribute to the existing EFR program in terms of scientific practice and knowledge and public awareness and use. The addition of ULTRAs to the existing EFRs represents a unique opportunity for long-term, standardized comparisons to understand how human activities modify ecological systems as well as respond to ecological change. For instance, the ULTRAs represent new opportunities for comparison with existing EFRs to examine ongoing changes in basic ecological patterns and processes, such as carbon sequestration, water yield, biodiversity, ecosystem productivity, and other ecosystem goods and services.

The ULTRAs can contribute to nascent EFR social science efforts for both on-site and regional EFR research. Although social science has not constituted a significant portion of on-site EFR research, examples can be found at some EFR sites (Charnley and Cerveny, Chap. 25, this book). Examples include (1) assisting in the development and application of socially acceptable and economically feasible natural resource management practices; (2) providing insight into how human activities, past and present, shape forest and rangeland ecosystems; (3) helping to solve resource management problems and to promote sustainable land-use practices; and (4) improving understanding of the social environment surrounding EFRs. These examples may be enhanced by interdisciplinary approaches emerging from the ULTRAs, including conceptual frameworks; novel approaches to long-term monitoring, experiments, comparisons, and modeling; innovative methods and data; and systems of practice with decision makers, educators, and diverse publics. Further, the adaptation of approaches from ULTRAs to on-site EFR activities can be relevant to EFR regional activities. This is an emerging and critical issue because EFRs are increasingly surrounded by growing urban and suburban sprawl (Charnley and Cerveny, Chapter 25, this book).

The ULTRAs can increase awareness of and support for existing EFRs. The EFR program is not well known to most Americans because EFRs are often in remote locations and because the connection between findings and applications from the EFR network and research implications is not often direct, immediate, or clear to most Americans. In contrast, the ULTRAs are located where most Americans live and ULTRA projects are more tightly coupled with the science findings and applications that directly and immediately benefit the majority of Americans.[1] Because of these connections, the ULTRAs have numerous and diverse opportunities to connect to the majority of Americans' perceptions, attitudes, and knowledge about their local, regional, and global environments. Further, knowledge about the ULTRAs can be a portal to increasing awareness of and support for the larger EFR network.

---

[1] Examples of recognition of the links between science findings and local benefits can be found from the BES, which was recognized through official proclamation for its service to the City of Baltimore by both Mayor Martin O'Malley (2004, 2006) and Mayor Sheila Dixon (2007, 2008).

27.4 Current Status of ULTRA Projects

There are currently several permutations for ULTRAs. The urban field stations initiated by the USDA Forest Service's individual Research Stations and its partners in New York City, Philadelphia, Baltimore, Chicago, and Los Angeles work together informally as a network. The Baltimore Field Station is also a key member of BES, which is an LTER project funded by NSF. BES participates in the LTER network and is considered a "cooperating EFR" in the EFR network.

In another type of ULTRA, projects are funded jointly by the Forest Service and NSF and selected through a competitive process. In April 2009, in anticipation of the full development of ULTRA projects based upon a competitive process, NSF and the Forest Service solicited proposals for exploratory ULTRA research—ULTRA-Ex projects—that would "identify and investigate topics and approaches that could advance both fundamental and applied knowledge regarding people and urban ecosystems," as stated in the NSF Program Announcement 2009. In the words of the program announcement, the intent of the ULTRA-Ex project awards is "to support research teams to conduct one or a limited number of related projects that will draw on and show promise of enhancing fundamental theory with respect to both human and biophysical systems as well as human–natural system interactions. Research teams will also generate knowledge about human–natural system interactions that can be used by government agencies, NGOs, businesses and individual citizens to improve the livability and sustainability of urban and urbanizing areas."

NSF and the Forest Service were surprised by the substantial response from the scientific community to the ULTRA-Ex solicitation, given the short time to prepare and submit proposals, limited project funding (US $ 300,000 total) and project length (2 years), and large expectations. More than 70 complete proposals were received. ULTRA-Ex proposals were reviewed in July 2009 for intellectual merit and broader impacts of the proposed project; the nature and scope of the partnerships; and the distribution of study areas in the USA by geographic location, size, age, and climatic zone. Funding was awarded to 21 proposals, with substantial diversity in terms of climate, size, and ecological, cultural, and economic histories. Additionally, some ULTRAs include urban gradients or multicity comparisons (Fig. 27.7).

ULTRA-Ex Projects (21)

- Hilo and Kailua-Kona, HI
- Los Angeles, CA
- Fresno/Clovis, CA
- Portland, OR and Vancouver, WA
- Tucson, AZ
- Albuquerque and Los Cruces, NM, and Phoenix, AZ (Southwest Gradient)
- New Orleans, LA
- Chicago, IL

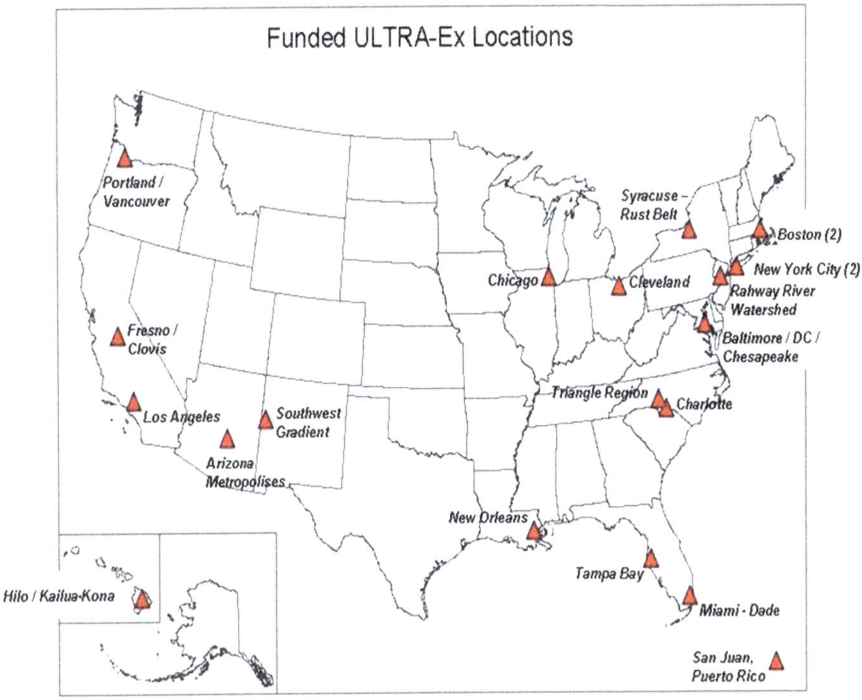

Fig. 27.7 21 ULTRA-Ex projects, located in diverse climatic, social, cultural, and economic landscapes

- Cleveland, OH
- San Juan, Puerto Rico
- Miami, FL
- Tampa Bay, FL
- Raleigh-Durham and Chapel Hill, NC (Triangle Region)
- Charlotte, NC
- Washington, D.C. and Baltimore, MD
- New Jersey Highlands, NJ (Rahway River Watershed)
- New York City, NY (2)
- Boston, MA (2)
- Syracuse, NY

It is hoped that a second solicitation for full ULTRAs will be made eventually. This call for proposals would not be limited to the existing 21 ULTRA-Ex projects, and some undetermined number of proposals would be funded on a funding cycle and level of funding similar to LTER sites. Fully funded ULTRAs would be selected based upon criteria similar to those for the ULTRA-Ex projects.

27.5 Conclusion

ULTRA projects have advanced scientific theory and practical knowledge about urban systems. This accomplishment is particularly important given the demographic trends for this century and a focus on adapting our cities to be more sustainable. In addition, ULTRA projects have two collateral effects that represent significant changes from the current state of academia and decision making. First, the establishment of ULTRA projects substantially legitimizes and recruits scientists and students to work on interdisciplinary topics, to focus on places where most humans live and will live, and to provide direct, practical benefits for decision making. Second, the ULTRAs provide decision makers with multi-scale, long-term, and systems-type knowledge about urban areas.

ULTRA projects can yield mutual benefits to the existing EFRs. Even one century after the establishment of the first Experimental Forest, there is still much to learn and apply from the EFRs to urban areas. This body of knowledge is related not only to comparing scientific understanding and methods of social–ecological systems in diverse wildland, rural, and urban contexts but also to the EFRs' culture of long-term research and addressing stakeholders' questions. At the same time, the explicit link between the ULTRAs and the EFRs can increase the public's knowledge about the Forest Service's investment in Experimental Forests and Ranges, the diverse types of ecosystems and conditions that exist in rural areas, and the scientific and practical benefits that the EFR network has provided for the past 100 years.

References

Batty M (2011) Building a science of cities. Cities 29(S1):9–16

Brand S (2006) "City Planet." strategy+business 42. http://www.strategy-business.com/press/16635507/06109. Accessed 10 April 2013

Brand S (2009) Whole earth discipline: an ecopragmatist manifesto. Viking Penguin, London, 316 p

Boone C, Cadenasso M, Grove JM (2009) Landscape, vegetation characteristics, and group identity in an urban and suburban watershed: why the 60s matter. Urban Ecosyst 13:255–271

Bormann FH, Likens GE (1979) Catastrophic disturbance and the steady state in northern hardwood forests. Am Sci 67(6):660–669

Carpenter SR (1998) The need for large-scale experiments to assess and predict the response of ecosystems to perturbation. In: Pace ML, Groffman PM (eds) Successes, limitations and frontiers in ecosystem science. Springer-Verlag, New York, pp 287–312

Chan S (2007) Considering the urban planet 2050. New York Times City Blog. Posted Dec. 4, 2007, 6:11 p.m. http://cityroom.blogs.nytimes.com/2007/12/04/considering-the-urban-planet-of-2050/?scp=1-b&sq=&st=nyt. Accessed 10 April 2013

City of New York (2007) plaNYC for a greener, greater New York City. http://nytelecom.vo.llnwd.net/o15/agencies/planyc2030/pdf/full_report_2007.pdf. Accessed 10 April 2013

Collins S, Swinton SM, Anderson CW, Benson BJ, Brunt J, Gragson TL, Grimm N, Grove JM, Henshaw D, Knapp AK, Kofinas G, Magnuson JJ, McDowell W, Melack J, Moore JC, Ogden L, Porter L, Reichman J, Robertson GP, Smith MD, Vande Castle J, Whitmer AC (2007) Integrated science for society and the environment: a strategic research initiative. LTER Network Office, Albuquerque, 35 p

Collins S, Carpenter SR, Swinton SM, Orenstein DE, Childers DL, Gragson TL, Grimm NB, Grove JM, Harlan SL, Kaye JP, Knapp AK, Kofinas GP, Magnuson JJ, McDowell WH, Melack JM, Ogden LA, Robertson GP, Smith MD, Whitmer AC (2011) An integrated conceptual framework for long-term social-ecological research. Front Ecol Environ 9(6):351–357

Couzin J (2008) Living in the danger zone. Science 319:748–749

Deutsch L, Jansson A, Troell M, Ronnback P, Folke C, Kautsky N (2000) The "ecological footprint": communicating human dependence on nature's work. Ecol Econ 32:351–355

Folke C, Jansson A, Larsson J, Costanza R (1997) Ecosystem appropriation by cities. Ambio 26:167–172

Galvin MF, Grove JM, O'Neil-Dunne JPM (2006) A report on Baltimore City's present and potential urban tree canopy. Maryland Department of Natural Resources, Forest Service, Annapolis, 17 p

Garreau J (1991) Edge city: life on the new frontier. Doubleday, New York

Grimm N, Faeth S, Golubiewski N, Redman C, Wu J, Bai X, Briggs J (2008) Global change and the ecology of cities. Science 319:756–760

Groffman PM, Crawford MK (2003) Denitrification potential in urban riparian zones. J Environ Qual 32:1144–1149

Groffman PM, Boulware NJ, Zipperer WC, Pouyat RV, Band LE, Colosimo MF (2002) Soil nitrogen cycling processes in urban riparian zones. Environ Sci Technol 36:4547–4552

Groffman PM, Bain DJ, Band LE, Belt KT, Brush GS, Grove JM, Pouyat RV, Yesilonis IC, Zipperer WC (2003) Down by the riverside: urban riparian ecology. Front Ecol Environ 1:315–321

Grove JM (2009) Cities: managing densely settled social-ecological systems. In: Chapin FSI, Kofinas G, Folke C (eds) Principles of ecosystem stewardship: resilience-based natural resource management in a changing world. Springer-Verlag, New York

Grove JM, Troy AR, O'Neil-Dunne JPM, Burch WR, Cadenasso ML, Pickett STA (2006) Characterization of households and its implications for the vegetation of urban ecosystems. Ecosystems 9:578–597

Grove JM, Pickett STA, Whitmer A, Cadenasso M (2013) Building an urban LTSER: the case of the Baltimore Ecosystem Study and the D.C./B.C. ULTRA-Ex Project. In: Singh SJ, Chertow M, Mirtl M, Schmid M (eds) Long term socio-ecological research: studies in society-nature across spatial and temporal scales. Springer-Verlag

Hough M (1984) City form and natural process: towards a new urban vernacular. Van Nostrand Reinhold, New York

Johnson S (2006) The ghost map: the story of London's most terrifying epidemic-and how it changed science, cities, and the modern world. Riverhead Books, New York

Katz B, Bradley J (1999) Divided we sprawl. Atl Month 284(6):26–42

Makse HA, Havlin S, Stanley HE (1995) Modeling urban growth patterns. Nature 377:608–612

Melosi MV (2000) The sanitary city: urban infrastructure in America from colonial times to the present. Johns Hopkins University, Baltimore

National Oceanic and Atmospheric Administration (2004) Population trends along the coastal United States: 1980–2008. Coastal Trends report series, US Department of Commerce, Washington, DC

O'Neil-Dunne JPM, Grove JM (2011) A report on the City of Philadelphia's existing and possible tree canopy. US Department of Agriculture, Forest Service, Northern Research Station, South Burlington, VT. http://www.fs.fed.us/nrs/utc/reports/UTC_Report_Philadelphia.pdf. Accessed 10 April 2013

Pickett STA, Cadenasso M (2007) Linking ecological and built components of urban mosaics: an open cycle of ecological design. Ecology 96(1):1–5

Pickett STA, Cadenasso ML (2009) Altered resources, disturbance, and heterogeneity: a framework for comparing urban and non-urban soils. Urban Ecosyst 12:23–44

Pickett STA, Cadenasso ML, Grove JM, Nilon CH, Pouyat RV, Zipperer WC, Costanza R (2001) Urban ecological systems: linking terrestrial ecological, physical, and socioeconomic components of metropolitan areas. Annu Rev Ecol Syst 32:127–157

Pickett STA, Cadenasso ML, Grove JM (2004) Resilient cities: meaning, models and metaphor for integrating the ecological, socio-economic, and planning realms. Landsc Urban Plan 69:369–384

Pickett STA, Belt KT, Galvin MF, Groffman P, Grove JM, Outen DC, Pouyat R, Stack WP, Cadenasso ML (2007) Watersheds in Baltimore, Maryland: understanding and application of integrated ecological and social processes. J Contemp Water Res Educ 136(June):44–55

Pickett STA, Boone CG, McGrath BP, Cadenasso ML, Childers DL, Ogden LA, McHale M, Grove JM (2013) Ecological science and transformation to the sustainable city. J Cities. http://dx.doi.org/10.1016/j.cities.2013.02.008. Accessed 10 April 2013

Raciti S, Galvin MF, Grove JM, O'Neil-Dunne J, Todd A (2006) Urban tree canopy goal setting: a guide for communities. US Department of Agriculture, Forest Service, Chesapeake Bay Program Office, Northeastern Area, State and Private Forestry, Annapolis, MD

Romolini M (2013) Governance of 21st century sustainable cities: examining stewardship networks in Baltimore & Seattle. Unpublished doctoral dissertation, University of Vermont, Burlington, VT

Romolini M, Grove JM (2010) Polycentric networks and resilience in urban systems: a comparison of Baltimore and Seattle. Available from Nature Precedings. http://dx.doi.org/10.1038/npre.2010.5222.1. Accessed 10 April 2013

Schaaf T, Zhao X, Keil G (1995) Towards a sustainable city: methods of urban ecological planning and its application in Tianjin, China. Urban System Consult GmbH, Berlin

Seto K, Sanchez-Rodriguez R, Fragkias M (2010) The new geography of contemporary urbanization and the environment. Ann Rev Environ Resour 35:167–194

Singh SJ, Chertow M, Mirtl M, Schmid M (eds) (2013) Long term socio-ecological research: studies in society-nature across spatial and temporal scales. Springer-Verlag, New York, 588 p

Spirn AW (1984) The granite garden: urban nature and human design. Basic Books, New York, 362 p

Svendsen ES, Campbell LK (2008) Urban ecological stewardship: understanding the structure, function and network of community-based urban land management. Cities Environ 1(1):1–32

Swanson FJ, Goodrich C, Moore KD (2008) Bridging boundaries: scientists, creative writers, and the long view of the forest. Front Ecol Environ 6(9):499–504

Thompson GE, Steiner FR (1997) Ecological design and planning. Wiley, New York, 348 p

Toepfer K (2005) From the desk of Klaus Toepfer, United Nations Under-Secretary-General and Executive Director, UNEP. Our Planet. 1

Troy AR, Grove JM, O'Neil-Dunne J, P M, Cadenasso M, Pickett STA (2007) Predicting patterns of vegetation and opportunities for greening on private urban lands. Environ Manag 40:394–412

Zhou W, Grove JM, Troy A, Jenkins JC (2009) Can money buy green? Demographic and socio-economic predictors of lawncare expenditures and lawn greenness in urban residential areas. Soc Nat Resour 22:744–760

Part IX
The Future

Chapter 28
The Future of USDA Forest Service Experimental Forests and Ranges

Peter Stine

Abstract Forest Service Research and Development is working to enhance the value of its network of Experimental Forests and Ranges (EFRs) by equipping them with modern data collection infrastructure to encourage cross-site and continental-scale monitoring, research, synthesis, and assessment. We envision a network of EFRs that shares resources across research sites, with managers and with scientists in other organizations. We anticipate focusing research across this network on the most pressing natural resource policy and management challenges of our time: natural disturbances and their interaction with human communities, sustainable production of forest and range goods and services, understanding and managing the effects of invasive species, understanding and coping with a changing environment, monitoring the pulse of our forest and range conditions, anticipating and addressing current and future land and resource management challenges, fully integrating urban natural resource research into our network, and optimizing the hierarchy of scales at which current and future natural resource challenges are addressed. As this vision is realized, the legacy of EFRs will be fully honored by sustained contributions to the well-being of the American people.

Keywords FS vision · Research networks · Environmental monitoring · Disturbance · Invasive species · Climate change · Forest management · Range management · Urban forestry · Cross-scale research

28.1 Introduction

Science provides us with the knowledge and insights that are critical to conserve and protect our precious natural resources for future generations to enjoy and utilize. As we move into a new and unprecedented era of land management and

P. Stine (✉)
John Muir Institute of the Environment, University of California at Davis, 1 Shields Ave, The Barn Rm 120A, Davis, CA 95616, USA
Phone: 530-334-0435
e-mail: pstine@fs.fed.us

Pacific Southwest Research Station, Research and Development, USDA Forest Service, Davis, CA, USA

D. C. Hayes et al. (eds.), *USDA Forest Service Experimental Forests and Ranges*, DOI 10.1007/978-1-4614-1818-4_28, © Springer New York 2014

conservation challenges, the American people will depend on the stewardship the Forest Service (FS) and other Federal agencies provide to deliver the information needed to meet those challenges. Our Experimental Forests and Ranges (EFRs) are the living laboratories that represent a rare and valuable asset which can serve the natural resource research needs of society for many years to come. The EFRs house scientific data from multiple areas of expertise and provide us with a wealth of data and information that can be used to answer scientific and policy questions through integrated analyses and synthesis. The network of FS EFRs across the country has four principal scientific advantages which can lead us into the future: *long-term, historical datasets of baseline environmental data, a diversity of expressly designated research areas strategically distributed across the USA, long-term study locations involving management and research partnerships, and the ability to conduct watershed-level field experiments where conditions can be manipulated for research purposes*. There is a critical need to equip these sites with the facilities needed to perform a broad range of high-quality scientific investigations, including deployment of modernized data collection infrastructure in the field for a subset (at least initially) of the EFRs where we stand to learn the most. These investments can provide increased opportunities for partners to work on the EFRs, increase the collaboration across EFRs, and mobilize an effort to synthesize scientific information at scales appropriate to the most important scientific and management questions.

28.2 Background

The legacy of FS EFRs is eloquently documented in the pages of this book. In recent years, Forest Service Research and Development has focused considerable effort on envisioning the most fertile future for these assets. Recently, we have recognized their potential as a network and what it will take to lead them as an integrated network into the future. New direction in our EFRs would support not only information technology and maintenance of our scientific and support infrastructure but also the collection of baseline data on several systems such as microclimate, stream water quality and flow, air quality, or carbon flux. Such data are instrumental in the full array of research themes that are carried out on EFRs with past, present, and future research investigations. Building upon national baseline data already collected by the Forest Inventory and Analysis (FIA) program, additional detailed data collected on EFRs would provide information essential to management and policy decisions of today and tomorrow.

In the twentieth century, the EFRs were established across the country to support long-term and manipulative experiments and monitoring across a variety of forest and range types. These studies were place-based to deal with the local natural resource questions of the day, and this role will continue to be important into the future. Fortuitously, EFRs also offer an ideal platform for Forest Service Research and Development, the National Forest System, and others to work in tandem to identify factors that drive current and future environmental change and to test management alternatives at regional to continental scales.

Our vision is that by modernizing data collection technology across the network, we will maximize the potential synergies of place-based and national or international research, of research to answer local management questions and to advance ecological theory. New technologies such as remote data sensors are creating new opportunities that will encourage scientists to work across organizational and disciplinary boundaries at common locations and, hence, leverage the scientific information gained from individual projects.

28.3 Opportunities

An assessment of EFRs in the USA indicates that there is a need for investment to enhance our capabilities across the network of EFRs and increase our capacity for a variety of existing and potential monitoring and research activities. EFRs are positioned to meet the most important and compelling forest and range management information needs in the near term and longer term. Some examples of compelling challenges that EFRs can address are provided in the following sections.

28.3.1 Persistent Disturbances as an Integral Part of Wild Ecosystems

Understanding and quantifying effects of disturbance agents (e.g., fire, insects, disease, timber harvest) and subsequent rates of recovery on hydrology and forest productivity are important for decision making and management. Human communities are concerned about the increasing intensity and frequency of fires, hurricanes, insect outbreaks, and spread of invasive species, and they seek answers from science. People and the natural resources that we depend on are found within a landscape that is subject to periodic disturbance from fire, wind, insect infestations, disease, and other forces that shape those landscapes. We cannot, nor should we try to, eliminate these forces; rather, we need to learn how to live with them. Disturbance factors have a substantial impact on ecosystem performance and the services ecosystems provide. We have made enormous strides in understanding how these factors affect different forest types, but we continue to need more information on how to manage forests and ranges in the face of impending or recent disturbance. With our array of EFRs, we are able to assess the ecological and economic effects of different methods of restoring forests and rangelands.

28.3.2 Providing the American People with the Valuable Goods and Services that Come from our Nation's Forests and Rangelands

Sustaining ecosystems goods and services into an uncertain future is critical. The urgency of the problems outlined in Sect. 28.1.3 requires a more synthetic response from science. Our forests and rangelands provide vital services for society and for environmental and human health. Freshwater, clean air, recreation, scenic values, and wildlife habitat are just some of the essential values that come from our forests and rangelands. Land managers need to understand how forest management affects the maintenance and delivery of these key ecosystem goods and services, such as the quantity and quality of surface and groundwater necessary to sustain growing human populations and ecosystem resilience during the next 100 years, maintenance or improvement of clean air through management of forests, the wide array of recreational opportunities and scenic values that Americans enjoy, habitat for fish and wildlife, sustainable production of wood and cellulose-based products, and carbon storage. EFRs provide a test bed for both baseline conditions and experimentation with management strategies that reveal insights into the response of key resources. We can take advantage of the distributed network of these sites to look at response to treatments at regional scales or across various kinds of environmental gradients.

28.3.3 Controlling or Eliminating the Effects of Invasive and Pest Species and the Impact on our Natural Resources

Understanding and addressing the spread of invasive plants and animals in our natural ecosystems is necessary for implementing the best management practices. Accelerating changes in our environment are also partly caused by invasive plants and animals, and society is especially concerned about invasive species when they are associated with disturbances or loss of ecosystem services. Invasive plant and animal species present an enormous challenge to maintenance of ecosystem health throughout the world. Forest and rangeland health, the provision of ecosystem services, and wood production are among the natural resources that could be negatively affected by invasive pests that have found their way into the USA. Monitoring these taxa offers an important and reliable scientific metric for detecting chronic, long-term changes in ecosystems. EFRs provide an effective platform for widespread monitoring and research strategies to assess changes in abundance and distribution, and controls of invasive species in forest and rangeland environments.

28.3.4 Understanding and Coping with a Changing Environment

Addressing the response of our forests and ranges to environmental change is critical to restore, retain, reforest, and maintain our natural systems. Society remains uncertain about the short- and long-term effects of a changing environment. We must learn how to cope with changes and uncertainty to sustain the productivity and stability of our ecosystems and the services they provide society. We need the capacity to understand the relative changes in ecosystems in response to changing climates. Changes are not uniform across geography and topography; thus, land managers require greater insight into the subtle to pronounced changes we are observing. We stand to learn much through links between observed climate station data and land-cover changes. EFRs are well situated to monitor change through collection of annual demographic data and to experiment with managing for adaptation to change. We can also experiment with strategies to mitigate greenhouse gas emissions through forest management strategies to sequester carbon.

28.3.5 Assessing the Pulse of our Forest and Range Ecosystem Conditions

Baseline monitoring of forest and range ecosystems is the foundation for our science-based management of natural resources. The public expects our leaders in land and resource management to be prepared to deal with the challenges and unknowns to conserve our natural resources. The distribution and abundance of EFR sites around the country give us a significant advantage for monitoring basic environmental parameters that have a vital influence on ecosystem performance. We will have the capacity to collect and manage the data on basic weather conditions (temperature, precipitation), stream water quantity and quality, air quality, and long-term, landscape-scale vegetation monitoring that enables us to link vegetation dynamics to the aforementioned watershed-scale environmental monitoring. Having these data collected using modern equipment and providing real-time accessibility will support a wide variety of research activities and provide land managers and the public with an accurate and precise tabulation of basic environmental conditions.

28.3.6 Anticipating and Addressing the Land and Resource Management Problems of the Present and Future

Support is necessary for long-term, cross-site research to assess forest and range variability in productivity and health of native ecosystems. The needs and desires of society change over time. The demands of the American public will evolve as conditions change and as we learn from our actions. A critical responsibility of our research and development capability is to forecast and anticipate what the land and resource management challenges will be in the near and longer-term future. We have established some long-term research and monitoring programs to examine key ecosystem functions such as soil productivity and forest regeneration, and to assess forest health and variability. This initiative will enable us to continue or initiate new long-term research addressing forest ecosystem response to proposed management treatments. With these data, combined with experimentation to test alternative forest management strategies, we can work with land managers to be prepared to address unforeseen problems.

28.3.7 Caring for the Land and Serving People Where They Live

Fully integrating long-term urban natural resource research sites into the EFR network is critical to understanding and managing the urban–wildland interface and the flow of services across this interface. More than 80% of the American people now live in urban areas, and Forest Service Research and Development has taken a lead role in understanding the natural resources of these areas and the relationships that people have with these natural resources. Continued success of the EFR network depends upon full integration of the sites and science of urban natural resource stewardship.

28.3.8 Optimizing the Hierarchy of Scales at Which Current and Future Natural Resource Challenges are Addressed

Integration of research on place-based challenges is important to managers within the EFR network and is key to long-term sustainability. Within the context of climatic variability and change, large-scale disturbances, and human needs, forest managers continue to need science-based solutions for today's problems that work in the specific forests they manage. A network of EFRs offers scientists and managers the opportunity to understand factors that explain responses to forest and ecosystem management, and how they vary at local, regional, national, or international scales. This book describes numerous historic efforts to understand these patterns, from species composition and productivity responses to uneven-aged management in the Lake States and Appalachia through broad-scale research about watershed responses to forest management across the country. Science at the cutting edge on EFRs will advance our understanding of these important distinctions. The EFRs of the future will need to be large enough to encompass and capture the ecological functions that drive forests (e.g. fire). We will need to explore means for expanding the spatial (and temporal) scales at which we study forest function and management activities.

Recognition of the EFRs as a system through implementation of national-level standards has tremendous opportunities. We now see the timely need for growing these activities and capitalizing on our 100+-year efforts with a more focused program of coordinated monitoring and research. We have partners in many different organizations that can enhance our prospects for succeeding in what must be a collaborative endeavor.

Demonstrating the value of the EFR network with these initial efforts would go a long way toward leveraging EFR use by individuals and small teams within Forest

Service Research and Development as well as partners at other land management or research institutions. For example, the FS is working to enhance EFR connections with the FIA program and with the National Earth Observation Network (NEON) to complement and enrich these collective investments.

28.4 Conclusion

It is through these collaborative programs that research and information from EFRs can be linked beyond small landscapes to regions and the continent. Scientists could use these sites to explore the robustness of their findings by comparing their results and conclusions across locations and habitat types. Once investments are made to increase the capacity and capabilities of the EFRs, the integrated information and data from the networked sites would be easily available for use by teams of researchers from other agencies, universities, nongovernmental organizations, and tribal communities. Increased use would allow scientists and managers access to integrated data and information to better design best management practices that take into account the considerable natural variability across ecosystem types. The 100 + year-old network of 80 experimental field sites across the USA represents an investment of immeasurable value. We offer that the potential for learning and provision of critical environmental knowledge merits our continued attention and investment.

Index

D. C. Hayes et al. (eds.), *USDA Forest Service Experimental Forests and Ranges*,
DOI 10.1007/978-1-4614-1818-4, © Springer New York 2014

D

E